Introduction to Quantum Mechanics

A Time-Dependent Perspective

INTRODUCTION TO QUANTUM MECHANICS

A Time-Dependent Perspective

David J. Tannor
Weizmann Institute of Science

UNIVERSITY SCIENCE BOOKS

Mill Valley, CA

University Science Books
www.uscibooks.com

Production Manager: Ann Knight
Manuscript Editor: Lee A. Young
Design: Mark Ong
Cover Design: George Kelvin
Illustrator: Lineworks
Compositor: Windfall Software, using ZzTEX
Proofreader: Jennifer McClain

Casebound ISBN : 978-1-891389-23-8
Softcover ISBN : 978-1-891389-99-3

Library of Congress Cataloging-in-Publication Data

Tannor, David Joshua, 1958–
    Introduction to quantum mechanics : a time-dependent perspective / David J. Tannor.
        p.   cm.
    Includes bibliographical references and index.
    ISBN 1-891389-23-8 (alk.paper)
1. Quantum theory.   I. Title.
QC174.12.T366 2006
530.12—dc22

                                                        2006043933

Printed in the United States of America
10 9 8 7 6 5 4 3 2

To Vivian,
my אשת חיל

Contents

Foreword

The importance of the time-dependent formulation of quantum mechanics is evident to theorists and experimentalists alike, but this wasn't always so. Even though Schrödinger's Time-Dependent Equation was the birth of modern quantum theory, and even though Schrödinger knew about the harmonic oscillator coherent states and their classical-like motion, attention quickly turned to the Time-Independent Schrödinder Equation. There were good reasons for this: physics was focused on line spectra and their predictions, and for this purpose the time-dependent formulation is indirect. Tradition quickly grow around this chronology, so that textbooks often begin with the Time-Dependent Schrödinder Equation or introduce it in Chapter 1, only to drop it like a hot potato for the remainder of the book, with the exception perhaps of a cameo appearance in the derivation of Fermi's Golden Rule.

Now the advent of explicitly time-dependent experiments, and especially the thrust toward many body systems (where computational and experimental requirements mean that no eigenstate can be found or measured, or ever needs to be measured, beyond at most the first few) make it imperative that students see time dependence put to good use as soon as possible. Indeed even in the heyday of line spectra the abandonment of the Time-Dependent Schrödinger Equation was far too wide a swing of the pendulum. It is enough of a shock for students of quantum mechanics that matter is a wave, and that Schrödinger's wave is a probability amplitude, and so on. Adding to that shock is that quantum physics was done with still, motionless, *stationary states*. Absurd questions like "How does the particle get past the node?" come up if one is divorced from time dependence. That teachers and textbook writers took this question seriously underscores the poverty of an education only about $H\psi = E\psi$. Using a time-dependent approach, some familiar intuition is still intact from the classical world, and the classical intuition we all have can be put to good use to understand quantum mechanics at a high level rather quickly.

The history of overemphasis on the Time-Independent Schrödinger Equation is there for anyone to see in the old textbooks. What is shocking is that it is there for everyone to see in most of the new ones too! Such "academic momentum" for textbook writers is a well-known phenomenon. Thankfully, David Tannor's book breaks free of this syndrome and takes a far more balanced approach, employing time-dependent and -independent approaches in the appropriate circumstances. The student of this text will come away well prepared to tackle today's explicitly time-dependent experiments and other stationary experiments with a strong link to a simple time-dependent picture via Fourier transform. This is the fun way to learn quantum mechanics applied to molecular dynamics.

All of the pillars of time-dependent methodology are here: wavepackets, correlation functions, semiclassical methods, and numerical methods. The treatment of strong fields, femtosecond multipulse control of reactions, photodissociation, and reactive scattering, all of which are treated in the latter half of the book, are made accessible banking on classical intuition and the preparation in time-dependent quantum and semiclassical methods in the first half.

My hope and expectation is that this book is not only a new and much needed vehicle for training the current generation of students, but also the impulse required to change the momentum of textbook writers of the future toward a balanced approach to quantum molecular dynamics.

The author says in his introduction essentially that this is the book I promised to write, but never did. He is half right: I promised to write a book, but the one I planned would not have been as good as this one, nor as comprehensive.

Eric Heller
Harvard University

Preface

Many people find time-dependent quantum mechanics the most interesting and understandable part of quantum mechanics. Yet it is the orphan child in the standard curriculum of physics and chemistry, both on the undergraduate and graduate levels. In physics courses one generally learns a little about *wavepackets*—solutions of the Time-Dependent Schrödinger Equation—but they are generally viewed as a curiosity, rarely compose more than a fraction of the course, and their relationship with the canonical time-independent topics in the standard course is generally disjointed. Traditional quantum chemistry courses start with the Time-Dependent Schrödinger Equation (TDSE); however, generally within half a lecture a separation of space and time variables is invoked and the rest of the course deals with the Time-Independent Schrödinger Equation (TISE). The latter then becomes treated as the fundamental equation, with little or no further mention of time in the course.

The Time-Dependent Schrödinger Equation is the truly fundamental equation of non-relativistic quantum mechanics in a way that the TISE is not. Many systems, most notably those with time-dependent Hamiltonians, simply cannot be treated using time-independent quantum mechanics. In contrast, there is no quantum phenomenon that cannot, in principle, be described using time-dependent quantum mechanics. It should therefore be possible to develop every major topic in quantum mechanics within a unified framework from the time-dependent point of view—including stationary states and energy resolved cross-sections. That is the approach of this book.

In addition to the above argument—that only a time-dependent approach has the potential for treating the many diverse topics in quantum mechanics within a single unifying framework—there are several additional compelling reasons to make the study of time-dependent quantum mechanics a major part of the standard curriculum. These reasons derive from three features of time-dependent quantum mechanics: its intuitive appeal, its convenience for semiclassical approximations, and its formal simplicity. It is worthwhile to expand on each of these aspects at greater length.

1. Time is a fundamental variable in virtually all areas of physics; the evolution of a particle or a wave with time is at the center of virtually all the fundamental equations of classical physics. Time is particulary important in applications of quantum mechanics to two of the most important processes in physical chemistry: photochemistry and reactive scattering. Any practicing organic chemist will tell you that a photochemical reaction proceeds from parent molecule to fragment as time progresses; and that in a chemical reaction, first reactants approach, then rearrange and then products separate, in that order. In the time-dependent wavepacket approach to reaction dynamics this aspect of a time sequence is preserved: the reactant wavepacket approaches the transition state, some of the wavepacket gets over the barrier and proceeds to products, while some is reflected back to the reactant

region. This fundamental description can be grasped by any bright undergraduate. In contrast, the time-independent theory of reactive scattering is technically quite formidable, and this intuitive picture of a chemical reaction progressing in time is lost.

2. Classical mechanics—Newton's laws—governs the dynamical behavior of macroscopic bodies all around us, and virtually everyone has a certain amount of intuition about the motion of classical trajectories. Traditionally, classical trajectories have played a central role in the study of chemical reaction dynamics and more recently in atomic physics. The central object in time-dependent quantum mechanics—the wavepacket—behaves very much like a classical trajectory. Its center, both in position and momentum, is often well described by a guiding classical trajectory, while its spatial width is simply a manifestation of the Heisenberg uncertainty principle. In other words, the correspondence principle between classical and quantum mechanics is fully operative in time-dependent quantum mechanics, often even at low energies, in contrast with the correspondence principle in time-independent quantum mechanics which is valid only at high quantum numbers. This provides a great aid to physical intuition and can be used as a starting point for developing a wide variety of semiclassical approximations.

3. In addition to being conceptually simpler for many processes, time-dependent quantum mechanics is actually formally simpler in many cases. Within time-dependent quantum mechanics, bound and free dynamics are treated formally in exactly the same way; in both cases a square integrable wavepacket describes the time-evolving state. This can lead to an enormous formal simplification. For example, the time-dependent theory of one- and two-photon absorption (or resonance Raman scattering) is completely unaffected by whether the excited eigenstates are bound or free. In contrast, the time-*independent* treatment employs square integrable wavefunctions to describe absorption to bound states, and non-normalizable scattering states to describe absorption to the continuum. In a similar vein, it is interesting to note that rigorous treatments of scattering theory all begin with time-dependent quantum mechanics because states are square normalizable; they then derive the more conventional and less intuitive time-independent expressions from there. In fact, in the treatment of scattering theory in Chapter 18 the fundamental formula for the S-matrix is expressed only in terms of square normalizable states. The widespread use of time-independent scattering theory apparently began because it provided the most convenient means to a numerical solution. There is reason to believe that all this has now changed: as a result of the remarkable strides in the numerical methods for solving the TDSE in the last fifteen years, time-dependent methods are now being used for state-of-the-art numerical work in reactive scattering. As another example of its formal advantages, using time-dependent quantum mechanics one can treat the interaction of a system with arbitrarily shaped time-dependent fields—either weak or strong—together with the internal time dependence of the system (e.g. electronic motion, phonon excitation or molecular vibrations), all within a single coherent framework. In contrast, time-independent quantum mechanics does not provide a convenient framework for treating the effect of arbitrarily shaped time-dependent fields.

This book has, therefore, three objectives:

1. To provide a unified treatment of quantum dynamical processes which keeps time-dependent quantum mechanics (TDQM) at center stage throughout. The formulation of virtually any experimental scattering or spectroscopic measurable—for time-dependent as well as continuous wave excitation—starting from a time-dependent approach provides a powerful connection between dynamics and experimental observables, and represents a new approach for a text.

2. To introduce a wide variety of applications of TDQM, in the text as well as in the exercises and the accompanying website (www.weizmann.ac.il/chemphys/tannor /Book), that have never been collected in one place before. These applications include many topics that are at the forefront of both experimental and theoretical research, including multiple pulse optical spectroscopy, control of molecular dynamics using specially designed laser pulse sequences, and the interpretation of spectra and scattering cross sections in terms of time correlation functions and features of the potential energy surface.

3. To present time-dependent quantum mechanics in a way that is accessible as early as possible in a student's training. This is attempted by developing the material in an intuitive and pictorial way, by interconnecting the material in different chapters such that the same phenomenon is viewed from multiple perspectives and with multiple formalisms, and via exercises and (classroom tested) computational projects which deepen and reinforce the understanding of the material in the text.

This book grew out of a one-semester graduate course (Quantum Mechanics III) which I taught in 1991 and 1994 at the University of Notre Dame, in 1993 at Columbia University and in 1996, 1998, 2000, 2003, and 2005 at the Weizmann Institute. Students who have had a course on the level of *Quantum Mechanics* by Cohen-Tannoudji et al. should be well prepared; with suitable selections the book may be appropriate for students who have studied from *Quantum Chemistry* by Ira Levine. Portions of the book can be used to supplement a course using *Molecular Reaction Dynamics* by Levine and Bernstein. It is hoped that the book will find a useful niche for students and researchers alike, interested in time-dependent quantum mechanics in general and in an introduction to the exciting new wavepacket approaches to molecular reaction dynamics in particular.

I gratefully acknowledge grants from the U.S. Office of Naval Research, the Israel Science Foundation, the U.S. National Science Foundation, the Sloan Foundation and the Birnbaum Foundation of South Bend. My thanks go out to the many colleagues and students at Notre Dame, Columbia and Weizmann who have made valuable suggestions and corrected errors in the text. Michael New typeset an early version of the book, starting with lecture notes from a course I gave at Columbia, and generated some of the figures. Einat Frishman, Efrat Rosenman, Anna Pomyalov and Yamit Maymon generated many of the other figures. Both Einat Frishman and Malki Cymbaliste were very helpful with the typesetting. The Matlab website (www.weizmann.ac.il/chemphys/tannor/Matlab_Programs) that accompanies the text was built by Yamit Maymon, Anna Pomyalov, and Shlomo Sklarz. Special thanks to them for an outstanding job. The Matlab programs in the website were written by Shlomo Sklarz, Efrat Rosenman, Einat Frishman, Yossi Elran, Jim Faeder, Yiftach Nevo, Nadav Katz, and Erez Boukobza.

The influence of Eric Heller on this book is enormous, as evidenced by the references to his work in most of the chapters. Had he published the book he had planned to there would probably be little need for this one. Ronnie Kosloff has also had an enormous impact on this work. He pioneered the numerical use of the Fourier method to solve the TDSE, a method which ushered in a new era in quantum dynamics calculations and which forms the basis for many of the computer programs on the accompanying website. He was also intimately involved with the original work on photon locking/heating and design of pulse sequences/optimal control which represent some of the most novel and interesting of the applications included here. Phil Pechukas taught me my first course in graduate quantum mechanics, and my treatment of linear algebra and some of the introductory material on wavepackets follow his presentation. Stuart Rice has contributed in numerous ways to the sections on pulse sequences and optimal control; other valued collaborators in this area include Pierre Gaspard, András Lörincz and my postdocs József Somlói and Vladimir Malinovsky. The research of my former students and postdocs, Sanjukta Das, Carl Williams, David Weeks, Sonya Garashchuk, Frank Grossmann and Daniela Kohen is the source of much of the material in the sections on the interaction representation, reactive scattering, semiclassical theory and the density operator, respectively. Vladimir Mandelshtam, Chris Meier, Ulrich Kleinekathöfer, Ignacio Martini, Maurice Schwartz, Jiri Vala, Shlomo Sklarz, Erez Boukobza, Yair Goldfarb, David Avisar, Danny Neuhauser, Uzy Smilansky, Doron Cohen, Nadav Katz, Adi Diner, Eli Pollak, Bob Wyatt, Aron Kuppermann, Jeremy Schiff, Jeff Cina, Paul Brumer, Jörn Manz, Daniel Lidar, Ilan Degani, Edriss Titi, Navin Khaneja, Ken Schafer, Bilha Segev, Tucker Carrington, Jr. and Kai Willner made many valuable comments on the manuscript. Lee Young did an expert job on copy editing. Jane Ellis and Bruce Armbruster of University Science Books have been enormously supportive and patient, and have demonstrated in every way a commitment to producing beautiful books. I am indebted to my father and father-in-law for their constant encouragement in this project, and my parents and sisters for their unending support. Finally, immeasurable thanks are due to my wife Vivian, and my four beautiful children Daniel, Philip, Clara and Shlomo, without whose constant companionship and support this book would never have been possible.

תושלב"ע

Structure of the Book and Teacher's Guide

This book is divided into three Parts: Part I, Pictures and Concepts; Part II, Formal Theory and Methods of Approximation; and Part III, Applications. Before getting into details, there are two major guidelines for using this book for teaching. First, for a one-semester course, the syllabus should be built almost entirely from Part I (Chapters 1–7) and Part III (Chapters 12–18), that is, most of the material in Part II (Chapters 8–11) should be skipped in a typical one-semester course. In many cases, the students will already have encountered the material in Chapter 8, on Linear Algebra and Quantum Mechanics. Chapters 9–11 are somewhat more specialized than the rest of the material in the book. The only portions of Part II that are essential for Part III are the interaction and Heisenberg pictures (Section 9.1), time-dependent perturbation theory (Section 9.2), and perhaps portions of Chapter 11 on numerical methods, depending on whether the students will be doing numerical projects. The second guiding principle is that wherever possible, the more advanced and/or more specialized material has been put at the end of each chapter. Thus, when designing a syllabus, a general rule of thumb would be to cut perhaps the last third from each chapter, although this is very chapter dependent and dependent on the interests of the class and the teacher.

Now to the contents of the various Parts. Part I, Pictures and Concepts, focuses on pictures and representations of the time evolving *wavepacket*. Wherever possible I have emphasized classical–quantum correspondence, in order to make the material as intuitive as possible. I have tried to present a wide variety of different handles on this correspondence, from the stationary phase method (Chapter 2) to analytic properties of Gaussian wavepackets (Chapter 3) , from Ehrenfest's theorem to the geometrical optics limit of quantum mechanics and wavepacket revivals (Chapter 4), to phase space pictures of the Wigner distribution (Chapter 5). In the later chapters of Part I, the concept of a wavepacket time-correlation function and its Fourier transform is introduced, first to obtain spectra and eigenfunctions (Chapter 6) and then to obtain reflection and transmission coefficients (Chapter 7). Chapter 6 contains an extensive, self-contained introduction to Fourier theory, with many pictorial illustrations. The concept of the Fourier transform of a time-correlation function will be a recurring theme under Applications (Part III); it appears there in the formulation of electronic absorption and emission spectroscopy (Chapter 14), Raman spectroscopy (Chapter 14), photodissociation (Chapter 17) and reactive scattering (Chapter 18). However, at this point the correlation function and the spectrum are viewed simply as an intrinsic property of the wavepacket, alternative *representations* of the wavepacket, independent of any particular application.

While Part I focuses on wavepacket evolution, Part II focuses on alternative representations of, and approximations to, the *evolution operator* or propagator. Of course, once the evolution operator is known the wavepacket which evolves from any initial state can be obtained immediately. Nevertheless, the isolation of the evolution operator and the study of its properties and approximations is more formal and more abstract than the concrete, readily visualized, time-evolving wavepacket. Several aspects of Part II deserve special mention. Chapter 8 is a self-contained introduction to linear algebra and the formalism of quantum theory. This chapter covers essentially the same material as in the corresponding section of *Quantum Mechanics* by Cohen-Tannoudji et al., perhaps a little more succinctly; it is included here to make the book self-contained, and accessible to students with an under-graduate course on the level of say *Quantum Chemistry* by I. Levine. This chapter could potentially be used for a half or a third of a course by itself. Logically, one could argue that this chapter should be taught at the beginning of the course, before Part I, since some of the mathematics used in Part I, such as Fourier transforms, Dirac δ functions and the operator $e^{-iHt/\hbar}$, is introduced without much discussion. However, I believe it is easier to draw students in with the visually appealing wavepacket dynamics, leaving a few unsettled mathematical questions, and then come to the full mathematical framework when the student is motivated.

Chapter 9 contains a discussion of dynamical representations and approximations, including the standard topics of the Heisenberg and interaction pictures and time-dependent perturbation theory. I have also included material that is not in standard textbook treatments, including the Magnus expansion, the Wei–Norman factorization, Floquet theory, the quantum adiabatic theorem and the geometric phase, and time-dependent variational methods. The interaction picture is used to derive the adiabatic theorem and its non-adiabatic corrections, leading to a heuristic derivation of the Landau–Zener formula.

Chapter 10 deals with semiclassical approximations. This chapter begins with a discussion of the classical action, and the first and second variations of the action. Then a brief, self-contained introduction to path integrals is given, followed by a derivation of the van Vleck–Gutzwiller (VVG) semiclassical limit of the propagator. The VVG propagator is then placed in the more general context of Miller's semiclassical algebra. Finally, the initial value representation (IVR) is introduced, along with the coherent state basis. These latter two representations allow a rigorous connection to be made between Gaussian wavepackets and the VVG propagator. To contrast the material in this chapter with the material on classical–quantum correspondence in Chapter 4, the focus here is the propagator and $\hbar \to 0$ approximations to it. Of course, there are connections with the earlier material on classical–quantum correspondence, in particular the stationary phase method, Gaussian wavepackets, and the geometrical optics limit of quantum mechanics, and these connections are noted.

Finally, Chapter 11 deals with numerical methods for solving the TDSE. I try to get to this material by the middle of the course because it provides a basis for understanding the computer programs that the students will use to carry out their projects in the second half of the course. Insofar as possible, the theory of numerical methods is developed in terms of projections of the infinite Hilbert space onto a finite Hilbert space. The mathematics of projection operators and unitary transformations, introduced in Chapter 8, is used here to integrate the treatment of numerical methods with the rest of the book. The guiding principle is that numerical methods should not be just a bag of tricks, or even based on some separate body of numerical analysis, but should emerge from the quantum mechanics itself as a well-controlled, and perhaps even optimal approximation.

Both semiclassical methods (Chapter 10) and numerical methods (Chapter 11) are areas of active research, with many new papers published each year. Unfortunately (or perhaps fortunately for the young reader!) there are just too many open questions in these areas to present a tidy package. I have tried to keep three goals in mind: to give an introduction to key concepts and general results (bearing in mind that this is a book about wavepackets, not semiclassical or numerical methods); to provide exposure to a wide variety of different methods and approaches (from path integration to Gaussian quadrature); and to develop the material wherever possible from the basic conceptual principles introduced earlier in the book (i.e. well-characterized or not-so-well-characterized approximations to the propagator). These chapters should leave the inquisitive reader with many questions, and perhaps novel ideas and connections. Each of these chapters contains many references to current research literature, and should set the ambitious student on his or her way.

Part III, Applications, deals in general terms with the different possible *projections* of the time-evolving wavepacket, and the recovery of both time- and energy-resolved observables. For completeness, I have included a chapter called Introduction to Molecular Dynamics (Chapter 12), which introduces the basic concepts of the Born–Oppenheimer approximation, potential energy surfaces, normal modes, transition state theory, and optical transitions in molecules, concepts that will appear in the subsequent applications. There is some discussion of molecular symmetry, with emphasis on understanding the molecular point group within the larger setting of the molecule's permutation-inversion group.

The applications *per se* begin in Chapter 13, starting with the interaction of a molecule with one or several ultrashort (femtosecond) laser pulses. This is one of the simplest applications of time-dependent perturbation theory (Chapter 9); at the same time it immediately introduces the reader to an extremely interesting and active area of research in chemical physics. The theoretical interpretation of "femtochemistry" experiments becomes immediately transparent by combining the formal expressions from first- and second-order time-dependent perturbation theory (Chapter 9) with the Born–Oppenheimer adiabatic basis (Chapter 12) and the concepts of wavepacket dynamics from Part I. Also included in this chapter is a discussion of nonlinear optical spectroscopy ($P^{(3)}$), both in the wavefunction and the density operator formulations. Within the context of $P^{(1)}$ we introduce linear response theory and derive the formula for the absorption spectrum.

Chapter 14 deals with the interaction of single frequency ("continuous wave" or CW) optical radiation with molecules. Experimental observables such as one-photon absorption and emission spectra and two-photon resonance Raman spectra are formulated in terms of the Fourier transform of a wavepacket time-correlation function at the optical excitation frequency. This powerful class of formulations has been used extensively in the last two decades to give physical insight into the width, structure, and intensities of experimental spectra in terms of the underlying wavepacket dynamics. The central formulas are derived here starting from the general first- and second-order perturbation theory expressions (Chapters 9 and 13), with the perturbations taken as single-frequency excitation. The first-order formula is shown to be identical to that derived using linear response theory in Chapter 13.

Chapter 15 deals with strong-field excitation, beginning with the two-level system. The importance of the two-level system is that it can be solved analytically almost completely, and hence provides intuition about more complicated systems. First, the Rabi solution for the time-dependent wavefunction is derived for CW excitation. Next, the optical Bloch equations for the density operator are formulated, along with the Feynman–Vernon–Hellwarth (FVH) geometrical picture. The FVH picture allows one to visualize the solution to the two-level system even for complicated pulse sequences where analytical solutions are not available. Although the Rabi solution and the FVH picture are treated quite well in several texts on nonlinear optics, they are developed here within the broader framework of wavepacket dynamics. The concept of dressed states is then introduced—states that are solutions of the Schrödinger equation in the presence of the field. This chapter also contains a discussion of stimulated Raman adiabatic passage (STIRAP), as well as a useful generalization of the optical Bloch equations from a two-level system to two electronic potential energy surfaces. Nonadiabatic corrections to STIRAP are calculated by applying the general non-adiabatic methodology of Chapter 9.

Chapter 16 deals with the very exciting new field of designing optical pulse sequences to create a desired chemical product. The first section, based on second-order perturbation theory, introduces a pump–dump sequence of two pulses to control reaction products. Next, a fascinating variational formulation is presented to find optimal pulse sequences that may have arbitrarily complicated time dependence. The close connection between this variational formulation and Optimal Control Theory, widely used in engineering applications, is made explicit. Additional topics in this chapter include controllability theory, femtosecond interfering pathway control, chirped excitation, and learning algorithms.

Chapters 17 and 18 deal with photodissociation and reactive scattering, respectively. Photodissociation is treated before reactive scattering for obvious reasons: photodissociation is a half-collision process and is simpler than the full reactive scattering process. Expressions for the partial and total photodissociation cross sections are derived in terms of wavepacket dynamics. The total cross section in particular is expressed in terms of the Fourier transform of a wavepacket correlation function, much like the expression for the one-photon absorption spectrum.

Although Chapter 17 builds heavily on the earlier chapters, two new ideas are required. The first is the concept of multiple degeneracy of the eigenstates of the asymptotic Hamiltonian. This degeneracy is concretized by viewing degenerate continuum states as superpositions of time evolving wavepackets with well-defined asymptotic boundary conditions, using the spectral method of Chapters 6 and 7. The second new concept is that of the Møller operators, closely related to the interaction picture of Chapter 9. By the end of Chapter 17 the reader will be familiar with much of the technical notation of scattering theory and its dynamical interpretation, and will be well poised to learn the material in Chapter 18.

The discussion of reactive scattering in Chapter 18 is in some sense the climax of the book. All the new elements introduced in Chapter 17 will be needed here, but doubled, once for reactants and once for products. Having learned about chemical arrangement channels (Chapters 12 and 16), having understood the structure of the asymptotic degeneracy and the concept of the Møller operators (Chapter 17), and having learned about the Fourier transform of wavepacket correlation functions (Chapters 6–7 and Chapter 14), these three major ideas are brought together. The full collision process is then expressed in a way that looks like a half collision in the reactant and the product arrangement channels, "matched" together in the transition state region via the Fourier transform of a wavepacket cross-correlation function. This formulation of reactive scattering has several attractive features: it highlights the role of the transition state region, it provides a time domain interpretation of the structure in the S-matrix element as a function of energy, and it manifestly illustrates the principle of detailed balance—the principle that if the identity of reactants and products is interchanged and time is reversed the identical value of the S-matrix element is obtained.

In the final section, related expressions are derived for the total reaction probability (i.e. the reaction probability summed over all initial or all final states) and the cumulative reaction probability at energy E (i.e. the reaction probability summed over both initial and final states).

The book contains a not-insignificant number of new derivations, ideas or discussion, that have not been published previously. For example, Chapter 2: the use of momentum space to calculate the center and width of the free particle wavepacket; Chapter 3: aspects of the discussion of saddle point vs. stationary phase integration; Chapter 4: the discussion of complex Bohmian dynamics and the classical limit; Chapter 5: the relationship of the purity of the density matrix, the uncertainty principle and positivity; Chapter 7: a) the formulation of reflection and transmission coefficients in terms of the Fourier transform of wavepacket correlation functions; b) the calculation of S-matrix elements via the explicit overlap of scattering eigenstates; Chapter 11: the use of projection operators to define pseudospectral methods, leading to a newly developed derivation of Gaussian quadrature; Chapter 12: a) the use of canonical transformations to obtain the G-matrix and skewed coordinates, resulting in a formula relating the G-matrix and the skew angle; b) a method for visualizing hyperspherical coordinates on a sphere.

Even when the material is not new, in many cases it has been synthesized from widely different sources. Examples of this synthesis are: Chapter 6: the unified treatment of Fourier transform, Fourier series and discrete Fourier transform Chapter 9: use of the structure of the equations of motion in the interaction picture to derive non-adiabatic corrections to the quantum adiabatic theorem, leading to a heuristic derivation of the Landau-Zener formula. The approach is taken up again in Chapter 15 to calculate nonadiabatic corrections to STIRAP; Chapter 10: integrated treatment of path integrals, the van Vleck propagator and Gaussian wavepackets; Chapter 13: unified treatment of femtochemistry, linear response theory and nonlinear spectroscopy, beginning from a wavepacket interpretation of time-dependent perturbation theory; Chapter 14: integrated treatment of absorption, emission, transition state spectroscopy and resonance Raman spectroscopy, with careful attention to the prefactors and the dimensional analysis; Chapter 15: synthesis of the Rabi solution, the FVH representation and the dressed states picture for strong fields; Chapter 16: integrated discussion of the pump-dump scheme for quantum control, interfering pathways, calculus of variations, optimal control theory, controllability and learning algorithms; Chapters 17–18: unified treatment of photodissociation and reactive scattering using wavepackets and Møller operators.

The book contains an extensive series of exercises, designed to help the student understand the interconnection between topics. In some cases the reader is asked, for pedagogical purposes, to solve the same problem using time-dependent and time-independent quantum mechanics. Getting the same answer using these two different approaches at once solidifies an understanding of the specific solution as well as the general relationship between the TISE and the TDSE. Wherever applicable, the reader is asked to apply the general principle of the chapter to the Gaussian wavepacket in the harmonic oscillator. In this way, the reader gains an appreciation of the connections between the chapters via the different perspectives offered on the same specific example. Accompanying the book is a library of Matlab programs to supplement the different chapters of the text (www.weizmann.ac.il/chemphys/tannor/Book). Virtually all these programs solve the TDSE numerically in one dimension, producing real time movies of the wavepacket while calculating the observables of interest. These programs serve two purposes. The first is the numerical verification of the analytical wavepacket results (e.g. for the harmonic oscillator). This provides the student with additional confidence in both the analytical solutions and the numerical methods. The second is to give the student a chance to explore a wide variety of applications of TDQM in a hands-on way, with convenient control of the input parameters. The applications include calculating wavepacket correlation functions and spectra, reflection and transmission coefficients, Wigner phase space movies of the evolving wavepacket, multiple pulse wavepacket excitation, STIRAP, coherent control, and the calculation of eigenstates from wavepacket dynamics via the spectral method, the relaxation method and by quantum adiabatic switching. The versatility of the time-dependent method given a short generic propagation routine is impressive indeed.

A word about notation. The usual designation of the position variable in this book is the letter x. However, in certain sections, where classical–quantum correspondence is being emphasized, the letter q is used instead for position. Where the letter q is used the reader will not have to look far to find the conjugate variable p. It is hoped that this notational nuance will heighten the reader's awareness of the points at which an underlying classical structure is manifesting itself in the quantum dynamics.

With this introduction, I welcome the reader to the wonderful world of time-dependent quantum mechanics!

Part I

Pictures and Concepts

Chapter 1

The Time-Dependent Schrödinger Equation

The central object of study in this book is the Time-Dependent Schrödinger Equation,

$$i\hbar \frac{\partial}{\partial t} \Psi(\mathbf{x}, t) = H \Psi(\mathbf{x}, t), \tag{1.1}$$

where $i = \sqrt{-1}$, $\hbar$ is Planck's constant divided by 2π, equal to 1.05450×10^{-27} erg sec, and H is the Hamiltonian operator:

$$H = -\frac{\hbar^2}{2m} \nabla^2 + V. \tag{1.2}$$

The potential function $V = V(\mathbf{x}, t)$ in general can depend on both coordinates and time; however, we will assume that $V = V(\mathbf{x})$, independent of time unless otherwise specified. Note that Eq. 1.2 is for a single particle and nonrelativistic. For simplicity we will work with the one-dimensional form of the Schrödinger equation unless otherwise specified, so that $\mathbf{x} \rightarrow x$ and $\nabla^2 \rightarrow \frac{\partial^2}{\partial x^2}$.

The function $\Psi(x, t)$, which satisfies this equation, is called the wavefunction. According to the conventional interpretation of quantum mechanics, $|\Psi(x, t)|^2 \, dx$ is the probability of finding the particle between x and $x + dx$, assuming that $\Psi(x, t)$ is normalized, that is, $\int_{-\infty}^{\infty} |\Psi(x, t)|^2 \, dx = 1$. $|\Psi(x, t)|^2$ is called the *probability density* for finding the particle at position x at time t. Given $\Psi(x, 0)$ if we can determine $\Psi(x, t)$ for all positions and all times we have a complete solution of the problem. The function $\Psi(x, t)$ is all that is in principle knowable about the position of a particle—before a measurement is made. The fact that there is a probability distribution associated with the outcome of the measurement of x is a manifestation of the uncertainty principle, to be discussed more fully below.

In introductory classes, the Time-Dependent Schrödinger Equation (which we will refer to by the abbreviation TDSE) is usually solved by a separation of variables in position and time. We will begin with this standard approach, but first offer a few remarks. The method of separation of variables leads naturally to *particular* solutions of the TDSE, which have the counterintuitive property of predicting time-independent observables. This point has been a conceptual pitfall for generations of students, who have been left to wonder where time disappeared in quantum mechanics. The resolution to this paradox is to note that the *general* solution of the TDSE is an arbitrary superposition of the particular solutions. Only by considering such superpositions can one observe time-dependent behavior in quantum mechanics and recover some measure of contact with our classical mechanical intuition.

A central tenet of this book is that these superposition solutions, or *wavepackets*, far from being anomalous, are both ubiquitous in nature as well as being conceptually more transparent than the particular solutions.

1.1   Separation of Variables and Reconstitution of the Wavepacket

The method of separation of variables assumes that solutions of the TDSE exist that can be written in the product form: $\Psi(x, t) = \psi(x)\chi(t)$. Substituting this trial form into Eq. 1.1 and dividing both sides by $\psi(x)\chi(t)$ yields

$$i\hbar\frac{\dot{\chi}(t)}{\chi(t)} = \frac{H\psi(x)}{\psi(x)}. \tag{1.3}$$

The left-hand side of Eq. 1.3 is a function only of time while the right-hand side is a function only of position. Thus both sides must be equal to a constant; call it E for energy. We then obtain two equations, one for $\psi(x)$ and one for $\chi(t)$:

$$i\hbar\dot{\chi}(t) = E\chi(t) \tag{1.4}$$

$$H\psi_E(x) = \left\{-\frac{\hbar^2}{2m}\frac{\partial^2}{\partial x^2} + V(x)\right\}\psi_E(x) = E\psi_E(x). \tag{1.5}$$

Equation 1.4 may be readily solved to yield

$$\chi(t) = \chi_0 e^{-\frac{i}{\hbar}Et}. \tag{1.6}$$

Equation 1.5 is the Time-*Independent* Schrödinger Equation (henceforth TISE) and depends on the specific form of the potential $V(x)$. This equation is an "eigenvalue" equation; the constant E is called the eigenvalue and the wavefunction $\psi_E(x)$ is called the eigenfunction or eigenstate. For bound potentials, physically meaningful solutions exist only at special, discrete values of E; for unbound potentials physically meaningful solutions exist over a continuous range of E. We will refer to the set of energies supporting physically meaningful solutions as the "spectrum." In most cases, Eq. 1.5 cannot be solved analytically. We may therefore write a solution to the Time-Dependent Schrödinger Equation as

$$\Psi(x, t) = \psi_E(x)\chi_0 e^{-\frac{i}{\hbar}Et}. \tag{1.7}$$

Note that χ_0 is an overall multiplicative factor, and as such may be absorbed into $\psi_E(x)$. The magnitude of the latter is determined by the Time-Independent Schrödinger Equation only to within an overall multiplicative factor, which is usually chosen so that the normalization condition,

$$\int_{-\infty}^{\infty} |\Psi(x, t)|^2 \, dx = 1,$$

is fulfilled, and therefore we may write without loss of generality

$$\Psi(x, t) = \psi_E(x)e^{-\frac{i}{\hbar}Et}. \tag{1.8}$$

Note that Eq. 1.8 has a rather trivial time dependence. If we consider the probability density $|\Psi(x, t)|^2$, determined by Eq. 1.8, we obtain

$$|\Psi(x, t)|^2 = (\psi_E(x)e^{-\frac{i}{\hbar}Et})^*\psi_E(x)e^{-\frac{i}{\hbar}Et} = |\psi_E(x)|^2, \tag{1.9}$$

independent of t! If the probability density does not change with time, there is no time dependence to where the particle is likely to be found—there is effectively no motion. Yet quantum mechanics has to be able to describe motion since the microworld it is intended to describe is constantly undergoing change. The resolution to this paradox is to realize that Eq. 1.7 is what is known as a "particular" solution of the TDSE. In general, any linear combination of particular solutions is also a solution of the TDSE. It is by forming linear combinations of particular solutions, each with its own characteristic time-dependent phase factor, that we obtain time dependence in the probability density. Consider the simplest example, a linear combination consisting of just two of the particular solutions:

$$\Psi(x, t) = a\psi_E(x)e^{-\frac{i}{\hbar}Et} + b\psi_{E'}(x)e^{-\frac{i}{\hbar}E't}. \tag{1.10}$$

Then the probability density is given by

$$|\Psi(x, t)|^2 = |a|^2|\psi_E(x)|^2 + |b|^2|\psi_{E'}(x)|^2 + 2\text{Re}\left\{a^*b\psi_E^*(x)\psi_{E'}(x)e^{-i\frac{(E'-E)t}{\hbar}}\right\}. \tag{1.11}$$

Note that Eq. 1.11 has three terms: one from ψ_E alone, one from $\psi_{E'}$ alone, and a third term that arises from an interference term between the two. This interference term is a result of having a superposition of eigenstates with different energies—a wavepacket. All the time dependence of $|\Psi(x, t)|^2$ is contained in this interference term, and hence we can remark quite generally that a *wavepacket*—a superposition of states having different energies—is required in order to have a time dependence in the probability density and in other observable quantities, such as the average position or average momentum of the particle.

We can write the *general* solution to the TDSE, expressed in terms of the particular solutions, as follows:

$$\Psi(x, t) = \sum_{n=1}^{\infty} a_n\psi_n(x)e^{-\frac{i}{\hbar}E_nt} \quad \text{(discrete spectrum)} \tag{1.12}$$

$$= \int_0^{\infty} a(E)\psi_E(x)e^{-\frac{i}{\hbar}Et}\, dE \quad \text{(continuous spectrum).} \tag{1.13}$$

(For systems with both a discrete and a continuous portion to their spectrum the general solution is a sum of Eq. 1.12 and Eq. 1.13.) The important lesson is that in order to get a nontrivial time dependence of any observable, after solving the TISE one must remember to "reconstitute" the wavepacket, that is, to superpose the solutions of the TISE, each with its own amplitude and phase factor.

1.2 Expectation Values

In quantum mechanics, all experimental observables are associated with the class of operators known as Hermitian operators. The average value of an experimental observable, given that the experiment is repeated many times using indistinguishable systems, is simply each possible outcome weighted by the probability of obtaining that outcome. This average is

Table 1.1 Some of the most common observables in quantum mechanics, the corresponding operators, and the effect of each of the operators on a wavefunction

Observable	Operator	Effect on a wavefunction
Position	$\hat{x}$	$x\Psi(x, t)$
Momentum	$\hat{p}$	$-i\hbar\frac{\partial}{\partial x}\Psi(x, t)$
Potential energy	$\hat{V}$	$V(x)\Psi(x, t)$
Kinetic energy	$\hat{T} = \frac{\hat{p}^2}{2m}$	$-\frac{\hbar^2}{2m}\frac{\partial^2}{\partial x^2}\Psi(x, t)$
Hamiltonian	$\hat{H} = \hat{T} + \hat{V}$	$\left\{-\frac{\hbar^2}{2m}\frac{\partial^2}{\partial x^2} + V(x)\right\}\Psi(x, t)$

known as the "expectation value," and may be represented in terms of the operator $\hat{A}$ that corresponds to the observable and the wavefunction $\Psi(x, t)$ as follows:

$$\langle A \rangle = \frac{\int_{-\infty}^{\infty} \Psi^*(x, t)\hat{A}\Psi(x, t)\,dx}{\int_{-\infty}^{\infty} \Psi^*(x, t)\Psi(x, t)\,dx}. \tag{1.14}$$

If $\Psi(x, t)$ is normalized, the expression for the expectation value takes the simpler form:

$$\langle A \rangle = \int_{-\infty}^{\infty} \Psi^*(x, t)\hat{A}\Psi(x, t)\,dx. \tag{1.15}$$

A full discussion of Hermitian operators, as well as a proof of Eq. 1.14, is deferred until Chapter 8. However, for our purposes here and the rest of Part I, the most commonly encountered Hermitian operators and their effect on wavefunctions of the form $\Psi(x, t)$ are summarized in Table 1.1.

Consider now the expectation value of $\hat{x}$ as a function of time:

$$\langle \hat{x} \rangle_t = \int \Psi^*(x, t)\,\hat{x}\,\Psi(x, t)\,dx. \tag{1.16}$$

For the wavefunction of Eq. 1.7,

$$\langle \hat{x} \rangle_t = \int \psi_E^*(x)e^{\frac{i}{\hbar}Et}\,x\,\psi_E(x)e^{-\frac{i}{\hbar}Et}\,dx$$

$$= \int \psi_E^*(x)\,x\,\psi_E(x)\,dx \tag{1.17}$$

$$= \langle \hat{x} \rangle_0$$

and thus the expectation value is independent of time.

Now consider the case that $\Psi(x, t)$ is a superposition of two different specific solutions, Eq. 1.10. Repeating the calculation of $\langle x \rangle_t$ using this wavefunction yields

$$\langle x \rangle_t = |a|^2 \int x|\psi_E(x)|^2\,dx + |b|^2 \int x|\psi_{E'}(x)|^2\,dx$$

$$+ 2\text{Re}\left\{a^*b \int x\psi_E^*(x)\psi_{E'}(x)\,dx\, e^{-i\frac{(E'-E)t}{\hbar}}\right\}. \tag{1.18}$$

As in the calculation of the probability density above, all the time dependence of $\langle \hat{x} \rangle_t$ is contained in the interference term. Quite generally, to obtain time dependence in *any* experimental observable it is necessary to have a superposition of states with different energies.

1.3 A Worked Example: Particle in Half a Box

As an example of an analytically solvable model, consider a particle in a box that extends from 0 to L. Assume the particle is in the lowest level of the box. At $t = 0$ the size of the box is suddenly expanded, so that the box extends from 0 to $2L$ as in Fig. 1.1. The object is to calculate the smallest period of time, τ, at which $\Psi(x, \tau) = \Psi(x, 0)$, and to draw a picture of $\Psi(x, t)$ at $t = \tau/2$.

At $t = 0$, $\Psi(x, 0)$ is a normalized eigenstate of the old box $[0, L]$:

$$\Psi(x, 0) = \sqrt{\frac{2}{L}} \sin\left(\frac{\pi x}{L}\right).$$

Note that the eigenfunctions $\psi_n(x)$ and eigenvalues E_n of the new box $[0, 2L]$ are

$$\psi_n(x) = \sqrt{\frac{1}{L}} \sin\left(\frac{n\pi x}{2L}\right) \qquad E_n = \frac{n^2\pi^2\hbar^2}{8mL^2} \quad n = 1, 2, \ldots$$

To calculate the time evolution of Ψ we expand $\Psi(x, 0)$ in a complete set of the *new* eigenfunctions, $\psi_n(x)$, since the time dependence of each of the new eigenfunctions is just a simple phase factor under the new Hamiltonian. The expansion takes the form

$$\Psi(x, 0) = \sum_n a_n \psi_n(x),$$

where

$$a_n = \int_0^L \Psi(x, 0)\, \psi_n^*(x)\, dx = \int_0^L \left\{ \sqrt{\frac{2}{L}} \sin\left(\frac{\pi x}{L}\right) \right\} \left\{ \sqrt{\frac{1}{L}} \sin\left(\frac{n\pi x}{2L}\right) \right\} dx.$$

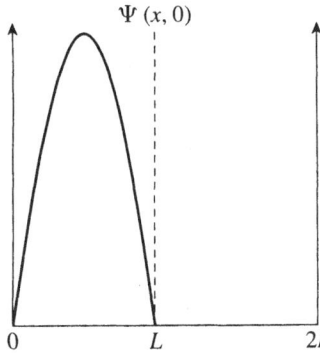

Figure 1.1 Initial conditions for particle in half a box

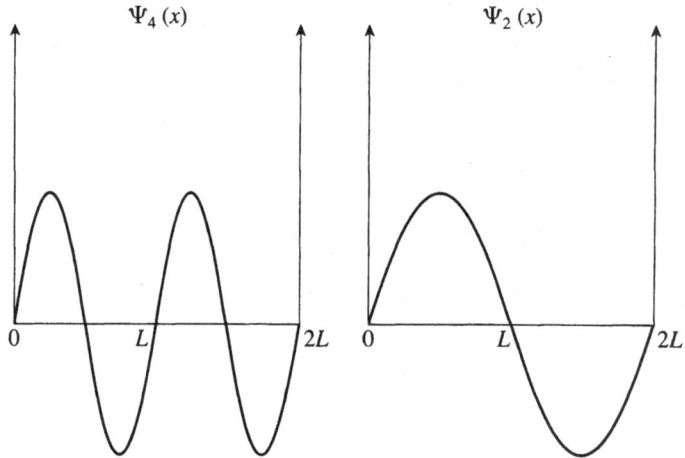

Figure 1.2 $\Psi(x, 0)$ is orthogonal to all $\psi_n(x)$ if n is even. This is because $\Psi(x, 0)$ vanishes outside of the range $[0, L]$, and in that range all the even eigenstates of the full box correspond, within a normalization factor, to excited eigenstates of the half box. This is shown in the left panel for $n = 4$. The one exception is the $n = 2$ state, shown in the right panel, which is essentially a replica of $\Psi(x, 0)$ in the range $[0, L]$. Note that we use capital Ψ for the time-evolving state and small ψ_n for the eigenstates.

Before doing this integral, notice that $a_n = 0$ for all even n, except $n = 2$, by symmetry (see Figure 1.2). Since $n = 2$ is the only state of its symmetry with nonvanishing coefficient, we calculate this overlap integral first:

$$a_2 = \int_0^L \left\{ \sqrt{\frac{2}{L}} \sin\left(\frac{\pi x}{L}\right) \right\} \left\{ \sqrt{\frac{1}{L}} \sin\left(\frac{\pi x}{L}\right) \right\} dx = \frac{1}{\sqrt{2}}.$$

To find a_n, n odd, we use the formula

$$\int_0^a \sin(bx) \sin(cx) \, dx = -\frac{1}{2} \left[\frac{\sin[(b+c)a]}{b+c} - \frac{\sin[(b-c)a]}{b-c} \right].$$

Setting $a = L$, $b = \frac{\pi}{L}$, $c = \frac{n\pi}{2L}$, we find

$$a_n = -\sqrt{2} \left[\frac{(-1)^{\frac{n+1}{2}}}{(n+2)\pi} + \frac{(-1)^{\frac{n-1}{2}}}{(n-2)\pi} \right]$$

$$= \frac{4\sqrt{2}(-1)^{\frac{n+1}{2}}}{(n+2)(n-2)\pi},$$

where we have used the fact that

$$\sin\left(\pi + \frac{n\pi}{2}\right) = (-1)^{\frac{n+1}{2}}; \qquad \sin\left(\pi - \frac{n\pi}{2}\right) = (-1)^{\frac{n-1}{2}}.$$

The numerical values for the first few overlap integrals are

$$a_1 = \frac{-4\sqrt{2}}{(3)(-1)\pi} = .600 \qquad a_5 = \frac{-4\sqrt{2}}{(7)(3)\pi} = -.086$$

$$a_2 = \frac{1}{\sqrt{2}} = .707 \qquad a_7 = \frac{4\sqrt{2}}{(9)(5)\pi} = .040$$

$$a_3 = \frac{4\sqrt{2}}{(5)\pi} = .360 \qquad a_9 = \frac{-4\sqrt{2}}{(11)(7)\pi} = -.023.$$

As a check, note that $\sum_{n=1}^{9} |a_n|^2 = .999 \cong 1$.

We now combine the eigenstates with their phase factors and their coefficients to analyze the time dependence of the wavepacket. Recall that

$$\Psi(x, t) = \sum_n a_n \sqrt{\frac{1}{L}} \sin\left(\frac{n\pi x}{2L}\right) e^{-i\frac{n^2\pi^2\hbar}{8mL^2}t}$$

$$= \sum_n a_n \sqrt{\frac{1}{L}} \sin\left(\frac{n\pi x}{2L}\right) e^{-in^2\omega_1 t},$$

where we have defined $\omega_1 = \frac{\pi^2\hbar}{8mL^2}$. To find the fundamental period, note that when $\omega_1 t = 2\pi$, then $n^2\omega_1 t$ is also an integer multiple of 2π. Therefore, at $t = \tau = \frac{2\pi}{\omega_1}$ the phase factor for *all* n is equal to 1, and so $\Psi(x, \tau) = \Psi(x, 0)$. Thus, $\tau = \frac{2\pi}{\omega_1} = \frac{16mL^2}{\hbar\pi}$ is the fundamental period.

Before drawing $\Psi\left(x, \frac{\tau}{2}\right)$, we ask: what makes $\Psi(x, 0)$ cancel on the right and not on the left? We know that $\sum_{n \text{ odd}} a_n\psi_n(x)$ must be symmetric (since it is the sum of symmetric functions); hence, we may infer that it must look as shown in Figure 1.3 to give cancellation on the right-hand side at $t = 0$.

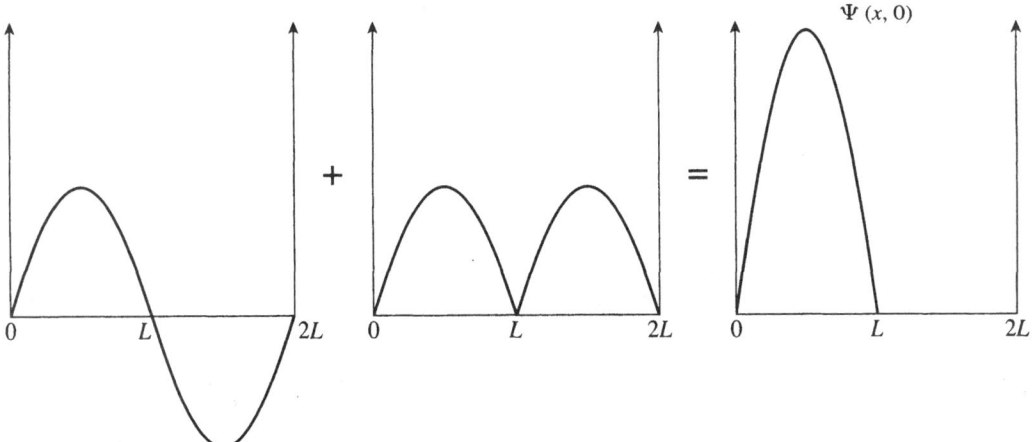

Figure 1.3 The component of ψ_2 (left panel) must be equal to that of all the ψ_n, n odd, put together (middle panel), to give precise cancellation in the amplitude of $\Psi(x, 0)$ on the right-hand side of the box at $t = 0$ (right panel). Note that the amplitude in the box on the right is twice that of the boxes on the left.

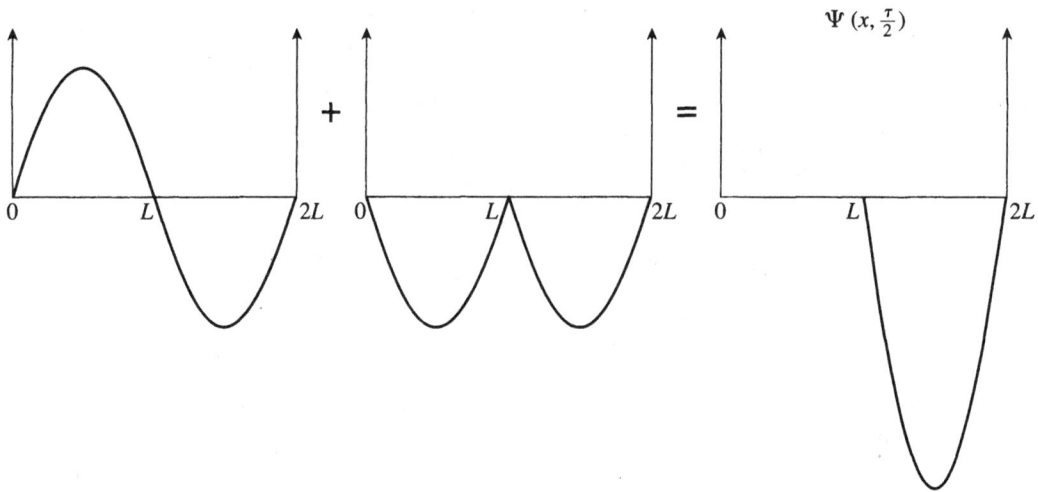

Figure 1.4 After half a period, the phase of ψ_2 is back to its original value (left), while the phase of all the ψ_n, n odd, has changed by (-1) (middle). This leads to complete cancellation of $\Psi(x, \frac{\tau}{2})$ on the *left-hand side of the box* (right panel). Note that the negative sign of the amplitude has no consequence for the probability density, which is proportional to $|\Psi|^2$. As in Figure 1.3, the amplitude in the box on the right is twice that of the boxes on the left.

We now ask what happens after half a period. When $t = \frac{\tau}{2}$,

$$\Psi(x, t) = \sum_n a_n \sqrt{\frac{1}{L}} \sin\left(\frac{n\pi x}{2L}\right) e^{-in^2\pi}.$$

For all odd n, $a_n e^{-in^2\pi} = -a_n$, while for $n = 2$, $a_n e^{-in^2\pi} = a_n$. Since all the odd ψ_n change sign and ψ_2 remains unchanged, $\Psi(x, t = \frac{\tau}{2})$ must look as shown in Figure 1.4, that is, the wavepacket is localized on the right-hand side of the box and cancels completely on the left-hand side.

Further Reading and Historical Notes

Schrödinger's original papers have been translated into English and collected (Schrödinger, 1928). It is interesting to note that Schrödinger started with the time-*independent* Schrödinger equation, and obtained the time-dependent Schrödinger equation only in the fourth paper in his series (Schrödinger, 1926a, Part IV). In (Schrödinger, 1926b), Schrödinger explores the time dependence of a displaced Gaussian in a harmonic potential—what would be called a "coherent state" in modern terminology, which we will discuss at length in Chapter 3. For a discussion of the introduction of time in Schrödinger's original papers and a provocative analysis, see Briggs (2000, 2001).

References

General

1. A. Messiah, *Quantum Mechanics*, vol. I (Wiley, New York, 1958).
2. C. Cohen-Tannoudji, B. Diu and F. Laloë, *Quantum Mechanics*, vol. I (Wiley, New York, 1977).
3. R. P. Feynman, R. B. Leighton and M. Sands, *The Feynman Lectures on Physics, vol. III: Quantum Mechanics* (Addison-Wesley, Reading, MA, 1965).
4. L. D. Landau and E. M. Lifshitz, *Quantum Mechanics, Nonrelativistic Theory* (Pergamon Press, Oxford, 1965).
5. L. I. Schiff, *Quantum Mechanics* (McGraw-Hill, New York, 1968).

Schrödinger's Original Papers (titles from English translation)

6. E. Schrödinger, Quantisation as a problem of proper values (Part I), Annalen der Physik 79, 361 (1926); (Part II), Annalen der Physik 79, 489 (1926); (Part III), Annalen der Physik 80, (1926); (Part IV), Annalen der Physik 81, 109 (1926).
7. E. Schrödinger, The continuous transition from micro- to macro-mechanics, Naturwissenschaften 28, 664 (1926).
8. E. Schrödinger, On the relation between the quantum mechanics of Heisenberg, Born, and Jordan, and that of Schrödinger, Annalen der Physik 79, (1926).
9. E. Schrödinger, *Collected Papers on Wave Mechanics*, translated from the second German edition by J. F. Shearer and W. M. Deans (Blackie and Son, London, 1928).

Articles

10. J. S. Briggs and J. M. Rost, Time dependence in quantum mechanics, Eur. Phys. J. D 10, 311 (2000).
11. J. S. Briggs and J. M. Rost, On the derivation of the time-dependent equation of Schrödinger, Foundations of Physics 31, 693 (2001).

Chapter 2

The Free-Particle Wavepacket

In this chapter we will investigate the simplest form of the TDSE, where the potential energy, $V(x)$, is independent of x. This is known as free-particle motion. For simplicity, we may take $V(x) = 0$. In this case we can write down a complete solution of the TDSE for an arbitrary initial state, and the time evolution of the wavepacket can be examined in detail. General expressions for the center and width of the wavepacket as a function of time will be obtained, expressions that will be seen to resemble closely the corresponding expressions in classical mechanics. The method of stationary phase integration, as well as the concepts of phase velocity, group velocity and dispersion, will be introduced. These methods and concepts are relevant beyond the TDSE, to wave equations in general.

The free-particle potential is a special case of a potential that is no higher than quadratic. The latter will form the subject of the next chapter, and it is well to ask, why treat the free particle separately? The answer is that although the next chapter deals with a broader class of potentials, it assumes that the initial state is a Gaussian function of coordinates. In the present chapter the initial state is allowed to be an arbitrary function of coordinates and an analytic solution is still possible because of the vanishing potential. In addition, the stationary phase methods introduced in this chapter will turn up again in Chapter 7 in the context of scattering from potential barriers. Nevertheless, the reader should bear in mind that the free particle is a special case of a potential that is no higher than quadratic and indeed the free particle will be revisited again in the next chapter, albeit with the assumption of a Gaussian initial state.

2.1   General Solution

Consider the free-particle potential, $V(x) = 0$. The Hamiltonian is then simply

$$H = -\frac{\hbar^2}{2m}\frac{\partial^2}{\partial x^2}.$$ (2.1)

We will construct general solutions of the Time-Dependent Schrödinger Equation, as in the previous chapter, by first finding particular solutions and then summing (integrating) to get the general solution. The spatial factor, $\psi(x)$, of the particular solution must satisfy the equation

$$H\psi(x) = E\psi(x).$$ (2.2)

The eigenfunctions and eigenvalues are

$$\psi(x) = e^{\pm ikx} \tag{2.3}$$

$$E = \frac{\hbar^2 k^2}{2m} = \frac{p^2}{2m}, \tag{2.4}$$

where we have defined $p \equiv \hbar k$ based on the analogy to the classical expression for the kinetic energy:

$$E_{\text{cl}} = \frac{1}{2} m v_{\text{cl}}^2 = \frac{p_{\text{cl}}^2}{2m}. \tag{2.5}$$

Note that the eigenvalues form a continuous spectrum—there is a physically acceptable solution for every value of k (and every positive value of E). Combining the spatial and temporal factors we obtain the particular solution:

$$\Psi(x, t) = e^{ikx} e^{-\frac{i}{\hbar} Et} = e^{i(kx - \frac{\hbar k^2}{2m} t)}, \tag{2.6}$$

where we now allow k to run from $-\infty$ to ∞.

Before analyzing how $\Psi(x, t)$ moves we discuss some of its static features in the context of wave equations in general. First, we fix t and vary x. Note that the wave will have the same value at x_2 and x_1 if $k(x_2 - x_1) = n2\pi$. Choosing $n = 1$ gives a unique definition to the wavelength $\lambda = x_2 - x_1$ and therefore $k\lambda = 2\pi$ or $k = 2\pi/\lambda$. Thus, k characterizes the number of radians (which is 2π times the number of cycles) per unit distance; k is called the "wavenumber." The definition of a wavenumber is a property of the solutions of any wave equation; what is unique to the Schrödinger equation is that $p = \hbar k = h/\lambda$. This is the famous de Broglie relationship that preceded wave mechanics. Its physical significance is that momentum is proportional to the density of spatial oscillations, that is, the number of oscillations per unit length.

We now fix x and vary t. We note that any wave of the form $e^{i\omega t}$ will have the same value at t_2 and t_1 if $\omega(t_2 - t_1) = n2\pi$. Choosing $n = 1$ defines the fundamental period, $\tau = t_2 - t_1$, and thus $\omega\tau = 2\pi$. Thus, ω characterizes the number of radians per unit time, and is called the "angular frequency"; it is a factor of 2π larger than the frequency ν, defined by $\nu = 1/\tau$. Again, the definition of an angular frequency is a property of the solutions of any wave equation; what is unique to the Schrödinger equation is that

$$E = \hbar\omega = h\nu. \tag{2.7}$$

This relation is due to Einstein and also preceded wave mechanics by many years.

Note that for the free particle,

$$\omega(k) = \frac{\hbar k^2}{2m}; \tag{2.8}$$

there is a simple functional relation between ω and k. This functional relationship is known as the "dispersion" relation. As in other wave equations, the dispersion relation $\omega(k)$ determines both the "phase velocity" and the "group velocity" of the wave:

$$\text{phase velocity} = \frac{\omega}{k} \tag{2.9}$$

$$\text{group velocity} = \frac{\partial \omega}{\partial k}. \tag{2.10}$$

It is the specific functional relationship between ω and k that distinguishes the motion of quantum waves from other waves that occur in nature.

To explore the concept of phase velocity, we rewrite Eq. 2.6 as

$$\Psi(x, t) = e^{ik(x - \frac{\hbar k}{2m} t)} \tag{2.11}$$

$$= e^{ik(x - vt)}.$$

As long as the quantity $x - vt = 0$, we can change the values of x and t to obtain a constant value of the wave amplitude. This leads to the condition $x = vt$. Since x increases as t increases, $\Psi(x, t)$ is a wave traveling to the right with a velocity $v = \frac{\hbar k}{2m} = \frac{\omega}{k}$ called the "phase velocity." Since the momentum of the wavepacket, p, is equal to $\hbar k$, the phase velocity is $p/2m$, which is one-half the classical expression for the velocity. To obtain classical behavior, we need to use a wavepacket and examine its "group velocity," $\frac{\partial \omega}{\partial k}$, as will be seen below.

From Eq. 1.13, we see that we may write the general solution for the free-particle wavepacket as:

$$\Psi(x, t) = \int_{-\infty}^{\infty} a(k) e^{i(kx - \frac{\hbar k^2}{2m} t)} \, dk. \tag{2.12}$$

The function $a(k)$ may be found from the $t = 0$ wavepacket through the use of the Fourier theorem. At $t = 0$, the wavepacket is

$$\Psi(x, 0) = \int_{-\infty}^{\infty} a(k) e^{ikx} \, dk. \tag{2.13}$$

Multiplying both sides of Eq. 2.13 by $\exp(-ik'x)$, integrating over x, and using the property of the Dirac delta function (see Appendix A)

$$\int_{-\infty}^{\infty} e^{-ik'x} e^{ikx} \, dx = 2\pi \delta(k - k'), \tag{2.14}$$

we find that

$$a(k) = \frac{1}{2\pi} \int_{-\infty}^{\infty} \Psi(x, 0) e^{-ikx} \, dx. \tag{2.15}$$

It is instructive to rewrite Eq. 2.12 as

$$\Psi(x, t) = \int_{-\infty}^{\infty} a(k, t) e^{ikx} \, dk, \tag{2.16}$$

where we have defined

$$a(k, t) = a(k) e^{-i \frac{\hbar k^2}{2m} t}. \tag{2.17}$$

Using Eq. 2.14 we find that

$$a(k, t) = \frac{1}{2\pi} \int_{-\infty}^{\infty} \Psi(x, t) e^{-ikx} \, dx. \tag{2.18}$$

Equations 2.16 and 2.18 display a symmetrical relationship between $\Psi(x, t)$ and $a(k, t)$. We will refer to $a(k, t)$ as the *k-representation* of the wavefunction.

▶ Exercise 2.1

    a. Use Eq. 2.14 to show that if $\Psi(x, 0)$ is normalized it follows that

$$\int_{-\infty}^{\infty} a^*(k, t) a(k, t) \, dk = \int_{-\infty}^{\infty} a^*(k) a(k) \, dk = \frac{1}{2\pi}. \tag{2.19}$$

    b. Combine the result of part a with Eq. 2.18 to show that if $\Psi(x, 0)$ is normalized so is $\Psi(x, t)$.

2.2   The Center of the Wavepacket

2.2.1   Calculation Using the k-Representation

We can now consider the time evolution of the center of the wavepacket,

$$\langle x \rangle_t = \frac{\int_{-\infty}^{\infty} \Psi^*(x, t) x \Psi(x, t) \, dx}{\int_{-\infty}^{\infty} \Psi^*(x, t) \Psi(x, t) \, dx}. \tag{2.20}$$

The evaluation of $\langle x \rangle_t$ can be done analytically in the k-representation. Using the fact that in the k-representation the position operator $\hat{x}$ is given by $i\frac{\partial}{\partial k}$ (see Chapter 8, Eq. 8.98), we have

$$\langle x \rangle_t = \int_{-\infty}^{\infty} a^*(k, t) i \frac{\partial}{\partial k} a(k, t) \, dk \bigg/ \int_{-\infty}^{\infty} a^*(k, t) a(k, t) \, dk \tag{2.21}$$

$$= 2\pi \int_{-\infty}^{\infty} \left(a(k) e^{-\frac{i\hbar k^2}{2m} t} \right)^* i \frac{d}{dk} \left(a(k) e^{-\frac{i\hbar k^2}{2m} t} \right) \, dk \tag{2.22}$$

$$= 2\pi \int_{-\infty}^{\infty} a^*(k) (\hbar k t / m) a(k) \, dk + \langle x \rangle_0 \tag{2.23}$$

$$= \frac{\langle p \rangle t}{m} + \langle x \rangle_0. \tag{2.24}$$

In obtaining Eq. 2.22 we have used Eqs. 2.17 and 2.19. The two terms in Eq. 2.23 come from the derivative in Eq. 2.22, where we have noted that at $t = 0$ Eq. 2.22 reduces to

$$\langle x \rangle_0 = 2\pi \int_{-\infty}^{\infty} a(k)^* i \frac{da(k)}{dk} \, dk. \tag{2.25}$$

Notice that in Eq. 2.23 $a(k)$ does not depend on time, and therefore $\langle p \rangle$ in Eq. 2.24 is time independent. Thus $\langle p \rangle_t = \langle p \rangle_0 = \langle p \rangle$, in agreement with the classical mechanics of the free particle.

▶ **Exercise 2.2**   Show that if $a(k)$ is real, $\langle x \rangle_0 = 0$.

Defining the "group velocity," v_g, as the velocity of the center of the wavepacket, we find that

$$v_g = \frac{d\langle x \rangle_t}{dt} = \frac{\langle p \rangle}{m}, \tag{2.26}$$

which is of course the analog of the classical relation $v_{cl} = \frac{p_{cl}}{m}$. We see that for the free particle the group velocity is independent of time, as in classical mechanics, and is completely determined by the distribution $a(k)$ at $t = 0$. That the correspondence between classical and quantum mechanics embodied in Eq. 2.24 survives *for an arbitrary distribution of $a(k)$* is surprising and represents a special case of Ehrenfest's theorem (see Chapter 4).

2.2.2   The Stationary Phase Method

We can obtain the same formula for the velocity of the center of the wavepacket from a stationary phase perspective. Equation 2.12 can be rewritten in the form

$$\Psi(x, t) = \int_{-\infty}^{\infty} a(k) e^{i\Phi(k,x;t)} \, dk, \tag{2.27}$$

where

$$\Phi(k, x; t) = kx - \frac{\hbar k^2}{2m} t. \tag{2.28}$$

We assume that $a(k)$ is a smooth function, peaked at $k = \langle k \rangle$. The phase factor in Eq. 2.27 is an oscillatory function, leading to an alternation in the sign of the integrand that tends to cancel out in the overall integral. To avoid the cancellation due to alternating sign, the oscillation in the phase should be minimal under the envelope $a(k)$. This leads to the so-called stationary phase condition, $d\Phi/dk|_{k=\langle k \rangle} = 0$. Applying the stationary phase condition to Eq. 2.28 leads to the equation

$$\langle x \rangle_t = \frac{\hbar \langle k \rangle t}{m} = \frac{\langle p \rangle t}{m}. \tag{2.29}$$

Note that Eq. 2.29 has precisely the form of the classical equation of motion for the position of a free particle (with $\langle x \rangle_0 = 0$; see Exercise 2.2). Since $\langle x \rangle_t$ defines the center of the wavepacket, we see that the center of the wavepacket moves in close agreement with the equations of classical mechanics.

The stationary phase argument can be made somewhat more precise (see Figure 2.1). We rewrite Eq. 2.12 as

$$\Psi(x, t) = e^{i\langle k \rangle x} \int_{-\infty}^{\infty} a(k) e^{i(k-\langle k \rangle)x} e^{-i\hbar[(k-\langle k \rangle)+\langle k \rangle]^2 t/2m} \, dk \tag{2.30}$$

$$= e^{i\langle k \rangle x} e^{-i\hbar\langle k \rangle^2 t/2m} \int_{-\infty}^{\infty} a(k) e^{-i\hbar(k-\langle k \rangle)^2 t/2m} e^{i(k-\langle k \rangle)(x-\langle x \rangle_t)} \, dk, \tag{2.31}$$

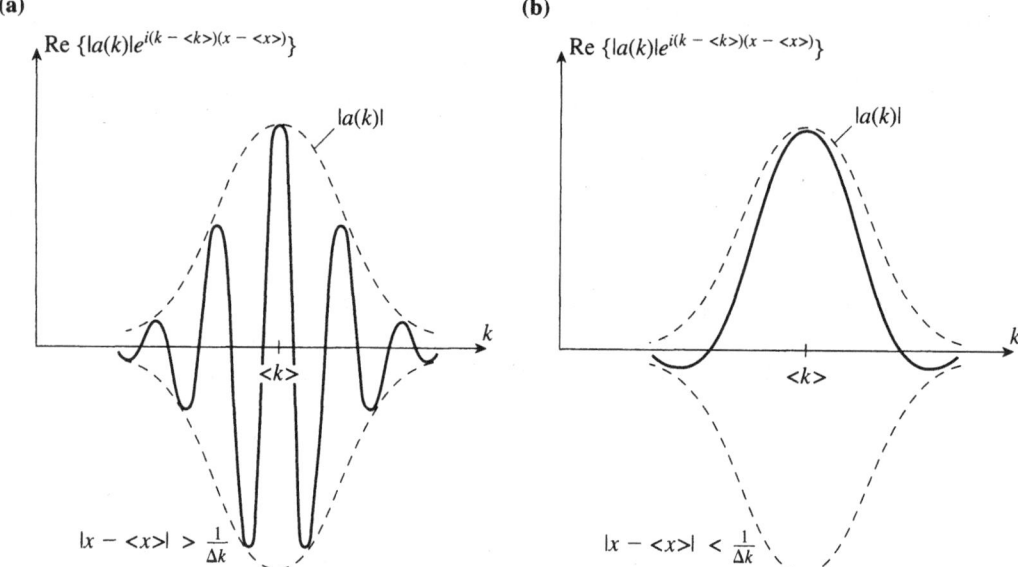

Figure 2.1 Pictorial explanation of the stationary phase method (cf. the last factor in the integral in Eq. 2.31). The figure shows both the amplitude and phase of a function of k whose integral yields $\psi(x)$. The phase factor depends parametrically on x. In (a), x is fixed at a value such that $|x - \langle x \rangle| > 1/\Delta k$, where Δk is the width of $a(k)$. In this case the function to be integrated oscillates several times within the interval Δk, and the integral is small. In (b), x is fixed at a value such that $|x - \langle x \rangle| < 1/\Delta k$, so that the function to be integrated undergoes less than one full oscillation within the interval Δk, and hence the integral is large. Adapted from Cohen-Tannoudji (1977).

where we have defined $\langle x \rangle_t \equiv \hbar \langle k \rangle t/m$. If $\hbar(\Delta k)^2 t/2m \ll 2\pi$, where Δk is the width of $a(k)$, then the factor $e^{-i\hbar(k-\langle k \rangle)^2 t/2m}$ can be taken as unity. When $|x - \langle x \rangle_t|$ is large, the integrand oscillates a large number of times within the interval Δk; this is depicted in Figure 2.1a, where it is seen that successive oscillations lead to a cancellation of the integral. On the other hand, if $x \approx \langle x \rangle_t$, the phase in the integrand in Eq. 2.31 is approximately constant, as depicted in Figure 2.1b, leading to a large value of $\Psi(x, t)$ for $x \approx \langle x \rangle_t$.

Note that the values of x for which $\Psi(x, t)$ is significant are those determined by the relation $\Delta k|x - \langle x \rangle_t| < 1$, so that the degree of oscillation of the integrand is not appreciable. This implies an uncertainty relation in position and momentum,

$$\Delta k \ \Delta x \approx 1. \tag{2.32}$$

2.3   The Dispersion of the Wavepacket

What is the time evolution of the width of the wavepacket? The width of the wavepacket is commonly measured by

$$\Delta x = [\langle x^2 \rangle - \langle x \rangle^2]^{1/2}. \tag{2.33}$$

A reasonable measure for the rate of change of the width is therefore the "dispersion," defined as

$$\frac{[(\Delta x)_t^2 - (\Delta x)_0^2]^{1/2}}{t}. \tag{2.34}$$

To calculate the second moments we again use the k-representation of the wavepacket:

$$\langle x^2 \rangle_t = 2\pi \int_{-\infty}^{\infty} \left(a(k) e^{-\frac{i\hbar k^2}{2m}t} \right)^* \left[-\frac{\partial^2}{\partial k^2} \left(a(k) e^{-\frac{i\hbar k^2}{2m}t} \right) \right] dk. \tag{2.35}$$

The second derivative is equal to

$$\left[\frac{1}{a} \frac{\partial^2 a}{\partial k^2} + \frac{2}{a} \frac{\partial a}{\partial k} \left(-i \frac{\hbar k t}{m} \right) + \left(-i \frac{\hbar t}{m} \right) + \left(-i \frac{\hbar k t}{m} \right)^2 \right] \left(a(k) e^{-\frac{i\hbar k^2}{2m}t} \right), \tag{2.36}$$

and so $\langle x^2 \rangle_t$ reduces to

$$\langle x^2 \rangle_t = -2\pi \int_{-\infty}^{\infty} a^*(k) a(k) \left[\frac{1}{a} \frac{\partial^2 a}{\partial k^2} + \frac{2}{a} \frac{\partial a}{\partial k} \left(-i \frac{\hbar k t}{m} \right) + \left(-i \frac{\hbar t}{m} \right) + \left(-i \frac{\hbar k t}{m} \right)^2 \right] dk. \tag{2.37}$$

If we assume that $a(k)$ is real ($a^* = a$), this allows us to integrate the second term by parts, with the result that the second and third terms cancel:

$$\langle x^2 \rangle_t = 2\pi \int_{-\infty}^{\infty} a^*(k) a(k) \left[-\frac{1}{a} \frac{\partial^2 a}{\partial k^2} + \frac{\hbar^2 k^2 t^2}{m^2} \right] dk. \tag{2.38}$$

At $t = 0$, this reduces to

$$\langle x^2 \rangle_0 = 2\pi \int_{-\infty}^{\infty} a^*(k) a(k) \left[-\frac{1}{a} \frac{\partial^2 a}{\partial k^2} \right] dk, \tag{2.39}$$

and thus,

$$\langle x^2 \rangle_t - \langle x^2 \rangle_0 = \frac{\hbar^2 t^2 \langle k^2 \rangle}{m^2}. \tag{2.40}$$

Finally,

$$(\Delta x)_t^2 - (\Delta x)_0^2 = (\langle x^2 \rangle_t - \langle x^2 \rangle_0) - (\langle x \rangle_t^2 - \langle x \rangle_0^2) \tag{2.41}$$

$$= \frac{\hbar^2 t^2 \langle k^2 \rangle}{m^2} - \left(\frac{\hbar t \langle k \rangle}{m} \right)^2 = \hbar^2 (\Delta k)^2 t^2 / m^2, \tag{2.42}$$

where we have used Eq. 2.40 with Eq. 2.24 and Exercise 2.2. Rearranging Eq. 2.42 we obtain

$$\frac{[(\Delta x)_t^2 - (\Delta x)_0^2]^{1/2}}{t} = \hbar \, \Delta k / m = \Delta p / m, \tag{2.43}$$

which is the desired formula for the rate of change of the width of the free-particle wavepacket with time.

Note that the assumption in the derivation of Eq. 2.43, that $a(k)$ is real, is not just to simplify the mathematics, but has a sound physical basis. If $a(k)$ is allowed to be complex, the width of the wavepacket may actually get smaller before it gets larger (cf. Exercises 3.8 and 5.6), and the LHS of Eq. 2.43 ceases to be a meaningful ratio. This problem is avoided by defining the $t = 0$ wavepacket to be maximally compact, implying that $a(k)$ be real.

We close this chapter by noting that the concepts of group velocity and dispersion are generally applicable to the solutions of any wave equation. The general definition of the group velocity is

$$v_g = \frac{\langle x \rangle_t - \langle x \rangle_0}{t} = \left\langle \frac{\partial \omega}{\partial k} \right\rangle, \tag{2.44}$$

and the general definition of the dispersion is

$$\text{dispersion} = \frac{[(\Delta x)_t^2 - (\Delta x)_0^2]^{1/2}}{t} = \left\langle \left(\frac{\partial \omega}{\partial k} \right)^2 \right\rangle - \left\langle \frac{\partial \omega}{\partial k} \right\rangle^2 \tag{2.45}$$

(see Exercises 2.4 and 2.5). Since for the Schrödinger equation $\omega(k) = \hbar k^2 / 2m$, we find that

$$v_g = \frac{\hbar \langle k \rangle}{m} = \frac{\langle p \rangle}{m}, \tag{2.46}$$

and

$$\text{dispersion} = \frac{\hbar}{m} \left(\langle k^2 \rangle - \langle k \rangle^2 \right)^{1/2} = \frac{\hbar}{m} \Delta k = \frac{\Delta p}{m}. \tag{2.47}$$

Problems

▶ **Exercise 2.3**   Define the scaled plane wave, $\psi_p(x)$, as follows:

$$\psi_p(x) = \frac{e^{ikx}}{\sqrt{2\pi\hbar}} = \frac{e^{ipx/\hbar}}{\sqrt{2\pi\hbar}}. \tag{2.48}$$

a.   Show that Eq. 2.48 describes a *momentum normalized* plane wave, in the sense that

$$\int_{-\infty}^{\infty} \psi_p^*(x) \psi_{p'}(x) \, dx = \delta(p - p'). \tag{2.49}$$

[Hint: use the property of δ-functions that $\delta(cx) = \frac{1}{|c|}\delta(x)$.]

b. $\Psi(x, t)$ can be expressed in terms of the momentum normalized states as follows:

$$\Psi(x, t) = \int_{-\infty}^{\infty} \tilde{\Psi}(p, t) \frac{e^{ipx/\hbar}}{\sqrt{2\pi\hbar}} \, dp. \tag{2.50}$$

Show that

$$\tilde{\Psi}(p, t) = \int_{-\infty}^{\infty} \Psi(x, t) \frac{e^{-ipx/\hbar}}{\sqrt{2\pi\hbar}} \, dx. \tag{2.51}$$

Note that the symmetry between $\Psi(x, t)$ and $\tilde{\Psi}(p, t)$ in Eqs. 2.50 and 2.51 is even more pronounced than that between $\Psi(x, t)$ and $a(k, t)$ in the analogous relations, Eqs. 2.16 and 2.18. The function $\tilde{\Psi}(p)$ is called the *p-representation* of $\Psi(x)$, and is discussed at length in Chapter 8.

▶ **Exercise 2.4**   Replace $a(k, t)$ in Eq. 2.17 with the more general form

$$a(k, t) = a(k)e^{-i\omega(k)t}, \tag{2.52}$$

which is valid for general wave equations, not just the TDSE. Show that the generalization of Eq. 2.26 is

$$v_g = \left\langle \frac{\partial \omega}{\partial k} \right\rangle. \tag{2.53}$$

▶ **Exercise 2.5**   Replace $a(k, t)$ in Eq. 2.17 with the more general expression, given by Eq. 2.52. Show that the generalization of Eq. 2.43 for arbitrary dispersion relations, $\omega(k)$, is

$$\frac{[(\Delta x)_t^2 - (\Delta x)_0^2]^{1/2}}{t} = \left\langle \left(\frac{\partial \omega}{\partial k}\right)^2 \right\rangle - \left\langle \frac{\partial \omega}{\partial k} \right\rangle^2 \tag{2.54}$$

Further Reading and Historical Notes

The concept of group velocity was developed by Lord Rayleigh (1877), extending earlier treatments by Hamilton and Stokes. Stationary phase integration and its relation to group velocity was discovered by William Thomson (later Lord Kelvin) (1887). The short book by Havelock (1914) is a gem; it provides a wonderful historical perspective on the development of the concept of group velocity and its applications to many types of wave phenomena. The book by Brillouin (1960) grapples with the question of group velocity versus signal velocity and is also an excellent reference for the history.

References

General

1. A. Messiah, *Quantum Mechanics*, vol. I (Wiley, New York, 1958).
2. C. Cohen-Tannoudji, B. Diu and F. Laloë, *Quantum Mechanics*, vol. I (Wiley, New York, 1977).
3. G. Strang, *Introduction to Applied Mathematics* (Wellesley-Cambridge Press, Wellesley, MA, 1986).

Group Velocity and Stationary Phase

4. Lord Rayleigh, On progressive waves, Proc. London Math. Soc. 9, 21 (1877); reprinted in *Scientific Papers by John William Strutt, Baron Rayleigh*, vol I, p. 322.
5. W. Thomson, On the waves produced by a single impulse in water of any depth, or in a dispersive medium, Phil. Mag. 23, 252 (1887); reprinted in *Mathematical and Physical Papers by Sir William Thomson, Baron Kelvin* (Cambridge University Press, 1910) vol. IV, p. 303.
6. T. H. Havelock, *The Propagation of Disturbances in Dispersive Media* (Cambridge University Press, 1914).
7. L. Brillouin, *Wave Propagation and Group Velocity* (Academic Press, New York, 1960).

Chapter 3

The Gaussian Wavepacket

In this chapter, we will look at the motion of wavepackets of the Gaussian form $\Psi(x) = \exp(-\alpha x^2)$ and its generalizations. If the potential energy function $V(x)$ is no higher than quadratic, an initial Gaussian wavefunction $\Psi(x, 0)$ will lead to a Gaussian $\Psi(x, t)$ for all times, a so-called "Gaussian wavepacket." Gaussian wavepackets are useful not only for potentials up to quadratic, but also as the basis for approximations for anharmonic potentials.

We will use the following formula for the integral of a generalized complex Gaussian many times in this chapter:

$$\int_{-\infty}^{\infty} e^{-ax^2 + ibx + ic} \, dx = \sqrt{\frac{\pi}{a}} e^{-b^2/4a} e^{ic} \quad \text{Re}(a) > 0. \tag{3.1}$$

3.1 The Gaussian Free Particle

We begin with the simplest case, a free-particle Gaussian wavepacket given at $t = 0$ by

$$\Psi(x, 0) = N e^{-\alpha_0 x^2}. \tag{3.2}$$

For generality, we will allow α_0 to be complex. The normalization factor, N, is determined by the condition that $\int_{-\infty}^{\infty} \Psi^*(x)\Psi(x) \, dx = 1$.

▶ Exercise 3.1   Use Eq. 3.1 to show that

$$N = \left(\frac{\alpha_0 + \alpha_0^*}{\pi} \right)^{1/4} = \left(\frac{2\text{Re}(\alpha_0)}{\pi} \right)^{1/4}. \tag{3.3}$$

To find an analytic expression for the time evolution of the wavepacket, we follow the same procedure as in the previous chapters. We expand the Gaussian in terms of the eigenstates of the free-particle Hamiltonian, e^{ikx}, with eigenvalues $E_k = \frac{\hbar^2 k^2}{2m}$:

$$\Psi(x, t) = \int_{-\infty}^{\infty} a(k) e^{ikx - i\frac{\hbar k^2}{2m}t} \, dk, \tag{3.4}$$

where the $a(k)$ are as yet unknown. At $t = 0$ the wavepacket is

$$\Psi(x, 0) = \int_{-\infty}^{\infty} a(k) e^{ikx} \, dk = N e^{-\alpha_0 x^2}. \tag{3.5}$$

Following the procedure in Eq. 2.15, and using Eqs. 3.5 and 3.1, the coefficients $a(k)$ are given by

$$a(k) = \frac{1}{2\pi} \int_{-\infty}^{\infty} \Psi(x, 0) e^{-ikx} \, dx = \frac{N}{2\pi} \sqrt{\frac{\pi}{\alpha_0}} e^{-k^2/4\alpha_0}. \tag{3.6}$$

Substituting Eq. 3.6 back into Eq. 3.4 we obtain

$$\Psi(x, t) = \frac{N}{2\pi} \sqrt{\frac{\pi}{\alpha_0}} \int_{-\infty}^{\infty} e^{-k^2/4\alpha_0} e^{-i(\hbar k^2/2m)\,t} e^{ikx} \, dk \tag{3.7}$$

$$= \frac{N}{\sqrt{1 + \frac{2i\hbar\alpha_0 t}{m}}} e^{-x^2/[4\{\frac{1}{4\alpha_0} + \frac{i\hbar t}{2m}\}]}. \tag{3.8}$$

Defining

$$\alpha_t = \frac{\alpha_0}{1 + \frac{2i\hbar\alpha_0 t}{m}} \tag{3.9}$$

$$\gamma_t = \frac{i\hbar}{2} \ln\left(1 + \frac{2i\hbar\alpha_0 t}{m}\right), \tag{3.10}$$

Eq. 3.8 may be rewritten as

$$\Psi(x, t) = N e^{-\alpha_t x^2 + i\gamma_t/\hbar}. \tag{3.11}$$

Here we see the important property of Gaussian wavepackets that when freely propagated, they remain Gaussian.

The time dependence of the free-particle Gaussian wavepacket can also be determined by a completely different method, by assuming that $\Psi(x, t) = \exp(-\alpha_t x^2 + i\gamma_t/\hbar)$ and then substituting this form into the Time-Dependent Schrödinger Equation. After equating powers of x, one obtains the above expressions for α_t and γ_t. We will demonstrate this technique by assuming a more general trial form,

$$\Psi(x, t) = N \exp\left(-\alpha_t(x - x_t)^2 + \frac{i}{\hbar} p_t(x - x_t) + \frac{i}{\hbar}\gamma_t\right). \tag{3.12}$$

In this expression, x_t, p_t, α_t, and γ_t are all (potentially) time-dependent parameters. The necessary derivatives are

$$\frac{\partial \Psi}{\partial x} = \left[-2\alpha_t(x - x_t) + \frac{i}{\hbar} p_t\right] \psi \tag{3.13}$$

$$\frac{\partial^2 \Psi}{\partial x^2} = \left[\left(-2\alpha_t(x - x_t) + \frac{i}{\hbar} p_t\right)^2 - 2\alpha_t\right] \psi \tag{3.14}$$

$$\frac{\partial \Psi}{\partial t} = \left[-\dot{\alpha}_t(x - x_t)^2 + 2\alpha_t(x - x_t)\dot{x}_t + \frac{i}{\hbar}(x - x_t)\dot{p}_t - \frac{i}{\hbar} p_t \dot{x}_t + \frac{i}{\hbar}\dot{\gamma}_t\right] \psi. \tag{3.15}$$

Using these derivatives in the Time-Dependent Schrödinger Equation and equating powers of $(x - x_t)$, one obtains the following differential equations governing the time evolution of the four wavepacket parameters:

$$\dot{\alpha}_t = -\frac{2i\hbar}{m}\alpha_t^2 \tag{3.16}$$

$$\dot{p}_t = 0 \tag{3.17}$$

$$\dot{x}_t = \frac{p_t}{m} \tag{3.18}$$

$$\dot{\gamma}_t = -\frac{p_t^2}{2m} + p_t\dot{x}_t - \frac{\hbar^2\alpha_t}{m}. \tag{3.19}$$

Note that in obtaining the equations for $\dot{p}$ and $\dot{x}$ we have equated separately the real and imaginary parts of the coefficients in the equation of order $(x - x_t)$. For the free particle, the classical Hamiltonian is simply $H_{cl} = p_{cl}^2/2m$. Hamilton's equations are then

$$\begin{aligned} \dot{p}_{cl} &= -\partial H_{cl}/\partial x_{cl} = 0 \\ \dot{x}_{cl} &= \partial H_{cl}/\partial p_{cl} = p_{cl}/m. \end{aligned} \tag{3.20}$$

By comparison, one sees that x_t and p_t evolve according to the classical Hamilton's equations and can be identified with the classical position, x_{cl}, and momentum, p_{cl}, respectively.

The equation for $\dot{\gamma}_t$ also has an intriguing interpretation in terms of classical mechanics. Recall that the classical Lagrangian is given by the difference between kinetic and potential energy (see Chapter 10):

$$L_{cl} = m\dot{x}^2/2 - V(x). \tag{3.21}$$

For the free particle, $V(x) = 0$, and the Lagrangian is simply $L_{cl} = m\dot{x}^2/2$. Therefore, the equation of motion for the parameter γ_t, Eq. 3.19, may be rewritten as

$$\dot{\gamma}_t = L_{cl} - \frac{\hbar^2\alpha_t}{m}. \tag{3.22}$$

Integrating Eqs. 3.16–3.19, one obtains the final expressions for the time dependence of the four wavepacket parameters:

$$p_t = p_0 \tag{3.23}$$

$$x_t = x_0 + \frac{p_0}{m}t \tag{3.24}$$

$$\alpha_t = \frac{\alpha_0}{1 + (2i\hbar\alpha_0 t/m)} \tag{3.25}$$

$$\gamma_t = \frac{p_0^2}{2m}t + \frac{i\hbar}{2}\ln\left(1 + \frac{2i\hbar\alpha_0 t}{m}\right), \tag{3.26}$$

where x_0 is the initial position of the wavepacket, p_0 is the initial momentum and α_0 is the initial width of the wavepacket. One should note that Eq. 3.25 is identical to Eq. 3.9 and that Eq. 3.26 is equivalent to Eq. 3.10, taking into account the term containing the initial momentum.

▶ **Exercise 3.2** Show that the classical Lagrangian can also be written as

$$L_{\text{cl}} = p\dot{x} - E. \tag{3.27}$$

Equation 3.27 will be used in Section 6.3.

3.2 General Properties of Gaussian Wavepackets

In the previous section we have seen the generalized Gaussian form, Eq. 3.12. In this section we will analyze some of the properties of such a Gaussian, in particular its center in both position and momentum, its width both in position and momentum, and position–momentum correlation. To emphasize that the discussion here has nothing to do with the time evolution of the Gaussian per se, but is a property only of the "anatomy" of the Gaussian, we will use the notation

$$\Psi(x) = N e^{-\alpha(x-x')^2 + ip'(x-x')/\hbar + i\gamma/\hbar}. \tag{3.28}$$

Here, α, x', p' and γ are parameters that do not depend on x. We assume that x' and p' are real, while α and γ may be complex. For complex γ, the normalization factor, N, takes the general form $N = (\frac{2\text{Re}\alpha}{\pi})^{1/4} e^{\text{Im}\gamma/\hbar}$.

▶ **Exercise 3.3** Show that if x' and p' are *not* real it is always possible to define a new set of variables α', γ', x'' and p'', such that x'' and p'' are real and the exponent in Eq. 3.28 is unchanged.

▶ **Exercise 3.4** Show that for the generalized Gaussian in Eq. 3.28, $\langle \hat{x} \rangle = x'$, $\langle \hat{p} \rangle = p'$. [Hint: to keep the algebra simple, change variables to $y = x - x'$.]

▶ **Exercise 3.5** Show that for the generalized Gaussian in Eq. 3.28,

$$\Delta x = \sqrt{\langle \hat{x}^2 \rangle - \langle \hat{x} \rangle^2} = \sqrt{\frac{1}{4\,\text{Re}[\alpha]}}$$

$$\Delta p = \sqrt{\langle \hat{p}^2 \rangle - \langle \hat{p} \rangle^2} = \left(2\hbar^2\alpha - \frac{\hbar^2\alpha^2}{\text{Re}(\alpha)} \right)^{1/2} = \frac{\hbar|\alpha|}{\sqrt{\text{Re}(\alpha)}}.$$

$$\left\langle \frac{\hat{x}\hat{p} + \hat{p}\hat{x}}{2} \right\rangle - \langle \hat{x} \rangle \langle \hat{p} \rangle = \frac{-\hbar}{2} \frac{\text{Im}(\alpha)}{\text{Re}(\alpha)}.$$

We now check that the uncertainty principle is satisfied by these relations. Using the formulas from the previous exercise we find

$$\Delta x(t)\, \Delta p(t) = \frac{\hbar}{2} \frac{|\alpha|}{\text{Re}\alpha} \geq \frac{\hbar}{2}. \tag{3.29}$$

For the free particle, we may use for α the explicit form α_t, Eq. 3.25. Assuming that α_0 is real we find that

$$\text{Re}[\alpha_t] = \frac{\alpha_0}{1 + 4\hbar^2\alpha_0^2 t^2/m^2} \qquad |\alpha_t| = \frac{\alpha_0}{(1 + 4\hbar^2\alpha_0^2 t^2/m^2)^{1/2}} \tag{3.30}$$

and therefore

$$\Delta x(t) = \frac{1}{2}\sqrt{\frac{1 + 4\hbar^2\alpha_0^2 t^2/m^2}{\alpha_0}} \tag{3.31}$$

$$\Delta p(t) = \hbar\sqrt{\alpha_0}. \tag{3.32}$$

The uncertainty relation for the wavepacket is therefore

$$\Delta x(t)\,\Delta p(t) = \frac{\hbar}{2}\sqrt{1 + \frac{4\hbar^2\alpha_0^2 t^2}{m^2}} \geq \frac{\hbar}{2}. \tag{3.33}$$

At $t = 0$, $\Delta x(t)\,\Delta p(t) = \hbar/2$ and the wavepacket is a minimum uncertainty wavepacket. For $t > 0$, $\Delta x(t)\,\Delta p(t)$ is always greater then $\hbar/2$, obeying the Heisenberg Uncertainty Principle.

▶ **Exercise 3.6** Substitute Eq. 3.7 into Eq. 2.51 to find the form of the free-particle Gaussian wavepacket in the momentum representation. Calculate the width $\Delta p(t)$ directly from the wavefunction in the momentum representation, and show consistency with Eq. 3.32.

▶ **Exercise 3.7** Show that Eqs. 3.31–3.32 for $\Delta x(t)$ and $\Delta p(t)$ are consistent with the general dispersion relations, Eqs. 2.45 and 2.47.

Figure 3.1 shows the time evolution of the free-particle wavepacket in space. The narrowing of the packet at negative times and spreading at positive times is clearly visible. Figure 3.2 shows the Δx of the packet as a function of time.

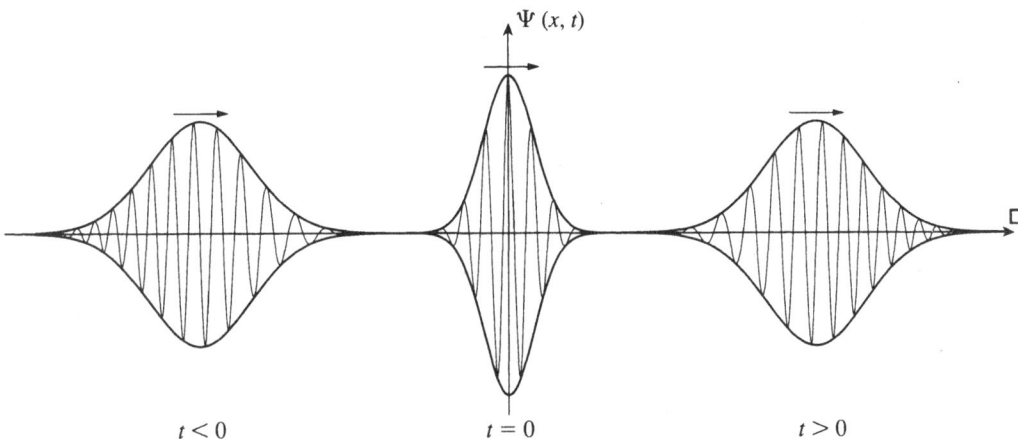

Figure 3.1 Time evolution of the free-particle Gaussian wavepacket. The envelope is the absolute value and the oscillatory curve in the interior is the real part of the wavepacket. The spacing between the oscillations is a measure of the average momentum. The central momentum is constant in time, while the central position changes at a uniform rate, $\langle x \rangle_t = \langle p \rangle_t t/m$. The width of the packet decreases until $t = 0$ and then increases: at $t = 0$ the wavepacket is a minimum uncertainty state, that is, $\Delta x \cdot \Delta p = \hbar/2$, while at earlier and later times $\Delta x \cdot \Delta p > \hbar/2$. Note that the height of the wavepacket changes in time in opposition to the width such that the norm of $\psi(x, t)$ is a constant.

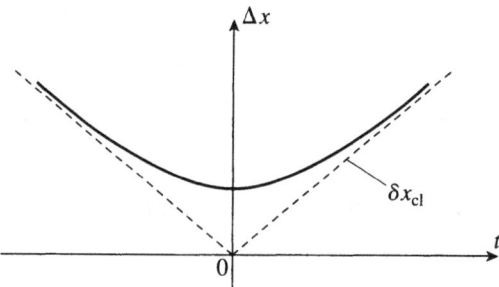

Figure 3.2 Variation in time of the width Δx of the wavepacket of Figure 3.1. For large t, Δx approaches the dispersion δx_{cl} of the positions of a swarm of classical particles that left $x = 0$ at $t = 0$ with a dispersion in velocity $\Delta p/m$.

▶ **Exercise 3.8** What is special about $t = 0$ that the wavepacket should be maximally compact at this time? What was it in our choice of initial conditions that led to this state, as opposed to, for example, the state that this evolves into at a later time? [Hint: see the discussion below Eq. 2.43.]

3.3   Gaussian in a Quadratic Potential

What happens to a Gaussian wavepacket in the presence of a potential? Substituting the *ansatz*

$$\Psi(x, t) = N \exp\left(-\alpha_t(x - x_t)^2 + \frac{i}{\hbar}p_t(x - x_t) + \frac{i}{\hbar}\gamma_t\right) \tag{3.34}$$

into the Time-Dependent Schrödinger Equation for a harmonic potential,

$$V(x) = \frac{1}{2}m\omega^2 x^2, \tag{3.35}$$

and equating equal powers of $(x - x_t)$ leads to the following equations for the wavepacket parameters (see Exercise 3.18):

$$\dot{\alpha}_t = -\frac{2i\hbar}{m}\alpha_t^2 + \frac{i}{2\hbar}m\omega^2 \tag{3.36}$$

$$\dot{x}_t = \frac{p_t}{m} \tag{3.37}$$

$$\dot{p}_t = -m\omega^2 x_t \tag{3.38}$$

$$\dot{\gamma}_t = \frac{p_t^2}{2m} - \frac{1}{2}m\omega^2 x_t^2 - \frac{\hbar^2}{m}\alpha_t. \tag{3.39}$$

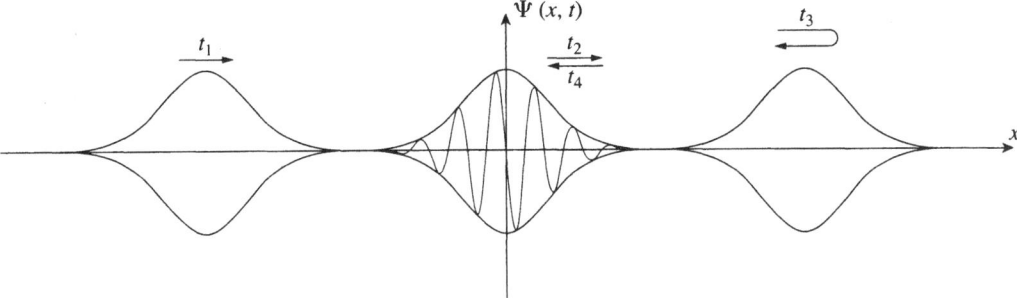

Figure 3.3 Gaussian wavepacket in a harmonic oscillator. The envelope is the absolute value and the oscillatory curve in the interior is the imaginary part of the wavepacket. The zero-point phase ($\omega t/2$) has been removed for clarity. Note that the average position and momentum change according to the classical equations of motion; the average momentum vanishes at the classical turning points and is maximum at the potential minimum. The initial width is the same as the width of the ground vibrational state of the oscillator: $\alpha_0 = m\omega/2\hbar$, where ω is the angular frequency of the oscillator. This is a so-called "coherent state," where the Gaussian moves without spreading: $\alpha_t = \alpha_0$.

Once again, x_t and p_t satisfy Hamilton's equations (cf. Eqs. 3.37–3.38), with solutions

$$x_t = x_0 \cos(\omega t) + (p_0/m\omega) \sin(\omega t) \tag{3.40}$$

$$p_t = p_0 \cos(\omega t) - m\omega x_0 \sin(\omega t). \tag{3.41}$$

Equation 3.36 may be integrated directly and after some algebraic manipulations, one finds that (Heller, 1975)

$$\alpha_t = a \left(\frac{\alpha_0 \cos \omega t + ia \sin \omega t}{i\alpha_0 \sin \omega t + a \cos \omega t} \right), \tag{3.42}$$

where $a = m\omega/2\hbar$. At $t = 0$, one has $\alpha_t = \alpha_0$ while for the special choice of $\alpha_0 = a$, one finds that $\alpha_t = a$ for all time. Since for $\alpha_0 = a = m\omega/2\hbar$ the wavepacket's width does not change, this state is called a "coherent state." It actually behaves quite like a classical particle since its average position and momentum follow Hamilton's equations and its width is a constant in time (see Figure 3.3).

If $\alpha_t \neq a$ the Gaussian is called a "squeezed state": the width spreads and contracts (or contracts and spreads) periodically in time (see Figure 3.4). The geometrical significance of the squeezed states will become clearer in Chapter 5, where the Wigner phase space representation is introduced.

▶ **Exercise 3.9**   The time-varying α_t leads to a time-varying $\Delta x(t)$. Calculate $\Delta x(t)$ as a function of α_0, ω and t, and make representative pictures.

▶ **Exercise 3.10**   Substitute Eq. 3.34 into Eq. 2.51 to obtain the wavepacket in momentum space. Show that for a coherent state the width of the wavepacket in momentum space is constant in time.

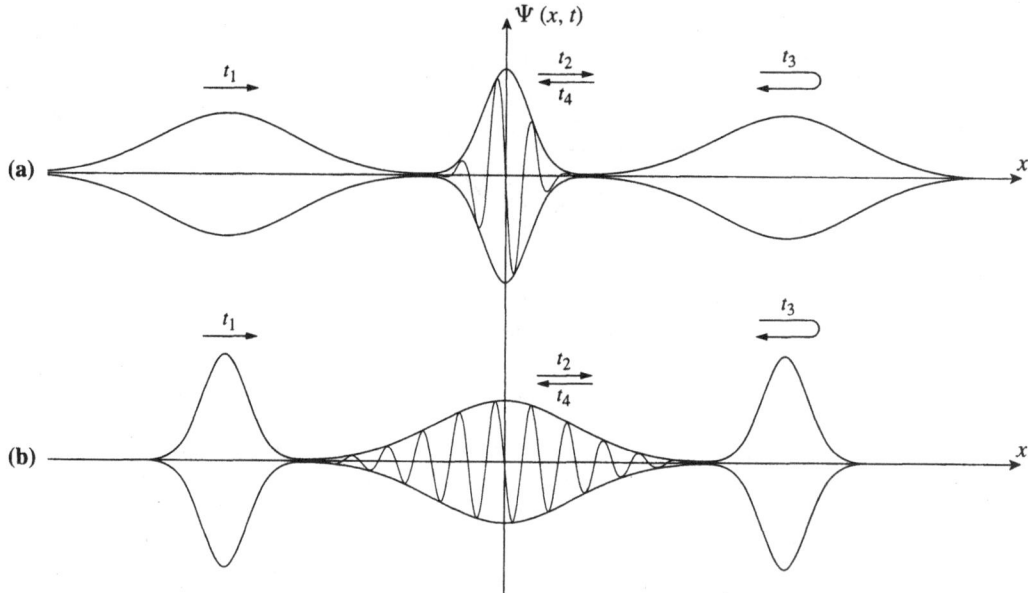

Figure 3.4 Same as Figure 3.3, but for intial Gaussians with different width parameters. (a) The initial Gaussian has width parameter $\alpha_0 < m\omega/2\hbar$. The center of the Gaussian still satisfies the classical equations of motion, however, the width first contracts and then spreads periodically in time, twice per period. (b) The initial Gaussian has width parameter $\alpha_0 > m\omega/2\hbar$. The center of the Gaussian again satisfies the classical equations of motion. Now the width first spreads and then contracts, twice per period.

The solution of Eq. 3.39 is slightly more complicated and deserves some extra discussion. Using the now known values of x_t, p_t, and α_t, it may be rewritten as

$$\dot{\gamma}_t = \frac{p_t^2}{2m} - \frac{1}{2}m\omega^2 x_t^2 - \frac{\hbar^2}{m}a\left(\frac{\alpha_0 \cos \omega t + ia \sin \omega t}{i\alpha_0 \sin \omega t + a \cos \omega t}\right). \tag{3.43}$$

Noting that the classical Lagrangian for the system is $L_{\rm cl} = \frac{p_t^2}{2m} - \frac{1}{2}m\omega^2 x_t^2$ and defining a new variable

$$z = i\alpha_0 \sin \omega t + a \cos \omega t \tag{3.44}$$

so that

$$-i\dot{z} = \omega\alpha_0 \cos \omega t + ia\omega \sin \omega t, \tag{3.45}$$

we can recast the equation of motion of γ_t in the form

$$\dot{\gamma}_t = L_{\rm cl} + \frac{ia\hbar^2}{m\omega}\frac{\dot{z}}{z}. \tag{3.46}$$

The second term may be integrated immediately to yield

$$\frac{i\hbar}{2}\ln\left(\frac{i\alpha_0 \sin \omega t + a \cos \omega t}{a}\right). \tag{3.47}$$

For the harmonic oscillator,

$$\int L_{cl} \, dt = \frac{p_t x_t - p_0 x_0}{2}, \tag{3.48}$$

an expression that is easily verified by differentiating both sides of the equation with respect to t, and using Eqs. 3.37 and 3.38. So, finally,

$$\gamma_t = \frac{p_t x_t - p_0 x_0}{2} + \frac{i\hbar}{2} \ln \left(\frac{i\alpha_0 \sin \omega t + a \cos \omega t}{a} \right). \tag{3.49}$$

For the coherent state, $\alpha_0 = a$, and γ_t reduces to

$$\gamma_t^{\text{coherent}} = \frac{p_t x_t - p_0 x_0}{2} - \frac{\hbar \omega t}{2}. \tag{3.50}$$

3.4   Reexamination of the Stationary Phase Method

We end this chapter with a discussion of the stationary phase evaluation of the Gaussian integral from k to x in Eq. 3.7. In the previous chapter we used the stationary phase procedure to derive an expression for the group velocity of the free particle. Here, we check the accuracy of the stationary phase integration against the exact solution of the Gaussian integral in Eq. 3.7. As we shall see, the stationary phase result is not exact, and in fact leads to singularities at specific times (e.g., for the free particle at $t = 0$).

Equation 3.7, modified to include a central momentum, $\hbar k_0$, can be written

$$\Psi(x, t) = \frac{N}{2\pi} \sqrt{\frac{\pi}{\alpha_0}} \int_{-\infty}^{\infty} e^{-(k-k_0)^2/4\alpha_0} e^{-i\frac{\hbar k^2}{2m}t} e^{ikx} \, dk. \tag{3.51}$$

Changing variables to $k' \equiv k - k_0$, we can rewrite this integral as

$$\Psi(x, t) = N' \int_{-\infty}^{\infty} e^{-k'^2/4\alpha_0} e^{-i\frac{\hbar(k'+k_0)^2}{2m}t} e^{i(k'+k_0)x} \, dk'$$

$$= N' \int_{-\infty}^{\infty} e^{-k'^2 \left(\frac{1}{4\alpha_0} + \frac{i\hbar t}{2m} \right)} e^{ik'\left(x - \frac{\hbar k_0}{m}t \right)} e^{ik_0\left(x - \frac{\hbar k_0}{2m}t \right)} \, dk'$$

$$= N' \left(\frac{4\pi\alpha_0}{1 + 2i\hbar\alpha_0 t/m} \right)^{1/2} e^{-\alpha_0 \left(x - \frac{\hbar k_0 t}{m} \right)^2 \frac{1 - 2i\hbar\alpha_0 t/m}{1 + 4\hbar^2 \alpha_0^2 t^2/m^2}} e^{ik_0\left(x - \frac{\hbar k_0}{2m}t \right)}, \tag{3.52}$$

where we have defined $N' \equiv \frac{N}{2\pi} \sqrt{\frac{\pi}{\alpha_0}}$. We now proceed to perform the integral in Eq. 3.51 by stationary phase. We begin by identifying the phase Φ as the imaginary part of the exponent of the integrand:

$$\Phi = kx - \frac{\hbar k^2 t}{2m}. \tag{3.53}$$

As in the previous chapter, the stationary phase condition is

$$\frac{d\Phi}{dk} \bigg|_{k_{sp}} = \left(x - \frac{\hbar k t}{m} \right) \bigg|_{k_{sp}} = 0 \tag{3.54}$$

or

$$k_{\text{sp}} = \frac{mx}{\hbar t},$$ (3.55)

which implies that

$$\Phi_{\text{sp}} = \Phi(k_{\text{sp}}) = \frac{mx^2}{2\hbar t}.$$ (3.56)

We will also need the result:

$$\Phi'' = \frac{d^2\Phi}{dk^2} = -\frac{\hbar t}{m}.$$ (3.57)

The stationary phase evaluation of the integral yields

$$\Psi(x, t) \approx N' \left(\frac{2\pi i}{\Phi''(k_{\text{sp}})} \right)^{1/2} e^{-(k_{\text{sp}} - k_0)^2/4\alpha_0} e^{i\Phi(k_{\text{sp}})}$$

$$= N' \left(\frac{2\pi}{i\hbar t/m} \right)^{1/2} e^{-\left(x - \frac{\hbar k_0 t}{m}\right)^2 / (4\hbar^2\alpha_0 t^2/m^2)} e^{i\frac{m}{2\hbar t}\left(\left[x - \frac{\hbar k_0}{2m}t\right] + \frac{\hbar k_0 t}{m}\right)^2}$$ (3.58)

$$= N' \left(\frac{2\pi}{i\hbar t/m} \right)^{1/2} e^{-\left(x - \frac{\hbar k_0 t}{m}\right)^2 \frac{1 - 2i\hbar\alpha_0 t/m}{4\hbar^2\alpha_0 t^2/m^2}} e^{ik_0\left(x - \frac{\hbar k_0 t}{m}\right)}.$$ (3.59)

Comparing Eq. 3.59 with Eq. 3.52, we see that the stationary phase result differs from the exact result only in the constant 1 missing from the denominator of both the prefactor and the Gaussian width parameter. At long times this constant difference is insignificant, being overwhelmed by the t (or t^2) term. However, at short times the difference is significant, and in fact at $t = 0$ the stationary phase result blows up! This is typical of stationary phase approximations: there are special times, t, at which all the amplitude piles up at one point in space, corresponding to times at which all the classical trajectories coalesce to a single value of coordinate. For the harmonic oscillator, such times occur twice per period, but for the free particle there is only one such time, at $t = 0$.

To gain further insight into stationary phase integration, note that the general stationary phase procedure is to expand the phase to second order in the integration variable, leading to a quadratic exponent. It therefore seems surprising that stationary phase integration does not give the exact result for a Gaussian integral. A moment's reflection reveals that it is because the real part of the quadratic exponent is not included in the evaluation of the stationary phase condition. One might well ask: why can't one improve on stationary phase integration, at least to the extent of gettting Gaussian integrals correctly, by including the real part of the exponent in determining the stationary phase point? If this is done, the stationary phase point moves off the real axis and into the complex plane. This generalization of the stationary phase method, known as saddle-point integration, in fact works beautifully. Saddle-point integration indeed gives the exact result for Gaussian integrals; in fact, one may view saddle-point integration as replacing the integrand by the closest Gaussian approximation and then evaluating the integral using the standard Gaussian integral formula.

The above example provides considerable insight into the potential advantages of a semiclassical method based on Gaussian dynamics, with the accuracy of saddle-point integration, compared to $x - p$ dynamics, with the accuracy of stationary phase integration. In particular, the saddle-point method avoids problems with the conjugate points in time. We will return to this point later in Chapter 10 dealing with semiclassical methods.

Problems

▶ **Exercise 3.11** Show that the product of two Gaussians is a third Gaussian centered between the first two.

▶ **Exercise 3.12** Find an expression for the integral of a product of two complex Gaussians of the form in Eq. 3.28. Note that this expression is symmetric in x' and p'.

The text was very sketchy on the derivation of many of the central formulas in this chapter. Exercises 3.13–3.18 fill in the details.

▶ **Exercise 3.13** Let $\Psi(x, 0) = Ne^{-\alpha_0 x^2}$. Find an expression for $\Psi(x, t)$ by using the Fourier integral theorem. Show that $\Psi(x, t)$ can be written in the form $\Psi(x, t) = Ne^{-\alpha_t x^2 + i\gamma_t/\hbar}$.

▶ **Exercise 3.14** An alternative approach to the previous problem is this: guess that $\Psi(x, t) = Ne^{-\alpha_t x^2 + i\gamma_t/\hbar}$. Substitute this form into the Time-Dependent Schrödinger Equation and, by matching equations in powers of x, find expressions for α_t and γ_t. Show that these expressions are equal to those derived in the previous problem.

▶ **Exercise 3.15** Take the more general trial form $\Psi(x, t) = Ne^{-\alpha_t(x-x_t)^2 + \frac{i}{\hbar}p_t(x-x_t) + \frac{i}{\hbar}\gamma_t}$, where x_t and p_t are real. Substitute this form into the TDSE and obtain expressions for α_t, x_t, p_t and γ_t.

▶ **Exercise 3.16** Find $\overline{x}(t)$ and $\overline{\Delta x^2}(t)$, $\overline{p}(t)$ and $\overline{\Delta p^2}(t)$ for Exercise 3.15 above. Show that the uncertainty principle is satisfied:

$$\Delta x(t)\,\Delta p(t) \geq \frac{\hbar}{2}. \tag{3.60}$$

▶ **Exercise 3.17** Take the form for the Gaussian wavepacket, $\Psi(x, t) = N \exp(-\alpha_t(x - x_t)^2 + \frac{i}{\hbar}p_t(x - x_t) + \frac{i}{\hbar}\gamma_t)$, and substitute it into the Time-Dependent Schrödinger Equation for the linear potential, $V(x) = -kx$. Obtain expressions for the parameters α_t, x_t, p_t and γ_t.

▶ **Exercise 3.18** Take the form for the Gaussian wavepacket, $\Psi(x, t) = N \exp(-\alpha_t(x - x_t)^2 + \frac{i}{\hbar}p_t(x - x_t) + \frac{i}{\hbar}\gamma_t)$, and substitute it into the Time-Dependent Schrödinger Equation for the harmonic oscillator potential, $V(x) = \frac{1}{2}m\omega^2 x^2$. Obtain expressions for the parameters α_t, x_t, p_t and γ_t. [Hint: to simplify the algebra, write $\frac{1}{2}m\omega^2 x^2 = \frac{1}{2}m\omega^2[(x - x_t) + x_t]^2$. To solve the differential equation for α_t and γ_t it is convenient to introduce the substitution $\alpha_t = \frac{c\dot{z}}{z}$, where c is a constant that gives the harmonic equation $\ddot{z} = -\omega^2 z$.] What happens when $\alpha_0 = \frac{m\omega}{2\hbar}$? In this case the wavepacket is sometimes called a "coherent state," a confusing but widely used label.

▶ **Exercise 3.19**   Adapt your solution from the previous exercise to the parabolic barrier potential, $V(x) = -1/2m\omega^2 x^2$, by making the replacement $\omega \to i\omega$. Draw a set of pictures of the wavepacket as a function of time, taking $\Psi(x, 0) = Ne^{-\alpha_0 x^2}$.

▶ **Exercise 3.20**   Consider the forced harmonic oscillator: $V(x) = \frac{1}{2}m\omega^2 x^2 - F(t)x$.

  a. Show that the wavepacket given by $\Psi(x, t) = N \exp(-\alpha_t(x - x_t)^2 + \frac{i}{\hbar}p_t(x - x_t) + \frac{i}{\hbar}\gamma_t)$ is a solution to this problem. Substitute this form into the Time-Dependent Schrödinger Equation and obtain expressions for x_t, p_t, α_t and γ_t.

  b. A convenient way of solving the differential equations for x_t and p_t is to define a complex quantity $z(t) = p_t + im\omega x_t$. Then $z(t) = e^{i\omega t}\int_{t_0}^{t} F(t')e^{-i\omega(t'-t_0)}\,dt' + e^{i\omega(t-t_0)}z(t_0)$. Differentiate $z(t)$ with respect to time and show that the resulting equation for $\dot{z}(t)$ is consistent with Hamilton's equations of motion for x_t and p_t.

  c. Solve for $z(t)$ for

   i. $F(t) = A\sin(\omega t)$

   ii. $F(t) = A$

   iii. $F(t) = At$.

   Draw pictures in the complex plane of $z(t)$ for all these cases.

  d. Show that in general, if $t(dF/dt) \ll F$, then $z(t) = iF(t)/\omega$ (taking $x_0 = p_0 = 0$). [Hint: integrate by parts.] This is the limit of adiabatic forcing: if the change in the Hamiltonian is slow enough, an initial eigenstate will stay an eigenstate for all time.

References

General

1. E. Schrödinger, Der stetige Übergang von der Mikro- zur Macromechanik, Naturwissenschaften 14, 664 (1926). English translation in *Collected Papers on Wave Mechanics* by E. Schrödinger (Blackie and Son, London, 1928).

2. D. ter Haar, *Problems in Quantum Mechanics* (Infosearch, London, 1960), problems 3.10–3.15.

3. E. J. Heller, Time-dependent approach to semiclassical dynamics, J. Chem. Phys. 62, 1544 (1975); E. J. Heller, Classical S-matrix limit of wavepacket dynamics, J. Chem. Phys. 65, 4979 (1976).

4. C. Cohen-Tannoudji, B. Diu and F. Laloë, *Quantum Mechanics*, vol. I (Wiley, New York, 1977).

Chapter 4

Correspondence between Classical and Quantum Dynamics

In this chapter we explore the connections between quantum and classical dynamics. The first part of the chapter deals with Ehrenfest's theorem, which provides a relationship between the expectation values $\langle \hat{x} \rangle_t$ and $\langle \hat{p} \rangle_t$ for a moving wavepacket and the classical values x_t and p_t. Ehrenfest's theorem is exact for potentials up to quadratic and is valid for arbitrary potentials at short enough times. The second part of the chapter deals with a formulation of the Time-Dependent Schrödinger Equation, which strongly resembles the classical time-dependent Hamilton–Jacobi equation; the latter is a lesser known but powerful formulation of classical mechanics. The third section deals with wavepacket fractional revivals—a quite general effect of wavepacket refocusing at times well beyond those for which Ehrenfest's theorem applies.

4.1 Ehrenfest's Theorem

In this section, q and p will everywhere refer to operators; to simplify the notation we will neglect the traditional hat to designate an operator. Also, the symbol q, rather than x, is used to indicate the coordinate operator, to highlight the symmetry of p and q.

Consider the time dependence of the expectation value of some operator $\langle A \rangle$. Since

$$\langle A \rangle = \int \Psi^* A \Psi \, d\tau, \tag{4.1}$$

then

$$\frac{d}{dt}\langle A \rangle = \left\langle \frac{d\Psi}{dt} | A\Psi \right\rangle + \left\langle \Psi | A \frac{d\Psi}{dt} \right\rangle + \left\langle \Psi | \frac{\partial A}{\partial t} \Psi \right\rangle. \tag{4.2}$$

From the Time-Dependent Schrödinger Equation we know that

$$i\hbar \frac{\partial \Psi}{\partial t} = H\Psi$$

$$i\hbar \frac{\partial \Psi^*}{\partial t} = -H\Psi^*,$$

so we can rewrite Eq. 4.2 as

$$\frac{d}{dt}\langle A \rangle = \frac{1}{i\hbar}\langle -H\Psi | A\Psi \rangle + \frac{1}{i\hbar}\langle \Psi A | H\Psi \rangle + \left\langle \Psi \left| \frac{\partial A}{\partial t} \Psi \right. \right\rangle$$

$$= \frac{1}{i\hbar}\langle \Psi | [A, H] | \Psi \rangle + \left\langle \frac{\partial A}{\partial t} \right\rangle. \tag{4.3}$$

If the operator A has no explicit time dependence, then the time evolution of its expectation value is described by

$$i\hbar\frac{d}{dt}\langle A \rangle = \langle \Psi | [A, H] | \Psi \rangle. \tag{4.4}$$

Specializing Eq. 4.4 to the position, q, and momentum, p, operators we obtain

$$i\hbar\frac{d}{dt}\langle q \rangle = \langle \Psi | [q, H] | \Psi \rangle \tag{4.5}$$

$$i\hbar\frac{d}{dt}\langle p \rangle = \langle \Psi | [p, H] | \Psi \rangle. \tag{4.6}$$

For Hamiltonians of the form $H = p^2/2m + V(q)$, the commutators are easily evaluated. The result is known as Ehrenfest's theorem:

$$\frac{d}{dt}\langle q \rangle = \frac{\langle p \rangle}{m} \tag{4.7}$$

$$\frac{d}{dt}\langle p \rangle = \left\langle -\frac{\partial V}{\partial q} \right\rangle. \tag{4.8}$$

It is important to notice that Eqs. 4.7 and 4.8 do not form a closed set of ordinary differential equations since we need to know the wavefunction in order to evaluate $\langle -\partial V/\partial q \rangle$. If Eq. 4.8 had been the more "classical,"

$$\frac{d}{dt}\langle p \rangle = -\frac{\partial V(\langle q \rangle)}{\partial \langle q \rangle}, \tag{4.9}$$

then given a knowledge of only $\Psi(q, 0)$ and Ehrenfest's theorem, one could predict the time evolution of $\langle q \rangle_t$ and $\langle p \rangle_t$.

Is it ever possible to use the classical equation, Eq. 4.9? Expand the derivative of the potential operator, $V'(q) = dV(q)/dq$, in Eq. 4.8 in an operator power series about the expectation value of q:

$$V'(q) = V'(\langle q \rangle) + (q - \langle q \rangle)V''(\langle q \rangle) + \frac{1}{2}(q - \langle q \rangle)^2 V'''(\langle q \rangle) + \cdots \tag{4.10}$$

The expectation value of $V'(q)$ may then be approximated by

$$\langle V'(q) \rangle \approx V'(\langle q \rangle) + \frac{1}{2}\chi V'''(\langle q \rangle), \tag{4.11}$$

where $\chi \equiv \langle q^2 \rangle - \langle q \rangle^2$. The first term of Eq. 4.11 is the classical term appearing in Eq. 4.9. The second term is the error in using the classical expression. This error term depends on the width of the wavepacket and the third derivative of the potential. For up to harmonic potentials, the third term in Eq. 4.10 vanishes and the classical Ehrenfest equations, Eqs. 4.7

and 4.9, are exact. For more general potentials, the expectation values of position and momentum follow the classical trajectories for compact wavefunctions (small width), for short times (little dispersion) or for nearly harmonic potentials.

Given that we have a good zeroth order estimate for the position of the center of the wavepacket, $\langle q \rangle$, and hence $V'''(\langle q \rangle)$, we could know when the classical and quantum expectations diverge if we knew χ. Note that χ is determined by the global shape of the wavefunction, and hence nominally requires solution of the full quantum equations of motion. However, we will now show that an approximate equation of motion for χ can be derived that only involves the solution of a small set of ordinary differential equations (ODEs). The equation of motion for χ may be obtained by applying Eq. 4.4:

$$
\begin{aligned}
\frac{d\chi}{dt} &= \frac{d}{dt}\{\langle q^2 \rangle - \langle q \rangle^2\} \\
&= \frac{1}{i\hbar}\{\langle [q^2, H] \rangle - 2\langle q \rangle \langle [q, H] \rangle\} \\
&= \frac{1}{m}\{\langle qp + pq \rangle - 2\langle q \rangle \langle p \rangle\}.
\end{aligned}
\tag{4.12}
$$

Since we cannot evaluate $\langle qp + pq \rangle$ without having the wavefunction in hand, we take the second derivative of χ:

$$
\begin{aligned}
\frac{d^2\chi}{dt^2} &= \frac{d}{dt}\left\{\frac{1}{m}(\langle qp + pq \rangle - 2\langle q \rangle \langle p \rangle)\right\} \\
&= \frac{1}{i\hbar}\frac{1}{m}\{\langle [qp, H] \rangle + \langle [pq, H] \rangle - 2\langle q \rangle \langle [p, H] \rangle - 2\langle p \rangle \langle [q, H] \rangle\}.
\end{aligned}
$$

The commutators may be calculated easily for time-independent Hamiltonians. If the Taylor expansion of the potential, Eq. 4.10, is then used and terms containing only $V'(\langle q \rangle)$ and $V''(\langle q \rangle)$ are retained, we obtain

$$
\frac{d^2\chi}{dt^2} \approx \frac{2}{m^2}\{\langle p^2 \rangle - \langle p \rangle^2 - mV''(\langle q \rangle)\chi\}.
\tag{4.13}
$$

We are not yet done since Eq. 4.13 contains $\langle p^2 \rangle$, which again nominally requires knowledge of the full shape of the wavepacket. However, note that $\langle H \rangle = \langle p^2 \rangle/2m + \langle V(q) \rangle$ is a conserved quantity. Moreover, the classical energy E_{cl} is conserved by the classical motion. If we make the approximation that

$$
E_{cl} = \frac{p_{cl}^2}{2m} + V_{cl} \approx \frac{\langle p \rangle^2}{2m} + V(\langle q \rangle),
\tag{4.14}
$$

then the quantity $\epsilon = \langle H \rangle - E_{cl}$, which is some sort of zero-point energy, is

$$
\epsilon = \frac{\langle p^2 \rangle}{2m} + \langle V(q) \rangle - \frac{\langle p \rangle^2}{2m} - V(\langle q \rangle).
\tag{4.15}
$$

Expanding $\langle V(q) \rangle$ as before, we see that

$$
\langle V(q) \rangle - V(\langle q \rangle) \approx \frac{1}{2}\chi V''(\langle q \rangle),
\tag{4.16}
$$

and so Eq. 4.15 reduces to

$$\epsilon = \frac{1}{2m}\{\langle p^2 \rangle - \langle p \rangle^2 + mV''(\langle q \rangle)\chi\}.$$ (4.17)

This is the key equation. Since ϵ is a conserved quantity it can be calculated once for all time, at $t = 0$ when the full shape of the wavefunction is known. This equation then allows us to eliminate the unknown quantity $\langle p^2 \rangle$ in terms of χ, that is, we can rewrite Eq. 4.13 as

$$\frac{d^2\chi}{dt^2} \approx \frac{4}{m}\{\epsilon - V''(\langle q \rangle)\chi\}.$$ (4.18)

We now have a set of three coupled ODEs that allow us to integrate the Ehrenfest equations, and to know when they are breaking down:

$$\frac{d}{dt}\langle q \rangle = \frac{\langle p \rangle}{m}$$

$$\frac{d}{dt}\langle p \rangle = -V'(\langle q \rangle) - \frac{1}{2}V'''(\langle q \rangle)\chi$$ (4.19)

$$\frac{d^2}{dt^2}\chi = \frac{4}{m}\{\epsilon - V''(\langle q \rangle)\chi\}.$$

To integrate this set of equations, we compute $\chi(0)$, $\dot{\chi}(0)$ and ϵ from the choice of $\Psi(x, 0)$. The value of $\ddot{\chi}$ can be computed from the classical trajectory.

4.2 Bohmian Mechanics and the Classical Limit

In this section, we will examine an alternative formulation of the TDSE, often called the hydrodynamic formulation. In this formulation, the TDSE is rewritten as two equations, one for the probability density and one for a phase function whose derivative gives the local velocity. The phase function is observed to obey an equation of motion almost identical to the Hamilton–Jacobi (HJ) equation of classical mechanics. The classical HJ equation, although it is a wave equation, is completely equivalent to Newton's laws, and therefore it is tempting to think of the quantum HJ equation as a natural way to pass to the classical limit. However, the quantum HJ equation has an extra term relative to the classical HJ equation, which is referred to as the "quantum potential." The quantum potential can be incorporated on the level of Newton's laws as an additional "quantum force," which was given a physical interpretation by David Bohm. Although Bohmian mechanics is completely equivalent to the Schrödinger equation, and recently efficient numerical methods have been developed for implementing the equations of motion, we discuss some difficulties in using the Bohmian viewpoint to pass to the classical limit, and present an alternative perspective.

4.2.1 Hydrodynamic Formulation of the Schrödinger Equation

We begin by writing the wavefunction in the following trial form:

$$\Psi(x, t) = A(x, t)e^{\frac{i}{\hbar}S(x,t)},$$ (4.20)

which is the product of a magnitude and a phase. Both $A(x, t)$ and $S(x, t)$ are considered to be real functions. Substituting this *ansatz* into the Time-Dependent Schrödinger Equation and separating it into its real and imaginary parts we obtain two equations:

$$\frac{\partial S}{\partial t} + \frac{1}{2m}\left(\frac{\partial S}{\partial x}\right)^2 + V = \frac{\hbar^2}{2mA}\frac{\partial^2 A}{\partial x^2} \tag{4.21}$$

$$\frac{\partial A}{\partial t} + \frac{1}{m}\frac{\partial A}{\partial x}\frac{\partial S}{\partial x} + \frac{A}{2m}\frac{\partial^2 S}{\partial x^2} = 0. \tag{4.22}$$

Multiplying both sides of Eq. 4.22 by $2A$ we obtain

$$\frac{\partial \rho(x, t)}{\partial t} + \frac{1}{m}\frac{\partial \rho}{\partial x}\frac{\partial S}{\partial x} + \frac{\rho}{m}\frac{\partial^2 S}{\partial x^2} = 0, \tag{4.23}$$

where we have defined ρ as

$$\rho \equiv A^2 = |\Psi(x, t)|^2. \tag{4.24}$$

Clearly, ρ is a probability density. Equations 4.21 and 4.22 (or 4.23) are called the hydrodynamic formulation of the Schrödinger equation.

By analogy to classical fluid dynamics, there are two approaches to solving the hydrodynamic equations. In the Eulerian approach, the equations are solved in a space-fixed frame, associated with $\partial/\partial t$. To pursue the Eulerian approach we rewrite Eq. 4.23 as

$$\frac{\partial \rho(x, t)}{\partial t} + \frac{\partial}{\partial x}\left(\rho\frac{1}{m}\frac{\partial S}{\partial x}\right) = \frac{\partial \rho(x, t)}{\partial t} + \frac{\partial \jmath(x, t))}{\partial x} = 0 \tag{4.25}$$

where

$$\jmath \equiv \rho\frac{1}{m}\frac{\partial S}{\partial x}. \tag{4.26}$$

If we identify $\frac{\partial S}{\partial x}$ with a momentum mv, then $\jmath = \rho v$ is seen to have the interpretation of a probability density times a velocity, which is a current, or flux. Equation 4.25 says that the change in probability density at point x is given by the difference between the flow in and the flow out at that point; this is just a statement of conservation of probability.

▶ **Exercise 4.1**

a. Consider the usual expression for the probability density, $\Psi^*(x, t)\Psi(x, t)$. Show that

$$\frac{\partial(\Psi^*\Psi)}{\partial t} = \frac{\partial}{\partial x}\text{Re}\left(\Psi^*\left(\frac{i\hbar}{m}\frac{\partial\Psi}{\partial x}\right)\right). \tag{4.27}$$

b. Equation 4.27 may be written in the form

$$\frac{\partial \rho'}{\partial t} + \frac{\partial \jmath'}{\partial x} = 0, \tag{4.28}$$

where $\rho' = \Psi^*\Psi$ is a probability density and $\jmath' = \text{Im}(\Psi^*(\frac{\hbar}{m}\frac{\partial\Psi}{\partial x}))$ is a current or flux. Substitute the *ansatz* for $\Psi(x, t)$, Eq. 4.20, into these expressions for ρ' and $\jmath'$, and show that $\rho' = \rho = A^2$ and $\jmath' = \jmath = \frac{A^2}{m}\frac{\partial S}{\partial x}$.

▶ **Exercise 4.2**   Show that in multidimensions, Eq. 4.25 takes the form

$$\frac{\partial \rho(\mathbf{x}, t)}{\partial t} + \nabla \cdot \mathbf{\mathcal{J}}(\mathbf{x}, t) = 0, \tag{4.29}$$

where

$$\mathbf{\mathcal{J}} \equiv \rho \frac{1}{m} \nabla S = \rho \mathbf{v}. \tag{4.30}$$

In the Lagrangian approach, the equations are solved in a reference frame moving with the fluid. From this perspective, x becomes a dependent variable, depending on t, and we see that

$$\frac{d\rho(x(t), t)}{dt} = \frac{\partial \rho}{\partial t} + \frac{\partial \rho}{\partial x} \frac{dx}{dt} = -\rho \frac{\partial v}{\partial x}, \tag{4.31}$$

where we have used Eq. 4.23 and the relation $v = \frac{dx}{dt} = \frac{1}{m} \frac{\partial S}{\partial x}$.

4.2.2   Bohmian Mechanics

We now turn to Eq. 4.21. Equation 4.21 can be written in the following suggestive way:

$$\frac{\partial S(x, t)}{\partial t} + \frac{1}{2m} \left(\frac{\partial S}{\partial x} \right)^2 + V(x, t) + Q(\rho; x, t) = 0. \tag{4.32}$$

Equation 4.21 is called the quantum Hamilton–Jacobi (HJ) equation; it is identical in form to the classical HJ equation,

$$\frac{\partial S(x, t)}{\partial t} + \frac{1}{2m} \left(\frac{\partial S}{\partial x} \right)^2 + V(x, t) = 0, \tag{4.33}$$

with the addition of the term Q, where

$$Q(\rho; x, t) = -\frac{\hbar^2}{2m} \frac{1}{A} \frac{\partial^2 A}{\partial x^2} = -\frac{\hbar^2}{2m} \rho^{-1/2} \frac{\partial^2 \rho^{1/2}}{\partial x^2}. \tag{4.34}$$

Q may be thought of as a time-dependent *quantum potential*. Notice that the quantum potential is a functional of the density distribution, ρ; in particular, Q is related to the curvature of the probability amplitude. Although Eq. 4.32 has its roots going back to de Broglie and Madelung in the 1920s, an attempt to give a physical interpretation to the quantum potential and the associated quantum force is associated with David Bohm in the early 1950s.

Both the classical and the quantum HJ equations are partial differential equations for the quantity $S(x, t)$. The classical HJ equation is completely equivalent to Newton's laws, and therefore can be solved by propagating classical trajectories. It is therefore natural to ask if the quantum HJ equation can also be solved by propagating *quantum* trajectories. The answer is yes, once one has calculated the quantum potential, whose determination is nonlocal. The propagation of these quantum trajectories, along with their interpretation, is known as Bohmian mechanics.

Taking the spatial derivative of Eq. 4.32 leads to the equation of motion:

$$m\left(\frac{\partial v}{\partial t} + v\frac{\partial v}{\partial x}\right) = -\frac{\partial}{\partial x}(V + Q). \tag{4.35}$$

Recognizing the LHS as the Lagrangian time derivative, $d/dt = \partial/\partial t + v\frac{\partial}{\partial x}$, of v, we find that

$$m\frac{dv}{dt} = f_{\text{cl}} + f_{\text{q}}, \tag{4.36}$$

where the classical force, f_{cl}, is given by $f_{\text{cl}} = -\frac{\partial V}{\partial x}$ and the quantum force, f_{q}, is given by $f_{\text{q}} = -\frac{\partial Q}{\partial x}$. Solving Eq. 4.36 is essentially the same as running classical trajectories, once the quantum force f_{q} is known. However, formally the quantum force is determined by nonlocal information; it requires knowing the full density distribution ρ in space, and can be sensitive to the behavior in the tails of the distribution. There are several key ideas involved in developing efficient algorithms to calculate the quantum force. The first idea is to solve for ρ in the Lagrangian frame, that is, in a frame that moves with the flow. Formally, Eq. 4.31 for ρ in the Lagrangian frame can be integrated to give (Lopreore, 1999)

$$\rho(t + \Delta t) = e^{-\frac{\partial v}{\partial x}\Delta t}\rho(t). \tag{4.37}$$

The second idea has to do with the numerical sampling of ρ; the idea is to use the trajectory locations calculated by solving Eq. 4.36 as a moving grid on which the equation for ρ is calculated.

Equations 4.32–4.36 describe the passage from the S function to the propagation of trajectories. It is interesting to consider the reverse passage as well—the reconstruction of the S function from trajectory information. Equation 4.31, together with Eq. 4.36, gives the density and the flux, but not the S function; yet, with a little extra work one can recover the S function, and hence the time-dependent wavepacket from the Bohmian trajectories. Since $\Psi(x, t) = A(x, t)e^{iS(x,t)/\hbar}$, we can write

$$A = \rho^{1/2} = e^{-\frac{1}{2}\frac{\partial v}{\partial x}}\rho(0), \tag{4.38}$$

where we have integrated Eq. 4.31. Furthermore, S can be calculated by the time integral of the quantum Lagrangian:

$$S(x(t)) = S(x(t_0)) + \int_{t_0}^{t} L_{\text{q}}(x(\tau))\, d\tau, \tag{4.39}$$

where $L_{\text{q}}(t)$ is the quantum Lagrangian at time t,

$$L_{\text{q}} = L_{\text{cl}} - Q = T - (V + Q). \tag{4.40}$$

Combining Eqs. 4.38–4.39 we obtain:

$$\Psi(x_j(t)) = \Psi(x_j(t_0)) \exp\left\{-\frac{1}{2}\int_{t_0}^{t}\left(\frac{\partial v}{\partial x}\right)_{x_j(\tau)} d\tau\right\} \times \exp\left\{\frac{i}{\hbar}\int_{t_0}^{t} L_{\text{q}}(x_j(\tau))d\tau\right\}, \tag{4.41}$$

where the index j indicates that all quantities are evaluated along the quantum trajectory of particle j. The full wavefunction, $\Psi(x, t)$, is the sum over all quantum trajectories with initial conditions determined by sampling the initial wavefunction (Wyatt, 1999).

4.2.3   The Classical Limit of the Quantum Hamilton–Jacobi Equation

It is tempting to use the quantum HJ equation as a starting point for analyzing the classical limit of quantum mechanics. Indeed, many standard textbooks present the quantum hydrodynamic equations and then argue that since the term on the RHS of Eq. 4.21 associated with the quantum potential, Q, goes as $\hbar^2$, that the classical limit is obtained by setting this term equal to zero. We now argue that this viewpoint is problematic, and present an alternative.

To begin with, we will analyze the size of the quantum potential, Q, for the case of a Gaussian wavepacket in a harmonic oscillator potential. Since this problem can be solved completely analytically (Section 3.3), it provides a good example to examine the relative size of all the terms. The Hamiltonian is given by

$$H = -\frac{\hbar^2}{2m}\frac{\partial^2}{\partial x^2} + \frac{1}{2}m\omega^2 x^2. \tag{4.42}$$

We choose as the wavepacket the coherent state from Section 3.3:

$$\Psi(x, t) = e^{-\alpha(x-x_t)^2 + \frac{ip_t}{\hbar}(x-x_t) + \frac{i\gamma_t}{\hbar}}, \tag{4.43}$$

with $\alpha = m\omega/2\hbar$. For this wavepacket, we can identify

$$A(x, t) = e^{-\alpha(x-x_t)^2} \tag{4.44}$$

$$S(x, t) = p_t(x - x_t) + \gamma_t. \tag{4.45}$$

Carrying out the necessary derivatives yields Eq. 4.21 in the explicit form

$$\dot{p}_t(x - x_t) + \dot{\gamma}_t - p_t\dot{x}_t + \frac{1}{2m}p_t^2 + V(x) = \frac{\hbar^2}{2m}(4\alpha^2(x - x_t)^2 - 2\alpha). \tag{4.46}$$

Using Eqs. 3.37–3.39, we may simplify Eq. 4.46 to

$$\frac{1}{2}m\omega^2(x - x_t)^2 = \frac{2\hbar^2}{m}\alpha^2(x - x_t)^2. \tag{4.47}$$

It is readily seen that if we take $\alpha = m\omega/2\hbar$, as for the coherent state in Section 3.3, the LHS and RHS of Eq. 4.46 are indeed equal. However, note that the $\hbar^2$ term on the right-hand side of the equation is no smaller than other terms we have been including! Specifically, Eq. 4.46 indicates that for a Gaussian wavepacket the quantum potential is an inverted parabola; since classical–quantum correspondence is exact for a Gaussian in a harmonic oscillator potential, one can well question the physical meaning of adding an inverted parabolic potential.

One may argue that the term on the RHS of Eq. 4.46 really *is* small if we take the limit $\hbar \to 0$ "properly." The form $\Psi = Ae^{iS/\hbar}$ presupposes that there is an $\hbar$ dependence only in the phase of Ψ, but not in its amplitude, while in Eq. 4.46 we have introduced an $\hbar$ dependence into the *envelope* of the Gaussian wavepacket through the relationship $\alpha = m\omega/2\hbar$. We will return to this objection at the end of this subsection, and take issue with it. For the present we simply note that there is another disturbing problem with claiming that the limit of the quantum potential $Q \to 0$ is the classical limit, and that is the inconsistency with respect to Ehrenfest's theorem. A key assumption in deriving Ehrenfest's theorem was that the wavepacket must be narrow relative to third derivatives of the potential (cf. Eq. 4.11). However, in neglecting the RHS of Eq. 4.21, the ratio $\frac{\partial^2 A}{\partial x^2}/A$ must be small, which is just a condition that the wavepacket be broad! From the Ehrenfest perspective, a "quantum correction" of the form $\frac{\partial^2 A}{\partial x^2}/A$ must be an artifact: narrow wavepackets do not in themselves signify a breakdown in classical–quantum correspondence, since the center of an arbitrarily narrow Gaussian in a harmonic potential moves according to classical mechanics. It is astonishing that many of the leading textbooks treat Ehrenfest's theorem and the classical limit based on Eq. 4.21 side by side, completely ignoring the contradictory nature of the underlying assumptions!

The paradoxes described above with viewing the quantum potential as a "correction" to classical mechanics prompts us to ask if we can reformulate Eqs. 4.21–4.22 so that at least for a Gaussian in a harmonic oscillator the $\hbar^2$ "quantum correction" term vanishes, or nearly vanishes. The key is to include the quadratic exponent of the Gaussian along with the phase in the definition of S, which in turn implies complex S. Consider the trial form,

$$\Psi(x,t) = e^{\frac{i}{\hbar}S(x,t)}, \tag{4.48}$$

but where we now allow $S(x,t)$ to be complex. Note that we now no longer have an envelope function, $A(x,t)$, since the envelope is included in the complex phase, $S(x,t)$. Substituting this *ansatz* into the Time-Dependent Schrödinger Equation we now obtain a single equation:

$$\frac{\partial S}{\partial t} + \frac{1}{2m}\left(\frac{\partial S}{\partial x}\right)^2 + V = \frac{i\hbar}{2m}\frac{\partial^2 S}{\partial x^2}. \tag{4.49}$$

Although Eq. 4.49 looks almost identical to Eq. 4.21, there is a profound difference: the quantum correction term now involves the second derivative of the phase, rather than the amplitude. For a Gaussian in a harmonic potential this quantum correction term takes the form

$$\frac{i\hbar}{2m}\frac{\partial^2 S}{\partial x^2} = -\frac{\hbar^2 \alpha_t}{m}. \tag{4.50}$$

Comparing the RHS of Eq. 4.50 with the RHS of Eq. 4.46, we see that all x dependence has now disappeared: at least for the harmonic potential, the "quantum correction" is independent of coordinate, and will contribute only a time-dependent phase. In fact, this quantum correction is identical to the peculiar nonclassical term that appears alongside the terms responsible for the classical action in the equation for the phase in Chapter 3 (Eq. 3.39)! There, the extra term played two roles: it led to a nonclassical phase factor—a time-dependent equivalent of the Maslov index that appears in Section 6.3—and it ensured conservation of wavepacket normalization. In all probability the quantum correction term plays the same role here. This points to a picture of quantum mechanics as classical mechanics with complex actions and additional, geometrical quantum phases. This contrasts with the Bohm interpretation of quantum mechanics, in which the action is real-valued, and the $\hbar^2$ term is viewed as giving rise to a peculiarly quantum force.

Notice that in adopting the *ansatz* $\Psi = e^{iS/\hbar}$, that is, incorporating the amplitude A into the phase factor, we have effectively defined the classical limit $\hbar \to 0$ as applying to both the amplitude and the phase. This reduces the size and character of the quantum potential dramatically in such a way that consistency with Ehrenfest's theorem is maintained. Since the phase is generally closely related to the classical action, this leads formally to the consideration of complex action integrals. Allowing S to be complex is reminiscent of our discussion in the previous chapter, where we showed that stationary phase integration of a Gaussian in a free-particle potential could be made exact by including the Gaussian envelope in the definition of the (complex) phase of the integrand. To the author's knowledge, Bohmian mechanics with complex actions has not yet been explored, although a related idea of time-dependent WKB theory with complex actions has been tested and gives impressive results (Boiron, 1998).

4.2.4   The Geometrical Optics Limit of Quantum Mechanics

In this section we describe an analogy between the classical limit of the hydrodynamic formulation of quantum mechanics and the geometical optics limit of wave optics. In the geometrical optics limit of wave optics, light follows rays that are perpendicular to the wavefronts representing constant optical path, or phase, of the electromagnetic field. Similarly, classical trajectories can be understood as rays that move perpendicular to the wavefronts representing constant values of Hamilton's characteristic function, S, which is the solution of the classical Hamilton–Jacobi equation.

To explore this idea more quantitatively, we return to Eq. 4.21. For a stationary state at energy E, $\partial S/\partial t = -E$ and Eq. 4.21 takes the form

$$\left(\frac{\partial S}{\partial x}\right)^2 = 2m(E - V) + \frac{\hbar^2}{A}\frac{\partial^2 A}{\partial x^2} \tag{4.51}$$

or in multidimensional notation

$$(\nabla S)^2 = 2m(E - V) + \hbar^2\frac{\nabla^2 A}{A}. \tag{4.52}$$

Equation 4.52 may be rewritten as

$$(\nabla S)^2 = \frac{\hbar^2}{\lambda^2}\left(1 + \lambda^2\frac{\nabla^2 A}{A}\right), \tag{4.53}$$

where we have defined a characteristic wavelength, λ, as

$$\lambda = \frac{\hbar}{\sqrt{2m(E - V(\mathbf{x}))}}. \tag{4.54}$$

If the condition

$$\lambda^2\frac{\nabla^2 A}{A} \ll 1 \tag{4.55}$$

is satisfied, Eq. 4.53 becomes

$$(\nabla S)^2 = \frac{\hbar^2}{\lambda^2}. \tag{4.56}$$

Since the velocities of the particles are proportional to ∇S, the trajectories of these particles are orthogonal to the surfaces of constant phase S, as shown in Figure 4.1. Equation 4.56

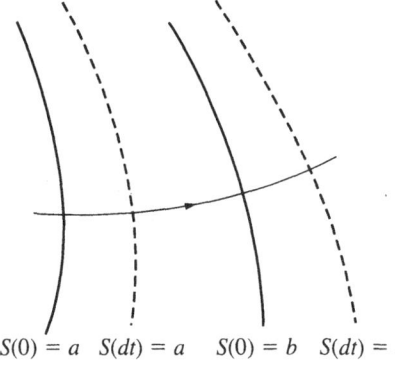

$S(0) = a$ $S(dt) = a$ $S(0) = b$ $S(dt) = b$

Figure 4.1 The motion of surfaces of constant S. These surfaces may be considered as wavefronts propagating in configuration space.

has the structure of the equation of the wavefronts in geometrical optics (Born, 1965; Messiah, 1958). In optics, S is called the "optical path" and is equivalent to the phase of the electric and magnetic field if polarization is neglected in Maxwell's equations. The role of $\hbar^2/\lambda^2$ is played by n^2, where n is the (position-dependent) index of refraction. In the limit of geometrical optics, the light rays move perpendicular to the wavefronts determined by constant optical path. There are two significant things about this limit. First, the path does not depend on the amplitude of the light, only on its phase. Second, once the amplitude of the light is given at one point along the path, the amplitude at all other points along the path is completely determined by geometrical considerations alone. One might say that the amplitude is "slaved" to the phase along the ray of light (see Exercise 4.9). Combining these considerations we find that the path of the light ray, as well as its amplitude, is completely decoupled from that of the neighboring rays. The analog in classical-limit quantum mechanics is that trajectory motion becomes decoupled from its neighbors, as opposed to fully quantum mechanical evolution where the change in amplitude and phase at each point depends in principle on the values at all other points. Put another way, quantum evolution is "global" while in the classical limit the evolution of both the phase and amplitude is "local." (Note that the analogy of the optical and the quantum mechanical case is not complete in that in the optical case the RHS depends only on the properties of the medium (the refractive index) whereas in the quantum mechanical case the RHS depends on the particular solution to the wave equation through the value of the energy E.) In the applied mathematics literature, the solution of a partial differential equation by solving ordinary differential equations along certain curves is known as the method of characteristics, and there is a rich literature on the subject (see, for example, Whitham (1974)).

▶ **Exercise 4.3**   How far can the geometrical optics analogy be taken if the wavefunction $\Psi(x, t)$ is not a stationary state?

Unfortunately, it is not generally the case that Eq. 4.55 is satisfied. Equation 4.55 corresponds to the neglect of the quantum force Q, which, as we discussed in the previous subsection, is often questionable. Indeed, the quantum force is precisely what makes quantum dynamics nonlocal. An alternative is to not neglect the LHS of Eq. 4.55 at all, but to add it to the classical potential as an additional quantum potential as discussed in

Section 4.2.2. One can still write a wavefront equation if one likes, but with a modified λ that now depends on the amplitude since the second term in Eq. 4.53 involves $\nabla^2 A$ (Holland, 1993). However, this viewpoint is now of limited utility, since the value of A is no longer determined just by the wavefronts of S and hence the evolution is fundamentally nonlocal as in the full quantum mechanics.

4.3   Fractional Revivals

Within the general context of classical–quantum correspondence falls the interesting topic of fractional revivals of wavepackets. Ehrenfest's theorem tells us that on sufficiently short time scales, the quantum expectation values $\langle x \rangle_t$ and $\langle p \rangle_t$ evolve in the same way as do the classical variables x_t and p_t. However, on longer time scales for anharmonic potentials Ehrenfest's theorem quite generally breaks down. It is a remarkable fact that on a still much longer time scale there is, in many cases of interest, an almost complete revival of the wavepacket, and a second Ehrenfest epoch. In between these full revivals are an infinite number of fractional revivals, which collectively have an interesting mathematical structure. The concept of an Ehrenfest time applies again to each of these fractional revivals (Averbukh, 1991).

Consider the preparation of a coherent superposition of bound states in an anharmonic oscillator or a Coulomb system. The initial wavepacket can be written as

$$\Psi(x, t) = \sum_n c_n \chi_n(x) e^{-i E_n t / \hbar}. \tag{4.57}$$

In the region of energy E corresponding to large quantum numbers n, we may expand E_n in a Taylor series about the central energy level, $E_{\bar{n}}$, to obtain

$$E_n = E_{\bar{n}} + \frac{dE}{d\bar{n}}(n - \bar{n}) + \frac{1}{2}\frac{d^2 E}{d\bar{n}^2}(n - \bar{n})^2 + \cdots. \tag{4.58}$$

In the limit of large n, we may make the identification $J = \bar{n}h$, where J is the classical "action" at energy E_n and h is Planck's constant. This is the Bohr semiclassical quantization condition, which will be discussed briefly in the next chapter. Substituting into Eq. 4.58 yields

$$E_n = E_{\bar{n}} + h\frac{dE}{dJ}(n - \bar{n}) + \frac{h^2}{2}\frac{d^2 E}{dJ^2}(n - \bar{n})^2 + \cdots. \tag{4.59}$$

This introduction of the classical variable, J, allows us to use the classical mechanical relation

$$\frac{dE}{dJ} = \nu_{\text{cl}} = \frac{1}{T_{\text{cl}}}, \tag{4.60}$$

where ν_{cl} is the classical frequency of motion and T_{cl} is the classical period at energy E_n (see, for example, Goldstein (1950), Ch. 9, Eq. (9-38)). Keeping only the first terms in the Taylor series expansion we find

$$E_n = E_{\bar{n}} + h\nu_{\text{cl}}(n - \bar{n}) + \frac{h^2}{2}\frac{d\nu_{\text{cl}}}{dJ}(n - \bar{n})^2 + \cdots \tag{4.61}$$

$$= E_{\bar{n}} + h(k/T_{\text{cl}} + k^2/T_{\text{rev}}), \tag{4.62}$$

where $k \equiv n - \bar{n} = 0, \pm 1, \pm 2, \ldots$, and we have defined

$$T_{\text{rev}} = \frac{2}{h} \left(\frac{d^2 E}{dJ^2} \right)^{-1} = \frac{2}{h} \left(v_{\text{cl}} \frac{dv_{\text{cl}}}{dE} \right)^{-1}, \tag{4.63}$$

where we have used the fact that

$$\frac{d^2 E}{dJ^2} = \frac{dv_{\text{cl}}}{dJ} = \frac{dv_{\text{cl}}}{dE} \frac{dE}{dJ} = \frac{dv_{\text{cl}}}{dE} v_{\text{cl}}. \tag{4.64}$$

Substituting only up to linear terms into Eq. 4.57 yields

$$\Psi(x, t) \approx \Psi_{\text{cl}}(x, t) \equiv \sum_k c_k \chi_k(x) e^{-2\pi i k t / T_{\text{cl}}}. \tag{4.65}$$

(We are neglecting the overall phase factor $e^{-iE_{\bar{n}}t/\hbar}$, which does not affect the spatial distribution). This series is perfectly periodic, and is identical to the series one obtains for a harmonic oscillator, where the levels are equally spaced. This approximation is valid for times

$$t \ll \frac{T_{\text{rev}}}{(\Delta n)^2}, \tag{4.66}$$

where Δn is the spread in quantum numbers with significant amplitude in the wavepacket. When $t \approx \frac{T_{\text{rev}}}{(\Delta n)^2}$ the quadratic terms in the Taylor series can no longer be neglected and we obtain the following expression for the wavefunction:

$$\Psi(x, t) = \sum_{k=-\infty}^{\infty} c_k \chi_k(x) e^{-2\pi i \left(k \frac{t}{T_{\text{cl}}} + k^2 \frac{t}{T_{\text{rev}}} \right)}. \tag{4.67}$$

At times

$$t = \frac{m}{n} T_{\text{rev}}, \tag{4.68}$$

where m and n are mutually prime integers, Eq. 4.67 can be written as

$$\Psi(x, t) = \sum_{k=-\infty}^{\infty} c_k \chi_k(x) e^{-2\pi i \left(k \frac{t}{T_{\text{cl}}} + \theta_k \right)}, \tag{4.69}$$

where

$$\theta_k = \left\{ \frac{m}{n} k^2 \right\} \tag{4.70}$$

is the fractional part of the argument. For example, $\{0.7\} = 0.7; \{10.3\} = 0.3$, and so on. The quantity θ_k is periodic in k. If we take $m = 1, n = 2$, the sequence obtained is $0, 1/2, 0, 1/2, \ldots$. If $m = 1, n = 3$, the sequence obtained is $0, 1/3, 1/3, 0, 1/3, 1/3$. The general rule is that the periodicity is n except when $\{\frac{n}{4}\} = 0$, in which case the periodicity is $\frac{n}{2}$.

Since the function $\theta_k = \{\frac{m}{n} k^2\}$ is discrete and periodic in k, so is the function $e^{-2\pi i \theta_k} = e^{-2\pi i \{\frac{m}{n} k^2\}}$. From the theory of the discrete Fourier transform (Section 6.2.4), the series $e^{-2\pi i \theta_k}$ can be expanded in a Fourier series with the number of terms equal to the periodicity—either n or $n/2$. With complete generality we may take the Fourier series as having n terms, since a series that has periodicity $n/2$ also has periodicity n. Therefore,

$$e^{-2\pi i \theta_k} = \sum_{s=0}^{n-1} a_s e^{-2\pi i k s/n} \qquad k = 0, 1, \ldots, n-1 \qquad (4.71)$$

with a_s given by

$$a_s = \frac{1}{n} \sum_{k=0}^{n-1} e^{-2\pi i \theta_k + 2\pi i k s/n}. \qquad (4.72)$$

Substituting Eq. 4.71 into Eq. 4.69 we obtain

$$\begin{aligned}
\Psi(x, t) &= \sum_{s=0}^{n-1} a_s \sum_{k=-\infty}^{\infty} c_k \chi_k(x) e^{-2\pi i (k \frac{t}{T_{cl}} + k \frac{s}{n})} \\
&= \sum_{s=0}^{n-1} a_s \sum_{k=-\infty}^{\infty} c_k \chi_k(x) e^{-2\pi i \frac{k}{T_{cl}} (t + \frac{s}{n} T_{cl})} \\
&= \sum_{s=0}^{n-1} a_s \Psi_{cl}(x, t + (s/n) T_{cl}).
\end{aligned} \qquad (4.73)$$

In obtaining the last line we have noted that the extra phase $e^{-2\pi i k \frac{s}{n} T_{cl}}$ corresponds to a shift in time by $\frac{s}{n} T_{cl}$. Thus, the composite wavepacket is made up of replicas of the $t = 0$ wavepacket, shifted by $\frac{s}{n}$ of the classical period, and weighted by the factor a_s. The number of replicas, r, is equal to the number of nonvanishing a_s.

Using Eq. 4.72 it can be shown that all nonzero a_s are of the same modulus, $|a_s| = \frac{1}{\sqrt{r}}$. If n is odd, the number of replicas, r, is equal to n. If n is even, the number of replicas r is equal to $n/2$, with all even a_s vanishing.

▶ **Exercise 4.4**   Show these properties of the a_s.

Solution   The terms in the expression for a_s in Eq. 4.72 are periodic in k with periodicity n. Therefore, as long as the sum spans n consecutive terms the result must be invariant to a shift in the index. Defining $k' = k + 1$ and $s' = (s + 2m) \pmod n$ we find

$$\begin{aligned}
a_{s'} &= \frac{1}{n} \sum_{k'=1}^{n} e^{-2\pi i \frac{m}{n} k'^2 + 2\pi i k' s'/n} \\
&= \frac{1}{n} \sum_{k=0}^{n-1} e^{-2\pi i \frac{m}{n} (k^2 + 2k + 1) + 2\pi i (k+1)(s+2m)/n} \\
&= e^{2\pi i \frac{m}{n}} e^{2\pi i \frac{s}{n}} \frac{1}{n} \sum_{k=0}^{n-1} e^{-2\pi i \frac{m}{n} k^2} e^{2\pi i k \frac{s}{n}} \\
&= e^{2\pi i \frac{m}{n}} e^{2\pi i \frac{s}{n}} a_s.
\end{aligned} \qquad (4.74)$$

Thus $|a_{s'}| = |a_s|$. If n is odd, n iterations of Eq. 4.74 show that all a_s have the same modulus. If n is even, starting with an odd s, $n/2$ iterations shows all odd $a_{s'}$ have the same modulus. Starting with an even s shows that all even a_s have the same modulus, but this modulus is zero. To see this, consider Eq. 4.72 for $s = 0$. Shifting the summation index by integer $n/2$ we find

$$a_0 = \frac{1}{n} \sum_{k'=0}^{n-1} e^{-2\pi i \frac{m}{n} k^2}$$

$$= \frac{1}{n} \sum_{k'=0}^{n-1} e^{-2\pi i \frac{m}{n} (k' + n/2)^2} = -a_0, \tag{4.75}$$

and therefore $a_0 = 0$. The same result obtains for all other a_s with even s. To find the numerical value of the a_s, we make use of the relation

$$\sum_s |a_s|^2 = 1, \tag{4.76}$$

which follows from Eq. 4.72. It follows that $a_s = 1/\sqrt{r}$ for those a_s that are nonzero, since all r nonzero values of a_s have the same magnitude.

The following are important special cases of the general numerology (cf. Eq. 4.68 and the paragraph preceding Exercise 4.4):

1. $t = T_{rev}$ $(m = n = 1; r = 1)$ $\implies \Psi(x, t) = \Psi_{cl}(x, t)$

2. $t = \frac{1}{2} T_{rev}$ $(m = 1, n = 2; r = 1) \implies \Psi(x, t) = \Psi_{cl}(x, t + \frac{1}{2} T_{cl})$

3. $t = \frac{1}{4} T_{rev}$ $(m = 1, n = 4; r = 2) \implies \Psi(x, t) = \frac{1}{\sqrt{2}} (e^{-i\frac{\pi}{4}} \Psi_{cl}(x, t) +$
   $$e^{i\frac{\pi}{4}} \Psi_{cl}(x, t + \frac{1}{2} T_{cl}))$$

4. $t = \frac{1}{3} T_{rev}$ $(m = 1, n = 3; r = 3)$ (also at $t = \frac{1}{6} T_{rev}, t = \frac{2}{3} T_{rev}, t = \frac{5}{6} T_{rev})$
   $$\implies \Psi(x, t) = \frac{1 + 2e^{-2\pi i/3}}{3} (\Psi_{cl}(x, t) + e^{\frac{2\pi i}{3}} \Psi_{cl}(x, t + \frac{1}{3} T_{cl}) + e^{\frac{2\pi i}{3}} \Psi_{cl}(x, t + \frac{2}{3} T_{cl}))$$

For example, Figure 4.2 shows the wavepacket fractional revival structure at representative times for an electron moving in a circular Kepler orbit. Fractional revivals have now been seen in numerous experiments on electronic wavepackets in Rydberg atoms and vibrational wavepackets in diatomic molecules.

Problems

▶ **Exercise 4.5**   Consider the generalized probability density, $\Psi_1^*(x, t)\Psi_2(x, t)$. Show that

$$\frac{\partial(\Psi_1^* \Psi_2)}{\partial t} = \frac{\partial}{\partial x} \left(\Psi_1^* \left(\frac{i\hbar}{2m} \frac{\partial \Psi_2}{\partial x} \right) - \Psi_2 \left(\frac{i\hbar}{2m} \frac{\partial \Psi_1^*}{\partial x} \right) \right). \tag{4.77}$$

Equation 4.77 may also be written in the form of a flux equation:

$$\frac{\partial \rho''}{\partial t} + \frac{\partial \mathcal{J}''}{\partial x} = 0, \tag{4.78}$$

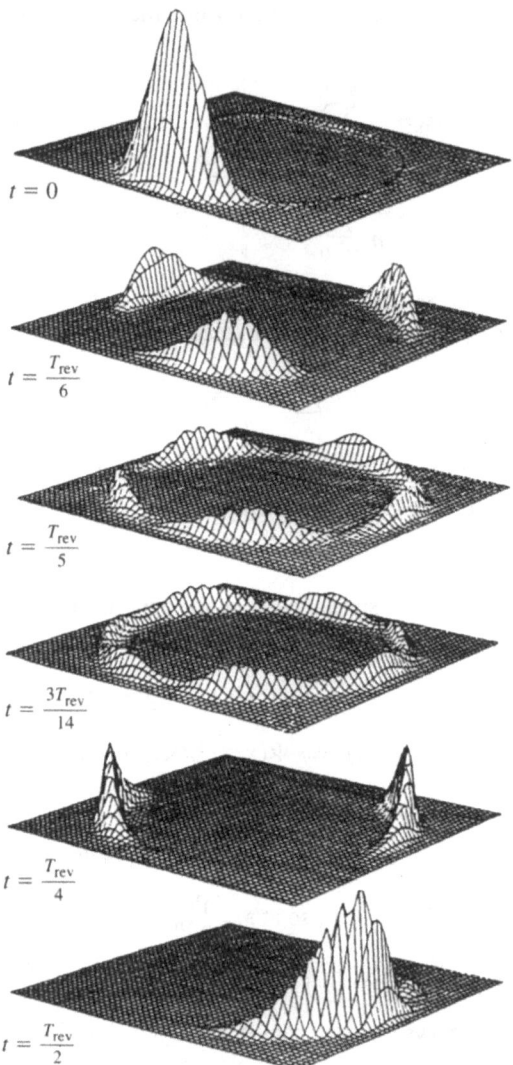

Figure 4.2 Evolution of the probability distribution in the x–y plane for the wavepacket of an electron moving in a circular Kepler orbit. The time t is measured in units of T_{rev}. The figure shows the absolute value of the wavepacket at $t = 0$ ($m = n = 0$), $t = T_{rev}/6$ ($m = 1, n = 6, r = 3$), $t = T_{rev}/5$ ($m = 1, n = 5, r = 5$), $t = 3T_{rev}/14$ ($m = 3, n = 14, r = 7$), $t = T_{rev}/4$ ($m = 1$, $n = 4, r = 2$) and $t = T_{rev}/2$ ($m = 1, n = 2, r = 1$). The position of the wavepacket replicas is somewhat different from that given in the equations following Eq. 4.76, due to a change in perspective from frame to frame. Adapted from Z. D. Goeta and C. R. Stroud, Jr., Phys. Rev. A 42, 6308 (1990).

where $\rho'' = \Psi_1^* \Psi_2$ is the generalized probability density and

$$\jmath'' = - \left(\Psi_1^* \left(\frac{i\hbar}{2m} \frac{\partial \Psi_2}{\partial x} \right) - \Psi_2 \left(\frac{i\hbar}{2m} \frac{\partial \Psi_1^*}{\partial x} \right) \right)$$

is a generalized flux.

▶ **Exercise 4.6** Consider the case where $\Psi_1(x, t) = \psi_1(x) e^{-i E_1 t/\hbar}$ and $\Psi_2(x, t) = \psi_2(x) e^{-i E_2 t/\hbar}$, where $H\psi_1 = E_1 \psi_1$ and $H\psi_2 = E\psi_2$. Show that in this case

$$\frac{\partial}{\partial x} \jmath'' = \frac{(E_1 - E_2)}{i\hbar} \Psi_1^* \Psi_2. \tag{4.79}$$

Equation 4.79 implies that if $E_2 = E_1$ (corresponding to a single eigenstate, as well as a pair of degenerate eigenstates), the generalized flux is independent of coordinate. This result is useful in the context of barrier scattering, where flux conservation under single energy conditions establishes relations between reflection and transmission coefficients (see Chapter 7).

▶ **Exercise 4.7** Consider the case where Ψ is a stationary state; then it must have the form

$$\Psi(x, t) = A(x, 0) e^{-\frac{i}{\hbar} E t}. \tag{4.80}$$

a. Equate this form with Eq. 4.20, to show that

$$\frac{\partial A}{\partial t} = 0 \qquad \frac{\partial S}{\partial t} = -E. \tag{4.81}$$

b. Set $\hbar \to 0$ (i.e., neglect the RHS of Eq. 4.21) to obtain

$$\frac{1}{2m} \left(\frac{\partial S}{\partial x} \right)^2 + V = E. \tag{4.82}$$

Show that there are two solutions to this equation, which differ by only a sign: $S(x, 0)$ and $-S(x, 0)$.

c. Show that a stationary state of the HJ equation must have the form

$$\Psi(x, t) = 2A(x, 0) \cos \left(\frac{S(x, 0)}{\hbar} \right) e^{-\frac{i}{\hbar} E t}. \tag{4.83}$$

▶ **Exercise 4.8** Consider the local kinetic energy, defined as

$$T_l = \text{Re} \left\{ \frac{\Psi^* (-\hbar^2/2m) \nabla^2 \Psi}{\Psi^* \Psi} \right\}. \tag{4.84}$$

Substitute $\Psi = A e^{iS/\hbar}$ into Eq. 4.84 and show that $T_l = T_f + T_s$ where the flow kinetic energy, T_f, is defined as

$$T_f = \frac{1}{2m} (\nabla S)^2 = \frac{1}{2} m v^2 \tag{4.85}$$

and the shape kinetic energy, T_s, is defined as

$$T_s = -\frac{\hbar^2}{2m} A^{-1} \nabla^2 A = Q. \tag{4.86}$$

▶ Exercise 4.9

a. Show that for a stationary state the continuity equation, Eq. 4.23, takes the multi-dimensional form

$$\nabla \cdot (\rho \nabla S) = 0 \tag{4.87}$$

b. Show that Eq. 4.87 implies that

$$\frac{\hbar}{\lambda} \frac{d\rho}{ds} + \rho \nabla^2 S = 0, \tag{4.88}$$

where s is the trajectory coordinate, perpendicular to the surfaces of constant S.

c. Equation 4.56 can be integrated to give

$$S(s) = S_0 + \int_0^s \frac{\hbar}{\lambda} ds \tag{4.89}$$

once S_0 and the energy E are specified. Combine Eqs. 4.88 and 4.89 to show that

$$\frac{\rho(s)}{\rho(0)} = \frac{\lambda(s)}{\lambda(0)}, \tag{4.90}$$

showing that the density is slaved to the local wavelength along the trajectory.

d. Use Eq. 4.90 to derive an expression for the quantum potential 4.34 in terms of derivatives of the local wavelength λ rather than the amplitude $A = \rho^{1/2}$.

▶ Exercise 4.10   The Morse potential is defined by

$$V(x) = D \left(1 - e^{-a(x-x_e)}\right)^2. \tag{4.91}$$

This is a bound potential that goes to ∞ as $x \to -\infty$ and goes to the dissociation energy D as $x \to \infty$. The potential has a minimum at $x = x_e$, and a harmonic frequency ω around the minimum given by $\frac{1}{2}m\omega^2 = Da^2$. The eigenvalues of the Hamiltonian $T + V$ are given by

$$E_n = \left(n + \frac{1}{2}\right)\hbar\omega - \left(n + \frac{1}{2}\right)^2 \frac{(\hbar\omega)^2}{4D}. \tag{4.92}$$

Classically, the energy of the Morse oscillator may be expressed in terms of the action, J, by

$$E = a \left(\frac{2D}{m}\right)^{1/2} \bar{J} - \frac{a^2}{2m} \bar{J}^2, \tag{4.93}$$

where $\bar{J} = J/2\pi$.

a. Draw a picture of the Morse potential. Verify the formula for the position of the minimum and the harmonic frequencies.

b. Use Eq. 4.60 to calculate the classical period, T_{cl}, as a function of E for the Morse oscillator. Show that at $E = 0$ the period is equal to the harmonic period, T_0, increasing to infinity as E goes to D, the dissociation energy.

c. Use Eq. 4.63 to calculate the revival time T_{rev} and the ratios T_{rev}/T_{cl} and T_{rev}/T_0. Discuss how the revival time is related to the other time scales in the system, as a function of a and E.

References

Ehrenfest's Theorem and Classical–Quantum Correspondence

1. P. Ehrenfest, Z. Phys. 45, 455 (1927).
2. P. A. M. Dirac, *The Principles of Quantum Mechanics* (Oxford, 1978).
3. A. Messiah, *Quantum Mechanics*, vol. I (Wiley, New York, 1958).
4. A. Peres, *Quantum Theory: Concepts and Methods* (Kluwer, 1995).
5. O. V. Prezhdo and Yu. V. Pereverzev, Quantized Hamilton dynamics, J. Chem. Phys. 113, 6557 (2000).

Bohmian Mechanics

6. L. de Broglie, Sur la possibilité de relier les phénomènes d'interference et de diffraction à la théorie des quanta de lumière, C. R. Acad. Sci. Paris 183, 447 (1926); L. de Broglie, La structure atomique de la matière et du rayonnement et la Mécanique ondulatoire, C. R. Acad. Sci. Paris 184, 273 (1927).
7. E. Madelung, Quantentheorie in hydrodynamischer Form, Z. Phys. 40, 322 (1926).
8. D. Bohm, A suggested interpretation of the quantum theory in terms of "hidden" variables. I, Phys. Rev. 85, 166 (1952); D. Bohm, A suggested interpretation of the quantum theory in terms of "hidden" variables. II, 85, 180 (1952).
9. P. R. Holland, *The Quantum Theory of Motion* (Cambridge, New York, 1993).
10. B. K. Dey, A. Askar and H. Rabitz, Multidimensional wavepacket dynamics within the fluid dynamical formulation of the Schrödinger equation, J. Chem. Phys. 109, 8770 (1998).
11. C. L. Lopreore and R. E. Wyatt, Quantum wavepacket dynamics with trajectories, Phys. Rev. Lett. 82, 5190 (1999).
12. R. E. Wyatt, Quantum wavepacket dynamics with trajectories: Wavefunction synthesis along quantum paths, Chem. Phys. Lett. 313, 189 (1999).
13. M. Boiron and M. Lombardi, Complex trajectory method in semiclassical propagation of wavepackets, J. Chem. Phys. 108, 3431 (1998).

Classical Limit of Quantum Mechanics

14. P. A. M. Dirac, *The Principles of Quantum Mechanics* (Oxford, 1978).
15. A. Messiah, *Quantum Mechanics*, vol. I (Wiley, New York, 1958).
16. H. Goldstein, *Classical Mechanics* (Addison-Wesley, New York, 1950).
17. A. Peres, *Quantum Theory: Concepts and Methods* (Kluwer, 1995).
18. R. P. Feynman and A. R. Hibbs, *Quantum Mechanics and Path Integrals* (McGraw-Hill, New York, 1965).

The Geometrical Optics Limit of Wave Optics

19. R. P. Feynman, R. B. Leighton and M. Sands, *The Feynman Lectures on Physics*, Vol. I (Addison-Wesley, Reading, MA, 1963), Chapter 26 and following.
20. M. Born and E. Wolf, *Principles of Optics: Electromagnetic Theory of Propagation, Interference and Diffraction of Light*, 3rd ed. (Pergamon, New York, 1965), Chapter 3.
21. G. B. Whitham, *Linear and Nonlinear Waves* (Wiley, New York, 1974).

Fractional Revivals

22. I. Averbukh and N. Perelman, Fractional revivals—Universality in the long-term evolution of quantum wavepacket beyond the correspondence principle dynamics, Phys. Lett. A, 139, 449 (1989).

23. I. Sh. Averbukh and N. F. Perelman, The dynamics of wavepackets of highly-excited states of atoms and molecules, Sov. Phys. Usp. 34, 572 (1991).

24. J. Parker and C. R. Stroud Jr., Coherence and decay of Rydberg wavepackets, Phys. Rev. Lett. 56, 716 (1986).

25. I. M. Suarez Barnes, M. Nauenberg, M. Nockleby and S. Tomsovic, Semiclassical theory of quantum propagation: The Coulomb potential, Phys. Rev. Lett. 71, 1961 (1993).

26. M. Mallalieu and C. Stroud, Semiclassical dynamics of circular-orbit Rydberg wavepackets, Phys. Rev. A 49, 2329 (1994).

27. M. Gruebele and A. H. Zewail, Femtosecond wavepacket spectroscopy: Coherences, the potential, and structural determination, J. Chem. Phys. 98, 883 (1993).

28. Ch. Meier and V. Engel, Electron kinetic energy distributions from multiphoton ionization of Na_2 with femtosecond laser pulses, Chem. Phys. Lett. 212, 691 (1993).

29. T. Baumert, V. Engel, C. R. Röttgermann, W. T. Strunz and G. Gerber, Femtosecond pump-probe study of the spreading and recurrence of a vibrational wavepacket in Na_2, Chem. Phys. Lett. 191, 639 (1992).

30. M. J. J. Vrakking, D. M. Villeneuve and A. Stolow, Observation of fractional revivals of a molecular wavepacket, Phys. Rev. A 54, R37 (1996).

31. C. Leichtle, I. Sh. Averbukh and W. P. Schleich, Multilevel quantum beats: An analytical approach, Phys. Rev. A 54 (1996).

Chapter 5

The Wigner Representation and the Density Operator

In previous chapters we focused primarily on the wavefunction in coordinate space, $\Psi(x)$. The main focus of this chapter will be a new representation of the wavefunction, called the Wigner function, given the symbol $f_W(p, q)$, where p and q are momentum and position, respectively (it is conventional to use q, rather than x, as the variable conjugate to p). We will see that this representation allows us to give a qualitative picture of the motion that corresponds closely to a classical description.

We will begin the chapter with a discussion of the utility of the p–q or phase space representation in classical mechanics. We then introduce the Wigner function, restricting our attention to pure state wavepackets and wavefunctions. These two sections are an integral part of the development of the book, in that they provide an important new perspective on wavepacket motion. Moreover, the Wigner function for pure states is developed without mentioning density operators, or using any operator calculus, and thus this part of the chapter is no more difficult than any of the earlier chapters.

The Wigner function is a particular representation of a more general object, the density operator. The full power of the density operator is in its ability to describe thermal states, and more generally, relaxation of a wavepacket to thermal equilibrium. In order to treat these topics we first devote a section to the density operator in its more abstract setting, and introduce the concept of mixed, or impure, states. Then, a qualitative discussion is given of equilibrium and relaxation of the density operator in the Wigner representation, although a discussion of the actual form of the relaxation equations is beyond the scope of this book. Finally, the Wigner representation of arbitrary operators is discussed, culminating in the (nondissipative) equations of motion for the Wigner distribution.

In principle, the density operator is more fundamental than its Wigner representation. For example, the Wigner representation is incapable of describing the important class of density operators corresponding to spin or two-level systems. Therefore, from a strictly formal viewpoint it would perhaps be more natural to begin with the density operator before introducing the Wigner distribution. However, the former is difficult to grasp in its general abstract setting, and the Wigner representation provides a convenient portal to its understanding. Moreover, the density operator requires fluency with the mathematics of Chapter 8. Indeed, with the exception of the phase space pictures of relaxation in the Wigner representation, Sections 5.3–5.4 dealing with the general density operator are more difficult than the rest of Part I of the book, and it is recommended that they be skipped on a first reading.

In Appendix B we introduce the concept of the *reduced* density matrix—the density matrix associated with a *subsystem* of a larger system. As is shown there, even if the larger system corresponds to a pure state, the reduced density matrix is generally characterized by a probabilistic mixture of pure states. In fact, the concept of the density operator is sometimes motivated in this way: that the probabilistic mixture associated with mixed states reflects the loss of information associated with the tracing out of part of a larger system. In this sense, Appendix B is an important complement to this chapter, but not necessary for the understanding of the material presented here.

5.1   The Concept of Phase Space

In classical mechanics, p and q are referred to as conjugate variables. The p and q coordinates, taken together, comprise the "phase space"; plotting orbits in the p–q, or phase plane, allows one to visualize motion in both p and q simultaneously. For example, Figure 5.1a shows phase space orbits for a harmonic oscillator, $H(p, q) = \frac{p^2}{2m} + \frac{1}{2}m\omega^2 q^2 = E$. Note that the equation for the energy is the usual quadratic function of position and momentum; however, when plotted as a function of p and q it is an ellipse. The ellipses are concentric, larger ellipses corresponding to higher energy orbits.

▶ **Exercise 5.1**   Consider a particle in a harmonic well, starting at $t = 0$ with $q = -q_0$ and $p = 0$. Sketch the location of the particle in phase space at $t = 0$, $\tau/4$, $\tau/2$, $3\tau/4$, and τ, where τ is a period of motion. Connect the points and show that the orbit is an ellipse. Show that one can scale the p and q coordinates to obtain a circular phase space orbit. Which way does the phase space orbit circulate, clockwise or counterclockwise?

Figure 5.1b shows phase space orbits for a pendulum, $H(p, \psi) = \frac{p^2}{2m} - mgl \cos(\psi) = E$, where the angle ψ is defined in Figure 5.1c and p is understood as the momentum conjugate to ψ. At low energies, the motion of the pendulum is the same as that of a harmonic oscillator; this can be seen by expanding the potential energy term up to quadratic order in ψ and is reflected in the elliptical curves at low energies in Figure 5.1b. At higher energies, the deviations from the harmonic oscillator become significant, reflected as distortions of the curves from ellipses. At a certain critical energy, $E = mgl$, the pendulum has enough energy to go over the top. At this energy and higher, the pendulum rotates with the same sense of ψ ad infinitum. There are thus two distinct types of motion possible for the pendulum, "librational," where $\dot{\psi}$ changes sign twice each period, and "rotational," where $\dot{\psi}$ maintains the same sign. The curve that divides rotational from librational motion is called the "separatrix." It corresponds to an orbit with exactly enough energy for the pendulum to stand on its head, but which takes an infinitely long time to get to that point. There are two distinct points that remain motionless for all time. One is at $E = 0$, where the pendulum hangs vertically; this point is called a stable, or elliptical, fixed point. The other is at $E = mgl$ with the pendulum standing on its head; this point is called an unstable, or hyperbolic, fixed point.

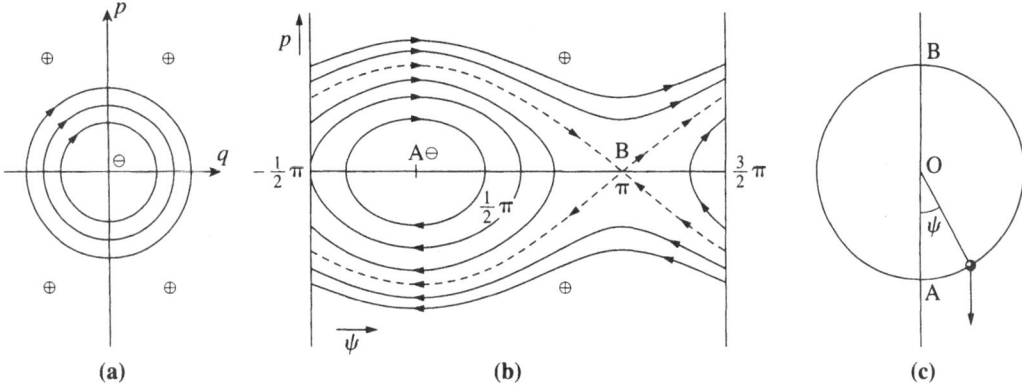

Figure 5.1 (a) Phase space contours of equal energy for the harmonic oscillator potential. The contours are ellipses, satisfying the equation $H(p, q) = \frac{p^2}{2m} + \frac{1}{2}m\omega^2 q^2 = E$. The symbols $\oplus(\ominus)$ indicate the direction of increasing (decreasing) energy. Circulation is always such that $\oplus$ is on the left. (b) Phase space contours of equal energy for the pendulum. The contours satisfy the equation $H(p, \psi) = \frac{p^2}{2m} - mgl\cos(\psi) = E$. At low energy, the motion is librational, similar to that of the harmonic oscillator; at sufficiently high energies the motion is rotational. The separatrix (dotted line) is the curve that divides these two distinct types of motion. (c) Diagram showing the definition of ψ as the angle of displacement of the pendulum from the negative vertical axis. Note that $\psi = \pi$ corresponds to the pendulum in the unstable position, "standing on its head." Adapted from Percival (1982).

▶ **Exercise 5.2**   Consider a pendulum, starting at $t = 0$ with $\psi = 0$ and $p = p_0$. Sketch the location of the particle in phase space at $t = 0$, $\tau/4$, $\tau/2$, $3\tau/4$, and τ, where τ is a period of motion. You should find three different regions, two corresponding to rotation and one to libration, depending on the value of p_0. Show that the direction of the rotational orbits in phase space correlate smoothly with those of the libration at the separatrix.

In semiclassical mechanics, the Bohr quantization rule is $J \equiv \int p\, dq = nh$, where J is called the classical "action." Note that $\int p\, dq$ has the geometrical interpretation of a phase space area. Indeed, the physical interpretation of the Bohr quantization rule is that quantum states come at intervals that correspond to increments in phase space area by an integer multiple of Planck's constant.

Phase space plays a central role in classical statistical mechanics: the single particle classical partition function in one dimension is given by

$$\frac{1}{h}\int_{-\infty}^{\infty}\int_{-\infty}^{\infty} dp\, dq\, e^{-\beta H(p,q)}. \tag{5.1}$$

We divide by h because this is the unit volume of a quantum state in phase space in the classical limit. More generally, the N-particle partition function in three dimensions contains a division by h^{3N}.

Phase space enters into classical mechanics in several other contexts. One is Liouville's theorem, which states that a distribution function in phase space may change shape in time but not change its phase space volume; this is simply a consequence of the conservation of probability.

5.2 The Wigner Representation of Wavepackets

5.2.1 Definition and Basic Properties

Given a wavepacket, $\Psi(x)$, we define the Wigner distribution function, $f_W(p, q)$, as

$$f_W(p, q) = \frac{1}{2\pi\hbar} \int_{-\infty}^{\infty} e^{\frac{i}{\hbar}ps} \left\langle q - \frac{s}{2} | \Psi \right\rangle \left\langle \Psi | q + \frac{s}{2} \right\rangle ds, \qquad (5.2)$$

where we are employing Dirac notation:

$$\left\langle q - \frac{s}{2} | \Psi \right\rangle = \Psi\left(q - \frac{s}{2}\right) \qquad (5.3)$$

$$\left\langle \Psi | q + \frac{s}{2} \right\rangle = \Psi^*\left(q + \frac{s}{2}\right). \qquad (5.4)$$

Defining new coordinates $x = q + \frac{s}{2}$ and $x' = q - \frac{s}{2}$ we note that

$$f_W(p, q) = \frac{1}{2\pi\hbar} \int_{-\infty}^{\infty} e^{\frac{i}{\hbar}p(x-x')} \langle x'|\Psi\rangle\langle\Psi|x\rangle \, ds. \qquad (5.5)$$

In words, the Wigner distribution function is constructed by first forming the product of $\Psi(x')$ with $\Psi^*(x)$, and then Fourier transforming along the difference coordinate, $s = x - x'$.

▶ **Exercise 5.3** To understand why the Fourier transform variable p has the physical meaning of a momentum, examine the structure of the quantity $\langle x'|\Psi\rangle\langle\Psi|x\rangle$ for the case where $\Psi(x) = Ne^{-\alpha(x-x_0)^2 + ip_0(x-x_0)/\hbar}$. This product is a function of the variables x and x' and is complex. Plot the real part of this product and discuss the wavelength of oscillations along the $s = x - x'$ coordinate. If the object is Fourier transformed along the $s = x - x'$ coordinate, where will its peak be in the Fourier variable, p?

The Wigner transform of a Gaussian wavepacket, $\Psi(x)$, is a Gaussian phase space distribution, $f_W(p, q)$. For example, Figure 5.2a shows the lowest eigenstate of the harmonic oscillator and its Gaussian Wigner function. The Wigner distribution has several important properties that suggest its interpretation as a probability distribution:

$$\int_{-\infty}^{\infty} f_W(p, q) \, dp = \langle q|\Psi\rangle\langle\Psi|q\rangle = |\Psi(q)|^2 \qquad (5.6)$$

$$\int_{-\infty}^{\infty} f_W(p, q) \, dq = \langle p|\Psi\rangle\langle\Psi|p\rangle = |\tilde{\Psi}(p)|^2 \qquad (5.7)$$

$$\int_{-\infty}^{\infty} \int_{-\infty}^{\infty} f_W(p, q) \, dp \, dq = 1, \qquad (5.8)$$

where $\tilde{\Psi}(p)$ is the momentum representation of the wavefunction $|\Psi\rangle$ (see Eq. 2.51).

▶ **Exercise 5.4** Derive Eqs. 5.6–5.8.

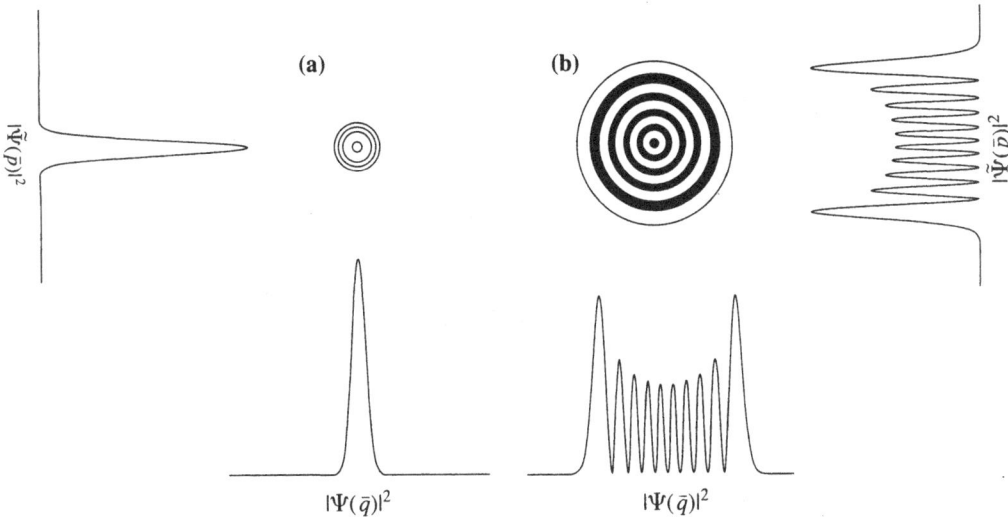

Figure 5.2 (a) Equal amplitude contours of the Wigner phase space distribution, $f_W(\bar{p}, \bar{q})$, corresponding to the lowest eigenstate of the harmonic oscillator. The contours represent a two-dimensional Gaussian with its maximum at $p = q = 0$. Shown below it is $\int_{-\infty}^{\infty} f_W(\bar{p}, \bar{q}) \, d\bar{p} = |\Psi(\bar{q})|^2$ and to the left is $\int_{-\infty}^{\infty} f_W(\bar{p}, \bar{q}) \, d\bar{q} = |\tilde{\Psi}(\bar{p})|^2$. Here and in the subsequent plots we use scaled coordinates, $\bar{p} \equiv \frac{1}{\sqrt{m\omega}} p$, $\bar{q} \equiv \sqrt{m\omega} q$, in order to get circular phase space contours. (b) Equal amplitude contours of the Wigner phase space distribution, $f_W(\bar{p}, \bar{q})$, corresponding to the ninth eigenstate of the harmonic oscillator. The dark regions correspond to negative values (see text). The maximum of the distribution is along the broad annulus near the outer extreme of the distribution. Shown below it is $\int_{-\infty}^{\infty} f_W(\bar{p}, \bar{q}) \, d\bar{p} = |\Psi(\bar{q})|^2$, and to the right is $\int_{-\infty}^{\infty} f_W(\bar{p}, \bar{q}) \, d\bar{q} = |\tilde{\Psi}(\bar{p})|^2$. These two functions, $|\Psi(\bar{q})|^2$ and $|\tilde{\Psi}(\bar{p})|^2$, have nodes but being true probabilities are nowhere negative.

However, the Wigner distribution function can take on negative as well as positive values, so that it cannot be interpreted strictly as a phase space probability density. For example, Figure 5.2b shows the Wigner distribution function corresponding to the ninth eigenstate of the harmonic oscillator, which has regions of negative amplitude. Note that these negative values never survive when the Wigner function is integrated over p or q, as can be seen from Eqs. 5.6–5.7. Despite the difficulty with interpreting it as a phase space probability density, the Wigner distribution provides a very useful starting point for comparing classical and quantum dynamics in phase space.

5.2.2   Phase Space Representation of a Gaussian in a Quadratic Potential

An important example of the dynamics of the Wigner distribution is that of the Gaussian wavepacket in the harmonic oscillator potential. This system was discussed at length in Chapter 3. Here we reexamine the solutions from that chapter using the Wigner phase space representation.

Consider the normalized wavepacket

$$\Psi(q, t) = \left(\frac{2\mathrm{Re}\alpha_t}{\pi}\right)^{1/4} e^{-\alpha_t(q-q_t)^2 + \frac{i}{\hbar} p_t(q-q_t) + \frac{i}{\hbar}\gamma_t}, \qquad (5.9)$$

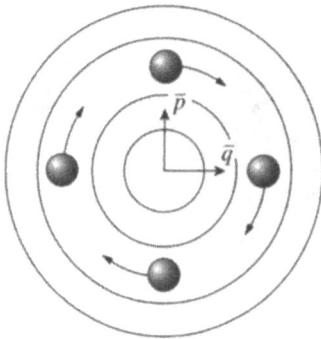

Figure 5.3 The motion of the Wigner distribution of a "coherent state," or minimum uncertainty wavepacket, superimposed on equal energy contours of the harmonic oscillator. $\Delta \bar{q} = \Delta \bar{p}$.

where we are taking γ_t to be real. As we saw in Chapter 3, the choice $\alpha_t = \frac{m\omega}{2\hbar}$ leads to a wavepacket whose width is constant in time. Substituting this value of α_t into Eq. 5.2 yields, after some algebra,

$$f_W(q, p) = \frac{1}{\pi \hbar} e^{-\frac{m\omega}{\hbar}(q-q_t)^2} e^{-\frac{1}{m\omega\hbar}(p-p_t)^2}. \qquad (5.10)$$

Figure 5.3 shows the motion of the Wigner distribution; note that the center of the distribution follows a circular orbit in phase space, entirely in accord with classical mechanics. (In this and subsequent figures we use scaled coordinates, $\bar{p} \equiv \frac{1}{\sqrt{m\omega}} p$, $\bar{q} \equiv \sqrt{m\omega} q$, in order to get circular, rather than elliptical, phase space orbits.) For this initial state, the width of the Gaussian is constant, both in $\bar{q}$ and $\bar{p}$, leading to a minimum uncertainty wavepacket ($\Delta \bar{p} \cdot \Delta \bar{q} = \Delta p \cdot \Delta q = \hbar/2$) at all times. This state is the "coherent state" defined in Section 3.3. In the more general case where

$$\alpha_t = a \left(\frac{\alpha_0 \cos \omega t + ia \sin \omega t}{i\alpha_0 \sin \omega t + a \cos \omega t} \right) \qquad (5.11)$$

and where

$$a = \frac{m\omega}{2\hbar}, \qquad (5.12)$$

the state is called a "squeezed state." The Wigner distribution is still a Gaussian:

$$f_W(q, p) = \frac{1}{\pi \hbar} e^{-\frac{2|\alpha_t|^2}{\operatorname{Re}\alpha_t}(q-q_t)^2} e^{-\frac{1}{2\hbar^2 \operatorname{Re}\alpha_t}(p-p_t)^2} e^{-\frac{2 \operatorname{Im}\alpha_t}{\hbar \operatorname{Re}\alpha_t}(q-q_t)(p-p_t)}, \qquad (5.13)$$

but the width, both in $\bar{p}$ and $\bar{q}$ changes in time, as shown in Figure 5.4.

▶ **Exercise 5.5**   Derive Eq. 5.13.

Figure 5.4 shows two examples of the evolution of squeezed states. In Figure 5.4a, $\Delta \bar{p}(0) > \Delta \bar{q}(0)$. As the distribution moves along the classical trajectory, it rotates rigidly in phase space, gaining uncertainty in $\bar{q}$ and losing uncertainty in $\bar{p}$, and then the opposite. In Figure 5.4b it is the reverse: $\Delta \bar{p}(0) < \Delta \bar{q}(0)$, the distribution gains uncertainty in $\bar{p}$ and loses it in $\bar{q}$, and then the opposite. Note that squeezed states in coordinate space require complex α, which leads to position–momentum correlation in the Wigner distribution. There

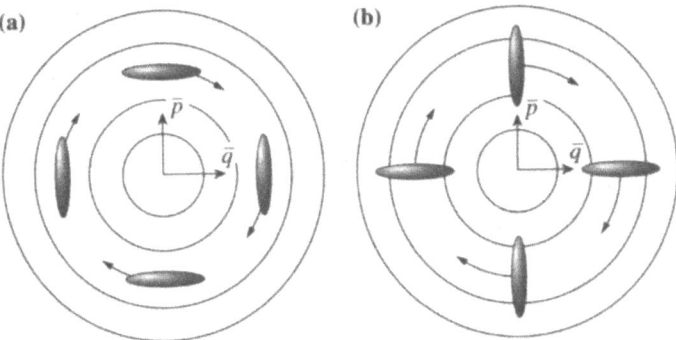

Figure 5.4 The motion of the Wigner distribution of two different "squeezed states," superimposed on equal energy contours of the harmonic oscillator. (a) $\Delta \bar{q}(t=0) < \Delta \bar{p}(t=0)$. (b) $\Delta \bar{q}(t=0) > \Delta \bar{p}(t=0)$. (The initial conditions are $\langle \bar{q} \rangle_0 < 0$, $\langle \bar{p} \rangle_0 = 0$.) The squeezed state is not a minimum uncertainty state in the conventional sense, since $\Delta \bar{p} \cdot \Delta \bar{q} > \hbar/2$, except at specific times. However, there is an (instantaneous) set of rotated phase space axes for which the squeezed state does have a minimum uncertainty product.

is nothing quantum mechanical per se about this rotation of the distribution function in phase space; a swarm of classical trajectories, each moving independently, would display the same overall rigid rotation of the distribution.

▶ **Exercise 5.6**   Calculate the Wigner distribution as a function of time for the free-particle Gaussian wavepacket. Sketch the shape of the Wigner distribution as a function of time. Use your results to answer Exercise 3.8.

▶ **Exercise 5.7**   Calculate the Wigner distribution as a function of time for a Gaussian wavepacket on a linear potential.

▶ **Exercise 5.8**   Calculate the Wigner distribution as a function of time for a Gaussian wavepacket on a parabolic potential at $t = 0$.

5.3   The Density Operator

To understand the Wigner phase space representation of quantum mechanics in its full generality, we must first introduce the density operator. The density operator takes some getting used to, for several reasons:

1. Mathematically, it is a different type of object than the wavefunctions we have been dealing with until now. It is an operator, and as such the mathematics is more abstract than that for wavefunctions.
2. Operators have representations, just as wavefunctions do; but because operators are to wavefunctions what matrices are to vectors, there are a greater number of possible

representations, each giving a different physical picture. For example, there are the (x, x'), (p, p'), (p, q) and (E, E') representations, just to name the common ones.

5.3.1 Definition of the Density Operator

The most natural starting point for introducing the density operator is to notice that Eq. 5.5 can be written as follows:

$$f_{\mathrm{W}}(p, q) = \frac{1}{2\pi\hbar} \int_{-\infty}^{\infty} e^{\frac{i}{\hbar} p(x-x')} \langle x' \{|\Psi\rangle\langle\Psi|\} x \rangle \, ds = \frac{1}{2\pi\hbar} \int_{-\infty}^{\infty} e^{\frac{i}{\hbar} p(x-x')} \langle x'|\rho|x\rangle \, ds,$$

(5.14)

where

$$x = q + \frac{s}{2} \qquad x' = q - \frac{s}{2}$$

(5.15)

and where we have defined the "pure" state density operator as

$$\rho \equiv |\Psi\rangle\langle\Psi|.$$

(5.16)

The most general definition of ρ includes both pure and "mixed" states:

$$\rho = \sum_i p_i |\Psi_i\rangle\langle\Psi_i|,$$

(5.17)

where $p_i \geq 0$ and $\sum p_i = 1$. This equation expresses ρ as an *incoherent* superposition of pure state density operators, $|\Psi_i\rangle\langle\Psi_i|$, where $|\Psi_i\rangle$ is a wavefunction but not necessarily an energy eigenstate. In Eq. 5.17, the p_i are the *probabilities* (not amplitudes) of finding the system in state $|\Psi_i\rangle$. Note that in addition to the usual probabilistic interpretation for finding the particle described by a particular wavefunction at a specified location, there is now a probability distribution for being in different states! Clearly, if one of the $p_i = 1$ and all the others are 0 the density operator corresponds to a pure state. Note that the states Ψ_i are assumed to be normalized but do not need to be orthogonal.

We want to show now that ρ is an operator and that ρ is a density. We will show the former now and the latter in the next section.

▶ **Exercise 5.9** Show that ρ is an operator.

Solution In Chapter 8, a full set of rules for manipulating objects of the form $|a\rangle\langle b|$ will be given. To see that $|a\rangle\langle b|$ is an operator, note that

$$\{|a\rangle\langle b|\}|c\rangle = |a\rangle\langle b|c\rangle = \{\langle b|c\rangle\}|a\rangle.$$

(5.18)

Since the object $\{|a\rangle\langle b|\}$ turns the vector $|c\rangle$ into another vector, $\{\langle b|c\rangle\}|a\rangle$, $\{|a\rangle\langle b|\}$ is an operator. We have used the fact that $\langle b|c\rangle$ is a scalar (as opposed to a vector or an operator), and hence its position in the expression can be moved freely. Note that operators are to wavefunctions what matrices are to vectors: an operator turns a wavefunction into another wavefunction, while a matrix turns a vector into another vector. It takes only one index to specify an element of a vector, but two to specify an element of a matrix. Similarly, it takes only one index to specify a wavefunction, for example, $\Psi(x) = \langle x|\Psi\rangle$, but two to specify an operator, $\langle x'|\rho|x\rangle$. Thus, the expression $\langle x'|\rho|x\rangle$ can be viewed as the (x', x) matrix element of ρ, where x and x' are continuous variables.

An important property of ρ is that $\text{Tr}(\rho)=1$ for all density operators.

▶ **Exercise 5.10** Demonstrate the foregoing statement.

Solution

$$\rho = \sum_i p_i |\Psi_i\rangle\langle\Psi_i| \tag{5.19}$$

$$\text{Tr}(\rho) = \sum_n \langle\psi_n|\{\sum_i p_i|\Psi_i\rangle\langle\Psi_i|\}|\psi_n\rangle = \sum_n\sum_i p_i\langle\Psi_i|\psi_n\rangle\langle\psi_n|\Psi_i\rangle = \sum_i p_i\langle\Psi_i|\Psi_i\rangle = 1. \tag{5.20}$$

Note that we have used the fact that the Ψ_i are normalized but have not assumed them to be orthogonal.

▶ **Exercise 5.11** Show that if the $|\Psi_i\rangle$ are orthogonal they are eigenfunctions of the ρ operator with eigenvalues p_i. (If the $|\Psi_i\rangle$ are not orthogonal, the eigenfunctions can be found by diagonalizing ρ in an orthogonal basis. Then ρ can be reexpressed in the new, orthogonal basis, $\{\Psi'_i\}$, as $\rho = \sum_i p'_i|\Psi'_i\rangle\langle\Psi'_i|$, where the $\sum_i p'_i = 1$.)

5.3.2 Why "Density"?

To see that ρ is a density we note that the expectation value or average value of an observable, A, corresponding to an operator $\hat{A}$ may be expressed in the compact form

$$\langle A\rangle_t = \text{Tr}(\rho\hat{A}). \tag{5.21}$$

▶ **Exercise 5.12** Show this.

Solution

$$\langle A\rangle_t = \langle\Psi(t)|\hat{A}|\Psi(t)\rangle = \text{Tr}(|\Psi(t)\rangle\langle\Psi(t)|\hat{A}) = \text{Tr}(\rho\hat{A}). \tag{5.22}$$

The proof of the second equality follows from the rules of operator manipulations that will be discussed extensively in Chapter 8. What is important for now is that this expression is the analog of the *classical mechanical* expression for the average of an observable, $A(p, q)$, over a classical density distribution, $\rho(p, q)$:

$$\langle A\rangle^{\text{cl}} = \int_{-\infty}^{\infty}\int_{-\infty}^{\infty} dp\, dq\, A(p, q)\rho(p, q). \tag{5.23}$$

Notice that in quantum mechanics, the trace operation plays the role of integration over p and q in classical mechanics.

▶ **Exercise 5.13**   Show that $\langle A \rangle = \text{Tr}(\rho \hat{A})$ for a general mixed state.

Solution

$$
\begin{aligned}
\langle A \rangle &= \sum_k p_k \langle \Psi_k | \hat{A} | \Psi_k \rangle \\
&= \sum_{m,n,k} p_k \langle \Psi_k | m \rangle \langle m | \hat{A} | n \rangle \langle n | \Psi_k \rangle \\
&= \sum_{m,n,k} p_k \langle n | \Psi_k \rangle \langle \Psi_k | m \rangle \langle m | \hat{A} | n \rangle \\
&= \sum_{m,n} \langle n | \rho | m \rangle \langle m | \hat{A} | n \rangle \\
&= \sum_n \langle n | \rho \hat{A} | n \rangle \\
&= \text{Tr}(\rho \hat{A}).
\end{aligned}
\tag{5.24}
$$

We have used the closure relation in Dirac notation, which will be discussed in Chapter 8. Note that the derivation never assumes that the Ψ_k are orthogonal, only that $\sum_k |\Psi_k\rangle\langle\Psi_k| = \mathbf{1}$.

The following table of analogies between properties of the classical density operator (classical phase space distribution function) and the quantum density operator shows the aptness of the term "density operator." The analogies will become even stronger below, when we will see quantum mechanical expressions involving phase space integrals, using the Wigner representation.

Classically		Quantum Mechanically
$\int \rho_{\text{cl}} \, dp \, dq = 1$	$\Longleftrightarrow$	$\text{Tr}(\rho) = 1$
$\langle A \rangle_{\text{cl}} = \int \rho_{\text{cl}} A_{\text{cl}} \, dp \, dq$	$\Longleftrightarrow$	$\langle A \rangle = \text{Tr}(\rho A)$
$\partial \rho_{\text{cl}} / \partial t = \{\rho_{\text{cl}}, H_{\text{cl}}\}$	$\Longleftrightarrow$	$\partial \rho / \partial t = \frac{i}{\hbar}[H, \rho]$
$\rho_{\text{cl}}^{\text{equil}} = \exp(-\beta H_{\text{cl}}) / \int \exp(-\beta H_{\text{cl}}) \, dp \, dq$	$\Longleftrightarrow$	$\rho^{\text{equil}} = \exp(-\beta H) / \text{Tr}(\exp(-\beta H))$
$S_{\text{cl}} = -k_B \int \rho_{\text{cl}} \ln \rho_{\text{cl}} \, dp \, dq$	$\Longleftrightarrow$	$S = -k_B \text{Tr}(\rho \ln \rho)$

The first two of these relationships we have already encountered; the remainder of the entries will be encountered below, either in the text or in exercises. Thus, we see that $|\Psi\rangle\langle\Psi|$ is an operator, and also plays the role of a density.

5.3.3 $\text{Tr}(\rho^2)$ as a Measure of Purity

As discussed above, if $p_n = 1$ for one particular Ψ_n and $p_n = 0$ for all others, ρ is a pure state; otherwise it is a mixed state. Thus, a measure of the coherence of the system is given by $\sum p_n^2$: $\sum p_n^2 = 1$ for a pure state and $\sum p_n^2 < 1$ for a mixed state. This same measure of coherence can be expressed as $\text{Tr}(\rho^2)$, as shown in the following exercise. The latter has the advantage of being basis independent.

▶ **Exercise 5.14**   Show that in general $\mathrm{Tr}(\rho^2) \leq 1$, and that for a pure state $\mathrm{Tr}(\rho^2) = 1$.

Solution

$$\mathrm{Tr}(\rho^2) = \sum_n \langle \psi_n | \left\{ \sum_{ij} p_i p_j | \Psi_i \rangle \langle \Psi_i | \Psi_j \rangle \langle \Psi_j | \right\} \psi_n \rangle \tag{5.25}$$

$$= \sum_{ij} p_i p_j \langle \Psi_j | \Psi_i \rangle \langle \Psi_i | \Psi_j \rangle \tag{5.26}$$

$$\leq \sum_{ij} p_i p_j \langle \Psi_i | \Psi_i \rangle \langle \Psi_j | \Psi_j \rangle \tag{5.27}$$

$$= \sum_i p_i \sum_j p_j \tag{5.28}$$

$$= 1, \tag{5.29}$$

where the second-to-last step follows from the Schwarz inequality. Note that we have not assumed orthogonality of the $\{\Psi_i\}$ in the proof. For a pure state, only one term in the double sum survives, and the Schwarz inequality becomes an equality. Thus, the proximity of $\mathrm{Tr}(\rho^2)$ to 1 indicates how close ρ is to a pure state.

▶ **Exercise 5.15**   In the previous exercise we saw that $\mathrm{Tr}(\rho^2) \leq 1$ for a general density operator. Below, we will see the result that

$$\mathrm{Tr}(\rho^2) = \frac{1}{2\pi\hbar} \int_{-\infty}^{\infty} \int_{-\infty}^{\infty} dp\, dq\, \rho_{\mathrm{W}}^2(p, q), \tag{5.30}$$

where

$$\rho_{\mathrm{W}}(p, q) = 2\pi\hbar f_{\mathrm{W}}(p, q). \tag{5.31}$$

Show that if the Wigner representation of the density operator is of the form

$$\rho_{\mathrm{W}}(q, p) = \frac{\hbar}{\Delta p \Delta q} e^{-(q-q_0)^2/2(\Delta q)^2} e^{-(p-p_0)^2/2(\Delta p)^2} \tag{5.32}$$

that the condition

$$\mathrm{Tr}(\rho^2) = \frac{1}{2\pi\hbar} \int_{-\infty}^{\infty} \int_{-\infty}^{\infty} dp\, dq\, \rho_{\mathrm{W}}^2(p, q) \leq 1$$

is equivalent to the uncertainty principle requirement:

$$\Delta p \cdot \Delta q \geq \frac{\hbar}{2}.$$

▶ **Exercise 5.16**   Show that $\mathrm{Tr}(\rho^2) > 1$ implies that at least one of the p_n in Eq. 5.17 must be negative, and hence there is a negative probability to be in one of the eigenstates. This has no physical meaning and is called "violation of positivity" of the density operator.

5.3.4 Energy Representation of the Density Operator

In Exercise 5.9 we saw that $\rho(x', x) = \langle x'|\rho|x\rangle$ can be viewed as the density operator in the coordinate representation. For pure states, this takes the form

$$\rho(x', x) = \langle x'|\Psi\rangle\langle\Psi|x\rangle. \tag{5.33}$$

By inserting two complete sets of states we can find the elements of the density operator in the energy basis:

$$\begin{aligned}
\rho(x', x) &= \langle x'|\Psi\rangle\langle\Psi|x\rangle \\
&= \sum_{m,n} \langle x'|\psi_m\rangle\langle\psi_m|\Psi\rangle\langle\Psi|\psi_n\rangle\langle\psi_n|x\rangle \\
&= \sum_{m,n} \psi_m(x')\,(a_m e^{-\frac{i}{\hbar}E_m t})\,(a_n^* e^{\frac{i}{\hbar}E_n t})\,\psi_n^*(x) \\
&= \sum_{m,n} \psi_m(x')\,\rho_{mn}\,\psi_n^*(x),
\end{aligned} \tag{5.34}$$

where we have identified the elements of the density operator in the energy representation, ρ_{mn}, as

$$\rho_{mn} = a_m a_n^* e^{\frac{i}{\hbar}(E_n - E_m)t}. \tag{5.35}$$

As an example, consider the two-level system with eigenvectors ψ_1 and ψ_2 and corresponding eigenvalues E_1 and E_2. The density operator generated from the pure state $\psi = a\psi_1 e^{-iE_1 t/\hbar} + b\psi_2 e^{-iE_2 t/\hbar}$, where $|a|^2 + |b|^2 = 1$, has the following form in the basis of energy eigenstates:

$$\rho_{mn} = \begin{pmatrix} |a|^2 & ab^* e^{-i(E_1 - E_2)t/\hbar} \\ a^* b e^{i(E_1 - E_2)t/\hbar} & |b|^2 \end{pmatrix}. \tag{5.36}$$

Note that the diagonal elements are real and positive, while the off-diagonal elements are complex. In the energy representation the diagonal elements are the *populations* in the different energy levels and the off-diagonal elements are sometimes called the *coherences* between the levels. In the absence of relaxation processes the populations do not change with time and hence the time dependence appears only in the off-diagonal matrix elements.

We turn now to an example of a mixed state. The thermal density matrix, $e^{-\beta H}/Q$, takes an extremely simple form in the energy representation:

$$\rho_{nm} = \langle\psi_n|\rho|\psi_m\rangle = \langle\psi_n|e^{-\beta H}/Q|\psi_m\rangle = \left(e^{-\beta E_n}/Q\right)\delta_{nm}, \tag{5.37}$$

in other words, the density operator is diagonal with elements $e^{-\beta E_n}/Q$. Here Q is the canonical partition function,

$$Q = \text{Tr}(e^{-\beta H}) = \sum_n e^{-\beta E_n},$$

which serves as a normalization factor for ρ. The variable $\beta = 1/k_B T$, where k_B is the Boltzmann constant. There are two kinds of processes required for an initially pure state to relax to an equilibrium state: the diagonal elements must redistribute to a Boltzmann distribution and the off-diagonal elements must decay to zero. The first of these processes is called population decay; in two-level systems this time scale is called T_1. The second of these processes is called dephasing, or coherence decay; in two-level systems there is a single time scale for this process called T_2. There is a well-known relationship in two-level systems that

$$\frac{1}{T_2} = \frac{1}{2T_1} + \frac{1}{T_2^*},$$ (5.38)

where T_2^* is the time scale for so-called pure dephasing. (See Exercise 5.29.) Equation 5.38 has the following significance: it is impossible to have population relaxation without dephasing. An alternative perspective on this relationship will be given below using the Wigner representation.

5.3.5   Equations of Motion for the Density Operator

A differential equation for the time evolution of the density operator may be derived by taking the time derivative of Eq. 5.17 and using the Time-Dependent Schrödinger Equation to replace the time derivative of the wavefunction with the Hamiltonian operating on the wavefunction. The result is the Liouville–von Neumann equation:

$$i\hbar \frac{\partial \rho}{\partial t} = [H, \rho].$$ (5.39)

▶ **Exercise 5.17**   Show that this equation holds, both for the pure state form of the density operator, $\rho = |\Psi\rangle\langle\Psi|$, as well as for the mixed state form, $\rho = \sum_n p_n |\Psi_n\rangle\langle\Psi_n|$.

Although this equation of motion applies to either pure or mixed states, it does not describe relaxation. For example, $\text{Tr}(\rho^2)$ is a constant of the motion under this equation, so there is no change in the purity of the state.

▶ **Exercise 5.18**   Show that $\text{Tr}(\rho^2)$ is a constant of motion under this equation.

This equation of motion is the quantum mechanical analog of the classical Liouville equation,

$$\partial \rho_{\text{cl}}/\partial t = \{H_{\text{cl}}, \rho_{\text{cl}}\},$$ (5.40)

where

$$\{A, B\} \equiv A \wedge B \equiv \frac{\partial A}{\partial q}\frac{\partial B}{\partial p} - \frac{\partial A}{\partial p}\frac{\partial B}{\partial q}$$ (5.41)

is called the Poisson bracket of A with B. In other words, the classical mechanical equation that corresponds to the quantum mechanical equation for the density operator is obtained by replacing the commutator by the Poisson bracket. At the end of the chapter the Poisson bracket will reappear (cf. Eqs. 5.59, 5.63). We will see that the equations of motion for the density operator in the Wigner representation, which are fully equivalent to Eq. 5.39, can be written in a form that is strikingly similar to that of Eq. 5.41.

5.4 Wigner Representation of the Density Operator

We now return to the Wigner representation of the density operator. The Wigner function for general densities is given by

$$f_W(p, q) = \frac{1}{2\pi\hbar} \int_{-\infty}^{\infty} e^{\frac{i}{\hbar} ps} \left\langle q - \frac{s}{2} \middle| \rho \middle| q + \frac{s}{2} \right\rangle ds \equiv \frac{1}{2\pi\hbar} \rho_W(p, q). \quad (5.42)$$

In the earlier sections of this chapter we examined the Wigner representation of wavepackets, that is, pure states only. We now explore the Wigner representation for more general densities (mixed states). The most interesting examples are the thermal equilibrium state and the process of relaxation—the approach to equilibrium—as it appears in phase space. Our plan is to begin with the equilibrium state, where we are on solid ground, and then give a qualitative discussion of relaxation by interpolating between an initial, pure state and a final, thermal state. We will avoid discussing the actual form of the dissipative equations of motion for the density operator or the Wigner distribution; however, an extensive list of references is given at the end of the chapter.

5.4.1 Wigner Function for Thermal Equilibrium

The density operator for a state at thermal equilibrium takes the form

$$\rho = \frac{e^{-\beta H}}{Q} \qquad Q = \text{Tr}(e^{-\beta H}), \quad (5.43)$$

where the normalization factor, Q, is called the canonical partition function. We will refer to this state below as the thermal density operator.

In the energy representation we have already seen that ρ takes the form

$$\langle \psi_n | \rho | \psi_m \rangle = \frac{e^{-\beta E_n}}{Q} \delta_{nm} \qquad Q = \sum_n e^{-\beta E_n}. \quad (5.44)$$

In the coordinate representation the thermal density matrix takes the form

$$\langle x | \rho | x' \rangle = \frac{\langle x | e^{-\beta H} | x' \rangle}{Q} \qquad Q = \int_{-\infty}^{\infty} \langle x | e^{-\beta H} | x \rangle \, dx. \quad (5.45)$$

To gain some insight into the form of the thermal density operator we examine the case of a harmonic oscillator potential, where the analytical solutions are known.

For the harmonic oscillator Hamiltonian, $H = \frac{p^2}{2m} + \frac{1}{2}m\omega^2 x^2$, the analytical expression for the normalized thermal density in the x-representation is (Feynman, 1972)

$$\frac{\langle x' | e^{-\beta H} | x \rangle}{Q} = \left(\frac{m\omega}{\pi\hbar} \tanh(f/2) \right)^{1/2} \exp\left\{ \frac{-m\omega}{2\hbar \sinh(f)} [(x^2 + x'^2) \cosh(f) - 2xx'] \right\}, \quad (5.46)$$

where $f = \hbar\omega/kT$. Transforming variables to $q = \frac{x+x'}{2}$, $s = x - x'$ we find that

$$\rho(q, s) = \left(\frac{m\omega}{\pi\hbar} \tanh(f/2) \right)^{1/2} \exp\left\{ \frac{-m\omega}{\hbar} \left[\tanh(f/2)q^2 + \frac{1}{4} \coth(f/2)s^2 \right] \right\}. \quad (5.47)$$

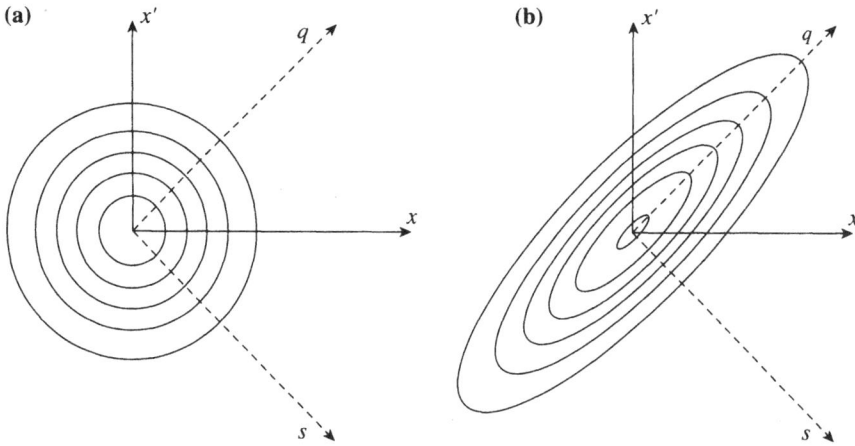

Figure 5.5 Equal probability contours of low-temperature and high-temperature density operators in the x, x' representation. (a) Pure state (0 temperature). (b) Intermediate temperature.

▶ **Exercise 5.19**   Derive Eq. 5.47 from Eq. 5.46. Verify that the width of $\rho(q, s)$, plotted in Figure 5.5, has the correct dependence on T.

Figure 5.5 shows a comparison of the harmonic oscillator thermal density operator at low temperature and high temperature in the x, x' representation. Note that for a pure state the coherence width is as large as the uncertainty width. As temperature increases, the width along q increases, signalling greater thermal fluctuations in the position variable, while the width in s decreases, signalling a reduction in coherence. The Wigner transform is simply the Fourier transform of $\rho(x, x')$ along the $s = x - x'$ coordinate. The analytic expression for the Wigner distribution for the harmonic oscillator thermal density is thus

$$\rho_W(p, q) = 2 \tanh(f/2) e^{-\tanh(f/2)(\frac{m\omega}{\hbar}q^2 + \frac{1}{m\omega\hbar}p^2)}. \tag{5.48}$$

▶ **Exercise 5.20**   Derive Eq. 5.48.

Figure 5.6 shows the Wigner distribution corresponding to these same low- and high-temperature density operators. Figure 5.6a is the Fourier transform of Figure 5.5a along the s coordinate. The width in s is maximally large, signifying full coherence, and its Fourier transform leads to the same value (Figure 5.5a). Similarly, Figure 5.6b is the Fourier transform of Figure 5.5b along the s coordinate. Since the distribution in s narrows with increasing temperature, the distribution in p broadens, while the width in q remains unchanged relative to Figure 5.5b. Note the spherical symmetry of the Wigner distribution in $\bar{q}$ and $\bar{p}$ for both low and high temperatures, a manifestation of the equipartition theorem in quantum statistical mechanics.

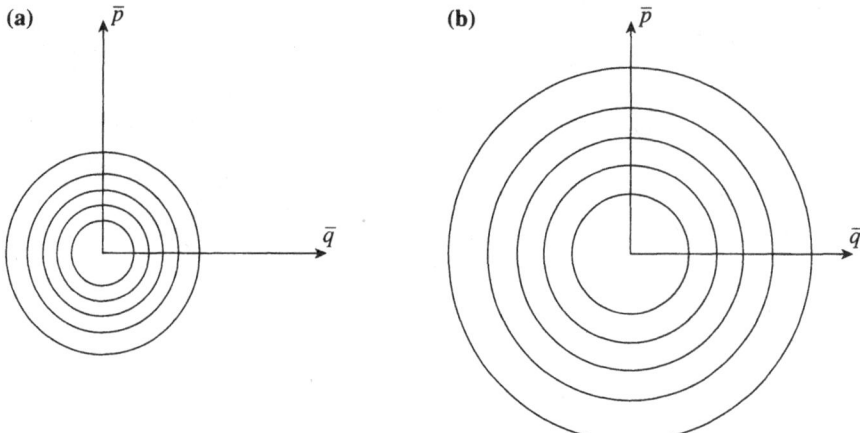

Figure 5.6 Equal probability contours of the Wigner distribution corresponding to the low- and high-temperature density operators. (a) Pure state (0 temperature). This figure is the Fourier transform of Figure 5.5a along the s coordinate. (b) Intermediate temperature. This figure is the Fourier transform of Figure 5.5b along the s coordinate.

5.4.2   Relaxation of the Wigner Function to Equilibrium

As mentioned at the beginning of this chapter, the density matrix is indispensible in describing quantum dissipative processes. The central issue in the theory of quantum dissipation, an equation of motion for ρ that includes relaxation induced by the environment, is beyond the scope of this book (for different approaches to this problem, see the end-of-chapter references). Nevertheless, using the Wigner distribution function, we can make several observations about different possible types of behavior of the density matrix. Figure 5.7a is a schematic picture of an initially pure state as it evolves toward thermal equilibrium in a harmonic oscillator potential. The state spirals in to the center, indicating energy loss, while the width evolves to the equilibrium thermal width of the Wigner distribution. In contrast, Figure 5.7b is a schematic picture of the same state experiencing pure dephasing with no energy loss. Note that the distribution becomes spread evenly on the original energy shell, without any radial motion. This is the quantum analog of an ensemble of classical oscillators, all with the same energy and starting with the same initial conditions, but by virtue of elastic collisions ending up with a uniform distribution of vibrational phases. Thus, we find that dephasing has a simple pictorial interpretation in phase space: it is a measure of the angle subtended by the distribution function in phase space.

Figure 5.7a can be obtained by starting in the energy representation and introducing a decay factor e^{-t/T_1} for the *nonequilibrium* component of each diagonal element and a decay factor of $e^{-t/2T_1}$ for each off-diagonal component, and then transforming to the Wigner representation. Figure 5.7b, in contrast, is obtained by letting T_1 go to infinity, and introducing a decay factor for the off-diagonal elements $e^{-t/|n-m|T_2}$.

An interesting observation is that it is impossible to have vibrational relaxation without accompanying dephasing: as the distribution in Figure 5.7a spirals into the center, without changing its width it automatically becomes more and more isotropic in phase until at equilibrium it subtends 2π. This is the phase space manifestation of the statement that

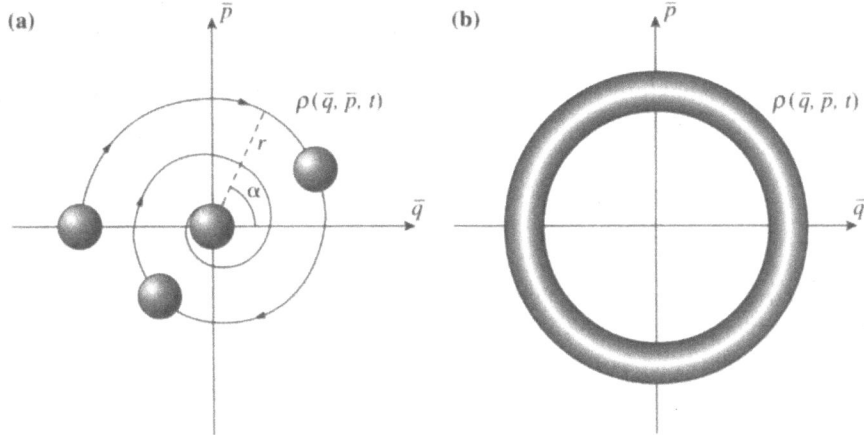

Figure 5.7 (a) Schematic evolution of the Wigner distribution in the presence of energy relaxation. Note the spiralling in to the equilibrium state. The energy relaxation process is normally accompanied by an increase in the width of the distribution (not shown). (b) Schematic evolution of the Wigner distribution, starting with the same initial conditions as in (a) but with only pure dephasing (i.e., no energy relaxation). Shown is the long-time distribution, which is spread out uniformly over the initial energy shell. Adapted from Kohen (1997).

$T_2 \leq 2T_1$. Even the factor of 2 in Eq. 5.38 comes out of this qualitative picture. We identify the inverse of the angle subtended with the radial distance of the center of the distribution function. Equating the time scale for the inverse angle to decay with T_2, we have

$$\frac{dr}{dt} = -\frac{1}{T_2} r. \tag{5.49}$$

Noting that the energy E goes as r^2,

$$E = \frac{\omega}{2}(\bar{p}^2 + \bar{q}^2) = \frac{\omega}{2} r^2, \tag{5.50}$$

we find that

$$\frac{dE}{dt} = -\frac{1}{T_1} E \implies \frac{dr^2}{dt} = -\frac{1}{T_1} r^2 = -\frac{2}{T_2} r^2, \tag{5.51}$$

where we have used Eq. 5.49. The RHS of Eq. 5.51 indicates that $\frac{1}{T_2} = \frac{1}{2T_1}$. Since the motion in the radial coordinate provides only a lower bound to the dephasing rate, we may write in general

$$\frac{1}{T_2} = \frac{1}{2T_1} + \frac{1}{T_2^*}, \tag{5.52}$$

identical to Eq. 5.38 for two-level systems in the energy representation.

5.4.3   Wigner Representation of an Arbitrary Operator

The prescription used to compute the Wigner distribution from the density operator can be used to construct a phase space representation of any quantum mechanical operator. The phase space, or Wigner, representation of any operator A is given by

$$A_W(p, q) = \int_{-\infty}^{\infty} e^{\frac{i}{\hbar}ps} \left\langle q - \frac{s}{2}|A|q + \frac{s}{2} \right\rangle ds. \tag{5.53}$$

▶ **Exercise 5.21**   Show that the Wigner transform of the $\hat{q}$ operator is $q_W(p, q) = q$, and of the $\hat{p}$ operator is $p_W(p, q) = p$. Show that the Wigner transform of the Hamiltonian, $H(\hat{q}, \hat{p}) = T(\hat{p}) + V(\hat{q})$, is $H_W(p, q) = H(p, q)$.

The Wigner distribution, f_W, is then related to the Wigner representation of the density operator, ρ_W, by

$$f_W = \frac{1}{2\pi\hbar} \rho_W. \tag{5.54}$$

The definition of the Wigner representation may be used to show that

$$\text{Tr}(\hat{A}\hat{B}) = \int_{-\infty}^{\infty}\int_{-\infty}^{\infty} A_W(p, q)B_W(p, q)\,dp\,dq, \tag{5.55}$$

where the "hats" have been used to emphasize that the quantities on the left-hand side of Eq. 5.55 are quantum mechanical operators.

▶ **Exercise 5.22**   Derive Eq. 5.55.

Equation 5.55 immediately leads to a key result of Wigner's approach, which is to express quantum mechanical expectation values as phase space integrals:

$$\langle A \rangle = \text{Tr}(\rho A)$$

$$= \frac{1}{2\pi\hbar} \int_{-\infty}^{\infty}\int_{-\infty}^{\infty} \rho_W(p, q)A_W(p, q)\,dp\,dq. \tag{5.56}$$

▶ **Exercise 5.23**   Derive Eq. 5.56.

An important special case, which we have used above, is

$$Tr(\rho^2) = \frac{1}{2\pi\hbar} \int_{-\infty}^{\infty}\int_{-\infty}^{\infty} \rho_W^2(p, q)\,dp\,dq. \tag{5.57}$$

Note that

$$(\hat{A}\hat{B})_W \neq \hat{A}_W\hat{B}_W; \tag{5.58}$$

only the traces are equal. The correct relationship is

$$(\hat{A}\hat{B})_W = A_W(p, q)e^{-\frac{i\hbar}{2}\Lambda}B_W(p, q), \tag{5.59}$$

where Λ is the Poisson bracket defined in Eq. 5.41. This somewhat cryptic-looking expression will be used below to discuss the time evolution of the Wigner distribution.

To go in the reverse direction, from a phase space function to a quantum operator, one uses Hermann Weyl's two-step prescription:

$$\alpha(\sigma, \tau) = \int \int e^{-\frac{i}{\hbar}(\sigma q + \tau p)} A_{\mathrm{W}}(p, q) \, dp \, dq$$

$$\hat{A}(\hat{p}, \hat{q}) = \int \int e^{-\frac{i}{\hbar}(\sigma \hat{q} + \tau \hat{p})} \alpha(\sigma, \tau) \, d\sigma \, d\tau.$$

Weyl's prescription does not remove the so-called operator ordering problem in passing from a classical to a quantum expression. The correspondence between quantum operators and phase space distribution functions given by Weyl's prescription, starting from the classical function, can be shown to be identical to that obtained by Wigner's prescription, which starts from the quantum operator.

5.4.4   Equations of Motion for the Wigner Distribution

Since the Wigner representation provides a path from quantum mechanical operators to classical-like phase space distribution functions, can we find an explicit equation of motion for the Wigner distribution function? Our starting point is the equation of motion of the density operator, written in the form

$$i\hbar \frac{\partial \rho}{\partial t} = H\rho - \rho H. \tag{5.60}$$

Taking the Wigner transform of Eq. 5.60 yields

$$i\hbar \frac{\partial \rho_{\mathrm{W}}}{\partial t} = (H\rho)_{\mathrm{W}} - (\rho H)_{\mathrm{W}}. \tag{5.61}$$

The "quantum" Liouville equation for the Wigner distribution is then

$$i\hbar \frac{\partial \rho_{\mathrm{W}}}{\partial t} = H_{\mathrm{W}} e^{-\frac{i\hbar}{2}\Lambda} \rho_{\mathrm{W}} - H_{\mathrm{W}} e^{\frac{i\hbar}{2}\Lambda} \rho_{\mathrm{W}} \tag{5.62}$$

or

$$\frac{\partial \rho_{\mathrm{W}}}{\partial t} = -\frac{2}{\hbar} H_{\mathrm{W}} \sin\left(\frac{\hbar \Lambda}{2}\right) \rho_{\mathrm{W}}. \tag{5.63}$$

Thus, the formal time evolution of the Wigner distribution is

$$\rho_{\mathrm{W}}(t) = e^{-\frac{2t}{\hbar} H_{\mathrm{W}} \sin(\frac{\hbar}{2}\Lambda)} \rho_{\mathrm{W}}(0). \tag{5.64}$$

To lowest order in $\hbar$, Eq. 5.63 is simply

$$\frac{d\rho_{\mathrm{W}}}{dt} = -H_{\mathrm{W}} \Lambda \rho_{\mathrm{W}} = \{H_{\mathrm{W}}, \rho_{\mathrm{W}}\}, \tag{5.65}$$

which is just the classical Liouville equation, 5.40. Equation 5.65 can be written in exponential form as

$$\rho_{\mathrm{W}}(t) = e^{-t H_{\mathrm{W}} \Lambda} \rho_{\mathrm{W}}(0). \tag{5.66}$$

▶ **Exercise 5.24**   Return to the table of classical–quantum correspondence at the end of Section 5.3, and add a third column giving the quantum expression in the Wigner representation. See how many entries in the third column you can figure out.

Problems

▶ **Exercise 5.25**   Calculate $\mathrm{Tr}(\rho)$ and $\mathrm{Tr}(\rho^2)$ for the 2×2 density matrix in Eq. 5.36. Then repeat, this time with the off-diagonal elements set equal to 0. Show that in the first case $\mathrm{Tr}(\rho^2) = 1$ (pure state) and in the second case $\mathrm{Tr}(\rho^2) < 1$ (mixed state).

▶ **Exercise 5.26**   Calculate the entropy

$$S = -\mathrm{Tr}(\rho \ln \rho) \tag{5.67}$$

for the pure state and mixed state in the above exercise.

▶ **Exercise 5.27**   Show explicitly that

$$\mathrm{Tr}(\rho^2) = \frac{1}{2\pi\hbar} \int_{-\infty}^{\infty} dp \int_{-\infty}^{\infty} dq \; \rho_{\mathrm{W}}^2(p, q)$$

for the equilibrium state of the harmonic oscillator potential. Calculate the LHS using both the energy representation for ρ, Eq. 5.44 and the x representation for ρ, Eq. 5.46, and the RHS using Eq. 5.48. Note that the higher the temperature, the farther ρ gets from a pure state; $\mathrm{Tr}(\rho^2)$ decreases with temperature. Plot $\mathrm{Tr}(\rho^2)$ as a function of T.

▶ **Exercise 5.28**   Calculate the entropy of the equilibrium state $\rho = e^{-\beta H}/Q$, and show that the result is the thermodynamic identity

$$S = \beta E + ln(Q). \tag{5.68}$$

Find explicit expressions for $E = \mathrm{Tr}(\rho H)$ and $Q = \mathrm{Tr}(e^{-\beta H})$ for the equilibrium state of the harmonic oscillator, using (a) the analytical expression for $\rho(E, E')$, (b) the analytical expression for $\rho_{\mathrm{W}}(p, q)$ and (c) the analytical expression for $\rho(x, x')$.

▶ **Exercise 5.29**   This exercise is to establish that under certain assumptions about the evolution of $\rho(t)$, $T_2 \leq 2T_1$ (see Section 5.3.4). We follow Pechukas (1994). Consider a two-level density matrix, and assume that

$$\rho(t) = \sum_n W_n \rho(0) W_n^\dagger, \tag{5.69}$$

where

$$W_n = \begin{pmatrix} a_n & c_n \\ d_n & b_n \end{pmatrix} \qquad W_n^\dagger = \begin{pmatrix} a_n^* & d_n^* \\ c_n^* & b_n^* \end{pmatrix}. \tag{5.70}$$

Furthermore, suppose that if

$$\rho(0) = \begin{pmatrix} p + \varepsilon & \delta \\ \delta^* & 1 - p - \varepsilon \end{pmatrix} \quad \text{and} \quad \rho(t \to \infty) = \begin{pmatrix} p & 0 \\ 0 & 1 - p \end{pmatrix} \quad (5.71)$$

that

$$\rho(t) = \begin{pmatrix} p + \alpha\varepsilon & \beta\delta \\ \beta\delta^* & 1 - p - \alpha\varepsilon \end{pmatrix}. \quad (5.72)$$

a. Show that $(c, a) = (d, b) = (c, d) = (d, a) = (b, c) = 0$, that $(c, c) + (b, b) = 1$, that $\beta = (a, b) = (b, a)$ and that $\alpha = (a, a) + (b, b) - 1$, where we use the notation $(u, v) = \sum_n u_n^* v_n$. Hint: the relaxation of the form Eq. 5.72 can be written as four independent conditions as follows:

$$\sum_n W_n \begin{pmatrix} p & 0 \\ 0 & 1 - p \end{pmatrix} W_n^\dagger = \begin{pmatrix} p & 0 \\ 0 & 1 - p \end{pmatrix} \qquad \sum_n W_n \begin{pmatrix} 1 & 0 \\ 0 & -1 \end{pmatrix} W_n^\dagger = \begin{pmatrix} \alpha & 0 \\ 0 & -\alpha \end{pmatrix}$$

$$\sum_n W_n \begin{pmatrix} 0 & 1 \\ 1 & 0 \end{pmatrix} W_n^\dagger = \begin{pmatrix} 0 & \beta \\ \beta & 0 \end{pmatrix} \qquad \sum_n W_n \begin{pmatrix} 0 & 1 \\ -1 & 0 \end{pmatrix} W_n^\dagger = \begin{pmatrix} 0 & \beta \\ -\beta & 0 \end{pmatrix}$$

b. Show that part a implies that $\beta \le \frac{\alpha+1}{2}$ [Hint: use the fact that $(a - b, a - b) = (a, a) + (b, b) - 2\,\mathrm{Re}(a, b) \ge 0$.]

c. Show that if the decay to equilibrium is exponential that part b implies that $T_2 \le 2T_1$.

d. Interpret the zero conditions in part a geometrically: the vectors c and d are orthogonal and the vectors a and b are both orthogonal to c and d. Use geometrical reasoning to establish the minimum value of n (dimension of the vector space) to obtain (i) any form of relaxation; (ii) $T_2 = 2T_1$; (iii) $T_2 \le 2T_1$. (Consider the possibility that c or d or both are zero.)

▶ **Exercise 5.30** Show that density operators satisfy a convexity property, that is, if ρ_1 and ρ_2 are density operators, then

$$\rho = p_1\rho_1 + p_2\rho_2 \quad (5.73)$$

is also a density operator, where $p_1 + p_2 = 1, 0 \le p_1, p_2 \le 1$.

▶ **Exercise 5.31** Recall that in the general definition of the density matrix,

$$\rho = \sum_{i=1}^n p_i |\psi_i\rangle\langle\psi_i|, \quad (5.74)$$

the $\{\psi_i\}$ need not be orthogonal. An important consequence of this fact is that the decomposition in Eq. 5.74 is not unique.

a. Consider the two ostensibly different density matrices,

$$\rho_1 = \frac{3}{4}|0\rangle\langle0| + \frac{1}{4}|1\rangle\langle1| \quad (5.75)$$

and

$$\rho_2 = \frac{1}{2}|a\rangle\langle a| + \frac{1}{2}|b\rangle\langle b|, \quad (5.76)$$

where

$$|a\rangle = \sqrt{\frac{3}{4}}|0\rangle + \sqrt{\frac{1}{4}}|1\rangle, \qquad |b\rangle = \sqrt{\frac{3}{4}}|0\rangle - \sqrt{\frac{1}{4}}|1\rangle. \qquad (5.77)$$

Show that these two density matrices are in fact equal and that in the $(|0\rangle, |1\rangle)$ representation both are given by the matrix

$$\rho_1 = \rho_2 = \begin{pmatrix} \frac{3}{4} & 0 \\ 0 & \frac{1}{4} \end{pmatrix}. \qquad (5.78)$$

b. The general rule for when two apparently different expressions for ρ are in fact indistinguishable may be expressed as follows. Defining

$$|\tilde{\psi}_i\rangle = \sqrt{p_i}|\psi_i\rangle, \qquad (5.79)$$

the sets $\{|\tilde{\psi}_i\rangle\}$ and $\{|\tilde{\phi}_j\rangle\}$ generate the same density matrix if and only if

$$|\tilde{\psi}_i\rangle = \sum_{j=1}^{n} u_{ij}|\tilde{\phi}_j\rangle, \qquad (5.80)$$

where u_{ij} are elements of a unitary matrix $(\sum_i |u_{ij}|^2 = \sum_j |u_{ij}|^2 = 1)$. Return to the example in part a, take

$$|\tilde{\psi}_1\rangle = \sqrt{\frac{3}{4}}|0\rangle, \qquad |\tilde{\psi}_2\rangle = \sqrt{\frac{1}{4}}|1\rangle \qquad (5.81)$$

and

$$|\tilde{\phi}_1\rangle = \frac{1}{\sqrt{2}}|a\rangle, \qquad |\tilde{\phi}_2\rangle = \frac{1}{\sqrt{2}}|b\rangle \qquad (5.82)$$

and verify that the transformation matrix is indeed unitary.

References

Phase Space

1. I. Percival and D. Richards, *Introduction to Dynamics* (Cambridge University Press, Cambridge, 1982).
2. H. Goldstein, *Classical Mechanics* (Addison-Wesley, New York, 1950).
3. V. I. Arnold, *Mathematical Methods of Classical Mechanics* (Springer-Verlag, New York, 1980).
4. A. J. Lichtenberg and M. A. Lieberman, *Regular and Stochastic Motion* (Springer-Verlag, New York, 1983).

Wigner Distribution

5. E. Wigner, On the quantum correction for thermodynamic equilibrium, Phys. Rev. 40, 749 (1932).

6. J. E. Moyal, Quantum mechanics as a statistical theory, Proc. Cambridge Philos. Soc. 45, 99 (1949).

7. M. S. Bartlett and J. E. Moyal, The exact transition probabilities of quantum-mechanical oscillators calculated by the phase-space method, Proc. Cambridge Philos. Soc. 45, 545 (1949).

8. K. Imre, E. Ozizmir, M. Rosenbaum, and P. F. Zweifel, Wigner method in quantum statistical mechanics, J. Math. Phys. 8, 1097 (1967).

9. E. J. Heller, Wigner phase space method: Analysis for semiclassical applications, J. Chem. Phys. 65, 1289 (1976).

10. M. V. Berry, Semiclassical mechanics in phase space: A study of Wigner's Function, Proc. Roy. Soc. London A 287, 237 (1977).

11. M. Hillery, R. F. O'Connell, M. O. Scully and E. P. Wigner, Distribution functions in physics: Fundamentals, Phys. Repts. 106, 121 (1984).

12. W. P. Schleich, *Quantum Optics in Phase Space* (Wiley-VCH, Berlin, 2001).

Density Operator

13. J. von Neumann, Wahrscheinlichkeitstheoritischer Aufbau der Quantenmechanik, Göttingen Nachrichten, p. 245 (1927); Thermodynamik quantenmechanischer Gesamtheiten, ibid, p. 273 (1927). Reprinted in *John von Neumann, Collected Works, vol. I* (Pergamon, New York, 1961).

14. L. Landau, Das Dämpfungsproblem in der Wellenmechanik, Z. Phys. 45, 430 (1927).

15. A. Messiah, *Quantum Mechanics*, vol. I (Wiley, New York, 1958).

16. R. P. Feynman, *Statistical Mechanics* (Addison-Wesley, New York, 1972).

17. A. Peres, *Quantum Theory: Concepts and Methods* (Kluwer, Dordrecht, 1993).

18. M. A. Nielsen and I. L. Chuang, *Quantum Computation and Quantum Information* (Cambridge University Press, Cambridge, 2000), pp. 98–108.

Relaxation of the Density Matrix: General

19. N. G. van Kampen, *Stochastic Processes in Physics and Chemistry* (North-Holland, Amsterdam, 1992), Chapter 17.

20. R. Zwanzig, *Nonequilibrium Statistical Mechanics* (Oxford University Press, New York, 2001).

21. R. Kubo, M. Toda and N. Hashitsume, *Statistical Physics II: Nonequilibrium Statistical Mechanics* (Springer, Berlin, 1995).

22. S. Nakajima, On quantum theory of transport phenomena: Steady diffusion, Prog. Theor. Phys. 20, 948 (1958); R. Zwanzig, Ensemble method in the theory of irreversibility, J. Chem. Phys. 33, 1338 (1960).

T_1–T_2

23. C. P. Slichter, *Principles of Magnetic Resonance*, 2nd ed. (Springer-Verlag, Berlin, 1978).

24. D. Oxtoby, Dephasing of molecular vibrations in liquids, Adv. Chem. Phys. 40, 1 (1979); D. Oxtoby, Vibrational population relaxation in liquids, Adv. Chem. Phys. 47, (Pt. 2) 487 (1981).

25. B. B. Laird, J. Budimir and J. L. Skinner, Quantum-mechanical derivation of the Bloch equations: Beyond the weak-coupling limit, J. Chem. Phys. 94, 4391 (1991); B. B. Laird and J. L. Skinner, T_2 can be greater than T_1 even at finite temperature, J. Chem. Phys. 94, 4405 (1991).

26. P. Pechukas, Reduced dynamics need not be completely positive, Phys. Rev. Lett. 73, 1060 (1994).

27. D. Kohen, *Phase Space Distribution Function Approach to Molecular Dynamics in Solution*, Ph.D. thesis, University of Notre Dame (1995).

Bloch–Redfield Theory

28. F. Bloch, Nuclear induction, Phys. Rev. 70, 460 (1946).

29. A. G. Redfield, IBM J. Res. Dev. 1, 19 (1957); The theory of relaxation processes, Adv. Magn. Reson. 1, 1 (1965).

30. K. Blum, *Density Matrix Theory and Applications* (Plenum Press, New York, 1981).

31. C. Cohen-Tannoudji, J. Dupont-Roc and G. Grynberg, *Atom-Photon Interactions* (Wiley, New York, 1992), Section IV.

32. J. M. Jean, R. A. Friesner and G. R. Fleming, Application of a multilevel Redfield theory to electron transfer in condensed phases, J. Chem. Phys. 96, 5827 (1992); W. T. Pollard, A. K. Felts and R. A. Friesner, The Redfield equation in condensed-phase quantum dynamics, Adv. Chem. Phys. 93, 77 (1996).

33. A. M. Walsh and R. D. Coalson, Redfield theory is quantitative for coupled harmonic oscillators, Chem. Phys. Lett. 198, 293 (1992).

34. D. Kohen and D. J. Tannor, Classical-quantum correspondence in the Redfield equation and its solutions, J. Chem. Phys. 107, 5141 (1997).

35. I. Kondov, U. Kleinekathöfer and M. Schreiber, Efficiency of different numerical methods for solving Redfield equations, J. Chem. Phys. 114, 1497 (2001).

36. V. May and O. Kühn, *Charge and Energy Transfer Dynamics in Molecular Systems* (Wiley-VCH, Weinheim, 1999).

Semigroup Theory

37. G. Lindblad, On the generators of quantum dynamical semigroups, Commun. Math. Phys. 48, 119 (1976).

38. V. Gorini, A. Kossakowski and E. C. G. Sudarshan, Completely positive dynamical semigroups of N-level systems, J. Math. Phys. 17, 821 (1976).

39. E. B. Davies, *Quantum Theory of Open Systems* (Academic, New York, 1976).

40. R. Alicki and K. Lendi, *Quantum Dynamical Semigroups and Applications* (Springer, New York, 1986).

41. P. Talkner, The failure of the quantum regression hypothesis, Annals of Physics (NY) 167, 390 (1986).

42. P. Pechukas, Reduced dynamics need not be completely positive, Phys. Rev. Lett. 73, 1060 (1994).

43. M. Berman, R. Kosloff and H. Tal-Ezer, Solution of the time-dependent Liouville-von Neumann equation: Dissipative evolution, J. Phys. A: Math. Gen. 25, 1283 (1992).

44. D.A. Lidar, Z. Bihary and K. B. Whaley, From completely positive maps to the quantum Markovian semigroup master equation, Chem. Phys. 268, 35 (2001).

Phase Space Approach

45. D. Kohen, C. C. Marston and D. J. Tannor, Phase space approach to theories of quantum dissipation, J. Chem. Phys. 107, 5236 (1997).

46. D. Kohen and D. J. Tannor, Classical-quantum correspondence in the Redfield equation and its solutions, J. Chem. Phys. 107, 5141 (1997).

47. G. Ashkenazi, U. Banin, A. Bartana, R. Kosloff and S. Ruhman, Quantum description of the impulsive photodissociation dynamics of I_3^- in solution, Adv. Chem. Phys., 100, 229–315 (1997).

Non-Markovian Approaches

48. Y. J. Yan and S. Mukamel, Semiclassical dynamics in Liouville space: Application to molecular electronic spectroscopy, J. Chem. Phys. 88, 5735 (1988).

49. S. Tsonchev and P. Pechukas, Binary collision model for Brownian motion, Phys. Rev. E 61, 6171 (2000).

50. A. Suarez, R. Silbey and I. Oppenheim, Memory effects in the relaxation of quantum open systems, J. Chem. Phys. 97, 5101 (1992).

51. C. Meier and D. J. Tannor, Non-Markovian evolution of the density operator in the presence of strong laser fields, J. Chem. Phys. 111, 3365–3376 (1999).

52. S. Jang, J. Cao and R. J. Silbey, Fourth-order quantum master equation and its Markovian bath limit, J. Chem. Phys. 116, 2705 (2002).

53. R. Baer and R. Kosloff, Quantum dissipative dynamics of adsorbates near metal surfaces: A surrogate Hamiltonian theory applied to hydrogen on nickel, J. Chem. Phys. 106, 8862 (1997).

Path Integral Approaches

54. R. P. Feynman, *Statistical Mechanics: A Set of Lectures* (Addison-Wesley, New York, 1972).

55. R. P. Feynman and F. L. Vernon, Jr., The theory of a general quantum system interacting with a linear dissipative system, Annals of Physics (NY) 24, 118 (1963).

56. A. O. Caldeira and A. J. Leggett, Path integral approach to quantum Brownian motion, Physica A 121, 587 (1983).

57. H. Grabert, P. Schramm and G.-L. Ingold, Quantum Brownian motion—The functional integral approach, Phys. Rep. 168, 115 (1988).

58. U. Weiss, *Quantum Dissipative Systems* (World Scientific, Singapore, 1993).

59. N. Makri, Time-dependent quantum methods for large systems, Annual Rev. Phys. Chem. 50, 167 (1999).

Stochastic Schrödinger Approaches

60. J. Dalibard, Y. Castin and K. Mølmer, Wavefunction approach to dissipative properties in quantum optics, Phys. Rev. Lett. 68, 580 (1992).

61. R. Dum, P. Zoller and H. Ritsch, Monte Carlo simulation of the atomic master equation for spontaneous emission, Phys. Rev. A 45, 4879 (1992).

62. H.-P. Breuer and F. Petruccione, Stochastic dynamics of quantum jumps, Phys. Rev. E 52, 428 (1995).

63. J. T. Stockburger and H. Grabert, Exact c-number representation of non-Markovian quantum dissipation, Phys. Rev. Lett. 88, 170407 (2002).

Chapter 6

Correlation Functions and Spectra

In this chapter we introduce the concept of a wavepacket time-correlation function, $C(t)$, and its Fourier transform, the wavepacket spectrum, $\sigma(\omega)$. Wavepacket correlation functions and their Fourier transforms will play a major role in our treatment of spectroscopy and reactive scattering in Part III of this book, Applications. However, for the time being we will present these quantities as intrinsic properties of any wavepacket, essentially as alternative *representations* of the packet. For example, the spectrum simply represents the absolute square of the energy components within the wavepacket; its definition does not require invoking any interaction with light. Similarly, the wavepacket correlation function, which is the Fourier transform of the spectrum, also has some sense of being a *time* representation of the wavepacket.

Because the concept of wavepacket correlation functions will appear in so many applications below, we use this opportunity to introduce many of the general features that will be exploited repeatedly below. In particular, since the correlation function and the spectrum are related via Fourier transform, an understanding of the time-domain approach to spectroscopy is greatly facilitated if one has a thorough grounding in Fourier transforms. Therefore, a significant fraction of this chapter is devoted to gaining experience with conjugate Fourier pairs, via analytic relations as well as via qualitative observations and general properties. The Fourier convolution theorem is introduced, which is the key to understanding multidimensional spectra.

Although the Fourier transforms described here are between the time and energy representations, the Fourier transform between position and momentum representations also plays a major role below, both in the formal theory of quantum mechanics (Chapter 8) and in the numerical methods (Chapter 11).

6.1 Spectra as Fourier Transforms of Wavepacket Correlation Functions

6.1.1 Formal Derivation

For a bound-state problem, we can expand a given wavepacket in the eigenfunctions of the problem:

$$\Psi(x, t) = \sum_n c_n \, \psi_n(x) \, e^{-\frac{i}{\hbar} E_n t}. \tag{6.1}$$

The "spectrum", $\sigma(\omega)$, is then given by

$$\sigma(\omega) = \sum_n |c_n|^2 \delta(\omega - \omega_n), \tag{6.2}$$

where $\omega_n = E_n/\hbar$; the spectrum is simply the sum of the absolute squares of the energy components of the wavepacket, positioned at the eigenvalue energies. We can also compute the spectrum from the Fourier transform of the wavepacket autocorrelation function:

$$\sigma(\omega) = \frac{1}{2\pi} \int_{-\infty}^{\infty} \langle \Psi(0)|\Psi(t)\rangle e^{i\omega t}\, dt. \tag{6.3}$$

To show that Eq. 6.3 is, indeed, equivalent to Eq. 6.2, we substitute Eq. 6.1 for $\Psi(x, t)$ into Eq. 6.3 and use the orthogonality of the eigenfunctions of H:

$$\sigma(\omega) = \frac{1}{2\pi} \int_{-\infty}^{\infty} \langle \Psi(0)|\Psi(t)\rangle e^{i\omega t}\, dt$$

$$= \frac{1}{2\pi} \int_{-\infty}^{\infty} \int_{-\infty}^{\infty} \left(\sum_m c_m^* \psi_m^*(x) \right) \left(\sum_n c_n e^{-\frac{i}{\hbar} E_n t} \psi_n(x) \right)\, dx\, e^{i\omega t}\, dt$$

$$= \frac{1}{2\pi} \int_{-\infty}^{\infty} \sum_{m,n} c_m^* c_n \delta_{mn} e^{-\frac{i}{\hbar} E_n t} e^{i\omega t}\, dt$$

$$= \frac{1}{2\pi} \sum_n |c_n|^2 \int_{-\infty}^{\infty} e^{-\frac{i}{\hbar} E_n t} e^{i\omega t}\, dt$$

$$= \sum_n |c_n|^2 \delta\left(\omega - \frac{E_n}{\hbar} \right)$$

$$= \sum_n |c_n|^2 \delta(\omega - \omega_n).$$

Equation 6.3 states that the spectrum, $\sigma(\omega)$, and the wavepacket correlation function, $C(t) = \langle \Psi(0)|\Psi(t)\rangle$, are Fourier transform pairs. As we show below, this allows many properties of the Fourier transform to be exploited in the intuitive interpretation of spectra.

▶ **Exercise 6.1**   We have adopted this definition of the spectrum to be normalized. Show that if the wavepacket is normalized ($\langle \Psi(0)|\Psi(0)\rangle = 1$), then the spectrum is normalized ($\int \sigma(\omega)d\omega = 1$).

6.1.2 Pictorial Treatment

The above result—that the spectrum of a system may be obtained from the Fourier transform of a wavepacket correlation function—is a powerful and general relationship. Although we derived this relationship under the assumption that the spectrum consists of a series of sharp (δ-function) lines, the result is actually much more general. To fully appreciate and exploit this relationship, it is necessary to gain some familiarity with Fourier transform pairs. Some of the more common examples are introduced in this subsection. A more

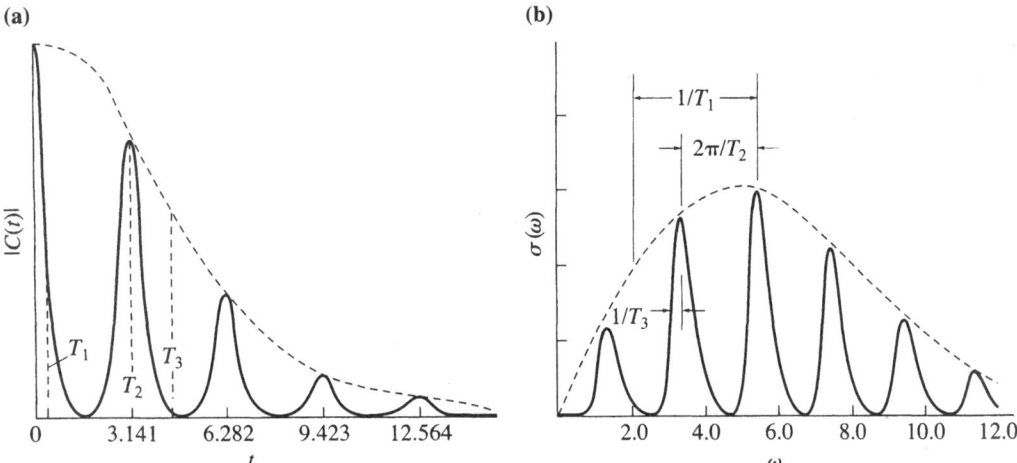

Figure 6.1 Schematic diagram showing three time scales in the autocorrelation function (a) and the corresponding three time scales in the energy spectrum (b). Adapted from (Heller, 1981).

complete description of the the most important properties of Fourier transforms are collected in Section 6.2 below.

Figure 6.1 shows a typical example of a wavepacket correlation function, $C(t)$, and its corresponding spectrum, $\sigma(\omega)$. The absolute value of the wavepacket correlation function $|C(t)| = |\langle\Psi|\Psi(t)\rangle|$ is shown in Figure 6.1a. At $t = 0$, $|\langle\Psi|\Psi(t)\rangle| = 1$ because Ψ is normalized. The subsequent behavior of $C(t)$ is characterized by three time scales. The first, T_1, corresponds to the initial decay of $C(t)$ as the wavepacket moves away from its initial location in phase space. After one period of motion, T_2, the wavepacket returns to its original phase space location, leading to a recurrence in $C(t)$. A series of such recurrences in $C(t)$ are seen in Figure 6.1a at multiples of the period of motion. There is a third time scale, T_3, on which the recurrences in $C(t)$ decay. This decay could have its physical origin in any number of sources, including anharmonicity, coupling of the wavepacket motion to other degrees of freedom, or irreversible decay due to indirect dissociation ("predissociation"), or radiationless or radiative decay. Figure 6.1b shows the spectrum, $\sigma(\omega)$, obtained by Fourier transforming $C(t)$. Each of the time scales in Figure 6.1a is manifested in a frequency scale in Figure 6.1b. The shortest time scale, T_1, manifests itself as the broadest scale in energy, $\omega_1 = 1/T_1$. The time scale for the recurrences, T_2, manifests itself as the spacing between features in frequency, $\omega_2 = 2\pi/T_2$. The time scale for the overall decay, T_3, corresponds to the width of the structures in the frequency domain, $1/T_3$. In principle, if there is no source of irreversible decay, there could be a time scale T_4 for later recurrences, which will ultimately resolve the spectrum into δ-functions.

We make two general observations on the correspondence between time-domain and frequency-domain structures. First, note that the shorter the scale of a feature in the time domain the broader is the corresponding feature in the energy domain, and vice versa; for example, the shortest time scale is T_1, which determines the broadest energy scale, $\omega_1 = 1/T_1$. Second, note that *decays* in the time domain correspond to *decays* (i.e., widths)

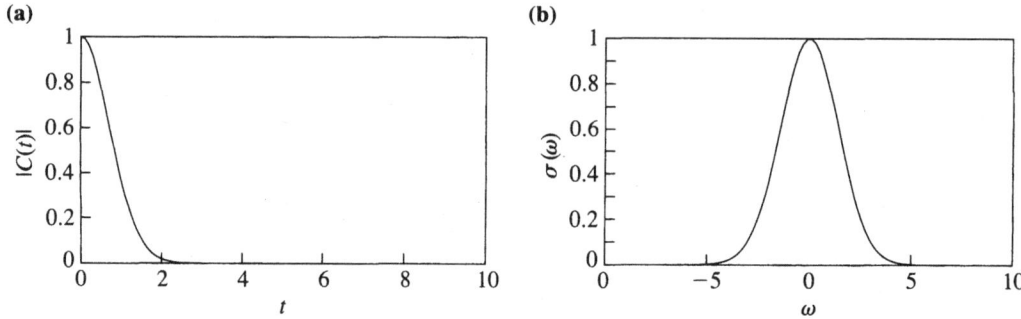

Figure 6.2 (a) Correlation function with a single Gaussian decay. (b) The spectrum corresponding to the correlation function in (a) is a single Gaussian peak. The spectrum has been scaled to unit magnitude.

in the energy domain, while *recurrences* in the time domain correspond to *recurrences* (i.e., spacings) in the frequency domain. For example, T_1 is a decay while $E_1 = 1/T_1$ is a width; T_2 is a recurrence while $E_2 = 2\pi/T_2$ is a spacing.

To explore Figure 6.1 in more detail, we will analyze the Fourier transform relationship, one time scale at a time. The shortest time scale decay is often close to Gaussian in shape. If we take the correlation function as a single Gaussian decay of unit width $C(t) = e^{-\alpha t^2}$, as shown in Figure 6.2a, then, since the Fourier transform of a Gaussian is another Gaussian, the spectrum, scaled to unit magnitude, will consist of a single Gaussian peak, $\sigma(\omega) = e^{-\frac{\omega^2}{4\alpha}}$, as shown in Figure 6.2b.

We now consider a series of perfect recurrences in the correlation function, as is characteristic of motion in a harmonic system (Figure 6.3a). If only a finite number of these peaks are included in the correlation function, the spectrum appears as in Figure 6.3b. The spacing of the peaks in the spectrum is inversely proportional to the period of the recurrences. The "ringing" of the spectrum (the oscillating positive and negative values) is the result of the sharp truncation of the time signal in the middle of the second recurrence (see Exercise 6.18). Finally, we consider the case when the correlation function is the product of a slow Gaussian decay and a sum of faster Gaussian peaks, as shown in Figure 6.4a. The spectrum retains the same qualitative peak structure as in Figure 6.3b, but is now smooth, with each spectral feature having a width inversely proportional to the decay of $C(t)$ (Figure 6.4b).

6.1.3   Survival Probabilities: P_{aa} and P_{ab}

In this section we discuss an additional interesting connection between the wavepacket correlation function and its spectrum. Define the wavepacket survival probability, P_{aa}, as follows:

$$P_{aa} \equiv \lim_{T \to \infty} \frac{1}{2T} \int_{-T}^{T} |\langle \Psi | \Psi(t) \rangle|^2 dt. \tag{6.4}$$

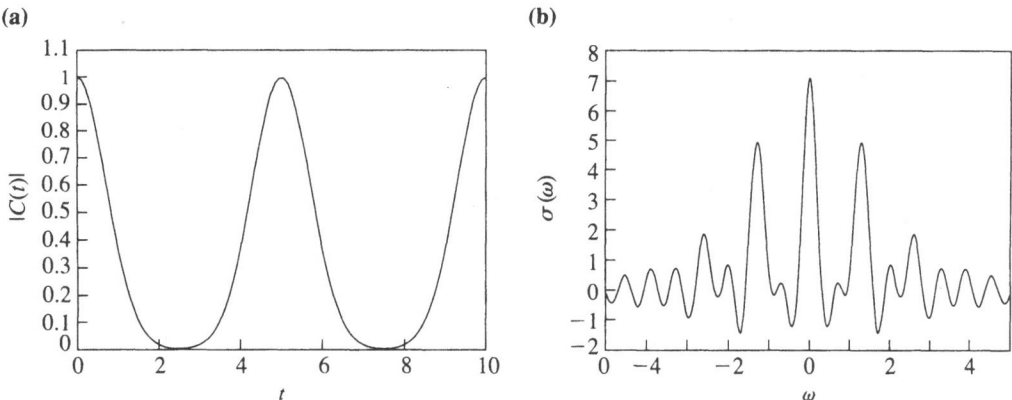

Figure 6.3 (a) Correlation function with a series of three recurring peaks. (b) The spectrum corresponding to the correlation function in (a) showing structure arising from the interference of the three peaks in the correlation function and "ringing" from the sharp truncation of the last of the peaks.

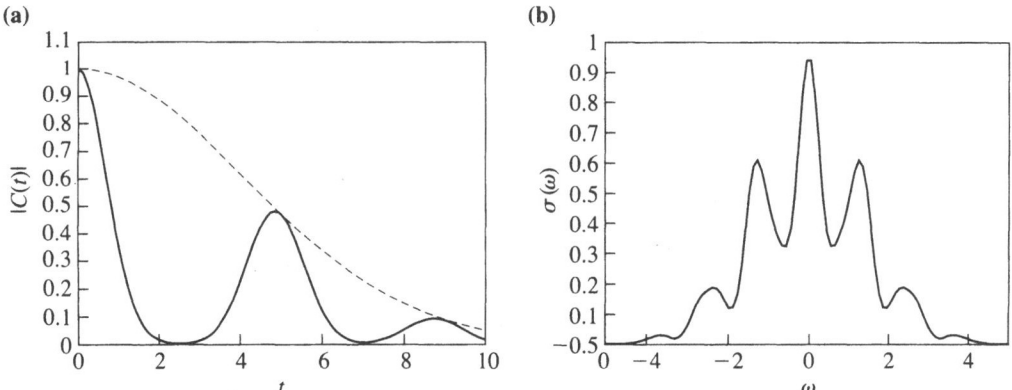

Figure 6.4 (a) Correlation function given by the product of a slow Gaussian decay and a series of three Gaussian peaks. The dashed line indicates the overall decay. (b) The spectrum corresponding to the correlation function in (a). Note that the structure remains but the ringing has disappeared.

This quantity has the physical interpretation of the time average of the *probability* that the wavepacket overlaps its initial state. This quantity can be rearranged as follows:

$$P_{aa} = \lim_{T\to\infty} \frac{1}{2T} \int_{-T}^{T} dt \, |\langle\Psi|e^{-iHt/\hbar}|\Psi\rangle\langle\Psi|e^{iHt/\hbar}|\Psi\rangle| \tag{6.5}$$

$$= \sum_{nm} c_n c_m^* \lim_{T\to\infty} \frac{1}{2T} \int_{-T}^{T} dt \, \langle\Psi|e^{-iE_n t/\hbar}|\psi_n\rangle\langle\psi_m|e^{iE_m t/\hbar}|\Psi\rangle \tag{6.6}$$

$$= \sum_{nm} c_n c_m^* \langle\Psi|\psi_n\rangle\langle\psi_m|\Psi\rangle \delta_{mn} \tag{6.7}$$

$$= \sum_{n} |c_n|^4, \tag{6.8}$$

where we have recognized that

$$\langle \Psi | \psi_n \rangle = c_n^* \quad \text{and} \quad \langle \psi_m | \Psi \rangle = c_m.$$

Thus, the wavepacket survival probability has a direct measure in the *square* of the peak intensities in the spectrum. Similarly, one can define

$$P_{ab} \equiv \lim_{T \to \infty} \frac{1}{2T} \int_{-T}^{T} |\langle \Psi_b | \Psi_a(t) \rangle|^2 dt, \tag{6.9}$$

which has the physical interpretation of the time-averaged probability the wavepacket Ψ_a overlaps the stationary wavepacket Ψ_b. Using steps that are parallel to those above for P_{aa}, one may show that

$$P_{ab} = \sum_n |c_n|^2 |d_n|^2, \tag{6.10}$$

where

$$\langle \Psi_a | \psi_n \rangle = c_n^* \quad \text{and} \quad \langle \psi_n | \Psi_b \rangle = d_n.$$

▶ Exercise 6.2   Derive Eq. 6.10.

6.2   General Properties of Fourier Transforms

6.2.1   Fourier Transform Pairs

We now turn to some general properties of Fourier transforms. The forward Fourier transform is defined as

$$\tilde{f}(\omega) = \int_{-\infty}^{\infty} f(t) e^{-i\omega t} \, dt \tag{6.11}$$

and the backward Fourier transform is defined as

$$f(t) = \frac{1}{2\pi} \int_{-\infty}^{\infty} \tilde{f}(\omega) e^{+i\omega t} \, d\omega. \tag{6.12}$$

▶ Exercise 6.3   Verify Eq. 6.12.

Solution   Substituting Eq. 6.11 into the RHS of Eq. 6.12, we obtain

$$\frac{1}{2\pi} \int_{-\infty}^{\infty} \tilde{f}(\omega) e^{+i\omega t} d\omega = \frac{1}{2\pi} \int_{-\infty}^{\infty} \left\{ \int_{-\infty}^{\infty} f(t') e^{-i\omega t'} dt' \right\} e^{+i\omega t} \, d\omega \tag{6.13}$$

$$= \frac{1}{2\pi} \int_{-\infty}^{\infty} \int_{-\infty}^{\infty} f(t') \{ e^{i\omega(t-t')} d\omega \} \, dt' \tag{6.14}$$

$$= \frac{1}{2\pi} \int_{-\infty}^{\infty} f(t') 2\pi \delta(t - t') \, dt' \tag{6.15}$$

$$= f(t). \tag{6.16}$$

Sometimes the forward and backward Fourier transforms are defined more symmetrically, namely, $\tilde{f}(\omega) = \frac{1}{\sqrt{2\pi}} \int_{-\infty}^{\infty} f(t) e^{-i\omega t} \, dt$, $f(t) = \frac{1}{\sqrt{2\pi}} \int_{-\infty}^{\infty} \tilde{f}(\omega) e^{i\omega t} \, d\omega$. We will not generally use this definition; as long as the forward and backward transforms are inverses, it makes no difference in the final result. Note that our definition of the spectrum, Eq. 6.3, is technically a *backward* Fourier transform of the wavepacket correlation function. This is simply a consequence of the opposite choices of convention used in the definition of the forward Fourier transform and in the evolution operator for the Time-Dependent Schrödinger Equation.

The exercises below explore several common Fourier transform pairs, as well as some general properties of Fourier transform pairs.

▶ **Exercise 6.4 (Fourier Transform Pairs)**

a. Show that the Fourier transform of a Gaussian, $N e^{-t^2/2\sigma_t^2}$, is a Gaussian, $N' e^{-\omega^2/2\sigma_\omega^2}$, where $\sigma_\omega = 1/\sigma_t$.

b. Show that the Fourier transform of an exponential, $e^{-\Gamma|t|}$, is a Lorentzian, $\frac{2\Gamma}{\omega^2 + \Gamma^2}$.

c. Show that the Fourier transform of the constant amplitude function $e^{i\omega_0 t}$ is the Dirac δ-function $\delta(\omega - \omega_0)$, times 2π.

▶ **Exercise 6.5 (Linearity)**   Show that the Fourier transform is linear, that is, $\widetilde{\{f_1(t) + f_2(t)\}} = \tilde{f}_1(\omega) + \tilde{f}_2(\omega)$. Use this property to find the Fourier transform of $a \cos(\omega_0 t) + b$.

▶ **Exercise 6.6 (Time Scaling)**   Show that the Fourier transform of $f(kt)$ is $\frac{1}{|k|} \tilde{f}(\frac{\omega}{k})$.

▶ **Exercise 6.7 (Time and Frequency Shifting)**

a. Show that the Fourier transform of $f(t) e^{i\omega_0 t}$ is $\tilde{f}(\omega - \omega_0)$. In words, multiplication of a function by a linear phase in time corresponds to a shift of the Fourier transform in frequency; similarly, a shift in time corresponds to a linear phase in frequency.

b. Use the property of linearity and time shifting to find the Fourier transform of the sum of three Gaussians, $e^{-\alpha(t+t_0)^2} + e^{-\alpha t^2} + e^{-\alpha(t-t_0)^2}$. Discuss the connection between your analytic result and Figure 6.4.

▶ **Exercise 6.8 (Complex Arithmetic Symmetries)**

a. Show that if the Fourier transform of $f(t)$ is $\tilde{f}(\omega)$, the Fourier transform of $f^*(t)$ is $(\tilde{f}(-\omega))^*$.

b. Show that if a function $f(t)$ is even ($f(t) = f(-t)$), its Fourier transform $\tilde{f}(\omega)$ is even ($\tilde{f}(\omega) = \tilde{f}(-\omega)$). Show that if it is real and even the FT is real and even; if it is imaginary and even the FT is imaginary and even.

c. Show that if a function $f(t)$ is odd ($f(t) = -f(-t)$), its Fourier transform $\tilde{f}(\omega)$ is odd ($\tilde{f}(\omega) = -\tilde{f}(-\omega)$). Show that if it is real and odd the FT is imaginary and odd; if it is imaginary and odd the FT is real and odd.

d. Combine the previous results to show that if $f(t) = f^*(-t)$ then $\tilde{f}(\omega)$ is real.

▶ **Exercise 6.9 (Fourier Inner Product Relation)**

   a. Let $\tilde{f}(\omega)$ be the FT of $f(t)$ and let $\tilde{g}(\omega)$ be the FT of $g(t)$. Show that

$$\int_{-\infty}^{\infty} f^*(t) g(t)\, dt = \frac{1}{2\pi} \int_{-\infty}^{\infty} \left(\tilde{f}(\omega)\right)^* \tilde{g}(\omega)\, d\omega. \qquad (6.17)$$

   b. Use Eq. 6.17 to derive Parseval's theorem,

$$\int_{-\infty}^{\infty} |f(t)|^2\, dt = \frac{1}{2\pi} \int_{-\infty}^{\infty} |\tilde{f}(\omega)|^2\, d\omega. \qquad (6.18)$$

▶ **Exercise 6.10 (Fourier Derivative Relation)**   Show that the Fourier transform of the derivative of a function, $F(t) \equiv \frac{d}{dt} f(t)$, is given by $\tilde{F}(\omega) = i\omega \tilde{f}(\omega)$, where $\tilde{f}(\omega)$ is the FT of $f(t)$.

Solution

$$\int_{-\infty}^{\infty} F(t) e^{-i\omega t}\, dt = \int_{-\infty}^{\infty} \left\{ \frac{d}{dt} f(t) \right\} e^{-i\omega t}\, dt$$

$$= -\int_{-\infty}^{\infty} f(t) \frac{d}{dt} e^{-i\omega t}\, dt$$

$$= \int_{-\infty}^{\infty} f(t) i\omega e^{-i\omega t}\, dt$$

$$= i\omega \tilde{f}(\omega) \equiv \tilde{F}(\omega), \qquad (6.19)$$

where we have done an integration by parts and neglected the boundary term, since $f(t)$ is assumed to vanish at $\pm\infty$.

The result of the previous exercise can be inverted, to calculate the derivative of a function over its entire domain in one simple and accurate procedure: (a) Fourier transform: $f(t) \to \tilde{f}(\omega)$, (b) Multiplication: $\tilde{f}(\omega) \to i\omega \tilde{f}(\omega) \equiv \tilde{F}(\omega)$, and (c) Inverse Fourier transform: $\tilde{F}(\omega) \to F(t) = \frac{d}{dt} f(t)$.

6.2.2   The Fourier Convolution Theorem

One of the most powerful properties of the Fourier transform is the *Fourier convolution theorem*:

$$\int_{-\infty}^{\infty} f(t) g(t) e^{-i\omega t}\, dt = \frac{1}{2\pi} \int_{-\infty}^{\infty} \tilde{f}(\omega') \tilde{g}(\omega - \omega')\, d\omega', \qquad (6.20)$$

where the LHS is the Fourier transform of the product of the two functions $f(t)$ and $g(t)$ and the RHS is the *convolution* of the Fourier transform of $\tilde{f}(\omega)$ and $\tilde{g}(\omega)$, where $\tilde{f}(\omega)$ and $\tilde{g}(\omega)$ are the Fourier transform of $f(t)$ and $g(t)$, respectively. The Fourier convolution theorem can be written more succinctly as

$$\widetilde{f(t)g(t)} = \tilde{f}(\omega) \otimes \tilde{g}(\omega), \qquad (6.21)$$

where tildes represent the Fourier transform of the function and the symbol $\otimes$ is used for the convolution.

▶ **Exercise 6.11**   Prove this theorem.

Solution

$$\int_{-\infty}^{\infty} f(t)g(t)e^{-i\omega t}\, dt$$

$$= \int_{-\infty}^{\infty} \left\{ \frac{1}{2\pi} \int_{-\infty}^{\infty} \tilde{f}(\omega')e^{i\omega' t}d\omega' \right\} \left\{ \frac{1}{2\pi} \int_{-\infty}^{\infty} \tilde{g}(\omega'')e^{i\omega'' t}d\omega'' \right\} e^{-i\omega t}\, dt \quad (6.22)$$

$$= \frac{1}{2\pi} \int_{-\infty}^{\infty} \int_{-\infty}^{\infty} \tilde{f}(\omega')\tilde{g}(\omega'')\delta(\omega - (\omega' + \omega''))\, d\omega'\, d\omega'' \quad (6.23)$$

$$= \frac{1}{2\pi} \int_{-\infty}^{\infty} \tilde{f}(\omega')\tilde{g}(\omega - \omega')\, d\omega' \quad (6.24)$$

The convolution of two functions has a pictorial interpretation as follows (see Figure 6.5): the function $\tilde{g}(\omega)$ is reflected around $\omega = 0$ ($\tilde{g}(\omega') \to \tilde{g}(-\omega')$); then the reflected function is translated to $-\infty$ and swept through to $+\infty$ ($\tilde{g}(\omega - \omega')$), where ω varies; at every value of ω the overlap integral with $\tilde{f}(\omega')$ is performed, integrating over the ω' variable. The result of the overlap, plotted as a function of ω, is the convolution of $\tilde{f}(\omega)$ and $\tilde{g}(\omega)$.

▶ **Exercise 6.12**   Derive the inverse form of the convolution theorem:

$$\int_{-\infty}^{\infty} \tilde{f}(\omega)\tilde{g}(\omega)e^{i\omega t}\, d\omega = 2\pi \int_{-\infty}^{\infty} f(t')g(t - t')\, dt'. \quad (6.25)$$

▶ **Exercise 6.13**   Show that $\tilde{f}(\omega) \otimes \tilde{g}(\omega) = \tilde{g}(\omega) \otimes \tilde{f}(\omega)$. Convolute a single sawtooth and a Gaussian pictorially and convince yourself that you get the same answer regardless of which function is stationary and which function is swept.

▶ **Exercise 6.14**   Show that the convolution of any function with a δ-function is the function itself.

▶ **Exercise 6.15**   Use the Fourier convolution theorem, set $g(t) = f^*(t)$ and set $\omega = 0$ on both sides to show Parseval's theorem:

$$\int_{-\infty}^{\infty} |f(t)|^2\, dt = \frac{1}{2\pi} \int_{-\infty}^{\infty} |\tilde{f}(\omega)|^2\, d\omega. \quad (6.26)$$

You will need the result of Exercise 6.8.

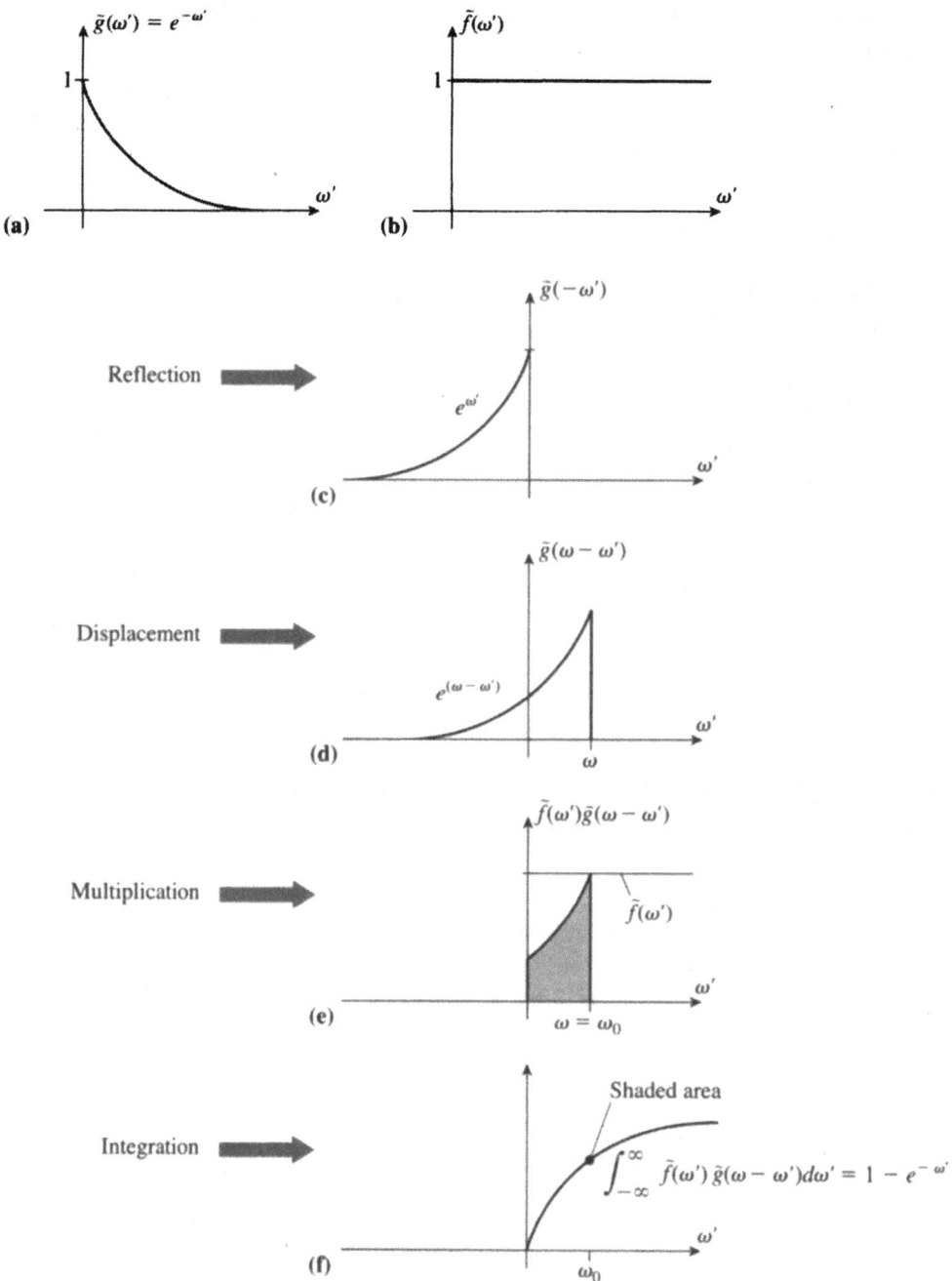

Figure 6.5 Pictorial interpretation of the process of convolution $f(\omega) \otimes g(\omega)$. The original functions are (a) $\tilde{g}(\omega) = e^{-\omega}$ and (b) $\tilde{f}(\omega) = H(\omega)$. The functions $\tilde{f}(\omega)$ and $\tilde{g}(\omega)$ are convolved by the following four steps: (c) reflection, (d) displacement, (e) multiplication, and (f) integration. Adapted from Brigham (1974).

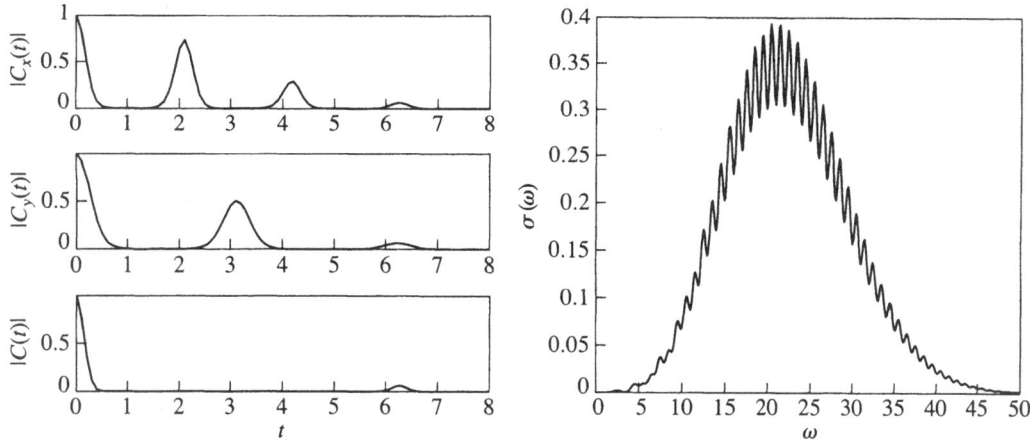

Figure 6.6 (a) Correlation function, $C(t)$, given by the product of two correlation functions, $C_x(t)$ and $C_y(t)$. If $C_x(t)$ has its first recurrence at 2τ and $C_y(t)$ has its first recurrence at 3τ, then $C(t) = C_x(t)C_y(t)$ will have its first recurrence at 6τ. (b) The spectrum $\sigma(\omega)$, given by the Fourier transform of the product spectrum $C(t)$. Note that $\sigma(\omega)$ has a low-resolution spacing of $\omega_0/6$, smaller than the fundamental frequency either in x or y!

An important consequence of the Fourier convolution theorem is that if a wavepacket correlation function can be factored, then the spectrum of the system is the convolution of the two factored spectra, that is, if

$$C(t) = \langle \Psi(x, y, 0) | \Psi(x, y, t) \rangle$$
$$= \langle \Psi_x(x, 0) | \Psi_x(x, t) \rangle \, \langle \Psi_y(y, 0) | \Psi_y(y, t) \rangle$$
$$= C_x(t) \, C_y(t),$$

then

$$\sigma(\omega) = \frac{1}{2\pi} \int_{-\infty}^{\infty} C_x(t)C_y(t)e^{i\omega t} \, dt = \int_{-\infty}^{\infty} \sigma_x(\omega')\sigma_y(\omega - \omega') \, d\omega', \qquad (6.27)$$

where

$$\sigma_x(\omega) = \frac{1}{2\pi} \int_{-\infty}^{\infty} C_x(t)e^{i\omega t} \, dt \qquad \sigma_y(\omega) = \frac{1}{2\pi} \int_{-\infty}^{\infty} C_y(t)e^{i\omega t} \, dt. \qquad (6.28)$$

An interesting example is if $C_x(t)$ has its first recurrence at 2τ and $C_y(t)$ has its first recurrence at 3τ, then $C(t) = C_x(t)C_y(t)$ will have its first recurrence at 6τ (see Figure 6.6a). The Fourier transform of $C(t)$, $\sigma(\omega)$ will have a low-resolution spacing of $\omega_0/6$, where $\omega_0 \equiv \frac{2\pi}{\tau}$ (see Figure 6.6b). Note that this interval is smaller than the fundamental frequency either in x or y!

▶ **Exercise 6.16**   By the convolution theorem, $\sigma(\omega) = \sigma_x(\omega) \otimes \sigma_y(\omega)$. Use this relation to gain a different perspective on where the spacing of $\omega_0/6$ comes from.

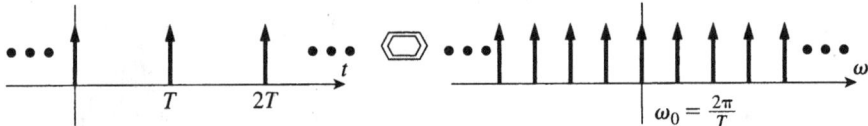

Figure 6.7 The Fourier transform of a "picket fence" time signal with spikes separated by T is a "picket fence" frequency spectrum with spikes separated by $\omega_0 = 2\pi/T$.

The convolution theorem is also useful when comparing an idealized, theoretical spectrum with a real spectrum, obtained from sampling with a finite number of points. This idea is developed in the next two exercises.

▶ **Exercise 6.17**   The Fourier transform of a "picket fence,"

$$S_T(t) = \sum_{n=-\infty}^{\infty} \delta(t - nT), \tag{6.29}$$

is another picket fence:

$$\omega_0 \tilde{S}_{\omega_0}(\omega) = \omega_0 \sum_{n=-\infty}^{\infty} \delta(\omega - n\omega_0) \tag{6.30}$$

(see Figure 6.7), where $\omega_0 = 2\pi/T$ (see Exercise 6.29). Multiply the picket fence function, $S_T(t)$, by a smooth function, $f(t)$, of width $\tau \gg T$ (for definiteness, consider a Gaussian, $f(t) = e^{-t^2/2\tau^2}$). The multiplication by $S(t)$ corresponds to discrete sampling of a continuous time signal. Use the Fourier convolution theorem to predict the error in the spectrum resulting from the discrete time sampling. This error is called "aliasing" and the spurious spectra are called "ghosts."

The previous exercise shows that *discrete sampling* in time corresponds to *periodicity* in frequency. Because of the reciprocity of the Fourier transform relationship, the converse must also be true: that periodicity in time implies a discrete spectrum in frequency. This important observation is the basis for the Fourier series expansion, which is developed in the next subsection. Using the picket fence Fourier transform pair, Fourier series emerge naturally as a special case of the Fourier transform when the function to be transformed is periodic.

▶ **Exercise 6.18**   Consider the window function, $W_T(t)$, defined as

$$W_T(t) = A(H(t + T) - H(t - T)), \tag{6.31}$$

where A is a constant and $H(t)$ is a Heaviside step-function:

$$H(t) = \begin{cases} 0, & \text{if } t < 0 \\ 1, & \text{otherwise.} \end{cases} \tag{6.32}$$

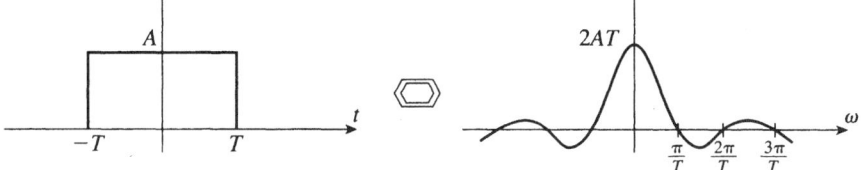

Figure 6.8 The Fourier transform of a rectangular box time signal.

Show that the Fourier transform of $W_T(t)$ is

$$\tilde{W}_T(\omega) = 2AT\frac{\sin(\omega T)}{\omega T} \qquad (6.33)$$

(see Figure 6.8). Combine this result with the Fourier convolution theorem to predict the error in the spectrum from the truncation of a time signal before it has decayed to zero. This effect is known as Gibbs's phenomenon or "ringing." Discuss the connection of this result with Figure 6.3.

▶ **Exercise 6.19**   Adapt the proof of the Fourier convolution theorem to show that

$$\int_{-\infty}^{\infty}\left\{\int_{-\infty}^{\infty} f(\tau)g(t+\tau)\,d\tau\right\} e^{-i\omega t}\,dt = \tilde{f}(-\omega)\tilde{g}(\omega). \qquad (6.34)$$

The term *correlation function* is sometimes used for the integral in brackets on the LHS.

6.2.3   Fourier Series

An important special case of Fourier transforms arises if the function is periodic in time. In this case, the Fourier transform, rather than consisting of a continuous set of frequencies, contains only discrete, evenly spaced frequencies. Although Fourier series are often developed independently of the Fourier integral, as we shall see the former emerge naturally from the latter using the picket fence Fourier pair that was introduced in Exercise 6.17.

Consider a function $f(t)$ that is periodic with period T,

$$f(t+T) = f(t). \qquad (6.35)$$

We shall show that its Fourier transform, $\tilde{f}(\omega)$, is a sequence of equidistant spikes,

$$\tilde{f}(\omega) = 2\pi \sum_{-\infty}^{\infty} \alpha_n \delta(\omega - n\omega_0), \qquad (6.36)$$

distance $\omega_0 = \frac{2\pi}{T}$ apart. Before proceeding with the proof, note that by applying an inverse Fourier transform to both sides of Eq. 6.36 we find that $f(t)$ can be written as a sum of exponentials,

$$f(t) = \sum_{n=-\infty}^{\infty} \alpha_n e^{in\omega_0 t}. \qquad (6.37)$$

Equation 6.37 is called a *Fourier series*; this series representation manifestly satisfies the periodicity requirement $f(t + T) = f(t)$ for $T = \frac{2\pi}{\omega_0}$. The constants α_n, which appear in both the discrete Fourier spectrum, Eq. 6.36, as well as in the Fourier series expansion of $f(t)$, Eq. 6.37, are given by

$$\alpha_n = \frac{1}{T} \int_{-T/2}^{T/2} f(t) e^{-in\omega_0 t} \, dt. \tag{6.38}$$

The α_n are equal, within a factor of $1/T$, to the Fourier integral of one unit cell of the function $f(t)$ at the discrete frequencies $\omega_n = n\omega_0 = 2\pi n/T$. This expression for the α_n underscores that the Fourier series is indeed a special case of the Fourier transform.

To prove Eq. 6.36, we begin by representing $f(t)$ as a sum of disjoint portions $f_0(t)$ in the different intervals:

$$f(t) = \sum_{-\infty}^{\infty} f_0(t + nT), \tag{6.39}$$

where

$$
\begin{aligned}
f_0(t) &= f(t) & |t| &< T/2 \\
&= 0 & |t| &> T/2.
\end{aligned}
\tag{6.40}
$$

Equation 6.40 may be expressed as a convolution:

$$f(t) = f_0(t) \otimes S_T(t), \tag{6.41}$$

where $S_T(t)$ is the picket fence from Exercise 6.17 and $\otimes$ designates convolution. Defining $\tilde{f}_0(\omega)$ as the Fourier integral of $f_0(t)$,

$$\tilde{f}_0(\omega) = \int_{-\infty}^{\infty} f_0(t) e^{-i\omega t} \, dt = \int_{-T/2}^{T/2} f(t) e^{-i\omega t} \, dt, \tag{6.42}$$

the Fourier convolution theorem gives

$$
\begin{aligned}
\tilde{f}(\omega) &= \tilde{f}_0(\omega) \omega_0 S_{\omega_0}(\omega) \\
&= \tilde{f}_0(\omega) \frac{2\pi}{T} \sum_{-\infty}^{\infty} \delta(\omega - n\omega_0) \\
&= \frac{2\pi}{T} \sum_{-\infty}^{\infty} \tilde{f}_0(n\omega_0) \delta(\omega - n\omega_0).
\end{aligned}
\tag{6.43}
$$

Equation 6.43 is of the form we sought, Eq. 6.36, with α_n given by Eq. 6.38. Taking the inverse Fourier transform of Eq. 6.38 we obtain Eq. 6.37, which may be written in the following alternative forms:

$$f(t) = \sum_{n=-\infty}^{\infty} f_0(t + nT) = \frac{1}{T} \sum_{n=-\infty}^{\infty} e^{in\omega_0 t} \tilde{f}_0(n\omega_0). \tag{6.44}$$

▶ **Exercise 6.20**   We derived Eq. 6.44 under the assumption that $f_0(t)$ equals zero for $|t| > T/2$. However, Eq. 6.44 actually holds for an arbitrary function $\phi(t)$ and its Fourier transform $\Phi(\omega)$, namely,

$$f(t) = \sum_{n=-\infty}^{\infty} \phi(t + nT) = \frac{1}{T} \sum_{n=-\infty}^{\infty} e^{in\omega_0 t} \Phi(n\omega_0), \qquad (6.45)$$

with $\omega_0 = \frac{2\pi}{T}$. Show this. Equation 6.45 is known as the Poisson sum formula.

Solution

$$\sum_{n=-\infty}^{\infty} \phi(t + nT) = \phi(t) \otimes S_T(t), \qquad (6.46)$$

where $S_T(t)$ is a picket fence function and $\otimes$ is the convolution. By the Fourier convolution theorem, the Fourier transform of Eq. 6.46 is

$$\tilde{f}(\omega) = \Phi(\omega) S_{\omega_0}(\omega) = \frac{2\pi}{T} \sum_{n=-\infty}^{\infty} \Phi(n\omega_0)\delta(\omega - n\omega_0). \qquad (6.47)$$

Taking the inverse Fourier transform of Eq. 6.47 yields Eq. 6.45. The only difference between the ordinary Fourier series, Eq. 6.44, and Eq. 6.45 is that $f_0(t)$ equals zero for $|t| > T/2$, whereas this is not true for $\phi(t)$. Taking $t = 0$ in Eq. 6.45 gives the following special form:

$$f(0) = \sum_{n=-\infty}^{\infty} \phi(nT) = \frac{1}{T} \sum_{n=-\infty}^{\infty} \Phi(n\omega_0). \qquad (6.48)$$

6.2.4   Discrete Fourier Transform

So far we have discussed separately the case in which the time signal is discrete and the case in which it is periodic. The former implies that the frequency spectrum is periodic, while the latter implies that the frequency spectrum is discrete (Fourier series). If the time series is *both* discrete and periodic, then the frequency spectrum also is discrete and periodic. In particular, let the time signal be spaced by Δt, with a periodicity of N points. This implies a period $T = N\Delta t$. By the Fourier series formula, the frequency spectrum will be discrete with spacing $\Delta\omega = \frac{2\pi}{T}$. Moreover, the discrete time signal with spacing Δt implies that the frequency spectrum must be periodic with period $\Omega = \frac{2\pi}{\Delta t} = \frac{2\pi N}{T} = N\Delta\omega$. Thus we have found the remarkable result that if a discrete time series is periodic with periodicity N, the discrete frequency spectrum is also periodic with periodicity N.

To proceed mathematically, we begin with Eq. 6.41 for a periodic function:

$$F(t) = f_0(t) \otimes S_T(t) = T \sum_{r=-\infty}^{\infty} f_0(t + rT). \qquad (6.49)$$

If the time signal is sampled discretely, the resulting function, $f(t) = F(t) \cdot S_{\Delta t}(t) = f_0(t) \otimes S_T(t) \cdot S_{\Delta t}$, can be written as

$$f(t) = T \sum_{r=-\infty}^{\infty} f_0(t + rT) \sum_{k=0}^{N-1} \delta(t - k\,\Delta t + rT)$$

$$= T \sum_{r=-\infty}^{\infty} \sum_{k=0}^{N-1} f_0(k\Delta t)\delta(t - k\,\Delta t + rT) \qquad (6.50)$$

Taking the Fourier transform of Eq. 6.50 we obtain

$$\tilde{f}(\omega) = \int_{-\infty}^{\infty} f(t)e^{-i\omega t}\,dt = T \sum_{r=-\infty}^{\infty} \sum_{k=0}^{N-1} f(k\,\Delta t)e^{-i\omega(k\Delta t - rT)} \qquad (6.51)$$

Evaluating this expression at $\omega = n\,\Delta\omega$ we find

$$\tilde{f}(n\,\Delta\omega) = T \sum_{r=-\infty}^{\infty} \sum_{k=0}^{N-1} f(k\,\Delta t)e^{-i2\pi nk/N}, \qquad (6.52)$$

where we have used the fact that $\Delta\omega\,\Delta t = \frac{2\pi}{N}$ and that $\Delta\omega\,T = 2\pi$. Since every replica with different r is equal, it is customary to write Eq. 6.52 in the form

$$\tilde{f}_n = \sum_{k=0}^{N-1} f_k e^{-i2\pi nk/N} \qquad n = 0, \pm 1, \pm 2, \ldots. \qquad (6.53)$$

where $\tilde{f}_n \equiv \tilde{f}(n\,\Delta\omega)$ and $t_k \equiv k\,\Delta t$. Equation 6.53 is called the discrete Fourier transform. The discrete Fourier transform will play an important role in the Fourier method in Chapter 11. There we discuss a remarkable algorithm for implementing the discrete Fourier transform, known as the Fast Fourier Transform, or FFT.

The inverse discrete Fourier transform is given by the relation

$$f_k = \frac{1}{N} \sum_{n=0}^{N-1} \tilde{f}_n e^{i2\pi nk/N} \qquad k = 0, \pm 1, \pm 2, \ldots \qquad (6.54)$$

To verify Eq. 6.54 we substitute Eq. 6.54 into Eq. 6.53:

$$\tilde{f}_n = \sum_{k=0}^{N-1} \left[\frac{1}{N} \sum_{r=0}^{N-1} \tilde{f}_r e^{2\pi irk/N} \right] e^{-2\pi ink/N} \qquad (6.55)$$

$$= \frac{1}{N} \sum_{r=0}^{N-1} \tilde{f}_r \left[\sum_{k=0}^{N-1} e^{2\pi irk/N} e^{-2\pi ink/N} \right] \qquad (6.56)$$

$$= \tilde{f}_n, \qquad (6.57)$$

where the last equality follows from the orthogonality relation

$$\sum_{k=0}^{N-1} e^{2\pi irk/N} e^{-2\pi ink/N} = N \text{ if } r = n \qquad (6.58)$$

$$= 0 \text{ otherwise.} \qquad (6.59)$$

The two sets of N numbers $\{f_k\}$ and $\{\tilde{f}_n\}$ are referred to as a discrete Fourier transform pair. Clearly, there is a symmetry between the forward and inverse discrete Fourier transform, reminiscent of the symmetry between the forward and reverse continuous Fourier transform. Strictly speaking, the discrete set of values $\{\tilde{f}_n\}$ and $\{f_k\}$ represent a Fourier transform pair only if they continue periodically over the infinite domain in both time and frequency. However, the truncation inherent in Eqs. 6.53 and 6.54 may be given two interpretations. The first is that the functions are still assumed to be periodic, and the truncation of the domain is simply for economy of notation (and computation), since there is no new information in the replica domains. The second interpretation is that the function f_k is a sampling of a function $f(t)$. If $\tilde{f}(\omega)$ has no frequency components higher than $\frac{\Omega}{2} = \frac{\pi}{\Delta t}$, then $\tilde{f}(\omega)$ is called a "bandwidth limited" function and it can be shown that all the information is contained in the discrete samples f_k. This is the content of the Nyquist (also called the Shannon) sampling theorem. Similarly, if $f(t)$ is completely contained within the bounds $[0, T]$, then $f(t)$ is called bandwidth limited and all the information is contained in the sampled function $\tilde{f}_n$. There are discrete analogs to virtually all of the relationships that hold for continuous Fourier transforms, including the symmetry properties and the convolution and correlation theorems. For further details, see Brigham (1974).

▶ **Exercise 6.21**   Show that

$$\sum_{k=0}^{N-1} |f_k|^2 = \frac{1}{N} \sum_{n=0}^{N-1} |\tilde{f}_n|^2. \tag{6.60}$$

This is the discrete version of Parseval's theorem, Eq. 6.26.

6.3   Eigenfunctions as Fourier Transforms of Wavepackets

Until now in this book we have been thinking of wavepackets as coherent superpositions of eigenstates. There is a fascinating converse relationship that we will explore in this section— that an eigenstate can be viewed as a superposition of wavepackets. To see this, consider the integral

$$\int_{-\infty}^{\infty} \Psi(x, t) e^{\frac{i}{\hbar} E_n t} \, dt = \int_{-\infty}^{\infty} e^{-\frac{i}{\hbar} H t} \Psi(x, 0) e^{\frac{i}{\hbar} E_n t} \, dt. \tag{6.61}$$

Expanding $\Psi(x, 0)$ in eigenfunctions of H, $\Psi(x, 0) = \sum_m a_m \psi_m(x)$, and substituting into the RHS of Eq. 6.61, yields

$$\int_{-\infty}^{\infty} \Psi(x, t) e^{\frac{i}{\hbar} E_n t} \, dt = \int_{-\infty}^{\infty} e^{-\frac{i}{\hbar} H t} \left(\sum_m a_m \psi_m(x) \right) e^{\frac{i}{\hbar} E_n t} \, dt$$

$$= \int_{-\infty}^{\infty} \sum_m a_m \psi_m(x) e^{-\frac{i}{\hbar} E_m t} e^{\frac{i}{\hbar} E_n t} \, dt$$

$$= 2\pi \hbar \sum_m a_m \psi_m(x) \delta(E_m - E_n) \propto \psi_n(x). \tag{6.62}$$

In the first step we brought the propagator inside the sum and used the relationship

$$e^{-\frac{iHt}{\hbar}}\psi_m(x) = e^{-\frac{i}{\hbar}E_m t}\psi_m(x). \qquad (6.63)$$

In the second step the integral relation for the Dirac δ-function was used.

Equation 6.62 is a remarkable formula, which for emphasis we rewrite as follows:

$$\psi_n(x) \propto \int_{-\infty}^{\infty} \overbrace{\Psi(x, t)}^{\text{moving wavepacket}} \overbrace{e^{\frac{i}{\hbar}E_n t}}^{\substack{\text{multiplied by phase factor}}} \overbrace{dt.}^{\text{integrated over time}} \qquad (6.64)$$

The LHS of the equation is the eigenstate of H with eigenvalue E_n: $H\psi_n = E_n\psi_n$. The RHS is the Fourier transform of the moving wavepacket, $\Psi(x, t)$, from time to energy at the eigenvalue energy E_n. If we recognize that the integral is simply the continuous limit of a sum, then the integral over time on the RHS can be viewed as the sum of wavepackets, $\Psi(x, t)$ at different times t, added with the phase factors, $e^{-iE_n t/\hbar}$. This is what we meant in the opening paragraph of this section when we said that an *eigenstate* (LHS) can be viewed as a superposition of *wavepackets* (RHS).

The basic concept is illustrated for the harmonic oscillator in Figure 6.9. Note how the wavepacket in different portions of its phase space orbit has different momenta: small momentum at the classical turning points and largest near the potential minimum. Momentum manifests itself as the local wave vector of the wavepacket, so that a high-momentum wavepacket has a highly oscillatory phase factor, while a low-momentum wavepacket has a nonoscillatory envelope. Adding these packets together produces an eigenfunction that has a small density of oscillations at the classical turning points and a large density of oscillations at the potential minimum; this corresponds to the well-known result that the local density of oscillations of an eigenfunction is in correspondence with the momentum of the classical trajectory at the same energy passing through that point.

In Figure 6.9, the average energy of the wavepacket was approximately equal to the energy of the eigenstate produced. This need not be the case: so long as the wavepacket contains any finite component of a particular eigenstate, that eigenstate can be produced by an appropriate superposition of wavepackets. Note that a single wavepacket can therefore generate many different eigenfunctions. If the eigenvalue spectrum is nondegenerate, the eigenfunction, $\psi_n(x)$, that is produced in a particular calculation is determined solely by the value of E_n that appears in the phase factor in Eq. 6.64. If E_n does not match any of the eigenvalues of H, the integral will give 0. Essentially, the production of the eigenfunction via superposition with phase $e^{iE_n t/\hbar}$ can be viewed as constructive interference of the component at energy E_n and destructive interference of the components at all other energies.

Equation 6.64 for the wavefunction ψ_n can be used for the numerical construction of eigenfunctions. One of the remarkable features of this formula is that although formally it requires the exact quantum mechanical $\Psi(x, t)$ at every time to obtain the wavefunction, in practice qualitatively accurate wavefunctions can be obtained with simple, approximate assumptions about the form of $\Psi(x, t)$. In particular, although the shape of the exact wavefunction changes with time and generally spreads, one obtains high-quality approximations to the wavefunction using a single Gaussian wavepacket whose center follows the classical trajectory and whose width is constant. The phase of the wavepacket involves two terms, the classical action and a time-dependent topological phase, which is the generalization of

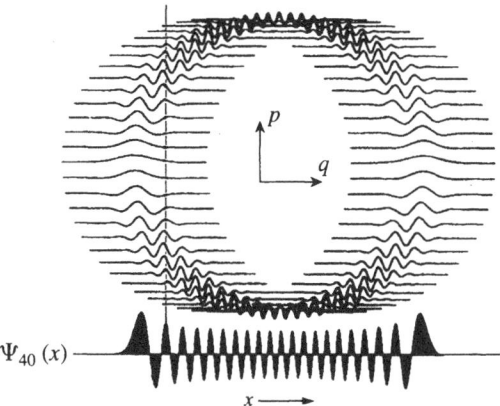

$\Psi_{40}(x)$

$x \longrightarrow$

Figure 6.9 The $n = 40$ harmonic oscillator eigenstate as a sum of Gaussian wavepackets. The upper portion of the plot shows a succession of snapshots of the real part of a Gaussian wavepacket at different times. The packets are positioned along the phase space orbit of the classical trajectory corresponding to the center of the packet. Note the high wave number where the classical momentum is highest, and the low wave number at the classical turning points. The lower portion of the plot represents the sum (or integral) of the contribution of all the packets at each value of x, producing the $n = 40$ eigenstate. Adapted from Heller (1991).

the factor $\omega t/2$ in Eq. 3.50 of Chapter 3, and which can be viewed as a continuously varying Maslov index.

Figure 6.10 shows the development of one of the highly excited eigenfunctions of the potential $H(x, y, p_x, p_y) = \frac{p_x^2}{2m} + \frac{p_y^2}{2m} + \frac{1}{2}m\omega_x^2 x^2 + \frac{1}{2}m\omega_y^2 y^2 - \lambda yx^2$ ($\omega_x = 1.1$, $\omega_y = 1.0$ and $\lambda = -0.042$) generated by forming a superposition of time-evolving Gaussian wavepackets. Both the energy E_n and the initial conditions for the trajectory were known in advance from a separate calculation. Figure 6.10 shows the wavefunction at different stages of formation—$\int_0^\tau \Psi(x, t)e^{\frac{i}{\hbar}E_n t}\, dt$ for different accumulation times, τ. Note that at short times the accumulated state looks like a superposition of all the places the moving wavepacket has visited. At longer times, however, the wavepacket returns to its initial location (in both position and momentum), and the ultimate nodal pattern of the wavefunction that is formed reflects the phase of the wavepacket as it returns in multiple passes; this phase in turn is determined by the classical action and topological phase of the wavepacket, as well as the energy at which the packet is Fourier transformed.

The success of the fixed-width Gaussians in generating eigenfunctions can be better understood by examining the explicit form of Eq. 6.64 with $\Psi(x, t)$ a Gaussian wavepacket:

$$\psi_n(x) \approx \int_{-\infty}^{\infty} N e^{-\alpha(x-x_t)^2 + i p_t x_t/\hbar + i\gamma_t/\hbar} e^{iE_n t/\hbar}\, dt. \tag{6.65}$$

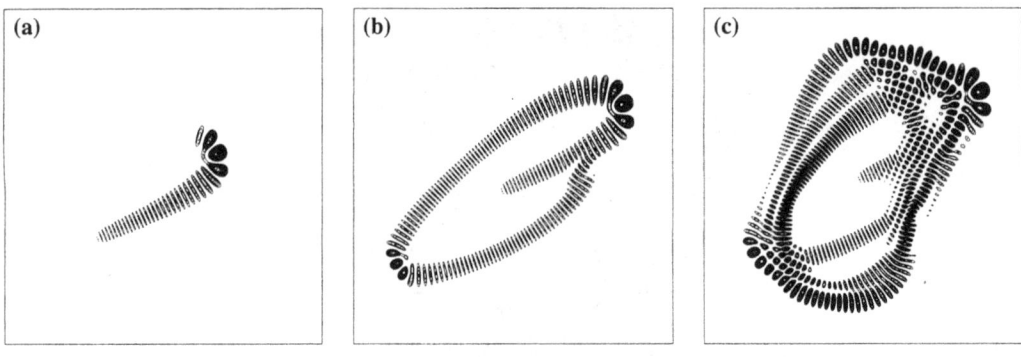

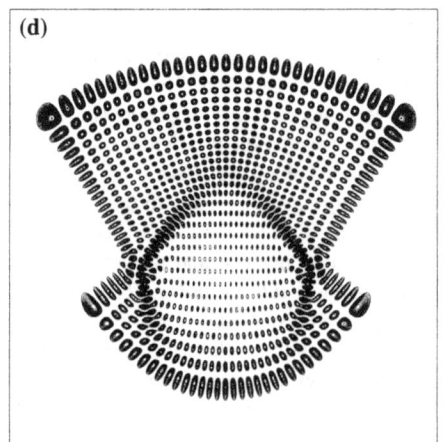

Figure 6.10 Construction of a semiclassical wavefunction for a two-dimensional Barbanis Hamiltonian, $H(x, y, p_x, p_y) = \frac{p_x^2}{2m} + \frac{p_y^2}{2m} + \frac{1}{2}m\omega_x^2 x^2 + \frac{1}{2}m\omega_y^2 y^2 - \lambda y x^2$ ($\omega_x = 1.1$, $\omega_y = 1.0$ and $\lambda = -0.042$). The wavefunction is generated from a single time-evolving Gaussian wavepacket, superposed at each instant of its evolution with the phase factor $e^{iE_n t/\hbar}$, where E_n is the energy of the desired eigenfunction. Both the energy E_n and the initial conditions for the wavepacket were known in advance from a separate calculation. The figure shows the real part of the accumulated wavefunction $\int_0^\tau \Psi(x, t)e^{\frac{i}{\hbar}E_n t}\, dt$ for various times: (a) $t = 2.5$, (b) $t = 7.5$, (c) $t = 20$, (d) $t = 230.3$ (one period of near return in both x and y). Based on Davis (1981).

Using Eq. 3.22 and Exercise 3.2 we have

$$\gamma_t = \int_0^t \{p_{t'}\dot{x}_{t'} - E\}\, dt' - \frac{\hbar\omega t}{2}, \tag{6.66}$$

where the last term comes from the replacement of α_t by the coherent state value $m\omega/2\hbar$. For anharmonic oscillators take ω to be the classical angular frequency at energy E, $\omega(E)$, which in general is different from the harmonic frequency near the bottom of the well. Taking the energy of the classical trajectory to be equal to the eigenvalue of interest, $E = E_n$, we have

$$\psi_n(x) \approx \int_{-\infty}^{\infty} Ne^{-\alpha(x-x_t)^2 + ip_t x_t/\hbar}e^{i\gamma_t'/\hbar}\, dt, \tag{6.67}$$

where

$$\gamma_t' \equiv \int_0^t p_{t'} \dot{x}_{t'} \, dt' - \frac{\hbar \omega t}{2}. \tag{6.68}$$

Let T be a period of classical motion. Then $p_{t+T} = p_t$ and $x_{t+T} = x_t$, and hence the Gaussian factor $e^{-\alpha(x-x_t)^2 + i p_t x_t / \hbar}$ in Eq. 6.67 is periodic with the period of classical motion. Whether the Gaussians will add constructively or destructively is determined by the phase factor γ_t': if $\gamma_{t+T}'/\hbar = \gamma_t'/\hbar + n2\pi$, then every time the Gaussian returns to a part of phase space it has visited previously it will add constructively with the Gaussians from the previous periods and an eigenfunction will build up. Thus we find the condition for eigenfunction build up:

$$\gamma_T' = \int_0^T p_{t'} \dot{x}_{t'} \, dt' - \pi \hbar = n2\pi \hbar, \tag{6.69}$$

where we have used the fact that $\omega T = 2\pi$. Rearranging Eq. 6.69 and using the relation $\int_0^T p_{t'} \dot{x}_{t'} \, dt' = \int_C p(x') \, dx'$, where C represents a closed circuit, we obtain

$$\int_C p(x') \, dx' = \left(n + \frac{1}{2} \right) h. \tag{6.70}$$

Equation 6.70 is the well-known Bohr quantization condition with n replaced by $n + \frac{1}{2}$. More generally, we may write

$$\int_C p(x') \, dx' = \left(n + \frac{\beta}{4} \right) h,$$

where β is known as the Maslov index, here equal to 2. The Maslov index is closely related to the Morse index described in Section 10.1.3, but the latter is formulated in terms of the dynamics of trajectories while the former is based purely on topological considerations in the phase space. Note how the Bohr quantization condition arises here in a context seemingly very different from the usual one, as the condition on the phase of the Gaussian wavepacket necessary to obtain constructive interference at successive periods.

The spectral formula for constructing wavefunctions from wavepackets, together with the success of substituting approximate, classical-like Gaussians for the exact wavepacket, implies that in general there is an underlying classical orbit that may be associated with a general eigenfunction. Some pictorial examples are given below. Figure 6.11 illustrates a family of wavefunctions with a "pretzel" type topology. The corresponding classical trajectories are shown in Figure 6.12. These trajectories live in a disjoint region of phase space known as a resonance zone, which is separated from the rest of the phase space by a separatrix. At the center of this resonance zone is a stable periodic trajectory (or "orbit") that makes 3 passes in the x direction for every 4 passes in the y direction; hence this is known as a 3:4 resonance zone. Figure 6.13 shows a "surface of section," which is a way of visualizing the motion in phase space and characterizing its topology. The surface of section is constructed by plotting p_x versus x for $y = 0$, $p_y > 0$, and p_y versus y for $x = 0$, $p_x > 0$. A single trajectory inside the resonance zone generates a point on the surface of section each time it passes through the defining surface; the locus of points generates a set of three ovals on one surface of section and two on the other.

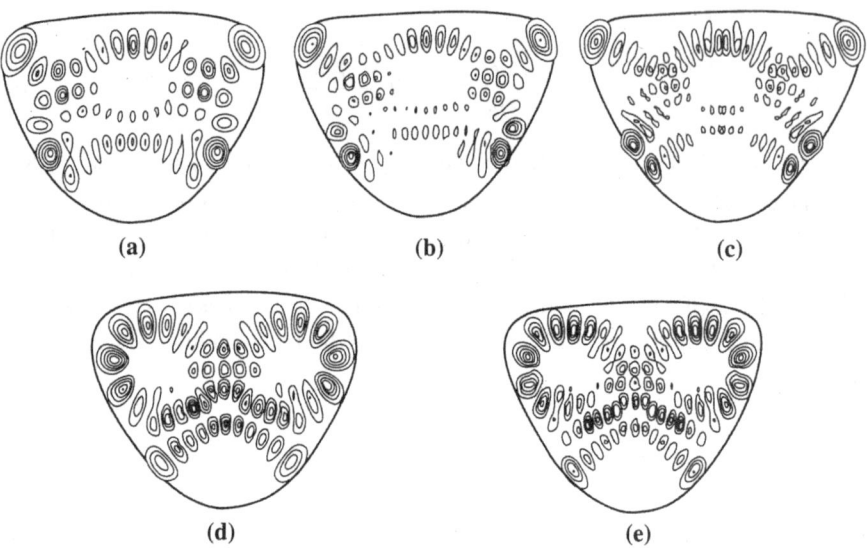

Figure 6.11 Semiclassical wavefunctions for a two-dimensional Barbanis potential, generated by forming a superposition of time-evolving Gaussian wavepackets. The pretzel-like topology reflects the resonance structure of the underlying classical mechanics, shown in the next figure. Adapted from DeLeon (1984).

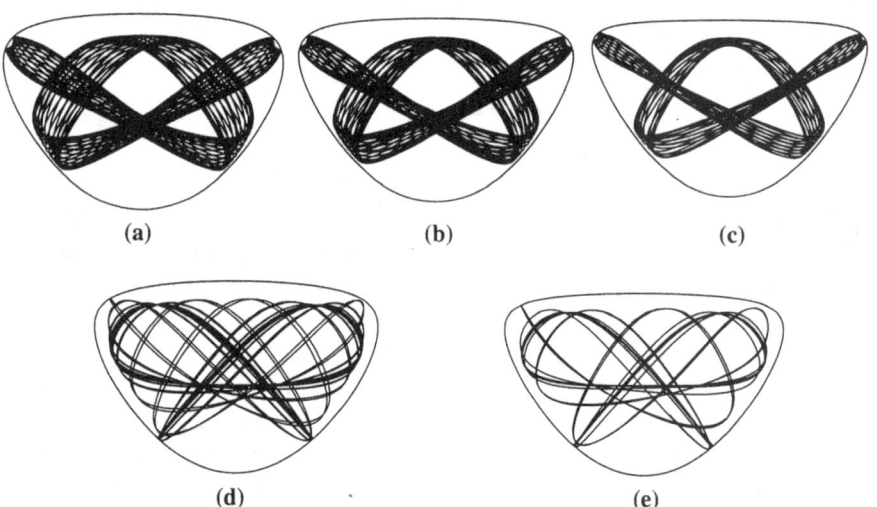

Figure 6.12 The classical orbits that generated the five wavefunctions in Figure 6.11. The pretzel-like shapes are indicative of a 3:4 classical resonance. Adapted from DeLeon (1984).

▶ **Exercise 6.22** The generalization of the Bohr quantization condition to multidimensions is known as EBK (Einstein–Brillouin–Keller) quantization. In two dimensions, for the trajectory topology shown in Figure 6.11, the EBK quantization conditions are

$$M_x \left(n_x + \frac{1}{2} \right) h = \int_C p_x \, dx \tag{6.71}$$

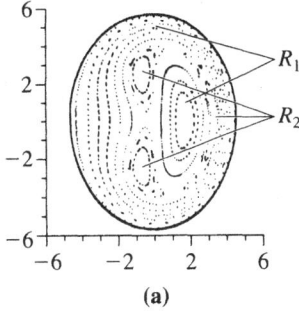
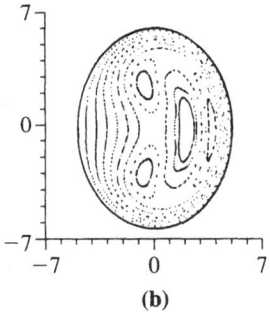
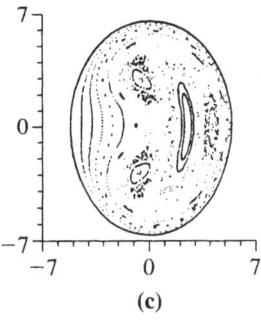

(a)   (b)   (c)

Figure 6.13 Classical surfaces of section (y versus p_y at $x = 0$, $p_x > 0$) for the same Hamiltonian that generated the trajectories in Figure 6.12. (a), (b), and (c) refer to increasing energy. R_1 and R_2 refer to classical resonance zones (R_2 is the 3:4 resonance zone where the trajectories of Figure 6.12 reside). Adapted from DeLeon (1984).

$$M_y \left(n_y + \frac{1}{2} \right) h = \int_C p_y \, dy, \tag{6.72}$$

where C refers to a closed path in phase space, the integers M_x and M_y refer to the number of windings that a particular trajectory makes in the x or y directions, and n_x and n_y are integer quantum numbers. Use the EBK quantization conditions to extend the reasoning about constructive interference of Gaussians to the multidimensional situation in Figure 6.11.

▶ **Exercise 6.23**   What determines which eigenstate is obtained if the energy, E_n, that is used in the spectral method is degenerate?

Figure 6.14 shows several representative wavefunctions for the stadium billiard potential. This system is known to be completely chaotic; if a surface of section for this system were plotted at any energy, it would not show any type of phase space dividing surface of the type in Figure 6.13; the surface of section looks like a scatter of dots. Yet, there is still some nonrandomness to many of the eigenfunctions (which has been termed "scarring"); this nonrandomness can be correlated, at least qualitatively, with unstable periodic orbits.

Figure 6.15 shows several representative wavefunctions for the lemon billiard potential. This system is known to have coexistence of chaotic and regular (i.e., quasi-periodic, of the type in Figures 6.10 and 6.11). Again, the close connection between features of the wavefunctions and unstable periodic classical orbits is striking.

Problems

▶ **Exercise 6.24**   If $C(t)$ is an autocorrelation function, it will generally satisfy $C(-t) = C^*(t)$. Show that this property can be used to save a factor of 2 in calculations.

▶ **Exercise 6.25**

a. Note that

$$C(2t) = \langle \Psi | e^{-2iHt/\hbar} | \Psi \rangle = \langle e^{+iHt/\hbar} \Psi | e^{-iHt/\hbar} \Psi \rangle = \langle \Psi^*(t) | \Psi(t) \rangle \tag{6.73}$$

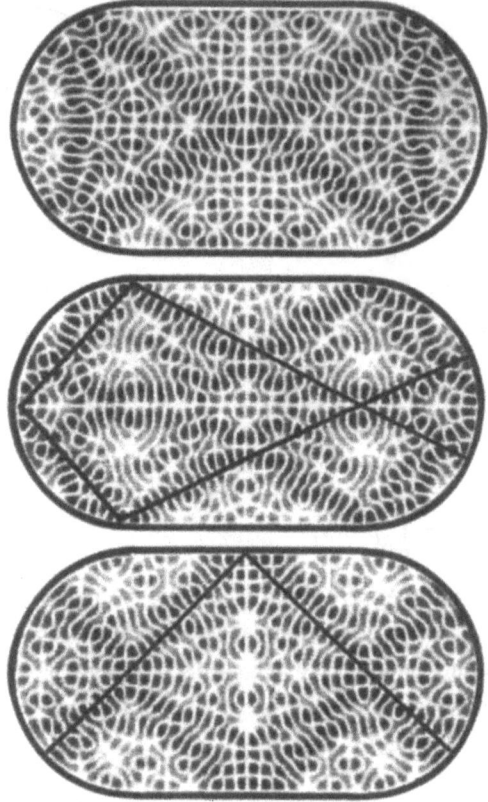

Figure 6.14 Typical "scarred" eigenstates of the stadium billiard. The scars, which are regions of high amplitude, are generated by periodic orbits, two of which are shown. Adapted from Heller (1993).

if Ψ is real. Discuss how this property can be used to save an additional factor of 2 in calculations.

b. Show that

$$C(t) = \langle \Psi(0) \mid \Psi(t) \rangle = \langle \Psi(\tau) \mid \Psi(t+\tau) \rangle,$$

that is, that the autocorrelation function is invariant with respect to a shift of the zero of time.

▶ **Exercise 6.26** Verify the formula that the spectrum is equal to the Fourier transform of a wavepacket autocorrelation function,

$$\sigma(E) = \frac{1}{2\pi} \int_{-\infty}^{\infty} \langle \Psi(0)|\Psi(t)\rangle e^{iEt/\hbar} \, dt, \tag{6.74}$$

by working out the formula analytically for the harmonic oscillator potential.

a. First construct the wavepacket autocorrelation function,

$$C(t) = \langle \Psi(0)|\Psi(t)\rangle = \int_{-\infty}^{\infty} \Psi^*(x,0)\Psi(x,t) \, dx, \tag{6.75}$$

using the coherent state wavepacket discussed in Chapter 3. After some rearrangement, the correlation function should be

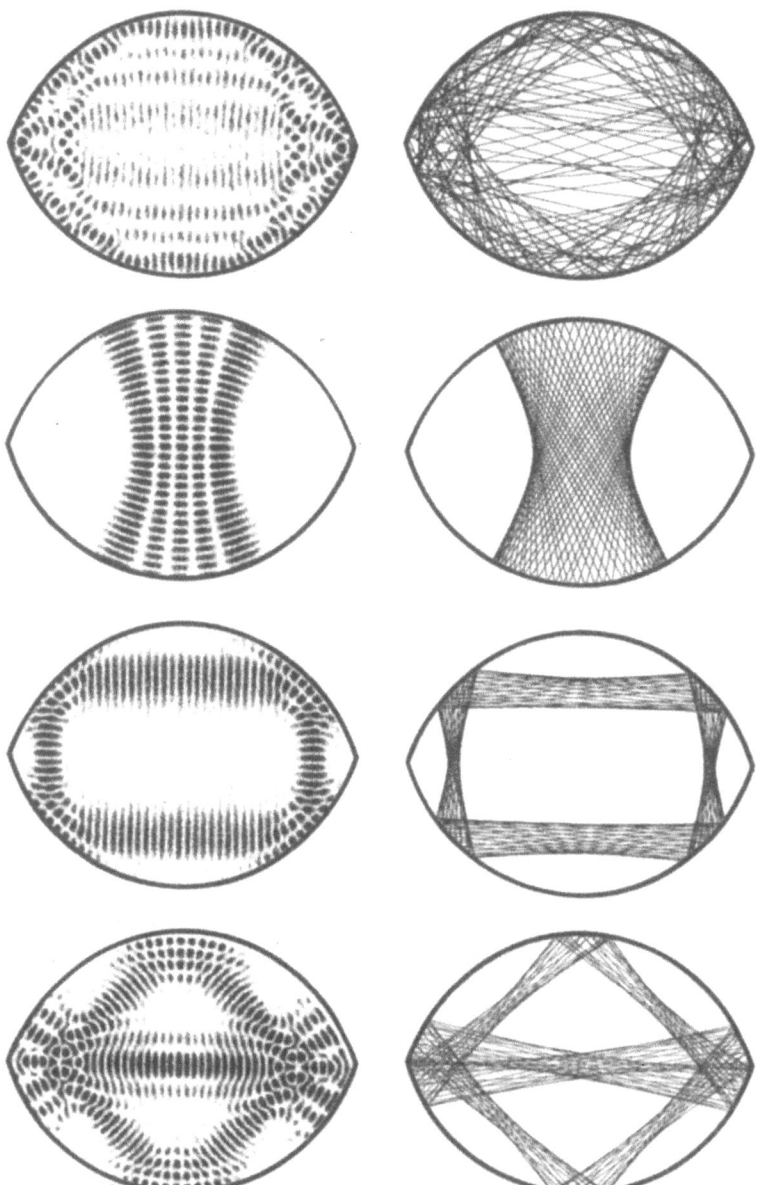

Figure 6.15 Quantum eigenstates (left) and classical orbits (right) for the "lemon" billiard system. This is a mixed dynamical system, having both chaotic motion (top two images) and quasi-periodic motion (bottom six images). Adapted from Heller (1993).

$$C(t) = e^{-\frac{\Delta^2}{2}\left(1-e^{-i\omega t}\right)-\frac{i}{2}\omega t}, \tag{6.76}$$

where $\Delta = \sqrt{m\omega/\hbar}x_0$ is the so-called dimensionless displacement and x_0 is the initial displacement of the center of the coherent state from the minimum of the harmonic potential (assume $p_0 = 0$).

b. Calculate the spectrum by performing the Fourier transform of the above result analytically. [Hint: consider the expression $e^{\frac{\Delta^2}{2}e^{-i\omega t}}$ to be of the form e^x. Expand in a Taylor's series and integrate term by term.] Your answer should agree with the standard expression for the spectrum, which is a set of δ-functions with a Poisson distribution of amplitudes:

$$\sigma(E) = \sum_{n=0}^{\infty} e^{-\frac{\Delta^2}{2}} \frac{\Delta^{2n}}{2^n n!} \delta\left(\frac{E}{\hbar} - \left(n + \frac{1}{2}\right)\omega\right) \tag{6.77}$$

▶ **Exercise 6.27** In this exercise we will analyze the origin of the short-time decay of the autocorrelation function in the previous exercise, by deriving the functional form of the decay two different ways.

a. First, expand the exponential in the exponent of Eq. 6.76 in a Taylor series, up to second order in t. This gives a Gaussian fit to the short-time behavior of $C(t)$.

b. Now, start from Eq. 6.75, substitute the coherent state $\Psi(x, t)$ from Chapter 3, expand x_t and p_t up to first order in t (take $p_0 = 0$) and finally do the integral over x. The answer should agree with the result from part a. Does the decay of $C(t)$ come from motion in x_t or p_t? Can you interpret this result using the Wigner phase space picture for the evolution of a coherent state, Figure 5.3, shown in the previous chapter?

▶ **Exercise 6.28** Verify the formula for an eigenstate in terms of an integral over wavepackets having the appropriate phases,

$$\frac{1}{2\pi} \int_{-\infty}^{\infty} \Psi(x, t) e^{\frac{i}{\hbar}Et} \, dt = \sum_n N_n \psi_n(x) \delta\left(\frac{E}{\hbar} - \frac{E_n}{\hbar}\right), \tag{6.78}$$

by working out the formula analytically for the harmonic oscillator potential. For this potential, $E_n = (n + \frac{1}{2})\hbar\omega$ and use the coherent state wavepacket discussed in Chapter 3.

a. First rearrange the coherent state wavepacket into the form

$$\Psi(X, t) = \left(\frac{m\omega}{\pi\hbar}\right)^{1/4} \exp\left(-\frac{X^2}{2} - \frac{\Delta^2}{4} + X\Delta e^{-i\omega t} - \frac{(\Delta e^{-i\omega t})^2}{4} - \frac{i\omega t}{2}\right), \tag{6.79}$$

where we have introduced the dimensionless coordinate $X = \sqrt{\frac{m\omega}{\hbar}} x$ and $\Delta = \sqrt{\frac{m\omega}{\hbar}} x_0$.

b. Do the integral over t. [Hint: identify part of your expression for $\Psi(x, t)$ with the generating function for the Hermite polynomials, $e^{2xz - z^2} = \sum_{n=0}^{\infty} H_n(x)\frac{z^n}{n!}$.] At energies $E \neq E_n$ you should obtain 0, while for $E = E_n$ you should obtain the well-known expression for the harmonic oscillator wavefunctions,

$$\psi_n(X) = \frac{1}{\sqrt{2^n n!}} H_n(X) e^{-\frac{X^2}{2}}, \tag{6.80}$$

times a Dirac δ-function, multiplied by a proportionality constant N_n. Compare the expression for N_n with the spectral intensity in Exercise 6.26b, and comment on its physical interpretation.

▶ **Exercise 6.29**   Show that the Fourier transform of a picket fence,

$$S_T(t) = \sum_{-\infty}^{\infty} \delta(t - nT), \tag{6.81}$$

is another picket fence,

$$\omega_0 S_{\omega_0}(\omega) = \omega_0 \sum_{-\infty}^{\infty} \delta(\omega - n\omega_0), \qquad \omega_0 = \frac{2\pi}{T} \tag{6.82}$$

(see Figure 6.7). [Hint: it suffices to show that the inverse transform of $(2\pi/T)S_{\omega_0}(\omega)$ equals $S_T(t)$:

$$\frac{1}{T} \sum_{-\infty}^{\infty} e^{in\omega_0 t} = S_T(t). \tag{6.83}$$

Consider the function

$$k_N(t) = \frac{1}{T} \sum_{-N}^{N} e^{in\omega_0 t}. \tag{6.84}$$

Clearly, $k_N(t)$ is periodic with period $T = \frac{2\pi}{\omega_0}$. Therefore, if we prove that in the interval $[-T/2, T/2]$ $k_N(t)$ tends to $\delta(t)$ as $N \to \infty$, it follows that

$$\lim_{N \to \infty} k_N(t) = S_T(t) \tag{6.85}$$

over the entire interval, which is equivalent to Eq. 6.83.]

Solution

$$k_N(t) = \frac{1}{T} \sum_{-N}^{N} e^{in\omega_0 t} = \frac{e^{i(N+1)\omega_0 t} - e^{-iN\omega_0 t}}{T(e^{i\omega_0 t} - 1)} \tag{6.86}$$

$$= \frac{\sin\left(N + \frac{1}{2}\right)\omega_0 t}{T \sin(\omega_0 t/2)} = \frac{\sin\left(N + \frac{1}{2}\right)\omega_0 t}{Tt} \frac{t}{\sin(\omega_0 t/2)}. \tag{6.87}$$

Note that the first factor satisfies the relation

$$\lim_{N \to \infty} \frac{\sin\left(N + \frac{1}{2}\right)\omega_0 t}{Tt} = \frac{\pi}{T}\delta(t), \tag{6.88}$$

and therefore,

$$\lim_{N \to \infty} k_N(t) = \frac{\pi}{T} \frac{t}{\sin(\omega_0 t/2)}\delta(t) = \frac{\pi}{T} \frac{t}{\sin(\omega_0 t/2)}\bigg|_{t=0} \delta(t) = \delta(t). \tag{6.89}$$

We have used the fact that the factor $\frac{t}{\sin(\omega_0 t/2)}$ is bounded in the region $[-T/2, T/2]$. To recapitulate, we have shown that $\lim_{N \to \infty} k_N(t) = \delta(t)$ in the interval $[-T/2, T/2]$. Combining this with the fact that $k_N(t)$ is periodic with period T completes the proof.

References

General

1. G. Arfken, *Mathematical Methods for Physicists*, 4th ed. (Academic, San Diego, 1995).

2. E. Oran Brigham, *The Fast Fourier Transform*, (Prentice Hall, Englewood Cliffs, NJ, 1974).

3. A. Papoulis, *The Fourier Integral and its Applications*, (McGraw-Hill, New York, 1962).

Spectra and Wavefunctions via Fourier Transforms

4. E. J. Heller, The semiclassical way to molecular spectroscopy, *Acc. Chem. Res.*, 14, 368 (1981).

5. M. J. Davis and E. J. Heller, Multidimensional wave functions from classical trajectories, J. Chem. Phys. 75, 3916 (1981).

6. N. DeLeon, M. J. Davis and E. J. Heller, Quantum manifestations of classical resonance zones, J. Chem. Phys. 80, 794 (1984).

7. E. J. Heller and S. Tomsovic, Postmodern quantum mechanics, Physics Today, 46, 38 (July 1993).

Alternative Papers on the Fourier Transform Approach to Spectroscopy

8. R. Zwanzig, Time correlation functions and transport coefficients in statistical mechanics, Annual Rev. Phys. Chem. 16, 67 (1965).

9. R. G. Gordon, Correlation functions for molecular motion, Adv. Mag. Resonance 3, 1 (1968).

10. M. Bixon and J. Jortner, Intramolecular radiationless transitions, J. Chem. Phys. 48, 715 (1968).

Chapter 7

One-Dimensional Barrier Scattering

In this chapter we discuss one-dimensional barrier scattering. The central dynamical event in barrier scattering is easily grasped: an incident wavepacket collides with a potential barrier and bifurcates into a transmitted and a reflected portion. The universality of this bifurcation process is at the heart of the formal theory of scattering, and we will use this dynamical event as a unifying theme to develop all the major ideas of this chapter.

To a certain extent, this chapter assumes that the reader has some minimal acquaintance with the standard treatments of barrier scattering, including the calculation of reflection and transmission coefficients for piecewise constant potentials and the use of conservation of flux arguments. In addition to method based on conservation of flux, we introduce several novel methods for calculating $R(E)$ and $T(E)$, two of which are time dependent and one of which is time independent.

The first method is based on the *autocorrelation functions* of the reflected and transmitted portions of the wavepacket after it has bifurcated. The Fourier transform of these autocorrelation functions, properly normalized by the incident wavepacket, give $R(E)$ and $T(E)$. Moreover, we show that $R(E)$ and $T(E)$ are simply related to the momentum representation of the reflected and transmitted portions of the wavepacket, again properly normalized. In the second method, we define four fundamental wavepackets—incoming from the left, outgoing to the left, incoming from the right and outgoing to the right. We show that the scattering or S-matrix can be expressed in terms of the Fourier transform of the *cross-correlation functions* of these four fundamental wavepackets, and we give the explicit relation between the S-matrix elements and $R(E)$ and $T(E)$. Corresponding to the four fundamental wavepackets are four fundamental eigenstates. In the third method, we show how the S-matrix can be calculated as the overlap of these scattering eigenstates. This is an interesting but delicate mathematical undertaking, since the scattering eigenstates are unnormalizable. However, after developing the rules for calculating the overlap of continuum states we show that we can obtain the orthogonality and normalization relations among the scattering eigenstates, as well as the symmetry relations among the reflection and transmission amplitudes, all within one coherent framework. Finally, we use the scattering eigenstates to reconstitute the wavepacket, showing how the wavepacket bifurcation behavior can be understood as emerging from a coherent superposition of the scattering eigenstates.

This material has a close relationship with three other parts of the book. It is the natural sequel to Chapter 2 on the free particle, since here, too, the wavepacket is a free particle at the beginning and the end of the process. The stationary phase arguments that were used there for the full wavepacket will be applied in this chapter to the reflected and transmitted portions of the wavepacket. The second connection is with Chapter 6 on correlation functions: by expressing the reflection and transmission coefficients in terms of the Fourier transform of time-correlation functions, we extend the approach of Chapter 6 from spectroscopy to barrier scattering. Finally, this chapter is to some extent an introduction to Chapter 18 on reactive scattering, where internal degrees of freedom of reactants and products will be added.

7.1   Wavepacket Formulation of Reflection and Transmission Coefficients

7.1.1   Autocorrelations of Bifurcated Wavepackets

Consider the asymmetric one-dimensional potential shown in Figure 7.1. The Hamiltonian is given by

$$H = -\frac{\hbar^2}{2m} \frac{\partial^2}{\partial x^2} + V(x), \quad \begin{cases} V(x) = 0 & x \to -\infty \\ V(x) = V_0 & x \to \infty. \end{cases} \tag{7.1}$$

Figure 7.1a shows an incoming wavepacket, incident from the left. The incoming wavepacket is defined such that as $t \to -\infty$ it has only positive momentum components. For convenience, we will refer below to the direction from which the packet is incident as "reactants," and the direction on the corresponding region on the other side of the barrier as "products." In Figure 7.1, the average energy of the wavepacket is slightly below the energy of the barrier maximum, and according to classical mechanics an incident particle would be reflected with unit probability. Quantum mechanically, the wavepacket bifurcates, with one portion reflected and one portion transmitted. This is seen during the collision in Figure 7.1b and after the collision, when the bifurcation is complete, in Figure 7.1c. We define $t = 0$ as a time just before the collision, when the incident wavepacket has approached the barrier but still consists entirely of positive momentum components. At time T, after the bifurcation is complete, we can define the amplitude on the product side as Ψ_T, and that on the reactant side as Ψ_R, and write

$$\Psi(x, T) = \Psi_T(x, T) + \Psi_R(x, T). \tag{7.2}$$

The bottom part of each figure shows the wavepacket in the momentum representation, at each of the three stages—before, during and after the collision. The bifurcation process is just as evident in the momentum representation as in the coordinate representation: the wavepacket before the collision is seen to have only positive momentum components; during the collision the momentum distribution becomes complicated, but after the collision it becomes fairly simple again, with both positive (transmitted) and negative (reflected) components. In the momentum representation we can write

$$\tilde{\Psi}(k) = \tilde{\Psi}_R(k) + \tilde{\Psi}_T(k). \tag{7.3}$$

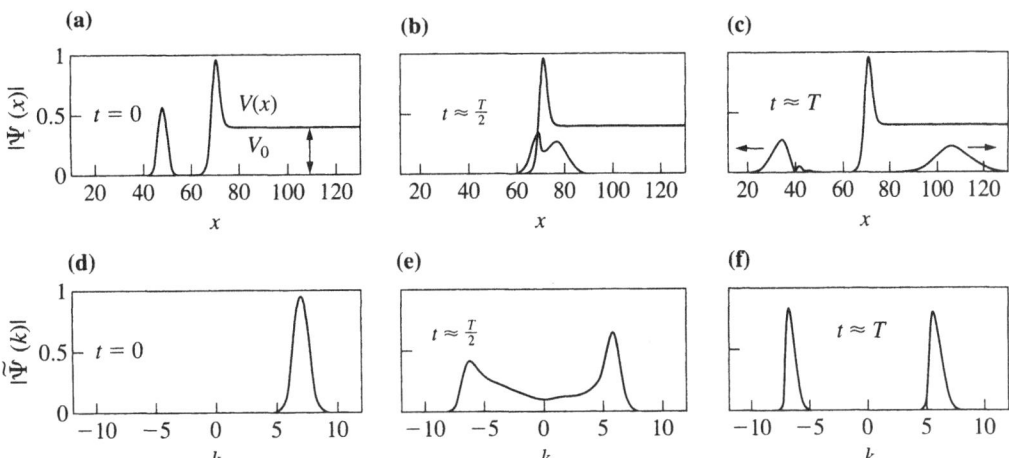

Figure 7.1 Potential energy barrier separating reactants (left) from products (right). (a) Initial wavepacket, incident from the left. The average energy of the wavepacket is slightly below the energy of the barrier height. (b) The wavepacket during the collision. (c) The wavepacket after the collision. The reflected $\left(\Psi_R(x, T)\right)$ and transmitted $\left(\Psi_T(x, T)\right)$ portions of the wavepacket are seen moving away from the region of the barrier. (d)–(f) The wavepacket in the momentum representation corresponding to the same times as in (a)–(c), respectively. Note the bifurcation of the wavepacket in momentum space, into a reflected portion $\left(\tilde{\Psi}_R(k)\right)$ and a transmitted portion $\left(\tilde{\Psi}_T(k)\right)$.

Consider the normalization condition applied to both sides of this equation:

$$\int_{-\infty}^{\infty} |\Psi(x, T)|^2 \, dx$$

$$= \int_{-\infty}^{\infty} |\Psi_T(x, T) + \Psi_R(x, T)|^2 \, dx \tag{7.4}$$

$$= \int_{-\infty}^{\infty} \left(|\Psi_T(x, T)|^2 + |\Psi_R(x, T)|^2 + 2\mathrm{Re}\{\Psi_T(x, T)^*\Psi_R(x, T)\} \right) dx \tag{7.5}$$

$$= \int_{-\infty}^{\infty} \left(|\Psi_T(x, T)|^2 + |\Psi_R(x, T)|^2 \right) dx, \tag{7.6}$$

where we have used the fact that the cross term in Ψ_T and Ψ_R vanishes. The vanishing of this overlap may be understood qualitatively by noting that these wavefunctions are in different regions of coordinate and momentum space. The rigorous proof that this overlap vanishes requires tools developed later in this chapter (see also the discussion in the next paragraph).

We now define the *spectrum* of the transmitted and reflected wavepackets. This is done by isolating the transmitted and the reflected wavepackets, constructing their individual autocorrelation functions, and Fourier transforming. Defining $\Psi \equiv \Psi(0)$, we write

$$\langle \Psi | \Psi(t) \rangle = \langle \Psi(T) | \Psi(T + t) \rangle$$

$$= \langle \Psi_R(T) + \Psi_T(T) | \Psi_R(T + t) + \Psi_T(T + t) \rangle$$

$$= \langle \Psi_R(T) | \Psi_R(T + t) \rangle + \langle \Psi_T(T) | \Psi_T(T + t) \rangle$$

$$= \langle \Psi_R | \Psi_R(t) \rangle + \langle \Psi_T | \Psi_T(t) \rangle. \tag{7.7}$$

In the first and last step we have used the fact that the correlation function is invariant with respect to the zero of time (see Exercise 6.25b). Moreover, we have neglected the cross-terms of the form $\langle \Psi_R(T)|\Psi_T(T+t)\rangle$. This is a stronger requirement than the neglect of $\langle \Psi_R(T)|\Psi_T(T)\rangle$ discussed above; it says that even if Ψ_T is propagated *backward* in time it will never develop any overlap with Ψ_R, even though they may have components located in the same region of coordinate space. This is a profound result that can be derived rigorously using tools we will develop later in this chapter (see Exercise 7.6c). The physics is that the orthogonality of the *destiny* of a pair of wavepackets at asymptotic positive times—outgoing to the left versus the right—guarantees their orthogonality at all other times, even if the two wavepackets evolve for different amounts of time. Similarly, orthogonality of the *history* of a pair of wavepackets at asymptotic negative times—incoming from the left versus the right—guarantees their orthogonality for all relative time evolutions.

Taking the time–energy Fourier transform of both sides of Eq. 7.7 we find that

$$\sigma(E) = \sigma_R(E) + \sigma_T(E), \tag{7.8}$$

where we have defined the spectrum of the incident wavepacket, the reflected wavepacket and the transmitted wavepacket using a self-evident notation. Dividing both sides of this equation by $\sigma(E)$ we find

$$1 = R(E) + T(E). \tag{7.9}$$

We have introduced the reflection coefficient, $R(E)$, and the transmission coefficient, $T(E)$, defined as

$$R(E) = \frac{\sigma_R(E)}{\sigma(E)}, \qquad T(E) = \frac{\sigma_T(E)}{\sigma(E)}. \tag{7.10}$$

The statement that $R(E)$ and $T(E)$ sum to 1 is simply a statement that at asymptotic times all particles must be either reflected or transmitted.

The whole procedure is shown graphically in Figure 7.2. Figure 7.2a shows the absolute value of the autocorrelation function for the incident wavepacket, $\langle \Psi|\Psi(t)\rangle$, the reflected

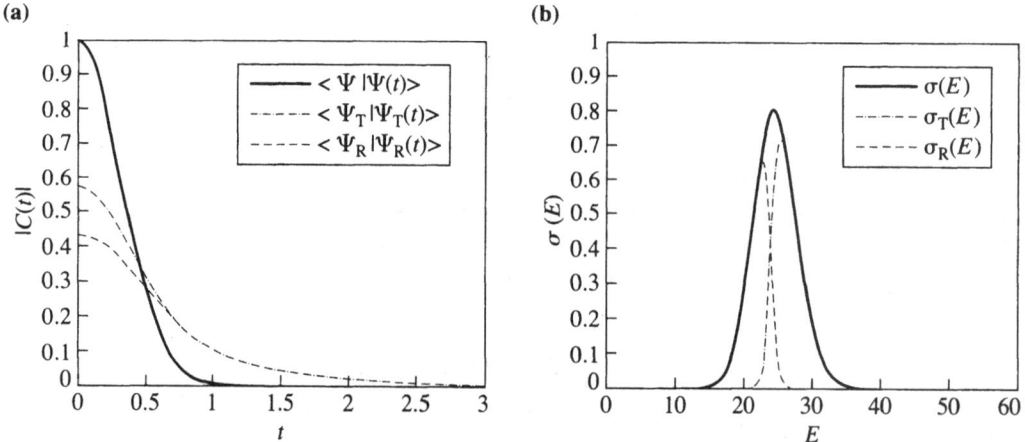

Figure 7.2 (a) Autocorrelation functions of the incident wavepacket, the reflected wavepacket and the transmitted wavepacket, $\langle \Psi|\Psi(t)\rangle$, $\langle \Psi_R|\Psi_R(t)\rangle$ and $\langle \Psi_T|\Psi_T(t)\rangle$, respectively. (b) Spectrum of the three wavepackets in (a), $\sigma(E)$, $\sigma_R(E)$ and $\sigma_T(E)$.

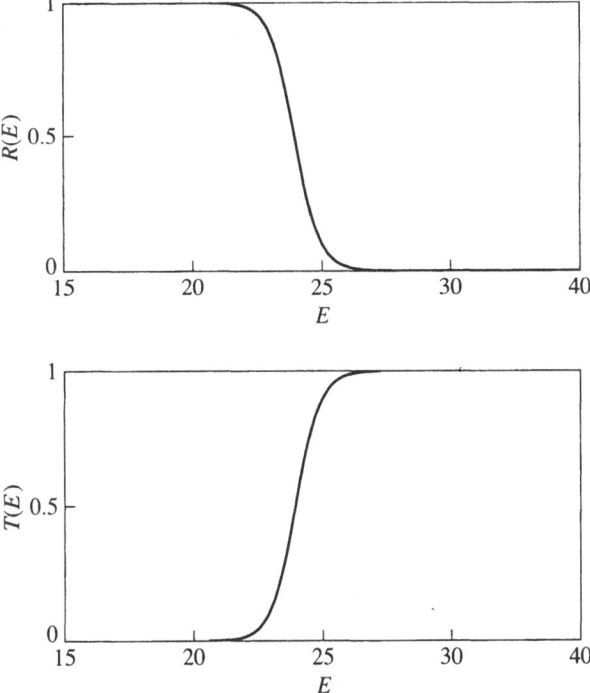

Figure 7.3 Reflection and transmission coefficients, given by the ratios of the reflected and transmitted spectra to the incident spectrum shown in Figure 7.2: $R(E) = \sigma_R(E)/\sigma(E)$ and $T(E) = \sigma_T(E)/\sigma(E)$.

wavepacket, $\langle \Psi_R | \Psi_R(t) \rangle$, and the transmitted wavepacket, $\langle \Psi_T | \Psi_T(t) \rangle$. Figure 7.2b shows the Fourier transform, or "spectrum," of each of these three autocorrelation functions. Note that the "spectrum" corresponding to the reflected and transmitted wavepacket sums precisely to that of the incident wavepacket (cf. Eq. 7.8). Figure 7.3 shows the reflection and transmission coefficients, $\sigma_R(E)$ and $\sigma_T(E)$ as a function of energy, from Eq. 7.10.

7.1.2 Projection of Bifurcated Wavepackets onto Plane Waves

We now continue discussing the wavepacket bifurcation process and present alternative formulas for the reflection and transmission coefficients in terms of the projection of the asymptotic wavepackets onto appropriate plane waves. We do the calculation first using *energy-normalized* plane waves, which in a sense are more natural for the problem. We then write the results in terms of *momentum-normalized* plane waves, where $R(E)$ and $T(E)$ are expressed directly in terms of the relevant wavepackets in k space but a density-of-states factor appears. We will need, however, to start with the definition of the momentum normalized plane waves.

Momentum- versus Energy-Normalized Plane Waves

We begin by defining an asymptotic TISE for reactants and products, in which the potential is taken as constant and equal to the asymptotic potential in the respective region. Since this just gives a free-particle Schrödinger equation, we may write down the solutions immediately

Table 7.1 Time-Independent Schrödinger Equation for reactants and products

	Asymptotic TISE	Solutions	Energy
Reactants	$H_0^\alpha \psi = -\frac{\hbar^2}{2m}\frac{\partial^2 \psi}{\partial x^2} = E\psi$	$\psi_{\pm k}(x) = \frac{e^{\pm ikx}}{\sqrt{2\pi}}$	$E = \frac{\hbar^2(\pm k)^2}{2m}$
Products	$H_0^\beta \psi = -\frac{\hbar^2}{2m}\frac{\partial^2 \psi}{\partial x^2} + V_0 = E\psi$	$\psi_{\pm\kappa}(x) = \frac{e^{\pm i\kappa x}}{\sqrt{2\pi}}$	$E = \frac{\hbar^2(\pm\kappa)^2}{2m} + V_0$

(see Table 7.1). The solutions given in Table 7.1 satisfy the normalization condition

$$\langle \psi_{\pm k}|\psi_{\pm k'}\rangle = \delta(k - k') \qquad \langle \psi_{\pm\kappa}|\psi_{\pm\kappa'}\rangle = \delta(\kappa - \kappa'). \tag{7.11}$$

We will refer to these states as momentum-normalized, despite the fact that we are using the variable k instead of $p = \hbar k$ (compare Exercise 2.3). Using the property $\delta(f(x)) = \sum_i \frac{1}{|f'(x_i)|}\delta(x - x_i)$, where $f(x_i) = 0$ (see Appendix A), we have

$$\delta(E(k) - E'(k')) = \frac{m}{\hbar^2|k|}\delta(k - k'). \tag{7.12}$$

With the aid of Eq. 7.12 we may define the energy-normalized plane waves as

$$\psi_{\pm k}(E) = \left(\frac{m}{2\pi\hbar^2|k|}\right)^{1/2} e^{\pm ikx} \qquad \langle \psi_{\pm k}(E)|\psi_{\pm k'}(E')\rangle = \delta(E - E'), \tag{7.13}$$

with analogous relations with k replaced by κ. To define a completeness relation in terms of the energy-normalized plane waves, we need to take into account the degeneracy of $\pm k$ at every E:

$$\int_0^\infty \left\{ |\psi_{k'}(E')\rangle\langle\psi_{k'}(E')| + |\psi_{-k'}(E')\rangle\langle\psi_{-k'}(E')| \right\} dE' = 1. \tag{7.14}$$

Equation 7.14 can also be written in terms of the κ states with the lower limit of the integral replaced by V_0, which are appropriate for expanding states in the product region of the potential.

The Three Spectral Projections

Consider now the formula for $\sigma(E)$:

$$\sigma(E) = \frac{1}{2\pi}\int_{-\infty}^\infty \langle\Psi|e^{-iHt/\hbar}|\Psi\rangle e^{iEt/\hbar}\,dt, \tag{7.15}$$

where Ψ is the incident wavepacket and $\sigma(E)$ is its energy spectrum. The key idea is to note that we can replace H by H_0^α in this equation. The argument for this replacement is as follows. The correlation function $\langle\Psi|\Psi(t)\rangle$ for a scattering process must decay on some finite time scale; although in certain cases, for example, when resonances are present, this time scale can be very long (see Section 7.6), it must be finite by the very definition of the scattering process. Noting that the correlation function is invariant with respect to a shift in time, we may take the zero of time far enough in the past such that the correlation function of the incident wavepacket will decay to zero before the wavepacket reaches the interaction potential. With this translation in time, the correlation function is determined entirely in the asymptotic region of reactants, and therefore it is permissible to replace H with H_0^α in the propagator. Noting that the states $\psi_k(E)$ are complete with respect to the incident wavepacket Ψ, we can write

$$\sigma(E) = \frac{1}{2\pi} \int_{-\infty}^{\infty} \langle \Psi | e^{-iH_0^\alpha t/\hbar} | \Psi \rangle e^{iEt/\hbar} \, dt \tag{7.16}$$

$$= \frac{1}{2\pi} \int_0^{\infty} \int_{-\infty}^{\infty} \langle \Psi | \psi_{k'}(E') \rangle \langle \psi_{k'}(E) | e^{-iE't/\hbar} | \Psi \rangle e^{iEt/\hbar} \, dt \, dE' \tag{7.17}$$

$$= \int_0^{\infty} |\langle \psi_{k'}(E') | \Psi \rangle|^2 \delta\left(E - E'\right) \, dE' = |\langle \psi_k(E) | \Psi \rangle|^2, \tag{7.18}$$

where we have used Eq. 7.14 and made use of the fact that Ψ does not have any projection onto the $-k'$ states.

Consider now the formula for $\sigma_R(E)$. The derivation is analogous to that of $\sigma(E)$. Again we replace H with H_0^α, except that now the argument is based on taking the time T after bifurcation sufficiently far into the future; moreover, the complete set of plane waves appropriate for Ψ_R is $\psi(-k)$. Thus

$$\sigma_R(E) = \int_{-\infty}^{\infty} \langle \Psi_R | \Psi_R(t) \rangle e^{iEt/\hbar} \, dt \tag{7.19}$$

$$= \int_0^{\infty} \int_{-\infty}^{\infty} \langle \Psi_R | \psi_{-k'}(E') \rangle \langle \psi_{-k'}(E') | e^{-iH_0^\alpha t/\hbar} | \Psi_R \rangle e^{iEt/\hbar} \, dt \, dE' \tag{7.20}$$

$$= |\langle \psi_{-k}(E) | \Psi_R \rangle|^2. \tag{7.21}$$

Again we have used Eq. 7.14, this time noting that Ψ_R does not have any projection onto the $+k'$ states.

Finally, consider the formula for $\sigma_T(E)$. The derivation is analogous to that of $\sigma_R(E)$, except that now H is replaced by H_0^β and the complete set of plane waves appropriate for Ψ_T is $\psi(\kappa)$:

$$\sigma_T(E) = \int_{-\infty}^{\infty} \langle \Psi_T | \Psi_T(t) \rangle e^{iEt/\hbar} \, dt \tag{7.22}$$

$$= \int_{V_0}^{\infty} \int_{-\infty}^{\infty} \langle \Psi_T | \psi_{\kappa'}(E') \rangle \langle \psi_{\kappa'}(E') | e^{-iH_0^\beta t/\hbar} | \Psi_T \rangle e^{iEt/\hbar} \, dt \, dE' \tag{7.23}$$

$$= |\langle \psi_\kappa(E) | \Psi_T \rangle|^2. \tag{7.24}$$

Again we have used Eq. 7.14, using the κ plane waves as a basis and noting that Ψ_T does not have any projection onto the $-\kappa'$ states.

Reflection and Transmission Coefficients
We now turn to the calculation of the reflection and transmission coefficients. Substituting Eqs. 7.21, 7.24 and 7.18 into Eq. 7.10 we obtain

$$R(E) = \frac{|\langle \psi_{-k}(E) | \Psi_R \rangle|^2}{|\langle \psi_k(E) | \Psi \rangle|^2} \qquad T(E) = \frac{|\langle \psi_\kappa(E) | \Psi_T \rangle|^2}{|\langle \psi_k(E) | \Psi \rangle|^2}. \tag{7.25}$$

The content of Eq. 7.25 is that $R(E)$ and $T(E)$ are the projections of the asymptotic wavepackets Ψ_R and Ψ_T onto the appropriate energy-normalized free-particle eigenstates at energy E, normalized by the coefficient at energy E in the incident wavepacket Ψ.

Equation 7.25 can be rewritten in terms of the ratio of projections onto k-normalized eigenstates:

$$R(E) = \frac{|\tilde{\Psi}_R(-k)|^2}{|\tilde{\Psi}(k)|^2} \qquad T(E) = \frac{k}{\kappa}\frac{|\tilde{\Psi}_T(\kappa)|^2}{|\tilde{\Psi}(k)|^2}, \qquad (7.26)$$

where we have used Eq. 7.13 and noted that $\tilde{\Psi}(k) = \langle \psi_k | \Psi \rangle$, and so on. The appeal of these expressions is that they provide a direct connection between $R(E)$ and $T(E)$, and $\tilde{\Psi}_R$, $\tilde{\Psi}_T$ and $\tilde{\Psi}$, which are the wavepackets Ψ_R, Ψ_T and Ψ in the momentum representation and calculated from the latter simply by performing Fourier transforms. The factor of $\frac{k}{\kappa}$ that appears in the transmission coefficient can be understood from a physical perspective, by noting that the density of states at energy E is given by $\rho(E(k)) \equiv \frac{dk}{dE}$, and $\rho(E(\kappa)) \equiv \frac{d\kappa}{dE}$. Using this relationship, Eq. 7.26 for $T(E)$ can be rewritten as

$$T(E) = \frac{|\tilde{\Psi}_T(\kappa)|^2 \rho(E(\kappa))}{|\tilde{\Psi}(k)|^2 \rho(E(k))}. \qquad (7.27)$$

Thus, the transmission coefficient $T(E)$ is the ratio of the probabilities of transmitted to incident wavepacket at value k *weighted by the density of k states at energy E in the corresponding region*. Note that the factor $\frac{k}{\kappa}$ is the reciprocal of the factor $\frac{\kappa}{k}$, which appears in the formula for the transmission coefficient appearing in many textbooks: $T(E) = \frac{\kappa}{k}|t(\kappa)|^2$. This latter relationship will be derived in Section 7.3.3 within a time-independent approach to scattering, and it will be seen to be entirely consistent with Eq. 7.26 for $T(E)$, after taking into account the precise definition of $t(\kappa)$ (Exercise 7.10).

7.2 Cross-Correlation Function Formulation of Barrier Scattering and the S-Matrix

7.2.1 The Wavepacket Correlation Matrix

We have seen how to calculate reflection and transmission coefficients from the propagation of an incoming wavepacket. What happens if the incoming wavepacket comes from the opposite direction? Is there a simple relationship between the transmission and reflection coefficients coming from the second direction relative to those coming from the first direction? The answer is that the transmission coefficient, $T(E)$, is the same in both directions. In fact, since $R(E) + T(E) = 1$, $R(E)$ is also the same in both directions (for the range of energies at which both asymptotic regions are energetically accessible).

In order to derive this result we develop a more symmetrical approach to the scattering process. We define four fundamental wavepackets, $\phi_\alpha^+, \phi_\alpha^-, \phi_\beta^+$ and ϕ_β^-, where α indicates the region to the left of the barrier, β to the right of the barrier. The superscript $+$ indicates incoming to the barrier at negative times; the superscript $-$ indicates outgoing from the barrier at positive times. These four fundamental wavepackets are, therefore, incoming from the left on the left, outgoing to the left on the left, incoming from the right on the right and outgoing to the right on the right, as shown in Figure 7.4.

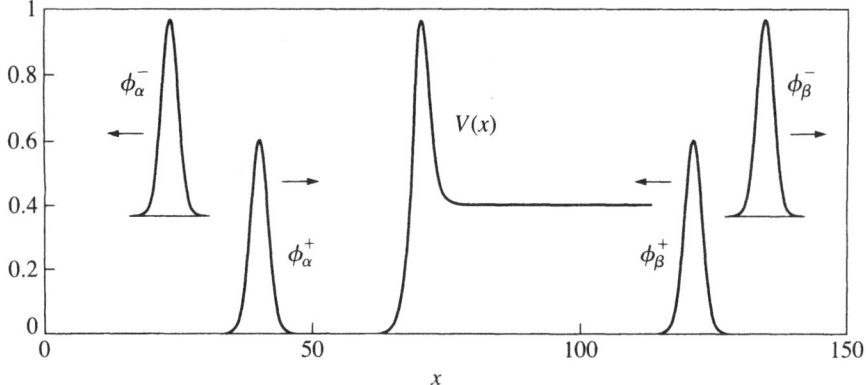

Figure 7.4 Pictures of the four fundamental wavepackets, incoming from the left on the left (ϕ_α^+), outgoing to the left on the left (ϕ_α^-), incoming from the right on the right (ϕ_β^+) and outgoing to the right on the right (ϕ_β^-). The vertical displacement of the wavepackets is arbitrary and was chosen for clarity.

We note that

$$\langle \phi_\alpha^+ | \phi_\alpha^+ \rangle = \langle \phi_\alpha^- | \phi_\alpha^- \rangle = \langle \phi_\beta^+ | \phi_\beta^+ \rangle = \langle \phi_\beta^- | \phi_\beta^- \rangle = 1 \qquad (7.28)$$

$$\langle \phi_\beta^+ | \phi_\alpha^+ \rangle = \langle \phi_\beta^- | \phi_\alpha^- \rangle = 0. \qquad (7.29)$$

The physical content of the first equation is simply normalization of the incoming and outgoing wavepackets. The content of the second equation is that an incoming wavepacket that is made up exclusively of components coming in from the left is orthogonal to an incoming wavepacket made up exclusively of components coming in from the right. Similarly, a final wavepacket outgoing to the right is orthogonal to a final wavepacket outgoing to the left. Again, the rigorous proof of this orthogonality will have to wait for the development of additional tools below. However, one can appreciate the basic reason for this orthogonality by examining the fundamental wavepackets in phase space. Figure 7.5 shows a phase space picture of the equipotentials of a one-dimensional barrier potential. The equipotentials are hyperbolae, with a separatrix dividing the phase space into different types of trajectories. The figure shows ϕ_α^+ (wavepacket incoming from the left) and ϕ_β^+ (wavepacket incoming from the right). As $t \to -\infty$ these packets are infinitely separated, and their overlap vanishes. Now consider propagating one of these packets, for example, ϕ_α^+, while keeping the other one fixed. Even at large positive times, when part of $\phi_\alpha^+(t)$ is located spatially in region β, the sign of its momentum is opposite to that of ϕ_β^+. This gives a qualitative picture of why the cross-correlation function of initial states with different initial histories vanishes even at later times.

▶ Exercise 7.1

a. Draw phase space pictures of ϕ_α^- and ϕ_β^-. Explain why their cross-correlation function is zero even at negative times.

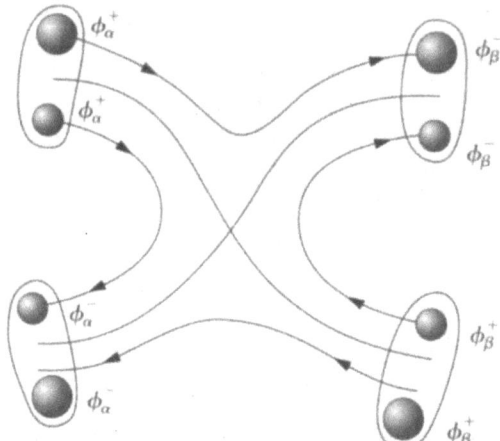

Figure 7.5 Phase space picture of the equipotential energy contours for a barrier potential. Overlaid are schematic phase space equiamplitude contours of ϕ_α^+, ϕ_β^+, ϕ_α^- and ϕ_β^- in the Wigner representation. Note that as $t \to -\infty$, ϕ_α^+ and ϕ_β^+ not only are separated in coordinate, but also have opposite signs of momentum. Similarly, as $t \to \infty$, ϕ_α^- and ϕ_β^- not only are separated in coordinate, but also have opposite signs of momentum.

b. Explain why the overlap of $\phi_\alpha^+(t)$ with $\phi_\beta^-(0)$ can be nonzero at positive times. Find three other cross-correlation functions that are generally nonzero, and describe the underlying phase space dynamics.

Thus, the only nontrivial overlaps are formed by combining $+$ and $-$ wavepackets, and there are four such combinations. We may define a 2×2 matrix of correlation functions:

$$\mathbf{C}(t) = \begin{pmatrix} C_{\alpha\alpha}(t) & C_{\alpha\beta}(t) \\ C_{\beta\alpha}(t) & C_{\beta\beta}(t) \end{pmatrix} = \begin{pmatrix} \langle\phi_\alpha^-|e^{-iHt/\hbar}|\phi_\alpha^+\rangle & \langle\phi_\beta^-|e^{-iHt/\hbar}|\phi_\alpha^+\rangle \\ \langle\phi_\alpha^-|e^{-iHt/\hbar}|\phi_\beta^+\rangle & \langle\phi_\beta^-|e^{-iHt/\hbar}|\phi_\beta^+\rangle \end{pmatrix}. \quad (7.30)$$

Note that the C-matrix is symmetric:

$$C_{\alpha\beta}(t) = C_{\beta\alpha}(t). \qquad (7.31)$$

To see this, we write

$$\langle\phi_\beta^-|e^{-iHt/\hbar}|\phi_\alpha^+\rangle = \langle\phi_\alpha^+|e^{iHt/\hbar}|\phi_\beta^-\rangle^* \qquad (7.32)$$

$$= \langle(\phi_\alpha^+)^*|e^{-iHt/\hbar}|(\phi_\beta^-)^*\rangle \qquad (7.33)$$

$$= \langle\phi_\alpha^-|e^{-iHt/\hbar}|\phi_\beta^+\rangle. \qquad (7.34)$$

Equation 7.34 is an expression of time-reversal symmetry: the amplitude to go from ϕ_β to ϕ_α is equal to that for going from ϕ_α to ϕ_β.

7.2.2 The S-Matrix

Suppose one is interested in the amplitude for making a transition from $\alpha \to \beta$ at a specific energy E. It seems quite plausible that this transition amplitude should be determined by the overlap between an incoming wavepacket from channel α with an outgoing wavepacket in channel β, that is, given by the function $C_{\beta\alpha}(t)$. However, calculating an overlap at just one instant of time would have the effect of averaging the dynamical information over all energies, since the reactant and product wavepackets have a broad spread in energies. We therefore anticipate that some kind of Fourier transform from time to energy will be required to filter the contribution from just a particular energy to the transition amplitude. We will now introduce the S-matrix, which is a measure of transition amplitude from $\alpha \to \beta$ at energy E, and we shall indeed see that its elements may be expressed in terms of the Fourier transform of the elements of the cross-correlation function matrix.

Scattering Eigenstates as Wavepacket Fourier Transforms

Before proceeding, we need to provide a little background about the scattering eigenstates. Since the latter will be discussed in detail in Section 7.3.1, in this section we will introduce only what is necessary for the presentation at hand. The scattering eigenstates are solutions of the TISE, $H\psi = E\psi$, with H given by Eq. 7.1. The reader may want to glance ahead at Figure 7.7, which shows the structure of the four fundamental scattering eigenstates. As we have seen in Chapter 6 (Section 6.3), the natural way to pass from a wavepacket to an eigenstate is via the spectral method. In the case of one-dimensional barrier scattering, there are eigenstates at every energy. In fact for $E > V_0$ the states are two-fold degenerate. Which of these degenerate states will be produced by the spectral method? The answer is that the *boundary conditions* of the wavepacket determine the boundary conditions of the eigenstate that it produces according to the correspondence

$$\phi_\alpha^+ \Longleftrightarrow \psi_\alpha^+ \qquad \phi_\alpha^- \Longleftrightarrow \psi_\alpha^- \qquad \phi_\beta^+ \Longleftrightarrow \psi_\beta^+ \qquad \phi_\beta^- \Longleftrightarrow \psi_\beta^-. \qquad (7.35)$$

Consider, for example, the Fourier transform at energy E of a wavepacket incident from channel α produces the eigenstate labelled $\psi_\alpha^+(E)$:

$$\psi_\alpha^+(E) \propto \int_{-\infty}^{\infty} e^{-iHt/\hbar} \phi_\alpha^+ e^{iEt/\hbar}\, dt. \qquad (7.36)$$

The eigenstates $\psi_\alpha^+(E)$, in turn, form a complete basis for the expansion of ϕ_α^+ (Exercise 7.2). Therefore we can write

$$\phi_\alpha^+ = \int_0^\infty \eta_\alpha(E)\psi_\alpha^+(E)\, dE, \qquad (7.37)$$

where we have defined $\eta_\alpha(E)$ as the coefficient of $\psi_\alpha^+(E)$ in ϕ_α^+. Substituting Eq. 7.37 into Eq. 7.36, we may obtain the proportionality constant of that equation:

$$\psi_\alpha^+(E) = \frac{(2\pi\hbar)^{-1}}{\eta_\alpha(E)} \int_{-\infty}^{\infty} e^{-iHt/\hbar} \phi_\alpha^+ e^{iEt/\hbar}\, dt. \qquad (7.38)$$

Similarly, we may write

$$\phi_\beta^- = \int_0^\infty \mu_\beta(E)\psi_\beta^-(E)\, dE \qquad (7.39)$$

and

$$\psi_\beta^-(E) = \frac{(2\pi\hbar)^{-1}}{\mu_\beta(E)} \int_{-\infty}^{\infty} e^{-iHt/\hbar} \phi_\beta^- e^{iEt/\hbar} \, dt. \tag{7.40}$$

▶ **Exercise 7.2** Prove that the eigenstates $\{\psi_\alpha^+\}$ form a complete basis for ϕ_α^+. [Hint: Suppose that the eigenstates $\{\psi_\alpha^+\}$ were not a complete basis for ϕ_α^+ and had to be supplemented by a linearly independent set, which we will call $\{\psi_\beta^+\}$:

$$\phi_\alpha^+ = \int_0^\infty \left\{ \eta_\alpha(E)\psi_\alpha^+(E) + \eta_\beta(E)\psi_\beta^+(E) \right\} dE. \tag{7.41}$$

What would the spectral method on ϕ_α^+ yield?]

S-Matrix Elements as Overlaps of Scattering Eigenstates

We are now ready to define the scattering or S-matrix. The elements of S are determined by the overlap integral of a $+$ eigenstate with a $-$ eigenstate. For example,

$$\langle \psi_\beta^-(E') | \psi_\alpha^+(E) \rangle = S_{\beta\alpha}(E)\delta(E - E') \tag{7.42}$$

with analogous definitions for $S_{\alpha\alpha}$, $S_{\alpha\beta}$ and $S_{\beta\beta}$. Since the scattering eigenstates are of infinite extent and non-normalizable in the usual sense, the overlap of two scattering eigenstates, when it is nonvanishing, is infinite. This is the meaning of the factor $\delta(E - E')$ in Eq. 7.42. However, we may define coefficients of the infinite overlap that are finite; these are the S-matrix elements.

We substitute Eq. 7.37 for $\psi_\alpha^+(E)$ together with Eq. 7.39 for $\psi_\beta^-(E')$ into the formula for the S-matrix, Eq. 7.42. We obtain a double integral over time and change variables to $t = t' - t''$ and $s = t' + t''$; the integral over s gives $\delta(E - E')$ and we obtain

$$\langle \psi_\beta^-(E') | \psi_\alpha^+(E) \rangle = \frac{(2\pi\hbar)^{-1}}{\mu_\beta^*(E')\eta_\alpha(E)} \int_{-\infty}^{\infty} \langle \phi_\beta^- | e^{-iHt/\hbar} | \phi_\alpha^+ \rangle e^{iEt/\hbar} dt \, \delta(E - E'). \tag{7.43}$$

Comparing with Eq. 7.42 we find that

$$S_{\beta\alpha}(E) = \frac{(2\pi\hbar)^{-1}}{\mu_\beta^*(E')\eta_\alpha(E)} \int_{-\infty}^{\infty} \langle \phi_\beta^- | e^{-iHt/\hbar} | \phi_\alpha^+ \rangle e^{iEt/\hbar} \, dt \tag{7.44}$$

$$= \frac{(2\pi\hbar)^{-1}}{\mu_\beta^*(E')\eta_\alpha(E)} \int_{-\infty}^{\infty} C_{\beta\alpha}(t) e^{iEt/\hbar} \, dt. \tag{7.45}$$

Note that the elements of the S-matrix are proportional to the Fourier transforms from time to energy of the elements of the C-matrix (correlation function matrix).

Equation 7.45 and its generalizations form the central core of Chapter 18 on reactive scattering. A more complete discussion of this formula is given there, and the reader who is so inclined may want to look at that chapter now. Here we will make just a few additional comments.

1. By expressing the S-matrix elements in terms of the Fourier transform of a time-correlation function, Eq. 7.45 extends the approach of Chapter 6 from spectroscopy to barrier scattering. Since the main features of the time-correlation function can

often be understood in terms of classical motion, this formulation gives an underlying dynamical interpretation to the S-matrix.

2. Equation 7.45 expresses the S-matrix in terms of square-normalizable wavepackets. The conventional formulation, which is treated in Section 7.4, expresses the S-matrix in terms of the non-normalizable scattering eigenstates. Although these two approaches are completely equivalent, the formulation of the S-matrix using only square-normalizable states has distinct advantages from a purely formal mathematical point of view.

▶ **Exercise 7.3**   Verify Eq. 7.43. [Hint: Define $E_+ \equiv (E + E')/2$, $E_- \equiv (E - E')/2$. You should obtain the factors $e^{iE_+t/\hbar}$ and $e^{iE_-s/\hbar}$. Note that there is a factor of 1/2 that comes from the Jacobian associated with the change of time variables that cancels the factor of 2 coming from $\delta(\frac{E-E'}{2})$.]

We may now assemble the various elements to form the scattering or S-matrix:

$$\mathbf{S}(E) = \begin{pmatrix} S_{\alpha\alpha}(E) & S_{\alpha\beta}(E) \\ S_{\beta\alpha}(E) & S_{\beta\beta}(E) \end{pmatrix} = \frac{1}{2\pi\hbar}\mu^{*-1}\tilde{\mathbf{C}}(E)\eta^{-1}, \tag{7.46}$$

where we have defined the matrices

$$\tilde{\mathbf{C}}(E) \equiv \int_{-\infty}^{\infty} \mathbf{C}(t)e^{iEt/\hbar}\,dt; \qquad \eta = \begin{pmatrix} \eta_\alpha & 0 \\ 0 & \eta_\beta \end{pmatrix}; \qquad \mu = \begin{pmatrix} \mu_\alpha & 0 \\ 0 & \mu_\beta \end{pmatrix}; \tag{7.47}$$

and we understand the Fourier transform of $\mathbf{C}(t)$ to be the matrix of the Fourier transform of the individual elements.

Reflection and Transmission Coefficients

As will be shown in Section 7.4 (Eq. 7.91 combined with Eqs. 7.70, 7.72), the S-matrix elements are related to the reflection and transmission coefficients, $R(E)$ and $T(E)$, as follows. For a wavepacket incident from the left,

$$R(E) = |S_{\alpha\alpha}(E)|^2 \qquad T(E) = |S_{\alpha\beta}(E)|^2, \tag{7.48}$$

while for a wavepacket incident from the right,

$$R'(E) = |S_{\beta\beta}(E)|^2 \qquad T'(E) = |S_{\beta\alpha}(E)|^2. \tag{7.49}$$

These equations provide us with an alternative to the method in Section 7.1 to calculate $R(E)$ and $T(E)$ from wavepacket dynamics. The cross-correlations, their Fourier transforms and the resulting reflection and transmission coefficients are shown in Figure 7.6.

▶ **Exercise 7.4**   Show that the S-matrix is symmetric, $S_{\alpha\beta} = S_{\beta\alpha}$. As we shall see in Section 7.4.2 it is also unitary, that is, $|S_{\alpha\alpha}|^2 + |S_{\alpha\beta}|^2 = 1$, although the proof does not follow as easily from Eq. 7.45.

The fact that $S_{\alpha\beta} = S_{\beta\alpha}$ leads immediately to the conclusion that the transmission coefficient $T(E)$ is equal for a wavepacket incident from the left and the right. The fact that $|S_{\alpha\beta}|^2 + |S_{\alpha\alpha}|^2 = |S_{\beta\alpha}|^2 + |S_{\beta\beta}|^2 = 1$ leads to the conclusion that $R(E)$ is the same for a

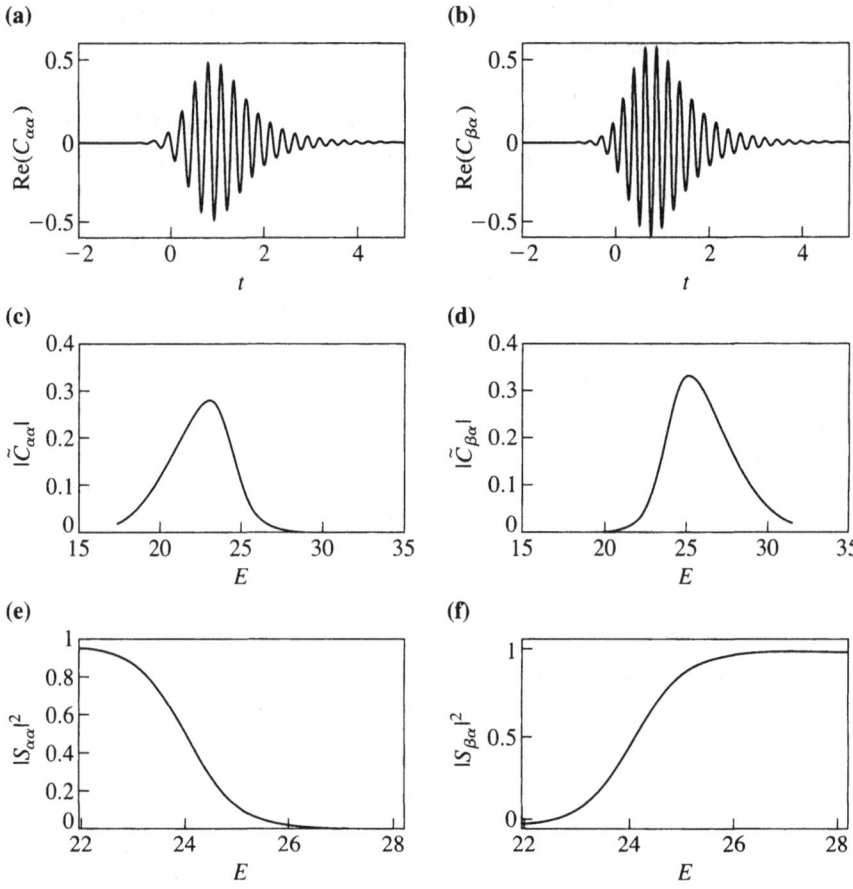

Figure 7.6 (a) Anti-autocorrelation function $(C_{\alpha\alpha}(t) = \langle\phi_\alpha^-|\phi_\alpha^+(t)\rangle)$. (b) Cross-correlation function $(C_{\beta\alpha}(t) = \langle\phi_\beta^-|\phi_\alpha^+(t)\rangle)$. (c) Absolute value of the Fourier transform of the correlation function in (a). (d) Absolute value of the Fourier transform of the correlation function in (b). (e) Normalized spectrum derived from the Fourier transform in (c), which gives the reflection coefficient. (f) Normalized spectrum derived from the Fourier transform in (d), which gives the transmission coefficient.

packet incident from the left and the right. Note that in general, $S_{\alpha\alpha} \neq S_{\beta\beta}$; only their absolute values squared are required to be equal. Furthermore, this equality of the reflection coefficients is valid only for energies for which both channels, α and β, are energetically accessible.

7.3   Scattering Theory Using Eigenstates

7.3.1   Asymptotic Behavior of the Scattering Eigenstates

We now return to the eigenstates of the scattering Hamiltonian, satisfying the equation

$$H\psi = E\psi. \tag{7.50}$$

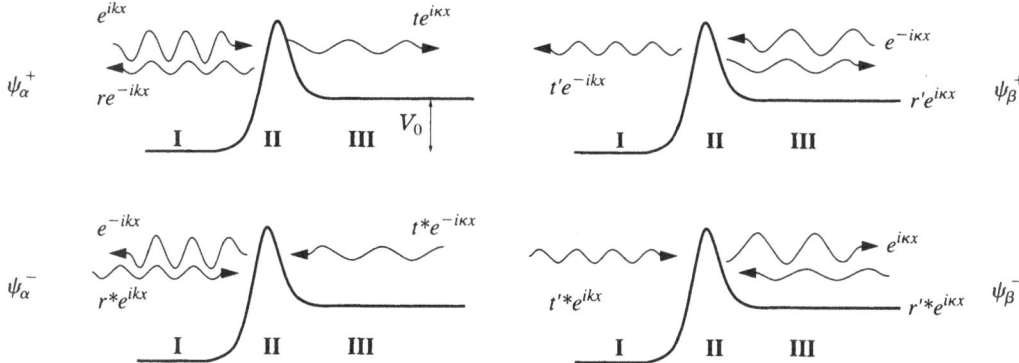

Figure 7.7 Pictures of the four fundamental eigenfunctions, ψ_α^+, ψ_β^+, ψ_α^- and ψ_β^-, corresponding to the four fundamental wavepackets in Figure 7.4. For the normalization of these states see the text.

Consider the potential in Figure 7.7. In region I, the potential is $V = 0$. The eigenstates behave like $e^{\pm ikx}$, where $E = \hbar^2 k^2/2m$. In region III, the potential is $V = V_0$. The eigenstates behave like $e^{\pm i\kappa x}$, where $E = \hbar^2 \kappa^2/2m + V_0$. Since the eigenstate is characterized by a single energy E, the asymptotic wave vectors k and κ must be related by the condition

$$\frac{\hbar^2 k^2}{2m} = \frac{\hbar^2 \kappa^2}{2m} + V_0 = E. \tag{7.51}$$

▶ **Exercise 7.5**

    a. Use Eq. 7.51 to show that $d\kappa/dk = k/\kappa$.

    b. Let $E' = \hbar^2 k'^2/2m = \hbar^2 \kappa'^2/2m + V_0$. Show that $\kappa^2 - \kappa'^2 = k^2 - k'^2$.

In region II the potential changes with distance and the eigenstates have no simple form; they are determined by the TISE and the condition of continuity of ψ and $\psi' = d\psi/dx$ at every point. The eigenfunctions are doubly degenerate for energies above V_0. In conventional scattering theory notation, the eigenstate with incoming boundary conditions from the reactants (here, from the left) is called $\psi_\alpha^+(k)$:

$$\psi_\alpha^+(k) = \begin{cases} e^{ikx} + re^{-ikx} & \text{region I} \\ te^{i\kappa x} & \text{region III}. \end{cases} \tag{7.52}$$

The degenerate partner of the state above is an eigenstate corresponding to incoming conditions from the right, called $\psi_\beta^+(\kappa)$:

$$\psi_\beta^+(\kappa) = \begin{cases} t'e^{-ikx} & \text{region I} \\ e^{-i\kappa x} + r'e^{i\kappa x} & \text{region III}. \end{cases} \tag{7.53}$$

Note the reversal of roles of region III and region I and the exchange of roles of κ and k with an accompanying change in direction: $k \to -\kappa$, $\kappa \to -k$.

    Since the eigenstates are degenerate, one can choose any linear combination of these two eigenstates and still have a solution of the TISE at the same energy. In particular the

Table 7.2 Asymptotic form of the four fundamental scattering eigenstates

State	Region I	Region III
$\psi_\alpha^+(k)$	$e^{ikx} + re^{-ikx}$	$te^{i\kappa x}$
$\psi_\alpha^-(k)$	$e^{-ikx} + r^*e^{ikx}$	$t^*e^{-i\kappa x}$
$\psi_\beta^+(\kappa)$	$t'e^{-ikx}$	$e^{-i\kappa x} + r'e^{i\kappa x}$
$\psi_\beta^-(\kappa)$	t'^*e^{ikx}	$e^{i\kappa x} + r'^*e^{-i\kappa x}$

complex conjugates of ψ_α^+ and ψ_β^+, which we will call ψ_α^- and ψ_β^-, respectively, can each be expressed as a linear combination of ψ_α^+ and ψ_β^+.

Consider first $\psi_\alpha^- \equiv \psi_\alpha^{+*}(k)$:

$$\psi_\alpha^-(k) = \begin{cases} e^{-ikx} + r^*e^{ikx} & \text{region I} \\ t^*e^{-i\kappa x} & \text{region III .} \end{cases} \tag{7.54}$$

The effect of complex conjugation is the same as replacing $k \to -k$ and $\kappa \to -\kappa$; since k(or κ) determines the momentum, changing the sign is equivalent to time reversal. The physical interpretation of the state ψ_α^- is that the reflected and transmitted waves at negative times combine to form a single outgoing wave at positive times, which proceeds to move away to the left of the barrier region—which is indeed the time reversal of the process that corresponds to ψ_α^+. This is a general rule, that complex conjugation is equivalent to time reversal in quantum mechanics.

Consider now $\psi_\beta^-(\kappa) \equiv \psi_\beta^{+*}(\kappa)$:

$$\psi_\beta^-(\kappa) = \begin{cases} t'^*e^{ikx} & \text{region I} \\ e^{i\kappa x} + r'^*e^{-i\kappa x} & \text{region III .} \end{cases} \tag{7.55}$$

The physical interpretation of the state ψ_β^- is that the reflected and transmitted wave at negative times combine to form a single outgoing incident wave at positive times, which proceeds to move away to the right of the barrier region. This is the time reversal of the process that corresponds to ψ_β^+. The asymptotic behavior of the four fundamental eigenstates is summarized in Table 7.2.

Note that we have adopted a convention in which the argument of $\psi_\alpha^\pm$ is written as k and the argument of $\psi_\beta^\pm$ is written as κ, independent of whether the wave is incoming or outgoing. This simplifies the notation; moreover, the sign of k (or κ) is completely determined by the $\pm$ superscript.

7.3.2 Momentum- versus Energy-Normalized Eigenstates

There is a one-to-one correspondence between the four fundamental scattering states (Section 7.3.1) and the four plane waves associated with incoming and outgoing solutions of H_0^α and incoming and outgoing solutions of H_0^β (Section 7.1.2). Explicitly, the correspondence is

$$\begin{aligned} \psi_\alpha^+(k) &\Longleftrightarrow \psi(+k) & \psi_\alpha^-(k) &\Longleftrightarrow \psi(-k) \\ \psi_\beta^+(\kappa) &\Longleftrightarrow \psi(-\kappa) & \psi_\beta^-(\kappa) &\Longleftrightarrow \psi(+\kappa). \end{aligned} \tag{7.56}$$

This correspondence may be appreciated immediately by inspection of Figure 7.7: the scattering eigenstates with channel label α are associated with the solutions of the TISE for H_0^α, while the scattering eigenstates with channel label β are associated with the solutions of the TISE for H_0^β. Incoming/outgoing $(\pm)$ labels on the scattering eigenstates correspond to left-moving/right-moving free-particle solutions $(\pm k)$ for channel α; and to right-moving/left-moving free-particle solutions $(\mp k)$ for channel β. In accordance with the momentum-normalization conditions for the plane waves, Eqs. 7.11 and its analogs, we scale the scattering eigenstates $\psi_\alpha^\pm(k)$, $\psi_\beta^\pm(\kappa)$ by the factor $1/\sqrt{2\pi}$ to match the multiplicative factor of the momentum-normalized plane wave in the corresponding region. It is a remarkable fact that the scattering eigenstates scaled in this way satisfy normalization equations that are precisely analogous to those of the corresponding free-particle eigenstates (cf. Eq. 7.11 and its analogs for $\psi(-k)$ and $\psi(\kappa)$):

$$\langle \psi_\alpha^\pm(k')|\psi_\alpha^\pm(k)\rangle = \delta(k - k') \qquad \langle \psi_\beta^\pm(\kappa')|\psi_\beta^\pm(\kappa)\rangle = \delta(\kappa - \kappa'); \qquad (7.57)$$

hence they are called the momentum-normalized scattering eigenstates. Furthermore, scattering eigenfunctions that are either incoming or outgoing to different directions obey the same orthogonality relations as the corresponding plane waves:

$$\langle \psi_\beta^+(\kappa')|\psi_\alpha^+(k)\rangle = \langle \psi(-\kappa')|\psi(+k)\rangle = 0;$$
$$\langle \psi_\beta^-(\kappa')|\psi_\alpha^-(k)\rangle = \langle \psi(+\kappa')|\psi(-k)\rangle = 0. \qquad (7.58)$$

Equations 7.57–7.58 are proved in Appendix C by calculating explicitly the overlap integrals between the scattering states. A much simpler proof, using the powerful method of Møller operators, is presented in Chapter 17. One heuristic explanation for these relations is to think of the scattering eigenstates as evolving from the corresponding plane wave while the barrier is turned on adiabatically. The quantum adiabatic theorem indicates that an eigenstate will evolve into an eigenstate under an adiabatically changing Hamiltonian. The norm of the wavefunction should not change during adiabatic evolution, and orthogonal states should remain orthogonal (see, for example, Wu (1962), section Q).

▶ **Exercise 7.6**   Show that the overlap will be zero for any superposition of eigenstates of the form ϕ_β^+ with ϕ_α^+, and ϕ_β^- with ϕ_α^-. This gives an orthogonality condition on entire families of incoming and outgoing wavepackets. This result is what gives the rigorous justification for neglecting the cross-terms in Eq. 7.7.

In Section 7.1.2 we discussed an alternative normalization of the plane waves, known as energy normalization. Applying the same scaling to the energy-normalized scattering eigenstates as we used for the energy-normalized plane waves we obtain

$$\psi_\alpha^\pm(E) = \left(\frac{m}{\hbar^2 k}\right)^{1/2} \psi_\alpha^\pm(k), \qquad \psi_\beta^\pm(E) = \left(\frac{m}{\hbar^2 \kappa}\right)^{1/2} \psi_\beta^\pm(\kappa). \qquad (7.59)$$

$\psi_\alpha^\pm(E)$ and $\psi_\beta^\pm(E)$ satisfy the normalization conditions

$$\langle \psi_\alpha^\pm(E')|\psi_\alpha^\pm(E)\rangle = \langle \psi_{\pm k'}(E')|\psi_{\pm k}(E)\rangle = \delta(E - E') \qquad (7.60)$$

$$\langle \psi_\beta^\pm(E')|\psi_\beta^\pm(E)\rangle = \langle \psi_{\pm \kappa'}(E')|\psi_{\pm \kappa}(E)\rangle = \delta(E - E') \qquad (7.61)$$

and hence are referred to as energy-normalized scattering eigenstates.

7.3.3 Reflection and Transmission Coefficients: The Method of Conservation of Flux

In this section, we will show the relationship between the amplitudes r and t in the scattering eigenstates, and the reflection and transmission coefficients, $R(E)$ and $T(E)$. In particular, we will show that

$$T(E) = \frac{\kappa}{k}|t(\kappa)|^2 \qquad R(E) = |r(k)|^2. \tag{7.62}$$

Moreover, we will derive relationships among the scattering amplitudes r, r', t and t', reflecting general relations between reflection and transmission for a wave incident from the left and a wave incident from the right. Our derivation is based on the principle of conservation of flux and generalized flux quantites in steady state. In Section 7.4.1 we give an alternative derivation of these same relations, showing that they emerge simply from the conditions of normalization and orthogonality of the scattering eigenstates.

Consider a steady-state situation corresponding to an infinite stream of incident particles being reflected and transmitted. The flux must be equal everywhere—otherwise there would be a buildup of amplitude at some location where the flux in was not equal to the flux out, which contradicts the assumption of a steady state. Classically, the flux is given by $\mathcal{J} = v\rho$, where v is the velocity of particles and ρ is their density. In quantum mechanics the flux corresponding to a wavefunction is given by the analogous formula

$$\mathcal{J} = \frac{-i\hbar}{2m}\left(\psi^*(x)\frac{d\psi(x)}{dx} - \psi(x)\frac{d\psi^*(x)}{dx}\right). \tag{7.63}$$

First, we prove that $d\mathcal{J}/dx = 0$; for an eigenstate the "flux" must be the same at all values of coordinate. We write

$$\psi^*(x)\frac{d^2\psi(x)}{dx^2} - \psi(x)\frac{d^2\psi^*(x)}{dx^2} = \psi^*(x)\left(-\frac{2m(E-V)\psi}{\hbar^2}\right)$$
$$- \psi(x)\left(-\frac{2m(E-V)\psi^*}{\hbar^2}\right) = 0, \tag{7.64}$$

where we have used the fact that

$$\frac{d^2\psi}{dx^2} = -\frac{2m}{\hbar^2}[E - V(x)]\psi(x), \tag{7.65}$$

which is simply a rewriting of $H\psi(x) = E\psi(x)$.

▶ **Exercise 7.7** Consider the generalized flux

$$\mathcal{J}_{ba} = \frac{-i\hbar}{2m}\left(\psi_b^*(x)\frac{d\psi_a(x)}{dx} - \psi_a(x)\frac{d\psi_b^*(x)}{dx}\right), \tag{7.66}$$

where ψ_a and ψ_b are two different wavefunctions. Show that $d\mathcal{J}_{ba}/dx = 0$, just as for the ordinary flux $\mathcal{J}$.

We now use the fact that $d\mathcal{J}/dx = 0$ to equate the flux to the left and right of the barrier. For clarity, we write the reflection and transmission amplitudes with an argument k

or κ, depending on whether the corresponding amplitude is located to the left or right of the barrier, respectively. Thus, for a wave incoming from the left we adopt the notation $r(k)$ and $t(\kappa)$, while for a wave incident from the right we adopt the notation $r'(\kappa)$ and $t'(k)$. Consider the flux corresponding to the density $(\psi_\alpha^+)^* \psi_\alpha^+$. To the left: $\psi(x) = e^{ikx} + r(k)e^{-ikx}$, $\psi^*(x) = e^{-ikx} + r^*(k)e^{ikx}$, and hence,

$$\psi^*(x)\frac{d\psi(x)}{dx} - \psi(x)\frac{d\psi^*(x)}{dx} = 2ik[1 - |r(k)|^2]. \tag{7.67}$$

To the right: $\psi(x) = t(\kappa)e^{i\kappa x}$, $\psi^*(x) = t^*(\kappa)e^{-i\kappa x}$, and so

$$\psi^*(x)\frac{d\psi(x)}{dx} - \psi(x)\frac{d\psi^*(x)}{dx} = i\kappa t^*(\kappa)t(\kappa) + i\kappa t(\kappa)t^*(\kappa) = 2i\kappa|t(\kappa)|^2. \tag{7.68}$$

Equating the LHS and RHS we obtain

$$|r(k)|^2 + \frac{\kappa}{k}|t(\kappa)|^2 = R(E) + T(E) = 1, \tag{7.69}$$

where we have defined

$$R(E) = |r(k)|^2 \qquad T(E) = \frac{\kappa}{k}|t(\kappa)|^2. \tag{7.70}$$

The equivalence of this definition of $R(E)$ and $T(E)$ with the definitions given at the beginning of the chapter, Eq. 7.26, will be established in Exercise 7.10. The factor κ/k is somewhat counterintuitive. It will be derived in Section 7.5 from stationary phase arguments.

A variety of other relationships among $r(k)$ and $t(\kappa)$, and $r'(\kappa)$ and $t'(k)$ (incident wave coming from the right) may be proved, by using the coordinate independence of the flux, and the generalized flux. Calculating the flux corresponding to $(\psi_\beta^+)^* \psi_\beta^+$, to the left and right of the barrier, we find that $k|t'(k)|^2 = k'[1 - |r'(\kappa)|^2]$, and hence that

$$|r'(\kappa)|^2 + \frac{k}{\kappa}|t'(k)|^2 = R'(E) + T'(E) = 1, \tag{7.71}$$

where we have defined

$$R'(E) = |r'(k)|^2 \qquad T'(E) = \frac{\kappa}{k}|t'(\kappa)|^2. \tag{7.72}$$

Equation 7.72 is the analog of Eq. 7.70 for a particle incident from the right. Next, we calculate the generalized flux corresponding to $(\psi_\beta^+)^* \psi_\alpha^+$, to the left and right of the barrier. We find that

$$kr(k)t'^*(k) = -\kappa t(\kappa)r'^*(\kappa). \tag{7.73}$$

Finally, calculating the generalized flux corresponding to $(\psi_\beta^-)^* \psi_\alpha^+$, to the left and right of the barrier, we find that

$$kt'(k) = \kappa t(\kappa). \tag{7.74}$$

Equations 7.69–7.74 are summarized in Table 7.3. The additional entries in the table, which take the form of overlap integrals, will be discussed in Section 7.4.

Combining Eq. 7.74 with Eqs. 7.70 and 7.72 we see that

$$T(E) = T'(E). \tag{7.75}$$

Table 7.3 Correspondence between generalized densities, reflection and transmission amplitudes, and overlap integrals

Generalized density	Relation between r,t,r',t'	Corresponding overlap
$(\psi_\alpha^+)^*\psi_\alpha^+$	$\lvert r \rvert^2 + \frac{\kappa}{k}\lvert t \rvert^2 = 1$	$\langle \psi_\alpha^+(k')\vert\psi_\alpha^+(k)\rangle = \delta(k-k')$
$(\psi_\beta^+)^*\psi_\beta^+$	$\lvert r' \rvert^2 + \frac{k}{\kappa}\lvert t' \rvert^2 = 1$	$\langle \psi_\beta^+(\kappa')\vert\psi_\beta^+(\kappa)\rangle = \delta(\kappa-\kappa')$
$(\psi_\beta^+)^*\psi_\alpha^+$	$kt'^*r + \kappa r'^*t = 0$	$\langle \psi_\beta^+(\kappa')\vert\psi_\alpha^+(k)\rangle = 0$
$(\psi_\beta^-)^*\psi_\alpha^+$	$kt' = \kappa t$	$\langle \psi_\beta^-(\kappa')\vert\psi_\alpha^+(k)\rangle = s_{\beta\alpha}\delta(k-k'(\kappa'))$

Equation 7.75, together with Eqs. 7.69 and 7.71, implies that

$$R(E) = R'(E). \tag{7.76}$$

The physical interpretation of Eq. 7.75 is that the probability for a particle incident from the right to be transmitted/reflected is identical with that of a particle incident from the left.

Equations 7.73–7.74 contain additional information not contained in Eqs. 7.75–7.76. In particular, Eq. 7.74 indicates that the *phase* of the transmission amplitude is identical whether the particle is incident from the left or the right. Similarly, Eq. 7.73 relates the *phase* of the reflection amplitude for a particle incident from the left and the right, although the relationship depends on the phase of the transmission amplitude.

7.4   Overlap Integrals of Scattering Eigenstates

In Section 7.3.2 we discussed the normalization and orthogonality relationships among the scattering eigenstates, based on the proof in Appendix C. In this section we show that these normalization and orthogonality relations lead to very general relationships among r, t, r' and t'. These relations are identical to, and in fact in one-to-one correspondence with, the relations derived by conservation of flux in Section 7.3.3. Nevertheless, it is conceptually appealing to show that these relations can be derived from the normalization and orthogonality conditions alone, without any need to invoke conservation of flux. In an analogous way, in Chapter 11 we show that just the conditions of normalization and orthogonality on a reduced subspace can be used as the foundation for deriving the theory of pseudospectral representation.

Our approach will involve calculating the overlap integrals among scattering states explicitly and comparing with the respective normalization or orthogonality condition. Calculating this overlap is a little subtle mathematically; in particular, we must learn how to combine sections of plane waves to obtain δ-functions. Nevertheless, the great number of relationships that come out from a couple of simple rules well justifies the effort. Moreover, by considering the overlap of scattering states of the form $\langle \psi_\beta^- \vert \psi_\alpha^+ \rangle$ within the same framework, we are led naturally to the S-matrix, which is discussed in Section 7.4.2. In a certain sense, all the results of this section are already contained in Appendix C (Eq. C.6): in the process of proving the normalization and orthogonality relations we obtain a very general formula that contains all the results of this section as special cases. Nevertheless,

Appendix C is somewhat abstract and it is worthwhile seeing how the individual relations come out one by one, as we derive them in this section.

7.4.1 Relationships among t, r, t' and r'

Consider the simplest example of a piecewise continuous potential, the potential step. We will focus on energies $E > V_0$, so that there are two degenerate eigenfunctions at each energy. Consider the overlap $\langle \psi_\alpha^+(k') | \psi_\alpha^+(k) \rangle$:

$$\langle \psi_\alpha^+(k') | \psi_\alpha^+(k) \rangle = \frac{1}{2\pi} \left\{ \int_{-\infty}^{0} (e^{ik'x} + r(k')e^{-ik'x})^* (e^{ikx} + r(k)e^{-ikx}) \, dx \right.$$

$$\left. + \int_{0}^{\infty} (t(\kappa')e^{i\kappa'x})^* (t(\kappa)e^{i\kappa x}) \, dx \right\}, \tag{7.77}$$

which according to Eq. 7.57 is equal to $\delta(k - k')$. We may group the terms in these integrals together according to whether they are "rotating" or "counterrotating":

$$\int_{-\infty}^{0} \left\{ e^{i(k-k')x} + r^*(k')r(k)e^{-i(k-k')x} \right\} dx + \int_{0}^{\infty} t^*(\kappa')t(\kappa)e^{i(\kappa-\kappa')x} \, dx \text{ (rotating)}$$

$$+ \int_{-\infty}^{0} \left\{ r^*(k')e^{i(k+k')x} + r(k)e^{-i(k+k')x} \right\} dx \text{ (counterrotating).} \tag{7.78}$$

In Appendix C we prove the remarkable result that the overlap integral between *any* two scattering eigenstates, for a single step or multiple steps, vanishes identically unless $k = k'$. In the case that $k = k'$, only the rotating terms need to be considered (see Appendix C), and we therefore shift our focus to these terms with $k = k'$. We will need the relations $\int_{-\infty}^{0} e^{-ikx} \, dx = \int_{0}^{\infty} e^{ikx} \, dx$ and $\text{Re} \int_{0}^{\infty} e^{i\kappa(k)x} \, dx = \text{Re} \int_{0}^{\infty} \frac{\kappa}{k} e^{ikx} \, dx$, which we will refer to as rules for combining partial delta functions. (These formulas are derived in Appendix A, Exercises A.6 and A.7.) Using these rules and setting $k = k'$ in the coefficients, the rotating terms can be rewritten as

$$\delta(k - k') = \frac{1}{2\pi} \left\{ \int_{-\infty}^{0} e^{i(k-k')x} \, dx + \int_{0}^{\infty} \left\{ |r|^2 e^{i(k-k')x} + \frac{\kappa}{k} |t|^2 e^{i(k-k')x} \right\} dx \right\}, \tag{7.79}$$

and thus

$$|r(k)|^2 + \frac{\kappa}{k} |t(\kappa)|^2 = 1. \tag{7.80}$$

Note that the limits of 0 in the integrals in Eq. 7.79 may be replaced by any finite values without changing the value of the $k = k'$ contribution (again, see Exercise A.7). Taking the first integral $(-\infty, a_L]$ and the second integral $[a_R, \infty)$, where a_L and a_R define the boundary of the scattering region, we see that the result in Eq. 7.80 depends only on the asymptotic coefficients of the scattering eigenstates and hence is valid for any form of the scattering potential in the intermediate region. This same reasoning may be applied to all the results in this section—that is, they depend only on the asymptotic coefficients of the scattering eigenstates—but for simplicity we will continue to write the integration limits as 0 instead of a_L and a_R. Equation 7.80 also comes out from a similar analysis on $\langle \psi_\alpha^- | \psi_\alpha^- \rangle$.

We next consider the overlap of $\langle \psi_\beta^+(\kappa') | \psi_\beta^+(\kappa) \rangle = \delta(\kappa - \kappa')$. Calculating the overlap in the two regions using the rules for manipulating partial δ-functions, we find that the rotating terms can be rewritten as

$$\frac{1}{2\pi} \left\{ \int_{-\infty}^{0} \left\{ |r'|^2 + \frac{k}{\kappa} |t'|^2 \right\} e^{-i(\kappa-\kappa')x} \, dx + \int_{0}^{\infty} \left\{ e^{-i(\kappa-\kappa')x} \right\} dx \right\}, \qquad (7.81)$$

and hence

$$|r'(\kappa)|^2 + \frac{k}{\kappa} |t'(k)|^2 = 1. \qquad (7.82)$$

Equation 7.82 also comes out from considering $\langle \psi_\beta^-(\kappa') | \psi_\beta^-(\kappa) \rangle$.

We now consider the overlap $\langle \psi_\beta^+(\kappa) | \psi_\alpha^+(k) \rangle = 0$. Again, using the rules for combining partial δ-functions we find that the rotating terms can be rewritten as

$$\frac{1}{2\pi} \left\{ \int_{-\infty}^{0} t'^* r e^{i(k-k')x} \, dx + \int_{0}^{\infty} \frac{\kappa}{k} t r'^* e^{i(k-k')x} \, dx \right\} \qquad (7.83)$$

and hence

$$k t'^* r + \kappa r'^* t = 0. \qquad (7.84)$$

The same relationship comes out from considering the other orthogonality relation,

$$\langle \psi_\beta^-(\kappa) | \psi_\alpha^-(k) \rangle = 0.$$

Finally, we consider the overlap $\langle \psi_\beta^-(\kappa) | \psi_\alpha^+(k) \rangle$. Overlaps of this form are neither zero nor simply δ-functions. They define the elements of the S-matrix, which is the subject of the next section. Using the rules for combining partial δ-functions, we find that the rotating terms can be rewritten as

$$\frac{1}{2\pi} \left\{ \int_{-\infty}^{0} t' e^{i(k-k')x} \, dx + \int_{0}^{\infty} \frac{\kappa}{k} t e^{i(k-k')x} \, dx \right\} \qquad (7.85)$$

and hence

$$k t' = \kappa t. \qquad (7.86)$$

To summarize, the normalization and orthogonality relations for the scattering eigenstates lead to relationships among t, r, t' and r' as summarized in Table 7.3, in one-to-one correspondence with the relationships derived by the method of conservation of flux (Section 7.3.3).

7.4.2 The S-Matrix

Consider now the overlap between a scattering eigenstate incoming from channel α and outgoing in channel β, $\langle \psi_\beta^- | \psi_\alpha^+ \rangle$. This overlap is not simply a δ-function, nor is it simply zero. Recall that for $E > V_0$ the TISE supports two eigenstates at every energy. This two-dimensional Hilbert space can be spanned in different ways, for example, using the orthogonal states $\{\psi_\alpha^+, \psi_\beta^+\}$ or using the orthogonal states $\{\psi_\alpha^-, \psi_\beta^-\}$. The first choice of basis partitions the Hilbert space according to the *history* of the particle, whether in the infinite past it was on the left or the right. The second choice partitions the Hilbert space according to the *destiny* of the particle, that is, whether in the infinite future it will be on the left or the right. However, the overlap $\langle \psi_\beta^- | \psi_\alpha^+ \rangle$ corresponds to neither of these partitionings: it represents the probability amplitude that the particle started on the left in the infinite past and will end on the right in the infinite future. This is an interesting, nontrivial quantity, which depends on the properties of the potential and the energy of the incident particle. As seen in Eq. 7.42, it defines an element of the scattering or S-matrix through the relationship

$$\langle \psi_\beta^-(E') | \psi_\alpha^+(E) \rangle \equiv S_{\beta\alpha} \delta(E - E'). \tag{7.87}$$

Similarly, we may consider

$$\langle \psi_\alpha^-(E') | \psi_\alpha^+(E) \rangle \equiv S_{\alpha\alpha} \delta(E - E'). \tag{7.88}$$

Analogous expressions define $S_{\alpha\beta}$ and $S_{\beta\beta}$. The four elements $S_{\alpha\alpha}$, $S_{\beta\alpha}$, $S_{\alpha\beta}$, $S_{\beta\beta}$ constitute the scattering or S-matrix. We have already come across the scattering matrix earlier in this chapter in the context of the time-dependent formulation.

Using the calculus of partial δ-functions introduced in the previous section, we may immediately derive relationships relating the S-matrix and the coefficients r, t, r', t'. Defining lowercase s as the coefficient of the δ-function in the k representation, we find that

$$
\begin{aligned}
\langle \psi_\beta^-(\kappa') | \psi_\alpha^+(k) \rangle &\equiv s_{\beta\alpha} \delta(k - k'(\kappa')) &&\implies t' = t\frac{\kappa}{k} = s_{\beta\alpha} \\
\langle \psi_\alpha^-(k') | \psi_\beta^+(\kappa) \rangle &\equiv s_{\alpha\beta} \delta(\kappa - \kappa'(k')) &&\implies t = t'\frac{k}{\kappa} = s_{\alpha\beta} \\
\langle \psi_\alpha^-(k') | \psi_\alpha^+(k) \rangle &\equiv s_{\alpha\alpha} \delta(k - k') &&\implies r = s_{\alpha\alpha} \\
\langle \psi_\beta^-(\kappa') | \psi_\beta^+(\kappa) \rangle &\equiv s_{\beta\beta} \delta(\kappa - \kappa') &&\implies r' = s_{\beta\beta}
\end{aligned}
\tag{7.89}
$$

(note that the first of the relationships in Eq. 7.89 was derived above as Eq. 7.86). To convert from s to S, two manipulations are necessary. First, each of the states in the overlap integral must be multiplied by its appropriate conversion from the k representation to the E representation. We multiply the scattering states $\psi_\alpha^+(k)$ and $\psi_\alpha^-(k)$ by $(m/\hbar^2 k)^{1/2}$ and the states $\psi_\beta^+(\kappa)$ and $\psi_\beta^-(\kappa)$ by $(m/\hbar^2 \kappa)^{1/2}$ (cf. Eq. 7.59). Second, we need the relations $\delta\left(E(k) - E'(k')\right) = \frac{m}{\hbar^2 k}\delta(k - k')$ and $\delta\left(E(\kappa) - E'(\kappa')\right) = \frac{m}{\hbar^2 \kappa}\delta(\kappa - \kappa')$ (cf. Eq. 7.12). Combining these relations gives

$$S_{\alpha\alpha} = s_{\alpha\alpha} \qquad S_{\beta\beta} = s_{\beta\beta} \qquad S_{\beta\alpha} = \sqrt{\frac{k}{\kappa}} s_{\beta\alpha} \qquad S_{\alpha\beta} = \sqrt{\frac{\kappa}{k}} s_{\alpha\beta}. \tag{7.90}$$

Assembling the elements of the S-matrix we have

$$S = \begin{pmatrix} S_{\alpha\alpha} & S_{\beta\alpha} \\ S_{\alpha\beta} & S_{\beta\beta} \end{pmatrix} = \begin{pmatrix} r & \sqrt{\frac{\kappa}{k}}t \\ \sqrt{\frac{k}{\kappa}}t' & r' \end{pmatrix}. \tag{7.91}$$

It is interesting to rewrite Eqs. 7.80, 7.82, 7.84 and 7.86 in terms of the elements of the S-matrix. Equation 7.86 implies that

$$S_{\alpha\beta} = S_{\beta\alpha}, \tag{7.92}$$

so the S-matrix is symmetric. Equations 7.80 and 7.82 take the form

$$|S_{\alpha\alpha}|^2 + |S_{\beta\alpha}|^2 = 1 = |S_{\alpha\beta}|^2 + |S_{\beta\beta}|^2, \tag{7.93}$$

thus the columns of the S-matrix are normalized vectors (similar relations can be shown for the rows). Finally, Eq. 7.84 takes the form

$$S_{\alpha\alpha}S_{\alpha\beta}^* + S_{\beta\alpha}S_{\beta\beta}^* = 0, \tag{7.94}$$

showing the orthogonality of the columns of the S-matrix. The last two properties taken together show that the S-matrix has orthonormalized columns (the same holds true for the rows), hence the S-matrix is unitary (see Section 8.2.4).

The S-matrix will be discussed in more detail in Chapters 17–18.

7.5   Reconstituting the Wavepacket from the Scattering Eigenstates

Now that we understand the form of the scattering eigenstates, we want to go in the other direction and reconstitute wavepackets as superpositions of these eigenstates. In particular, we ask the following question: how do these unnormalizable scattering eigenstates combine to give the wavepacket picture of approach, collision and ultimately reflection and transmission? In this section we consider simple scattering potentials, and focus on the effect of different asymptotic potentials for the reflected and transmitted wavepacket. In the next section we focus on more complicated potentials that support quasibound states, or "resonances," where interesting time delay effects come into play.

Consider a superposition of eigenstates of different energies, all of which have boundary conditions corresponding to a particle incident from the left, so we have a superposition of $\psi_\alpha^+(k)$ eigenstates:

$$\Psi(x, t) = \int_{-\infty}^{\infty} dk \, a(k)\psi_\alpha^+(k)e^{-iEt/\hbar} = \int_{-\infty}^{\infty} dk \, a(k)\psi_\alpha^+(k)e^{-i\hbar k^2 t/2m}, \tag{7.95}$$

where $a(k)$ is a smooth envelope function peaked at $k = k_0$. Substituting Eq. 7.52 for $\psi_\alpha^+(k)$, we obtain

$$\Psi(x, t) = \int_{-\infty}^{\infty} dk \, a(k)(e^{ikx} + re^{-ikx})e^{-i\hbar k^2 t/2m} \quad \text{region I} \tag{7.96}$$

$$= \int_{-\infty}^{\infty} dk \, a(k)te^{i\kappa x}e^{-i\hbar k^2 t/2m} \quad \text{region III.} \tag{7.97}$$

For simplicity, we revert here to the unnormalized form of $\psi_\alpha^+(k)$, absorbing the normalization factor into $a(k)$. According to the wavepacket picture, as $t \to -\infty$ the wavepacket is entirely to the left of the barrier (only e^{ikx} should contribute); similarly for $t \to +\infty$ only re^{-ikx} should contribute to the left of the barrier and only $te^{i\kappa x}$ should contribute to the right of the barrier. This suggests that if we consider the two terms in the integral in Eq. 7.96 separately, we can identify the first term with the incident wavepacket, $\Psi_I(x, t)$, and the second term with the reflected wavepacket, $\Psi_R(x, t)$:

$$\Psi(x, t) = \int_{-\infty}^{\infty} dk \, a(k)e^{ikx}e^{-i\hbar k^2 t/2m} + \int_{-\infty}^{\infty} dk \, a(k)r(k)e^{-ikx}e^{-i\hbar k^2 t/2m} \quad \text{region I} \tag{7.98}$$

$$= \Psi_I(x, t) + \Psi_R(x, t). \tag{7.99}$$

What is the mathematical justification for this picture? Stationary phase analysis (see Chapter 2) shows that the first term on the RHS is centered at $x_t = \hbar k_0 t / m$, while the second term is centered at $x_t = -\hbar k_0 t / m$. Thus, for $t \to -\infty$, the first term on the RHS is indeed on the left of the barrier (region I), while the second term is to the right of the barrier (region III), where the region I form of the solution does not apply. Thus, for large negative times only the first term contributes to the integral, consistent with the identification of this term with the incident wavepacket. As $t \to +\infty$ the first term is to the right of the barrier, where the region I solution is not physical, while the second term is to the left of the barrier, where the region I form of the solution applies, consistent with the identification of the second term with the reflected wavepacket.

Similarly, we would like to identify the solution to the right of the barrier with the transmitted wavepacket, $\Psi_T(x, t)$:

$$\Psi_T(x, t) = \int_{-\infty}^{\infty} dk \, a(k) t(\kappa) e^{i\kappa(k)x} e^{-i\hbar k^2 t / 2m} \quad \text{region III.} \tag{7.100}$$

Note that we have left the integration variable as k, as in region I, but since the plane waves in region III are characterized by wave vector κ, we need to view κ as a function of k with the functional relationship determined by Eq. 7.51. We now expand the function $\kappa(k)$ in a Taylor series around the value $k = k_0$, which characterizes the center of the wavepacket:

$$\kappa(k) = \kappa(k_0) + (k - k_0) \left. \frac{d\kappa}{dk} \right|_{k=k_0} + \cdots = \kappa_0 + (k - k_0) \frac{k_0}{\kappa_0}. \tag{7.101}$$

With this expansion, the transmitted wavepacket can be rewritten as

$$\Psi_T(x, t) \approx \int_{-\infty}^{\infty} dk \, a(k) t(\kappa) e^{i\kappa_0 x} e^{i[(k-k_0)\frac{k_0}{\kappa_0}x - \hbar k^2 t / 2m]}. \tag{7.102}$$

Stationary phase analysis indicates that this function is centered at $x_t = \hbar \kappa_0 t / m$, where $\frac{\hbar^2 \kappa_0^2}{2m} + V_0 = \frac{\hbar^2 k_0^2}{2m} = E_0$, thus the transmitted wavepacket propagates at the velocity predicted by classical mechanics on the basis of energy conservation.

▶ **Exercise 7.8**   Derive the stationary phase expression for x_t.

Solution   The phase, ϕ, is given by

$$\phi = \kappa_0 x + (k - k_0) \frac{k_0}{\kappa_0} x - \frac{\hbar k^2 t}{2m}. \tag{7.103}$$

The stationary phase condition is

$$\left. \frac{d\phi}{dk} \right|_{k=k_0} = \frac{k_0}{\kappa_0} x_t - \frac{\hbar k_0 t}{m} = 0 \Longrightarrow x_t = \frac{\hbar \kappa_0 t}{m}. \tag{7.104}$$

Note that if higher-order terms in $(k - k_0)$ had been kept in the Taylor series expansion, Eq. 7.101, they would in any event have vanished at the stationary phase point $k = k_0$.

As $t \to -\infty$, the function is centered on the left of the barrier, where the solution is not valid. However, for $t \to +\infty$ the function is centered to the right of the barrier, and thus contributes. Thus, the stationary phase analysis allows us to recover the intuitive wavepacket picture: at negative times the particle is impinging from the left while at positive times there is a reflected and a transmitted portion of the wavepacket.

Comparing the expression for the transmitted wavepacket, Eq. 7.102, with that for the incident wavepacket (first term in Eq. 7.98), we see that

$$|\Psi_T(x, t)| \approx t(\kappa_0) \left| \Psi_I\left(\frac{k_0}{\kappa_0}x, t\right) \right|, \quad t > 0, \tag{7.105}$$

where we have assumed that $t(\kappa)$ is a smooth function of κ and therefore k, and can be replaced in the integral by its value at κ_0. Equation 7.105 is a remarkable relationship, which suggests that the transmitted wavepacket is a close replica of the incident wavepacket, on a contracted length scale of $\frac{\kappa_0}{k_0}$. For example, the width of the transmitted packet relative to that of the incident packet is $\Delta x_T = \frac{\kappa_0}{k_0}\Delta x_I$, and similarly the velocity is $v_T = \frac{\kappa_0}{k_0}v_I$, consistent with the expressions for x_t for the transmitted and incident wavepackets, given above. Also, due to the contraction of the transmitted packet the transmission coefficient is seen to be approximately given by the product of two factors, $T(E_0) = \frac{\kappa_0}{k_0}|t(\kappa_0)|^2$. This gives a new insight into the origin of the factor $\frac{\kappa}{k}$ that we have seen several times already: the amplitude factor of $|t(\kappa_0)|^2$ must be multiplied by the spatial contraction before it can be interpreted as a transmission probability. Note that in a similar vein, one may show that

$$|\Psi_R(x, t)| \approx r(k_0)|\Psi_I(-x, t)|, \quad t > 0, \tag{7.106}$$

thus the reflected wavepacket is approximately a mirror image of the incident wavepacket, travelling in the opposite direction and diminished in intensity by the factor $r(k_0)$.

Equations 7.105–7.106 require some additional comment. Note that at $t > 0$ Eq. 7.99 for $\Psi_I(t)$ is not valid. That expression at $t > 0$ corresponds to the incident wavepacket continuing to propagate for $t > 0$ under the same free-particle potential it experienced at $t < 0$. Although such a wavepacket does not exist physically, Eqs. 7.105–7.106 express the reflected and transmitted wavepackets in terms of the functional form that this "ghost" wavepacket would take if it did continue to propagate unperturbed.

7.6   Resonances and Time Delay

One of the most dramatic effects in scattering theory is that of resonances—sudden changes in phase shifts and/or reflection/transmission coefficients as a function of energy. This sharp dependence of the reflection and transmission amplitude on energy generally corresponds to a matching of the energy of the free incoming or outgoing wave with an almost bound, or "quasibound" state of the scattering potential.

From a time-dependent point of view, portions of the wavepacket are reflected or transmitted immediately, while other portions, with energy contents approximately equal to that of the quasibound state, spend a non-negligible amount of time in the region of the interaction potential before reaching the asymptotic region of reactants or products. The closer the quasibound state is to being truly bound, the longer the time delay and the sharper the resonance in energy, by the time–energy uncertainty principle.

We will explore this phenomenon now by first looking at the scattering eigenstates in the vicinity of a resonance. We will see how there are abrupt changes in the phase of the scattering eigenstates in the vicinity of the energy of a quasibound state.

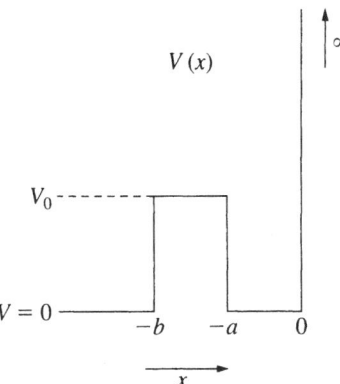

$V(x)$

V_0 ------

$V = 0$

$-b$ $-a$ 0

x

Figure 7.8 Potential energy function supporting scattering resonances. If V_0 is sufficiently high, the potential supports quasibound states, or "resonances" at energies around those of the bound states for an infinite well in the region $[-a, 0]$.

Consider the potential shown in Figure 7.8. There is only one acceptable eigenstate at each energy, since the wavefunction must vanish at $x = 0$. This solution takes the form

$$\psi_k(x) = -\sin(kx) \qquad -a \le x \le 0 \tag{7.107}$$

$$\psi_k(x) = -A \sin[k(x + b) - \chi] \quad x \le -b, \tag{7.108}$$

where A is an amplitude factor and χ is a phase shift, and both are functions of k.

We now proceed to write the expression for $\psi_k(x)$ in the form

$$\psi_k(x) = e^{ikx} + re^{-ikx}.$$

To do this, we first rewrite Eq. 7.108 in terms of complex exponentials:

$$\psi_k(x) = -A \frac{e^{ik(x+b)}e^{-i\chi} - e^{-ik(x+b)}e^{i\chi}}{2i} \tag{7.109}$$

We next divide both sides of this expression by $-\frac{A}{2i}e^{ikb}e^{-i\chi}$ to obtain

$$\psi_k(x) = e^{ikx} - e^{-2ikb}e^{2i\chi}e^{-ikx}. \tag{7.110}$$

Comparing Eq. 7.110 with Eq. 7.52 allows us to identify

$$r = -e^{-2ikb}e^{2i\chi}. \tag{7.111}$$

Note that $R = |r|^2 = 1$, as it must be: there is no transmission through an infinite barrier.

To explore the phenomenon of resonance, consider the following schematic pictures of the scattering eigenstate as a function of energy, as the wave number k passes from below to above the first resonance at $k = \pi/a$. Note that the eigenstate grows sharply in amplitude in the region of the potential well and then decreases again (Figure 7.9a, c). This is accompanied by a sharp change in the phase, $\chi(k)$, over this narrow energy range (Figure 7.9b). This abrupt change in the eigenfunction as a function of energy can be readily understood from the following considerations. First, the wavefunction and its first derivative must be continuous everywhere the potential is finite, and therefore at the points $x = -a$ and $x = -b$. Moreover, consideration of the *curvature* of the wavefunction shows that in the region $-b < x < -a$ the function is made up of an exponentially growing and an exponentially decaying solution. Under very special circumstances, when the value of the function is very small at the point $x = -a$, the exponentially decaying solution satisfies

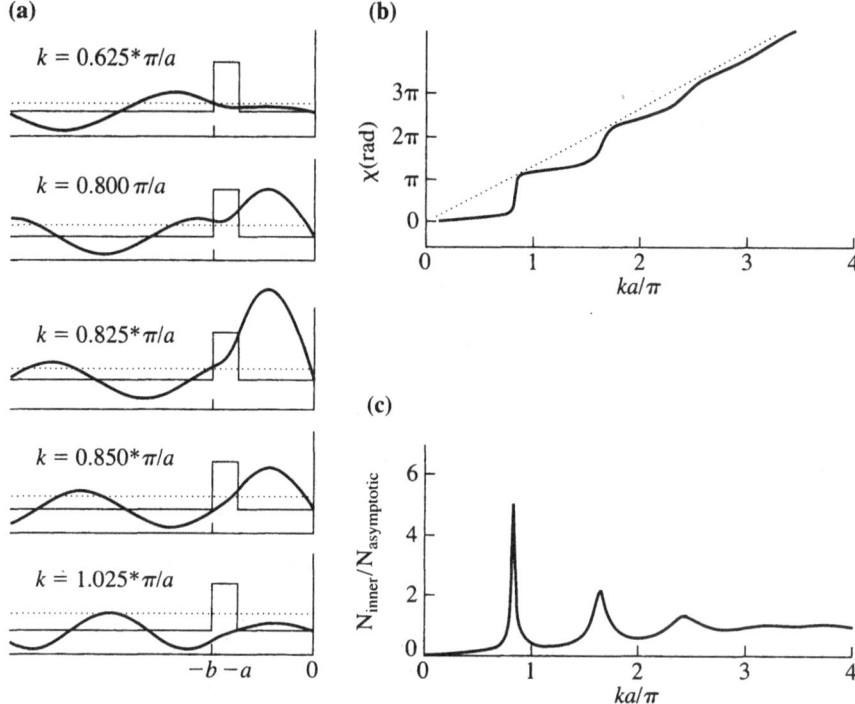

Figure 7.9 (a) The scattering eigenfunction as the energy is varied from below the energy of the first bound state to above the energy of the first bound state of the corresponding potential with infinite walls. At particular energies, the eigenstate grows sharply in amplitude in the region of the potential well and then decreases again, indicating a resonance. (b) Phase of the eigenfunction at position $-b$ as a function of k. Note the sharp changes of phase every time k passes through an integer multiple of $k_{\text{res}} \equiv \pi/b$. (c) Ratio of the square magnitude of the wavefunction in the well region compared with the asymptotic region. Note the sharp peaks at integer multiples of k_{res}, consistent with the behavior of the wavefunction in (a).

the matching condition of the function and its first derivative at $x = -a$, and the amplitude inside the well is much larger than outside the well—this is the signature of the resonance. It occurs at energies close to bound state energies of the corresponding well because the bound states would have to satisfy the condition $\psi \approx 0$ at $x = -a$ at the eigenvalue energies. The rapid change in phase stems from the fact that since the value of the function at $x = -a$ is small at energies near a resonance, and since the value changes continuously, it passes through 0. This causes ψ to change sign in or near the region $-b < x < -a$ and ultimately the function that satisfies the continuity conditions at $x = -b$ has the opposite sign.

Finally, we construct a wavepacket out of a linear superposition of eigenstates:

$$\Psi(x, 0) = \int_{-\infty}^{\infty} dk\, a(k)\psi_k(x), \tag{7.112}$$

where we assume that $a(k)$ is a smooth function, sharply peaked at k_0. Substituting Eq. 7.110 for $\psi_k(x)$, we have

$$\Psi(x, 0) = \int_{-\infty}^{\infty} dk\, a(k)\{e^{ikx} - e^{-2ikb}e^{2i\chi}e^{-ikx}\} \tag{7.113}$$

and hence

$$\Psi(x, t) = \int_{-\infty}^{\infty} dk \, a(k)\{e^{ikx}e^{-i\hbar k^2 t/2m} - e^{-2ikb}e^{2i\chi}e^{-ikx}e^{-i\hbar k^2 t/2m}\}. \quad (7.114)$$

As discussed above, the two terms in the integral correspond to the incident and the reflected wavepackets, respectively, and satisfy two different stationary phase (SP) conditions. For the incident wavepacket the SP condition is $\frac{d\phi_1}{dk}\Big|_{k=k_0} = 0$, where $\phi_1 = kx - \hbar k^2 t/2m$, and thus the center of the wavepacket is given by

$$x_t = \frac{\hbar k_0 t}{m}. \quad (7.115)$$

Identifying $\frac{\hbar k_0}{m} = v_0$ with the classical velocity of a particle moving to the right, we see that the center of the incident wavepacket is moving to the right at the classical velocity. Note that for $t > 0$ this gives $x_t > 0$. As discussed before, there is no physical significance to this result: the scattering eigenstates are defined to take on finite values only for $x < 0$.

For the reflected wavepacket we find that the SP condition is $\frac{d\phi_R}{dk}\Big|_{k=k_0} = 0$, where $\phi_R = -kx - 2kb + 2\chi - \hbar k^2 t/m$, and thus

$$x_t = -2b - \frac{\hbar k_0 t}{m} + 2\frac{d\chi}{dk}. \quad (7.116)$$

For a classical particle with instantaneous reflection off the wall at $x = -b$, the position would be given by $x_t = -2b - v_0 t$. Identifying $v_0 = \frac{\hbar k_0}{m}$ with the velocity of the corresponding classical particle, we may write

$$x_t = -2b - v_0(t - \tau), \quad (7.117)$$

where

$$\tau = \frac{2}{v_0}\frac{d\chi}{dk}\Big|_{k=k_0}. \quad (7.118)$$

If k_0 is far away from a resonance, $\frac{d\chi}{dk}\Big|_{k=k_0} \approx 0$ and hence $\tau \approx 0$. However, if k_0 is close to a resonance, $\frac{d\chi}{dk}\Big|_{k=k_0}$ can be large and hence τ can be very large.

▶ **Exercise 7.9**   Show that if the width of the resonance is Δk, there is associated with the resonance an energy width ΔE, such that $\Delta E \, \tau \approx h$.

Solution   If the width of the first resonance is Δk, then $(\frac{d\chi}{dk})_{k_0}$ is approximately

$$\frac{\Delta\chi}{\Delta k} \approx \frac{\pi}{\Delta k}, \quad (7.119)$$

where k_0 is in the middle of the first resonance. Therefore,

$$\tau \approx \frac{2\pi}{v_0 \, \Delta k} = \frac{2m\pi}{\hbar k_0 \, \Delta k}. \quad (7.120)$$

Now

$$\Delta E = \Delta \left(\frac{\hbar^2 k^2}{2m} \right) \approx \frac{\hbar^2 k_0 \, \Delta k}{m}, \tag{7.121}$$

so

$$\Delta E \, \tau \approx \left(\frac{\hbar^2 k_0 \, \Delta k}{m} \right) \left(\frac{2m\pi}{\hbar k_0 \, \Delta k} \right) = 2\pi \hbar = h. \tag{7.122}$$

Problems

▶ **Exercise 7.10**   Show that the equation for $T(E)$, Eq. 7.26, is consistent with Eq. 7.69, that is,

$$T(E) = \frac{k}{\kappa} \frac{|\tilde{\Psi}_T(\kappa)|^2}{|\tilde{\Psi}(k)|^2} = \frac{\kappa}{k} |t(\kappa)|^2,$$

despite the apparent reversal of k and κ in the prefactor. [Hint: First show that

$$|\tilde{\Psi}_T(\kappa)|^2 = |\langle \psi(\kappa) | \Psi_T \rangle|^2 = |\langle \psi_\beta^-(\kappa) | \Psi_T \rangle|^2 = |\langle \psi_\beta^-(\kappa) | \Psi \rangle|^2. \tag{7.123}$$

Then insert the set of states $\psi_\alpha^+(k')$, which is complete with respect to the incident wavepacket Ψ and use the first of the equations in 7.89. You will need to use the result of Exercise 7.6a and b both in the forward and reverse direction.]

▶ **Exercise 7.11**   Show that the asymptotic form of the four fundamental eigenfunctions can be reexpressed in terms of plane waves with coefficients given by S-matrix elements:

Region I   Region III

$$\psi_\alpha^+(E) \sqrt{\frac{m}{\hbar^2 k}} e^{ikx} + S_{\alpha\alpha} \sqrt{\frac{m}{\hbar^2 k}} e^{-ikx} \qquad S_{\beta\alpha} \sqrt{\frac{m}{\hbar^2 \kappa}} e^{i\kappa x} \tag{7.124}$$

$$\psi_\beta^+(E) S_{\alpha\beta} \sqrt{\frac{m}{\hbar^2 k}} e^{-ikx} \qquad \sqrt{\frac{m}{\hbar^2 \kappa}} e^{-i\kappa x} + S_{\beta\beta} \sqrt{\frac{m}{\hbar^2 \kappa}} e^{i\kappa x} \tag{7.125}$$

The outgoing states are given by complex conjugation. Note the general pattern: in this representation of the scattering eigenstates, the S-matrix elements serve as the coefficients of "energy-normalized" plane waves; however, the normalization factor is not uniform, but is determined in each region by the local k-vector.

▶ **Exercise 7.12**   Derive the following expressions for the reflection and transmission coefficients, which are slight variations on Eqs. 7.45–7.49. These expressions exploit the fact that $R(E)$ and $T(E)$ involve only the absolute value squared of the S-matrix elements

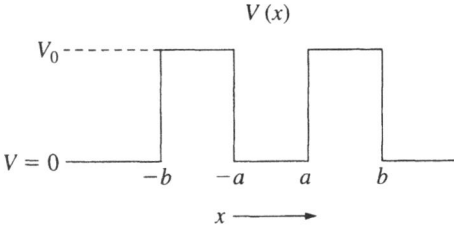

Figure 7.10 Symmetric potential with two barriers separating reactants (left) from products (right).

and hence only the magnitude of the normalization constants η and μ, and therefore can be determined from the power spectrum of the reactant (product) wavepacket.

$$R(E) = |S_{\alpha\alpha}(E)|^2 = \frac{\int_{-\infty}^{\infty} \langle \phi_\alpha^- | e^{-iHt/\hbar} | \phi_\alpha^+ \rangle e^{iEt/\hbar}\, dt}{\int_{-\infty}^{\infty} \langle \phi_\alpha^+ | e^{-iHt/\hbar} | \phi_\alpha^+ \rangle e^{iEt/\hbar}\, dt} \tag{7.126}$$

and

$$T(E) = |S_{\beta\alpha}(E)|^2 =$$
$$\frac{\int_{-\infty}^{\infty} \langle \phi_\beta^- | e^{-iHt/\hbar} | \phi_\alpha^+ \rangle e^{iEt/\hbar}\, dt}{|\int_{-\infty}^{\infty} \langle \phi_\alpha^+ | e^{-iHt/\hbar} | \phi_\alpha^+ \rangle e^{iEt/\hbar}\, dt|^{1/2}\, |\int_{-\infty}^{\infty} \langle \phi_\beta^+ | e^{-iHt/\hbar} | \phi_\beta^+ \rangle e^{iEt/\hbar}\, dt|^{1/2}} . \tag{7.127}$$

▶ **Exercise 7.13**   The formulation of the S-matrix in terms of the Fourier transform of a wavepacket cross-correlation function (Section 7.2.2) is a useful approach for seeing the connection between time delay and resonances. The time delay for a state-to-state transition may be defined as follows (Smith, 1960):

$$\tau_{\beta\alpha}(E) = \mathrm{Re}\left(-i\hbar (S_{\beta\alpha})^{-1} \frac{d}{dE} S_{\beta\alpha} \right) . \tag{7.128}$$

a.   Use the cross-correlation function formulation of the S-matrix to show that, with a judicious choice for ϕ_α^+ and ϕ_β^-, $\tau(E)$ can be written as

$$\tau_{\beta\alpha}(E) = \mathrm{Re}\left[\frac{\int_{-\infty}^{\infty} \langle \phi_\beta^- | e^{-iHt/\hbar} | \phi_\alpha^+ \rangle e^{iEt/\hbar}\, t\, dt}{\int_{-\infty}^{\infty} \langle \phi_\beta^- | e^{-iHt/\hbar} | \phi_\alpha^+ \rangle e^{iEt/\hbar}\, dt} \right] . \tag{7.129}$$

b.   Discuss why $\tau_{\beta\alpha}$ gives an average measure of time delay, and analyze the behavior of $\tau_{\beta\alpha}$ as a function of energy in the presence of resonances.

c.   Show the relationship between Eq. 7.128 and Eq. 7.118.

▶ **Exercise 7.14**   Go through an analysis of the phase shifts, resonances and time delays for the potential shown in Figure 7.10.

As opposed to the potential in Figure 7.8, this potential admits transmission. Show that the phase shifts described in the previous example lead to dramatic changes in the transmission coefficient as a function of energy, including energies below V_0 for which there is complete transmission. [Hint: Because the potential is even it is possible to construct eigenstates with even or odd symmetry around $x = 0$:

$$\psi_1(x, k) = A_1 \sin[k(x - b) + \chi_1] \qquad x > b$$
$$\psi_2(x, k) = A_2 \sin[k(x - b) + \chi_2] \qquad x > b$$

$$\psi_1(x, k) = A'_1 \cos(kx) \quad -b < x < b$$

$$\psi_2(x, k) = A'_2 \sin(kx) \quad -b < x < b$$

$$\psi_1(x, k) = A_1 \sin[k(-x - b) + \chi_1] \quad x < -b$$

$$\psi_2(x, k) = -A_2 \sin[k(-x - b) + \chi_2] \quad x < -b.$$

Reexpress these functions in terms of complex exponentials, and form a linear combination to get rid of the e^{-ikx} terms.]

References

General

1. A. Messiah, *Quantum Mechanics*, vol. I (Wiley, New York, 1958).
2. C. Cohen-Tannoudji, B. Diu and F. Laloë, *Quantum Mechanics*, vol. I (Wiley, New York, 1977).
3. G. C. Schatz and M. A. Ratner, *Quantum Mechanics in Chemistry* (Prentice Hall, Englewood Cliffs, NJ, 1993), Chapter 7.
4. E. Merzbacher, *Quantum Mechanics* (Wiley, New York, 1970).
5. L. D. Landau and E. M. Lifshitz, *Quantum Mechanics, Nonrelativistic Theory* (Pergamon Press, Oxford (1965).
6. T.-Y. Wu and T. Ohmura, *Quantum Theory of Scattering* (Prentice Hall, Englewood Cliffs, NJ, 1962).
7. L. I. Schiff, *Quantum Mechanics* (McGraw-Hill, New York, 1968).

Time Delay in Scattering Theory

8. L. Eisenbud, *The Formal Properties of Nuclear Collisions*, dissertation, Princeton, June 1948.
9. D. Bohm, *Quantum Theory* (Prentice Hall, New York, 1951).
10. E. P. Wigner, Lower limit for the energy derivative of the scattering phase shift, Phys. Rev. 98, 145 (1955).
11. F. T. Smith, Lifetime matrix in collision theory, Phys. Rev. 118, 349 (1960).
12. D. Wardlaw, P. Brumer and T. A. Osborn, Channel specific spectral properties of time delays for bimolecular collisions, J. Chem. Phys. 76, 4916 (1982).
13. E. Pollak and W. H. Miller, New physical interpretation for time in scattering theory, Phys. Rev. Lett. 53, 115 (1984).

Wavepacket Correlation Function Formulation of Scattering Theory

14. D. J. Tannor and D. E. Weeks, Wavepacket correlation function formulation of scattering theory: The quantum analog of classical S-matrix theory, J. Chem. Phys. 98, 3884 (1993).
15. S. Garashchuk and D. J. Tannor, Correlation function formulation for the state-selected total reaction probability, J. Chem. Phys. 109, 3028 (1998).
16. D. J. Tannor and S. Garashchuk, Semiclassical calculation of chemical reaction dynamics via wavepacket correlation functions, Annu. Rev. Phys. Chem. 51, 553 (2000).

Part II

Formal Theory and Methods
of Approximation

Chapter 8

Linear Algebra and Quantum Mechanics

This chapter contains a brief review of the mathematical structure of quantum mechanics. The starting point is the requirement that all physically allowed wavefunctions in quantum mechanics must be square normalizable. It is shown that the set of all square normalizable wavefunctions has the mathematical properties of a linear vector space: add any two members with any finite coefficients and the sum is also a member of the space. This mathematical structure allows us to exploit fully an analogy between quantum wavefunctions and ordinary (real) vectors in three-dimensional Euclidean space. All the concepts of linear independence, subspaces, dot (or inner) product, orthogonality, normalization and completeness, which are intuitively clear in Euclidean space, have their analogs with wavefunctions. For wavefunctions, the dimensionality of the linear vector space (the "Hilbert space") is infinite, with each point in coordinate space (or any alternative basis) representing one of the orthogonal directions. Thus, the inner product in quantum mechanics is a sum (integral) over the product of two wavefunctions at all points in coordinate space.

The concept of operators is then introduced. Operators map one wavefunction onto another wavefunction in the Hilbert space, in the same way that a matrix maps one vector into another vector in a finite linear vector space. The three main kinds of operators, Hermitian, unitary and projection operators, are described. Next, the concept of different bases for both the wavefunctions and operators is developed. Starting with discrete bases, we show how a basis set leads to the representation of a differential operator as a matrix. Similarly, the representation of a wavefunction in terms of its components in a discrete basis is easily understood by appeal to the analogous idea in Euclidean space. Finally, we turn to continuous representations. We focus on the wavefunction both in the position and momentum representations, and compare the structure with the representation in a discrete basis.

8.1   Linear Vector Spaces

The square integral of a wavefunction in quantum mechanics is the probability density of finding the particle somewhere in space. Thus, the only physically acceptable wavefunctions are those that are square normalizable (for shorthand called L^2):

$$\int_{-\infty}^{\infty} |\psi(x)|^2 \, dx < +\infty.$$

The set of all square normalizable wavefunctions has the properties of a linear vector space (LVS). This simple-sounding statement is actually very profound and has many important consequences for quantum mechanics. We begin this section, therefore, with a review of the properties of a linear vector space.

A linear vector space (LVS) is a set of vectors $u, v, \ldots, \in S$ such that if $u, v \in S$ so is any linear combination $\alpha u + \beta v$.

In a real LVS α, β must be real numbers; in a complex LVS α, β may be complex. Here are some examples:

1. the set of real numbers
2. the set of complex numbers
3. the set of vectors in three-dimensional Euclidean space
4. the set of L^2 quantum states.

To prove that the set of all L^2 wavefunctions comprises an LVS, we have to show that if ψ_1 and ψ_2 are square integrable so is $\alpha \psi_1 + \beta \psi_2$. The proof is straightforward. Consider:

$$|\alpha \psi_1 + \beta \psi_2|^2 = \left(\alpha^* \psi_1^* + \beta^* \psi_2^*\right)\left(\alpha \psi_1 + \beta \psi_2\right) \tag{8.1}$$

$$= |\alpha|^2 |\psi_1|^2 + |\beta|^2 |\psi_2|^2 + \alpha^* \beta \psi_1^* \psi_2 + \alpha \beta^* \psi_1 \psi_2^* \tag{8.2}$$

$$\leq |\alpha|^2 |\psi_1|^2 + |\beta|^2 |\psi_2|^2 + 2|\alpha||\beta||\psi_1||\psi_2| \tag{8.3}$$

$$\leq |\alpha|^2 |\psi_1|^2 + |\beta|^2 |\psi_2|^2 + |\alpha||\beta| \left(|\psi_1|^2 + |\psi_2|^2\right). \tag{8.4}$$

The last inequality holds since

$$\left(|\psi_1| - |\psi_2|\right)^2 \geq 0,$$

which implies that

$$|\psi_1|^2 + |\psi_2|^2 \geq 2|\psi_1||\psi_2|.$$

The dimension of the "square normalizable wavefunction space" is infinite. It is called a Hilbert space.

A linear manifold is a subset of an LVS that is itself an LVS.

A linear manifold is a special kind of subspace. Here are some examples:

1. 0 (the empty set)
2. S (the original LVS)
3. the set of vectors αu, where α is any complex number.

Table 8.1 Comparison of properties of inner (dot) products for real vectors and wavefunctions

Properties	Real vectors	Wavefunctions
1. $(u, v) = (v, u)^*$	$u \cdot v = v \cdot u$	$\int \phi^*(x)\psi(x)\, dx = \left\{ \int \psi^*(x)\phi(x)\, dx \right\}^*$
2. $(u, u) \geq 0$	$\|u\| = (u, u)^{1/2}$	$\int \psi^*(x)\psi(x)\, dx = \int |\psi(x)|^2\, dx \geq 0$
3. $\left(u, av_1 + bv_2\right)$	$u \cdot \left(av_1 + bv_2\right) = au \cdot v_1 + bu \cdot v_2$	$\int \phi^*(x)\left\{\alpha\psi_1(x) + \beta\psi_2(x)\right\} dx$
$\quad = a(u, v_1) + b(u, v_2)$		$\quad = \alpha \int \phi^*(x)\psi_1(x)\, dx + \beta \int \phi^*(x)\psi_2(x)\, dx$

The linear manifold essentially divides the complete LVS into two parts, the part in the manifold and the part outside. Any vector in the full LVS can be written as $v = v_M + v_\perp$, where v_M is that component in the linear manifold and $v_\perp$ is the part outside.

An important feature of a linear vector space is that one can define an inner product or dot product on the space, which is a measure of both the length and linear dependence of two vectors. The inner product of two vectors, u and v, is written (u, v), and is defined by the following three properties:

1. $(u, v) = (v, u)^*$
2. $(u, u) \geq 0$
3. $\left(u, av_1 + bv_2\right) = a(u, v_1) + b(u, v_2)$

Table 8.1 compares the properties of the inner product as they appear for real vectors and for wavefunctions.

▶ **Exercise 8.1**   Show that $(au_1 + bu_2, v) = a^*(u_1, v) + b^*(u_2, v)$.

▶ **Exercise 8.2**   Show that for wavefunctions the above equality takes the form

$$\left(a\phi_1 + b\phi_2, \psi\right) = \left(\psi, a\phi_1 + b\phi_2\right)^*$$
$$= \left[a\left(\psi, \phi_1\right) + b\left(\psi, \phi_2\right)\right]^* = a^*\left(\phi_1, \psi\right) + b^*\left(\phi_2, \psi\right).$$

Note that the inner product is a measure of the overlap of two *vectors* in Euclidean space, and of the overlap of two *wavefunctions* in the Hilbert space.

The normalization, or "norm," of a vector is a measure of its length, and is defined as $\|u\| = (u, u)^{1/2}$. For wavefunctions this definition takes the form

$$\|\psi\| = \left\{ \int_{-\infty}^{\infty} |\psi(x)|^2\, dx \right\}^{1/2}.$$

A vector, v, is called "normalized" if $\|v\| = 1$. Two vectors u and v are called "orthogonal" if $(u, v) = 0$. For wavefunctions this takes the form $\int_{-\infty}^{\infty} \phi^*(x)\psi(x)\, dx = 0$. This is a generalization of the idea of perpendicular vectors in three-dimensional space. A set of vectors is called "orthonormal" if each vector is normalized, and the vectors are pairwise orthogonal.

From the definition of the inner product several inequalities, of essentially geometrical nature, follow:

1. Schwarz inequality: $|(u, v)| \leq ||u|| \, ||v||$
2. Triangle inequality: $||u + v|| \leq ||u|| + ||v||$.

These properties take the following form when applied to wavefunctions:

1. $|\int_{-\infty}^{\infty} \phi^*(x)\psi(x) \, dx| \leq \left\{ \int_{-\infty}^{\infty} |\phi(x)|^2 \, dx \right\}^{1/2} \left\{ \int_{-\infty}^{\infty} |\psi(x)|^2 \, dx \right\}^{1/2}$
2. $\left\{ \int_{-\infty}^{\infty} |\phi(x) + \psi(x)|^2 \, dx \right\}^{1/2} \leq \left\{ \int_{-\infty}^{\infty} |\phi(x)|^2 \, dx \right\}^{1/2} + \left\{ \int_{-\infty}^{\infty} |\psi(x)|^2 \, dx \right\}^{1/2}$

▶ **Exercise 8.3** Explain the geometrical meaning of the Schwarz and triangle inequalities in three-dimensional space, and describe what theorems in geometry and trigonometry they express.

▶ **Exercise 8.4** Derive the Schwarz inequality.

Solution Consider the quantity $(u - \alpha v, u - \alpha v)$. This quantity must be greater than 0 for any α, since it is the square of the norm of a vector. The key to the derivation is to consider the particular choice, $\alpha = \frac{(v,u)}{(v,v)}$. Expanding the inner product, and substituting in the definition of α, we obtain

$$(u, u) - \alpha(u, v) - \alpha^*(v, u) + (\alpha v, \alpha v) = ||u||^2 - \frac{(v, u)(u, v)}{||v||^2}$$
$$- \frac{|(u, v)|^2}{||v||^2} + \frac{|(u, v)|^2}{||v||^4}||v||^2 \geq 0. \tag{8.5}$$

The second and fourth terms cancel; rearranging the remaining terms gives

$$|(u, v)|^2 \leq ||u||^2 ||v||^2 \tag{8.6}$$

or

$$|(u, v)| \leq ||u|| \, ||v||. \tag{8.7}$$

▶ **Exercise 8.5** Show that the Heisenberg uncertainty relation follows from the Schwarz inequality.

Solution Consider the quantity

$$\int_{-\infty}^{\infty} x \frac{\partial}{\partial x} [\psi^*(x)\psi(x)] \, dx. \tag{8.8}$$

Integrating by parts we obtain

$$x|\psi(x)|^2 |_{-\infty}^{+\infty} - \int_{-\infty}^{\infty} |\psi(x)|^2 \, dx. \tag{8.9}$$

For ψ normalized, the second term equals -1; the first term vanishes since, if ψ is square integrable, $|\psi(x)|^2$ must go to zero faster than $1/x$ as $x \to \pm\infty$. Therefore,

$$1 = \left| \int_{-\infty}^{\infty} x \frac{\partial}{\partial x} [\psi^* \psi] \, dx \right| = \left| \int_{-\infty}^{\infty} x \frac{\partial \psi^*}{\partial x} \psi \, dx + \int_{-\infty}^{\infty} \psi^* x \frac{\partial \psi}{\partial x} \, dx \right| \qquad (8.10)$$

$$\leq \left| \int_{-\infty}^{\infty} \frac{\partial \psi^*}{\partial x} x \psi \, dx \right| + \left| \int_{-\infty}^{\infty} (x\psi)^* \frac{\partial \psi}{\partial x} \, dx \right|$$

$$= \left| \left(\frac{\partial \psi}{\partial x}, x\psi \right) \right| + \left| \left(x\psi, \frac{\partial \psi}{\partial x} \right) \right| \qquad (8.11)$$

$$\leq 2 \left\| \frac{\partial \psi}{\partial x} \right\| \, \|x\psi\|. \qquad (8.12)$$

Writing out explicitly the expression for the norm we obtain

$$1 \leq 2 \left[\int_{-\infty}^{\infty} \psi^* \frac{\hat{p}^2}{\hbar^2} \psi \, dx \right]^{1/2} \left[\int_{-\infty}^{\infty} \psi^* \hat{x}^2 \psi \, dx \right]^{1/2}, \qquad (8.13)$$

which implies that $\Delta p \, \Delta x \geq \frac{\hbar}{2}$.

▶ **Exercise 8.6**   Derive the triangle inequality.

Solution

$$0 \leq (u + v, u + v) = \|u + v\|^2 \qquad (8.14)$$

$$= (u, u) + (v, u) + (u, v) + (v, v) \qquad (8.15)$$

$$\leq \|u\|^2 + \|v\|^2 + 2|(u, v)| \qquad (8.16)$$

$$\leq \|u\|^2 + \|v\|^2 + 2\|u\|\|v\| \qquad (8.17)$$

$$= (\|u\| + \|v\|)^2, \qquad (8.18)$$

where we have used the Schwarz inequality. Taking the square root of both sides gives

$$\|u\| + \|v\| \geq \|u + v\|, \qquad (8.19)$$

the desired result.

8.2   Operators: Mapping a Wavefunction to Another Wavefunction

Operators take a vector to another vector in the linear vector space: $Av = v'$, $\|v'\| \leq \infty$. In general, v' does not point in the same direction as (i.e., is not proportional to) v. However, there are special cases where Av is proportional to v, in which case v is called an eigenvector of A. Eigenvectors will be discussed extensively below.

Linear algebra is the study of linear operators. A linear operator is defined by the relation

$$A\left(av_1 + bv_2\right) = aAv_1 + bAv_2.$$

▶ **Exercise 8.7**   Consider the operator A such that $Av = (v, v)v$. Check whether A is a linear operator.

Products of operators are defined as operating from right to left: $BA\psi(x) = B(A\psi(x))$. Powers of operators are simply repeated action of the fundamental operator: $A^n = AA \ldots A$. More generally, functions of operators are defined in terms of their Taylor series expansions, and hence in terms of sums of powers of the fundamental operator.

▶ **Exercise 8.8**   Prove that if A and B are linear operators then BA is also a linear operator.

▶ **Exercise 8.9**   Prove that if A is a linear operator so is e^A.

In quantum mechanics an operator maps an L^2 wavefunction onto another L^2 wavefunction. Here are some examples of commonly used operators in quantum mechanics:

1. multiplication: $A\psi(x) = x\psi(x)$ where x is a number
2. differentiation: $A\psi(x) = \frac{d}{dx}\psi(x)$
3. translation: $A\psi(x) = e^{a\frac{d}{dx}}\psi(x) = \psi(x + a)$
4. scaling or dilation: $A\psi(x) = e^{ax\frac{d}{dx}}\psi(x) = \psi(e^a x)$
5. evolution: $A\psi(x) = e^{-iHt/\hbar}\psi(x) = \psi(x, t)$
6. projection:

   a. $A\psi(x) = h(x)\psi(x)$ $\begin{cases} h(x) = 0 & x < 0 \\ h(x) = 1 & x \geq 0 \end{cases}$ (8.20)

   b. $A\psi(x) = \frac{1}{\sqrt{2}}\left[\psi(x) + \psi(-x)\right]$

▶ **Exercise 8.10**   Show that all the operators above take any L^2 function into another L^2 function.

▶ **Exercise 8.11**   Show that each of the above operators is linear.

▶ **Exercise 8.12**   Draw pictures of the action of each of the operators on a wavefunction.

The "translation," or "shift," operator $e^{a\frac{d}{dx}}$ is so called because it translates the wavefunction by an amount $-a$ without changing its shape: $e^{a\frac{d}{dx}}\psi(x) = \psi(x + a)$. The proof of this relationship provides a nice illustration of the use of Taylor series expansions of operators:

$$A\psi(x) = e^{a\frac{d}{dx}}\psi(x) = \left(1 + a\frac{d}{dx} + \frac{1}{2!}a^2\frac{d^2}{dx^2} + \frac{1}{3!}a^3\frac{d^2}{dx^3} + \cdots\right)\psi(x).$$

But the RHS is also the Taylor series expansion of $\psi(x + a)$ around $x = 0$; hence the two expressions are equal.

The "scaling" or "dilation" operator squeezes the wavefunction. The proof provides a nice illustration of another method for manipulating operators that appear in the exponent. Define

$$f(a, x) = e^{ax\frac{d}{dx}}\psi(x).$$

Differentiating both sides of this equation with respect to a we obtain a differential equation for f:

$$\frac{\partial f}{\partial a} = x\frac{\partial f}{\partial x} \qquad f(0, x) = \psi(x).$$

The reader may easily check that the solution is

$$f = \psi(e^a x).$$

A variant of the scaling operator that preserves the norm is

$$A\psi(x) = e^{a(x\frac{d}{dx} + \frac{d}{dx}x)}\psi(x) = e^a\psi(e^{2a}x).$$

We leave the proof to the reader.

The "evolution" operator $e^{-iHt/\hbar}$ is so named because it generates the evolution of $\psi(x, t = 0)$ into $\psi(x, t)$. To prove this, consider the Time-Dependent Schrödinger Equation: $i\hbar\frac{\partial \psi}{\partial t} = H\psi$. Separating variables and integrating both sides we obtain

$$\int_{\psi(0)}^{\psi(t)}\frac{d\psi}{\psi} = \int_0^t\frac{1}{i\hbar}H\,dt' \Longrightarrow \ln\frac{\psi(t)}{\psi(0)} = \frac{-i}{\hbar}Ht.$$

This implies that

$$\psi(x, t) = e^{-iHt/\hbar}\psi(x, 0),$$

and hence $e^{-iHt/\hbar}$ maps $\psi(x, t = 0)$ into $\psi(x, t)$. Because $||\psi(t)|| = ||\psi(0)||$ (conservation of probability), the evolution operator is also a norm-preserving operator.

▶ **Exercise 8.13**   Show that the translation and evolution operators preserve the norm of $\psi(x)$. Show that the second variant of the dilation operator preserves the norm but the first form does not.

Projection operators are sufficiently important that we devote a separate section to them below. As we shall see there, the defining feature of a projection operator, P, is the property $P^2 = P$.

▶ **Exercise 8.14**   Show that the two projection operators cited above satisfy the relationship $A^2\psi = A\psi$.

Much of linear algebra deals with the solutions to the equation $Av = \lambda v$. This is known as an eigenvalue equation. The vectors v are known as eigenvectors, and the quantity λ is known as an eigenvalue. The word "eigen" is a German word meaning "self": when A operates on an arbitrary vector v, mapping it onto a new vector v', in the vast majority of cases the new vector is not proportional to the old vector. Only in special cases is the new vector proportional to the old vector, and the search for these special vectors, and the proportionality constant, is one of the main goals of linear algebra.

▶ **Exercise 8.15** Find eigenfunctions and eigenvalues for each of the operators enumerated preceding Exercise 8.10.

An operator A has a spectrum of eigenvalues, which may be finite or infinite, discrete or continuous. If there are two or more eigenvectors with the same eigenvalue the spectrum is called "degenerate": specification of a degenerate eigenvalue is not sufficient to determine the eigenfunction. The portion of the LVS associated with a particular degenerate eigenvalue is a linear manifold, sometimes called the degenerate manifold.

▶ **Exercise 8.16** Show that any linear combination of eigenvectors in the degenerate manifold is also an eigenvector.

▶ **Exercise 8.17** Show that if v is an eigenfunction of A, v is also an eigenfunction of $f(A)$, where $f(A)$ is any linear function of A; if $Av = \lambda v$, $f(A)v = f(\lambda)v$. Use this result to show that if $H\psi_n = E_n\psi_n$ then $e^{-iHt/\hbar}\psi_n = e^{-iE_nt/\hbar}\psi_n$.

8.2.1 Adjoint of an Operator

Given an operator A, its adjoint, $A^\dagger$, is defined as follows:

$$(u, Av) = \left(A^\dagger u, v\right).$$

In terms of wavefunctions this relation is

$$\int_{-\infty}^{\infty} \phi^*(x)(A\psi(x))\, dx = \int_{-\infty}^{\infty} (A^\dagger\phi(x))^*\psi(x)\, dx.$$

An adjoint is sometimes called a Hermitian conjugate. To find $A^\dagger$ for a given operator A one begins with the expression for the inner product, and then must ingeniously rearrange it to get rid of any operator operating on ψ, moving all operators onto ϕ. Here are some examples.

1. Consider the operator $A\psi(x) = x\psi(x)$. This is an easy example, but we will write out each step carefully as a template for the procedure in more complicated cases.

$$\int_{-\infty}^{\infty} \phi^*(x)(A\psi(x))\, dx = \int_{-\infty}^{\infty} \phi^*(x)(x\psi(x))\, dx$$

$$= \int \left(x\phi^*(x)\right)\psi(x)\, dx \qquad (8.21)$$

(replace the operator on ψ by an operator
that operates on ϕ^*)

$$= \int \left(x^*\phi(x)\right)^*\psi(x)\, dx \qquad (8.22)$$

(remove the * outside of the parentheses
containing the operator)

$$= \int (x\phi(x))^*\psi(x)\, dx \qquad (8.23)$$

(simplify the operator; x^* can be replaced
by x, since x is real)

$$= \int \left(A^\dagger\phi(x)\right)^*\psi(x)\, dx \qquad (8.24)$$

(identify what appears in parentheses
as $A^\dagger\phi$ by definition).

Hence, $A^\dagger\phi(x) = x\phi(x)$.

2. We now follow the same steps for the more complicated case, $A\psi(x) = i\frac{\partial}{\partial x}\psi(x)$:

$$\int_{-\infty}^{\infty} \phi^*(x)(A\psi(x))\, dx = \int_{-\infty}^{\infty} \phi^*(x)\left(i\frac{\partial}{\partial x}\psi(x)\right)\, dx \qquad (8.25)$$

$$= \phi^*(x)(i\psi(x))|_{-\infty}^{\infty} - \int_{-\infty}^{\infty} \left(i\frac{\partial}{\partial x}\phi^*(x)\right)\psi(x)\, dx \qquad (8.26)$$

$$= -\int_{-\infty}^{\infty} \left((i)^*\frac{\partial}{\partial x}\phi(x)\right)^*\psi(x)\, dx \qquad (8.27)$$

$$= \int_{-\infty}^{\infty} \left(i\frac{\partial}{\partial x}\phi(x)\right)^*\psi(x)\, dx \qquad (8.28)$$

$$= \int_{-\infty}^{\infty} \left(A^\dagger\phi(x)\right)^*\psi(x)\, dx. \qquad (8.29)$$

In Eq. 8.26 we integrated by parts; this is the crucial trick to get the operator off ψ and onto ϕ. Notice that the boundary term vanished: it must be 0 at both limits if ψ and ϕ are both square integrable. Hence, $A^\dagger\phi(x) = i\frac{\partial}{\partial x}\phi(x)$.

Notice that in both examples $A = A^\dagger$, and hence both operators, x and $i\frac{\partial}{\partial x}$, belong to the class of Hermitian, or self-adjoint, operators.

▶ **Exercise 8.18** Find the adjoint of each of the operators enumerated preceding Exercise 8.10.

▶ **Exercise 8.19** If the adjoint of A is $A^\dagger$ and the adjoint of B is $B^\dagger$, find an expression for the adjoint of AB.

8.2.2 Hermitian Operators

An important class of operators in quantum mechanics is the Hermitian operators. These operators are their own adjoint: $A^\dagger = A$. This leads to the property that inner products of the form (u, Au) are real. Since measurements in quantum mechanics correspond to inner products of this form, and since all physical measurements are real, all observables in quantum mechanics are associated with Hermitian operators. Here we describe the main properties of Hermitian operators.

1. The eigenvalues of a Hermitian matrix are real. Proof:
   The basic eigenvalue equation can be written $Av = \lambda v$. Now,

   $$(v, Av) = (v, \lambda v) = \lambda(v, v).$$

   But,

   $$(v, Av) = \left(A^\dagger v, v\right) = (Av, v) = (\lambda v, v) = \lambda^*(v, v)$$

   $$\implies \lambda = \lambda^*.$$

2. Given two distinct eigenvalues of a Hermitian operator, the corresponding eigenfunctions must be orthogonal. Proof:

   $$\text{If } Av_1 = \lambda_1 v_1, \qquad Av_2 = \lambda_2 v_2, \qquad \text{with } \lambda_1 \neq \lambda_2,$$

   then

   $$(v_2, Av_1) = (v_2, \lambda_1 v_1) = \lambda_1 (v_2, v_1).$$

   But

   $$(v_2, Av_1) = \left(A^\dagger v_2, v_1\right) = (Av_2, v_1) = \lambda_2^* (v_2, v_1)$$

   $$= \lambda_2 (v_2, v_1) \text{ since } \lambda_2 \text{ must be real.}$$

   This implies that

   $$\lambda_1 (v_2, v_1) = \lambda_2 (v_2, v_1)$$

   or

   $$(\lambda_1 - \lambda_2)(v_2, v_1) = 0.$$

   Since $(\lambda_1 - \lambda_2) \neq 0$ by assumption, (v_2, v_1) must equal 0.

▶ **Exercise 8.20** Show that the Hamiltonian operator, $H = \frac{-\hbar^2}{2m}\frac{\partial^2}{\partial x^2} + V(x)$, is Hermitian.

▶ **Exercise 8.21** Which of the operators enumerated preceding Exercise 8.10 are Hermitian?

▶ **Exercise 8.22** If A is Hermitian and B is Hermitian, under what conditions is AB Hermitian?

▶ **Exercise 8.23** Let $\phi = \sum_i a_i v_i$, where $A v_i = \lambda_i v_i$. Show that

$$\frac{(\phi, A\phi)}{(\phi, \phi)} = \sum_i p_i \lambda_i, \tag{8.30}$$

where

$$p_i \equiv \frac{|a_i|^2}{\sum_i |a_i|^2}. \tag{8.31}$$

A central tenet of quantum theory is the following three-part statement:

1. Every observable is associated with a Hermitian operator, A.
2. The only possible outcomes of measuring the observable associated with A are the eigenvalues of A, $\{\lambda_i\}$.
3. Given a system in state $\phi = \sum_i a_i v_i$, the probability of obtaining the outcome λ_i is given by p_i, Eq. 8.31.

Thus, the RHS of Eq. 8.30 is the average value, or expectation value, of measuring A. Writing the expectation value as $\langle A \rangle$, we find that

$$\frac{(\phi, A\phi)}{(\phi, \phi)} = \langle A \rangle. \tag{8.32}$$

8.2.3 Inverse of an Operator

The inverse of an operator A, A^{-1}, is defined as

$$A u = v \implies u = A^{-1} v$$

for all u and v in the space. Multiplying both sides of the equation to the right of the arrow by A gives

$$A u = (A A^{-1}) v;$$

multiplying both sides of the equation to the left of the arrow by A^{-1} gives

$$(A^{-1} A) u = A^{-1} v.$$

Comparing these equations with the original equation gives

$$A A^{-1} = A^{-1} A = 1.$$

Not all operators have inverses. Consider a situation in which an operator A brings two different intial vectors to the same final vector:

$$A u_1 = v \tag{8.33}$$

$$A u_2 = v. \tag{8.34}$$

In this case, we have no way to decide whether $A^{-1}v = u_1$ or u_2, and hence we say that the inverse does not exist. An instructive way of expressing the condition for the nonexistence of the inverse is obtained by subtracting Eq. 8.34 from Eq. 8.33:

$$A\left(u_1 - u_2\right) = 0.$$

Thus A kills some nonzero vector.

▶ Exercise 8.24

    a. Consider an operator that maps every eigenstate of the harmonic oscillator onto the state below it:

$$A\psi_n = \psi_{n-1}, \tag{8.35}$$

    with the rule $A\psi_0 = 0$. Does this operator have an inverse?

    b. Now consider an operator that maps every eigenstate of the harmonic oscillator onto the state above it:

$$A\psi_n = \psi_{n+1}. \tag{8.36}$$

    Does this operator have an inverse? [Hint: one needs to define the domain and the range of the mapping.]

8.2.4 Unitary Operators

Another important class of operators in quantum mechanics is that of the unitary operators. Unitary operators are defined by the property

$$A^{-1} = A^\dagger.$$

Here we describe several important additional properties of unitary operators.

1. Unitary operators are norm-preserving. Proof:

$$
\begin{aligned}
(Au, Au) &= \left(A^\dagger(Au), u\right) \\
&= \left(A^{-1}(Au), u\right) \\
&= \left(\left(A^{-1}A\right)u, u\right) \\
&= (u, u)\,.
\end{aligned}
$$

2. Eigenvalues of unitary operators have magnitude 1. Proof: First note that $Uv = \lambda v \Rightarrow U^{-1}v = \frac{1}{\lambda}v$. Then,

$$(v, Uv) = (v, \lambda v) = \lambda(v, v). \tag{8.37}$$

But,

$$(v, Uv) = \left(U^\dagger v, v\right) = \left(U^{-1}v, v\right) = \left(\frac{1}{\lambda}v, v\right) = \frac{1}{\lambda^*}(v, v). \tag{8.38}$$

Therefore,

$$\lambda = \frac{1}{\lambda^*} \implies |\lambda|^2 = 1 \implies |\lambda| = 1. \tag{8.39}$$

3. Eigenvectors of unitary operators with distinct eigenvalues are orthogonal. Proof:

$$\left(v_2, \, U v_1\right) = \lambda \left(v_2, \, v_1\right).$$

But,

$$\left(v_2, \, U v_1\right) = \left(U^\dagger v_2, \, v_1\right) = \left(U^{-1} v_2, \, v_1\right)$$

$$= \left(\frac{1}{\lambda_2} v_2, \, v_1\right) = \frac{1}{\lambda_2^*} \left(v_2, \, v_1\right) = \lambda_2 \left(v_2, \, v_1\right).$$

Since, by assumption, $\lambda_2 - \lambda_1 \neq 0 \implies \left(v_2, \, v_1\right) = 0$.

There is an interesting relationship between Hermitian and unitary operators: if A is Hermitian, e^{iA} is unitary. Proof:

$$\left(e^{iA}\right)^{-1} = e^{-iA} = 1 + (-iA) + \frac{(-iA)^2}{2!} + \cdots$$

$$= 1 + (iA)^\dagger + \left[\frac{(iA)^\dagger}{2!}\right]^2 + \cdots = \left(1 + iA + \frac{(iA)^2}{2} + \cdots\right)^\dagger$$

$$= \left(e^{iA}\right)^\dagger$$

The translation and evolution operators encountered above are examples of unitary operators:

1. $A = e^{a\frac{\partial}{\partial x}} = e^{i\frac{a}{\hbar}\left(-i\hbar\frac{\partial}{\partial x}\right)} = e^{\frac{ia\hat{p}}{\hbar}}$ (recall that $\hat{p}$ is Hermitian)
2. $A = e^{-iHt/\hbar}$ (recall that the Hamiltonian H is Hermitian)

▶ **Exercise 8.25**   Which of the operators enumerated preceding Exercise 8.10 are unitary?

▶ **Exercise 8.26**   If A is unitary and B is unitary, what are the conditions for AB to be unitary?

▶ **Exercise 8.27**   Are there any Hermitian operators that are also unitary? If so, what are the allowed eigenvalues of the operator?

8.2.5   Projection Operators

Projection operators project a vector onto a manifold of the linear vector space. An example of this in three-dimensional space is the projection of an arbitrary vector onto the x-axis:

$$P\left(v_1, \, v_2, \, v_3\right) = \left(v_1, \, 0, \, 0\right).$$

More generally, we can express the action of a projection operator in terms of linear manifolds. P discards that portion of the vector that is not inside the manifold:

$$\phi = \phi_M + \phi_\perp$$
$$P\phi = \phi_M.$$

From this simple expression follow the key properties of a projection operator:

1. $P^2 = P$. This follows because $P(P\phi_M) = P\phi_M = \phi_M$ and $P(P\phi_\perp) = P(0) = 0 = P(\phi_\perp)$. Adding these two equations gives $P(P\phi) = P\phi$.

2. P is Hermitian. Proof:

$$(u, Pv) = (u_M + u_\perp, P(v_M + v_\perp)) = (u_M + u_\perp, v_M) = (u_M, v_M)$$
$$(Pu, v) = (P(u_M + u_\perp), v_M + v_\perp) = (u_M, v_M + v_\perp) = (u_M, v_M).$$

But, $(u, Pv) = (P^\dagger u, v)$ by the definition of the adjoint. Therefore, $P^\dagger = P$.

3. The eigenvalues of P are either 0 or 1. Proof:
    Let $Pv = \lambda v$. Then,

$$(v, Pv) = \lambda(v, v) = (Pv, Pv) = |\lambda|^2 (v, v).$$

Therefore,

$$|\lambda|^2 = \lambda.$$

The only numbers for which this equation is satisfied are 0 and 1. Note that the earlier, defining equations we used for projection operators can in fact be written in the form of eigenvalue equations, with eigenvalue 1 or 0 depending on whether ϕ is inside or outside the manifold onto which P projects:

$$P\phi_M = \phi_M = 1\phi_M$$
$$P\phi_\perp = 0 = 0\phi_\perp.$$

▶ **Exercise 8.28**   Show that the two operators referred to as projection operators preceding Exercise 8.10 indeed satisfy the condition $P^2 = P$.

   a. $A\psi(x) = h(x)\psi(x)$, where $h(x)$ is the Heaviside function

$$h(x) = 1 \text{ if } x \geq 0$$
$$h(x) = 0 \text{ if } x < 0.$$

Then,

$$A^2\psi(x) = A(A\psi(x)) = A(h(x)\psi(x))$$
$$= h(x)(h(x)\psi(x))$$
$$= h(x)\psi(x)$$

since

$$h(x)h(x) = h(x).$$

b. $A\psi(x) = \frac{1}{2}\left[\psi(x) + \psi(-x)\right] \equiv \psi_s(x)$. This operator has the effect of symmetrizing $\psi(x)$ about the value $x = 0$; hence the subscript ψ_s.

$$A^2\psi(x) = A(A\psi(x)) = A\left(\frac{1}{2}\left[\psi(x) + \psi(-x)\right]\right) \tag{8.40}$$

$$= \frac{1}{2}\left[\left\{\frac{1}{2}\left[\psi(x) + \psi(-x)\right]\right\} + \left\{\frac{1}{2}\left[\psi(-x) + \psi(x)\right]\right\}\right] \tag{8.41}$$

$$= \frac{1}{2}\left[\psi(x) + \psi(-x)\right] = \psi_s(x) = A\psi(x). \tag{8.42}$$

Again, $A^2 = A$, indicating that A is a projection operator.

▶ **Exercise 8.29**   Are there any projection operators that are also unitary?

▶ **Exercise 8.30**   If P_1 is a projection operator and P_2 is a projection operator, what is the condition that $P_1 P_2$ be a projection operator?

8.2.6   Commutators of Operators

In general operators do not commute:

$$AB \neq BA.$$

The commutator of two operators A and B is defined as

$$[A, B] = AB - BA.$$

The commutator is itself an operator. In many cases its generic form can be simplified. For example, consider the commutator of the $\hat{p}$ and $\hat{x}$ operators, $\left[-i\hbar\frac{d}{dx}, x\right]$. To calculate the commutator, we proceed according to the following steps:

1. Put in a wavefunction $\psi(x)$ for the commutator to act on:

$$\left[-i\hbar\frac{d}{dx}, x\right]\psi(x) = -i\hbar\frac{d}{dx}\left[x(\psi(x))\right] - x\left(-i\hbar\frac{d}{dx}\psi(x)\right) \tag{8.43}$$

$$= -i\hbar x\frac{d}{dx}\psi(x) - i\hbar\frac{dx}{dx}(\psi(x)) + i\hbar x\frac{d}{dx}\psi(x). \tag{8.44}$$

2. Cancel like terms:

$$\left[-i\hbar\frac{d}{dx}, x\right]\psi(x) = -i\hbar\psi(x). \tag{8.45}$$

3. Remove $\psi(x)$:

$$\left[-i\hbar\frac{d}{dx}, x\right] = -i\hbar. \tag{8.46}$$

▶ **Exercise 8.31**   Find the commutator $\left[e^{a\frac{d}{dx}}, x\right]$.

▶ **Exercise 8.32**   Show that $[A, B] = -[B, A]$.

▶ **Exercise 8.33**

    a. Show that $[AB, C] = A[B, C] + [A, C]B$.

    b. Show that $[A, BC] = B[A, C] + [A, B]C$.

▶ **Exercise 8.34**   Show the Jacobi relation

$$[A, BC] + [B, CA] + [C, AB] = 0. \tag{8.47}$$

▶ **Exercise 8.35**   Verify the "conjugation-commutation" formula

$$e^{A}Be^{-A} = B + [A, B] + \frac{1}{2!}[A, [A, B]] + \frac{1}{3!}[A, [A, [A, B]]] + \cdots \tag{8.48}$$

by expanding the LHS in a Taylor series and matching with the RHS term by term.

A useful relation involving commutators is the Baker–Hausdorff relation,

$$e^{A+B} = e^{A}e^{B}e^{-\frac{1}{2}[A,B]}, \tag{8.49}$$

valid if $[A, B]$ commutes with A and B (see Exercise 9.8). A generalization of this relation known as the Zassenhaus relation will appear in the context of the interaction picture (Exercise 9.25); the Magnus expansion in Section 9.3.1 can also be viewed as a generalization of this relation to the case where an integral, rather than a sum, of operators appears in the exponent.

    Commutators enter in several ways in quantum mechanics. First, if two operators do not commute, their corresponding observables cannot be measured simultaneously. Perhaps the most important example is the noncommutation of $\hat{p}$ and $\hat{x}$, which gives rise to the Heisenberg uncertainty relation, $\Delta x\, \Delta p \geq \hbar/2$. Conversely, if $[A, B] = 0$ one can find a set of common eigenfunctions. This leads to the idea of a complete set of commuting observables, which gives an unambiguous labelling for a quantum state.

▶ **Exercise 8.36**   Show that the Heisenberg uncertainty principle can be derived from the commutation relation between x and p.

Solution

$$\langle [x, p] \rangle = -i\hbar \left\{ \int_{-\infty}^{\infty} \psi^{*} x \frac{\partial}{\partial x} \psi\, dx - \int_{-\infty}^{\infty} \psi^{*} \frac{\partial}{\partial x}(x\psi)\, dx \right\} \tag{8.50}$$

$$= -i\hbar \left\{ \int_{-\infty}^{\infty} x\psi^{*} \frac{\partial \psi}{\partial x}\, dx + \int_{-\infty}^{\infty} x \frac{\partial \psi^{*}}{\partial x} \psi\, dx \right\} \tag{8.51}$$

$$= -i\hbar \int_{-\infty}^{\infty} x \frac{\partial}{\partial x} \left[\psi^{*}(x)\psi(x) \right] dx. \tag{8.52}$$

From the Schwarz inequality we have

$$-\int_{-\infty}^{\infty} x \frac{\partial}{\partial x} \left[\psi^{*}(x)\psi(x) \right] dx \leq 2 \left\| \frac{\partial \psi}{\partial x} \right\| \ \| x\psi \| \tag{8.53}$$

(cf. Exercise 8.5); hence

$$\langle [x, p] \rangle = i\hbar \leq 2i\,\Delta p\,\Delta x \tag{8.54}$$

or

$$\Delta p\,\Delta x \geq \frac{\hbar}{2}. \tag{8.55}$$

Of particular importance are operators that commute with the Hamiltonian operator, H. Finding an operator that commutes with H may simplify the problem of finding eigenvalues and eigenfunctions of H. If $[A, H] = 0$, and if one can find easily the eigenvalues and eigenfunctions of A, one can use these eigenfunctions as a "basis" (see the next section) to solve for the eigenvalues and eigenfunctions of H. This partitions the problem of finding eigenvalues and eigenfunctions of H into separate, smaller problems associated with each different eigenvalue of A. A well-known example is the hydrogen atom, where H, L^2, L_z and s commute, allowing one to find common eigenfunctions of all four of these operators. The eigenfunction is then labelled by the indices $\{n, l, m, s\}$, which gives it a unique specification. Another example, of great importance to chemistry, involves the point group operators of molecules with some symmetry. These operators commute with the Hamiltonian, and their eigenfunctions are found "trivially" through the use of appropriate symmetry projection operators. Using the eigenfunctions of the symmetry operators as a basis partitions the problem of finding eigenvalues and eigenfunctions of H into separate, smaller problems associated with each different eigenvalue of the symmetry operator.

Commutators are also closely related to invariance principles. If an operator commutes with $\hat{p}$ it is translationally invariant; if it commutes with the Hamiltonian operator, H, it is time invariant—it is a constant of the motion. Conversely, if an operator does not commute with H, it cannot be time invariant. Indeed, we saw in Chapter 5 that the equation of motion of the density operator is proportional to $[H, \rho]$; thus the noncommutation of ρ with H is responsible for the time evolution. Moreover, in the context of quantum control (Chapter 16), we shall see that the ability to control a system governed by a Hamiltonian H_0 requires the existence of a control Hamiltonian H_1 such that $[H_0, H_1] \neq 0$; if $[H_0, H_1] = 0$ then H_1 cannot steer the system to somewhere other than where it would evolve under H_0.

8.3   Discrete Basis Sets: The Bridge between Operators–Wavefunctions $\Longleftrightarrow$ Matrices–Vectors

8.3.1   Introduction

Consider a set of orthonormal vectors $\{\phi_i\}$ that span the LVS:

$$\int \phi_i^*(x)\phi_j(x)\,dx = 1 \text{ if } i = j$$

$$\int \phi_i^*(x)\phi_j(x)\,dx = 0 \text{ if } i \neq j. \tag{8.56}$$

For shorthand, we write

$$\int \phi_i^*(x)\phi_j(x)\,dx = \delta_{ij},\tag{8.57}$$

where we have introduced the Kronecker delta symbol, δ_{ij}. Now consider some arbitrary wavefunction ψ. ψ can be expanded in the complete set $\{\phi_i\}$:

$$\psi(x) = \sum_{i=0}^{\infty} a_i\phi_i(x).$$

How does one find a_j? (Note that we use the symbol j for a particular coefficient; j is one of the allowed values of i.) The trick is to left multiply by ϕ_j^* and integrate over x: by orthogonality this will eliminate any component of ψ that is not along ϕ_j.

$$\begin{aligned}
\int_{-\infty}^{\infty} \phi_j^*\psi(x)\,dx &= \int_{-\infty}^{\infty} \phi_j^*(x)\left\{\sum_{i=0}^{\infty} a_i\phi_i(x)\right\}dx \\
&= \sum_{i=0}^{\infty} a_i \int_{-\infty}^{\infty} \phi_j^*(x)\phi_i(x)\,dx \\
&= \sum_{i=0}^{\infty} a_i\delta_{ij} = a_j.
\end{aligned}$$

Now note that we can do a similar manipulation on $A\psi$:

$$A\psi(x) = \sum_{i=0}^{\infty} c_i\phi_i(x)$$

$$c_j = \int_{-\infty}^{\infty} \phi_j^*(x)(A\psi(x))\,dx = \int_{-\infty}^{\infty} \phi_j^*(x)\left(A\left\{\sum_{i=0}^{\infty} a_i\phi_i(x)\right\}\right)dx$$

$$= \sum_{i=0}^{\infty}\left\{\int_{-\infty}^{\infty} \phi_j^*(x)\left(A\phi_i(x)\right)dx\right\}a_i.\tag{8.58}$$

It is instructive to think of Eq. 8.58 as a matrix equation of the form

$$c_j = \sum_{i=0}^{\infty} A_{ji}a_i,$$

where A_{ji} is the expression in braces,

$$A_{ji} \equiv \int_{-\infty}^{\infty} \phi_j^*(x)\left(A\phi_i(x)\right)dx.$$

In other words, the expression in braces depends on two indices, i and j; hence the values of the integral as these two indices are varied can be stored as elements of a matrix whose rows run over $\{\phi_j\}$ and whose columns run over $\{\phi_i\}$. Each entry is a (complex) number.

$$\begin{pmatrix} c_1 \\ c_2 \\ c_3 \\ \vdots \end{pmatrix} = \begin{pmatrix} A_{11} & A_{12} & A_{13} & \cdots \\ A_{21} & A_{22} & A_{23} & \cdots \\ A_{31} & A_{32} & A_{33} & \cdots \\ \vdots & \vdots & \vdots & \ddots \end{pmatrix} \begin{pmatrix} a_1 \\ a_2 \\ a_3 \\ \vdots \end{pmatrix}$$

$$= \begin{pmatrix} A_{11}a_1 + A_{12}a_2 + A_{13}a_3 + \cdots \\ A_{21}a_1 + A_{22}a_2 + A_{23}a_3 + \cdots \\ A_{31}a_1 + A_{32}a_2 + A_{33}a_3 + \cdots \\ \vdots \end{pmatrix}.$$

▶ **Exercise 8.37**   Express the Time-Dependent Schrödinger Equation in matrix form.

Solution   Starting with the TDSE,

$$i\hbar \frac{\partial \Psi}{\partial t}(x, t) = H\Psi(x, t), \tag{8.59}$$

we left multiply by $\phi_j^*(x)$ and integrate over x:

$$i\hbar \int_{-\infty}^{\infty} \phi_j^*(x) \left(\frac{\partial \Psi}{\partial t}(x, t) \right) dx = \int_{-\infty}^{\infty} \phi_j^*(x) \{H\Psi(x, t)\} dx$$

$$i\hbar \frac{\partial}{\partial t} \int_{-\infty}^{\infty} \phi_j^*(x) \Psi(x, t) dx = \int_{-\infty}^{\infty} \phi_j^*(x) \{H\Psi(x, t)\} dx$$

$$i\hbar \frac{\partial}{\partial t} \int_{-\infty}^{\infty} \phi_j^*(x) \left\{ \sum_{i=0}^{\infty} a_i(t)\phi_i(x) \right\} dx = \int_{-\infty}^{\infty} \phi_j^*(x) \left\{ H \sum_{i=0}^{\infty} a_i(t)\phi_i(x) \right\} dx$$

$$i\hbar \frac{d}{dt} \sum_{i=0}^{\infty} a_i(t) \int_{-\infty}^{\infty} \phi_j^*(x)\phi_i(x) dx = \sum_{i=0}^{\infty} a_i(t) \int_{-\infty}^{\infty} \phi_j^*(x) \left(H\phi_i(x) \right) dx.$$

In matrix form, this last equation reduces to

$$i\hbar \sum_{i=0}^{\infty} S_{ji} \dot{a}_i(t) = \sum_{i=0}^{\infty} H_{ji} a_i(t), \tag{8.60}$$

where **S** is called the *overlap matrix*, with elements $S_{ji} = \int_{-\infty}^{\infty} \phi_j^*(x)\phi_i(x)\,dx$. If the $\{\phi_i\}$ are orthonormal then $S_{ji} = \delta_{ij}$ and Eq. 8.60 gives

$$i\hbar \frac{da_j(t)}{dt} = \sum_{i=0}^{\infty} H_{ji}a_i(t). \tag{8.61}$$

▶ **Exercise 8.38**   Express the Time-*Independent* Schrödinger Equation in matrix form.

Solution   Starting with the TISE,

$$H\Psi_n = E_n\Psi_n, \tag{8.62}$$

we left multiply by $\phi_j^*(x)$ and integrate over x:

$$\int_{-\infty}^{\infty} \phi_j^*(x) H\Psi_n(x)\,dx = E_n \int_{-\infty}^{\infty} \phi_j^*(x)\Psi_n(x)\,dx. \tag{8.63}$$

Expanding $\Psi_n = \sum_{i=0}^{\infty} a_{in}\phi_i(x)$ we obtain

$$\sum_{i=0}^{\infty} a_{in} \int_{-\infty}^{\infty} \phi_j^*(x) H\phi_i(x)\, dx = E_n \sum_{i=0}^{\infty} a_{in} \int_{-\infty}^{\infty} \phi_j^*(x)\phi_i(x)\, dx \qquad (8.64)$$

or

$$\sum_{i=0}^{\infty} a_{in} H_{ji} = \sum_{i=0}^{\infty} E_n a_{in} S_{ji}, \qquad (8.65)$$

where the S_{ji} are defined as in Exercise 8.37. If the $\{\phi_i\}$ are orthonormal, Eq. 8.65 reduces to

$$\sum_{i=0}^{\infty} a_{in} H_{ji} = \sum_{i=0}^{\infty} E_n a_{in} \delta_{ij} = E_n a_{jn}. \qquad (8.66)$$

In matrix form, Eq. 8.65 is $(\mathbf{HU})_{jn} = (\mathbf{SUE})_{jn}$, or

$$\mathbf{HU} = \mathbf{SUE}, \qquad (8.67)$$

where $\mathbf{E}$ is a diagonal matrix and we have defined $U_{ij} = a_{ij}$. Equation 8.67 is usually referred to as the generalized eigenvalue problem. In matrix form, Eq. 8.66 takes the form $(\mathbf{HU})_{jn} = (\mathbf{UE})_{jn}$, or

$$\mathbf{HU} = \mathbf{UE}. \qquad (8.68)$$

This is the usual form of the matrix eigenvalue problem. The solution of matrix eigenvalue problems of the form in Eq. 8.68 will be discussed in Section 8.3.2, while the solution of the generalized eigenvalue problem, Eq. 8.67, is treated in Exercises 8.79 and 8.80.

Adjoint of a Matrix and Hermitian Matrices

The adjoint of a matrix is given by the complex conjugate of its transpose. Proof:

$$\left(A^\dagger\right)_{ij} = \int \phi_i^*(x) \left(A^\dagger \phi_j(x)\right) dx$$

$$= \int \left(A^\dagger \phi_j(x)\right) \phi_i^*(x)\, dx$$

$$= \left[\int \left(A^\dagger \phi_j(x)\right)^* \phi_i(x)\, dx\right]^*$$

$$= \left[\int \phi_j^*(x) \left(A\phi_i(x)\right) dx\right]^*$$

$$= A_{ji}^*.$$

For a Hermitian matrix, $A^\dagger = A$ and hence

$$A_{ij} = A_{ji}^*,$$

thus the complex conjugate of the transpose is equal to the matrix itself. For diagonal elements

$$A_{ii} = A_{ii}^*,$$

that is, the diagonal elements of a Hermitian matrix must be real.

▶ **Exercise 8.39**   Verify that $\begin{pmatrix} 2 & i \\ -i & 1 \end{pmatrix}$ is a Hermitian matrix.

Inverse of a Matrix and Unitary Matrices

Given a matrix A, its inverse, A^{-1}, is defined by the relation

$$AA^{-1} = A^{-1}A = 1,$$

where **1** is the identity matrix

$$\mathbf{1} = \begin{pmatrix} 1 & 0 & 0 & \cdots \\ 0 & 1 & 0 & \cdots \\ 0 & 0 & 1 & \cdots \\ \vdots & \vdots & \vdots & \ddots \end{pmatrix}.$$

A general procedure for finding the inverse of a matrix is to transform the matrix into diagonal form (see below), invert the diagonal form, and transform back. In quantum mechanics, the main interest in inverses is with regard to unitary matrices, where the inverse is simply the adjoint:

$$U_{ij}^{-1} = U_{ij}^\dagger = U_{ji}^*.$$

From this relation, the following further property of unitary matrices can be derived:

$$\left(U^{-1}U\right)_{ik} = \delta_{ik} = \left(U^\dagger U\right)_{ik} = \sum_j U_{ij}^\dagger U_{jk}.$$

But $\sum_j U_{ij}^\dagger U_{jk} = \sum_j U_{ji}^* U_{jk}$, which is just the inner product of two columns of the matrix U. Since this must equal δ_{ik}, we conclude that the columns of a unitary matrix must be orthonormal vectors.

▶ **Exercise 8.40**   Show that the same conclusions hold for the rows of a unitary matrix.

▶ **Exercise 8.41**   Consider the following matrix:

$$\begin{pmatrix} \cos\frac{\theta}{2}e^{-i\phi/2} & \sin\frac{\theta}{2}e^{i\phi/2} \\ -\sin\frac{\theta}{2}e^{-i\phi/2} & \cos\frac{\theta}{2}e^{i\phi/2} \end{pmatrix}.$$

Show that it is unitary (calculate its adjoint and show that it is equal to its inverse). Then show that its columns and rows are orthonormal vectors.

8.3.2   Diagonalizing Matrices

Diagonalizing a matrix refers to finding a unitary transformation that puts the matrix in diagonal form. The diagonal elements will then be the eigenvalues, and the columns of the unitary transformation will be the eigenvectors. Generally in quantum mechanics one is interested in diagonalizing Hermitian matrices. The eigenvalues are then real, and represent the possible values of the physical observable that corresponds to the operator; the state of the system at the time of the measurement is the eigenvector associated with the measured eigenvalue. If the eigenvalue spectrum is degenerate, that is, if any of the eigenvalues occurs more than once in the spectrum, then that measurement is not sufficient to specify completely which state the system is in. The state of the system has then been determined to within a degenerate manifold of the Hilbert space, and other measurements must be performed, *simultaneously*, on observables that are not codegenerate but yet commute with the original operator. In Section 8.3.1 (Exercise 8.38) we saw that the TISE can be reformulated as a matrix eigenvalue equation, which accounts for the chief application of diagonalizing matrices in quantum mechanics.

We begin with a review of the method for calculating eigenvalues and eigenvectors, using as an example a 2×2 matrix. For matrices larger than 2×2 (or certainly 3×3) a computer is recommended. Consider the matrix

$$A = \begin{pmatrix} 2 & 1 \\ 1 & 2 \end{pmatrix}.$$

The eigenvalue equation is

$$\begin{pmatrix} 2 & 1 \\ 1 & 2 \end{pmatrix} \begin{pmatrix} v_1 \\ v_2 \end{pmatrix} = \lambda \begin{pmatrix} v_1 \\ v_2 \end{pmatrix}.$$

Solving this equation consists of finding all possible λ, and for each λ a vector (v_1, v_2). The associated eigenvector is determined within an overall multiplicative constant (try multiplying the vector (v_1, v_2) by a constant c and you will see that the equation is unchanged). By convention, the eigenvector is put into normalized form, with the phase of the first element real and positive. To begin, we subtract the right-hand side from the left to obtain

$$\begin{pmatrix} 2 - \lambda & 1 \\ 1 & 2 - \lambda \end{pmatrix} \begin{pmatrix} v_1 \\ v_2 \end{pmatrix} = 0.$$

Since we are interested in cases where the vector $(v_1, v_2) \neq 0$, the only way this equation can hold is if the rank of the new matrix is less than two, that is, the rows (or columns) are linearly dependent. The condition for this is

$$\begin{vmatrix} 2 - \lambda & 1 \\ 1 & 2 - \lambda \end{vmatrix} = 0, \tag{8.69}$$

where

$$\begin{vmatrix} a & b \\ c & d \end{vmatrix} = ad - bc$$

is called the determinant. Expanding the determinant we find

$$(2 - \lambda)(2 - \lambda) - 1 = \lambda^2 - 4\lambda + 3 = 0 \tag{8.70}$$

or

$$\lambda_{1,2} = \frac{4 \pm \sqrt{16 - 12}}{2} = \frac{4 \pm 2}{2} = 3, 1.$$

Equation 8.70 is known as the "characteristic equation" for the matrix A.

▶ **Exercise 8.42**   Substitute these values of λ back into Eq. 8.69 one at a time and show that both the rows and columns are linearly dependent.

▶ **Exercise 8.43**   Consider the matrix

$$\begin{pmatrix} a & b \\ c & d \end{pmatrix}.$$

Show that if either the rows or the columns are linearly dependent then the determinant vanishes.

▶ **Exercise 8.44**   The Cayley–Hamilton theorem says that every matrix satisfies its own characteristic equation (see Exercise 8.78). Verify this for the matrix $A = \begin{pmatrix} 2 & 1 \\ 1 & 2 \end{pmatrix}$.

We now pick the eigenvalues one at a time, and solve for the corresponding eigenvector:

$$\begin{pmatrix} 2 & 1 \\ 1 & 2 \end{pmatrix} \begin{pmatrix} 1 \\ a \end{pmatrix} = 1 \begin{pmatrix} 1 \\ a \end{pmatrix}. \tag{8.71}$$

Rather than using (v_1, v_2) we have written $(1, a)$. To understand why we can do this, note that Eq. 8.71 is really two equations, but they are linearly dependent (in general, for an $N \times N$ matrix only $N - 1$ equations would be linearly independent). Thus, any one of the vector elements can be chosen at random, so long as the others are chosen commensurately. Ultimately, we will normalize the vector, and the initial form $(1, a)$ will be moot. The choice $(1, a)$ simplifies the intermediate algebra and gives the conventional choice of phase to the first element of the eigenvector. Equating components of the vector on the left- and right-hand side we obtain

$$2 + a = 1 \Longrightarrow a = -1$$
$$1 + 2a = a \Longrightarrow a = -1.$$

It is clear that these two equations are linearly dependent. Substituting $a = -1$ we obtain

$$v_1 = \begin{pmatrix} 1 \\ -1 \end{pmatrix} \longrightarrow \begin{pmatrix} \frac{1}{\sqrt{2}} \\ \frac{-1}{\sqrt{2}} \end{pmatrix},$$

where the right arrow implies after normalization. Turning to the second eigenvalue we find

$$\begin{pmatrix} 2 & 1 \\ 1 & 2 \end{pmatrix} \begin{pmatrix} 1 \\ a \end{pmatrix} = 3 \begin{pmatrix} 1 \\ a \end{pmatrix},$$

which implies

$$2 + a = 3 \Longrightarrow a = 1$$
$$1 + 2a = 3a \Longrightarrow a = 1.$$

Substituting $a = 1$ we find

$$v_2 = \begin{pmatrix} 1 \\ 1 \end{pmatrix} \longrightarrow \begin{pmatrix} \frac{1}{\sqrt{2}} \\ \frac{1}{\sqrt{2}} \end{pmatrix}.$$

▶ **Exercise 8.45** Verify that each of the two eigenvectors solves the eigenvector equation, with its correct eigenvalue.

▶ **Exercise 8.46** Arrange the two normalized eigenvectors as columns of a 2×2 matrix, and show that the matrix is a unitary matrix, U.

▶ **Exercise 8.47** Show explicitly for the 2×2 matrix that $AU = U\Lambda$, but $AU \neq \Lambda U$.

▶ **Exercise 8.48** Show explicitly for the 2×2 matrix that $U^{-1}AU = \Lambda$:

$$\underbrace{\begin{pmatrix} \frac{1}{\sqrt{2}} & -\frac{1}{\sqrt{2}} \\ \frac{1}{\sqrt{2}} & \frac{1}{\sqrt{2}} \end{pmatrix}}_{\text{unitary}} \underbrace{\begin{pmatrix} 2 & 1 \\ 1 & 2 \end{pmatrix}}_{\text{Hermitian}} \underbrace{\begin{pmatrix} \frac{1}{\sqrt{2}} & \frac{1}{\sqrt{2}} \\ -\frac{1}{\sqrt{2}} & \frac{1}{\sqrt{2}} \end{pmatrix}}_{\text{unitary}} = \underbrace{\begin{pmatrix} 1 & 0 \\ 0 & 3 \end{pmatrix}}_{\text{diagonal}}$$

This exercise gives a slightly different perspective on the diagonal structure of Λ: when the matrix A operates on each column of U, it returns that column times the eigenvalue. When the inner product of that column is taken with the rows of U^{-1}, orthogonality guarantees that the off-diagonal elements must vanish, and normalization guarantees that the inner product of the diagonal elements equals 1 times the corresponding eigenvalue.

▶ **Exercise 8.49** The trace of a matrix is defined as the sum of its diagonal elements. Verify that the trace of A is the same in the diagonal form as in the original form.

▶ **Exercise 8.50** Show that the determinant of A is the same in the diagonal form as in the original form.

In general, the trace of a matrix is invariant to a unitary transformation. Here is the proof:

$$\begin{aligned}
\text{Tr}\left(U^\dagger A U\right) &= \sum_{ijk} \left(U^\dagger\right)_{ij} (A)_{jk} (U)_{ki} \\
&= \sum_{ijk} (U)^*_{ji} \left(U^{-1}\right)_{ki} A_{jk} \\
&= \sum_{jk} \delta_{jk} A_{jk} \\
&= \sum_{j} A_{jj} \\
&= \text{Tr}(A).
\end{aligned}$$

The trace of a product of matrices is invariant to the order of the product. Here is the proof:

$$\begin{aligned}
\text{Tr}(AB) &= \sum_{i} (AB)_{ii} = \sum_{ij} A_{ij} B_{ji} \\
&= \sum_{ij} B_{ji} A_{ij} = \sum_{j} (BA)_{jj} = \text{Tr}(BA).
\end{aligned}$$

▶ **Exercise 8.51**   Show that

$$\mathrm{Tr}(ABC) = \mathrm{Tr}(CAB) = \mathrm{Tr}(BCA).$$

Use this result to give a simpler proof that $\mathrm{Tr}(U^\dagger AU) = \mathrm{Tr}(A)$ than that given above.

Every unitary transformation can be thought of as a change in basis, and every change in basis as a unitary transformation. The elements of the unitary matrix, U_{ij}, are given by (u'_i, u_j), where $\{u\}$ is the old basis and $\{u'\}$ is the new basis. If one takes two sets of orthonormal bases for an N-dimensional vector space (that is, two different sets of orthonormal column vectors of length N, each set containing N vectors), then an $N \times N$ matrix can be constructed from the $N \times N$ combinations of inner products of the basis vectors. This $N \times N$ matrix will be unitary. A special case is the transformation to the basis that diagonalizes a matrix, where the old basis is $(1, 0, 0, \ldots)$, $(0, 1, 0, \ldots)$, $\ldots$, and the new basis is the eigenvectors.

▶ **Exercise 8.52**   Consider again the matrix $A = \begin{pmatrix} 2 & 1 \\ 1 & 2 \end{pmatrix}$. Using the eigenvectors $(\frac{1}{\sqrt{2}}, \frac{1}{\sqrt{2}})$, $(\frac{1}{\sqrt{2}}, \frac{-1}{\sqrt{2}})$ derived in the text, construct all combinations of inner products with the vectors $(1, 0)$ and $(0, 1)$ and verify that the unitary matrix that diagonalizes A is obtained.

▶ **Exercise 8.53**   The determinant of a product of matrices is the product of the determinants. Also, the determinant of a unitary matrix is 1. Using these properties, show that the determinant of a matrix is unchanged by a unitary transformation.

▶ **Exercise 8.54**   Evaluate $\mathrm{Tr}\left(e^{-\beta H}\right)$ for the harmonic oscillator by representing the operator $e^{-\beta H}$ in the basis of harmonic oscillator eigenstates.

8.3.3   Projection Matrices

In the same way that we defined projection operators above, we may define projection matrices by the condition that P be a Hermitian matrix and

$$P^2 = P. \tag{8.72}$$

▶ **Exercise 8.55**   Show that

$$\frac{1}{2}\begin{pmatrix} 1 & 1 \\ 1 & 1 \end{pmatrix} \quad \text{and} \quad \frac{1}{2}\begin{pmatrix} 1 & -1 \\ -1 & 1 \end{pmatrix} \tag{8.73}$$

are projection matrices.

Like a projection operator, a projection matrix kills vectors that are not in the subspace and does not affect vectors in the subspace onto which they project:

$$v \equiv v_\mathrm{M} + v_\perp \tag{8.74}$$

$$P v_\mathrm{M} = v_\mathrm{M} \tag{8.75}$$

$$P v_\perp = 0. \tag{8.76}$$

▶ **Exercise 8.56** Find vectors v_M and $v_\perp$ for each of the projection matrices in the previous exercise.

It is possible to form projection operators from vectors by taking their *outer product* with themselves:

$$\begin{pmatrix} v_1 \\ v_2 \end{pmatrix} (v_1, v_2)^* = \begin{pmatrix} v_1 v_1^* & v_1 v_2^* \\ v_2 v_1^* & v_2 v_2^* \end{pmatrix}.$$

Such projection operators project onto the one-dimensional vector space spanned by the vector used to form the outer product, and kill any vector perpendicular to this space.

▶ **Exercise 8.57** In Section 8.3.2 above, we used the example of the matrix

$$A = \begin{pmatrix} 2 & 1 \\ 1 & 2 \end{pmatrix} \tag{8.77}$$

to illustrate finding the eigenvalues and eigenvectors of a matrix.

a. Use the two eigenvectors of the matrix A,

$$\begin{pmatrix} \frac{1}{\sqrt{2}} \\ \frac{1}{\sqrt{2}} \end{pmatrix} \quad \text{and} \quad \begin{pmatrix} \frac{1}{\sqrt{2}} \\ -\frac{1}{\sqrt{2}} \end{pmatrix}, \tag{8.78}$$

to construct projection matrices.

b. Show that the matrix A may be represented as a linear combination of its projection matrices:

$$A = \lambda_1 P_1 + \lambda_2 P_2, \tag{8.79}$$

where λ_i is the ith eigenvalue and P_i is the projection operator formed from the outer product of the ith eigenvector. Show that $P_1 + P_2 = \mathbf{1}$ and that $P_1 P_2 = P_2 P_1 = 0$.

▶ **Exercise 8.58** In diagonalizing the matrix A earlier in this chapter we obtained the equation

$$\begin{pmatrix} 2-\lambda & 1 \\ 1 & 2-\lambda \end{pmatrix} \begin{pmatrix} v_1 \\ v_2 \end{pmatrix} = 0, \tag{8.80}$$

where $\lambda = 3$ or 1. Show that, for a given choice of λ, the matrix in Eq. 8.80 is proportional to the projection matrix corresponding to the eigenvector associated with the other value of λ, and find the proportionality constant.

Solution Consider first the case that $\lambda = \lambda_1 = 3$. Equation 8.80 may be written as $A - \lambda_1(P_1 + P_2)$, since $P_1 + P_2 = \mathbf{1}$. From the previous exercise, $A = \lambda_1 P_1 + \lambda_2 P_2$ and therefore $A - \lambda_1(P_1 + P_2) = (\lambda_2 - \lambda_1)P_2$. Similarly, $A - \lambda_2(P_1 + P_2) = (\lambda_1 - \lambda_2)P_1$.

8.3.4 Commutators of Matrices

The commutator of two matrices, A and B, is defined as

$$[A, B] = AB - BA. \tag{8.81}$$

For example, if

$$A = \begin{pmatrix} 1 & 0 \\ 0 & -1 \end{pmatrix} \quad \text{and} \quad B = \begin{pmatrix} 0 & i \\ -i & 0 \end{pmatrix}, \tag{8.82}$$

$$[A, B] = AB - BA = 2 \begin{pmatrix} 0 & i \\ i & 0 \end{pmatrix}.$$

▶ **Exercise 8.59**   The Pauli matrices are defined as

$$\sigma_X = \begin{pmatrix} 0 & 1 \\ 1 & 0 \end{pmatrix}, \quad \sigma_Y = \begin{pmatrix} 0 & -i \\ i & 0 \end{pmatrix}, \quad \sigma_Z = \begin{pmatrix} 1 & 0 \\ 0 & -1 \end{pmatrix}. \tag{8.83}$$

Calculate the commutator of each pair of Pauli matrices, and show that the result is proportional to another Pauli matrix.

If a set of matrices commutes it is possible to find a set of common eigenvectors for all the matrices in the set.

▶ **Exercise 8.60**   Consider the matrices

$$A = \begin{pmatrix} 1 & 0 & 0 \\ 0 & -1 & 0 \\ 0 & 0 & -1 \end{pmatrix}, \quad B = \begin{pmatrix} 1 & 0 & 0 \\ 0 & 0 & 1 \\ 0 & 1 & 0 \end{pmatrix}. \tag{8.84}$$

Show that these matrices commute and find a set of common eigenvectors for them.

▶ **Exercise 8.61**   Calculate the matrix elements of the operators $\hat{x}$ and $\hat{p}$ in the basis set of the harmonic oscillator eigenstates. Then calculate the commutator of these matrices and compare with the commutator of the operators $\hat{x}$ and $\hat{p}$.

8.3.5   Matrices as Representations of Operators: Dirac Notation

We now introduce a shorthand notation for matrix manipulation, called Dirac notation. This notation is especially useful for representing changes of basis. A very powerful result called the "closure" relation, which expresses completeness of a basis, will emerge naturally in Dirac notation. In the next section, the more abstract concept of continuous basis sets will be introduced, and it is there that the full power of Dirac notation will become apparent: a unified notational scheme for changing between any representations, discrete or continuous.

We take as a definition of Dirac notation

$$A_{ji} = \int \phi_j^*(x) \left(A\phi_i(x) \right) dx = \langle \phi_j | A | \phi_i \rangle.$$

The object $\langle \phi_j |$ is known as the bra and $|\phi_i\rangle$ is known as the ket; together they bra-ket (bracket) the operator A. Setting $A = \mathbf{1}$ we find

$$\int \phi_j^*(x) \psi_i(x) \, dx = \langle \phi_j | \psi_i \rangle,$$

where by convention the two vertical bars are merged. Thus, the scalar product of two vectors in Dirac notation is actually similar to our earlier notation (u, v). Note that

$$A_{ji} = \int \phi_j^*(x) \left(A\phi_i(x) \right) dx = \langle \phi_j | A | \phi_i \rangle = \langle \phi_j | A\phi_i \rangle .$$

In other words, the equation $\int \phi_j^*(x) \left(A\phi_i(x) \right) dx$ may be viewed either as a matrix element of the operator A or as an inner product of ϕ with $A\psi$. The significance of this observation is that the operator in the center in Dirac notation can be brought into the ket to make a new ket, whose inner product is then taken with the bra.

Note that since

$$(v, Au) = (A^\dagger v, u) = (u, A^\dagger v)^*$$

it follows that

$$\langle \phi | A | \psi \rangle = \langle \phi | A\psi \rangle = \langle A^\dagger \phi | \psi \rangle = \langle \psi | A^\dagger \phi \rangle^* = \langle \psi | A^\dagger | \phi \rangle^*.$$

We now give some examples of the use of Dirac notation.

1. We return to the equation for the Time-Dependent Schrödinger Equation, $i\hbar \frac{\partial \psi}{\partial t} = H\psi$, in matrix form (cf. Exercise 8.37):

$$i\hbar \frac{d}{dt} a_j(t) = \sum_{i=0}^{\infty} H_{ji} a_i(t).$$

In Dirac notation this takes the form

$$i\hbar \frac{d}{dt} \langle \phi_j | \psi \rangle = \langle \phi_j | H\psi \rangle = \sum_i \langle \phi_j | H \{ | \phi_i \rangle \langle \phi_i | \} \psi \rangle.$$

▶ **Exercise 8.62**   Verify this.

Comparing these two forms of the equation in Dirac notation leads to the identification

$$\sum_i | \phi_i \rangle \langle \phi_i | = \mathbf{1}.$$

This latter equation is called the "closure" relation, and is a statement of the completeness of the basis set $\{\phi_i\}$ and the invariance of the scalar product to a change in basis. One of the great simplifying features of Dirac notation is the ability to insert the identity operator $\mathbf{1}$ in this form wherever desired, referring to any complete basis.

For nonorthogonal bases the closure relation takes the form

$$\sum_{ij} | \phi_i \rangle \left(S^{-1} \right)_{ij} \langle \phi_j | = \mathbf{1}, \tag{8.85}$$

where $\mathbf{S}$ is the overlap matrix, with elements $S_{ij} \equiv \langle \phi_i | \phi_j \rangle$ (see Exercise 8.81).

2. As a second example, we reexamine the equations for diagonalizing a matrix in Dirac notation. Recall that we had the generic matrix equation

$$U^{-1}AU = A'.$$

In Dirac notation this takes the form

$$\langle i|U^{-1}AU|j\rangle = \langle i|A'|j\rangle,$$

where we use $|i\rangle$, $|j\rangle$ for $|\phi_i\rangle$, $|\phi_j\rangle$, and so on. Taking $\{i, j\}$ to be the old basis, the significance of the operators U, U^{-1} is that they transform the operator into a new operator that is diagonal in the old basis. An alternative interpretation is obtained by writing

$$\langle i|U^{-1}AU|j\rangle = \langle i'|A|j'\rangle,$$

where $|j'\rangle = |Uj\rangle$, $|i'\rangle = |Ui\rangle$. From this perspective, unitary operators rotate the original basis vectors to a basis in which A is diagonal. Inserting the closure relationship twice we obtain

$$\sum_{kl} \langle i|U^{-1}|l\rangle\langle l|A|k\rangle\langle k|U|j\rangle = \sum_{kl} \langle i'|l\rangle\langle l|A|k\rangle\langle k|j'\rangle, \qquad (8.86)$$

which leads to the identification

$$\langle k|U|j\rangle = \langle k|j'\rangle$$

and so on. This equation indicates that the matrix of U in the original representation (LHS) can be constructed by forming the inner product of the original basis $\{l, k\}$ with the basis that diagonalizes A, $\{i', j'\}$ (RHS).

3. As a third example, we use Dirac notation to show that $\mathrm{Tr}(A)$ is invariant to a change in basis.

$$\begin{aligned}
\mathrm{Tr}(A) &= \sum_i \langle\psi_i|A|\psi_i\rangle \\
&= \sum_{ijk} \langle\psi_i|\phi_j\rangle\langle\phi_j|A|\phi_k\rangle\langle\phi_k|\psi_i\rangle \\
&= \sum_{ijk} \langle\phi_k\{|\psi_i\rangle\langle\psi_i|\}\phi_j\rangle\langle\phi_j|A|\phi_k\rangle \\
&= \sum_{jk} \langle\phi_k|\phi_j\rangle\langle\phi_j|A|\phi_k\rangle \\
&= \sum_k \langle\phi_k|A|\phi_k\rangle .
\end{aligned}$$

The elements $\langle\phi_k|\psi_i\rangle$ (and $\langle\psi_i|\phi_j\rangle$) define a unitary matrix, whose elements are the projection of each of the new basis functions onto each of the old basis functions. This is the analog in Hilbert space of a change of axes (or equivalently, a change of unit vectors) in Euclidean geometry. Rewriting the above equations in matrix form we obtain

$$\text{Tr}(A) = \sum_i A_{ii}^{\psi} = \sum_i (U^\dagger A^\phi U)_{ii} = \sum_k A_{kk}^{\phi}, \qquad (8.87)$$

where we have used the notation A^ψ and A^ϕ to express the representation of the operator A in the two different sets of bases.

▶ **Exercise 8.63** Use Dirac notation to show that $\text{Tr}(AB) = \text{Tr}(BA)$.

8.4 · Continuous Basis Sets

One of the great milestones in the history of quantum mechanics was the development by Dirac of a unified symbolic treatment of Schrödinger's wave mechanics and Heisenberg's matrix mechanics. Essentially, Dirac's insight was that the quantum formalism of matrices as representations of operators could be applied using the continuous bases, x and p, just as well as using any discrete basis set. We will begin by showing this using ordinary wavefunction notation, and then show how simply it arises, and can be used, using Dirac notation.

8.4.1 Momentum Representation

Table 8.2 examines four of the important characteristics satisfied by any basis. The first column states the property in words; the second column shows the now familiar equation for this property in a discrete basis. The third column shows the four equations as they apply in the momentum representation, the first three of which we have already encountered in the first section of this book. The important point here is that if we identify the basis functions $\{\phi_i(x)\}$ with $\frac{e^{ipx/\hbar}}{\sqrt{2\pi\hbar}}$ and the expansion coefficients $\{a_i\}$ with the $\tilde{\psi}(p)$, we see that these familiar equations for expressing the wavefunction in the momentum representation follow all the patterns of a basis set expansion. Note that the discrete (but infinite) index i is replaced by the continuous variable p in these expressions; that is the key.

Table 8.2 Fundamental relations in the momentum representation compared with a discrete basis representation

	Discrete basis representation	Momentum representation
Expansion in a basis	$\psi(x) = \sum_{i=0}^{\infty} a_i \phi_i(x)$	$\psi(x) = \frac{1}{\sqrt{2\pi\hbar}} \int_{-\infty}^{\infty} \tilde{\psi}(p) e^{ipx/\hbar} \, dp$
Orthonormality relation	$\int_{-\infty}^{\infty} \phi_i^*(x)\phi_j(x) \, dx = \delta_{ij}$	$\frac{1}{2\pi\hbar} \int_{-\infty}^{\infty} e^{i(p'-p)x/\hbar} \, dx = \delta\left(p - p'\right)$
Expansion coefficient	$a_i = \int_{-\infty}^{\infty} \phi_i^*(x)\psi(x) \, dx$	$\tilde{\psi}(p) = \frac{1}{\sqrt{2\pi\hbar}} \int_{-\infty}^{\infty} \psi(x) e^{-ipx/\hbar} \, dx$
Completeness relation	$\langle \psi | \psi \rangle = 1 = \sum_i |a_i|^2$	$\langle \psi | \psi \rangle = 1 = \int_{-\infty}^{\infty} dp \, |\tilde{\psi}(p)|^2$

The last relation can be rewritten as $\int_{-\infty}^{\infty} |\psi(x)|^2 \, dx = \int_{-\infty}^{\infty} |\tilde{\psi}(p)|^2 \, dp$, and is called Parseval's relation. Here is its proof:

$$\int_{-\infty}^{\infty} dp \, |\tilde{\psi}(p)|^2$$

$$= \int_{-\infty}^{\infty} dp \left\{ \frac{1}{\sqrt{2\pi\hbar}} \int_{-\infty}^{\infty} \psi(x') e^{-ipx'/\hbar} \, dx' \right\}^* \left\{ \frac{1}{\sqrt{2\pi\hbar}} \int_{-\infty}^{\infty} \psi(x) e^{-ipx/\hbar} \, dx \right\} \quad (8.88)$$

$$= \int_{-\infty}^{\infty} \int_{-\infty}^{\infty} \psi^*(x')\psi(x) \left\{ \int_{-\infty}^{\infty} \frac{e^{ip(x'-x)/\hbar}}{2\pi\hbar} dp \right\} dx \, dx' \quad (8.89)$$

$$= \int_{-\infty}^{\infty} \int_{-\infty}^{\infty} \psi^*(x')\psi(x) \left\{ \delta(x'-x) \right\} dx \, dx' \quad (8.90)$$

$$= \int_{-\infty}^{\infty} \psi^*(x)\psi \, dx = \int_{-\infty}^{\infty} |\psi(x)|^2 \, dx. \quad (8.91)$$

Parseval's relation was already encountered in Exercises 6.9 and 6.15. Note the symmetric form of Parseval's relation obtained here, due to the symmetric definition of the forward and backward Fourier transform (see Table 8.2).

8.4.2   Position Representation

Table 8.3 examines the same four characteristics of an arbitrary basis. The first column again shows the equation for this property in a discrete basis, while the second column shows four equations that follow trivially from properties of the Dirac δ-function. The importance of the equations in the last column is not in any novel physics; rather it is in the revealing of a consistent algebra if the basis functions $\{\phi_i(x)\}$ are identified with $\delta(x - x')$ and the expansion coefficients $\{a_i\}$ with $\psi(x')$. Again, note that the discrete index i is replaced by the continuous variable x' in these expressions.

▶ **Exercise 8.64**   Prove the second entry of the array by using one of the two limiting expressions for a δ-function:

$$\text{a. } \delta(x) = \lim_{h \to 0} f(x) \quad f(x) = 1/h \ (-h/2 < x < h/2) \quad (8.92)$$

$$f(x) = 0 \ (x < -h/2; x > h/2) \quad (8.93)$$

$$\text{b. } \delta(x) = \lim_{h \to 0} f(x) \quad f(x) = \sqrt{h/\pi} \, e^{-hx^2}. \quad (8.94)$$

For pedagogical purposes, we will now prove the third and fourth entries in Table 8.3. Although these relations are trivial from one point of view, we wish to highlight the essential parallel of this representation to the others by deriving the relations in a way that

Table 8.3 Fundamental relations in the position representation compared with a discrete basis representation

Discrete basis representation	Position representation
$\psi(x) = \sum_{i=0}^{\infty} a_i \phi_i(x)$	$\psi(x) = \int_{-\infty}^{\infty} \psi(x')\delta(x - x')\, dx'$
$\int_{-\infty}^{\infty} \phi_i^*(x)\phi_j(x)\, dx = \delta_{ij}$	$\int_{-\infty}^{\infty} \delta(x - x')\delta(x - x'')\, dx = \delta(x' - x'')$
$a_i = \int_{-\infty}^{\infty} \phi_i^*(x)\psi(x)\, dx$	$\psi(x') = \int_{-\infty}^{\infty} \psi(x)\delta(x - x')\, dx$
$\langle \psi|\psi \rangle = 1 = \sum_i |a_i|^2$	$\langle \psi|\psi \rangle = 1 = \int_{-\infty}^{\infty} \psi^*(x')\psi(x')\, dx'$

is clumsy, but analogous to the derivations for the other representations, in that the basis and the components are written explicitly.

$$\int_{-\infty}^{\infty} \psi(x)\delta(x - x')\, dx = \int_{-\infty}^{\infty} \left\{ \int_{-\infty}^{\infty} \psi(x'')\delta(x - x'')\, dx'' \right\} \delta(x - x')\, dx$$

$$= \int_{-\infty}^{\infty} \psi(x'')\delta(x' - x'')\, dx''$$

$$= \psi(x')$$

$$\int_{-\infty}^{\infty} \psi^*(x)\psi(x)\, dx = \int_{-\infty}^{\infty} \left\{ \int_{-\infty}^{\infty} \psi^*(x'')\delta(x - x'')\, dx'' \right\} \left\{ \int_{-\infty}^{\infty} \psi(x')\delta(x - x')\, dx' \right\} dx$$

$$= \int_{-\infty}^{\infty} \int_{-\infty}^{\infty} \psi^*(x'')\psi(x') \left\{ \int_{-\infty}^{\infty} \delta(x - x')\delta(x - x'')\, dx \right\} dx'\, dx''$$

$$= \int_{-\infty}^{\infty} \int_{-\infty}^{\infty} \psi^*(x'')\psi(x') \left\{ \delta(x' - x'') \right\} dx'\, dx''$$

$$= \int_{-\infty}^{\infty} \psi^*(x')\psi(x')\, dx'$$

8.4.3 Dirac Notation for Continuous Basis Sets

If we now collect our results from the previous sections, we can make a table of alternative expressions for $\psi(x)$. Using the concept of continuous basis sets, we now understand that all three of these expressions are expansions of $\psi(x)$ in different basis sets. The basis set and the component along each basis vector are shown in Table 8.4.

We may proceed to one more level of abstraction. Underlying $\psi(x)$ is an abstract vector $|\psi\rangle$, of which $\psi(x)$ itself is only one representation. To see this more clearly, consider the

Table 8.4 Comparison of the discrete basis, the momentum, and the position representations

	Discrete	Momentum	Position
Expansion of $\psi(x)$	$\sum_i^{\infty} a_i \phi_i(x)$	$\int_{-\infty}^{\infty} \tilde{\psi}(p) \frac{e^{ipx/\hbar}}{\sqrt{2\pi\hbar}}\, dp$	$\int_{-\infty}^{\infty} \psi(x_0)\delta(x - x_0)\, dx_0$
Basis functions	$\phi_i(x)$	$\frac{e^{ipx/\hbar}}{\sqrt{2\pi\hbar}}$	$\delta(x - x_0)$
Components of the basis	a_i	$\tilde{\psi}(p)$	$\psi(x_0)$

Table 8.5 Expressions for the inner product in different representations

Expression for $\langle \phi | \psi \rangle$	Equivalent expression in Dirac notation
$\sum_{i=1}^{\infty} b_i^* a_i$	$\sum_{i=1}^{\infty} \langle \phi | \phi_i \rangle \langle \phi_i | \psi \rangle$
$\int_{-\infty}^{\infty} \phi^*(x) \psi(x)\, dx$	$\int_{-\infty}^{\infty} \langle \phi | x \rangle \langle x | \psi \rangle\, dx$
$\int_{-\infty}^{\infty} \tilde{\phi}^*(p) \tilde{\psi}(p)\, dp$	$\int_{-\infty}^{\infty} \langle \phi | p \rangle \langle p | \psi \rangle\, dp$

following three expressions for the inner product $\langle \phi | \psi \rangle$, associated with each of the three representations above:

$$\langle \phi | \psi \rangle = \sum_{i=1}^{\infty} b_i^* a_i = \int_{-\infty}^{\infty} \phi^*(x) \psi(x)\, dx = \int_{-\infty}^{\infty} \tilde{\phi}^*(p) \tilde{\psi}(p)\, dp,$$

where b_i is the expansion coefficient of ϕ associated with ϕ_i. These three expressions can be brought into analogous forms if we make the identification

$$\delta(x - x_0) \Longrightarrow \langle x_0 |$$

$$\frac{e^{ipx/\hbar}}{\sqrt{2\pi\hbar}} \Longrightarrow \langle p |.$$

With this identification, we may take the inner product with the abstract vector $| \psi \rangle$:

$$\langle x_0 | \psi \rangle = \int \psi(x) \delta(x - x_0)\, dx = \psi(x_0)$$

$$\langle p | \psi \rangle = \int \psi(x) \frac{e^{ipx/\hbar}}{\sqrt{2\pi\hbar}}\, dx = \tilde{\psi}(p).$$

It is clear that the subscript 0 on x_0 is only an artifact of our treatment in which we began with an axiomatic definition of the inner product in the x representation, and used x_0 to distinguish this variable from x. In reality, x_0 is a dummy variable, and we can now dispense with the subscript; the final result, of great significance, is the pair of identifications

$$\langle x | \psi \rangle = \psi(x)$$

$$\langle p | \psi \rangle = \tilde{\psi}(p).$$

Table 8.5 shows the three expressions for the inner product and their analogous structure in Dirac notation.

We now take notice of the three different expressions for the identity operator reflected in these three expressions:

$$\sum_{i=1}^{\infty} | \phi_i \rangle \langle \phi_i | = \mathbf{1}$$

$$\int_{-\infty}^{\infty} | x \rangle \langle x |\, dx = \mathbf{1}$$

$$\int_{-\infty}^{\infty} | p \rangle \langle p |\, dp = \mathbf{1}.$$

Table 8.6 Important special cases of the closure relation for x

Position space notation	Dirac notation
$\int_{-\infty}^{\infty} \delta(x - x')\delta(x - x'')\, dx = \delta(x' - x'')$	$\int_{-\infty}^{\infty} \langle x''|x\rangle\langle x|x'\rangle\, dx = \langle x''|x'\rangle$
$\int_{-\infty}^{\infty} \left(\frac{e^{ip'x/\hbar}}{\sqrt{2\pi\hbar}}\right)\left(\frac{e^{ip''x/\hbar}}{\sqrt{2\pi\hbar}}\right)^{*} dx = \delta(p' - p'')$	$\int_{-\infty}^{\infty} \langle p''|x\rangle\langle x|p'\rangle\, dx = \langle p''|p'\rangle$
$\int_{-\infty}^{\infty} \delta(x - x')\frac{e^{ip'x/\hbar}}{\sqrt{2\pi\hbar}}\, dx = \frac{e^{ip'x'/\hbar}}{\sqrt{2\pi\hbar}}$	$\int_{-\infty}^{\infty} \langle x'|x\rangle\langle x|p'\rangle\, dx = \langle x'|p'\rangle$

The first relationship we have seen before: it is the closure relationship in a discrete basis. The latter two equations are closure relationships in the continuous variables x and p. As in the case of the discrete closure relationship, these representations of unity can be inserted wherever desired.

We illustrate the closure relationship for x on some important special cases in Table 8.6. For comparison, the same relationships are written in conventional form; it is apparent that Dirac notation makes these relationships trivial to derive and exposes the underlying unity of representation. Summarizing the key relationships, we have

$$\langle x'|x\rangle = \delta(x - x'); \qquad \langle p'|p\rangle = \delta(p - p'); \qquad \langle x|p\rangle = \frac{1}{\sqrt{2\pi\hbar}}e^{ipx/\hbar}. \quad (8.95)$$

As an example, we show the invariance of the trace to a transformation from a discrete basis representation to the x-representation:

$$\mathrm{Tr}(A) = \sum_i \langle \phi_i|A|\phi_i\rangle$$

$$= \sum_i \int dx \int dx' \langle \phi_i|x\rangle\langle x|A|x'\rangle\langle x'|\phi_i\rangle$$

$$= \sum_i \int dx \int dx' \langle x'|\phi_i\rangle\langle \phi_i|x\rangle\langle x|A|x'\rangle$$

$$= \int dx \int dx' \langle x'|x\rangle\langle x|A|x'\rangle$$

$$= \int dx \int dx' \, \delta(x - x')\langle x|A|x'\rangle$$

$$= \int dx \, \langle x|A|x\rangle.$$

Two other important relationships involve the action of the operators $\hat{x}$ and $\hat{p}$ on the kets $|x\rangle$ and $|p\rangle$:

$$\hat{x}|x\rangle = x|x\rangle; \qquad \hat{p}|p\rangle = p|p\rangle. \quad (8.96)$$

▶ **Exercise 8.65**   Prove these relationships by using the definition of $|x\rangle$ and $|p\rangle$.

▶ **Exercise 8.66**   Rewrite the three expansions for $\psi(x)$ in Table 8.4 in Dirac notation, and show how they represent different insertions of unity in Dirac notation:

a. $\psi(x) = \sum_i a_i \psi_i(x)$

b. $\psi(x) = \frac{1}{\sqrt{2\pi\hbar}} \int_{-\infty}^{\infty} \tilde{\psi}(p) e^{ipx/\hbar} \, dp$

c. $\psi(x) = \int_{-\infty}^{\infty} \psi(x_0) \delta(x - x_0) \, dx_0.$

▶ **Exercise 8.67**   Rewrite the three expressions for the coefficients of the basis functions in the previous problem using Dirac notation:

a. $a_i = \int_{-\infty}^{\infty} \phi_i^*(x) \psi(x) \, dx$

b. $\tilde{\psi}(p) = \frac{1}{\sqrt{2\pi\hbar}} \int_{-\infty}^{\infty} \psi(x) e^{-ipx/\hbar} \, dx$

c. $\psi(x_0) = \int_{-\infty}^{\infty} \psi(x) \delta(x - x_0) \, dx.$

▶ **Exercise 8.68**   Prove Parseval's relation, $\int_{-\infty}^{\infty} dp \, |\tilde{\psi}(p)^2| = \int_{-\infty}^{\infty} dx \, |\psi(x)^2|$, using Dirac notation.

▶ **Exercise 8.69**   Consider the expression $\langle \phi_j | \phi_i \rangle = \delta_{ij}$, where ϕ_i and ϕ_j are orthonormal members of the harmonic oscillator basis. Insert a complete set of x states and interpret the orthogonality relation graphically.

▶ **Exercise 8.70**   Consider the expression $\langle x' | x \rangle = \delta(x - x')$ and insert a complete set of harmonic oscillator eigenstates. Interpret the resulting orthogonality relation graphically.

We are now in a position to prove two of the most important, but most subtle, relations in quantum mechanics. From an early stage we have implicitly identified $\langle x | \hat{H} | \psi \rangle = -\frac{\hbar^2}{2m} \frac{\partial^2}{\partial x^2} \psi(x) + V(x)\psi(x)$. This follows from the identification of $\langle x | V(\hat{x}) | \psi \rangle = V(x)\psi(x)$ and $\langle x | \hat{p} | \psi \rangle = -i\hbar \frac{\partial}{\partial x} \psi(x)$, relations that we now prove.

$$\langle x | V(\hat{x}) | \psi \rangle = \langle \psi | V(\hat{x}) | x \rangle^*$$
$$= \langle \psi | V(x) | x \rangle^* \text{(recall that if } Av = \lambda v, \ f(A)v = f(\lambda)v)$$
$$= (\langle \psi | x \rangle V(x))^*$$
$$= V(x)\psi(x).$$

Turning to the second relationship, we have

$$\langle x | \hat{p} | \psi \rangle = \int \langle x | p \rangle \langle p | \hat{p} | \psi \rangle \, dp$$

$$= \int \frac{e^{ipx/\hbar}}{\sqrt{2\pi\hbar}} p\tilde{\psi}(p) \, dp$$

$$= \int -i \frac{\partial}{\partial x} \left\{ \frac{1}{\sqrt{2\pi\hbar}} \tilde{\psi}(p) e^{ipx/\hbar} \right\} dp$$

$$= -i \frac{\partial}{\partial x} \psi(x), \tag{8.97}$$

where in the last step we have used the fact that the order of integration and differentiation can be interchanged.

Finally, we wish to derive a formula for the reciprocal expression, $\langle p|\hat{x}|\psi\rangle$. Proceeding in the analogous way we find

$$\langle p|\hat{x}|\psi\rangle = \int \langle p|x\rangle\langle x|\hat{x}|\psi\rangle \, dx$$

$$= \int \frac{e^{-ipx/\hbar}}{\sqrt{2\pi\hbar}} x\psi(x) \, dx$$

$$= \int i\frac{\partial}{\partial p} \left\{ \frac{1}{\sqrt{2\pi\hbar}}\psi(x)e^{-ipx/\hbar} \right\} dx$$

$$= i\frac{\partial}{\partial p}\tilde{\psi}(p). \qquad (8.98)$$

Equation 8.98 was used in Chapter 2, to obtain Eqs. 2.24 and 2.35.

8.4.4   Uncapped Bras and Kets: General Properties of Dirac Notation

Dirac notation is at the same time the most abstract and the most powerful system of notation in quantum mechanics. We conclude this chapter by collecting together the various rules associated with the manipulation of the symbols in Dirac notation.

1. In general, state vectors are represented in Dirac notation as kets: $u, v \rightarrow |u\rangle, |v\rangle$.

2. Operators operating on bras and kets may be manipulated as follows:

$$A|u\rangle = |Au\rangle; \qquad \langle v|A = \langle A^\dagger v|.$$

3. The general rules for exchanging bras and kets in Dirac notation are

   a. $c \rightarrow c^*$ (c is a number)
   b. $|\psi\rangle \rightarrow \langle\psi|$
   c. $\langle\psi| \rightarrow |\psi\rangle$
   d. $A \rightarrow A^\dagger$
   e. reverse order of operators.

▶ **Exercise 8.71**   Find the bra that corresponds to the ket $c^*\langle u|AB|v\rangle CD|w\rangle$.

Solution $\langle w|D^\dagger C^\dagger\langle v|B^\dagger A^\dagger|u\rangle c.$

---

4. The object $|u\rangle\langle v|$ is an operator. To see this, note that

$$\{|u\rangle\langle v|\}|w\rangle = |u\rangle\langle v|w\rangle = \langle v|w\rangle|u\rangle,$$

   hence $|u\rangle\langle v|$ maps the vector $|w\rangle$ into a vector proportional to $|u\rangle$. Anything that maps a vector into a vector is an operator. (We have already seen that $\sum_i |u_i\rangle\langle u_i| = 1$, the identity operator.)

5. The adjoint of $|u\rangle\langle v|$ is $|v\rangle\langle u|$. To see this, consider

$$\langle a|\{|u\rangle\langle v|\}b\rangle = \langle a|u\rangle\langle v|b\rangle = \{\langle b|v\rangle\langle u|a\rangle\}^*.$$

Recall that

$$\langle a|A|b\rangle = \langle A^\dagger a|b\rangle = \langle b|A^\dagger a\rangle^* = \langle b|A^\dagger|a\rangle^*.$$

We therefore identify $A^\dagger$ with $|v\rangle\langle u|$.

▶ **Exercise 8.72**   Find the adjoint of $\{c\langle u|A|v\rangle|w\rangle\langle\psi|\}$.

Solution $|\psi\rangle\langle w|\langle v|A^\dagger|u\rangle c^*$.

---

6. The object $|u\rangle\langle u|$ is a fundamental projection operator if $||u|| = \langle u|u\rangle = 1$. To see this, note that $\{|u\rangle\langle u|\}|w\rangle = \langle u|w\rangle|u\rangle$, that is, $|u\rangle\langle u|$ maps the vector $|w\rangle$ onto a vector proportional to $|u\rangle$; the proportionality coefficient is the projection of w onto u, $\langle u|w\rangle$. To check that $|u\rangle\langle u|$ is a projector, note that

$$\{|u\rangle\langle u|\}^2 = |u\rangle\langle u|\{|u\rangle\langle u|\} = |u\rangle\langle u|.$$

▶ **Exercise 8.73**   Show that $\sum_i |u_i\rangle\langle u_i|$ is a projection operator if and only if $\langle u_i|u_j\rangle = \delta_{ij}$. In the case that the sum over i ranges over a complete basis we recover the closure relationship, $\sum_i |u_i\rangle\langle u_i| = \mathbf{1}$, which is effectively a complete projection.

Problems

▶ **Exercise 8.74**   In this exercise we consider the form of the unitary matrix, **U**, that diagonalizes a general 2×2 Hermitian matrix, **V**. Show that for any Hermitian $\mathbf{U}^\dagger\mathbf{V}\mathbf{U} = \mathbf{\Lambda}$, where

$$\mathbf{V} = \begin{pmatrix} V_1 & V_{12} \\ V_{21} & V_2 \end{pmatrix} \quad \mathbf{U} = \begin{pmatrix} \cos\frac{\theta}{2}e^{-i\phi/2} & -\sin\frac{\theta}{2}e^{-i\phi/2} \\ \sin\frac{\theta}{2}e^{i\phi/2} & \cos\frac{\theta}{2}e^{i\phi/2} \end{pmatrix} \quad \mathbf{\Lambda} = \begin{pmatrix} \lambda_1 & 0 \\ 0 & \lambda_2 \end{pmatrix}. \quad (8.99)$$

The angles θ and ϕ are defined by

$$\sin\theta = \frac{2|V_{21}|}{D^{1/2}} \quad \cos\theta = \frac{V_1 - V_2}{D^{1/2}} \quad V_{21} = |V_{21}|e^{i\phi} = V_{12}^*, \quad (8.100)$$

where $D = 4|V_{21}|^2 + (V_1 - V_2)^2$. The elements of the diagonal matrix, $\mathbf{\Lambda}$, are the two eigenvalues of **V**:

$$\lambda_{1,2} = \frac{1}{2}(V_1 + V_2) \pm \frac{1}{2}\sqrt{(V_1 - V_2)^2 + 4|V_{12}|^2}. \quad (8.101)$$

[Hint: Do all the matrix multiplications using the above representation of U. At the end, use the trigonometric half-angle relations $\sin^2\frac{\theta}{2} = \frac{1}{2}(1 - \cos\theta)$, $\cos^2\frac{\theta}{2} = \frac{1}{2}(1 + \cos\theta)$ and $\sin\frac{\theta}{2}\cos\frac{\theta}{2} = \frac{1}{2}\sin\theta$.] The angles θ and ϕ have a geometrical representation closely connected to the Bloch sphere, discussed in Section 15.2 (see Cohen-Tannoudji (1977), Complement BIV, CIV).

▶ Exercise 8.75

    a. Use a Taylor series expansion to show that

$$e^{\mathbf{A}} = \mathbf{U} e^{\mathbf{\Lambda}} \mathbf{U}^{-1}, \tag{8.102}$$

        where

$$\mathbf{A} = \mathbf{U} \mathbf{\Lambda} \mathbf{U}^{-1} \tag{8.103}$$

        and $\mathbf{\Lambda}$ is a diagonal matrix.

    b. Show that in general

$$f(\mathbf{A}) = \mathbf{U} f(\mathbf{\Lambda}) \mathbf{U}^{-1} \tag{8.104}$$

        provided that the Taylor series of $f(x)$ exists.

    c. Give a 2×2 representation for $e^{\mathbf{A}}$ with $\mathbf{A} = \begin{pmatrix} 2 & 1 \\ 1 & 2 \end{pmatrix}$.

▶ Exercise 8.76

    a. Use the result of Exercise 8.75 to show that

$$\det \left(e^{\mathbf{A}} \right) = e^{\text{Tr} \mathbf{A}}. \tag{8.105}$$

    b. Verify Eq. 8.105 explicitly for $\mathbf{A} = \begin{pmatrix} 2 & 1 \\ 1 & 2 \end{pmatrix}$.

▶ Exercise 8.77

    a. Use a Taylor series expansion, together with the Cayley–Hamilton theorem (Exercise 8.44) to prove that

$$f(\mathbf{A}) = \sum_{n=0}^{\infty} a_n \mathbf{A}^n = \sum_{n=0}^{N-1} a'_n \mathbf{A}^n, \tag{8.106}$$

        where $\mathbf{A}$ is a matrix of dimension $N \times N$. [Hint: the Cayley–Hamilton theorem can be interpreted as a statement about the linear dependence of powers of a matrix of finite dimension.]

    b. Consider $f(\mathbf{A}) = e^{\mathbf{A}}$ with $\mathbf{A} = \begin{pmatrix} 2 & 1 \\ 1 & 2 \end{pmatrix}$. Find the sets of coefficients $\{a_n\}$ and $\{a'_n\}$ for the two expansions in part a.

▶ **Exercise 8.78**   An operator that satisfies the property

$$A^\dagger A = AA^\dagger \tag{8.107}$$

is called "normal." All normal operators admit a decomposition of the form (cf. Eq. 8.79)

$$A = \sum_{i=1}^{N} \lambda_i P_i, \tag{8.108}$$

where the P_i are complete ($\sum_{i=1}^{N} P_i = 1$) and orthonormal ($P_i P_j = P_i \delta_{ij}$).

a. Show that Hermitian and unitary operators are normal operators.

b. Use the decomposition in Eq. 8.108 to prove the Cayley–Hamilton theorem (cf. Exercises 8.44 and 8.78). (Note that the Cayley–Hamilton theorem applies even to matrices that do not admit a decomposition of the form in Eq. 8.108, but the proof for general matrices is more involved.)

c. Show that the matrix $\begin{pmatrix} 1 & 1 \\ 0 & 1 \end{pmatrix}$ is not normal, but still satisfies the Cayley–Hamilton theorem.

▶ **Exercise 8.79**   In Exercise 8.38 we encountered the generalized eigenvalue problem, which we saw could be written in the form (Eq. 8.67)

$$\mathbf{HU} = \mathbf{SUE}. \tag{8.109}$$

Show that this equation can be transformed into the form

$$\mathbf{H'U'} = \mathbf{U'E} \tag{8.110}$$

by an appropriate transformation.

Solution   Let $\mathbf{T}$ be the unitary transformation that diagonalizes $\mathbf{S}$,

$$\mathbf{TST}^\dagger = \mathbf{s}. \tag{8.111}$$

Then Eq. 8.67 can be rewritten

$$\mathbf{THT}^\dagger \mathbf{TU} = \mathbf{TST}^\dagger \mathbf{TUE} = \mathbf{sTUE}. \tag{8.112}$$

Left multiplying both sides by the diagonal matrix $\mathbf{s}^{-1/2}$ we obtain

$$\mathbf{s}^{-1/2}\mathbf{THT}^\dagger \mathbf{s}^{-1/2}\mathbf{s}^{1/2}\mathbf{TU} = \mathbf{s}^{1/2}\mathbf{TUE} \tag{8.113}$$

or

$$\mathbf{H'U'} = \mathbf{U'E}, \tag{8.114}$$

where we have defined

$$\mathbf{H'} = \mathbf{s}^{-1/2}\mathbf{THT}^\dagger \mathbf{s}^{-1/2}, \qquad \mathbf{U'} = \mathbf{s}^{-1/2}\mathbf{TU}. \tag{8.115}$$

The generalized eigenvalue problem has an interesting geometrical interpretation. Consider two generalized ellipsoids, one corresponding to $\mathbf{S}$ and one to $\mathbf{H}$. In general, it is impossible to find joint principal axes for both ellipsoids. However, one may perform a principal axis transformation on **S** and then scale the length of the **S** axes, deforming the S-ellipsoid into a sphere. The same two-part transformation that turns **S** into a sphere must then be applied to the **H**-ellipsoid, which in general remains an ellipsoid. Now, a principal axis transformation is applied to the deformed **H**-ellipsoid; the same transformation must be applied to the deformed S-ellipsoid, but since the latter is a sphere, all directions correspond to principal axes.

▶ Exercise 8.80

    a. In Exercise 8.79, take

$$\mathbf{H} = \begin{pmatrix} 1 & 0 \\ 0 & 1 \end{pmatrix} \quad \text{and} \quad \mathbf{S} = \begin{pmatrix} 2 & 1 \\ 1 & 2 \end{pmatrix}. \tag{8.116}$$

        Find the eigenvalues and eigenvectors of Eq. 8.67 by solving Eq. 8.114. Note that one must back-transform the solutions of Eq. 8.114 to obtain the eigenvectors of the original Eq. 8.67.

    b. It is possible to solve Eq. 8.67 directly via the equation

$$|\mathbf{H} - E\mathbf{S}| = 0. \tag{8.117}$$

        Using the same H and S as in part a, calculate the eigenvalues and eigenvectors using Eq. 8.117.

▶ Exercise 8.81   Show that for nonorthogonal bases the closure relation takes the form

$$\sum_{ij} |\phi_i\rangle \left(S^{-1}\right)_{ij} \langle\phi_j| = \mathbf{1}, \tag{8.118}$$

where **S** is the *overlap matrix*, with elements $S_{ij} \equiv \langle\phi_i|\phi_j\rangle$.

Solution   Apply the operator on the LHS of Eq. 8.118 to an arbitrary state $|\Psi\rangle = \sum_k \alpha_k|\phi_k\rangle$, where $\{\phi_k\}$ is a nonorthogonal basis. Then

$$\sum_{ij} |\phi_i\rangle \left(S^{-1}\right)_{ij} \langle\phi_j|\Psi\rangle = \sum_{ijk} \alpha_k|\phi_i\rangle \left(S^{-1}\right)_{ij} \langle\phi_j|\phi_k\rangle \tag{8.119}$$

$$= \sum_{ijk} \alpha_k|\phi_i\rangle \left(S^{-1}\right)_{ij} S_{jk} \tag{8.120}$$

$$= \sum_{ik} \alpha_k|\phi_i\rangle \left(S^{-1}S\right)_{ik} \tag{8.121}$$

$$= \sum_{ik} \alpha_k|\phi_i\rangle \delta_{ik} = \sum_k \alpha_k|\phi_k\rangle = |\Psi\rangle. \tag{8.122}$$

Since the operator on the LHS of Eq. 8.118 returns the original state $|\Psi\rangle$ for any $|\Psi\rangle$ it must be equal to the identity operator.

References

General

1. P. A. M. Dirac, *The Principles of Quantum Mechanics* (Oxford, 1978).
2. C. Cohen-Tannoudji, B. Diu and F. Laloë, *Quantum Mechanics*, vol. I (Wiley, New York, 1977).
3. G. Strang, *Linear Algebra and Its Applications*, 3rd ed. (Harcourt Brace Jovanovich, San Diego, 1986).
4. E. Merzbacher, *Quantum Mechanics* (Wiley, New York, 1970).

Linear Algebra in Quantum Chemistry

5. I. N. Levine, *Quantum Chemistry*, 5th ed. (Prentice Hall, Upper Saddle River, NJ, 2000).
6. D. A. McQuarrie, *Quantum Chemistry* (University Science Books, Sausalito, 1983).
7. M. D. Fayer, *Elements of Quantum Mechanics* (Oxford, New York, 2001).
8. N. Moiseyev, *Quantum Mechanics: A Chemistry Perspective* (Michlol, Haifa, 1996; in Hebrew).
9. A. Szabo and N. S. Ostlund, *Modern Quantum Chemistry: Introduction to Advanced Electronic Structure Theory*, revised 1st ed. (McGraw-Hill, New York, 1982).

Chapter 9

Approximate Solutions of
the Time-Dependent Schrödinger Equation

This chapter will introduce a variety of methods for approximating the Time-Dependent Schrödinger Equation and its solutions. In general, the quality of an approximation depends on the choice of representation with which one chooses to work, and a good choice of representation is essential for an accurate approximation. Therefore, this chapter will begin with a discussion of some alternative representations of the Time-Dependent Schrödinger Equation, specifically with a discussion of the Heisenberg and interaction pictures. Although these are not approximations in themselves, the interaction picture in particular provides an alternative starting point that is useful for a perturbation series expansion. The chapter continues with a description of the Magnus expansion, which provides an expression for the evolution operator for time-dependent Hamiltonians as the exponential of a generally infinite sum of operators. The infinite sum in the exponent can be truncated at any desired order, giving an approximation to the propagator. An alternative representation of the propagator for time-dependent Hamiltonians, due to Wei and Norman, is also discussed. The Wei–Norman formulation, based on the assumption of a closed Lie algebra, represents the propagator as a product of exponentials of operators, as opposed to the exponential of a sum in the Magnus expansion. The chapter proceeds with a discussion of certain limiting cases of time-dependent Hamiltonians, specifically slowing varying Hamiltonians (adiabatic dynamics) and periodic Hamiltonians. Normally, the evolution under time-dependent Hamiltonians requires a more complicated description than for time-independent Hamiltonians. However, in these two special cases, the dynamics is well described using one or a small number of time-dependent eigenstates. All the above topics, loosely speaking, involve choosing the right "representation," which then leads very naturally to a good approximation by keeping just a small number of terms or a small number of basis functions. The chapter concludes with a discussion of time-dependent variational principles and the time-dependent self-consistent field approach, which follows naturally.

It may be worthwhile to contrast this chapter with Chapter 11 on numerical methods. Numerical methods, after all, also provide approximations to the true wavefunction. In fact, a good choice of representation is essential for an efficient numerical method. Many of the topics in this chapter have, in fact, been used in conjunction with numerical methods, and there are probably many unexplored connections between topics in these two chapters that warrant further thought. However, the key distinguishing feature of the methods of this chapter is that one or a small number of terms in the expansion provides a useful representation of the solution. In contrast, numerical methods are typically designed to be efficient with hundreds of basis functions or iterations, and normally give a meaningless representation of the wavefunction when only one or two terms or iterations are retained.

9.1   The Schrödinger, Heisenberg and Interaction Pictures

In the previous chapter (Section 8.2) we encountered the evolution operator, $e^{-\frac{i}{\hbar}Ht}$, which maps $\Psi(0)$ into $\Psi(t)$:

$$\Psi(t) = e^{-\frac{i}{\hbar}Ht}\Psi(0). \tag{9.1}$$

The evolution operator is unitary (see Section 8.2.4) and is sometimes given the symbol

$$U(t) = e^{-iHt/\hbar}. \tag{9.2}$$

(If the Hamiltonian is time dependent, $H = H(t)$, the evolution operator is still unitary but the expression on the RHS has to be modified; see Section 9.3.1). Differentiating both sides of Eq. 9.2 with respect to time we obtain

$$i\hbar\dot{U}(t) = HU(t), \qquad U(0) = \mathbf{1}, \tag{9.3}$$

indicating that U satisfies the Time-Dependent Schrödinger Equation. Equation 9.3 is in fact more general than Eq. 9.2, since it is valid even for time-dependent Hamiltonians, $H = H(t)$. The evolution operator is often called the *propagator*.

Consider the expectation value of an operator A as a function of time:

$$\begin{aligned}
\langle A\rangle_t &= \langle\Psi(t)|A|\Psi(t)\rangle \\
&= \langle e^{-\frac{i}{\hbar}Ht}\Psi(0)|A|e^{-\frac{i}{\hbar}Ht}\Psi(0)\rangle.
\end{aligned} \tag{9.4}$$

This is called the Schrödinger picture: operators are stationary but expectation values may evolve in time since the state vector Ψ evolves in time.

We now turn to the Heisenberg picture. In the Heisenberg picture, the wavefunction Ψ_H is defined to be

$$\Psi_H(t) = e^{\frac{i}{\hbar}Ht}\Psi_S(t). \tag{9.5}$$

Since $\Psi_S(t) = \exp(-iHt/\hbar)\Psi_S(0)$, we see that $\Psi_H(t) = \Psi_S(0) = \Psi_H(0)$; in the Heisenberg picture the state vector is constant in time! Operators in the Heisenberg picture, A_H, are related to their counterparts in the Schrödinger picture, A, by the relation

$$A_H(t) = e^{\frac{i}{\hbar}Ht}\,A\,e^{-\frac{i}{\hbar}Ht}. \tag{9.6}$$

The constancy of the state vectors in the Heisenberg picture is compensated by the time dependence of the operators, with the result that the expectation value of any operator in the Heisenberg picture is the same as in the Schrödinger picture:

$$\begin{aligned}
\langle A\rangle_t &= \langle\Psi_S(t)|A|\Psi_S(t)\rangle \\
&= \langle\Psi_S(t)|e^{-\frac{i}{\hbar}Ht}e^{\frac{i}{\hbar}Ht}\,A\,e^{-\frac{i}{\hbar}Ht}e^{\frac{i}{\hbar}Ht}|\Psi_S(t)\rangle \\
&= \langle\Psi_H(0)|A_H(t)|\Psi_H(0)\rangle.
\end{aligned}$$

Since this is true for any state vector, it implies that the two formalisms give the same result for any observable.

▶ **Exercise 9.1**   Show that the Hamiltonian in the Heisenberg representation H_H is

$$H_H = \frac{p_H^2}{2m} + V(x_H) = H, \tag{9.7}$$

where H is the Hamiltonian in the Schrödinger representation. Are the two Hamiltonians H_H and H equal term by term?

It is interesting to consider the form of the equations of motion governing the time evolution of operators in the Heisenberg picture. Differentiating the definition of A_H leads to

$$i\hbar \frac{d}{dt} A_H = [A_H, H]. \tag{9.8}$$

In general, if the operator A has some explicit time dependence, then

$$i\hbar \frac{d}{dt} A_H = [A_H, H] + i\hbar\, e^{\frac{i}{\hbar}Ht} \frac{\partial A}{\partial t} e^{-\frac{i}{\hbar}Ht} \tag{9.9}$$

$$= [A_H, H] + i\hbar \frac{\partial A_H}{\partial t}. \tag{9.10}$$

Now specialize Eq. 9.8 to the operators p and q. Noting that $[p, H] = -i\hbar dH/dq$ and $[q, H] = i\hbar p/m$, we have $[p_H, H] = -i\hbar(dH/dq)_H$ and $[q_H, H] = i\hbar p_H/m$. Using Eq. 9.7 we obtain

$$\frac{dp_H}{dt} = -\frac{\partial H}{\partial q_H} = -\frac{\partial V}{\partial q_H} \tag{9.11}$$

$$\frac{dq_H}{dt} = \frac{\partial H}{\partial p_H} = \frac{p_H}{m}. \tag{9.12}$$

These equations are the operator analog to Hamilton's equations in classical mechanics. Although in the Heisenberg picture the degree of classical–quantum correspondence would seem to be complete, recall that operators themselves are not measured, only the corresponding expectation values. The degree to which the expectation values, that is, the average of the operator over the wavefunction of the system, satisfy equations of motion identical to Hamilton's equations is the subject Ehrenfest's theorem, which was discussed in Chapter 4. In fact, Eqs. 9.11–9.12 served as the starting point for the derivation there.

Having discussed the Schrödinger and Heisenberg picture, we now turn to the interaction picture. It is often the case that the full Hamiltonian, H, can be partitioned into $H = H_0 + V$, where the time evolution generated by H_0 is known. We define the interaction picture wavefunction $\Psi_I(t)$ as

$$\Psi_I(t) = e^{\frac{i}{\hbar}H_0 t} \Psi_S(t) = e^{\frac{i}{\hbar}H_0 t} e^{-\frac{i}{\hbar}Ht} \Psi_S(0), \tag{9.13}$$

where $\Psi_S(t)$ is the Schrödinger picture wavefunction that we have been using all along. If "most" of the time evolution is generated by H_0, then the forward propagation under

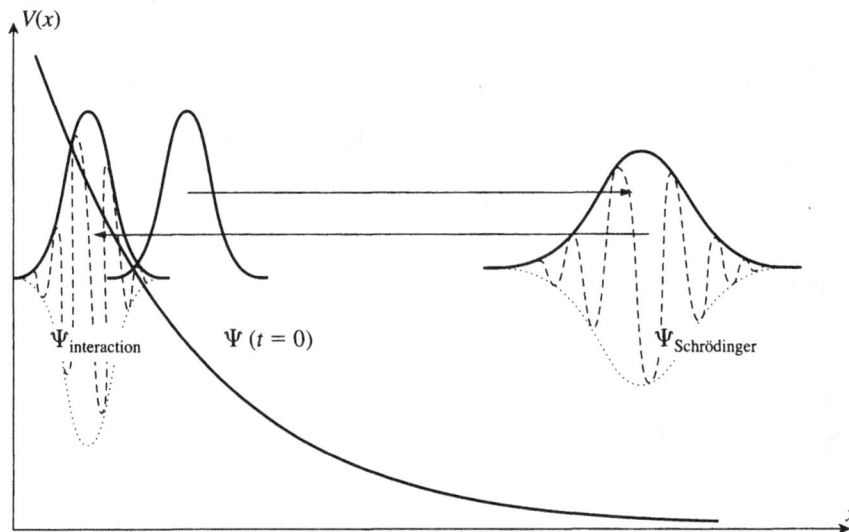

Figure 9.1 Qualitative diagram showing the relationship between the Schrödinger and interaction wavefunctions on a purely repulsive potential. The initial Schrödinger wavefunction starts from rest on the repulsive wall, and proceeds to the asymptotic region of the potential. The dashed line represents the real part of the wavefunction, corresponding to the momentum that the wavefunction develops as it accelerates down the potential. If H_0 is taken as the kinetic energy operator, the interaction wavefunction corresponds to propagation of the Schrödinger wavefunction backward in time under free-particle evolution (see Eq. 9.13). Note that the interaction wavefunction is much closer to the initial state, both in its width and its average position, than the time-evolving Schrödinger wavefunction, because the wavepacket spreading and translation are largely reversed by the backward propagation. Reprinted from D. J. Tannor et al., J. Chem. Phys. 95, 1721 (1991).

H should almost cancel with the backward propagation under H_0, and Ψ_I should undergo much less evolution than in the Schrödinger picture. This is shown in Figure 9.1.

▶ **Exercise 9.2**   If $V \to V_0$ as $x \to \infty$, and if the Schrödinger wavepacket is completely unbound, then $\Psi_I(t)$ reaches a limit (i.e., stops evolving) as $t \to \infty$. Explain why. This property is closely tied to the definition of the Møller operators, which will play an important role in Chapters 17–18.

We can obtain the equation of motion for the interaction picture wavefunction by simply substituting it into the Time-Dependent Schrödinger Equation. After multiplying the result by $\exp(i H_0 t / \hbar)$, one obtains

$$i\hbar \frac{\partial \Psi_I(t)}{\partial t} = [e^{\frac{i}{\hbar} H_0 t} \, V \, e^{-\frac{i}{\hbar} H_0 t}]\Psi_I(t) \tag{9.14}$$

$$= H_I(t)\Psi_I(t), \tag{9.15}$$

where we have defined the interaction picture Hamiltonian, H_I, as

$$H_I(t) = e^{\frac{i}{\hbar} H_0 t} \, V \, e^{-\frac{i}{\hbar} H_0 t}. \tag{9.16}$$

Note the structural similarity between Eq. 9.15 and the original Time-Dependent Schrödinger Equation, although in the interaction picture the Hamiltonian, $H_I(t)$, is in general time dependent.

Now suppose we have a set of functions $\{\psi_n\}$ such that $H_0\psi_n = E_n^0\,\psi_n$. Then we can expand the interaction picture wavefunction:

$$\Psi_I(t) = \sum_n a_n(t)\,\psi_n. \tag{9.17}$$

When this *ansatz* is substituted in the equation of motion, and some algebra is performed, one obtains a set of differential equations for the time evolution of the expansion coefficients:

$$i\hbar\dot{a}_m(t) = \sum_n V_{mn} e^{\frac{i}{\hbar}(E_m - E_n)t} a_n(t), \tag{9.18}$$

where

$$V_{mn} = \int \psi_m^* \, V \, \psi_n \, dx. \tag{9.19}$$

In the interaction picture, the $\{a_m(t)\}$ are relatively slowly varying so a large time step may be used to integrate the equations of motion. Defining

$$H_{mn}(t) = V_{mn} e^{\frac{i}{\hbar}(E_m - E_n)t}, \tag{9.20}$$

we obtain a matrix representation of the Time-Dependent Schrödinger Equation in the interaction picture,

$$i\hbar\dot{a}_m(t) = \sum_n H_{mn}(t)a_n(t). \tag{9.21}$$

The interaction picture can be seen as intermediate between the Schrödinger picture and the Heisenberg picture. In the Schrödinger picture it is the wavefunction that evolves, in the Heisenberg picture it is the operator that evolves, and in the interaction picture both the state vector and the operator contain part of the time evolution. The relationship of the three pictures is summarized by the following equalities:

$$\langle A \rangle_t = \langle \Psi_S(t)|A_S|\Psi_S(t)\rangle \qquad \text{(Schrödinger)}$$

$$= \langle \Psi_S(0)|e^{\frac{i}{\hbar}Ht} A_S \, e^{-\frac{i}{\hbar}Ht}|\Psi_S(0)\rangle$$

$$= \langle \Psi_H|A_H(t)|\Psi_H\rangle \qquad \text{(Heisenberg)} \tag{9.22}$$

$$= \langle \Psi_S(0)|\underbrace{e^{\frac{i}{\hbar}Ht}e^{-\frac{i}{\hbar}H_0 t}}\,\underbrace{e^{\frac{i}{\hbar}H_0 t} A_S \, e^{-\frac{i}{\hbar}H_0 t}}\,\underbrace{e^{\frac{i}{\hbar}H_0 t}e^{-\frac{i}{\hbar}Ht}}|\Psi_S(0)\rangle$$

$$= \langle \Psi_I(t)|A_I(t)|\Psi_I(t)\rangle. \qquad \text{(Interaction)}$$

▶ **Exercise 9.3**   Show that the generalization of Eq. 9.14 to time-dependent Hamiltonians,

$$H(t) = H_0(t) + V(t), \tag{9.23}$$

is

$$i\hbar\frac{\partial \Psi_I(t)}{\partial t} = U_0^{-1}(t)H(t)U_0(t) - i\hbar\dot{U}_0(t) \tag{9.24}$$

$$\equiv H_I(t)\Psi_I(t), \tag{9.25}$$

where

$$\Psi_{\mathrm{I}}(t) = U_0^{-1}(t)\Psi(t); \tag{9.26}$$

$U_0(t)$ is the evolution operator generated by $H_0(t)$.

▶ **Exercise 9.4**   Consider the periodic Hamiltonian, $H(t)$,

$$H(t) = \begin{pmatrix} E_a & -\mu_{ab}Ee^{i\omega t} \\ -\mu_{ba}Ee^{-i\omega t} & E_b \end{pmatrix}. \tag{9.27}$$

Find an interaction picture that transforms this Hamiltonian into a time-independent Hamiltonian, H'. Diagonalize H', and use the eigenvalues and eigenvectors to represent the evolution of the state vector with initial conditions $(1, 0)$ in the original basis.

9.2   Time-Dependent Perturbation Theory

The interaction picture, described in the previous section, is formally exact. In this section we introduce time-dependent perturbation theory (TDPT), which uses the same partitioning of the Hamiltonian into $H_0 + V$ to develop a series of successive approximations to the evolving wavefunction. Like the interaction picture, TDPT is useful only if the dynamics under H_0 are known. Moreover, TDPT is not recommended unless $V \ll H_0$ in some sense, in other words unless V can be regarded as a perturbation on H_0. In fact, if V is too large, TDPT will not converge. Below we will present two derivations of TDPT. The first employs an ordering parameter to collect terms of the same order in the perturbation. In the second derivation different terms in the perturbation expansion emerge naturally from an expansion of the propagator in the interaction picture.

9.2.1   Derivation Using an Ordering Parameter λ

Suppose we are interested in the time evolution generated by a Hamiltonian of the form $H = H_0 + H_1$, where H_0 is time independent and H_1 need not be. Further assume that we can compute the time evolution generated by H_0. If H_1 is "small enough," then it should act as a small perturbation on the time evolution generated by H_0. We introduce a parameter, λ, which will be used simply to organize terms that are in similar orders of the perturbation:

$$H = H_0 + \lambda H_1. \tag{9.28}$$

In the end, λ will be set equal to one. We can expand the wavefunction of interest in a power series in λ:

$$\Psi(t) = \Psi^{(0)}(t) + \lambda\Psi^{(1)}(t) + \lambda^2\Psi^{(2)}(t) + \cdots = \sum_n \lambda^n \Psi^{(n)}(t). \tag{9.29}$$

Of course, we do not know *a priori* that this expansion will converge, but we will assume it does. Substituting the expansion into the Time-Dependent Schrödinger Equation and equating terms with like powers of λ leads to a series of equations for the terms in the expansion, shown here up to second order:

$$i\hbar \frac{\partial}{\partial t} \Psi^{(0)} = H_0 \Psi^{(0)} \tag{9.30}$$

$$i\hbar \frac{\partial}{\partial t} \Psi^{(1)} = H_0 \Psi^{(1)} + H_1 \Psi^{(0)} \tag{9.31}$$

$$i\hbar \frac{\partial}{\partial t} \Psi^{(2)} = H_0 \Psi^{(2)} + H_1 \Psi^{(1)}. \tag{9.32}$$

The first equation, Eq. 9.30, is simply the Time-Dependent Schrödinger Equation for H_0 and serves as a consistency check for our method. We can solve Eqs. 9.30–9.32 to obtain the correction terms $\Psi^{(0)}$, $\Psi^{(1)}$ and $\Psi^{(2)}$ in Eq. 9.29:

$$\Psi^{(0)}(t) = e^{-\frac{i}{\hbar} H_0 (t - t_0)} \, \Psi^{(0)}(t_0) \tag{9.33}$$

$$\Psi^{(1)}(t) = \frac{1}{i\hbar} e^{-\frac{i}{\hbar} H_0 t} \int_{t_0}^{t} e^{\frac{i}{\hbar} H_0 t'} H_1(t') \, \Psi^{(0)}(t') \, dt' + e^{-\frac{i}{\hbar} H_0 (t - t_0)} \, \Psi^{(1)}(t_0) \tag{9.34}$$

$$\Psi^{(2)}(t) = \frac{1}{i\hbar} e^{-\frac{i}{\hbar} H_0 t} \int_{t_0}^{t} e^{\frac{i}{\hbar} H_0 t'} H_1(t') \, \Psi^{(1)}(t') \, dt' + e^{-\frac{i}{\hbar} H_0 (t - t_0)} \, \Psi^{(2)}(t_0). \tag{9.35}$$

One can easily confirm that Eqs. 9.33–9.35 solve Eqs. 9.30–9.32, respectively, by taking the derivatives of the former. We can rewrite Eq. 9.35 by making use of Eq. 9.34:

$$\Psi^{(2)}(t) = \frac{1}{(i\hbar)^2} \int_{t_0}^{t} dt' \int_{t_0}^{t'} dt'' \, e^{-\frac{i}{\hbar} H_0 (t - t')} H_1(t') e^{-\frac{i}{\hbar} H_0 (t' - t'')} H_1(t'') \Psi^{(0)}(t'')$$

$$+ \frac{1}{i\hbar} \int_{t_0}^{t} dt' \, e^{-\frac{i}{\hbar} H_0 (t - t')} H_1(t') e^{-\frac{i}{\hbar} H_0 (t' - t_0)} \Psi^{(1)}(t_0)$$

$$+ e^{-\frac{i}{\hbar} H_0 (t - t_0)} \Psi^{(2)}(t_0) \tag{9.36}$$

where $\Psi^{(n)}(t_0)$ is the initial value of $\Psi^{(n)}$. Usually, we will assume that $\Psi(t_0) = \Psi^{(0)}(t_0)$ and so $\Psi^{(1)}(t_0) = \Psi^{(2)}(t_0) = 0$. Then

$$\Psi^{(1)}(t) = \frac{1}{i\hbar} \int_{t_0}^{t} dt' \, e^{-\frac{i}{\hbar} H_0 (t - t')} \, H_1(t') \, e^{-\frac{i}{\hbar} H_0 (t' - t_0)} \Psi^{(0)}(t_0) \tag{9.37}$$

$$\Psi^{(2)}(t) = \frac{1}{(i\hbar)^2} \int_{t_0}^{t} dt' \int_{t_0}^{t'} dt'' \, e^{-\frac{i}{\hbar} H_0 (t - t')} H_1(t') e^{-\frac{i}{\hbar} H_0 (t' - t'')} H_1(t'') e^{-\frac{i}{\hbar} H_0 (t'' - t_0)} \Psi^{(0)}(t_0). \tag{9.38}$$

When the perturbation is on for a finite time one may take $t_0 = 0$, defining the zero of time to be before the perturbation takes effect. In Chapter 13 we will see examples involving excitation with ultrashort laser pulses. In other situations, the perturbation is steady state so that in principle it has been in effect since the infinite past. In this case it is appropriate to choose $t_0 \to -\infty$. In Chapter 14 we will see examples of this $t_0 \to -\infty$ limit in the context of continuous wave (CW) electronic absorption and resonance Raman spectroscopy.

9.2.2   Derivation from the Interaction Picture

Equations 9.37–9.38 can also be derived using the interaction picture. Instead of finding corrections to the wavefunction, we will find corrections to the propagator, $U(t, t_0)$. We will associate with the unperturbed Hamiltonian, H_0, a propagator, $U^{(0)}(t, t_0)$, which satisfies the Time-Dependent Schrödinger Equation:

$$i\hbar \frac{\partial}{\partial t} U^{(0)}(t, t_0) = H_0 U^{(0)}(t, t_0). \tag{9.39}$$

Recall the definition of the interaction picture wavefunction (Eq. 9.13):

$$\Psi_{\mathrm{I}}(t) = e^{\frac{i}{\hbar} H_0 t} \Psi_{\mathrm{S}}(t) = e^{\frac{i}{\hbar} H_0 t} e^{-\frac{i}{\hbar} H t} \Psi_{\mathrm{S}}(0). \tag{9.40}$$

We can therefore define an interaction picture propagator, $U_{\mathrm{I}}(t, t_0)$ as

$$U_{\mathrm{I}}(t, t_0) = e^{\frac{i}{\hbar} H_0(t-t_0)} e^{-\frac{i}{\hbar} H(t-t_0)} = U^{(0)\dagger}(t, t_0)\, U(t, t_0). \tag{9.41}$$

The interaction picture propagator also satisfies a Time-Dependent Schrödinger Equation of the form

$$i\hbar \frac{\partial}{\partial t} U_{\mathrm{I}}(t, t_0) = H_{\mathrm{I}}(t) U_{\mathrm{I}}(t, t_0), \tag{9.42}$$

where

$$H_{\mathrm{I}}(t) = e^{\frac{i}{\hbar} H_0 t} H_1(t) e^{-\frac{i}{\hbar} H_0 t} \tag{9.43}$$

is the interaction picture Hamiltonian. Eq. 9.42 may be solved (exactly) iteratively yielding

$$U_{\mathrm{I}}(t, t_0) = 1 + \sum_{n=1}^{\infty} U_{\mathrm{I}}^{(n)}(t, t_0), \tag{9.44}$$

where

$$U^{(n)}(t, t_0) = \frac{1}{(i\hbar)^n} \int_{t_0}^{t} d\tau_n \int_{t_0}^{\tau_n} d\tau_{n-1} \cdots \int_{t_0}^{\tau_2} d\tau_1\, H_{\mathrm{I}}(\tau_n) H_{\mathrm{I}}(\tau_{n-1}) \cdots H_{\mathrm{I}}(\tau_1) U_{\mathrm{I}}(\tau_1, t_0) \tag{9.45}$$

and the integrals are performed subject to the constraint that $t > \tau_n > \tau_{n-1} > \cdots > \tau_1 > t_0$. To construct a perturbation series from Eqs. 9.44 and 9.45, we replace $U_{\mathrm{I}}(\tau_1, t_0)$ with 1, which is the same as saying that $H = H_0$. After making this substitution, and using Eq. 9.41, we can now write down a perturbation series for the Schrödinger picture propagator, $U(t, t_0)$:

$$U(t, t_0) = U^{(0)}(t, t_0) + \sum_{n=1}^{\infty} U^{(n)}(t, t_0), \tag{9.46}$$

where

$$U^{(n)}(t, t_0) = \frac{1}{(i\hbar)^n} \int_{t_0}^{t} d\tau_n \int_{t_0}^{\tau_n} d\tau_{n-1} \cdots \int_{t_0}^{\tau_2} d\tau_1\, U^{(0)}(t, \tau_n) H_1(\tau_n) U^{(0)}(\tau_n, \tau_{n-1})$$
$$\times H_1(\tau_{n-1}) \cdots U^{(0)}(\tau_2, \tau_1) H_1(\tau_1) U^{(0)}(\tau_1, t_0) \tag{9.47}$$

and the integrals are still subject to the constraint that $t > \tau_n > \tau_{n-1} > \cdots > \tau_1 > t_0$. Substituting

$$U^{(0)}(t, t') = e^{-\frac{i}{\hbar}H_0(t-t')} \tag{9.48}$$

into Eqs. 9.46–9.47 and applying the result to $\Psi^{(0)}(t_0)$ recovers our previous result for the time-dependent perturbative corrections to the wavefunction (Eqs. 9.37–9.38).

9.3   The Magnus Expansion and Wei–Norman Factorization

9.3.1   The Magnus Expansion

We have already encountered in several places the integral representation of the Schrödinger equation. For a time-independent Hamiltonian, H, the Schrödinger equation

$$i\hbar \frac{\partial}{\partial t} \Psi(x, t) = H\Psi(x, t) \tag{9.49}$$

can be rewritten as

$$\Psi(x, t) = e^{-iHt/\hbar}\Psi(x, 0). \tag{9.50}$$

The same does not hold true if the Hamiltonian is time dependent. One might imagine that the following simplest generalization holds:

$$\Psi(x, t) = e^{-i\int_0^t H(t')dt'/\hbar}\Psi(x, 0), \tag{9.51}$$

but this is not correct. To see why, we Taylor series expand Eq. 9.51 to get

$$\Psi(x, t) = \Psi(x, 0) - i/\hbar \int_0^t H(t')\, dt'\, \Psi(x, 0)$$

$$- \frac{1}{2\hbar^2} \int_0^t \int_0^t H(t')H(t'')\, dt'\, dt''\, \Psi(x, 0) + \cdots \tag{9.52}$$

Note that this expression indicates that $H(t')$ always operates after $H(t'')$, regardless of whether $t' > t''$ or not, whereas the differential form of the Time-Dependent Schrödinger Equation indicates that the Hamiltonian at different times must operate on the wavefunction in chronological order. A formal way of writing the integral representation for time-dependent Hamiltonians is

$$\Psi(x, t) = \mathrm{T}e^{-i\int_0^t H(t')dt'/\hbar}\Psi(x, 0), \tag{9.53}$$

where T is the so-called time-ordering operator, which by definition orders operators in the Taylor series chronologically. However, this is just a shorthand notation; what one would like is to express the propagator in the form

$$U(t, 0) = e^{A(t)}, \qquad A(t) = A_1(t) + A_2(t) + A_3(t) + \cdots. \tag{9.54}$$

This is what the Magnus expansion does.

The form of the first few A's is as follows:

$$A_1 = \frac{1}{i\hbar} \int_0^t dt_1 H(t_1) \tag{9.55}$$

$$A_2 = -\frac{1}{2} \left(\frac{1}{i\hbar}\right)^2 \int_0^t dt_2 \int_0^{t_2} dt_1 [H(t_1), H(t_2)] \tag{9.56}$$

$$A_3 = -\frac{1}{6} \left(\frac{1}{i\hbar}\right)^3 \int_0^t dt_3 \int_0^{t_3} dt_2 \int_0^{t_2} dt_1 [H(t_1), [H(t_2), H(t_3)]] + [[H(t_1), H(t_2)], H(t_3)]$$

$$\tag{9.57}$$

and so forth. Note that the leading term, A_1, is indeed of the form anticipated in Eq. 9.51. Furthermore, note that the correction terms, A_2 and higher, involve averages of the commutator of $H(t_1)$ with $H(t_2)$, and so forth. This is consistent with the observation that the error in using the A_1 term is only that the action of $H(t)$ is not correctly time ordered; it is thus to be expected that the corrections involve commutators, which may be viewed as subtracting off the difference between the correct and the incorrect time orderings.

There are several derivations of the Magnus formula. We follow that of Pechukas and Light (Pechukas, 1966), giving only the key components of the derivation. For more details, the reader is referred to their original paper. The strategy is to find an equation for the time derivative of A (which we will denote in this section by A') in terms of $H(t)$. To this end, we consider the time derivative of $U = e^A$:

$$(e^A)' = A' + [(AA' + A'A)/2!] + [(AAA' + AA'A + A'AA)/3!] + \cdots. \tag{9.58}$$

Defining $C_A X = [A, X]$, $C_A^2 X = C_A(C_A X) = [A, [A, X]]$, and so on, one may show that

$$(e^A)' = \left\{ \frac{e^{C_A} - 1}{C_A} A' \right\} e^A, \tag{9.59}$$

where, as usual, e^{C_A} is defined in terms of its Taylor series expansion. Since $U' = \frac{H}{i\hbar} U$, Eq. 9.59 can be inverted to give

$$A' = \left\{ \frac{C_A}{e^{C_A} - 1} \right\} \frac{H}{i\hbar} \tag{9.60}$$

$$= \left\{ 1 - \frac{C_A}{2} + \sum_{n=1}^{\infty} \frac{(-1)^{n+1} B_n C_A^{2n}}{(2n)!} \right\} \frac{H}{i\hbar}, \tag{9.61}$$

where the B's are the Bernoulli numbers, $B_1 = \frac{1}{6}$, $B_2 = \frac{1}{30}$, and so on. Writing $A = A_1 + A_2 + A_3 + \cdots$, the terms A_n may be obtained by integrating Eq. 9.61 term by term. The leading term in Eq. 9.61, A_1, is seen to be of the form in Eq. 9.55. Higher-order A_n are obtained by iteration; it is readily seen that the general term A_n will be a sum of $1, \ldots n$-fold integrals of $1, \ldots n$-fold commutators of $H/i\hbar$, as indicated above.

Since the higher-order terms in the Magnus expansion involve sums of integrals, it turns out that there are many equivalent ways of expressing this sum, and hence apparently very different looking expressions for the higher-order terms in the Magnus expansion. Salzman (1985) has given a straightforward procedure for calculating higher-order terms in the Magnus expansion, although the form of his terms looks very different from those above. Salzman begins with the expression

$$\frac{\partial}{\partial t} U = VU, \tag{9.62}$$

where $V \equiv H/i\hbar$. Note that

$$U(t, 0) = 1 + I_1 + I_2 + I_3 + \cdots, \tag{9.63}$$

where the I's are time-ordered integrals of products of V's, for example,

$$I_1 = \int_0^t dt_1 V(t_1)$$

$$I_2 = \int_0^t dt_2 \int_0^{t_2} dt_1 V(t_2) V(t_1)$$

$$I_3 = \int_0^t dt_3 \int_0^{t_3} dt_2 \int_0^{t_2} dt_1 V(t_3) V(t_2) V(t_1). \tag{9.64}$$

Equating Eq. 9.63 with the assumed form of the Magnus expansion, Eq. 9.54, yields

$$1 + I_1 + I_2 + I_3 + I_4 + \cdots$$
$$= 1 + (A_1 + A_2 + A_3 \cdots)/1! + (A_1 + A_2 + A_3 + \cdots)^2/2!$$
$$+ (A_1 + A_2 + A_3 + \cdots)^3/3! + (A_1 + A_2 + A_3 + \cdots)^4/4! + \cdots. \tag{9.65}$$

Matching terms with equal powers of V on both sides of Eq. 9.65 (taking care that the A's of different order in V do not necessarily commute), one finds

$$I_1 = A_1$$
$$I_2 = A_2 + A_1^2/2!$$
$$I_3 = A_3 + (A_2 A_1 + A_1 A_2)/2! + A_1^3/3!$$
$$I_4 = A_4 + (A_3 A_1 + A_1 A_3 + A_2^2)/2! + (A_2 A_1^2 + A_1 A_2 A_1 + A_1^2 A_2)/3! + A_1^4/4! \tag{9.66}$$

and so forth. These equations can be inverted to yield

$$A_1 = I_1$$
$$A_2 = I_2 - I_1^2/2$$
$$A_3 = I_3 - (I_2 I_1 + I_1 I_2)/2 + I_1^3/3$$
$$A_4 = I_4 - (I_3 I_1 + I_1 I_3 + I_2^2)/2 + (I_2 I_1^2 + I_1 I_2 I_1 + I_1^2 I_2)/3 - I_1^4/4 \tag{9.67}$$

and so on.

▶ **Exercise 9.5**   Show explicitly for A_2 that Salzman's form and the Magnus form are equivalent.

Solution   Using the notation $V = H/i\hbar$ (cf. Eq. 9.62) we have

$$A_2 = -\frac{1}{2} \int_0^t dt_2 \int_0^{t_2} dt_1 [V(t_1), V(t_2)]. \tag{9.68}$$

Replace the double integral by

$$\int_0^t dt_2 \int_0^t dt_1\, \theta_{21}, \tag{9.69}$$

where $\theta_{21} \equiv \theta(t_2 - t_1)$ is the Heaviside step-function ($\theta(t) = 0,\ t < 0;\ \theta(t) = 1,\ t \geq 0$). Write out the commutator to obtain

$$A_2 = -\frac{1}{2}\left(\int_0^t dt_2 \int_0^t dt_1\, \theta_{21} V(t_1) V(t_2) - \int_0^t dt_2 \int_0^t dt_1\, \theta_{21} V(t_2) V(t_1)\right). \tag{9.70}$$

The labelling of the dummy variables of integration is immaterial, so the labels on t_1 and t_2 in the first term can be exchanged. Moreover, the integration order can be exchanged. Performing these two operations we obtain

$$A_2 = -\frac{1}{2}\left(\int_0^t dt_2 \int_0^t dt_1 (\theta_{12} V(t_2) V(t_1) - \theta_{21} V(t_2) V(t_1)\right). \tag{9.71}$$

Noting that $\theta_{12} = 1 - \theta_{21}$ in the first term, we find

$$A_2 = -\frac{1}{2}\left(\int_0^t dt_2 \int_0^t dt_1\, (V(t_2) V(t_1) - 2\theta_{21} V(t_2) V(t_1)\right). \tag{9.72}$$

This expression is readily seen to be equal to Eq. 9.56.

The Magnus expansion is an elegant alternative to time-dependent perturbation theory, since it can be truncated at any order in the expansion and it still gives a unitary expression for the propagator. First-order Magnus is quite accurate for sudden perturbations, while there is some evidence that second-order Magnus is accurate for adiabatically changing Hamiltonians. Issues of convergence of the Magnus expansion are difficult to prove in general, but numerical evidence shows that in representative examples the series does not converge in the Schrödinger picture but does in the interaction picture (Salzman, 1986). Recently, there has been great interest in the Magnus expansion among applied mathematicians, with progress in developing general methods to sum the Magnus series (Iserles, 2000).

9.3.2   Wei–Norman Factorization

An alternative to the Magnus formulation of the propagator has been given by Wei and Norman (1964), in which the propagator is represented as a product of exponentials. There are a number of reasons to consider this form. First, as pointed out in the original paper by Wei and Norman, the Magnus representation of the propagator $U(t)$ holds only in the neighborhood of $t = 0$ ($U(t) = 1$). By contrast, there exists a product of exponentials form that is valid for all t provided that the operators in the exponent form a closed algebra with fairly general properties. Second, the Lie algebraic framework that is used to derive the Wei–Norman propagator anticipates the Lie algebraic analysis of controllability discussed in Section 16.2.4. A third motivation is provided by quantum computing, where much current effort is being expended in designing desired unitary transformation in systems with a $2^{\otimes N}$ tensor product structure (tensor product of N two-level systems). There is a general theorem that any $2^{\otimes N} \times 2^{\otimes N}$ unitary matrix can be decomposed into unitary operators that act on only one or two of the two-level systems at a time (one- and two-"qubit" operations). Despite this general theorem, there are no general algorithms at present for actually finding such decompositions. The Wei–Norman factorization is almost certain to be of value in the quest for such one- and two-qubit decompositions.

We start with the equation of motion for the propagator

$$\frac{\partial U(t)}{\partial t} = \frac{H(t)}{i\hbar} U(t) \qquad U(0) = 1 \tag{9.73}$$

and we assume that

$$\frac{H(t)}{i\hbar} = \sum_{i=1}^{n} a_i(t) X_i, \quad n \text{ finite}, \tag{9.74}$$

where the X_i are operators that are independent of time and that in general do not commute. We define an *algebra* generated by the X_i under the commutation product $[X_i, X_j] = X_i X_j - X_j X_i$; such an algebra is known as a Lie algebra. The dimension of the algebra is defined to be the number of linearly independent operators produced by evaluating all possible commutators and commutators of commutators in an iterative fashion. We assume that the algebra is of finite dimension N, a property that is guaranteed if $H(t)$ (and therefore $U(t)$) are finite matrices. Wei and Norman showed that if $U(t)$ is a solution of Eq. 9.73, $U(t)$ can be represented in the form

$$U(t) = \prod_{i=1}^{N} e^{g_i(t) X_i} \qquad (U^{-1} = \prod_{i=N}^{1} e^{-g_i(t) X_i}). \tag{9.75}$$

The first n members of the set $\{X_i, i = 1, \ldots, N\}$ are the same as in Eq. 9.74, but we add an additional $N - n$ members to complete the algebra.

To prove Eq. 9.75, we need an intermediate result. We begin with the conjugation-commutation formula derived in Exercise 8.35:

$$e^A X e^{-A} = X + [A, X] + [A, [A, X]]/2! + [A, [A, [A, X]]]/3! + \cdots . \tag{9.76}$$

Let $X_1, X_2, \cdots, X_N$ be a basis for the Lie algebra with the multiplication table $[X_i, X_j] = \sum_{k=1}^{N} \gamma_{ij}^k X_k$, $i, j = 1, 2, \cdots, N$. Then repeated application of Eq. 9.76 gives

$$\left(\prod_{j=1}^{r} e^{g_j X_j} \right) X_i \left(\prod_{j=r}^{1} e^{-g_j X_j} \right) = \sum_{k=1}^{N} \xi_{ki}^{(r)} X_k, \tag{9.77}$$

where $\xi_{ki}^{(r)} = \xi_{ki}(g_1, \ldots, g_r)$ and r can take on any value between 1 and N. With Eq. 9.77 in hand, we turn to the proof of Eq. 9.75. Taking $U(t)$ to be of the form in Eq. 9.75 we have

$$\frac{dU}{dt} = \sum_{i=1}^{N} \dot{g}_i(t) \left(\prod_{j=1}^{i-1} e^{g_j X_j} \right) X_i \left(\prod_{j=i}^{N} e^{g_j X_j} \right) \tag{9.78}$$

$$= \frac{H(t)}{i\hbar} U(t) = \sum_{i=1}^{N} a_i(t) X_i U(t), \tag{9.79}$$

where without loss of generality we have replaced n by N in the upper bound of the sum on the RHS. Substituting Eq. 9.75 into the RHS of Eq. 9.79 and postmultiplying by U^{-1} we obtain

$$\sum_{i=1}^{N} a_i(t) X_i = \sum_{i=1}^{N} \dot{g}_i(t) \left(\prod_{j=1}^{i-1} e^{g_j X_j} \right) X_i \left(\prod_{j=i-1}^{1} e^{-g_j X_j} \right). \tag{9.80}$$

Substituting Eq. 9.77 with $r = i - 1$ into Eq. 9.80 we may write

$$\sum_{i=1}^{N} a_i(t) X_i = \sum_{i=1}^{N} \sum_{k=1}^{N} \dot{g}_i(t) \xi_{ki} X_k, \tag{9.81}$$

where we write ξ_{ki} instead of $\xi_{ki}^{(i-1)}$ since the superscript is superfluous. Changing the summation index on the LHS from i to k we may match coefficients of X_k to obtain

$$\overbrace{\begin{pmatrix} a_1 \\ a_2 \\ \vdots \\ a_N \end{pmatrix}}^{a} = \overbrace{\begin{pmatrix} \xi_{11} & \cdots & \xi_{1N} \\ & \vdots & \\ \xi_{N1} & \cdots & \xi_{NN} \end{pmatrix}}^{\xi} \overbrace{\begin{pmatrix} \dot{g}_1 \\ \dot{g}_2 \\ \vdots \\ \dot{g}_N \end{pmatrix}}^{\dot{g}}. \tag{9.82}$$

It is shown by Wei (1964) that ξ is guaranteed to be invertible in the neighborhood of $t = 0$, and therefore

$$\frac{dg}{dt} = \xi^{-1} a = f(a, g), \qquad g(0) = 0. \tag{9.83}$$

This completes the proof of Eq. 9.75.

▶ **Exercise 9.6**   Using the notation $C_A X = [A, X]$, $C_A^2 X = [A, [A, X]]$ as in the previous section, show that Eq. 9.76 can be written as

$$e^A X e^{-A} = e^{C_A} X. \tag{9.84}$$

▶ **Exercise 9.7**   Using the result of Exercise 9.6, show that Eq. 9.80 can be rewritten

$$\sum_{i=1}^{N} a_i(t) X_i = \sum_{i=1}^{N} \dot{g}_i(t) \left(\prod_{j=1}^{i-1} e^{g_j X_j} \right) X_i \left(\prod_{j=i-1}^{1} e^{-g_j X_j} \right)$$

$$= \sum_{i=1}^{N} \dot{g}_i(t) \left(\prod_{j=1}^{i-1} e^{g_j C_{X_j}} \right) X_i. \tag{9.85}$$

▶ **Exercise 9.8**   The Baker–Hausdorff formula states that

$$e^{A+B} = e^A e^B e^{-\frac{1}{2}[A,B]} \tag{9.86}$$

if $[A, B]$ commutes with A and B. Derive this formula as a special case of the Wei–Norman factorization.

▶ Exercise 9.9   Let

$$\frac{\partial U(t)}{\partial t} = \frac{H(t)}{i\hbar} U(t) = (a_0(t)A_0 + a_1(t)A_1 + a_2(t)A_2)U(t) \tag{9.87}$$

with the Lie multiplication table

$$[A_1, A_2] = A_0; \qquad [A_1, A_0] = 2A_1; \qquad [A_2, A_0] = -2A_2. \tag{9.88}$$

Setting

$$U(t) = e^{g_0(t)A_0} e^{g_1(t)A_1} e^{g_2(t)A_2} \tag{9.89}$$

derive the differential equations for g_0, g_1 and g_2 (Wei, 1964).

Solution   Using the notation of Exercise 9.6 we obtain

$$a_0 A_0 + a_1 A_1 + a_2 A_2 = \dot{g}_0 A_0 + \dot{g}_2(e^{g_0 C_{A_0}})A_1 + \dot{g}_2(e^{g_0 C_{A_0}} e^{g_1 C_{A_1}})A_2 \tag{9.90}$$

(cf. Eq. 9.80). Moreover,

$$(e^{g_0 C_{A_0}})A_1 = A_1 - 2g_0 A_1 + \frac{(2g_0)^2}{2!}A_1 - \cdots = e^{-2g_0}A_1 \tag{9.91}$$

$$(e^{g_0 C_{A_0}} e^{g_1 C_{A_1}})A_2 = (e^{g_0 C_{A_0}})(A_2 + g_1 A_0 + g_1^2 A_1) = e^{2g_0}A_2 + g_1 A_0 + g_1^2 e^{-2g_0}A_1 \tag{9.92}$$

(cf. Eq. 9.77). Hence g_0, g_1, g_2 satisfy the equations

$$\begin{pmatrix} a_0 \\ a_1 \\ a_2 \end{pmatrix} = \begin{pmatrix} 1 & 0 & g_1 \\ 0 & e^{-2g_0} & g_1^2 e^{-2g_0} \\ 0 & 0 & e^{2g_0} \end{pmatrix} \begin{pmatrix} \dot{g}_0 \\ \dot{g}_1 \\ \dot{g}_2 \end{pmatrix}. \tag{9.93}$$

9.4   Adiabatic Dynamics and the Geometrical Phase

The topic of this section is the special case of Hamiltonians that change slowly with time, or "adiabatically." The wavefunction evolving under such a Hamiltonian satisfies some quite general properties. The three main results we will discuss are the quantum adiabatic theorem, nonadiabatic corrections and the geometric phase.

9.4.1   Quantum Adiabatic Theorem

Consider a quantum system with a discrete level structure. The Time-Dependent Schrödinger Equation reads

$$i\hbar \frac{\partial}{\partial t} \Psi(x, t) = H(x, t)\Psi(x, t), \tag{9.94}$$

where

$$H(t) = \begin{pmatrix} E_a & V_{ab}(t) & \cdots \\ V_{ba}(t) & E_b & \cdots \\ \cdots & \cdots & \cdots \end{pmatrix} \tag{9.95}$$

$$\Psi(t) = \begin{pmatrix} \Psi_a \\ \Psi_b \\ \cdots \end{pmatrix}. \tag{9.96}$$

If the off-diagonal couplings are slowly varying, some general statements can be made. Define the unitary transformation that diagonalizes the instantaneous $H(t)$:

$$U^{-1}(t)H(t)U(t) = D(t) \tag{9.97}$$

$$U^{-1}\Psi \equiv \Psi'. \tag{9.98}$$

The TDSE can be written as

$$i\hbar \frac{\partial}{\partial t}\left(U\Psi'(t)\right) = i\hbar\left(U(t)\frac{\partial\Psi'}{\partial t} + \frac{\partial U(t)}{\partial t}\Psi'\right) = H(t)U(t)\Psi'(t) \tag{9.99}$$

or

$$i\hbar \frac{\partial}{\partial t}\Psi'(t) = D(t)\Psi'(t) - i\hbar U^{-1}(t)\frac{\partial U(t)}{\partial t}\Psi'(t). \tag{9.100}$$

If $H(t)$ is slowly varying so will be $U(t)$ and therefore also $U^{-1}(t)$. In this case the second term on the RHS may be neglected. This is the adiabatic approximation. The adiabatic theorem in quantum mechanics states that if $\Psi(0)$ is an eigenfunction of $H(0)$, and $H(t)$ is a slowly varying function of time, then $\Psi(t)$ will evolve in such a way as to remain an eigenfunction of $H(t)$ for all time, in the limit of infinitely slowly varying $H(t)$. The reader is referred to Messiah (1958) for a very careful derivation, but the essence is already captured in Eq. 9.100.

The quantum adiabatic theorem may be exploited to calculate eigenfunctions for complicated potentials, V, by starting with an eigenfunction of a simple potential V_0 and slowly "switching" on the difference potential, $\Delta V \equiv V - V_0$, that is, $H(t) = H_0 + \lambda(t)\,\Delta V$, where $\lambda(t)$ varies slowly from 0 to 1. This switching process is illustrated in Figure 9.2: V_0 is chosen as the harmonic oscillator potential, $V_0 = \frac{1}{2}kx^2$, and V is the double-well potential, $V = k_1 x^4 - k_2 x^2$. The initial state may be chosen as any of the harmonic oscillator eigenstates; by switching on the difference $V - V_0$ slowly, the eigenfunction of the nth eigenfunction of the harmonic oscillator evolves into the nth eigenfunction of the double-well potential. The eigenvalues as a function of the switching parameter yield a "correlation diagram" between the spectrum of H_0 and the spectrum of H (Figure 9.2c). Note the pairwise coalescence of eigenvalues in the double-well system, corresponding to the quasi-degenerate symmetric and antisymmetric doublets of a double well.

9.4.2 Nonadiabatic Corrections: The Landau–Zener Formula

In simple cases it is possible to evaluate the nonadiabatic correction (the second term on the RHS of Eq. 9.100) explicitly. Consider the 2×2 Hamiltonian

$$\mathbf{H}(t) = \begin{pmatrix} E_a(t) & V_{ab}(t) \\ V_{ba}(t) & E_b(t) \end{pmatrix}. \tag{9.101}$$

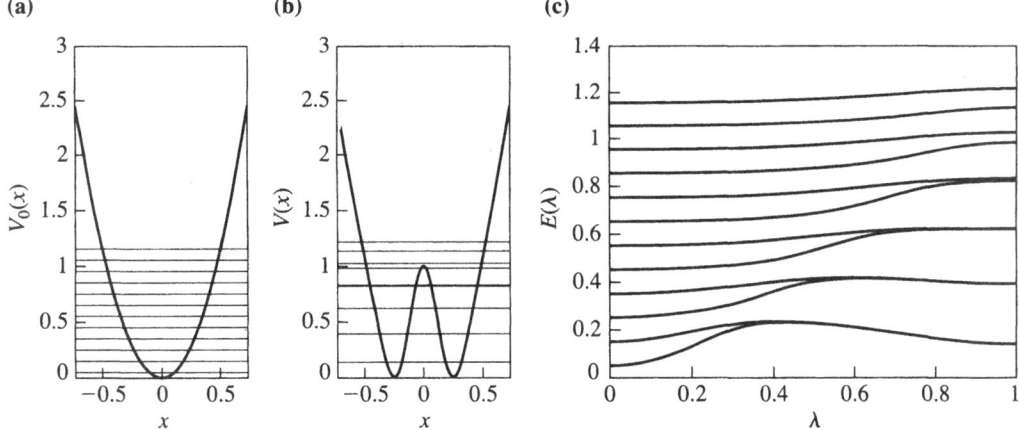

Figure 9.2 (a) Energy levels of the initial potential, $V_0(x)$ (harmonic oscillator). (b) Energy levels of the final potential $V(x)$ (double well). (c) The correlation diagram of energy levels, E, as a function of a switching parameter, λ, that changes the potential adiabatically from a harmonic oscillator to a double well. An initial state that is an eigenstate of the oscillator will remain an eigenstate of the switching potential and ultimately of the double well if the switching time $T \to \infty$.

Using the general form for the unitary matrix that diagonalizes a 2×2 Hamiltonian matrix (see Exercise 8.74) we have

$$\mathbf{U}^{-1}(t)\mathbf{H}(t)\mathbf{U}(t) = \mathbf{D}(t), \qquad (9.102)$$

where

$$\mathbf{U}(t) = \begin{pmatrix} \cos(\frac{\theta}{2})e^{-i\phi/2} & -\sin(\frac{\theta}{2})e^{-i\phi/2} \\ \sin(\frac{\theta}{2})e^{i\phi/2} & \cos(\frac{\theta}{2})e^{i\phi/2} \end{pmatrix} \qquad \mathbf{D}(t) = \begin{pmatrix} E_- & 0 \\ 0 & E_+ \end{pmatrix} \qquad (9.103)$$

with

$$E_\pm = \frac{1}{2}(E_a + E_b) \pm \frac{1}{2}\sqrt{(E_a - E_b)^2 + 4|V_{ab}|^2} \qquad (9.104)$$

$$\tan\theta = \frac{2|V_{ab}|}{E_a - E_b} \quad 0 \le \theta < \pi \qquad (9.105)$$

$$V_{ba} = |V_{ba}|e^{i\phi} \quad 0 \le \phi < 2\pi. \qquad (9.106)$$

Substituting Eq. 9.103 into Eq. 9.100 we obtain

$$i\hbar \frac{d}{dt}\begin{pmatrix} \Psi_- \\ \Psi_+ \end{pmatrix} = \begin{pmatrix} E_- & 0 \\ 0 & E_+ \end{pmatrix}\begin{pmatrix} \Psi_- \\ \Psi_+ \end{pmatrix}$$

$$+ \left\{ -\frac{i\hbar\dot{\theta}}{2}\begin{pmatrix} 0 & -1 \\ 1 & 0 \end{pmatrix} - \frac{\hbar\dot{\phi}}{2}\begin{pmatrix} \cos\theta & -\sin\theta \\ -\sin\theta & -\cos\theta \end{pmatrix} \right\}\begin{pmatrix} \Psi_- \\ \Psi_+ \end{pmatrix}, \qquad (9.107)$$

where we have defined

$$\begin{pmatrix} \Psi_- \\ \Psi_+ \end{pmatrix} = \mathbf{U}^{-1}\begin{pmatrix} \Psi_a \\ \Psi_b \end{pmatrix}. \qquad (9.108)$$

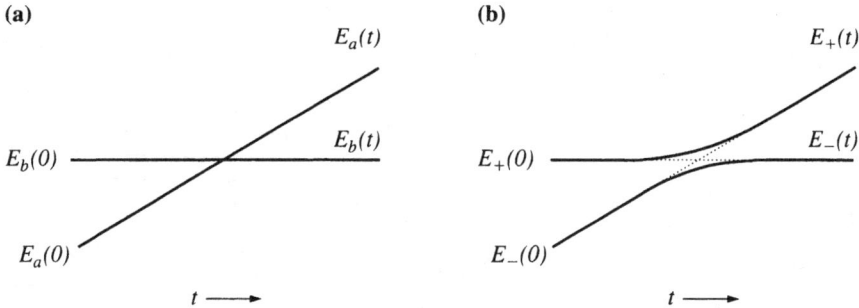

Figure 9.3 Time-dependent energy levels in a two-level system, corresponding to the scenario discussed in the text. (a) The bare energy levels, E_a and E_b, pass through a degeneracy at some intermediate time. (b) Due to the coupling between E_a and E_b, the eigenvalues of the coupled system, E_+ and E_-, are never actually degenerate.

The condition for the nonadiabatic coupling to be small is (see Exercise 9.10)

$$\left[\left(\hbar \dot{\theta} \right)^2 + \left(\hbar \dot{\phi} \right)^2 \sin^2 \theta \right] \ll (E_+ - E_-)^2 = \left(\hbar \dot{\phi} \cos \theta \right)^2 . \tag{9.109}$$

▶ **Exercise 9.10**   Equation 9.107 can be rewritten in the form

$$i\hbar \frac{d}{dt} \begin{pmatrix} \Psi_- \\ \Psi_+ \end{pmatrix} = H'(t) \begin{pmatrix} \Psi_- \\ \Psi_+ \end{pmatrix} .$$

a.   Find $H'(t)$ and its eigenvalues $E'_\pm(t)$ [You may use Eq. 9.104 with appropriate substitutions.]

b.   Adiabaticity is the condition that the $E'_\pm(t)$ are not significantly different from the diagonal elements of $H'(t)$. Show that this leads to the inequality in Eq. 9.109.

Consider the simple case

$$E_a(t) = E_b + \dot{E}t \tag{9.110}$$

$$V_{ab} = V_{ba} = V \tag{9.111}$$

(see Figure 9.3). To calculate $\dot{\theta}$ we use Eq. 9.105:

$$\frac{d}{dt}(\tan \theta) = -\frac{2V}{(E_a - E_b)^2} \dot{E} = \sec^2 \theta \, \dot{\theta}$$

$$= (1 + \tan^2 \theta) \, \dot{\theta} = \left(1 + \frac{4V^2}{(E_b - E_a)^2} \right) \dot{\theta} = \frac{(E_+ - E_-)^2}{(E_b - E_a)^2} \dot{\theta} \tag{9.112}$$

and therefore

$$\dot{\theta} = -\frac{2V\dot{E}}{(E_+ - E_-)^2} . \tag{9.113}$$

Since $\phi = \dot{\phi} = 0$, the adiabaticity condition is

$$\hbar |\dot{\theta}| = \frac{2\hbar V |\dot{E}|}{(E_+ - E_-)^2} \ll \frac{|E_+ - E_-|}{2} \quad \text{or} \quad |E_+ - E_-|^3 \gg 4\hbar V |\dot{E}|. \quad (9.114)$$

At the point of closest crossing, $E_a = E_b$, hence $E_+ - E_- = 2V$, so the adiabaticity condition is

$$2V^2/(\hbar |\dot{E}|) \gg 1. \quad (9.115)$$

Given that as $t \to -\infty$ all the amplitude starts in $E_a \approx E_-$, where will the amplitude be as $t \to \infty$? This problem was addressed by Landau (1932) and Zener (1932) in a different form. They considered the case of two potential curves with spatial dependence; taking the potentials to be linear in the region of the crossing we may write $E_a(x) = F_a x$ and $E_b(x) = F_b x$, where $E_a(x)$ and $E_b(x)$ are the potential curves and F_a and F_b are the slopes. The Landau–Zener formula expresses the transition probability between two crossing *diabatic* states as

$$P = 1 - p, \qquad p = \exp\left[-\frac{2\pi V^2}{\hbar v |F_a - F_b|} \right], \quad (9.116)$$

where v is the speed of passing through the crossing and $F_a - F_b$ is the difference in slopes of the diabatic potentials at the crossing. Identifying

$$|F_a - F_b| v = \left| \frac{dE}{dx} \right| \left| \frac{dx}{dt} \right| = |\dot{E}|, \quad (9.117)$$

we obtain

$$P = 1 - p, \qquad p = \exp\left[-\frac{2\pi V^2}{\hbar |\dot{E}|} \right]. \quad (9.118)$$

Clearly, Eq. 9.118 is consistent with the adiabatic criterion, Eq. 9.115: if $V^2/(\hbar |\dot{E}|) \gg 1$, $p \approx 0$ and $P \approx 1$, indicating that the crossing of amplitude between *diabatic* states is nearly complete. Similarly, if $V^2/(\hbar |\dot{E}|) \ll 1$ we have $p \approx 1$ and $P \approx 0$, indicating that the amplitude remains on the diabatic state on which it originated. In physical terms, for strong coupling of diabatic states and slow passage through the crossing region, the amplitude follows the adiabatic state. For weak coupling of the diabatic states and rapid passage through the crossing region, the system follows the diabatic state.

9.4.3 The Geometrical Phase

There is an additional intriguing aspect of adiabatically changing Hamiltonians to which we now devote extensive discussion. The adiabatic theorem states that the system will be in an eigenstate of the instantaneous Hamiltonian, but says nothing about the *phase*. Indeed, if the Hamiltonian is changed adiabatically in such a way that the circuit described by the parameters (we are necessarily dealing with more than one-dimensional parameter space) encloses a value of the parameters for which the spectrum of the Hamiltonian is degenerate, then the phase of the eigenstate upon return will have a purely geometrical portion, in addition to the usual dynamical portion. This additional phase is called the geometrical phase, or "Berry" phase, after the mathematical physicist who gave the general theory, although the effect was known and understood in a variety of special contexts prior to Berry's work. In this section we follow closely Berry's seminal paper (Berry, 1984).

Consider the Time-Dependent Schrödinger Equation

$$i\hbar|\dot{\Psi}\rangle = H(\mathbf{R}(t))|\Psi(t)\rangle, \tag{9.119}$$

where the Hamiltonian depends on a set of parameters, $\mathbf{R} = (X, Y, \ldots)$, which vary with time. At any instant in time there is an instantaneous basis of eigenstates of $H(\mathbf{R}(t))$, which we may call $\{n(\mathbf{R}(t))\}$:

$$H(\mathbf{R}(t))|n(\mathbf{R}(t))\rangle = E_n(\mathbf{R}(t))|n(\mathbf{R}(t))\rangle. \tag{9.120}$$

Note that there is an arbitrariness to the phase of $|n(\mathbf{R}(t))\rangle$ but we assume it to be single valued for a given value of $\mathbf{R}$, so that if $\mathbf{R}(T) = \mathbf{R}(0)$ then $|n(\mathbf{R}(T))\rangle = |n(\mathbf{R}(0))\rangle$.

If the system is prepared in state $|n(\mathbf{R}(0))\rangle$ at time 0, and if $H(\mathbf{R}(t))$ is varied adiabatically, the quantum adiabatic theorem states that the system will stay in the eigenstate $|n(\mathbf{R}(t))\rangle$, which correlates with $|n(\mathbf{R}(0))\rangle$. Thus, $|\Psi(t)\rangle$ can be written

$$|\Psi(t)\rangle = \exp\left\{\frac{-i}{\hbar}\int_0^t dt' E_n(\mathbf{R}(t'))\right\} e^{i\gamma_n(t)}|n(\mathbf{R}(t))\rangle; \tag{9.121}$$

that is, the dynamical portion of the phase factor is given by the energy of the eigenvalue associated with $|n(\mathbf{R})\rangle$, but we allow for the possibility of an additional, purely geometrical phase, $\gamma_n(t)$. Substituting Eq. 9.121 into Eq. 9.119 gives

$$\dot{\gamma}_n(t) = i\langle n(\mathbf{R}(t))|\nabla n(\mathbf{R}(t))\rangle \cdot \dot{\mathbf{R}}(t), \tag{9.122}$$

where $\nabla n(\mathbf{R}) \equiv \nabla_{\mathbf{R}} n(\mathbf{R})$. Let the parameter set $\mathbf{R}(t)$ trace out a circuit in parameter space such that $\mathbf{R}(T) = \mathbf{R}(0)$. Integrating Eq. 9.122 and substituting into Eq. 9.121, the total phase change of $|\Psi\rangle$ accumulated along the circuit C is obtained:

$$|\Psi(T)\rangle = e^{i\gamma_n(C)} e^{-\frac{i}{\hbar}\int_0^T dt\, E_n(R(t))}|\Psi(0)\rangle, \tag{9.123}$$

where

$$\gamma_n(C) = i\oint_C \langle n(\mathbf{R})|\nabla n(\mathbf{R})\rangle \cdot d\mathbf{R} \tag{9.124}$$

is the geometrical phase. The fact that $\langle n(\mathbf{R})|n(\mathbf{R})\rangle = 1$ for all $\mathbf{R}$ implies that

$$\text{Re}(\langle n(\mathbf{R})|\nabla n(\mathbf{R})\rangle) = 0, \tag{9.125}$$

so that $\langle n(\mathbf{R})|\nabla n(\mathbf{R})\rangle$ is purely imaginary.

▶ **Exercise 9.11**   Verify Eqs. 9.122 and 9.125.

Solution   Taking the derivative of Eq. 9.121 we obtain

$$|\dot{\Psi}(t)\rangle = \left(-\frac{i}{\hbar}E_n(R(t)) + i\dot{\gamma}_n(t) + \nabla n \cdot \dot{\mathbf{R}}\right)|\Psi(t)\rangle = -\frac{i}{\hbar}H(\mathbf{R}(t))|\Psi(t)\rangle. \tag{9.126}$$

Using Eq. 9.120, the first term in the middle expression cancels the RHS (note that the Hamiltonian operator depends on $\mathbf{R}$ but does not operate on this variable) and hence

$$\dot{\gamma}_n(t)|n(\mathbf{R}(t))\rangle = i\nabla n(\mathbf{R}) \cdot \dot{\mathbf{R}}|n(\mathbf{R}(t))\rangle. \tag{9.127}$$

Applying the bra $\langle n(\mathbf{R})|$ to both sides, and using the normalization condition on $|n(\mathbf{R})\rangle$ for all $\mathbf{R}$ gives Eq. 9.122. To show Eq. 9.125, note that

$$\int n^*(x;\mathbf{R})n(x;\mathbf{R})\,dx = 1 \tag{9.128}$$

and thus

$$\nabla_{\mathbf{R}} \int n^*(x;\mathbf{R})n(x;\mathbf{R})\,dx = 0 \tag{9.129}$$

$$= \int \nabla_{\mathbf{R}} n^*(x;\mathbf{R})n(x;\mathbf{R})\,dx + \int n^*(x;\mathbf{R})\nabla_{\mathbf{R}} n(x;\mathbf{R})\,dx \tag{9.130}$$

$$= 2\mathrm{Re}\left(\int n^*(x;\mathbf{R})\nabla_{\mathbf{R}} n(x;\mathbf{R})\,dx\right), \tag{9.131}$$

which is the desired result.

It is by no means obvious from Eq. 9.124 that $\gamma_n(C)$ should be unique: there is an arbitrary overall phase to $|n(\mathbf{R})\rangle$ and a somewhat more complicated arbitrary phase to $|\nabla n(\mathbf{R})\rangle$, since the latter involves components of all the basis functions $|n(\mathbf{R})\rangle$. We therefore seek an alternative expression in which it is apparent that any arbitrariness in the phase of the basis set $|n(\mathbf{R})\rangle$ disappears. From Stokes's theorem, the line integral in Eq. 9.124 can be replaced by a surface integral:

$$\gamma_n(C) = i\int\int_C d\mathbf{S} \cdot \nabla \times \langle n|\nabla n\rangle \tag{9.132}$$

$$= i\int\int_C d\mathbf{S} \cdot \langle \nabla n| \times |\nabla n\rangle \tag{9.133}$$

$$= i\int\int_C d\mathbf{S} \cdot \sum_{m \neq n} \langle \nabla n|m\rangle \times \langle m|\nabla n\rangle. \tag{9.134}$$

$\mathbf{S}$ denotes the element of area in $\mathbf{R}$ space; the exclusion in the summation is because $\langle n|\nabla n\rangle$ is purely imaginary and $\gamma_n(C)$ must be real. From Eq. 9.120 one can show that

$$\langle m|\nabla n\rangle = \langle m|\nabla H|n\rangle/(E_n - E_m), \quad m \neq n. \tag{9.135}$$

Thus we obtain our final expression for the phase:

$$\gamma_n(C) = -\int\int_C d\mathbf{S} \cdot \mathbf{V}_n(\mathbf{R}), \tag{9.136}$$

where

$$\mathbf{V}_n(\mathbf{R}) \equiv \mathrm{Im} \sum_{m \neq n} \frac{\langle n(\mathbf{R})|\nabla H(\mathbf{R})|m(\mathbf{R})\rangle \times \langle m(\mathbf{R})|\nabla H(\mathbf{R})|n(\mathbf{R})\rangle}{(E_m(\mathbf{R}) - E_n(\mathbf{R}))^2}. \tag{9.137}$$

From Eq. 9.136, $\gamma_n(C)$ is zero if a circuit retraces itself and encloses no area. Clearly, any arbitrariness in phase of $|n(\mathbf{R})\rangle$ and $|m(\mathbf{R})\rangle$ cancels; that is, the symmetric appearance of these states in the bra and ket ensures the overall independence of $\gamma_n(C)$ on any phase convention for these states. Equation 9.137 was first derived by Mead and Truhlar in 1979, assuming infinitesimal circuits, in the context of molecules in the Born–Oppenheimer approximation.

▶ **Exercise 9.12**   In deriving Eq. 9.133 we have used the relation

$$\nabla_{\mathbf{R}} \times (n^*(\mathbf{R})\nabla_{\mathbf{R}}n(\mathbf{R})) = \nabla_{\mathbf{R}}n^*(\mathbf{R}) \times \nabla_{\mathbf{R}}n(\mathbf{R}). \tag{9.138}$$

Using the definition of ∇ and $\nabla\times$ show, for each component X, Y, Z, that the two expressions are equal.

▶ **Exercise 9.13**   Verify Eq. 9.135.

Solution   Applying $\nabla_{\mathbf{R}}$ to both sides of Eq. 9.120, we obtain

$$\nabla_{\mathbf{R}}(H(\mathbf{R})|n(\mathbf{R})\rangle) = \nabla_{\mathbf{R}}(E_n(\mathbf{R})|n(\mathbf{R})\rangle). \tag{9.139}$$

Thus,

$$\nabla_{\mathbf{R}}H(\mathbf{R})\,|n(\mathbf{R})\rangle + H(\mathbf{R})\nabla_{\mathbf{R}}|n(\mathbf{R})\rangle = \nabla_{\mathbf{R}}E_n(\mathbf{R})|n(\mathbf{R})\rangle + E_n(\mathbf{R})\nabla_{\mathbf{R}}|n(\mathbf{R})\rangle. \tag{9.140}$$

Applying the bra $\langle m(\mathbf{R})|$ to both sides of the equation yields

$$\langle m(\mathbf{R})|\nabla_{\mathbf{R}}H(\mathbf{R})|n(\mathbf{R})\rangle + \langle m(\mathbf{R})|H(\mathbf{R})\nabla_{\mathbf{R}}|n(\mathbf{R})\rangle$$
$$= \langle m(\mathbf{R})|\nabla_{\mathbf{R}}E_n(\mathbf{R})|n(\mathbf{R})\rangle + \langle m(\mathbf{R})|E_n(\mathbf{R})\nabla_{\mathbf{R}}|n(\mathbf{R})\rangle. \tag{9.141}$$

Recognizing that $\langle m|H = \langle m|E_m$, $\langle m|n\rangle = 0$ for $m \neq n$, and rearranging terms we find that

$$\langle m|\nabla n\rangle = \langle m|\nabla H|n\rangle/(E_n - E_m), \quad m \neq n, \tag{9.142}$$

which is the desired result.

Equations 9.136–9.137 indicate that the contributions to $\gamma_n(C)$ will be dominated by states m in near degeneracy. In the most common situation the degeneracy involves only two states, which we will denote + and –, with $E_+(\mathbf{R}) \geq E_-(\mathbf{R})$. Near a point of degeneracy, $\mathbf{R}^*$, the leading term for $\mathbf{V}_+(\mathbf{R})$ is:

$$\mathbf{V}_+(\mathbf{R}) = \text{Im}\frac{\langle +(\mathbf{R})|\nabla H(\mathbf{R}^*)| - (\mathbf{R})\rangle \times \langle -(\mathbf{R})|\nabla H(\mathbf{R}^*)| + (\mathbf{R})\rangle}{(E_+(\mathbf{R}) - E_-(\mathbf{R}))^2}. \tag{9.143}$$

It follows from the definition of the cross product that $\mathbf{V}_-(\mathbf{R}) = -\mathbf{V}_+(\mathbf{R})$, and therefore that $\gamma_-(C) = -\gamma_+(C)$. Without loss of generality we can take $E_\pm(\mathbf{R}^*) = 0$ and $\mathbf{R}^* = 0$; the Hamiltonian $H(\mathbf{R})$ can then be written in a standard form:

$$\frac{1}{2}\begin{pmatrix} Z & X - iY \\ X + iY & -Z \end{pmatrix}. \tag{9.144}$$

▶ **Exercise 9.14**   Show that the eigenvalues of this Hamiltonian are $E_+(\mathbf{R}) = -E_-(\mathbf{R})$ $= \frac{1}{2}(X^2 + Y^2 + Z^2)^{1/2} = \frac{1}{2}R$.

▶ **Exercise 9.15**   Show that $\nabla H = \frac{1}{2}\bar{\sigma}$, that is, $(\nabla H)_X = \frac{1}{2}\sigma_X$, $(\nabla H)_Y = \frac{1}{2}\sigma_Y$, $(\nabla H)_Z$ $= \frac{1}{2}\sigma_Z$, where the components, σ_X, σ_Y and σ_Z of $\bar{\sigma}$ are the Pauli spin matrices (see Eq. 8.83 for their definition).

▶ **Exercise 9.16**   If the z-axis is taken to be along $\mathbf{R}$, the action of the Pauli spin operators on the states $|-(\mathbf{R})\rangle$ and $|+(\mathbf{R})\rangle$ is given by the usual relations:

$$\sigma_X|\pm\rangle = |\mp\rangle, \qquad \sigma_Y|\pm\rangle = \pm i|\mp\rangle, \qquad \sigma_Z|\pm\rangle = \pm|\pm\rangle. \tag{9.145}$$

Substituting into Eq. 9.143 the result of Exercise 9.15, and using Eq. 9.145, show that

$$V_{X+} = \mathrm{Im}(\langle+|\sigma_Y|-\rangle\langle-|\sigma_Z|+\rangle)/2R^2 = 0 \tag{9.146}$$

$$V_{Y+} = \mathrm{Im}(\langle+|\sigma_Z|-\rangle\langle-|\sigma_X|+\rangle)/2R^2 = 0 \tag{9.147}$$

$$V_{Z+} = \mathrm{Im}(\langle+|\sigma_X|-\rangle\langle-|\sigma_Y|+\rangle)/2R^2 = 1/2R^2. \tag{9.148}$$

Because of the isotropy of spin, the result of Exercise 9.16 generalizes, for arbitrary orientations of $\mathbf{R}$, to

$$\mathbf{V}_+(\mathbf{R}) = -\mathbf{V}_-(\mathbf{R}) = \mathbf{R}/2R^3. \tag{9.149}$$

Equation 9.149 combined with Eq. 9.136 has the form of an expression for the flux through C of a magnetic field arising from a monopole of strength $-\frac{1}{2}$ located at the degeneracy. A result from the theory of electricity and magnetism is that this flux is just the product of the monopole strength times the solid angle that C subtends at the degeneracy (see Figure 9.4a), that is, $\gamma_C = -\frac{1}{2}\Omega(C)$, and therefore

$$e^{i\gamma_\pm(C)} = e^{\mp\frac{1}{2}i\Omega(C)}. \tag{9.150}$$

For real Hamiltonians Eq. 9.150 simplifies. Since the circuit is now confined to the $X - Z$ plane, the solid angle subtended is $\Omega = \pm 2\pi$ if the degeneracy is enclosed; $\Omega = 0$ otherwise. Thus,

$$e^{i\gamma_\pm(C)} = \begin{cases} -1, & \text{if } C \text{ encircles the degeneracy} \\ 0 & \text{otherwise.} \end{cases} \tag{9.151}$$

This result implies a sign change of real wavefunctions as a degeneracy is encircled. This is illustrated in Figure 9.4b, which shows a particular eigenfunction of a triangle as its shape is taken in a circuit around the degenerate equilateral geometry. When the triangle returns to its original shape (at the top of the figure) dark has changed to light and vice versa, signalling the adiabatic sign change.

The result of the adiabatic sign change for real wavefunctions encircling a degeneracy was known to Herzberg and Longuet–Higgins as early as 1963, in the context of degenerate electronic states with conical intersections. As the electronic wavefunction is taken in a circuit in the parameter space of nuclear displacements, the electronic wavefunction will have changed sign when the nuclear displacements return to their original values if the circuit

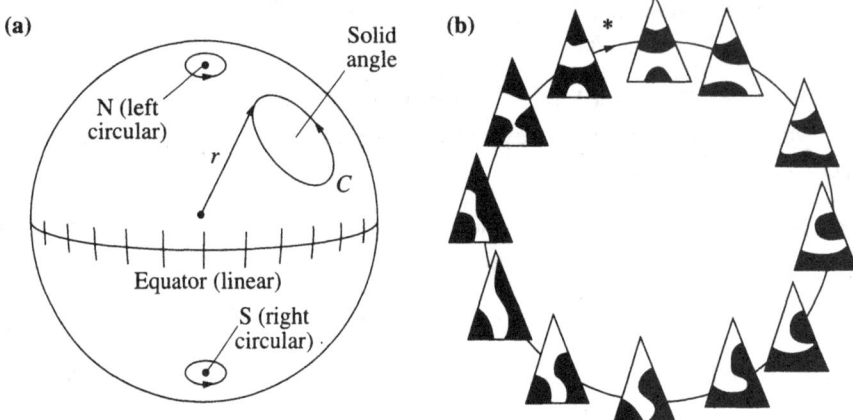

Figure 9.4 (a) As the vector $\mathbf{R}(t)$ sweeps out a circuit C in parameter space, the vector $\frac{\mathbf{R}}{|\mathbf{R}|} = \mathbf{r}(t)$ traces out a closed circuit on the unit sphere whose area is equal to the solid angle subtended by C at the center of the sphere. The phase change associated with a circuit C of the $\mathbf{R}$-vector is just half this solid angle. (b) Illustration of the circuit C in the space of boundary shapes of triangular drums, surrounding a shape for which the vibrational mode is degenerate. The dark and light areas of the vibrating drums correspond to the conventional + and − labels. Note that the triangles on either side of the arrow (at the *) differ only by a sign change—a phase of 180°. Adapted from Berry (1990).

encloses a degeneracy. Note that if one considers the *total* wavefunction—the product of nuclear and electronic wavefunctions—the result must be single valued, and hence must return to its original value when the original coordinates are restored. Since the electronic wavefunction changes sign, this implies that the *nuclear* wavefunction must also change sign as a result of the adiabatic transformation of the displacements. This sign change in the nuclear wavefunction shows up a half-integral quantum number in the coordinate for *pseudorotation*: the coordinate that encloses the degeneracy. This is described in more detail in Section 12.2.3. In the context of time-dependent calculations, the wavepacket describing nuclear motion will pick up a geometric phase of π upon encircling the degeneracy and returning to its original location.

9.5 Periodic Hamiltonians and Floquet Theory

Another special case of time-dependent Hamiltonians is that of Hamiltonians that are periodic in time. In this case it is also possible to make certain general statements about the solution of the Time-Dependent Schrödinger Equation. The theory of these systems is known as Floquet theory. The application of Floquet theory to quantum mechanical systems was first introduced by Shirley (1965).

Consider the usual Time-Dependent Schrödinger Equation,

$$i\hbar \frac{\partial}{\partial t} \Psi(x, t) = H(x, t)\Psi(x, t), \tag{9.152}$$

where the Hamiltonian is periodic in time with period T, that is, $H(x, t + T) = H(x, t)$. We assume the trial form

$$\Psi_\lambda(x, t) = e^{-i\varepsilon_\lambda t/\hbar} \Phi_\lambda(x, t). \tag{9.153}$$

Substituting into the Time-Dependent Schrödinger Equation and regrouping terms we find

$$H_F(x, t)\Phi_\lambda(x, t) = \varepsilon_\lambda \Phi_\lambda(x, t), \tag{9.154}$$

where

$$H_F(x, t) \equiv \left\{ H(x, t) - i\hbar \frac{\partial}{\partial t} \right\} \tag{9.155}$$

is called the Floquet Hamiltonian and $\Phi_\lambda(x, t)$ is called a Floquet eigenstate. Equation 9.154 is an eigenvalue equation in the two variables, (x, t), with time-independent eigenvalues given by the Floquet energies, ε_λ. The key result of Floquet theory is that the Floquet eigenstates, $\Phi_\lambda(x, t)$, are periodic in time with the same period T as the Hamiltonian. Equation 9.153 represents a particular solution of the TDSE. The general solution is given by

$$\Psi(x, t) = \sum_\lambda a_\lambda e^{-i\varepsilon_\lambda t/\hbar} \Phi_\lambda(x, t). \tag{9.156}$$

In order to develop some intuition for the Floquet eigenstates we examine some limiting examples. First, if the Hamiltonian is time independent, $H(x, t) = H(x)$, then the Floquet states are just the eigenfunctions of $H(x)$, and the Floquet eigenvalues are just the eigenvalues of $H(x)$. Second, if the Hamiltonian is changing adiabatically with no change to the energy levels, as in the case of a harmonic oscillator whose minimum is shifted slowly and periodically, the Floquet eigenstates are just the adiabatic eigenstates and the Floquet eigenvalues are just the time-independent eigenvalues of $H(x, t)$. Third, if the Hamiltonian is changing adiabatically but with eigenvalues that do change with time, again the Floquet eigenfunctions are the adiabatic eigenfunctions; the time-independent Floquet eigenvalues, ε_λ, are given by the sum of two terms:

$$\varepsilon_\lambda = \int_{-\infty}^{\infty} \Phi_\lambda(x, t) H(x, t) \Phi_\lambda(x, t)\, dx - i \int_{-\infty}^{\infty} \Phi_\lambda(x, t) \frac{\partial}{\partial t} \Phi_\lambda(x, t)\, dx. \tag{9.157}$$

The second term, involving a derivative with respect to time, compensates for the adiabatic change in the eigenvalues of $H(x, t)$ to ensure the constancy of the Floquet eigenvalue, ε_λ. The behavior of the Floquet eigenfunctions can deviate dramatically from that of the adiabatic eigenfunctions when there is resonance, that is, when $H(x, t)$ has frequency components that promote transitions between the adiabatic eigenfunctions; indeed, this is where Floquet theory is most useful. The case of resonance will be discussed below, after we introduce an alternative formulation of the formalism in terms of basis sets in space and time.

Note that it is possible to regroup the factors in Eq. 9.153, as follows:

$$\Psi_\lambda(x, t) = e^{-i\varepsilon_\lambda t/\hbar} \Phi_\lambda(x, t) \tag{9.158}$$

$$= e^{-i(\varepsilon_\lambda + n\hbar\omega)t/\hbar} e^{in\omega t} \Phi_\lambda(x, t) \tag{9.159}$$

This repartitioning corresponds to a series of new Floquet eigenvalues, $\varepsilon_\lambda + n\hbar\omega$, with corresponding eigenfunctions, $\Phi_{\lambda n}(x, t) \equiv e^{in\omega t}\Phi_\lambda(x, t)$. If n is an integer, $\Phi_{\lambda n}(x, t)$ will be periodic in t so long as $\Phi_\lambda(x, t)$ is. Yet the physical state $\Psi_\lambda(x, t)$ is unchanged, and thus the Floquet eigenvalues associated with distinct physical states are defined only modulo $\hbar\omega$.

It is convenient to define a composite Hilbert space in both position and time, $R \oplus T$ (Sambe, 1973). The spatial part is spanned by square-integrable functions in configuration space, while the temporal part is spanned by the complete, orthonormal set of functions $e^{in\omega t}$, where $n = 0, \pm 1, \pm 2, \ldots$. The Floquet eigenstates, $\Phi_{\lambda n}(x, t)$, satisfy the orthonormality condition

$$\langle\langle\Phi_{\kappa n}|\Phi_{\nu m}\rangle\rangle \equiv \frac{1}{T}\int_0^T dt \int_{-\infty}^\infty dx \, \Phi_{\kappa n}^*(x, t)\Phi_{\nu m}(x, t) = \delta_{\kappa\nu}\delta_{nm} \qquad (9.160)$$

and form a complete set in $R \oplus T$:

$$\sum_{\kappa n} |\Phi_{\kappa n}\rangle\rangle\langle\langle\Phi_{\kappa n}| = \mathbf{1}, \qquad (9.161)$$

where we have introduced the double bra-ket notation to denote the inner product over both x and t. Note that the vectors $\Phi_{\kappa n}(x, t)$ and $\Phi_{\kappa n'}(x, t)$, differing only in n, are distinct vectors in the Hilbert space $R \oplus T$, and all values of n are required for the complete expansion of an arbitrary function. In other words, a single value of n, such as $n = 0$, is sufficient for expanding the physically evolving wavefunction, $\Psi(x, t)$, but is not sufficient for expanding an arbitrary wavefunction in $R \oplus T$.

We now turn to the representation of the Floquet Hamiltonian in an arbitrary complete basis. As we shall see, using a complete basis in both position and time provides a representation of the Hamiltonian that is time independent, allowing the application of virtually all the techniques and theorems of time-independent Hamiltonians to periodic time-dependent Hamiltonians. We introduce the notation $|\alpha n\rangle\rangle \equiv |\alpha\rangle|n\rangle$, where $|\alpha\rangle$ are atomic or molecule eigenstates that solve $H(x)\alpha(x) = E_\alpha^0\alpha(x)$ and $|n\rangle$ are Fourier vectors that satisfy $\langle t|n\rangle = e^{in\omega t}$. Since both $\Phi_\lambda(x, t)$ and $H(x, t)$ are periodic in time, they may be expanded in a Fourier series:

$$\int_{-\infty}^\infty \beta^*(x)\Phi_\lambda(x, t)\,dx = \sum_{m=-\infty}^\infty \phi_{\beta\lambda}^{(m)}e^{im\omega t} \qquad (9.162)$$

and

$$\int_{-\infty}^\infty \alpha^*(x)H(x, t)\beta(x)\,dx = \sum_{n=-\infty}^\infty H_{\alpha\beta}^{(n)}e^{in\omega t}. \qquad (9.163)$$

On substituting these expansions into the Schrödinger equation, one obtains an infinite set of relations for $\phi_{\alpha\lambda}^{(n)}$:

$$\sum_{\beta m}\langle\langle\alpha n|\hat{H}_{\mathrm{F}}|\beta m\rangle\rangle\phi_{\beta\lambda}^{(m)} = \varepsilon_\lambda\phi_{\alpha\lambda}^{(n)}, \qquad (9.164)$$

where $\hat{H}_{\mathrm{F}}$ is the time-independent Floquet Hamiltonian with matrix elements

$$\langle\langle\alpha n|\hat{H}_{\mathrm{F}}|\beta m\rangle\rangle = H_{\alpha\beta}^{(n-m)} + n\hbar\omega\delta_{\alpha\beta}\delta_{nm}. \qquad (9.165)$$

	$n = 2$	$n = 1$	$n = 0$	$n = -1$	$n = -2$	
	$A + 2\omega\mathbf{1}$	B	0	0	0	$n' = 2$
	B	$A + \omega\mathbf{1}$	B	0	0	$n' = 1$
$[H_F] =$	0	B	A	B	0	$n' = 0$
	0	0	B	$A - \omega\mathbf{1}$	B	$n' = -1$
	0	0	0	B	$A - 2\omega\mathbf{1}$	$n' = -2$

Figure 9.5 Structure of the Floquet Hamiltonian matrix. The matrix **A** on the diagonal consists of matrix elements of the time-averaged Hamiltonian. Typically, **A** is itself diagonal, with entries E_i, the eigenenergies of the bare system Hamiltonian. In each diagonal block, the matrix **A** is repeated with a shift of $n\hbar\omega\mathbf{1}$. The matrix **B** contains the Fourier component at frequency ω of the coupling between the basis functions that is induced by the periodic Hamiltonian.

The typical structure of this matrix is illustrated in Figure 9.5. The matrix **A** is given by

$$A_{\alpha\beta} = \langle\langle\alpha 0|H(x, t)|\beta 0\rangle\rangle = \frac{1}{T}\int_0^T \langle\alpha|H(x, t)|\beta\rangle dt = \langle\alpha|\bar{H}(x)|\beta\rangle, \quad (9.166)$$

where $\bar{H}(x)$ signifies the time average of $H(x, t)$. For the common case that $H(x, t) = H_0(x) + V(x, t)$ and $\bar{V}(x) = 0$, the time-averaged Hamiltonian is just the bare system Hamiltonian, $H_0(x)$. The basis functions $\{\alpha, \beta\}$ may be chosen to be eigenstates of $\bar{H}(x)$, in which case **A** is diagonal. The matrix **B** is given by

$$B_{\alpha\beta} = \langle\langle\alpha 0|H(x, t)|\beta 1\rangle\rangle = \frac{1}{T}\int_0^T \langle\alpha|H(x, t)|\beta\rangle e^{i\omega t} dt. \quad (9.167)$$

The Floquet matrix in Figure 9.5 corresponds to the common case of a Hamiltonian with time dependence $\cos(\omega t)$, so that blocks with $|n - m| > 1$ vanish.

In addition to the matrix **A**, the diagonal blocks contain a shift by $n\hbar\omega\mathbf{1}$. The eigenstates of $\bar{H}(x)$, shifted by multiples of $\hbar\omega$, are sometimes called "field-dressed states." The off-diagonal blocks, which arise from Fourier components of $H(x, t)$, supply the coupling necessary to mix the field-dressed states to produce the "dressed states" (cf. Section 15.3); this mixing is most important when the states dressed by different values of n are quasi-degenerate. Normally, in the dressed state treatment only a 2×2 block within Figure 9.5 is considered, with diagonal elements $|\alpha 1\rangle\rangle$ and $|\beta 0\rangle\rangle$, where $E_\alpha + \hbar\omega \approx E_\beta$. In the full Floquet theory, the molecular states are coupled also by higher-order interaction through quasidegeneracies of the form $|\alpha n\rangle\rangle$ and $|\beta n - 1\rangle\rangle$. Nevertheless, in cases of isolated resonances, the reduced 2×2 treatment may contain much of the physics and can be diagonalized, often with quite accurate results.

▶ **Exercise 9.17**   Return to Exercise 9.4 and interpret the time-independent matrix that appears there, along with its eigenvalues and eigenvectors, in terms of Floquet theory and dressed states. How many Floquet states are there? What are their eigenvalues?

Shirley has given an elegant interpretation of the matrix structure in Figure 9.5. The physical source of the time periodicity of the system Hamiltonian has its origin in the interaction with some degrees of freedom that are not explicitly treated. If those additional degrees of freedom are treated explicitly, the joint system will have a Hamiltonian structure of the form in Figure 9.5. In the commonly encountered case where the periodic change in the system Hamiltonian arises from interaction with an electromagnetic field, the Hamiltonian structure in Figure 9.5 corresponds to a treatment in which the field is quantized. The basis functions of the joint system then become labelled by a quantum number for the molecule and a quantum number for the field (number of photons in the field). Within this interpretation, the quasidegeneracy described in the previous paragraph corresponds to the statement that material state α plus n photons is quasi-degenerate with material state β plus m photons within a treatment in which both the matter and light are quantized.

What happens when the matrix in Figure 9.5 is diagonalized? Formally, the number of eigenvalues of the Floquet matrix in Figure 9.5 is infinite. However, we have seen above that if ε_λ is an eigenvalue, so is $\varepsilon_\lambda + n\hbar\omega \equiv \varepsilon_{\lambda n}$, for any integer n. Similarly, although the eigenvectors are distinct in the larger space of $R \oplus T$, they are interrelated through the periodicity relationship

$$\langle\langle \alpha, n + p | \Phi_{\lambda, m+p} \rangle\rangle = \langle\langle \alpha, n | \Phi_{\lambda, m} \rangle\rangle. \tag{9.168}$$

Since the basis functions in the bra differ by a simple phase factor, $e^{ip\omega t}$, from those in the ket, once they are combined with their associated Floquet eigenvalue they become generators for identical dynamics. Thus, the formally infinite number of eigenvalues and eigenfunctions obtained by diagonalizing the infinite Floquet matrix in fact reduces to N physically distinct eigenvalues and eigenfunctions, where N is the number of basis functions of $\bar{H}(x)$, due to the above periodicity relations.

As mentioned above, the matrix representation of the Floquet Hamiltonian in the joint Hilbert space $R \oplus T$ provides a representation of periodic Hamiltonians that is compatible with virtually all the techniques and theorems of time-independent Hamiltonians. One of the remarkable consequences of this result is that one may write a closed-form expression for the evolution operator in terms of the eigenvalues and eigenfunctions of the Floquet Hamiltonian. This expression for the propagator is an attractive alternative to the Magnus expansion because its basis set character allows for well-controlled convergence properties. Specifically, the propagator is given by the following expression (Shirley, 1965):

$$U_{\beta\alpha}(t, t_0) = \sum_n \sum_{\lambda, l} \langle\langle \beta n | \Phi_{\lambda l} \rangle\rangle e^{-i\varepsilon_{\lambda l}(t - t_0)/\hbar} \langle\langle \Phi_{\lambda l} | \alpha 0 \rangle\rangle e^{in\omega(t - t_0)} \tag{9.169}$$

$$= \sum_n \langle\langle \beta n | e^{-i\hat{H}_F(t - t_0)/\hbar} | \alpha 0 \rangle\rangle e^{in\omega(t - t_0)}. \tag{9.170}$$

Equation 9.170 is consistent with two interpretations. First, it is the amplitude of a system initially in molecular state α at time t_0 evolving to state β at time t according to the time-dependent Hamiltonian, $H(x, t)$. Alternatively, it can be interpreted as the amplitude of a system initially in Floquet state $|\alpha 0\rangle\rangle$ at time t_0 evolving to Floquet state $|\beta n\rangle\rangle$ at time t according to the time-independent Floquet Hamiltonian, $\hat{H}_F$, summed over n with weighting factors $e^{in\omega t}$. Note that in the simplest case of one-photon absorption in a dressed-state picture, $n = -1$, since the photon number decreases upon absorption; the phase factor becomes $e^{-i\omega t}$, consistent with the contribution of the photon energy, $\hbar\omega$, to the forward-in-time evolution of the dressed state.

The numerical diagonalization of the large Floquet matrix can be inconvenient, and other numerical approaches to Floquet theory have been developed. It can be shown (Milfield, 1983) that for time-periodic Hamiltonians,

$$U(t + T, 0) = U(t, 0)U(T, 0), \tag{9.171}$$

from which it follows that

$$U(nT, 0) = U(T, 0)^n. \tag{9.172}$$

Thus, the time evolution operator over one period, $U(T, 0)$, provides all the information ever needed for the long-time propagation of a system with a periodic Hamiltonian. In a truncated basis set, $U(T, 0)$ may be diagonalized by some unitary transformation Z, yielding

$$U(T, 0) = Ze^{-iDT/\hbar}Z^\dagger, \tag{9.173}$$

where D is a diagonal matrix of the time-independent Floquet eigenvalues.

It stands to reason that the unitary time evolution operator obtained in Floquet theory should be in agreement with the unitary time evolution operator one would obtain for the same time-dependent Hamiltonian using the Magnus expansion and keeping all the terms. Indeed, in an important paper, Lior has definitively established the equivalence of these two approaches (Lior, 1992).

We close this section with a remarkable generalization of Floquet theory, which allows the use of time-independent methods for *any* time-dependent Hamiltonian, not necessarily periodic: the so-called (t, t') formalism (Peskin, 1993; Pfeifer, 1983). In this approach the time-dependent solution

$$\Psi(x, t) = U(t, t_0)\Psi(x, t_0) \tag{9.174}$$

for the explicitly Time-Dependent Schrödinger Equation is written as

$$\Psi(x, t) = \Psi(x, t', t)|_{t'=t}, \tag{9.175}$$

where

$$\Psi(x, t', t) = e^{-\frac{i}{\hbar}H_F(x,t')(t-t_0)}\Psi(x, t', t_0) \tag{9.176}$$

with

$$H_F = H(x, t') - i\hbar\frac{\partial}{\partial t'}. \tag{9.177}$$

▶ **Exercise 9.18** Prove Eqs. 9.175 and 9.176.

Solution Using Eq. 9.176 we have

$$i\hbar\frac{\partial}{\partial t}\Psi(x, t', t) = H_F(x, t')e^{-iH_F(x,t')(t-t_0)/\hbar}\Psi(x, t', t_0) \tag{9.178}$$

$$= -i\hbar\frac{\partial}{\partial t'}\Psi(x, t', t) + H(x, t')\Psi(x, t', t). \tag{9.179}$$

Hence,

$$i\hbar \left(\frac{\partial}{\partial t} + \frac{\partial}{\partial t'} \right) \Psi(x, t', t) = H(x, t')\Psi(x, t', t). \tag{9.180}$$

Since we are interested in t' only on the contour $t' = t$, where $\frac{\partial t'}{\partial t} = 1$, we find that

$$\frac{\partial \Psi(x, t', t)}{\partial t'}\bigg|_{t'=t} + \frac{\partial \Psi(x, t', t)}{\partial t}\bigg|_{t'=t} = \frac{\partial \Psi(x, t', t)}{\partial t}, \tag{9.181}$$

which, combined with Eq. 9.180, completes the proof.

9.6 Variational Principles and the Time-Dependent Self-Consistent Field Approximation

In this section we will discuss the use of variational principles to generate equations of motion for an approximate wavefunction. Many different possible equations of motion may be derived from variational principles. The common thread is that the functional form for the wavefunction is assumed, but a restricted set of parameters within the function may be varied. Variational principles lead to equations of motion for these parameters such that the difference between the approximate wavefunction and the exact wavefunction is minimized (Dirac, 1978; Frenkel, 1934; McLachlan, 1964).

Specifically, in this section we will derive two very different forms of variationally optimal equations of motion, originating from very different representations of the wavefunction. The first is a fairly obvious set of equations for propagation of a wavefunction represented in a finite basis; its counterpart, with an infinite set of basis functions, was derived in Chapter 8. The second approximate equations of motion are the time-dependent self-consistent field (TDSCF) equations (Hartree, 1927; Dirac, 1930). This is the counterpart of the time-independent self-consistent field equations, in which the wavefunction is taken by construction to be a separable product of one-dimensional (or more generally, lower-dimensional) wavefunctions. Given the constraint of this form for the wavefunction, the variational principle provides an optimal form for the approximate equations of motion.

A wide variety of different equations of motion can be traced back to variational principles. Some of this variety originates from the many alternative representations possible to describe the wavefunction; each representation will lead to a different form of equation for the adjustable parameters. Moreover, one may quantify the difference between the exact and the approximate wavefunction in a variety of different ways: in terms of the real or the imaginary part of the error, the integral over space of the absolute value of the difference or the absolute value squared, and so forth. Finally, an important point is that the variational formulations described in this section are local in time—they minimize the difference between the "exact" and approximate wavefunction at time $t + \Delta t$, assuming implicitly that the wavefunction at time t is exact; however, in practice, if the wavefunction is propagated forward in time iteratively the wavefunction at time t is known only imperfectly, and thus the optimization of parameters is being performed by comparing with an imperfect reference. An alternative is to develop variational formulations that are global—those that minimize the difference between the exact and approximate wavefunction over the entire time interval. Some of the numerical methods described in Chapter 11 involve global representations for the propagator, and satisfy variational criteria defined over the entire time interval. We can only conclude that variational formulations are probably as rich and diverse as time-dependent quantum mechanics itself, and there is much remaining for the reader to explore!

We will meet variational equations again in the next chapter, in the context of path integration, and again in Part III, in Chapter 16 on control of chemical reactions. Moreover, as mentioned above, variational methods are closely connected with some of the numerical methods that we will discuss in Chapter 11. Finally, it is worth noting that variational formulations of spectra and scattering cross sections are possible, formulations that formally circumvent the evaluation of the time-dependent wavefunction.

Our starting point is the Time-Dependent Schrödinger Equation:

$$i\hbar \frac{\partial \Psi}{\partial t} = H\Psi. \tag{9.182}$$

For some small time increment, call it τ, we can write

$$\Psi(t + \tau) = \Psi(t) - \frac{i}{\hbar}\theta(t)\tau, \tag{9.183}$$

where $\theta(t)$ is

$$\theta(t) = i\hbar \frac{\partial \Psi}{\partial t}. \tag{9.184}$$

If Ψ were an exact solution to the Time-Dependent Schrödinger Equation then $\theta = H\Psi$. Since this is not true in general, we will let θ vary. We can therefore define a variational functional $\mathfrak{I}$,

$$\mathfrak{I} = \int |\theta(t) - H\Psi(t)|^2 \, d\mathcal{V}, \tag{9.185}$$

and minimize it subject to variations in θ. We have defined $d\mathcal{V}$ as the volume $dq_1 dq_2 \ldots dq_N$, where N is the number of degrees of freedom. The functional $\mathfrak{I}$ is called the McLachlan variational functional. Its first variation with respect to θ is simply

$$\delta\mathfrak{I} = \int \delta\theta^*(\theta(t) - H\Psi(t)) \, d\mathcal{V} + \int (\theta(t) - H\Psi(t))^* \, \delta\theta \, d\mathcal{V}. \tag{9.186}$$

Note that the McLachlan functional is local in time: it optimizes $\frac{\partial \Psi}{\partial t}$ at time t, given $\Psi(t)$. This has two implications. First, minimizing the error locally in time at every time step in general is not equivalent to minimizing the global error over the full time interval. Second, at intermediate times t in the propagation, $\Psi(t)$ in general will not be the exact wavefunction, due to the approximate propagation in the previous steps. Thus, the McLachlan functional is very different from what are in principle better measures of the true error: the time-local functional

$$\mathfrak{I}_1(t) = \int |\Psi_{\text{ex}}(\mathbf{q}, t) - \Psi_{\text{calc}}(\mathbf{q}, t)|^2 \, d\mathcal{V} \tag{9.187}$$

or the time-global functional

$$\mathfrak{I}_2 = \iint |\Psi_{\text{ex}}(\mathbf{q}, t) - \Psi_{\text{calc}}(\mathbf{q}, t)|^2 \, d\mathcal{V} \, dt, \tag{9.188}$$

where $\Psi_{\text{ex}}(\mathbf{q}, t)$ is the *exact* wavefunction and $\Psi_{\text{calc}}(\mathbf{q}, t)$ is the *calculated* wavefunction at time t. Unfortunately, these alternative measures are not generally available since they require knowledge of the exact wavefunction, and therefore we proceed with the discussion of the McLachlan functional.

We can use the McLachlan variational functional to obtain the form of the Time-Dependent Schrödinger Equation in a finite basis. Choose a set of N functions $\{\Phi_n\}$ and expand Ψ and θ in them:

$$\Psi(t) = \sum_{n=1}^{N} c_n(t)\Phi_n \qquad (9.189)$$

$$\theta(t) = \sum_{n=1}^{N} d_n(t)\Phi_n. \qquad (9.190)$$

For completeness, introduce another set of functions $\{\phi_\alpha\}$ such that the union, $\{\Phi_n\} \bigcup \{\phi_\alpha\}$, is complete. In this basis, the part of the integrand of the McLachlan variational functional that multiples $\delta\theta^*$ is

$$\theta - H\Psi = \sum_{n=1}^{N} \left(d_n(t) - \sum_{m=1}^{N} H_{mn}c_m(t) \right) \Phi_n - \sum_{m=1}^{N} \sum_{\alpha} H_{\alpha m}c_m(t)\phi_\alpha. \qquad (9.191)$$

The first variation in $\mathfrak{I}$ is therefore

$$\delta\mathfrak{I} = \sum_{n=1}^{N} \delta d_n^* \left(d_n - \sum_{m=1}^{N} H_{mn}c_m \right) + \text{c.c.} = 0. \qquad (9.192)$$

We immediately obtain the equation

$$d_n(t) = \sum_{m=1}^{N} H_{mn}c_m(t). \qquad (9.193)$$

From Eqs. 9.184, 9.189 and 9.190 we see that $d_n(t) = i\hbar\,\dot{c}_n(t)$ and so Eq. 9.193 becomes

$$i\hbar\,\dot{c}_n(t) = \sum_{m=1}^{N} H_{mn}\,c_m(t) \qquad (9.194)$$

which is the representation of the Time-Dependent Schrödinger Equation in a finite basis.

The McLachlan variational functional is also at the heart of the time-dependent self-consistent field method for solving multidimensional problems. Suppose we wish to approximate a many-particle wavefunction as a product of single-particle or single-coordinate wavefunctions:

$$\Psi(q_1, q_2, \ldots, q_N, t) = \psi_1(q_1, t)\psi_2(q_2, t) \cdots \psi_N(q_N, t). \qquad (9.195)$$

For the Hamiltonian describing two-body interactions,

$$H = H_0 + V = \sum_{i=1}^{N} H_i(q_i) + \sum_{i=1}^{N} \sum_{j>i}^{N} V_{ij}(q_i, q_j), \qquad (9.196)$$

what is the optimal choice of $\{\psi_i\}$? As before, we can vary each of the one-particle wavefunctions independently, namely:

$$\psi_i(q_i, t + \tau) = \psi_i(q_i, t) - \frac{i}{\hbar}\theta_i(q_i, t)\tau. \qquad (9.197)$$

The change, θ, in the many-body wavefunction is then given to first order in τ by

$$\theta(q_1, q_2, \ldots, q_N, t) = \sum_{i=1}^{N}\{\psi_1(q_1, t)\cdots\theta_i(q_i, t)\cdots\psi_N(q_N, t)\}. \qquad (9.198)$$

Equation 9.186 takes the form

$$\delta\mathfrak{I} = \int \delta\theta^*\chi\,d\mathcal{V} + \text{c.c.}, \qquad (9.199)$$

where

$$\chi = \theta - H\Psi$$

$$= \sum_{i=1}^{N}\{\psi_1(q_1, t)\cdots[\theta_i(q_i, t) - H_i\psi_i(q_i, t)]\cdots\psi_N(q_N, t)\}$$

$$- \sum_{i=1}^{N}\sum_{j>i=1}^{N}\{\psi_1(q_1, t)\cdots[V_{ij}(q_i, q_j)\psi_i(q_i, t)\psi_j(q_j, t)]\cdots\psi_N(q_N, t)\}.$$

Setting $\delta\mathfrak{I} = 0$ leads to a set of N equations for the N unknown ψ_i:

$$i\hbar\frac{\partial\psi_i}{\partial t} = (H_i + G_i + \lambda_i)\psi_i, \qquad (9.200)$$

where

$$G_i(q_i, t) = \sum_{j\neq i=1}^{N}\int \psi_j^*(q_j, t)V_{ij}(q_i, q_j)\psi_j(q_j, t)\,dq_j \qquad (9.201)$$

$$\lambda_i(t) = \sum_{k\neq i=1}^{N}\langle H_k - \epsilon_k\rangle + \sum_{k\neq i}^{N}\sum_{j>k\neq i}^{N}\langle V_{kj}(q_k, q_j)\rangle \qquad (9.202)$$

$$\epsilon_i(t) = i\hbar\int \psi_i^*\frac{\partial\psi_i}{\partial t}\,dq_i \qquad (9.203)$$

and where the brackets stand for averages.

▶ **Exercise 9.19**   Verify Eqs. 9.200–9.203 for the case $N = 3$ by writing out the explicit expressions for $\delta\theta^*$ and χ and setting the coefficients of $\delta\theta_1$, $\delta\theta_2$ and $\delta\theta_3$ independently equal to zero.

The following general relations are satisfied by the ϵ's, λ's and G's:

$$\epsilon_i = \langle H_i + G_i + \lambda_i\rangle \qquad \sum_{i=1}^{N}\epsilon_i = \langle H\rangle \qquad \sum_{i=1}^{N}\lambda_i = -\langle V\rangle. \qquad (9.204)$$

▶ **Exercise 9.20** Verify these relations by checking them explicitly for the case $N = 3$.

The presence of the λ_i's is inconvenient as they depend on integrals over all the coordinates. However, the λ_i's are arbitrary in the sense that changing any pair of them does not change the solution, as long as the sum $\sum \lambda_i$ remains constant. So instead of solving Eq. 9.200, we may solve

$$i\hbar \frac{\partial \phi_i}{\partial t} = (H_i + G_i)\phi_i \tag{9.205}$$

and then construct the final solution as

$$\Psi(q_1, q_2, \ldots, q_n, t) = \left(\prod_{i=1}^{N} \phi_i(q_i, t) \right) e^{-\frac{i}{\hbar} S(t)}, \tag{9.206}$$

where

$$S(t) = \int \sum_i \lambda_i(t')\, dt' = - \int \langle V(t') \rangle\, dt'. \tag{9.207}$$

Equations 9.200 with Eq. 9.195 (or Eq. 9.205 with Eq. 9.206) together constitute the time-dependent Hartree (TDH) or time-dependent self-consistent field (TDSCF) method. The name time-dependent Hartree generally implies that all the particles in the system are identical, and hence all H_i and V_{ij} have the same form.

▶ **Exercise 9.21**

a. Show that if the individual orbitals in the TDH method, $\{\phi_i\}$, are normalized at $t = 0$ they are normalized at all times.

b. Show that if H is independent of time, the total energy defined as $\sum_i H_i + G_i/2$ is conserved in the TDH method, but the energy in the individual modes is not. Thus, the TDH allows for the description of time-dependent energy transfer.

c. Show that the individual orbitals in TDH do not in general retain orthogonality unless one includes the so-called self-interaction term, that is, unless $G_i \rightarrow \tilde{G}_i$, where $\tilde{G}_i$ includes the term $j = i$ in Eq. 9.201.

d. The TDH equations of motion can be modified to incorporate antisymmetric exchange symmetry of identical particles as follows (Dirac, 1930):

$$i\hbar \frac{\partial \phi_i}{\partial t} = (H_i + \tilde{G}_i)\phi_i - \sum_j K_j \phi_j, \tag{9.208}$$

where the exchange part of the Hamiltonian is given by

$$K_j = \int \psi_j^*(q) V_{ji}(q, q_i) \psi_i(q)\, dq. \tag{9.209}$$

Equation 9.208 goes by the name of the time-dependent Hartree-Fock (TDHF) equation (Fock, 1930; Dirac, 1930). Show that the TDHF equation removes the problematic self-interaction term, yet still maintains the time-dependent orthogonality of the orbitals.

e. Which of the properties in **a.–d.** apply more generally to the TDSCF method, where H_i and V_{ij} may be different for different particles?

Solution

a. Consider

$$\frac{d}{dt}\langle\phi_i(t)|\phi_i(t)\rangle = \left\langle\frac{\partial\phi_i}{\partial t}\Big|\phi_i(t)\right\rangle + \text{c.c.} \tag{9.210}$$

$$= \frac{1}{-i\hbar}\langle[H_i + G_i]\phi_i(t)|\phi_i(t)\rangle + \text{c.c.} \tag{9.211}$$

$$= \frac{1}{-i\hbar}\langle\phi_i(t)|H_i + G_i|\phi_i(t)\rangle + \text{c.c.} = 0. \tag{9.212}$$

In the last line we have noted that since the first term is purely imaginary the c.c. is its negative.

b. Define the energy per mode as

$$\varepsilon_i(t) = \langle\phi_i(t)|H_i + \frac{1}{2}\sum_{j\neq i}\langle\phi_j(t)|V_{ij}|\phi_j(t)\rangle|\phi_i(t)\rangle, \tag{9.213}$$

with the total energy

$$\varepsilon = \sum_i \varepsilon_i. \tag{9.214}$$

Then

$$\frac{d\sum_i \varepsilon_i}{dt} = \sum_i\left\langle\frac{\partial\phi_i(t)}{\partial t}\Big|H_i + \frac{1}{2}\sum_{j\neq i}\langle\phi_j(t)|V_{ij}|\phi_j(t)\rangle|\phi_i(t)\rangle + \text{c.c.}\right.$$
$$+ \frac{1}{2}\sum_i\left\langle\phi_i(t)\Big|\sum_{j\neq i}\left\langle\frac{\partial\phi_j(t)}{\partial t}\Big|V_{ij}|\phi_j(t)\rangle\right|\phi_i(t)\right\rangle + \text{c.c.} \tag{9.215}$$

Taking the potential to be symmetric in the indices i and j the second term can be rewritten

$$\frac{1}{2}\sum_j\sum_{i\neq j}\left\langle\frac{\partial\phi_j(t)}{\partial t}\Big|\langle\phi_i(t)|V_{ji}|\phi_i(t)\rangle|\phi_j(t)\rangle + \text{c.c.}\right. \tag{9.216}$$

Noting that i and j are dummy indices we have

$$\frac{d\sum_i \varepsilon_i}{dt} = \sum_i\left\langle\frac{\partial\phi_i(t)}{\partial t}\Big|H_i + \sum_{j\neq i}\langle\phi_j(t)|V_{ij}|\phi_j(t)\rangle|\phi_i(t)\rangle + \text{c.c.}\right. \tag{9.217}$$

$$= \sum_i\left\langle\frac{\partial\phi_i(t)}{\partial t}\Big|H_i + G_i|\phi_i(t)\rangle + \text{c.c.}\right. \tag{9.218}$$

$$= \frac{1}{-i\hbar}\sum_i\langle\phi_i(t)|[H_i + G_i]^2|\phi_i(t)\rangle + \text{c.c.} = 0. \tag{9.219}$$

Note that the proof relied on interchanging the double summation over i and j and hence there is no reason that the individual $d\varepsilon_i/dt$'s should vanish. This implies that although the total energy is conserved in TDH, the energy in the individual modes is not.

c.

$$\frac{d}{dt}\langle\phi_i|\phi_j\rangle = \langle\dot\phi_i|\phi_j\rangle + \langle\phi_i|\dot\phi_j\rangle$$

$$= \frac{1}{-i\hbar}\langle\phi_i|H_i + G_i|\phi_j\rangle + \frac{1}{i\hbar}\langle\phi_i|H_j + G_j|\phi_j\rangle.$$

Note that for identical particles, $H_i = H_j$, but

$$G_i = \sum_{k\neq i=1}^{N}\langle\phi_k(t)|V_{ik}|\phi_k(t)\rangle \neq G_j = \sum_{k\neq j=1}^{N}\langle\phi_k(t)|V_{jk}|\phi_k(t)\rangle.$$

Hence, $\frac{d}{dt}\langle\phi_i|\phi_j\rangle = 0$. However, including the self-energy term we have

$$\tilde G_i = \sum_{k=1}^{N}\langle\phi_k(t)|V_{ik}|\phi_k(t)\rangle = \tilde G_j,$$

and hence $\frac{d}{dt}\langle\phi_i|\phi_j\rangle = 0$. We leave the proof of **d** and **e** to the reader.

▶ **Exercise 9.22**   If the range of the interaction between particles is much shorter than the extent of the wavefunctions, one may write

$$V_{ij} \approx g\delta(q_i - q_j). \tag{9.220}$$

Substitute this form for V_{ij} into the TDH equation and set all the orbitals to be identical. Show that one obtains the Gross–Pitaevskii equation (Gross, 1961; Pitaevskii, 1961),

$$i\hbar\frac{\partial\phi_i}{\partial t} = (H_i + g(N-1)|\phi_i|^2)\phi_i. \tag{9.221}$$

Note that Eq. 9.221 is of the form of a nonlinear Schrödinger equation.

Problems

▶ **Exercise 9.23**   Using the conjugation-commutation formula (Exercise 8.35),

$$e^A Be^{-A} = B + [A, B] + \frac{1}{2!}[A, [A, B]] + \frac{1}{3!}[A, [A, [A, B]]] + \cdots, \tag{9.222}$$

find an expression for the interaction picture Hamiltonian, $H_I(t)$, for

   a. $H_0 = p^2/2m$ and $V(x) = -kx$,
   b. $H_0 = p^2/2m$ but arbitrary $V(x)$, and
   c. $H_0 = p^2/2m + 1/2m\omega^2x^2$ and arbitrary $V(x)$.

▶ **Exercise 9.24**   Show that for the interaction picture Hamiltonian, $H_I(t)$, with $H_0 = p^2/2m$ and $V(x) = -kx$, the time-dependent interaction wavepacket is given by $\Psi_I(x, t) = N\exp(-\alpha_t(x - x_t)^2 + \frac{i}{\hbar}p_t(x - x_t) + \frac{i}{\hbar}\gamma_t)$. Substitute this form for the wavefunction into the interaction picture equations of motion and obtain the following expressions for the wavepacket parameters:

$$x_t = x_0 - \frac{\alpha_0}{2m}t^2 \tag{9.223}$$

$$p_t = p_0 + \alpha_t \tag{9.224}$$

$$\alpha_t = \alpha_0 \tag{9.225}$$

$$\gamma_t = -\frac{\alpha_0^2}{6m}t^3 + \alpha_0 x_0 t + \gamma_0. \tag{9.226}$$

▶ **Exercise 9.25**   The Baker–Hausdorff formula states (see Exercise 9.8)

$$e^{A+B} = e^A e^B e^{-\frac{1}{2}[A,B]}. \tag{9.227}$$

This formula is valid only if $[A, B]$ commutes with A and B. The Zassenhaus formula provides a generalization of the Baker–Hausdorff formula when the latter assumption does not hold:

$$e^{A+B} = e^A e^B \prod_{n=2}^{\infty} e^{c_n(A,B)}, \tag{9.228}$$

where

$$c_2(A, B) = -\frac{1}{2}[A, B] \tag{9.229}$$

$$c_3(A, B) = -\frac{1}{3}[[A, B], B] - \frac{1}{6}[[A, B], A]. \tag{9.230}$$

Use the Zassenhaus formula to expand $\exp(-\frac{i}{\hbar}Ht)$, where $H = p^2/2m - kx$. Show that the four factors in the expansion can be related to a shift of the wavepacket in coordinate space, a shift in momentum space, a wavepacket spreading factor, and a coordinate-independent phase factor.

▶ **Exercise 9.26**   Consider the "nested" interaction wavefunction

$$\Psi_I'(t) = e^{-\frac{i}{\hbar}\langle P \rangle R} e^{\frac{i}{\hbar}H_0 t} \Psi_S(t), \tag{9.231}$$

where $H_0 = p^2/2m$ and $\langle P \rangle$ is assumed to be independent of time.

a. Find the Hamiltonian, H_I', such that Ψ_I' obeys an equation of motion of the same structure as the Time-Dependent Schrödinger Equation:

$$i\hbar\frac{\partial \Psi_I'(t)}{\partial t} = H_I'(t)\Psi_I'(t). \tag{9.232}$$

b. Show that H_I' can be manipulated into the form

$$H_I' = e^{\frac{i}{\hbar}H_0 t} V\left(R + \frac{\langle P \rangle}{m}t\right) e^{-\frac{i}{\hbar}H_0 t} = V\left(R + \frac{(P + \langle P \rangle)}{m}t\right) \tag{9.233}$$

▶ **Exercise 9.27**   Derive the TDSCF equations for the two-dimensional harmonic potential, $H = p_x^2/2 + p_y^2/2 + kx^2 + ky^2 + \lambda xy$. Calculate the correlation function and the spectrum for the TDSCF motion. How does it compare with the exact solution? How does it compare with the energies obtained from time-independent SCF?

References

General

1. P. A. M. Dirac, *The Principles of Quantum Mechanics* (Oxford, 1978).
2. A. Messiah, *Quantum Mechanics*, vols. I and II (Wiley, New York, 1958).

Interaction Picture: Quantum Mechanics

3. S. Das and D. J. Tannor, Time dependent Quantum mechanics with picosecond time steps: Application to the predissociation of HeI$_2$, J. Chem. Phys. 92, 3403 (1990).
4. D. J. Tannor, A. Besprozvannaya and C. J. Williams, Nested interaction representations in time dependent quantum mechanics, J. Chem. Phys. 96, 2998 (1992).
5. J. Z. H. Zhang, Interaction representation in time-dependent quantum scattering: Elimination of finite boundary reflection, Chem. Phys. Lett. 160, 417 (1989); J. Z. H. Zhang, New method in time-dependent quantum scattering theory: Integrating the wave function in the interaction picture, J. Chem. Phys. 92, 324 (1990); J. Z. H. Zhang, Multichannel quantum wavepacket propagation in the interaction picture: Application to gas-surface scattering, Comput. Phys. Commun. 63, 28 (1991).
6. D. H. Zhang and J. Z. H. Zhang, Time dependent treatment of vibrational predissociation within the golden rule approximation, J. Chem. Phys. 95, 6449 (1991).
7. M. Founargiotakis and J. C. Light, A split interaction representation for quantum correlation functions of dissociative systems, J. Chem. Phys. 93, 633 (1991).

Interaction Picture: Semiclassical

8. R. T. Skodje, On the use of the interaction picture in classical mechanics, Chem. Phys. Lett. 109, 221 (1984); R. T. Skodje, Gaussian wavepacket dynamics expressed in the classical interaction picture, Chem. Phys. Lett. 109, 227 (1984).
9. J. D. Kress and A. E. DePristo, Semiclassical Gaussian wavepacket dynamics for collinear reactive scattering, J. Chem. Phys. 89, 2886 (1988).
10. N. E. Hendricksen and E. J. Heller, Gaussian wavepacket dynamics and scattering in the interaction picture, Chem. Phys. Lett. 148, 567 (1988).
11. J. Shao and N. Makri, Forward-backward semiclassical dynamics in the interaction representation, J. Chem. Phys. 113, 3681 (2000).

Exponential Operators: Magnus Expansion and Wei–Norman Factorization

12. P. Pechukas and J. C. Light, On the exponential form of time-displacement operators in quantum mechanics, J. Chem. Phys. 44, 3897 (1966).
13. W. R. Salzman, An alternative to the Magnus expansion in time-dependent perturbation theory, J. Chem. Phys. 82, 822 (1985).

14. W. R. Salzman, Convergence of Magnus and Magnus-like expansions in the Schrödinger representation, J. Chem. Phys. 85, 4605 (1986).

15. A. Lior, Equivalence between dynamical averaging methods of the Schrödinger equation: Average Hamiltonian, secular averaging and van Vleck transformation, Chem. Phys. Lett. 199, 383 (1992).

16. A. Iserles, H. Z. Munthe-Kaas, S. P. Nørsett and A. Zanna, Lie-group methods, Acta Numerica, pp. 215–365 (2000).

17. W. Magnus, On the exponential solution of differential equations for a linear operator, Commun. Pure Appl. Math. 7, 649 (1954).

18. R. M. Wilcox, Exponential operators and parameter differentiation in quantum physics, J. Math. Phys. 8, 962 (1967).

19. J. Wei and E. Norman, On global representations of the solutions of linear differential equations as a product of exponentials, Proc. Amer. Math. Soc. 15, 327 (1964).

20. C. Altafini, Parameter differentiation and quantum state decomposition for time varying Schrödinger equations, Reports Math. Phys. 52, 381 (2003).

Quantum Adiabatic Theorem and the Geometric Phase

21. A. Messiah, *Quantum Mechanics*, vol. II (Amsterdam: North Holland, 1962).

22. M. V. Berry, Quantal phase factors accompanying adiabatic changes, Proc. R. Soc. Lond. A 392, 45 (1984).

23. M. V. Berry, Anticipations of the Geometric Phase, Physics Today, December, 1990.

24. A. Shapere and F. Wilczek, eds. *Geometrical Phases in Physics* (World Scientific, Singapore, 1989).

25. Y. Aharonov and D. Bohm, Significance of electromagnetic potentials in the quantum theory, Phys. Rev. 115, 485 (1959).

26. G. Herzberg and H. C. Longuet-Higgins, Intersection of potential energy surfaces in polyatomic molecules, Discuss. Faraday Soc. 35, 77 (1963).

27. C. A. Mead and D. G. Truhlar, On the determination of Born-Oppenheimer nuclear motion wave functions including complications due to conical intersections and identical nuclei, J. Chem. Phys. 70, 2284 (1979).

28. G. Delacrétaz, E. R. Grant, R. L. Whetten, L. Wöste, and J. W. Zwanziger, Fractional quantization of molecular pseudorotation in Na_3, Phys. Rev. Lett. 56, 2598 (1986).

29. A. Kuppermann and Y.-S. M. Wu, The geometrical phase effect shows up in chemical reactions, Chem. Phys. Lett. 205, 577 (1993).

30. Z. Wu, Evolution of systems with a slowly changing Hamiltonian, Phys. Rev. A 40, 2184 (1989).

Landau–Zener Formula

31. L. D. Landau, Zur Theorie der Energieübertragung II, Physik. Z. Sowjetunion 2, 46 (1932); English translation: A theory of energy transfer. II. In *Collected Papers of L. D. Landau*, ed. D. ter Haar (Gordon and Breach, New York, 1965).

32. C. Zener, Non-adiabatic crossing of energy levels, Proc. Roy. Soc. London A137, 696 (1932).

33. E. C. G. Stückelberg, Theorie der unelastischen Stösse zwischen Atomen, Helv. Phys. Acta 5, 369 (1932).

Floquet Theory

34. J. H. Shirley, Solution of the Schrödinger equation with a Hamiltonian periodic in time, Phys. Rev. 138, 979 (1965).

35. H. Sambe, Steady states and quasienergies of a quantum-mechanical system in an oscillating field, Phys. Rev. A 7, 2203 (1973).

36. S. I. Chu, Recent developments in semiclassical Floquet theories for intense-field multiphoton processes, Adv. At. Mol. Phys. 21, 197 (1985).

37. K. F. Milfield and R. E. Wyatt, Study, extension, and application of Floquet theory for quantum molecular systems in an oscillating field, Phys. Rev. A 27, 72 (1983).

38. U. Peskin, R. Kosloff and N. Moiseyev, The solution of the time-dependent Schrödinger equation by the (t, t') method: The use of global polynomial propagators for time-dependent Hamiltonians, J. Chem. Phys. 100, 8849 (1994).

39. M. Grifoni and P. Hänggi, Driven quantum tunneling, Phys. Rep. 304, 229 (1998).

40. U. Peskin and N. Moiseyev, The solution of the time dependent Schrödinger equation by the (t, t') method: Theory, computational algorithm and applications, J. Chem. Phys. 99, 4590 (1993).

41. P. Pfeifer and R. D. Levine, A stationary formulation of time-dependent problems in quantum mechanics, J. Chem. Phys. 79, 5512 (1983).

Time-Dependent Variational Principles

42. P. A. M. Dirac, Note on exchange phenomena in the Thomas atom, Proc. Cambridge Philos. Soc. 26, 376 (1930).

43. J. Frenkel, *Wave Mechanics* (Clarendon, Oxford, 1934), p. 253.

44. A. D. McLachlan, A variational solution of the time-dependent Schrödinger equation, Mol. Phys. 8, 39 (1964).

45. E. J. Heller, Time dependent variational approach to semiclassical dynamics, J. Chem. Phys. 64, 63 (1976).

46. S.-Y. Lee and E. J. Heller, Exact time-dependent wavepacket propagation: Application to the photodissociation of methyl iodide, J. Chem. Phys. 76, 3035 (1982).

47. R. D. Coalson and M. Karplus, Extended wavepacket dynamics: Exact solution for collinear atom, diatomic molecule scattering Chem. Phys. Lett. 90, 301 (1982).

48. R. C. Brown, *Time-Dependent Methods for Radiative and Non-radiative Electronic Transitions in Polyatomic Molecules*, Ph.D. thesis, UCLA, 1983.

49. R. T. Skodje and D. G. Truhlar, Localized Gaussian wavepacket methods for inelastic collisions involving anharmonic oscillators, J. Chem. Phys. 80, 3123 (1984).

50. S. Sawada, R. Heather, B. Jackson, and H. Metiu, A strategy for time dependent quantum mechanical calculations using a Gaussian wavepacket representation of the wave function, J. Chem. Phys. 83, 3009 (1985).

51. M. Ben-Nun and T. J. Martinez, A multiple spawning approach to tunneling dynamics, J. Chem. Phys. 112, 6113 (2000); M. Ben-Nun, J. Quenneville and T. J. Martinez, Ab initio multiple spawning: Photochemistry from first principles quantum molecular dynamics, J. Phys. Chem. 104, 5161 (2000).

52. D. Shalashilin and M. S. Child, Time dependent quantum propagation in phase space, J. Chem. Phys. 113, 10028 (2000); D. Shalashilin and M. S. Child, Description of tunneling with the help of coupled frozen Gaussians, J. Chem. Phys. 114, 9296 (2001); D. Shalashilin and M. S. Child, Multidimensional quantum propagation with the help of coupled coherent states, J. Chem. Phys. 115, 5367 (2001).

Time-Dependent Self-Consistent Field Method

53. D. R.Hartree, The wave mechanics of an atom with a non-coulomb central field. I. Theory and methods; II. Some results and discussion, Proc. Cambridge Philos. Soc. 24, 89, 111 (1927).

54. V. Fock, Näherungsmethode zur Lösung des quantenmechanischen Mehrkörperproblems, Zeit. für Phys. 61, 126 (1930). English translation: An approximate method for solving the quantum many-body problem, in V. A. Fock, *Selected Works: Quantum Mechanics and Quantum Field Theory* (CRC Press, 2004).

55. E. P. Gross, Structure of a quantized vortex in boson systems, Nuovo Cimento 20, 454 (1961).

56. L. P. Pitaevskii, Vortex lines in an imperfect bose gas, Zh. Eksp. Teor. Fiz. 40, 646 (1961) [Sov. Phys. JETP 13, 451 (1961)].

57. P. A. M. Dirac, Note on exchange phenomena in the Thomas atom, Proc. Cambridge Philos. Soc. 26, 376 (1930).

58. *Time-Dependent Hartree Fock and Beyond*, ed. K. Goeke and P.-G. Reinhard, Lecture Notes in Physics, vol. 71 (Springer, Berlin, 1982).

59. S. Levit, Time-dependent mean-field approximation for nuclear dynamical problems, Phys. Rev. C 21, 1594 (1980); S. Levit, J. W. Negele and Z. Paltiel, Phys. Rev. C 21, 1603 (1980).

60. R. B. Gerber, V. Buch and M. A. Ratner, Time-dependent self-consistent field approximation for intramolecular energy transfer. I. Formulation and application to dissociation of van der Waals molecules, J. Chem. Phys. 77, 3022 (1982).

61. R. Kosloff, A. D. Hammerich and M. A. Ratner, Time dependent quantum mechanical calculation of the dissociation dynamics of the cluster He_n-I_2, in *Large Finite Systems: Proceedings of the Twentieth Jerusalem Symposium of Quantum Chemistry and Biochemistry*, ed. J. Jortner and B. Pullman (Reidel, Dordrecht, 1987).

62. N. Makri and W. H. Miller, Time-dependent self-consistent field (TDSCF) approximation for a reaction coordinate coupled to a harmonic bath: Single and multiple configuration treatments, J. Chem. Phys. 87, 5781 (1987).

63. J. Campos-Martinez and R. D. Coalson, Adding configuration interaction to the time-dependent Hartree grid approximation, J. Chem. Phys. 93, 4740 (1990).

64. U. Manthe, H.-D. Meyer and L. S. Cederbaum, Wavepacket dynamics within the multiconfiguration Hartree framework: General aspects and application to NOCl, J. Chem. Phys. 97, 3199 (1992).

65. M. H. Beck, A. Jäckle, G. A. Worth and H.-D. Meyer, The multiconfiguration time-dependent Hartree (MCTDH) method: A highly efficient algorithm for propagating wavepackets, Physics Reports 324, 1 (2000).

Chapter 10

Path Integration, the van Vleck Propagator and Semiclassical Mechanics

This chapter deals with semiclassical mechanics. Semiclassical mechanics refers to a range of methodologies in which the trajectories obey classical mechanics, but the dynamical information is used to construct quantum mechanical amplitudes, rather than probabilities. By adding the amplitudes before calculating the absolute value squared to obtain the probabilities, one can observe interference effects between classical trajectories that travel from the same initial point to the same final point by different routes. The construction of quantum amplitudes based on the underlying classical dynamics is sometimes described picturesquely as placing "quantum mechanical flesh on classical bones." In this chapter we will pay particular attention to the propagator, and its semiclassical approximations, that is, the use of classical trajectory information to approximate the quantum propagator.

Despite many beautiful results, the subject of semiclassical mechanics is vast, difficult and incomplete. Here we focus on several of the main concepts and perspectives. We will begin with a discussion of the path integral formulation of quantum mechanics. Since the classical action integral and its variations play a central role in this formulation, we have added a preliminary section summarizing the key features of this quantity. After introducing the path integral formulation we derive the van Vleck–Gutzwiller semiclassical propagator, which arises very naturally as an approximation to the full quantum path integral. Again, the classical action integral and its variations are central, but now the sum is over only *classical* paths, not over all *virtual* paths as in the quantum expression. We next introduce Miller's semiclassical algebra: the concept that matrix elements of unitary operators in quantum mechanics can be approximated quite generally by a sum over classical paths, each weighted by an amplitude times a phase, $Ae^{i\phi}$. The phase derives from the generating function of the appropriate *classical* canonical transformation, while the amplitude is related to the second derivative of the generating function with respect to the appropriate initial and final variables. Probabilities are calculated as the absolute value squared of this sum, and therefore generally have cross-terms that lead to interference contributions. The term semiclassical "algebra" refers to the fact that matrix elements obtained by successive canonical transformations give the same result as those obtained in a single global canonical transformation, if all integrals are performed by stationary phase. The van Vleck–Gutzwiller propagator is then seen to be a special case, where the canonical transformation is the dynamical transformation of the coordinate variable and the generator for the transformation is the classical action. Finally, we turn to the links with wavepacket propagation. A variety of approximate wavepacket propagation techniques have been proposed over the years, especially by Heller. A particularly simple method expands an arbitrary initial state in a basis of coherent states (i.e., Gaussians), and then propagates each coherent state according to classical mechanics with a frozen width parameter. A very gratifying connection with the van Vleck–Gutzwiller (VVG) expression was derived by Herman and Kluk (HK), who started with the VVG expression and rigorously derived the frozen Gaussian method. Their derivation led to the discovery of a curious new amplitude prefactor that can be viewed as a generalizaton of the VVG prefactor, but is free from caustic singularities. Numerical results with this technique are significantly better than with other wavepacket techniques, or other semiclassical techniques in general. Thus, the inclusion of the central formulas of this approach is highly appropriate in the framework of this chapter and this book as a whole, due to the close relation to the VVG formula, the inherent wavepacket character of the derivation and the high quality of the results.

10.1   The Classical Action

The classical action integral, or Hamilton's principal function, S, plays a major role in virtually every time-dependent semiclassical theory. We have already seen this quantity in Part I, both in the context of Gaussian wavepackets (Chapter 3) and in the classical limit of quantum mechanics (Chapter 4). In fact, the material in this chapter will have many connections with those results, and provides a more complete framework for understanding them.

10.1.1   Lagrange's Equations of Motion

Before proceeding to the classical action, we will briefly review the Lagrangian formulation of classical mechanics. The central object, the Lagrangian L, is defined as

$$L = T - V, \tag{10.1}$$

where T is the kinetic energy and V is the potential energy. Lagrange's equations of motion are given in general by

$$\frac{d}{dt}\left(\frac{\partial L}{\partial \dot{q}}\right) - \frac{\partial L}{\partial q} = 0. \tag{10.2}$$

We will be interested in the most common situation in which

$$T = \frac{m\dot{q}^2}{2} \tag{10.3}$$

(m time independent) and $V = V(q)$ is assumed to be a function of coordinates q only. In this case Lagrange's equations reduce to

$$m\frac{d^2q}{dt^2} + \frac{\partial V}{\partial q} = 0, \tag{10.4}$$

which is immediately seen to be the usual expression for Newton's second law.

The momentum is defined in Lagrangian mechanics as

$$p \equiv \frac{\partial L}{\partial \dot{q}} = m\dot{q}. \tag{10.5}$$

This formula shows a simple proportionality between p and $\dot{q}$, which belies the generality of Lagrangian mechanics. Combining Eqs. 10.4 and 10.5 yields

$$\frac{dp}{dt} + \frac{\partial V}{\partial q} = 0, \tag{10.6}$$

which is a more general expression of Newton's second law that applies to time-dependent masses, motion in magnetic fields and so forth. Newton himself used this more general formulation, which, by identifying the momentum as a fundamental object, to some extent anticipates Hamilton's formulation of mechanics.

Having defined the momentum, p, the Hamiltonian, $H(p, q)$, is related to the Lagrangian by a Legendre transform:

$$H(p, q) = \dot{q}p - L = \dot{q}(p)\frac{\partial L(\dot{q}(p), q, t)}{\partial \dot{q}} - L(\dot{q}(p), q, t). \tag{10.7}$$

▶ **Exercise 10.1**   Show that for a Lagrangian given by $L = T - V = \frac{m\dot{q}^2}{2} - V$, Eq. 10.7 gives the usual quantity, $H = T + V$.

▶ **Exercise 10.2**   Show that Hamilton's equations,

$$\dot{p} = -\frac{\partial H}{\partial q} \tag{10.8}$$

$$\dot{q} = \frac{\partial H}{\partial p}, \tag{10.9}$$

are fully equivalent to Lagrange's equations.

10.1.2   Hamilton's Principal Function and Its First Variation

The previous section dealt with the equations of motion for *trajectories* in terms of the Lagrangian. We now distinguish between classical trajectories and *paths*. By a "path" we mean a *schedule* of (q, t) with the endpoints (q', t'), (q'', t'') fixed. Each path completely determines the velocity $\dot{q}$ at all times, since $\dot{q}(t) = \lim_{\Delta t \to 0} \frac{q(t+\Delta t)-q(t)}{\Delta t}$, that is, two values of q nearby in time determine the velocity at the midpoint in time. As a result, one can assign values to $L(\dot{q}, q, t)$ along any path, regardless of whether that path is consistent with Newton's laws (or equivalently, Lagrange's equations). There are an infinite number of paths from (q', t') to (q'', t''), although all but a set of measure zero are *virtual* trajectories, in the sense that they don't satisfy Newton's laws.

Hamilton's principal function, S, is the time integral of the Lagrangian defined for each of these paths, regardless of whether it satisfies Newton's laws:

$$S[q(t)](q'', t'', q', t') = \int_{t'}^{t''} L(\dot{q}, q, t) \, dt \tag{10.10}$$

(we follow Goldstein and most of the semiclassical literature in using the notation S, although Gutzwiller and others use W). Note that S is a function of the endpoints $(q'', t''; q', t')$, but is a *functional* of $q(t)$. Hamilton's variational principle states that for a classical *trajectory*, namely, a path that *does* satisfy Lagrange's equations,

$$\delta S[q_{\text{cl}}(t)](q'', t'', q', t') = \delta \int_{t'}^{t''} L(\dot{q}_{\text{cl}}, q_{\text{cl}}, t) \, dt = 0, \tag{10.11}$$

where we have used the shorthand

$$\delta S[q_{\text{cl}}(t)] \equiv \delta S[q(t)]|_{q(t)=q_{\text{cl}}(t)} \tag{10.12}$$

and so on. The variation here must be applied with care. Recall that the only paths that are being considered are those with the fixed endpoints $q'(t')$ and $q''(t'')$. Thus the variation is over schedules of the path that connects between these two endpoints. Note that there may be more than one classical path—or none!—that connects q' at time t' with q'' at time t''. This is because of the double-ended boundary conditions. If instead of specifying $q'(t')$ and $q''(t'')$, one were to specify *initial conditions*, $q'(t')$ and $p'(t')$, the classical path would be uniquely determined, but it might not lead to q'' at t''. Typically, the number of classical root trajectories grows rapidly as the time interval $t'' - t'$ increases.

▶ Exercise 10.3

    a. Write down the Lagrangian for a classical trajectory with free-particle motion.

    b. Show that for the classical free particle

$$S[q_{\text{cl}}(t)](q'', t'', q', t') = \frac{m(q'' - q')^2}{2(t'' - t')}. \tag{10.13}$$

▶ Exercise 10.4   Show that for a classical trajectory (i.e., $S = S[q_{\text{cl}}(t)]$)

$$a. \left. \frac{\partial S}{\partial q''} \right|_{q',\tau} = p''; \qquad b. \left. \frac{\partial S}{\partial q'} \right|_{q'',\tau} = -p'; \qquad c. \left. \frac{\partial S}{\partial \tau} \right|_{q',q''} = -E,$$

where $\tau \equiv t'' - t'$.

Solution

    a. We need to show that $\frac{\partial S}{\partial q''} = p''$, where

$$S = \int_{t'}^{t''} (p\dot{q} - H) \, dt. \tag{10.14}$$

We begin by writing out the full functional dependence of S:

$$S(q'', q', t'', t') = \int_{t'}^{t''} \left\{ p(q'', q', t)\dot{q}(q'', q', t) \right.$$
$$\left. -H(p(q'', q', t), q(q'', q', t)) \right\} dt. \tag{10.15}$$

Thus,

$$
\left(\frac{\partial S}{\partial q''}\right)_{q'} = \int_{t'}^{t''} \left\{ \left(\frac{\partial p}{\partial q''}\right)_{q'} \dot{q} + p \left(\frac{\partial \dot{q}}{\partial q''}\right)_{q'} \right.
$$
$$
\left. - \left(\frac{\partial H}{\partial p}\right)_{q} \left(\frac{\partial p}{\partial q''}\right)_{q'} - \left(\frac{\partial H}{\partial q}\right)_{p} \left(\frac{\partial q}{\partial q''}\right)_{q'} \right\} dt \qquad (10.16)
$$

$$
= p \left(\frac{\partial q}{\partial q''}\right)_{q'}\Bigg|_{t'}^{t''} + \int_{t'}^{t''} \left\{ -\dot{p} \left(\frac{\partial q}{\partial q''}\right)_{q'} - \left(\frac{\partial H}{\partial q}\right)_{p} \left(\frac{\partial q}{\partial q''}\right)_{q'} \right\} dt \quad (10.17)
$$

$$
= p'' \left(\frac{\partial q''}{\partial q''}\right)_{q'} - p' \left(\frac{\partial q'}{\partial q''}\right)_{q'} \qquad (10.18)
$$

$$
= p''. \qquad (10.19)
$$

b. The proof for p' follows similarly.

c. We need to show that

$$
\frac{\partial S(q', q'', \tau)}{\partial \tau} = -E(q', q'', \tau), \qquad (10.20)
$$

where

$$
S = \int_0^\tau (p\dot{q} - E)dt = \int_0^\tau \left(p(q', q'', \tau; t)\, \dot{q}(q', q'', \tau; t) - E(q', q'', \tau) \right) dt.
$$
$$
(10.21)
$$

Note the functional dependence; this is of paramount importance to get the correct answer. The independent variables are taken to be q', q'' and τ. Changing τ but leaving q' and q'' unchanged leads to a different trajectory that extremizes the action, and a different value for S, p', p'' and E. It is in this sense that E is a function of τ (but not t—energy is conserved along the trajectory!).

$$
\frac{\partial S}{\partial \tau} = (p\dot{q} - E)|_\tau + \int_0^\tau \left\{ \frac{\partial p}{\partial \tau}\dot{q} + p\frac{\partial \dot{q}}{\partial \tau} - \frac{\partial E}{\partial \tau} \right\} dt \qquad (10.22)
$$

$$
= (p\dot{q} - E)|_\tau + \int_0^\tau \left\{ \frac{\partial p}{\partial \tau}\frac{\partial H}{\partial p} + \frac{\partial H}{\partial q}\frac{\partial q}{\partial \tau} - \frac{\partial E}{\partial \tau} \right\} dt + p\frac{\partial q}{\partial \tau}\Bigg|_0^\tau \quad (10.23)
$$

$$
= p\frac{\partial q}{\partial t}\Bigg|_\tau - E + \int_0^\tau \left\{ \frac{\partial H(p, q)}{\partial \tau} - \frac{\partial E}{\partial \tau} \right\} dt + p\frac{\partial q}{\partial \tau}\Bigg|_0^\tau \qquad (10.24)
$$

$$
= -E + p(\tau)\left(\frac{\partial q}{\partial t} + \frac{\partial q}{\partial \tau}\right)\Bigg|_\tau - p(0)\frac{\partial q(t)}{\partial \tau}\Bigg|_0. \qquad (10.25)
$$

In the second line we have integrated by parts with respect to t, and then used Hamilton's equations, Eqs. 10.8–10.9. In the fourth line we have equated the Hamiltonian with the energy, and hence cancelled their derivatives with respect to τ. Proceeding with the derivation, the last term in Eq. 10.25 vanishes by the boundary condition at $t = 0$: $q(0) = q'$. The second and third terms cancel because of the boundary condition $q(t; \tau)|_{t=\tau} = q''$. The proof of the latter requires some care. Recall that q is being viewed as a function of t and τ, and therefore, for fixed q'', along the line $t = \tau$:

$$\delta q(t, \tau)|_{t=\tau} = 0 = \left(\frac{\partial q}{\partial t}\right)_\tau \delta t + \left(\frac{\partial q}{\partial \tau}\right)_t \delta \tau = \left[\left(\frac{\partial q}{\partial t}\right)_\tau + \left(\frac{\partial q}{\partial \tau}\right)_t\right] \delta \tau. \quad (10.26)$$

In physical terms, if the final velocity is positive ($\left(\frac{\partial q}{\partial t}\right)_\tau > 0$), increasing the time interval τ with fixed endpoints q' and q'' will slow down the trajectory and therefore lead to a decrease in the value of $q(t)$ for a given t ($\left(\frac{\partial q}{\partial \tau}\right)_t < 0$). In conclusion, the only term on the RHS of Eq. 10.25 that survives is the term $-E$, which is the desired result.

▶ **Exercise 10.5**   Show that if H does not depend explicitly on time, for paths where $\delta S = 0$ it follows that $\frac{dH}{dt} = 0$.

▶ **Exercise 10.6**   Show that the statement $\delta S = 0$ is equivalent to Lagrange's equations of motion,

$$\frac{d}{dt}\frac{\partial L}{\partial \dot{q}} - \frac{\partial L}{\partial q} = 0. \quad (10.27)$$

Solution

$$\delta S = \delta \int_{t'}^{t''} L(\dot{q}, q, t)dt = \int_{t'}^{t''} \left\{\frac{\partial L}{\partial \dot{q}}\delta\dot{q} + \frac{\partial L}{\partial q}\delta q\right\} dt = 0. \quad (10.28)$$

Integrating the first term on the RHS by parts and combining with the second term, we obtain

$$\frac{\partial L}{\partial \dot{q}}\delta q\Big|_{t'}^{t''} - \int_{t'}^{t''} \left\{\frac{d}{dt}\frac{\partial L}{\partial \dot{q}} - \frac{\partial L}{\partial q}\right\} \delta q \, dt = 0. \quad (10.29)$$

The first term on the LHS vanishes since $\delta q(t') = \delta q(t'') = 0$. The integral must be 0 for any variation δq, hence the quantity in braces must be 0. We recognize this as Lagrange's equation.

It may be worth working this same exercise out a little more explicitly. Take $L = \frac{1}{2}m\dot{q}^2 - V(q)$. Then

$$\delta S = S[q_0 + \delta q] - S[q_0] \quad (10.30)$$

$$= \int_{t'}^{t''} \left\{\frac{1}{2}m(\dot{q}_0 + \dot{\delta q})^2 - V(q_0 + \delta q)\right\} dt - S[q_0] \quad (10.31)$$

$$= \int_{t'}^{t''} \left\{\frac{1}{2}m\dot{q}_0^2 + m\dot{q}_0\dot{\delta q} - V(q_0) - V'(q_0)\delta q)\right\} dt - S[q_0] \quad (10.32)$$

$$= \int_{t'}^{t''} \left\{m\dot{q}_0\dot{\delta q} - V'(q_0)\delta q\right\} dt \quad (10.33)$$

$$= \int_{t'}^{t''} \left\{-m\ddot{q}_0\delta q - V'(q_0)\delta q\right\} dt + m \dot{q}_0\delta q\Big|_{t'}^{t''} \quad (10.34)$$

$$= 0. \quad (10.35)$$

In the third line we have neglected terms proportional to $\dot{\delta q}^2$ and δq^2. In the fifth line we integrated by parts. The boundary term is 0 since the variation in q is constrained to be 0 at the endpoints of the path. Thus we conclude that

$$m\ddot{q}_0 + V'(q_0) = 0, \tag{10.36}$$

which is just Newton's second law.

10.1.3   The Second Variation

The statement is sometimes made that the classical trajectory follows the path of least action, but this statement is not precise. In analogy with the ordinary calculus, the condition $\delta S = 0$ guarantees that S is an extremum. But is S a maximum, a minimum or a saddle point? To clarify this question, we note that even for one-dimensional dynamics the variations $\delta S[q(t)]$ are with respect to an infinite dimensional function space. If the change in S is negative for all variations in this infinite dimensional space, S is a maximum; if the change in S is positive for all variations, S is a minimum. If the change in S is 0 for first-order changes, but positive for some second-order changes and negative for other second-order changes, S is a generalized saddle point in the infinite dimensional space.

Consider the path

$$q(t) = q_0(t) + \delta q(t), \tag{10.37}$$

where $q_0(t)$ corresponds to a classical trajectory. We assume throughout the fixed endpoints (q'', t'', q', t'). Then S can be expanded in terms of various orders of δq:

$$S = S_0 + \delta S_0 + \frac{1}{2}\delta^2 S_0 + \cdots, \tag{10.38}$$

where

$$S_0 = \int_{t'}^{t''} L(\dot{q}_0, q_0, t)\, dt \tag{10.39}$$

$$\delta S_0 = \delta \int_{t'}^{t''} L(\dot{q}_0, q_0, t)\, dt = 0 \tag{10.40}$$

$$\delta^2 S_0 = \delta^2 \int_{t'}^{t''} L(\dot{q}_0, q_0, t)\, dt. \tag{10.41}$$

This third term, the *second* variation of S around the classical trajectory, determines the character of the extremum, just as the second derivative in the ordinary calculus determines whether one has a maximum, a minimum or a saddle point. To obtain an explicit expression for $\delta^2 S_0$ we write

$$\delta^2 S_0 = \delta S(q_0 + \delta q) - \delta S(q_0) \tag{10.42}$$

$$= \int_{t'}^{t''} \left\{ -m(\ddot{q}_0 + \ddot{\delta q}) - V'(q_0 + \delta q) \right\} \delta q\, dt - \int_{t'}^{t''} \left\{ -m\ddot{q}_0 - V'(q_0) \right\} \delta q\, dt \tag{10.43}$$

$$= \int_{t'}^{t''} \left\{ -m\ddot{\delta q} - V''(q_0)\delta q \right\} \delta q\, dt. \tag{10.44}$$

Thus, the second variation can be written

$$\delta^2 S = \int_{t'}^{t''} \delta q(t) \left\{ -V''(q_0(t)) - m\frac{d^2}{dt^2} \right\} \delta q(t)\, dt. \tag{10.45}$$

To make further progress it is convenient to expand $\delta q(t)$ in a basis of orthonormal eigenfunctions of $\Lambda \equiv -V''(q_0(t)) - m\frac{d^2}{dt^2}$. That is, we define a set of basis functions in time, $\{u_n(t)\}$, such that

$$\Lambda u_n(t) = \lambda_n u_n(t) \qquad u_n(t') = u_n(t'') = 0. \tag{10.46}$$

Then we may expand $\delta q(t)$ in a complete basis of the $u_n(t)$:

$$\delta q(t) = \sum_n a_n u_n(t) \tag{10.47}$$

and it follows that

$$\delta^2 S = \sum_n \lambda_n a_n^2. \tag{10.48}$$

The operator Λ has an infinite number of eigenvalues.

▶ **Exercise 10.7** Find the eigenvalues of Λ for the free particle and show that all eigenvalues are positive.

For short time $\tau \equiv t'' - t'$, the motion is essentially free particle–like and hence all the eigenvalues of Λ are positive; this implies that for small τ the path of extremal action is a minimum. However, for larger τ one or more of the eigenvalues generally becomes negative. It can be shown that, as a function of τ, once an eigenvalue becomes negative it never again becomes positive (see Figure 10.1). Thus, for longer time intervals, the paths of extremal action are neither minima or maxima, but saddle points, heading closer toward the character of a maximum as the time interval increases. The number of negative eigenvalues, ν, is known as the Morse index.

To gain an additional perspective on Eq. 10.45, consider the classical equations of motion for $q(t) = q_0(t) + \delta q(t)$. We now imagine, somewhat differently from before, that the quantity $\delta q(t)$ is fixed to be 0 at $t = t'$ but not at $t = t''$; its value at time t is now not to be chosen freely, but to be determined by propagating the classical equations of motion with initial conditions $q' = q(t') = q_0(t')$ and $p' = p(t') = m\dot{q}_0(t') + m\delta\dot{q}(t')$. It is possible to solve for $\delta q(t)$ directly. Substituting the expression $q(t) = q_0(t) + \delta q(t)$ into Newton's equations and cancelling the lowest-order terms yields

$$m\ddot{\delta q} + \frac{\partial^2 V}{\partial q^2} = -\Lambda\, \delta q = 0 \tag{10.49}$$

Equation 10.49 is known as the Jacobi equation and δq is known as the Jacobi field. This equation must be solved with initial conditions

$$\delta q(t') = 0 \tag{10.50}$$

$$m\delta\dot{q}(t') = \delta p'. \tag{10.51}$$

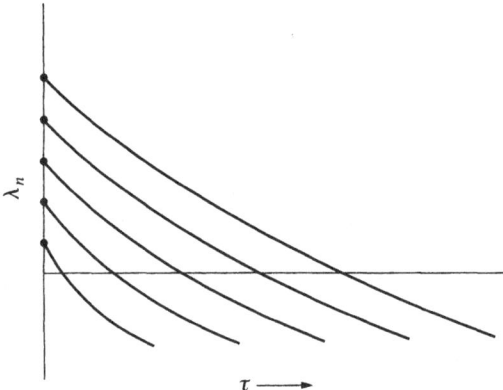

Figure 10.1 Schematic figure showing the eigenvalues $\{\lambda_n\}$ of the second variation operator, as a function of time. All the eigenvalues are positive at short times. As time progresses the eigenvalues may become negative; once they become negative they can never again become positive. The number of negative eigenvalues at time $\tau = t'' - t'$ is known as the Morse index ν.

The variation in initial momenta leads to a spread of trajectories. The central trajectory, $q_0(t)$, by definition satisfies the condition $\delta q(t'') = 0$; the neighboring trajectories may be described by the equation

$$\delta q'' = N \delta p', \tag{10.52}$$

where

$$N = \left(\frac{\partial q''}{\partial p'}\right)_{q'} = \left(-\frac{\partial^2 S}{\partial q'' \partial q'}\right)^{-1}. \tag{10.53}$$

▶ **Exercise 10.8**

a.   Write Eq. 10.52 as

$$\delta q(t) = N(t) \delta p', \tag{10.52'}$$

where $N(t) = \left(\frac{\partial q(t)}{\partial p'}\right)_{q'}$. Substitute Eq. 10.52′ into Eq. 10.49 to show that $N(t)$ satisfies the Jacobi equation.

b.   Differentiate Lagrange's equation (Eq. 10.2) with respect to p, employing the chain rule with respect to q and $\dot{q}$. Show that $N(t)$ satisfies the Jacobi equation.

Levit and Smilansky proved a profound relationship between the product of the eigenvalues of Λ and the partial derivative, N, namely that

$$\prod_{n=1}^{\infty} \frac{\lambda_n}{\lambda_n^0} = \frac{N}{N^0} = \frac{Nm}{\tau}, \tag{10.54}$$

where λ_n^0, N^0 refer to the free-particle dynamics and in obtaining the last expression we have used Eq. 10.13. Both the LHS and the two RHS expressions of Eq. 10.54 depend on the

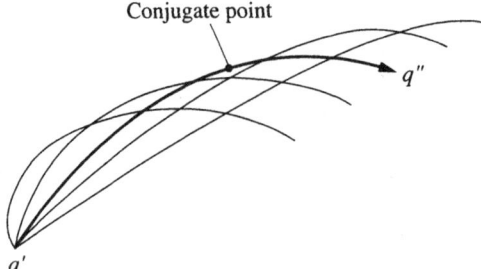
Conjugate point

Figure 10.2 Fan of trajectories originating in the point q' and intersecting one another to form a caustic. The main (heavy) trajectory touches the caustic in the point conjugate to q'.

time interval $\tau = t'' - t'$. For short time intervals, the swarm of trajectories behaves as a set of free particles. However, for longer time intervals there can be special times at which the swarm reconverges; the trajectories coalesce to a point, called a conjugate point. At these instants N vanishes, because changes in the initial momentum p' lead to the same final coordinate, q''. What happens to the Morse index at these values of τ? Since the two RHS expressions of Eq. 10.54 go to zero, the LHS must as well. But since, as described above, the eigenvalues of Λ all start out positive, and once they become negative they never come back, it follows that the Morse index increases precisely at these times. Specifically, the number of negative eigenvalues of the operator Λ is precisely equal to the number of times a conjugate point was reached, known as the Morse index. This is the one-dimensional statement of Morse's theorem. The Morse index ν will enter below in the phase of the semiclassical propagator (Eq. 10.80), leading to a discontinuous change in the latter every time a conjugate point is reached.

In multidimensions, the variation in initial momenta leads to a *fan* of trajectories. N becomes a matrix with elements

$$N_{ij} = \left(\frac{\partial q_i''}{\partial p_j'} \right)_{p_1', p_2', \ldots, q_1', q_2', \ldots} , \tag{10.55}$$

where p_j' is excluded from the variables held constant on the RHS. At the conjugate point the fan of trajectories intersect one another to form a *caustic* (see Figure 10.2). At this instant in time the matrix $\mathbf{N}$ becomes singular, that is, one or more of its rows or columns becomes linearly dependent. We may order the times for which this decrease in rank happens, $t_1 < t_2 < t_3 = t_4 < \ldots$, where decrease in rank by n is represented by n distinct entries with an equal sign. The multidimensional version of Morse's theorem states that the number of negative eigenvalues of Λ is equal to the number of times t_n in the list.

It is useful to define a second matrix, $\mathbf{M}$, with elements

$$M_{ij} = -\frac{\partial^2 S}{\partial q_i' \partial q_j''}. \tag{10.56}$$

It can be shown (see Exercise 10.9) that

$$\mathbf{M} = \mathbf{N}^{-1}. \tag{10.57}$$

In multidimensions, Eq. 10.54 can now be written

$$\prod_{n=1}^{\infty} \frac{\lambda_n}{\lambda_n^0} = \frac{\det(\mathbf{N})}{\det(\mathbf{N}^0)} = \frac{\det(\mathbf{M}^0)}{\det(\mathbf{M})}, \tag{10.58}$$

where det $\mathbf{M}^0$ refers to the free-particle dynamics and we have used the property that $(\det \mathbf{M}) = \det(\mathbf{N}^{-1}) = (\det \mathbf{N})^{-1}$. The quantity $\det \mathbf{M} = (\det \mathbf{N})^{-1}$ determines the density of trajectories in the fan. The square root of this quantity will emerge below as the amplitude of the semiclassical propagator. Note that the matrix $\mathbf{M}$ is manifestly symmetric and therefore the matrix $\mathbf{N} = \mathbf{M}^{-1}$ must be symmetric as well.

▶ **Exercise 10.9**   Show that $\mathbf{M} = \mathbf{N}^{-1}$.

Solution   For the one-dimensional case the proof is trivial, since

$$N^{-1} = \left(\frac{\partial p'}{\partial q''}\right)_{q'} = -\frac{\partial^2 S}{\partial q' \partial q''} = M. \tag{10.59}$$

For the general multidimensional case the situation is more subtle, since the independent variables of N are p', q' and the independent variables of M are q', q'', and hence the variables that are held fixed when derivatives are taken are different. Consider, for example, the two-dimensional case. The matrix $\mathbf{M}$ is given by

$$\mathbf{M} = \begin{pmatrix} -\dfrac{\partial^2 S}{\partial q'_1 \partial q''_1} & -\dfrac{\partial^2 S}{\partial q'_1 \partial q''_2} \\[2mm] -\dfrac{\partial^2 S}{\partial q'_2 \partial q''_1} & -\dfrac{\partial^2 S}{\partial q'_2 \partial q''_2} \end{pmatrix} = \begin{pmatrix} -\dfrac{\partial p'_1}{\partial q''_1} & -\dfrac{\partial p'_1}{\partial q''_2} \\[2mm] -\dfrac{\partial p'_2}{\partial q''_1} & -\dfrac{\partial p'_2}{\partial q''_2} \end{pmatrix}. \tag{10.60}$$

We need to show that

$$\mathbf{MN} = \begin{pmatrix} -\dfrac{\partial p'_1}{\partial q''_1} & -\dfrac{\partial p'_1}{\partial q''_2} \\[2mm] -\dfrac{\partial p'_2}{\partial q''_1} & -\dfrac{\partial p'_2}{\partial q''_2} \end{pmatrix} \begin{pmatrix} -\dfrac{\partial q''_1}{\partial p'_1} & -\dfrac{\partial q''_1}{\partial p'_2} \\[2mm] -\dfrac{\partial q''_2}{\partial p'_1} & -\dfrac{\partial q''_2}{\partial p'_2} \end{pmatrix} = \begin{pmatrix} 1 & 0 \\ 0 & 1 \end{pmatrix}. \tag{10.61}$$

Using the chain rule,

$$(\mathbf{MN})_{11} = \left(\frac{\partial p'_1}{\partial q''_1}\right)_{q''_2 q'_1 q'_2} \left(\frac{\partial q''_1}{\partial p'_1}\right)_{q'_1 q'_2 p'_2} + \left(\frac{\partial p'_1}{\partial q''_2}\right)_{q''_1 q'_1 q'_2} \left(\frac{\partial q''_2}{\partial p'_1}\right)_{q'_1 q'_2 p'_2}$$

$$= \left(\frac{\partial p'_1}{\partial p'_1}\right)_{q'_1 q'_2 p'_2} = 1, \tag{10.62}$$

where we have viewed p'_1 as a function of the independent variables $\{q'_1 q'_2 p'_1 p'_2\}$, namely, $p'_1(q''_1(q'_1 q'_2 p'_1 p'_2), q''_2(q'_1 q'_2 p'_1 p'_2), q'_1, q'_2)$. By similar analysis,

$$(\mathbf{MN})_{12} = \left(\frac{\partial p'_1}{\partial p'_2}\right)_{q'_1 q'_2 p'_1} = 0; \quad (\mathbf{MN})_{21} = \left(\frac{\partial p'_2}{\partial p'_1}\right)_{q'_1 q'_2 p'_2} = 0; \quad (\mathbf{MN})_{22} = \left(\frac{\partial p'_2}{\partial p'_2}\right)_{q'_1 q'_2 p'_1} = 1.$$

$$\tag{10.63}$$

The general multidimensional case proceeds in a similar way.

10.2 Path Integration

In earlier chapters we have devoted a great deal of attention to the evolution operator, or propagator, $U = e^{-iHt/\hbar}$. In this section we will explore a new perspective on the propagator, due to Feynman, with close connections to classical mechanics (Feynman, 1965). The new perspective expresses the propagator matrix element $\langle x'|e^{-iHt/\hbar}|x\rangle$ as the sum over all *paths* from x to x' in time t, weighted by the phase factor $e^{iS/\hbar}$. In the classical limit, $S/\hbar \to \infty$, this phase factor is rapidly varying and the contribution from nearby paths cancels out, except in the region where $\frac{\delta S}{\delta x} = 0$, just where S is stationary with respect to small changes in the path. But this is just the necessary condition for a classical path, according to Hamilton's principle! This, then, explains how classical mechanics arises from quantum mechanics in the classical limit.

The picture of summing over paths from x to x' is conveniently understood as a two-step process. In the first step, we divide the time t into N pieces of time $\Delta t = t/N$, and insert N complete sets of states $\int_{-\infty}^{\infty} |x_i\rangle\langle x_i|\, dx_i$, $i = 1, \ldots, N$. In the second step, a set of these short time segments are strung together, with the terminal value of x_i from one segment of time serving as the initial value of x_i for the next segment of time. Hamilton's principal function, S, for the combined path is the sum of Hamilton's principal function for each of the segments, and the complex exponential of the total S gives the amplitude for a single path; the full propagator is then the sum of the amplitudes over all such paths. Inserting N complete sets in position space, the matrix element of the quantum mechanical propagator is given by

$$\langle x'|e^{-iHt/\hbar}|x\rangle = \int dx_1 dx_2 \ldots dx_N \langle x'|e^{-iH\Delta t/\hbar}|x_N\rangle \langle x_N|e^{-iH\Delta t/\hbar}|x_{N-1}\rangle$$

$$\ldots \langle x_3|e^{-iH\Delta t/\hbar}|x_2\rangle\langle x_2|e^{-iH\Delta t/\hbar}|x_1\rangle\langle x_1|e^{-iH\Delta t/\hbar}|x\rangle. \quad (10.64)$$

Consider now one of the short time propagators, such as $\langle x_2|e^{-iH\Delta t/\hbar}|x_1\rangle$. In the limit $\Delta t \to 0$ (i.e., the number of partitions $N \to \infty$) we can neglect the commutator of the $\hat{p}$ and $\hat{V}$ operators and write

$$e^{-iH\Delta t/\hbar} \approx e^{-iT\Delta t/\hbar} e^{-V\Delta t/\hbar}, \quad (10.65)$$

where T is the kinetic energy operator and V is the potential energy operator, that is, the commutator of the two factors on the RHS vanishes as $\Delta t \to 0$. Substituting the explicit form for T and V we find

$$\langle x_2|e^{-iH\Delta t/\hbar}|x_1\rangle \approx \langle x_2|e^{-i\hat{p}^2\Delta t/2m\hbar} e^{-iV(\hat{x})\Delta t/\hbar}|x_1\rangle \quad (10.66)$$

$$= \int\int\int \langle x_2|p\rangle\langle p|e^{-i\hat{p}^2\Delta t/2m\hbar}|p'\rangle\langle p'|x\rangle\langle x|e^{-iV(\hat{x})\Delta t/\hbar}|x_1\rangle dp\, dp'\, dx \quad (10.67)$$

$$= \int \frac{e^{\frac{ipx_2}{\hbar}}}{\sqrt{2\pi\hbar}} e^{-ip^2\Delta t/2m\hbar} \frac{e^{\frac{ipx_1}{\hbar}}}{\sqrt{2\pi\hbar}} e^{-iV(x_1)\Delta t/\hbar} dp \quad (10.68)$$

$$= \frac{1}{2\pi\hbar} e^{-iV(x_1)\Delta t/\hbar} \int_{-\infty}^{\infty} e^{-ip^2\Delta t/2m\hbar} e^{ip(x_2-x_1)/\hbar} dp \quad (10.69)$$

$$= \left(\frac{m}{2\pi i\hbar\,\Delta t}\right)^{1/2} e^{-iV(x_1)\Delta t/\hbar} e^{im\frac{(x_2-x_1)^2}{2\Delta t^2}\Delta t}, \quad (10.70)$$

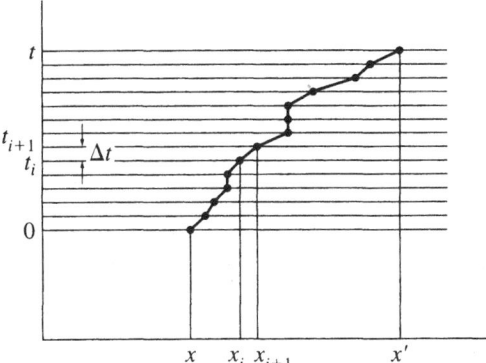

Figure 10.3 Illustration of the "sum over paths" in path integration. A path is defined by slicing time into small intervals, Δt, and specifying the value of the coordinate x_i at each time t_i. All sequences of x_i versus t_i are admissible, with no requirement that the sequence of x_i values satisfy any equation of motion; in particular, each x_i can run over $(-\infty, \infty)$. The sum over paths refers to a sum over all sequences of (x_i, t_i). The path integral is defined as the limit of the sum over paths as Δt approaches 0.

where we have assumed that $V(x)$ is continuous and smooth. In the limit $\Delta t \to 0$, $\frac{(x_2-x_1)^2}{\Delta t^2} \to \dot{x}^2$, leading to

$$\left(\frac{m}{2\pi i \hbar\, \Delta t}\right)^{1/2} e^{i(\frac{m\dot{x}^2}{2} - V(x))\Delta t/\hbar} = \left(\frac{m}{2\pi i \hbar\, \Delta t}\right)^{1/2} e^{iL\Delta t/\hbar} \qquad (10.71)$$

We have applied the operator $V(\hat{x})$ to x, rather than x'. As long as Δt tends to 0, the difference is negligible, since the difference in the potential can be made arbitrarily small. Note the appearance of the Lagrangian, $L(x, \dot{x}) = \frac{m\dot{x}^2}{2} - V(x)$, which is a fundamental quantity in classical mechanics. The appearance of the Lagrangian here will be presented from a different and more general perspective in Section 10.4.

The second step in achieving the picture of independent "paths" is to string together these short time segments into a connected thread from x to x' in time t. We now combine N of the above short time factors, corresponding to a preselected set of N intermediate positions $x_1, \ldots, x_N$ (cf. Figure 10.3). Combining the short time factors yields the amplitude contribution from a single path:

$$\langle x'|e^{-iHt/\hbar}|x\rangle_{\text{path}} = \left(\frac{m}{2\pi i \hbar\, \Delta t}\right)^{1/2} e^{i\sum_i L_i \Delta t/\hbar}. \qquad (10.72)$$

In the limit $\Delta t \to 0$, the sum can be replaced by an integral, and we recognize the exponent in Eq. 10.72 as $S(x', x, t) = \int_0^t L\, dt'$, where $S(x', x, t)$ is known as Hamilton's principal function in classical mechanics, or sometimes called just the "action." To complete the expression for the propagator, we note that the above expression is the contribution for one particular path—a set of intermediate values of x. The full expression for the propagator contains N integrals, an integral over the value of x at each of the intermediate times. Thus, the complete expression for the propagator is

$$\langle x' | e^{-iHt/\hbar} | x \rangle = \left(\frac{m}{2\pi i \hbar t} \right)^{1/2} \sum_{\text{all paths}} e^{iS(x,x',t)/\hbar}$$

$$\equiv \left(\frac{m}{2\pi i \hbar t} \right)^{1/2} \int \mathcal{D}[x(t)] e^{iS(x,x',t)/\hbar}, \tag{10.73}$$

where $\mathcal{D}[x(t)]$ represents an integral over all paths. Note that in the path integral formulation all paths are given precisely equal weight, what Feynman calls "the democracy of histories." As discussed above, in the classical limit $S/\hbar \to \infty$ the phase factor is rapidly varying and the contribution from nearby paths cancels out, except in the region where $\frac{\delta S}{\delta x} = 0$, that is, where S is stationary with respect to small changes in the path. But this condition is just Hamilton's principle in classical mechanics, which provides a necessary condition on classical trajectories. This then accounts for the emergence of classical paths from amidst the sea of all possible quantum paths.

Although we have worked out the normalization factor for the case of the free particle, a careful analysis shows that the normalization factor is unaffected by the inclusion of the potential energy. Hence, Eq. 10.73 is the correct expression for arbitrary potentials.

▶ **Exercise 10.10**   It is possible to write an alternative expression for the path integral in the basis of eigenfunctions of the operator Λ defined at Eq. 10.45. The sum over all paths can be written equivalently as the sum over all possible values of the a_n, the coefficients of the basis functions in the variation, as $N \to \infty$, that is,

$$\langle x' | e^{-iHt/\hbar} | x \rangle = \lim_{N \to \infty} \left(\frac{m}{2\pi i \hbar t} \right)^{1/2} \int e^{iS(x,x',t)/\hbar} \prod_{n=1}^{N} da_n \left(\frac{\lambda_n^{(0)}}{2\pi i \hbar} \right)^{1/2}. \tag{10.74}$$

Work out the path integral for the free particle using this representation, and show that we have indeed given the correct normalization factor.

10.3   The van Vleck Propagator

10.3.1   The Semiclassical Limit of the Path Integral

Since, in the classical limit, the main contribution to $\langle x' | e^{-iHt/\hbar} | x \rangle$ comes from classical paths, it should be possible to use the path integral expression to develop a semiclassical approximation to the propagator. Thus,

$$\langle x' | e^{-iHt/\hbar} | x \rangle = \left(\frac{m}{2\pi i \hbar t} \right)^{1/2} \sum_{\text{all paths}} e^{iS(x,x',t)/\hbar}$$

$$\approx \left(\frac{m}{2\pi i \hbar t} \right)^{1/2} \sum_{\text{classical paths}} \int \mathcal{D} e^{i(S + \delta S + \frac{\delta^2 S}{2})/\hbar}. \tag{10.75}$$

Note that we have approximated the full path integral by a sum of path integrals with quadratic functional variations around the classical paths. Although from the bookkeeping point of view there is a multiple counting of each path, with the multiplicity equal to the number of classical paths, the exponentially rapid decay of the contribution away from a classical path makes this multiple counting irrelevant in most cases. We now use the results obtained in Section 10.1. For classical paths, $\delta S = 0$. To treat the second variation it is again simplest to work in the basis that diagonalizes the stability operator Λ, described in the beginning of this chapter:

$$\delta^2 S = \sum_n \lambda_n a_n^2. \tag{10.76}$$

Using the expansion Eq. 10.75 along with Eq. 10.76, we see that the integral Eq. 10.74 is now a multidimensional Gaussian integral and can be performed exactly. The result is

$$\langle x'|e^{-iHt/\hbar}|x\rangle^{\mathrm{SC}} = \sum_{\text{all } classical \text{ paths}} \left(\frac{m}{2\pi i\hbar t}\right)^{1/2} \lim_{N\to\infty} \prod_{n=1}^{N} \left(\frac{\lambda_n^{(0)}}{\lambda_n}\right)^{1/2}\Bigg|_{x_{\mathrm{cl}}} e^{iS/\hbar} \tag{10.77}$$

$$= \sum_{\text{all } classical \text{ paths}} \left(\frac{m}{2\pi i\hbar t}\right)^{1/2} \left(\frac{N^{(0)}(t)}{N(t)}\right)^{1/2}\Bigg|_{x_{\mathrm{cl}}} e^{iS/\hbar} \tag{10.78}$$

$$= \sum_{\text{all } classical \text{ paths}} \left(\frac{-\partial^2 S/\partial x'\partial x}{2\pi i\hbar}\right)^{1/2} e^{iS/\hbar}, \tag{10.79}$$

where the superscript SC indicates that the matrix element is evaluated semiclassically, and we have used the result of Levit and Smilansky, Eq. 10.54, in its one-dimensional form. In n dimensions, Eqs. 10.78–10.79 take the form

$$\langle \mathbf{x}'|e^{-iHt/\hbar}|\mathbf{x}\rangle^{\mathrm{SC}} = \sum_{\text{all } classical \text{ paths}} \left(\frac{m}{2\pi i\hbar t}\right)^{n/2} \left(\frac{\det \mathbf{N}^{(0)}(t)}{\det \mathbf{N}(t)}\right)^{1/2}\Bigg|_{\mathbf{x}_{\mathrm{cl}}} e^{iS/\hbar} \tag{10.78$'$}$$

$$= \sum_{\text{all } classical \text{ paths}} \left(\frac{1}{2\pi i\hbar}\right)^{n/2} \left[\det\left(\frac{-\partial^2 S}{\partial \mathbf{x}'\partial \mathbf{x}}\right)\right]^{1/2} e^{iS/\hbar}. \tag{10.79$'$}$$

Equation 10.79$'$, is known as the van Vleck propagator, and the prefactor is known as the van Vleck determinant. Note that the sign of the van Vleck determinant, and therefore the phase of the semiclassical propagator, is determined by the number of negative λ_n's in Eq. 10.77, which is just the Morse index ν. This number, in turn, is just the number of conjugate times encountered between 0 and t. Following Gutzwiller, the phase of the prefactor can be factored out from the amplitude, giving the following expression for the semiclassical propagator:

$$\langle x'|e^{-iHt/\hbar}|x\rangle^{\mathrm{SC}} = \sum_{\text{all } classical \text{ paths}} \left|\frac{-\partial^2 S/\partial x'\partial x}{2\pi i\hbar}\right|^{1/2} e^{iS/\hbar - i\frac{\nu\pi}{2}}, \tag{10.80}$$

where the magnitude and phase of the propagator are written explicitly and we have returned to one dimension. As discussed above, there may be times when the prefactor of the exponential diverges. These singular points are precisely the times when the Morse index changes.

Equation 10.80 contains a sum over all classical paths. The contribution from multiple classical paths in general leads to interference, since each path contributes with a different phase. This lies at the heart of the statement that classical mechanics plus quantum superposition provides a qualitative description of quantum interference phenomena.

▶ **Exercise 10.11**   Show that Eq. 10.80 reduces to the expression in Exercise 10.10 if $V(x)$ is a constant.

Note the following relations governing the prefactor in the semiclassical propagator:

$$-\frac{\partial^2 S}{\partial x' \partial x} = \left(\frac{\partial p}{\partial x'}\right)\bigg|_x = -\left(\frac{\partial p'}{\partial x}\right)\bigg|_{x'}. \tag{10.81}$$

The first equality, in particular, gives an interesting perspective on the significance of this prefactor. This expression can be rewritten as

$$\left(\frac{\partial p}{\partial x'}\right)\bigg|_x = \left(\frac{\partial x'}{\partial p}\right)\bigg|_x^{-1}. \tag{10.82}$$

Thus, it emerges that the prefactor is the inverse of the sensitivity of the final position to the initial momentum. The more sensitive this quantity is the larger is $(\partial x'/\partial p)_x$ and hence the smaller is the weight from this trajectory; thus, if the final position is very sensitive to initial momentum, trajectories that differ slightly in initial momentum do not focus closely. On the other hand, there are times when an entire range of different initial momenta lead to the same final position: these are times of extreme focusing, and are the conjugate points described earlier. At these times $(\frac{\partial x'}{\partial p})|_x$ goes to zero and the van Vleck prefactor blows up. The behavior of $\left(\frac{\partial x'}{\partial p}\right)\big|_x$ as a function of (x', t) can be understood from Figure 10.4, which provides a phase space perspective.

The quantity $(\frac{\partial p}{\partial x'})\big|_x$ is an element of the linear stability matrix, **T** (also called the "monodromy" matrix), defined as

$$\mathbf{T} = \begin{pmatrix} \frac{\partial p'}{\partial p}\big|_x & \frac{\partial p'}{\partial x}\big|_p \\ \frac{\partial x'}{\partial p}\big|_x & \frac{\partial x'}{\partial x}\big|_p \end{pmatrix}. \tag{10.83}$$

▶ **Exercise 10.12**   Show that the determinant of the **T** matrix is equal to 1.

▶ **Exercise 10.13**

a. Show that for the harmonic oscillator

$$\mathbf{T} = \begin{pmatrix} \cos(\omega t) & -m\omega \sin(\omega t) \\ \sin(\omega t)/m\omega & \cos(\omega t) \end{pmatrix}. \tag{10.84}$$

b. Show that for the harmonic oscillator Eq. 10.84 can be rewritten as

$$\mathbf{T} = \exp\left[\begin{pmatrix} 0 & -m\omega^2 \\ 1/m & 0 \end{pmatrix} t\right]. \tag{10.85}$$

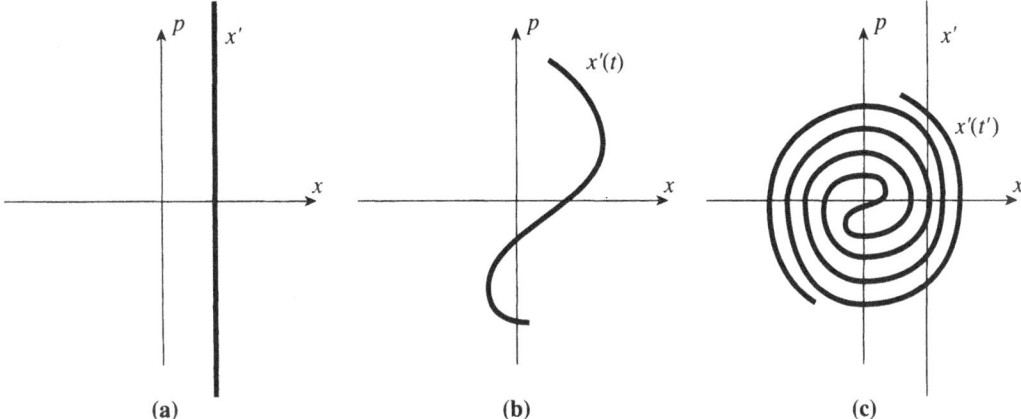

Figure 10.4 Phase space interpretation of the quantity $(\frac{\partial x'}{\partial p})|_x$, which enters into the van Vleck prefactor. (a) At $t = 0$, the state $\langle x'|x(0)\rangle = \delta(x - x')$ is a vertical line with equal probability for all p. (b) As time progresses, $\langle x'|x(t)\rangle = \langle x'|e^{-iHt/\hbar}|x\rangle$ evolves; the local slope is $\frac{\partial x'}{\partial p}|_x^{-1}$. When a portion of $x(t)$ develops a vertical tangent $\frac{\partial p}{\partial x'}|_x = \infty$ and hence $\frac{\partial x'}{\partial p}|_x = 0$, a whole range of nearby initial p lead to the same final x, giving a singular burst in the amplitude of the van Vleck propagator. (c) At long times, especially in chaotic systems, $\langle x'|x(t)\rangle$ becomes very filamented, and quite generally has singularities at all times. Adapted from Heller (1991).

▶ Exercise 10.14

    a. Show that for the parabolic barrier

$$\mathbf{T} = \begin{pmatrix} \cosh(\omega t) & m\omega \sinh(\omega t) \\ \sinh(\omega t)/m\omega & \cosh(\omega t) \end{pmatrix}. \tag{10.86}$$

    b. Show that Eq. 10.86 can be rewritten as

$$\mathbf{T} = \exp\left[\begin{pmatrix} 0 & m\omega^2 \\ 1/m & 0 \end{pmatrix} t\right]. \tag{10.87}$$

    c. Find the eigenvalues of $\mathbf{T}$, $\{\lambda_i\}$, and calculate the Liapunov exponents

$$\sigma_i = \lim_{t \to \infty} \frac{1}{t} \ln(\lambda_i(t)). \tag{10.88}$$

    The Liapunov exponents are a measure of the rate of separation of nearby trajectories. For typical trajectories at long times the rate of separation is determined by the largest Liapunov exponent.

▶ Exercise 10.15 Show that

$$\frac{d}{dt}\mathbf{T} = \mathbf{KT}, \tag{10.89}$$

where

$$\mathbf{K} = \begin{pmatrix} 0 & -V'' \\ m^{-1} & 0 \end{pmatrix} \tag{10.90}$$

with the initial conditions

$$\mathbf{T}(0) = \begin{pmatrix} 1 & 0 \\ 0 & 1 \end{pmatrix}. \tag{10.91}$$

Solution

$$\frac{d}{dt} \begin{pmatrix} \delta p_t \\ \delta x_t \end{pmatrix} = \begin{pmatrix} 0 & -V'' \\ m^{-1} & 0 \end{pmatrix} \begin{pmatrix} \delta p_t \\ \delta x_t \end{pmatrix} \equiv \mathbf{K} \begin{pmatrix} \delta p_t \\ \delta x_t \end{pmatrix}. \tag{10.92}$$

Using

$$\begin{pmatrix} \delta p_t \\ \delta x_t \end{pmatrix} = \mathbf{T} \begin{pmatrix} \delta p_0 \\ \delta x_0 \end{pmatrix}, \tag{10.93}$$

we have

$$\frac{d}{dt} \mathbf{T} \begin{pmatrix} \delta p_0 \\ \delta x_0 \end{pmatrix} = \mathbf{KT} \begin{pmatrix} \delta p_0 \\ \delta x_0 \end{pmatrix}. \tag{10.94}$$

▶ **Exercise 10.16**   Since the propagator is a solution of the TDSE, it is perhaps not surprising that the van Vleck propagator is a solution of the semiclassical TDSE, the coupled equations for the amplitude and phase in Chapter 4 (Eqs. 4.21 and 4.25–4.26), with the neglect of the $\hbar^2$ term. Show that this is indeed the case.

Solution   The phase of the wavefunction is the classical action, which satisfies the Hamilton–Jacobi equation. The amplitude of the wavefunction satisfies the continuity equation, as we now show. The continuity equation can be written

$$\frac{\partial A^2}{\partial t} = -\frac{\partial}{\partial x} \left(A^2 \frac{p}{m} \right), \tag{10.95}$$

which means in our current notation

$$\frac{\partial A^2}{\partial t_2} = -\frac{\partial}{\partial x_2} \left(A^2 \frac{p_2}{m} \right)_{x_1}, \tag{10.96}$$

where $A = A(x, t)$, $p_2 = \frac{\partial S(x_2, x_1, t)}{\partial x} = p_2(x_2, x_1, t)$ and $t = t_2 - t_1$. The claim is that $A^2 = (\frac{\partial p_1}{\partial x_2})_{x_1} = -\frac{\partial^2 S(x_2, x_1, t)}{\partial x_1 \partial x_2}$ is a solution. Substituting this trial form into Eq. 10.96, we find

$$\frac{\partial}{\partial t_2} A^2 = \frac{\partial}{\partial t_2} \left(\frac{\partial p_1}{\partial x_2} \right)_{x_1} \tag{10.97}$$

$$= -\frac{\partial^2}{\partial x_1 \partial x_2} \frac{\partial S}{\partial t_2} = \frac{\partial^2}{\partial x_1 \partial x_2} H(x_2, p_2(x_2, x_1))$$

$$= \frac{\partial}{\partial x_2} \left(\left(\frac{\partial H}{\partial x_1} \right)_{x_2} \right)_{x_1} = \frac{\partial}{\partial x_2} \left(\left(\frac{\partial H}{\partial p_2} \right)_{x_2} \left(\frac{\partial p_2}{\partial x_1} \right)_{x_2} \right)_{x_1}$$

$$= \frac{\partial}{\partial x_2} \left(\left(\frac{p_2}{m} \right) \left(-\frac{\partial p_1}{\partial x_2} \right)_{x_1} \right)_{x_1} = -\frac{\partial}{\partial x_2} \left(A^2 \frac{p_2}{m} \right)_{x_1}. \tag{10.98}$$

10.3.2   The Initial Value Representation of the van Vleck Propagator

An important variation on the van Vleck propagator is the so-called initial value representation (IVR). Consider a correlation function between two square integrable wavefunctions, Φ_i and Φ_f:

$$C_{fi}(t) = \langle \Phi_f | e^{-iHt/\hbar} | \Phi_i \rangle = \int dx_1 \int dx_2 \Phi_f(x_2)^* \langle x_2 | e^{-iHt/\hbar} | x_1 \rangle \Phi_i(x_1). \quad (10.99)$$

Using the VVG expression for the propagator, we obtain

$$C_{fi}(t) = \int dx_1 \int dx_2 \Phi_f^*(x_2) \left[2\pi \hbar \left| \frac{\partial x_2(x_1, p_1)}{\partial p_1} \right| \right]^{-1/2} e^{iS(x_2, x_1)/\hbar - i\nu\pi/2} \Phi_i(x_1). \quad (10.100)$$

We now change integration variables from x_2 to p_1. Since x_1 is constant inside the x_2 integral, we can write

$$dx_2 = dp_1 \left| \frac{\partial x_2(x_1, p_1)}{\partial p_1} \right|, \quad (10.101)$$

provided $\frac{\partial x_2(x_1, p_1)}{\partial p_1} \neq 0$. Substituting Eq. 10.101 into Eq. 10.100, we obtain

$$C_{fi}(t) = \int dx_1 \int dp_1 \Phi_f^*(x_2(x_1, p_1)) \left[\left| \frac{\partial x_2(x_1, p_1)}{\partial p_1} \right| / 2\pi \hbar \right]^{1/2} e^{iS(x_1, p_1(x_1, x_2))/\hbar - i\nu\pi/2} \Phi_i(x_1).$$

$$(10.102)$$

All trajectory-related quantities now appear as functions only of the initial conditions (x_1, p_1). There is the additional attractive feature that the Jacobian factor, $|\partial x_2/\partial p_1|_{x_1}$, now appears in the *numerator* in the integrand, rather than the denominator. Since the semiclassical propagator is most in error near classical turning points (the zeros of this Jacobian), the turning points (or caustics, in multidimensions) now enter as *zeroes* in the integrand, rather than as singularities. Also, because when the Jacobian passes through zero the Maslov index ν changes discontinuously, the fact that the integrand of this equation is zero at such points means that the integrand itself is continuous even at these caustics.

The root search problem that would result if one evaluated the integral in Eq. 10.100 by stationary phase has thus been replaced with an integral over the phase space of initial conditions, Eq. 10.102. Note that Eq. 10.100, and thus Eq. 10.102, formally has a fractional error of order $\hbar$ (because the integrand does) and is thus *in principle* no more accurate than the stationary phase approximation to the integrand. In practice, however, it may be more accurate.

10.3.3   Matrix Elements of the van Vleck Propagator with Gaussian States

We now return to Eq. 10.99:

$$C_{fi}(t) = \langle \Phi_f | e^{-iHt/\hbar} | \Phi_i \rangle = \int dx_1 \int dx_2 \Phi_f(x_2)^* \langle x_2 | e^{-iHt/\hbar} | x_1 \rangle \Phi_i(x_1), \quad (10.103)$$

and take the initial and final states, Φ_i and Φ_f, as Gaussians:

$$\Phi_i(x_1) = N_i e^{-\alpha_i(x_1-x_i)^2+ip_i(x_1-x_i)/\hbar}, \qquad \Phi_f(x_2) = N_f e^{-\alpha_f(x_2-x_f)^2+ip_f(x_2-x_f)/\hbar}. \quad (10.104)$$

Using the VVG expression for the propagator, we obtain

$$C_{fi}(t) = \int dx_1 \int dx_2 N_f^* e^{-\alpha_f^*(x_2-x_f)^2-ip_f(x_2-x_f)/\hbar}$$

$$\sum_{\text{classical paths}} \left[2\pi\hbar \left|\frac{\partial x_2(x_1, p_1)}{\partial p_1}\right|\right]^{-1/2} e^{iS(x_2,x_1)/\hbar-i\nu\pi/2} N_i e^{-\alpha_i(x_1-x_i)^2+ip_i(x_1-x_i)/\hbar}.$$

$$(10.105)$$

Expanding the action to second order in the coordinates, we obtain

$$S(x_1, x_2; \tau) = S(x_i, x_f, \tau) + p_f(\tau)(x_2-x_f) - p_i(\tau)(x_1-x_i)$$

$$+ \frac{1}{2}\left[\frac{\partial^2 S}{\partial x_f^2}(x_2-x_f)^2 + \frac{\partial^2 S}{\partial x_i^2}(x_1-x_i)^2 + \frac{\partial^2 S}{\partial x_i \partial x_f}(x_1-x_i)(x_2-x_f)\right] + \cdots,$$

$$(10.106)$$

where we have used the classical relations $\partial S/\partial x_i = -p_i(\tau)$ and $\partial S/\partial x_f = p_f(\tau)$. Here, $p_i(\tau) = p_i(x_i, x_f; \tau)$ and $p_f(\tau) = p_f(x_i, x_f; \tau)$ are the initial and final momenta, respectively, of a classical trajectory starting from x_i and ending in x_f at time τ. Inserting the expansion Eq. 10.106 into Eq. 10.105 and integrating over x_1 and x_2, we obtain

$$C_{fi}(t) = \sum_{\text{classical paths}} N(\tau) e^{iS(\tau)/\hbar} e^{-\frac{1}{4}[\Delta p(\tau)]\cdot A^{-1}\cdot[\Delta p(\tau)]}, \quad (10.107)$$

where

$$A = \alpha - \frac{iS_2}{2\hbar}, \qquad \Delta p(\tau) = [p_f(\tau) - p_f, p_i(\tau) - p_i], \quad (10.108)$$

$$S_2 = \begin{bmatrix} \partial^2 S/\partial x_f^2 & \partial^2 S/\partial x_f \partial x_i \\ \partial^2 S/\partial x_f \partial x_i & \partial^2 S/\partial x_i^2 \end{bmatrix} \qquad \alpha = \begin{bmatrix} \alpha_f^* & 0 \\ 0 & \alpha_i \end{bmatrix}. \quad (10.109)$$

Equation 10.107 will be useful in Section 16.5 (see Exercise 16.13).

10.4   The Propagator as a Unitary Transformation

Miller has given a beautiful, general framework for semiclassical theory in which semiclassical matrix elements are calculated by performing the integrals using stationary phase integration (Miller, 1974). An important feature of this classical–quantum correspondence is the internal consistency of a *semiclassical algebra*. That is to say, if the matrix element $\int dQ \langle q'|U_1|Q\rangle\langle Q|U_2|q\rangle$ is evaluated by using the stationary phase expression for each matrix element, and performing the integration over the intermediate variable Q by stationary phase, one should obtain the same result as if the matrix element $\langle q'|U_1U_2|q\rangle$ were evaluated by stationary phase in one step. We will illustrate this "composition" property below for the important example of successive operations of the semiclassical propagator (Exercise 10.21).

Miller's approach focuses on matrix elements of unitary operators. Most interesting quantities in quantum mechanics are transition amplitudes from one state to another, such as Franck–Condon factors, S-matrix elements and so on, or transformation elements from one basis to another, as in the case of coordinate representations of eigenstates, and therefore can be expressed as matrix elements of unitary operators. For definiteness, consider the transition between one value of coordinate and another, which can be written as $\langle q|Q \rangle$. Writing the matrix element in polar form, $Ae^{i\phi}$, Miller shows that in the classical limit, $\phi = F_1(q, Q)/\hbar$, while $A = \left[\frac{-\partial^2 F_1(q,Q)/\partial q \partial Q}{2\pi i\hbar} \right]^{1/2}$, where $F_1(q, Q)$ is the classical generating function from (q, p) to (Q, P), employing one old and one new coordinate (see below). Analogous relations hold for each of the other three combinations of one old variable (q or p) and one new variable (Q or P), with $F_1(q, Q)$ being replaced by $F_2(q, P)$, $F_3(p, Q)$ or $F_4(p, P)$ and appropriate sign changes made (see Table 10.1 for the definition of the generating functions).

Note the similarity in structure of the general semiclassical matrix element and the van Vleck–Gutzwiller expression for the propagator. That is because the expression $\langle x'|e^{-iHt/\hbar}|x \rangle$ is the matrix element of a unitary operator. This connection will be made explicit below. Thus, the semiclassical limit of $\langle x'|e^{-iHt/\hbar}|x \rangle$ as expressed by the VVG propagator can be put into a general context of the semiclassical limit of unitary operators in a very elegant way. We begin with a review of classical canonical transformations. We follow closely the treatment of Miller (1974), which is a wonderful introduction to canonical transformations as well as to semiclassical algebra. The reader is referred to the original treatment for additional details.

10.4.1 Classical Canonical Transformations

Within the Hamiltonian formulation of classical mechanics, a change of variables $(p, q) \rightarrow (P, Q)$ is performed via a canonical transformation—a transformation ensuring that Hamilton's equations hold in the new variables (P, Q) just as they do in the old variables (p, q). Canonical transformations preserve volume in phase space. By the same token, a change of variables is accomplished in quantum mechanics via a unitary transformation, which is norm preserving. As we shall now discuss, there is a deep correspondence between these two.

We begin with a review of canonical transformations in classical mechanics. A canonical transformation is a transformation from an old set of variables, p and q, satisfying Hamilton's equations,

$$\frac{dp}{dt} = -\frac{\partial H(p, q)}{\partial q}$$

$$\frac{dq}{dt} = \frac{\partial H(p, q)}{\partial p},$$

to a new set of variables, P and Q, which also satisfy Hamilton's equations:

$$\frac{dP}{dt} = -\frac{\partial H(P(p, q), Q(p, q))}{\partial Q}$$

$$\frac{dQ}{dt} = \frac{\partial H(P(p, q), Q(p, q))}{\partial P}.$$

(These conditions are valid for a canonical transformation that is independent of time, which is the case of interest to us.) If we take (p, q) to be the independent variables, then the dependent variables are expressed as $P(p, q)$, $Q(p, q)$, and $H(P(p, q), Q(p, q)) = H(p, q)$. The condition that P and Q be canonical variables is equivalent to the condition that the determinant of the transformation from $(p, q) \rightarrow (P, Q) = 1$:

$$\left| \frac{\partial(P, Q)}{\partial(p, q)} \right| \equiv \det \begin{bmatrix} (\frac{\partial P}{\partial p})_q & (\frac{\partial P}{\partial q})_p \\ (\frac{\partial Q}{\partial p})_q & (\frac{\partial Q}{\partial q})_p \end{bmatrix} = 1.$$

These equations provide a way to test if a transformation is canonical, but do not provide a useful way to construct a canonical transformation. For that, it is necessary to choose one old variable and one new variable as the independent variables; there are four combinations for this choice. For example, consider (q, Q) as the independent variables, with $p(q, Q)$ and $P(q, Q)$ taken as functions of them. Then a necessary and sufficient condition that the transformation $(p, q) \rightarrow (P, Q)$ be canonical is that

$$-\frac{\partial p(q, Q)}{\partial Q} = \frac{\partial P(q, Q)}{\partial q}.$$

To satisfy this equation it is convenient to introduce an auxiliary function of the independent variables, the "generating function," $F(q, Q)$. If an $F(q, Q)$ is found such that

$$\frac{\partial F(q, Q)}{\partial q} = p(q, Q) \quad \text{and} \quad -\frac{\partial F(q, Q)}{\partial Q} = P(q, Q) \tag{10.110}$$

the transformation will be canonical. It is generally quite easy to find a suitable function F that satisfies these two conditions, which is what makes this technique useful. (Note that Eq. 10.110 takes a different form for each of the four combinations of one old and one new variable.)

The first equation immediately gives $p(q, Q)$, which must be inverted to give $Q(q, p)$. The second equation then gives $P(q, Q(q, p))$. This inversion in general has more than one solution, and this is intimately tied in with the existence of multiple solutions to the classical mechanics whose superposition leads to interference in the quantum mechanics.

In practice, one generally knows from the beginning *one* of the new variables as a function of the old variables, in this case $Q(q, p)$. The first part of the procedure is therefore circular:

1. Invert $Q(q, p)$ to get $p(q, Q)$;
2. Integrate $\int p(q, Q) \, dq$ to get $F(q, Q)$ to within an unknown function $C(Q)$;
3. Take

$$\frac{\partial F(q, Q)}{\partial q} = p(q, Q) \tag{10.111}$$

and invert to get $Q(q, p)$. This completes the circular part of the procedure.
4. Finally, take

$$-\frac{\partial F(q, Q)}{\partial Q} = P(q, Q) \tag{10.112}$$

and substitute $Q(q, p)$ to get $P(q, Q(q, p))$.

Table 10.1 The defining relationships for the four types of generating functions. Note that on the RHS of the first column the dependent variables are assumed to be expressed in terms of the independent ones.

Generator definition	Derivative relation (I)	Derivative relation (II)
$F_1(q, Q)$	$\frac{\partial F_1(q,Q)}{\partial q} = p(q, Q)$	$-\frac{\partial F_1(q,Q)}{\partial Q} = P(q, Q)$
$F_2(q, P) = F_1(q, Q) + PQ$	$\frac{\partial F_2(q,P)}{\partial q} = p(q, P)$	$\frac{\partial F_2(q,P)}{\partial P} = Q(q, P)$
$F_3(p, Q) = F_1(q, Q) - pq$	$-\frac{\partial F_3(p,Q)}{\partial p} = q(p, Q)$	$-\frac{\partial F_3(p,Q)}{\partial Q} = P(p, Q)$
$F_4(p, P) = F_1(q, Q) + PQ - pq$	$-\frac{\partial F_4(p,P)}{\partial p} = q(p, P)$	$\frac{\partial F_4(p,P)}{\partial P} = Q(p, P)$

Clearly, one may define four types of F functions, depending on which of the two old variables (q, p) and which of the two new variables (Q, P) are taken to be the independent variables. Thus, we have $F_1(q, Q)$, $F_2(q, P)$, $F_3(p, Q)$, $F_4(p, P)$. The defining relationships for the four types of generating functions are collected in Table 10.1. Note that the new coordinates and momenta may not bear any simple relationship to the conventional concepts of coordinate and momenta; in fact, there is a canonical transformation that interchanges the role of coordinates and momenta. It is therefore worth emphasizing that the defining factor in determining whether a new variable is a coordinate or a momentum is whether it satisfies the equation $\dot{U} = \partial H / \partial V$ or $\dot{U} = -\partial H / \partial V$.

▶ **Exercise 10.17** Find an F_2 generator such that the new variables (Q, P) are identical to the old variables (q, p). This is an example of a point transformation, a transformation such that the new coordinates are functions only of the old coordinates and not the old momenta.

▶ **Exercise 10.18** Find an F_1 generator such that the new variables (Q, P) are identical to the old variables (q, p), but with the roles of position and momentum reversed: $(q, p) \rightarrow (-P, Q)$.

10.4.2 Canonical Transformations and Classical–Quantum Correspondence

As discussed above, quite generally a matrix element of a unitary operator in quantum mechanics can be expressed in the classical limit in terms of the appropriate classical generating function for the corresponding classical canonical transformation. For example, consider the matrix element of the form $\langle q|U|q'\rangle = \langle q|Q\rangle$; this is a complex number and can be written in polar form as $Ae^{i\phi}$. In the classical limit, $\phi = F_1(q, Q)/\hbar$, where $F_1(q, Q)$ is the classical generating function for the transformation $(q, p) \rightarrow (Q, P)$, which employs one old and one new coordinate, and $A = \left[\frac{-\partial^2 F_1(q,Q)/\partial q \partial Q}{2\pi i \hbar}\right]^{1/2}$. With minor variations, the same types of formulas apply for every combination of old and new variable, involving the corresponding classical generating function. We now discuss four interesting examples, representing some of the variety of combinations of old and new variables that are possible.

A. The Fundamental Matrix Element: $\langle q | p \rangle$

The fundamental starting point for the semiclassical algebra is the object $\langle q|p \rangle$, which is taken *by assumption* to be

$$\langle q|p \rangle = \frac{1}{\sqrt{2\pi i \hbar}} e^{ipq/\hbar}.$$

It follows that

$$\langle p|q \rangle = \frac{1}{\sqrt{-2\pi i \hbar}} e^{-ipq/\hbar}.$$

As discussed by Dirac, these transformations are equivalent to assuming the commutation relation $[\hat{q}, \hat{p}] = i\hbar$, and are thus equivalent to the uncertainty principle. These relations are equivalent to taking the generator of the canonical transformation to be $F_2 = pq = Pq$ or $F_3 = -pq = -pQ$, that is, the new P equals the old p or the new Q equals the old q. This is the identity canonical transformation. Note that $\frac{\partial^2 F_2}{\partial p \partial q} = 1$ and $\frac{\partial^2 F_3}{\partial p \partial q} = -1$, which is consistent with the form of the prefactors.

B. Semiclassical Eigenstates: $\langle q|E \rangle \Longleftrightarrow \langle q|P \rangle$

The coordinate dependence of the energy eigenstate, $\langle q|E \rangle$, can be regarded as the matrix element of a unitary transformation between the eigenstate basis and the coordinate basis. The associated canonical transformation from $(p, q) \to (P, Q)$ is such that the new momentum $P = E = p^2/2m + V(q)$. The associated generating function, $F_2(q, P)$, satisfies the following equations:

$$\frac{\partial F_2(q, P)}{\partial q} = p(q, P), \qquad \frac{\partial F_2(q, P)}{\partial P} = Q(q, P).$$

From the first equation we see that $F_2(q, P) = \int p(q, P)\, dq + C(P)$. From Eq. 10.111 we have $p(q, P) = \pm(2m(P - V(q)))^{1/2}$, leading to $F_2(q, P) = \pm \int (2m(P - V(q)))^{1/2}\, dq$, where we may neglect $C(P)$ without loss of generality. This is the solution to the Hamilton–Jacobi equation in classical mechanics. Note that the multivalued inversion (the $\pm$ sign of the square root in this case) will lead to more than one classical contribution.

We can now find Q from the relation

$$\frac{\partial F_2(q, P)}{\partial P} = Q(q, P) = \pm \int \left(\frac{2}{m}(P - V(q)) \right)^{-1/2} dq. \qquad (10.113)$$

To within an additive constant, Q is the time, t.

▶ **Exercise 10.19**   Show that the RHS of Eq. 10.113 is indeed the time, and that retaining $C(P)$ just corresponds to a shift in time.

The required second derivative of F_2 is given by

$$\frac{\partial^2 F_2}{\partial q \partial P} = \pm \left(\frac{2}{m}(P - V(q)) \right)^{-1/2} = \pm \frac{1}{v}.$$

The final expression for the energy normalized wavefunction is

$$\langle q|E\rangle = (2\pi\hbar)^{-1/2} \left(\frac{2}{m} \left[E - V(q) \right] \right)^{-1/4}$$

$$\times \left(e^{-i\pi/4 + i\int dq\ (2m[E-V(q)])^{1/2}} + e^{i\pi/4 - i\int dq\ (2m[E-V(q)])^{1/2}} \right) \qquad (10.114)$$

$$= (2\pi\hbar)^{-1/2} v^{-1/2} (e^{-i\pi/4 + i\int p\ dq} + e^{i\pi/4 - i\int p\ dq}). \qquad (10.115)$$

This is the standard semiclassical (so-called WKB) expression for the wavefunction. The two terms emerge from the two roots in the inversion of the equations involving the generating function. The phases $e^{\pm i\pi/4}$ are well known in WKB theory, and emerge here automatically from careful attention to the phase of the prefactor.

C. The Semiclassical Propagator: $\langle x'|e^{-iHt/\hbar}|x\rangle \Longleftrightarrow \langle q_2|q_1(t)\rangle$

We now come to the important example of the propagator. As discussed above, the general semiclassical approach should lead to the van Vleck–Gutzwiller propagator directly. We now show this.

In classical mechanics, the coordinates and momenta at time t_1, (q_1, p_1), are related to the coordinates and momenta at time t_2, (q_2, p_2), by a canonical transformation. The generating function depending on the old coordinate, $q_1(0)$, and the new coordinate, $q_2 = q_1(t)$, is of the type $F_1(q, Q)$ and is given by the classical action, $S(q_2, q_1, t)$. Thus,

$$\langle q_2|q_1(t)\rangle = \left(\frac{-\partial^2 S/\partial q_1 \partial q_2}{2\pi i\hbar} \right)^{1/2} e^{iS(q_2, q_1, t)/\hbar}, \qquad (10.116)$$

exactly the expression obtained in the previous section (Eq. 10.79). In the present context we gain a different perspective on this expression, as arising from the properties of the F_1 generator for the canonical transformation from q to $q(t)$. Again, the nonuniqueness of the classical trajectories going from $q_1(0)$ to $q_2(t)$ can be viewed as a manifestation of the nonuniqueness of the functional inversion of the equation $\frac{\partial F_1(q,Q)}{\partial q} = p(q, Q)$ to find $Q(q, p)$.

▶ Exercise 10.20

a. For the free particle, $q_2 = q_1 + \frac{p_1 \Delta t}{m}$. Setting $(P, Q) = (p_1, q_1)$ and $(p, q) = (p_2, q_2)$ (note the counterintuitive assignment of old and new variables), use the first generating function equation, $\frac{\partial F_1(q,Q)}{\partial q} = p(q, Q)$, to show that $F_1(q, Q) = \frac{(\frac{1}{2}q^2 - qQ)m}{\Delta t} + C(Q)$, where $C(Q)$ is a so-far undetermined function of Q. Use the second generating function relation, $-\frac{\partial F_1(q,Q)}{\partial Q} = P(q, Q)$, to show that $C(Q) = \frac{m}{2\Delta t} Q^2$, so that $F_1(q, Q) = \frac{m}{2\Delta t} (q - Q)^2$.

b. Identifying $F_1(q, Q)$ with $S(q_2, q_1, t)$, use Eq. 10.116 to evaluate the semiclassical transition amplitude.

The following equations follow immediately from the derivative relations for the generating function:

$$\frac{\partial S(q_2, q_1)}{\partial q_2} = p_2$$

$$\frac{\partial S(q_2, q_1)}{\partial q_1} = -p_1$$

and thus

$$-\frac{\partial^2 S}{\partial q_2 \partial q_1} = \left(\frac{\partial q_2}{\partial p_1}\right)^{-1}\bigg|_{q_1},$$

which as discussed in Section 10.1 is a measure of the sensitivity of final position to initial momentum.

▶ **Exercise 10.21**   In ordinary quantum mechanics the propagator for two time steps obeys a composition property:

$$\int_{-\infty}^{\infty} \langle x'|e^{-iHt_2/\hbar}|x_1\rangle\langle x_1|e^{-iHt_1/\hbar}|x\rangle dx_1 = \langle x'|e^{-iH(t_1+t_2)/\hbar}|x\rangle. \tag{10.117}$$

Show that the property of composition holds for the van Vleck–Gutzwiller propagator as well if the integral over the value of the coordinate at the intermediate time t_1 is performed by stationary phase. What we need to prove is

$$\int_{-\infty}^{\infty}{}^{SP} \langle x'|e^{-iHt_2/\hbar}|x_1\rangle^{SC}\langle x_1|e^{-iHt_1/\hbar}|x\rangle^{SC}dx_1 = \langle x'|e^{-iH(t_1+t_2)/\hbar}|x\rangle^{SC}. \tag{10.118}$$

Solution   Using Eq. 10.80, we need to show

$$\int_{-\infty}^{\infty}{}^{SP} \left(\sum \left|\sqrt{\frac{-\frac{\partial^2 S(x',x_1,t_2)}{\partial x' \partial x_1}}{2\pi i\hbar}}\right| e^{iS(x',x_1,t_2)/\hbar - i\frac{v_2\pi}{2}}\right)\left(\sum \left|\sqrt{\frac{-\frac{\partial^2 S(x_1,x,t_1)}{\partial x_1 \partial x}}{2\pi i\hbar}}\right| e^{iS(x_1,x,t_1)/\hbar - i\frac{v_1\pi}{2}}\right)dx_1$$

$$= \left(\sum \left|\sqrt{\frac{-\frac{\partial^2 S(x',x,t)}{\partial x' \partial x}}{2\pi i\hbar}}\right| e^{iS(x',x,t)/\hbar - i\frac{v\pi}{2}}\right), \tag{10.119}$$

where the sum is over all classical paths. This equality reduces to two conditions:

$$a. \quad S(x', x, t) = S(x', \tilde{x}_1, t_2) + S(\tilde{x}_1, x, t_1) \tag{10.120}$$

and

$$b. \quad \left(-\frac{\partial^2 S}{\partial x' \partial x_1}\right)^{1/2}_{x_1 = \tilde{x}_1} \left(-\frac{\partial^2 S}{\partial x_1 \partial x}\right)^{1/2}_{x_1 = \tilde{x}_1} \left(\frac{\partial^2 S(x', x_1)}{\partial x_1^2} + \frac{\partial^2 S(x_1, x)}{\partial x_1^2}\right)^{-1/2}_{x_1 = \tilde{x}_1}$$

$$= \left(-\frac{\partial^2 S}{\partial x' \partial x}\right)^{1/2}, \tag{10.121}$$

where $\tilde{x}_1$ is the stationary phase value of x_1. The second condition may be written equivalently as

$$\left(-\frac{\partial p'(x', x_1)}{\partial x_1}\right)^{1/2} \left(\frac{\partial p(x_1, x)}{\partial x_1}\right)^{1/2} \left(-\frac{\partial p_1(x', x_1)}{\partial x_1} + \frac{\partial p_1(x_1, x)}{\partial x_1}\right)^{-1/2}$$

$$= \left(-\frac{\partial p'(x', x)}{\partial x}\right)^{1/2}. \tag{10.122}$$

(Note that we must specify carefully the arguments of the quantity whose partial derivative is taken; when a cross partial derivative is taken the arguments are automatically assumed to be the two variables with respect to which the derivatives are taken.)

We begin with the first condition. The stationary phase condition is that

$$\left.\frac{\partial S(x', x_1)}{\partial x_1}\right|_{x_1=\tilde{x}_1} + \left.\frac{\partial S(x_1, x)}{\partial x_1}\right|_{x_1=\tilde{x}_1} = 0 \tag{10.123}$$

or

$$-p_1(x', \tilde{x}_1) + p_1(\tilde{x}_1, x) = 0. \tag{10.124}$$

This stationary phase condition determines $\tilde{x}_1$ as a function of the fixed endpoints, x and x'. Thus, we may write

$$-p_1(x', x_1(x, x')) + p_1(x_1(x, x'), x) = 0, \tag{10.125}$$

where here and henceforth we take $x_1 = \tilde{x}_1$. The continuity of the momentum at coordinate x' is the condition that the two "subtrajectories," from x to x_1 in time t_1 and from x_1 to x' in time t_2, combine to form one single trajectory from x to x' in time $t = t_1 + t_2$. The action integrals for the two subtrajectories add, giving Eq. 10.120. Note that action integrals always add for subtrajectories; however, if $x_1 \neq \tilde{x}_1$ the composite action integral will not correspond to a classical trajectory and hence will not have a vanishing first variation. This is shown schematically in Figure 10.5.

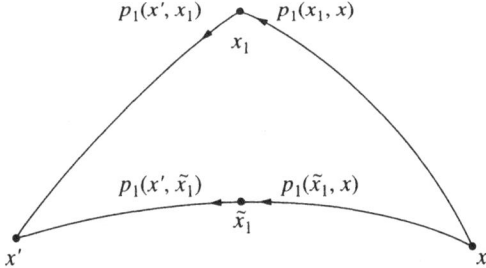

Figure 10.5 Schematic figure showing the significance of the continuity of the first derivative of the action at the stationary phase point, $\tilde{x}_1$. The first derivative of the action is the momentum; a continuous first derivative implies that the final momentum of one segment of the path is equal to the initial momentum for the next segment of the path. This is just the condition that the union of the path segments correspond to a bona fide trajectory. Adapted from Berry (1972).

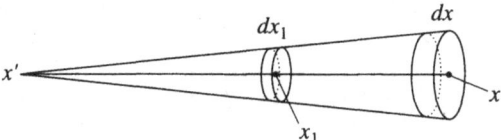

Figure 10.6 Schematic figure showing the geometric significance of setting $\frac{\partial}{\partial x}(\frac{\partial S}{\partial x_1}) = 0$. Adapted from Berry (1972).

We turn now to establishing the second condition. We take the derivative of Eq. 10.125 with respect to x for fixed x' and set the result to 0:

$$-\frac{\partial p_1(x', x_1)}{\partial x_1}\left(\frac{\partial x_1}{\partial x}\right)_{x'} + \frac{\partial p_1(x_1, x)}{\partial x_1}\left(\frac{\partial x_1}{\partial x}\right)_{x'} + \frac{\partial p_1}{\partial x} = 0. \qquad (10.126)$$

The physical justification for setting the derivative to zero is important to understand. Eq. 10.125 views the intermediate position $x_1 = x_1(x, x')$ as a "floating" variable, always satisfying the stationary phase condition. Thus, when we allow the value of x to change, x_1 changes accordingly to maintain the stationary phase condition, Eq. 10.125. With this understanding for the way in which x_1 varies, the RHS of Eq. 10.125 is maintained at 0 as x is varied, and hence $\frac{\partial}{\partial x}$ of the LHS is equal to 0. This is shown schematically in Figure 10.6. Equation 10.126 gives

$$-\frac{\partial p_1(x', x_1)}{\partial x_1} + \frac{\partial p_1(x_1, x)}{\partial x_1} = -\frac{\partial p_1(x_1, x)/\partial x}{\partial x_1(x', x)/\partial x}. \qquad (10.127)$$

Substituting this equation into Eq. 10.122, we obtain

$$\left(-\frac{\partial p'(x', x)}{\partial x_1}\right)^{1/2}\left(\frac{\partial p(x_1, x)}{\partial x_1}\right)^{1/2}\left(-\frac{\partial p_1(x_1, x)/\partial x}{\partial x_1(x', x)/\partial x}\right)^{-1/2}$$

$$= \left(-\frac{\partial p'(x', x_1)}{\partial x}\right)^{1/2}\left(\frac{\partial p(x_1, x)}{\partial x_1}\right)^{1/2}\left(-\frac{\partial p_1(x_1, x)}{\partial x}\right)^{-1/2} \qquad (10.128)$$

$$= \left(-\frac{\partial p'(x', x)}{\partial x}\right)^{1/2}. \qquad (10.129)$$

In Eq. 10.128, the last two factors cancel since in both cases the quantity in parentheses is equal to $-\frac{\partial^2 S(x_1, x)}{\partial x_1 \partial x}$. Because of the condition that the momentum be continuous, the double sum appearing on the LHS of Eq. 10.119 reduces to a single sum. Moreover, for each different classical trajectory that satisfies the stationary phase condition, Eq. 10.123, the Morse index is just $\nu = \nu_1 + \nu_2$. This completes the derivation.

▶ **Exercise 10.22** Show that the VVG propagator satisfies time reversal symmetry; that is, show that propagation forward for time t followed by backward propagation by the same time t gives the identity operator.

D. Semiclassical S-Matrix Elements: $\langle \mathbf{n}'|e^{-iHt/\hbar}|\mathbf{n}\rangle \Longleftrightarrow \langle p|p(t)\rangle$

In the quantum theory of scattering, which governs all molecular collisions as well as chemical reactions, the central object is the S-matrix—the probability amplitude for making a transition from a particular internal state of reactants, $\mathbf{n}$, to a particular internal state of products, $\mathbf{n}'$, at fixed total energy E. This subject will be discussed at length in Chapter 18. Here, however, we note that the S-matrix is a unitary matrix, and thus it is natural to apply the quite general treatment we have been using above to obtain semiclassical expressions for the S-matrix elements.

Schematically, the scattering amplitude is of the form $\langle \mathbf{n}'|e^{-iHt/\hbar}|\mathbf{n}\rangle$. To discuss the semiclassical representation of this matrix element we first need to say a few words about classical action–angle variables, which were used briefly in Chapter 4. In classical mechanics there is a canonical transformation, $(p, q) \rightarrow (J, \theta)$, where J is a constant of the motion; it immediately follows that $H(J, \theta)$ is independent of θ, since $\dot{J} = -dH/d\theta = 0$, while $\dot{\theta} = dH/dJ = \omega$. The quantity J is called the classical *action* and θ is the classical *angle*. The classical action variables are the analog of good quantum numbers in quantum mechanics: operators that commute with the Hamiltonian. These values are quantized in semiclassical mechanics; hence we use the notation $\mathbf{n}$ instead of J.

The classical action is a generalized momenta variable; hence, the matrix element of interest is of the form $\langle p|p(t)\rangle$. This object is similar in spirit to the $\langle q|q(t)\rangle$ transformation above, but since it involves momentum, the governing generating function is of the type F_4. Specifically,

$$\langle p_2|e^{-iHt/\hbar}|p_1\rangle = \left(-\frac{\partial^2\phi(p_2, p_1)/\partial p_1\partial p_2}{2\pi i\hbar}\right)^{1/2} e^{i\phi(p_2, p_1)/\hbar} \qquad (10.130)$$

$$= (-2\pi i\hbar(\partial p_2/\partial q_1)_{p_1}))^{-1/2} e^{i\phi(p_2, p_1)/\hbar}, \qquad (10.131)$$

where

$$\phi(p_2, p_1) = \int_{t_1}^{t_2} dt\{-q(t)\dot{p}(t) - H[p(t), q(t)]\} \qquad (10.132)$$

and $t = t_2 - t_1$.

The S-matrix describes the amplitude for state-to-state transitions between specific quantum states of collision partners. It will be discussed in much greater detail in Chapter 18. Here we present without proof a convenient formula for the S-matrix in terms of the propagator (Miller, 1974):

$$S_{\mathbf{n}_2,\mathbf{n}_1}(E) = -\lim_{R_1, R_2 \to \infty} \left(\frac{\hbar^2 k_1 k_2}{\mu^2}\right)^{1/2} e^{-ik_1 R_1 - ik_2 R_2} \int_0^\infty dt e^{iEt/\hbar}\langle R_2\mathbf{n}_2|e^{-iHt/\hbar}|R_1\mathbf{n}_1\rangle.$$

$$(10.133)$$

Here R_1 and R_2 are the values of the relative translational coordinate of the fragments before and after the collision, respectively, and μ is the reduced mass in this coordinate. Similarly, $\mathbf{n}_1$ and $\mathbf{n}_2$ are labels for the quantized internal state of the fragments before and after the collision. The $|\mathbf{n}\rangle$ states are eigenfunctions of the internal Hamiltonian of the fragments with $N - 1$ quantum numbers corresponding to the $N - 1$ internal degrees of freedom. (For simplicity we specialize here to nonreactive collisions, although it is straightforward to extend the approach to the reactive case.) The classical coordinates and momenta that correspond to the matrix elements in Eq. 10.133 are the ordinary translational coordinate and momentum, R and P, and the action–angle variables, $\{n_i\}$ and $\{q_i\}$, $i = 1, \ldots, N - 1$, for the internal degrees of freedom. The action variables $\mathbf{n}$ are the classical counterpart of the quantum numbers for the internal degrees of freedom, and are a set of generalized momenta. The classical Hamiltonian function is given in terms of these variables by

$$H(P, R, \mathbf{n}, \mathbf{q}) = \frac{P^2}{2\mu} + \epsilon(\mathbf{n}) + V(R, \mathbf{n}, \mathbf{q}),$$

where $\epsilon(\mathbf{n})$ is the energy associated with the semiclassically quantized wavefunction for the internal degrees of freedom. The quantum numbers $\{\mathbf{n}\}$ are constant before and after the collision, since $V \to 0$ as $R_1, R_2 \to \infty$, and hence

$$\dot{\mathbf{n}}(t) = -\frac{\partial H}{\partial \mathbf{q}} \to 0.$$

In terms of the canonical variables $(P, R, \mathbf{n}, \mathbf{q})$ the matrix elements of the propagator, Eq. 10.133, are in a mixed representation: a coordinate representation with respect to translation and a momentum representation with respect to the internal degrees of freedom. This causes no difficulty, however, and in the classical limit Eq. 10.133 becomes

$$S_{\mathbf{n}_2, \mathbf{n}_1}(E) = -\lim_{R_1, R_2 \to \infty} \sum_{\text{classical trajectories}} \left(-\frac{P_1 P_2}{\mu^2}\right)^{1/2} e^{i/\hbar(P_1 R_1 - P_2 R_2)}$$

$$\times \int_0^\infty dt \left[\frac{\partial^2 \phi(R_2\mathbf{n}_2, R_1\mathbf{n}_1; t)/\partial(R_2\mathbf{n}_2)\partial(R_1\mathbf{n}_1)}{(-2\pi i\hbar)^N}\right]^{1/2} e^{i/\hbar[Et + \phi(R_2\mathbf{n}_2, R_1\mathbf{n}_1; t)]},$$

$$\tag{10.134}$$

where the action integral in the mixed representation is

$$\phi(R_2\mathbf{n}_2, R_1\mathbf{n}_1; t) = \int_{t_1}^{t_2} dt' \left[P(t')\dot{R}(t') - \mathbf{q}(t')\dot{\mathbf{n}}(t') - H\right]$$

and we have used the definitions

$$k_1 = -P_1/\hbar, \qquad k_2 = P_2/\hbar.$$

Equation 10.134 can be simplified by evaluating the time integral by stationary phase and expressing the action integral, ϕ, in terms of the new phase Φ:

$$\phi(R_2\mathbf{n}_2, R_1\mathbf{n}_1; t) \equiv \Phi(\mathbf{n}_2, \mathbf{n}_1; E) + P_2 R_2 - P_1 R_1 - Et. \tag{10.135}$$

With these modifications, Eq. 10.134 can be written as

$$S_{\mathbf{n}_2, \mathbf{n}_1}(E) = i \left[\frac{\partial^2 \Phi(\mathbf{n}_2, \mathbf{n}_1; E)/\partial \mathbf{n}_2 \partial \mathbf{n}_1}{(-2\pi i\hbar)^{N-1}}\right]^{1/2} e^{i\Phi(\mathbf{n}_2, \mathbf{n}_1; E)/\hbar}. \tag{10.136}$$

We refer the reader to Miller (1974) for details of the derivation.

The sum over classical paths has the elegant interpretation of a sum over classical trajectories that have quantized values of the action in both the reactant and product arrangement channels, but connect smoothly across the transition state in time interval t. We will return to this picture in Chapter 18, where we will present a rigorous quantum wavepacket approach to reactive scattering with much of the same spirit.

In all the preceding examples we have suppressed the explicit dependence on the Morse index. The amplitude prefactor as written has an unspecified phase; if the absolute value of the amplitude prefactor were written instead, it would be necessary to give the phase factor separately, with the Morse index ν calculated as specified in the first section of this chapter.

10.5   Gaussian Wavepackets and the van Vleck Propagator

Despite the beauty of the van Vleck–Gutzwiller form for the propagator, it suffers from several problems in actual dynamical applications:

1. There is the need to solve a double-ended boundary value problem for the root trajectories. This is significantly more difficult than solving an initial value problem.

2. If there are nearby root trajectories the simple procedure of adding their contributions and squaring the sum is no longer accurate; a so-called uniformizing procedure is required, which essentially goes beyond the quadratic expansion around the classical path in the derivation of the semiclassical propagator.

3. Probability amplitudes diverge at certain times, when the denominator in the prefactor goes to zero.

4. The Morse (or Maslov) indices are nontrivial to calculate.

In recent years, these problems have been largely overcome by two modifications to the VVG propagator. The first is the use of the initial value representation, which avoids the search for root trajectories, and also effectively avoids the problem with nearby roots. The second modification is to change the representation of the semiclassical propagator from coordinate (or momentum) to the "coherent state" representation, which is a compromise between the two. The coherent state representation avoids the problem with divergence of the probability amplitude, and also makes the Maslov index trivial to calculate. The combined method is the culmination of an approach introduced by Heller in 1981 on heuristic grounds, known as "frozen Gaussian" propagation. In 1984, Herman and Kluk formally derived the frozen Gaussian method starting with the VVG propagator, by introducing a coherent state basis and changing over to an initial value representation (Herman, 1984). Their derivation revealed a complex amplitude factor that had not been in Heller's treatment, a factor that is a natural generalization of the VVG prefactor but is free from caustic singularities and takes care of the Maslov index automatically.

The outline of the following sections is as follows. First, the coherent state representation will be introduced. Then, the path integral formula in the coherent state representation, as well as its classical limit, will be described. Finally, the frozen Gaussian method will be discussed. The Herman–Kluk prefactor will be derived from the classical limit of the coherent state propagator, by going over to an initial value representation. The rigorous connection between Gaussian wavepacket propagation and the VVG semiclassical propagator is one of the great achievements in semiclassical theory in recent years, and has led to what is presently probably the most accurate and efficient semiclassical method for molecular systems.

10.5.1 The Coherent State Representation

The coherent state basis is a basis of nonorthogonal Gaussians all with the same width, but centered at different locations in either q or p. The basis functions take the form

$$g_{p',q';\gamma}(x) = \left(\frac{\gamma}{\pi}\right)^{1/4} e^{-\gamma(x-q')^2/2+ip'(x-q'/2)/\hbar} \equiv \langle x|p', q'; \gamma\rangle, \qquad (10.137)$$

for all real p' and q' and we take the width parameter γ to be real. Among the basic properties of the coherent states are

$$\langle p\rangle = \langle p', q'; \gamma|\hat{p}|p', q'; \gamma\rangle = p', \qquad (10.138)$$

$$\langle q\rangle = \langle p', q'; \gamma|\hat{q}|p', q'; \gamma\rangle = q', \qquad (10.139)$$

independent of γ. Thus, the parameters (q', p') determine the average position and momentum of the state $g_{p',q';\gamma}(x)$. As q' and p' vary, the centers of the Gaussians vary continuously over the entire range of the p–q plane. Note that this set of Gaussians differs from the Gaussian basis set used in electronic structure calculations, since these Gaussians are complex. The complex Gaussian has a coordinate dependent phase that can describe rapidly oscillating portions of a wavefunction with one or several basis functions. There are other useful relationships between coherent states. It is useful to define

$$z \equiv \frac{1}{\sqrt{2}}\left(\frac{q}{b} + i\frac{p}{c}\right), \qquad (10.140)$$

where $b = \sqrt{\hbar/m\omega}$ and $c = \sqrt{\hbar m\omega}$, and ω is related to γ through the equation $\gamma = m\omega/\hbar$. Then the overlap of two coherent states is given by

$$\langle z''|z'\rangle \equiv \langle p'', q''; \gamma|p', q'; \gamma\rangle = e^{-\frac{1}{2}(|z''|^2+|z'|^2)+z''^*z'}. \qquad (10.141)$$

▶ **Exercise 10.23** Compare this formula with the formula for the overlap of two Gaussians in Chapter 3. Show that the overlap has a simple physical interpretation—that the overlap of two Gaussians decreases exponentially with the square of the phase space distance between their centers.

The time evolution of a Gaussian in a harmonic oscillator was studied in Chapter 3. Recall that if the Gaussian is a coherent state (the width parameter, $\alpha = \frac{m\omega}{2\hbar} = \frac{\gamma}{2}$, where ω is the frequency of the oscillator) the center of the Gaussian in both p and q evolves according to the classical equations of motion. This result is expressed concisely in terms of the coherent state notation z:

$$g_z(x, t) = g_{z(t)}(x) = \left(\frac{\gamma}{\pi}\right)^{1/2} e^{-\gamma(x-q_t)^2/2+ip_t(x-q_t/2)/\hbar}, \qquad (10.142)$$

where $z(t) \equiv z(0)e^{-i\omega t}$, $q_t \equiv \sqrt{2}b\,\mathrm{Re}(z(t))$ and $p_t = \sqrt{2}c\,\mathrm{Im}(z(t))$.

▶ **Exercise 10.24** Show that q_t and p_t are the *classical* values of q and p, given initial conditions q_0, p_0 at time 0.

▶ **Exercise 10.25**   Substitute Eq. 10.142 into Eq. 10.141, identifying $g_z(x, t)$ with z' and $g_z(x, 0)$ with z''. Show that, aside from the zero-point phase, the result agrees with the Gaussian time correlation derived in Eq. 6.76.

From Eq. 10.141 it is clear that the coherent states are not mutually orthogonal, although if the centers of the coherent states are well separated the overlap is virtually zero. Although the coherent states are nonorthogonal they are complete; in fact they are *overcomplete* in the sense that there are many ways of choosing complete subsets among them. Despite this flexibility in choosing subsets, the most useful completeness relationship for the coherent states is one that involves the infinite, continuous set of all phase space centers:

$$\frac{1}{2\pi\hbar} \int \int dp' \, dq' \, |p', q'; \gamma\rangle\langle p', q'; \gamma| = \frac{1}{\pi} \int \int d^2z \, |z\rangle\langle z| = 1, \quad (10.143)$$

where d^2z signifies integration over the whole complex plane:

$$d^2z = d\left(\frac{q}{\sqrt{2b}}\right) d\left(\frac{p}{\sqrt{2c}}\right) = \frac{dq \, dp}{2\hbar}. \quad (10.144)$$

▶ **Exercise 10.26**   Take the matrix element $\langle x'|x\rangle$ and insert the above completeness relation for the coherent states. Do the phase space integral explicitly and show that the result is $\delta(x' - x)$, as it must be.

The above coherent state completeness relationship has an interesting interpretation in terms of even coverage of phase space. Any completeness relationship in quantum mechanics may be viewed as generating a complete phase space coverage. Whereas the usual completeness forms, $\int |x\rangle\langle x| \, dx$ and $\int |p\rangle\langle p| \, dp$, correspond to inserting *lines* in phase space whose area (according to a well-defined limiting procedure) is h (cf. Figure 10.7), the coherent state completeness relationship corresponds to inserting *ellipses* of area h (cf.

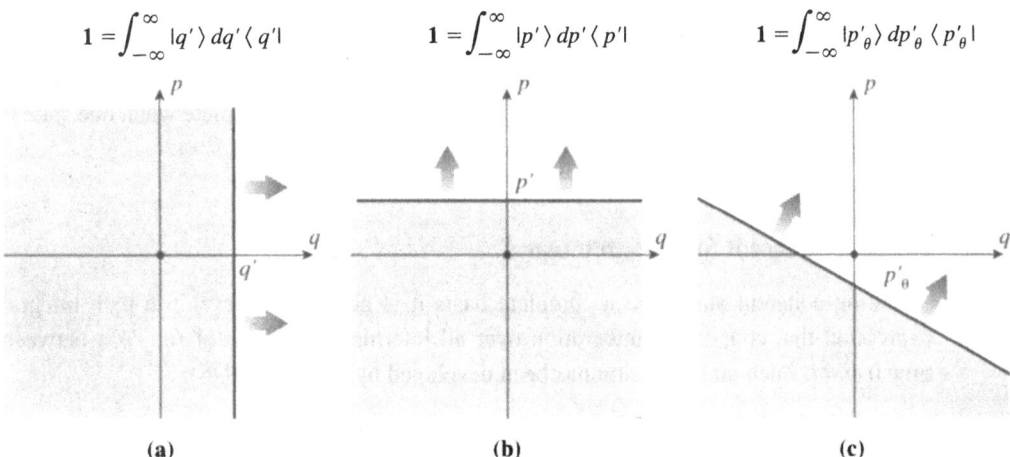

Figure 10.7 Phase space representation of the completeness relation: (a) $\int |q'\rangle\langle q'| dq' = 1$, (b) $\int |p'\rangle\langle p'| \, dp' = 1$ and (c) a mixed representation. Adapted from Heller (1977).

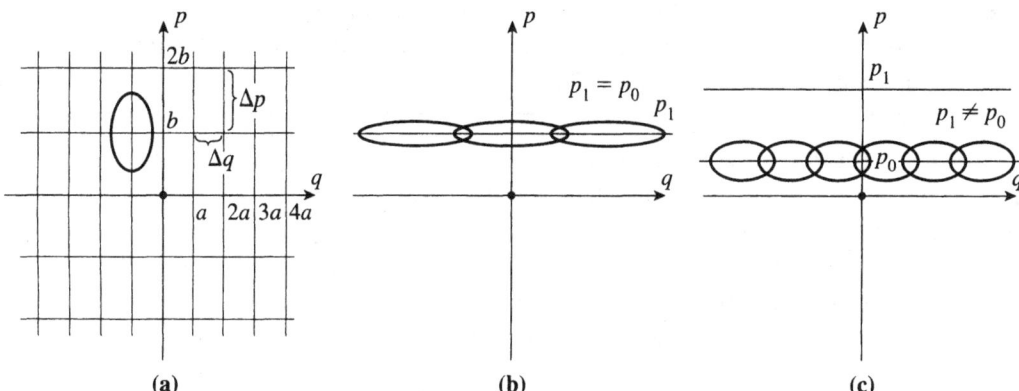

Figure 10.8 (a) Phase space representation of the coherent state completeness relation $\frac{1}{2\pi\hbar}\int dp'\,dq'$ $|p',q';\gamma\rangle\langle p',q';\gamma| = \mathbf{1}$. Note that the completeness relation calls for a continuous distribution of Gaussians centered at every q',p'. However, because the above set is so vastly overcomplete, placing one Gaussian at each phase space point on an evenly spaced grid is also complete, if the product of grid spacings in p and q satisfy $\Delta p\,\Delta q = h$. Other complete subsets of the coherent states include a continuous line in q' or in p'. (b) Use of a line of coherent states all with centers at $p = p_0$ to describe the state $|p_0\rangle$. (c) Use of a line of coherent states all with centers at $p = p_0$ to describe the state $|p_1\rangle$, $p_1 \neq p_0$. Adapted from Davis (1979).

Figure 10.8). In either case, the basis functions satisfy the uncertainty principle; however, in the first case all the uncertainty is found in x or p, while in the coherent state representation the uncertainty is evenly distributed between the two. This *compromise-in-uncertainty* state generally displays well-behaved numerics, while the x and p kets, because of their infinite extent in one coordinate or the other, can lead to pathologies when dynamical or numerical approximations are made.

The interpretation of the coherent state completeness relationship in terms of phase space coverage has an interesting footnote. If one divides the phase space into regular cells of area S and places a coherent state at the center of each cell, this gives a subset of the complete set of coherent state, known as the von Neumann lattice (von Neumann, 1931). It can be shown (Perelomov, 1971; Bargmann, 1971) that (a) if $S < 2\pi\hbar$ this set is overcomplete, and remains so when a finite number of states are removed; (b) if $S > 2\pi\hbar$ the subset is not complete; (c) if $S = 2\pi\hbar$ the subset is complete; it remains complete when one state is removed but becomes incomplete when any two states are removed.

10.5.2 Coherent State Path Integral

Since the coherent states are a complete basis it is possible to develop a path integral expression that consists of integration over all intermediate values of $(q_{t'}, p_{t'})$ between time 0 and t. Such an expression has been developed by Klauder (1978):

$$\langle z''|e^{-iH(t''-t')/\hbar}|z'\rangle \equiv \langle p'',q''|e^{-iH(t''-t')/\hbar}|p',q'\rangle$$

$$= \lim_{\epsilon\to 0} N \int e^{i\int_{t'}^{t''}[\frac{1}{2}(p\dot{q}-\dot{p}q)+\frac{1}{4}i\epsilon(\dot{p}^2+\dot{q}^2)-H(p,q)]dt}\mathcal{D}p\mathcal{D}q, \quad (10.145)$$

where $\mathcal{D}p\mathcal{D}q \equiv \prod_t dp(t)dq(t)$, $H(p,q) \equiv \langle p,q|H|p,q\rangle$ and N is a normalization factor fixed by the composition law,

$$
\begin{aligned}
&\langle p''', q''', t''' | p', q', t'\rangle \\
&\qquad = \int \langle p''', q''', t''' | p'', q'', t''\rangle\langle p'', q'', t'' | p', q', t'\rangle \frac{dp''dq''}{2\pi\hbar}.
\end{aligned}
\tag{10.146}
$$

The semiclassical limit of the coherent state propagator has been derived by a number of workers. It was derived by Klauder (1979) without the normalization factor in the form

$$
\langle z'' | e^{-iH(t''-t')/\hbar} | z'\rangle \approx e^{\frac{1}{2}i(q''\bar{p}''-p''\bar{q}''+\bar{q}'p'-\bar{p}'q')+i\int_{t'}^{t''}[\frac{1}{2}(\dot{\bar{p}}\bar{q}-\dot{\bar{p}}\bar{q})-H(\bar{p},\bar{q})]dt},
\tag{10.147}
$$

where $\bar{q}(t)$ and $\bar{p}(t)$ are (generally) complex solutions of the usual classical equations

$$
\dot{\bar{q}} = \frac{\partial H(\bar{p},\bar{q})}{\partial\bar{p}}
\tag{10.148}
$$

$$
\dot{\bar{p}} = -\frac{\partial H(\bar{p},\bar{q})}{\partial\bar{q}}
\tag{10.149}
$$

with boundary conditions

$$
\frac{\bar{q}(t')}{b} + i\frac{\bar{p}(t')}{c} = \frac{q'}{b} + i\frac{p'}{c}
\tag{10.150}
$$

$$
\frac{\bar{q}(t'')}{b} - i\frac{\bar{p}(t'')}{c} = \frac{q''}{b} - i\frac{p''}{c}.
\tag{10.151}
$$

We emphasize that the endpoints q', q'' and p', p'' are real, while the endpoints $\bar{q}(t')$, $\bar{p}(t')$, $\bar{q}(t'')$, $\bar{p}(t'')$ and hence the trajectories $\bar{q}(t)$, $\bar{p}(t)$ are generally complex. Complex $\bar{q}(t)$, $\bar{p}(t)$ are required in order to satisfy what would otherwise be an overdetermined set of boundary conditions: given (q', p'), and admitting only real trajectories, $\bar{q}(t)$ and $\bar{p}(t)$ are completely determined; hence it is not generally possible to satisfy the terminal conditions $\bar{q}(t'') = q''$, $\bar{p}(t'') = p''$. However, allowing $\bar{q}(t)$ and $\bar{p}(t)$ to be complex makes it generally possible to satisfy the conditions given by Eqs. 10.150–10.151 simultaneously.

An equivalent expression for the semiclassical coherent state propagator, including the normalization factor, was derived independently by Heller, by positing the extension of Miller's semiclassical algebra to eigenstates of non-Hermitian operators (Heller, 1977). Heller's approach was recast in an elegant way by Weissman (1982), who showed that the classical generating functions for the canonical transformation for $(q, p) \rightarrow (Q, P)$ may be associated with the unitary transformation in quantum mechanics from $|q\rangle$ and $|p\rangle$ to the coherent states $|u\rangle$ and $|v\rangle$, where

$$
Q = (\hbar/2i)^{1/2}\left(\frac{q}{b} - i\frac{p}{c}\right), \qquad P = (\hbar/2i)^{1/2}\left(\frac{q}{b} + i\frac{p}{c}\right).
\tag{10.152}
$$

In Weissman's expression, the final expression has the same structure as the van Vleck propagator; that is, the phase is the generating function for the canonical transformation and the prefactor is the square root of the second derivative of the generating function with respect to one initial and one final variable, yielding the general structure we have seen in Section 10.4. Below we will use a slightly modified version of Weissman's expression derived by de Aguiar and Baranger (referenced in Grossmann (1998); see also Baranger (2001)) that does not include the $(\frac{\hbar}{i})^{1/2}$ prefactor in the new variables in Eq. 10.152. This latter convention will be convenient in what follows, although as we shall see this leads to a form of Hamilton's equations, as well as relationships among the derivatives of the generating function, which have an extra factor of i.

Following the notation of de Aguiar and Baranger, we may write the semiclassical expression for the matrix element of the propagator as

$$\langle z''|e^{-iHt/\hbar}|z'\rangle^{SC} = \sum_j \sqrt{\frac{i}{\hbar}\left|\frac{\partial^2 S_j^c}{\partial u'\partial v''}\right|}\, e^{\frac{i}{\hbar}S_j^c(v'',u',t)-\frac{1}{2}(|v''|^2+|u'|^2)}. \quad (10.153)$$

Here, the complex action of the jth trajectory is given by

$$S_j^c(v'', u', t) = \int_0^t \left[\frac{i\hbar}{2}(v_j(t')\dot{u}_j(t') - \dot{v}_j(t')u_j(t')) - H(u_j(t'), v_j(t'))\right] dt'$$

$$-\frac{i\hbar}{2}(v''u_j(t) + u'v_j(0)). \quad (10.154)$$

The variables

$$u_j = \frac{1}{\sqrt{2}}\left(\frac{q_j}{b} + i\frac{p_j}{c}\right), \qquad v_j = \frac{1}{\sqrt{2}}\left(\frac{q_j}{b} - i\frac{p_j}{c}\right) \quad (10.155)$$

satisfy a modified form of Hamilton's equations:

$$i\hbar\dot{u} = \frac{\partial H}{\partial v} \qquad i\hbar\dot{v} = -\frac{\partial H}{\partial u}. \quad (10.156)$$

The variables u and v each have one endpoint fixed:

$$u(0) \equiv u' = z' = \frac{1}{\sqrt{2}}\left(\frac{q'}{b} + i\frac{p'}{c}\right), \quad (10.157)$$

$$v(t) \equiv v'' = z''^* = \frac{1}{\sqrt{2}}\left(\frac{q''}{b} - i\frac{p''}{c}\right), \quad (10.158)$$

while the other endpoint, $u_j(t)$ or $v_j(0)$, is determined by the dynamics. The Hamiltonian H is obtained by substituting the inverse of the transformation formula, Eq. 10.152, into $H(p, q)$, and thus is generally complex. Note that the sum is over all possible complex trajectories satisfying the boundary conditions given by Eqs. 10.157–10.158.

▶ **Exercise 10.27** Show that the derivatives of the complex action satisfy the following generating function type relationship:

$$\frac{\partial S^c}{\partial v''} = -i\hbar u(t), \qquad \frac{\partial S^c}{\partial u'} = -i\hbar v(0). \quad (10.159)$$

▶ Exercise 10.28 Show that

$$\frac{i}{\hbar} \frac{\partial^2 S^c}{\partial u' \partial v''}\Bigg|_{q_t, p_t} = \frac{\partial u(t)}{\partial u'}\Bigg|_{q_t, p_t} = \left(\frac{\partial u(t)}{\partial p'} \frac{\partial p'}{\partial u'} + \frac{\partial u(t)}{\partial q'} \frac{\partial q'}{\partial u'} \right)\Bigg|_{q_t, p_t}$$

$$= \frac{1}{2} \left(\frac{\partial p_t}{\partial p'} + \frac{\partial q_t}{\partial q'} - im\omega \frac{\partial q_t}{\partial p'} - \frac{1}{im\omega} \frac{\partial p_t}{\partial q'} \right). \quad (10.160)$$

Equation 10.160 will figure prominently in the next section as the stability prefactor in the Heller–Herman–Kluk–Kay propagator.

10.5.3 Frozen Gaussians and the Heller–Herman-Kluk–Kay (HHKK) Propagator

In 1981 Heller proposed the following simple semiclassical-type method for propagating wavepackets based on coherent states (Heller, 1981):

1. Decompose the initial wavepacket into a set of coherent states. For simplicity, the coherent states may be chosen such that their centers form an evenly spaced grid in the phase space.

2. Propagate each of these coherent states approximately, letting the centers move in time according to classical mechanics; the width is kept fixed at its initial value, hence the name "frozen Gaussians." The central classical trajectory may be called the "guiding" trajectory.

3. Take the phase of the wavepacket to be the classical action integral along the path of the guiding classical trajectory. Because the coherent states have a fixed width the method was relatively simple to implement. Moreover, it proved to be accurate, relative to available methods at that time.

Heller was well aware that a Gaussian wavepacket is not an accurate description of the wavepacket in an anharmonic potential, and, as we saw in Chapter 3, even in a harmonic potential the analytical Gaussian solution does not generally have a fixed width. Why then the fixed width assumption? Why not let the width vary in some physically motivated way, for instance as determined by an instantaneous locally quadratic potential around the guiding classical trajectory? (This procedure, which had previously been tested by Heller (1975, 1976), he subsequently called "thawed Gaussians.") The reasons are twofold: First, the fixed width assumption is the simplest possible prescription, and simplicity is vital to have a general, flexible and well-controlled algorithm. The second reason is that although the individual Gaussians may not provide an accurate description of the wavepacket motion, collectively they may do much better. For example, in a harmonic potential, although the individual frozen Gaussians have fixed width, collectively they can describe the spreading and contracting discussed in detail in Chapter 3. Moreover, the swarm of frozen Gaussians can show position–momentum correlation, which is a key feature of the analytical solution (see Figure 10.9). Since the individual frozen Gaussians evolve according to local equations of motion, their collective behavior never deviates wildly from the form of the ensemble of classical trajectories; in contrast, the width in the thawed Gaussian approach is governed by a single global number and can become highly nonphysical. Finally, it happens quite generally in full quantum wavepacket calculations with multiple arrangement channels or

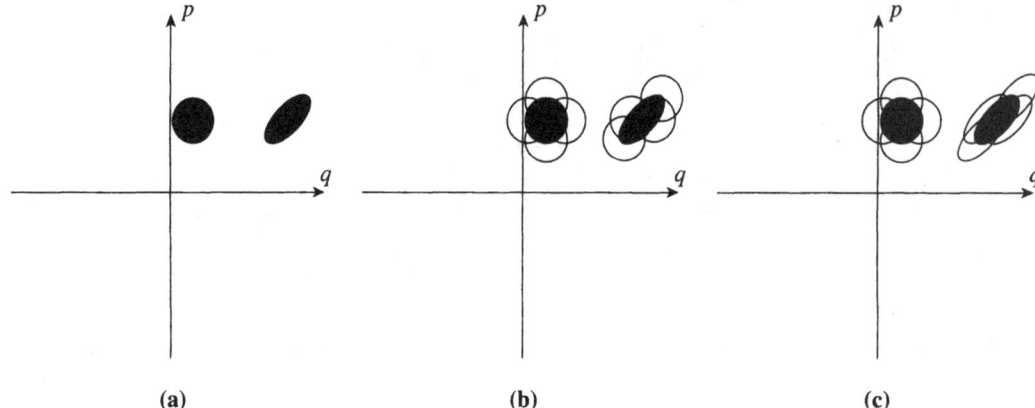

(a)   **(b)**   **(c)**

Figure 10.9 Phase space diagram showing how a swarm of frozen (nonspreading) Gaussians can describe wavepacket spreading and position–momentum correlation. (a) Initial state (circle) and time-evolved state (ellipse) in phase space. (b) Decomposition of the initial state in terms of a set of nonspreading Gaussians. Although the individual Gaussians do not spread, collectively their evolution accounts for wavepacket spreading, position–momentum correlation and even bifurcation. (c) Decomposition of the initial state in terms of a set of *spreading* Gaussians. The collective dynamics is qualitatively or quantitatively the same as that of the nonspreading Gaussians. Adapted from Heller (1981).

electronic states that an initially localized wavepacket will bifurcate. It is quickly perceived that a swarm of frozen Gaussians can describe this bifurcation process since the Gaussians evolve independently, while a single thawed Gaussian, by virtue of its functional form, inherently fails to describe bifurcation.

In retrospect, one recognizes in Heller's approach two key features. First, its simplicity stems from the forward propagation of trajectories and the avoidance of any type of root search. This suggests a close relationship to initial value representations of the propagator. Second, the use of coherent states leads exclusively to Gaussian integrals in determining spectra and cross sections, and thereby avoids mathematical singularities that often plague semiclassical techniques. Since the coherent states form a perfectly legitimate basis set in quantum mechanics, this suggests that the frozen Gaussian approach could be derived from the van Vleck propagator in the initial value representation using a coherent state basis.

The connection between the frozen Gaussians and the van Vleck propagator was proposed and proved by Herman and Kluk in 1984, just three years after Heller's initial publication on the frozen Gaussian method. Beginning with the VVG propagator, they inserted a complete set of (nonorthogonal) Gaussian basis functions. In addition, they transformed from the double-ended boundary value representation to an initial value representation. They succeeded in recovering the frozen Gaussian approximation, but discovered an additional prefactor, which is an analog of the stability matrix element in the usual VVG propagator but was absent in the earlier frozen Gaussian approach. With the addition of this prefactor the frozen Gaussian approach is rigorously semiclassical, norm preserving within the stationary phase approximation (empirically the norm is preserved much better than in the earlier frozen Gaussian approach) and it is exact for harmonic potentials.

The expression for the semiclassical propagator derived by Herman and Kluk in N dimensions takes the form

$$K^{SC}(\mathbf{x}'', t; \mathbf{x}', 0) = \frac{1}{(2\pi\hbar)^N} \int \int d\mathbf{p}' \, d\mathbf{q}' \, R_{\mathbf{p}'\mathbf{q}'t} e^{iS(\mathbf{p}'\mathbf{q}'t)/\hbar} g_\gamma(\mathbf{q}_t, \mathbf{p}_t, \mathbf{x}'') g_\gamma^*(\mathbf{q}', \mathbf{p}', \mathbf{x}'),$$

(10.161)

where

$$g_\gamma(\mathbf{q}', \mathbf{p}', \mathbf{x}') = \left(\frac{\gamma}{\pi}\right)^{N/4} \exp\left(-\frac{\gamma}{2}(\mathbf{x}' - \mathbf{q}')^2 + \frac{i}{\hbar}\mathbf{p}'(\mathbf{x}' - \mathbf{q}')\right),$$

(10.162)

$$g_\gamma(\mathbf{q}_t, \mathbf{p}_t, \mathbf{x}'') = \left(\frac{\gamma}{\pi}\right)^{N/4} \exp\left(-\frac{\gamma}{2}(\mathbf{x}'' - \mathbf{q}_t)^2 + \frac{i}{\hbar}\mathbf{p}_t(\mathbf{x}'' - \mathbf{q}_t)\right),$$

(10.163)

and

$$R_{\mathbf{p}'\mathbf{q}'t} = \sqrt{\det\left(\frac{1}{2}\left(\frac{\partial \mathbf{p}_t}{\partial \mathbf{p}'} + \frac{\partial \mathbf{q}_t}{\partial \mathbf{q}'} - i\gamma\hbar\frac{\partial \mathbf{q}_t}{\partial \mathbf{p}'} + \frac{i}{\hbar\gamma}\frac{\partial \mathbf{p}_t}{\partial \mathbf{q}'}\right)\right)}.$$

(10.164)

Here, $\mathbf{q}_t$ and $\mathbf{p}_t$ are the coordinates and momenta at time t of a classical trajectory started with initial conditions $\mathbf{p}'$ and $\mathbf{p}'$ at time zero. The integration goes over all initial values $(\mathbf{q}', \mathbf{p}')$. $S_{\mathbf{p}'\mathbf{q}'t}$ is the classical action,

$$S_{\mathbf{p}', \mathbf{q}', t} = \int_0^t [\mathbf{p}_{t'}\dot{\mathbf{q}}_{t'} - H(\mathbf{p}_{t'}, \mathbf{q}_{t'}, t')] \, dt'.$$

(10.165)

The derivative of a vector with respect to a vector in Eq. 10.164 implies a matrix of partial derivatives of components of one vector with respect to components of another vector. Note that the prefactor has the form of a generalized van Vleck determinant.

In 1994, Kay provided a unified framework for comparing and deriving new initial value representations of semiclassical dynamics, and gave a simplified procedure for calculating the Maslov (Morse) index in the Heller–Herman–Kluk approach (Kay, 1994). In particular, as shown by Kay, if the phase of Eq. 10.164 is chosen such that $R_{\mathbf{p}\mathbf{q}t}$ is a continuous function of time, the Maslov (Morse) index is automatically taken into account. We will refer to the final expression for the propagator as the Heller–Herman–Kluk–Kay or HHKK propagator, to acknowledge the major contributions of Heller and Kay, as well as Herman and Kluk, to the development of the expression.

▶ Exercise 10.29 Perhaps the most straightforward way to derive the HHKK propagator is to insert two complete sets of coherent states into the (x'', x') expression for the propagator. If the semiclassical expression, Eq. 10.153, is used for the propagator between initial and final coherent states and *final* phase space variables are integrated via saddle point integration while retaining the integral over initial phase space variables, the HHKK propagator is obtained (Grossmann, 1998). Derive the one-dimensional HHKK propagator this way.

Solution Write

$$\langle x''|e^{-iHt/\hbar}|x'\rangle = \int \frac{d^2z''}{\pi} \int \frac{d^2z'}{\pi} \langle x''|z''\rangle\langle z''|e^{-iHt/\hbar}|z'\rangle\langle z'|x'\rangle.$$

(10.166)

After substitution of Eq. 10.153 for the semiclassical coherent state propagator, the part of the exponent in the integrand of Eq. 10.166 that depends on the final phase space variables is denoted by

$$\sigma_j \equiv \frac{i}{\hbar} S_j^c(v'', u', t) - \frac{1}{2}|z''|^2 - \frac{1}{2b^2}(x'' - \sqrt{2}bz'')^2 + \frac{z''}{2}(z'' - z''^*) \quad (10.167)$$

$$= \frac{i}{\hbar} S_j^c(v'', u', t) - \frac{1}{2}|z''|^2 - \frac{1}{2b^2}(x'' - q'')^2 + \frac{i}{bc} p'' \left(x'' - \frac{q''}{2} \right). \quad (10.168)$$

The saddle point condition, $\partial \sigma_j / \partial z''^* = 0$, leads to the relation

$$u_j(t) - u'' = 0, \quad (10.169)$$

where we have used Eq. 10.159 for $\partial S^c / \partial v''$, and written u'' for z'' since they are defined by the same equation. Equation 10.169 can be rewritten as

$$\frac{1}{b}(q_j(t) - q'') + \frac{i}{bc}(p_j(t) - p'') = 0. \quad (10.170)$$

Since $q = \text{Re}(q) + i\text{Im}(q)$ and $p = \text{Re}(p) + i\text{Im}(p)$, Eq. 10.170 can be rewritten as

$$\frac{\text{Re}(q_j(t))}{b} - \frac{\text{Im}(p_j(t))}{c} = \frac{q''}{b} \quad (10.171)$$

$$\frac{\text{Re}(p_j(t))}{c} + \frac{\text{Im}(q_j(t))}{b} = \frac{p''}{c}. \quad (10.172)$$

But according to the boundary conditions given in Eq. 10.158, we have

$$\frac{\text{Re}(q_j(t))}{b} + \frac{\text{Im}(p_j(t))}{c} = \frac{q''}{b} \quad (10.173)$$

$$\frac{\text{Re}(p_j(t))}{c} - \frac{\text{Im}(q_j(t))}{b} = \frac{p''}{c}. \quad (10.174)$$

Therefore, we conclude that $\text{Im}(q_j(t)) = \text{Im}(p_j(t)) = 0$. This implies that the saddle point conditions are satisfied only for trajectories that are real throughout their time evolution. For the case of real trajectories there exists only a single trajectory starting from an initial point in phase space, and the phase factor in Eq. 10.153 reduces to

$$\left(S^c + \frac{i\hbar}{2}(|v''|^2 + |u'|^2) \right)\Big|_{q_t, p_t} = \int_0^t [p_{t'}\dot{q}_{t'} - H] dt' - \frac{1}{2}qp \Big|_0^t. \quad (10.175)$$

As shown in Eq. 10.160, at the stationary phase point the radicand of the prefactor takes the form

$$\frac{i}{\hbar} \left| \frac{\partial^2 S}{\partial u' \partial v''} \right|_{q_t, p_t} = \left| \frac{\partial u(t)}{\partial u'} \right|_{q_t, p_t} = \left| \frac{1}{2} \left(\frac{\partial p_t}{\partial p'} + \frac{\partial q_t}{\partial q'} - im\omega \frac{\partial q_t}{\partial p'} - \frac{1}{im\omega} \frac{\partial p_t}{\partial q'} \right) \right|, \quad (10.176)$$

which is immediately recognized as the generalized stability prefactor of the HHKK propagator.

The remaining task is the Gaussian integration using the second derivatives of the exponent. By differentiating Eq. 10.168 we find that

$$\frac{\partial^2 \sigma}{\partial q''^2}\bigg|_{q_t, p_t} = -\frac{3}{2b^2}, \qquad \frac{\partial^2 \sigma}{\partial p''^2}\bigg|_{q_t, p_t} = -\frac{1}{2c^2}, \qquad \frac{\partial^2 \sigma}{\partial p''^2 \partial q''^2}\bigg|_{q_t, p_t} = -\frac{i}{2bc}. \quad (10.177)$$

Together with the factor of $1/(2\pi\hbar)$ appearing in the volume element of the final phase space integration, this integration gives unity. The final result is then

$$\langle x'' | e^{-iHt/\hbar} | x' \rangle = \int \frac{dp' dq'}{\pi} \langle x'' | z_t \rangle \left| \frac{1}{2} \left(\frac{\partial p_t}{\partial p'} + \frac{\partial q_t}{\partial q'} - im\omega \frac{\partial q_t}{\partial p'} - \frac{1}{im\omega} \frac{\partial p_t}{\partial q'} \right) \right|^{1/2}$$

$$\times e^{\frac{i}{\hbar} \int_0^t [p_{t'} \dot{q}_{t'} - H] dt' - \frac{i}{2\hbar}(q_t p_t - q' p')} \langle z' | x' \rangle \quad (10.178)$$

which is the one-dimensional version of the HHKK expression for the propagator, Eq. 10.161.

▶ Exercise 10.30   Show that if one does the integral over dp_0 and dq_0 in the HHKK propagator one recovers the VVG propagator.

The above exercise illustrates an important general principle: the VVG propagator is in some sense the lowest-level semiclassical approximation. It is obtained from the full path integral expression by performing the integral over *every* time slice between t' and t'' by stationary phase. However, it is possible to leave one or more integrals expressed exactly, giving an expression that is in principle closer to the full quantum mechanical expression. This is what the initial value representation does.

As a final comment, the general form for the semiclassical time-correlation function of the two states ψ_A, ψ_B is

$$C_{AB}^{SC}(t) = \int \int d\mathbf{x}' d\mathbf{x}'' \psi_B^*(\mathbf{x}'', 0) K^{SC}(\mathbf{x}'', t; \mathbf{x}', 0) \psi_A(\mathbf{x}', 0).$$

Substituting the HK expression for the propagator into this equation leads to the useful formula

$$C_{AB}^{SC}(t)$$
$$= \int \int d\mathbf{x}' d\mathbf{x}'' \psi_B^*(\mathbf{x}'', 0) \int \int \frac{d\mathbf{p}_0 d\mathbf{q}_0}{(2\pi\hbar)^N} R_{pqt} e^{iS_{pqt}/\hbar} g_\gamma(\mathbf{q}_t, \mathbf{p}_t, \mathbf{x}'') g_\gamma^*(\mathbf{q}_0, \mathbf{p}_0, \mathbf{x}') \psi_A(\mathbf{x}', 0).$$

$$(10.179)$$

Recent numerical results using the HHKK propagator show it to be essentially free from all the problems with semiclassical methods described above, and to give impressive accuracy. Applications include CO_2 photodissociation (Walton, 1995), $H + H_2$ reactive scattering (Garashchuk, 1996) and a variety of system-bath calculations (Shao, 2000; Miller, 2001; Ovchinnikov, 2001).

References

Classical Mechanics

1. M. Gutzwiller, *Chaos in Classical and Quantum Mechanics* (Springer-Verlag, New York, 1990).
2. H. Goldstein, *Classical Mechanics* (Addison-Wesley, Reading, MA 1950).
3. I. C. Percival and D. Richards, *Introduction to Dynamics* (Cambridge University Press, 1982).
4. W. H. Miller, Classical-limit quantum mechanics and the theory of molecular collisions, Adv. Chem. Phys. 25 69 (1974).

Path Integration

5. P. A. M. Dirac, *The Principles of Quantum Mechanics* (Oxford, 1978).
6. R. P. Feynman and A. R. Hibbs, *Quantum Mechanics and Path Integrals* (McGraw-Hill, New York, 1965).
7. L. S. Schulman, *Techniques and Applications of Path Integrals* (Wiley, New York, 1981).

Semiclassical Propagator

8. J. H. van Vleck, The correspondence principle in the statistical interpretation of quantum mechanics, Proc. Natl. Acad. Sci. 14, 178 (1928).
9. M. C. Gutzwiller, Phase-integral approximation in momentum space and the bound states of an atom, J. Math. Phys. 8, 1979 (1967).
10. P. Pechukas, Time-dependent semiclassical scattering theory. I. Potential scattering, Phys. Rev. 181, 166 (1969); P. Pechukas, Time-dependent semiclassical scattering theory. II. Atomic collisions, Phys. Rev. 181, 174 (1969).
11. S. Levit and U. Smilansky, A theorem on infinite products of eigenvalues of Sturm-Liouville type operators, Proc. Am. Math. Soc. 65, 299 (1977).
12. S. Levit and U. Smilansky, A new approach to Gaussian path integrals and the evaluation of the semiclassical propagator, Ann. Phys. 103, 198 (1977).
13. M. V. Berry and K. E. Mount, Semiclassical approximations in wave mechanics, Rep. Prog. Phys. 35, 315 (1972).
14. W. H. Miller, Classical-limit quantum mechanics and the theory of molecular collisions, Adv. Chem. Phys. 25, 69 (1974).
15. M. S. Child, *Semiclassical Mechanics with Molecular Applications* (Clarendon, Oxford, 1991).
16. E. J. Heller, Phase space interpretation of semiclassical theory, J. Chem. Phys. 67, 3339 (1977).
17. E. J. Heller, Wavepacket dynamics and quantum chaology, in *Les Houches, Session LII, Chaos and Quantum Physics*, eds. M. J. Giannoni et al. (Elsevier, 1991).
18. R. G. Littlejohn, The van Vleck formula, Maslov theory, and phase space geometry, J. Stat. Phys. 68, 7 (1992).
19. S. Tomsovic and E. J. Heller, Semiclassical dynamics of chaotic motion: Unexpected long time accuracy, Phys. Rev. Lett. 67, 664, (1991); S. Tomsovic and E. J. Heller, Long-time semiclassical dynamics of chaos: The stadium billiard, Phys. Rev. E 47, 282 (1993).
20. E. J. Heller and S. Tomsovic, Postmodern quantum mechanics, Physics Today (July, 1993).
21. F. Grossmann and E. J. Heller, A semiclassical correlation function approach to barrier tunneling, Chem. Phys. Lett. 241, 45 (1995).

Morse and Maslov Index

22. M. Morse, *Variational Analysis* (Wiley, New York, 1973).

23. J. W. Milnor, *Morse Theory* (Princeton University Press, Princeton, NJ, 1963).

24. V. P. Maslov, *Théorie des Perturbations et Méthodes Asymptotiques* (Paris, Dunod, 1972).

25. V. I. Arnold, *Mathematical Methods of Classical Mechanics*, (Springer, New York, 1978).

26. R. G. Littlejohn and J. M. Robbins, New way to compute Maslov indices, Phys. Rev. A 36, 2953 (1987).

Initial Value Representation

27. W. H. Miller, Classical S-matrix: Numerical application to inelastic collisions, J. Chem. Phys. 53, 3578 (1970); W. H. Miller and T. F. George, Analytic continuation of classical mechanics for classically forbidden collision processes, J. Chem. Phys. 56, 5668 (1972). W. H. Miller, Comment on: Semiclassical time evolution without root searches, J. Chem. Phys. 95, 9428 (1991) and references therein.

28. E. J. Heller, Cellular dynamics: A new semiclassical approach to time-dependent quantum mechanics, J. Chem. Phys. 94, 2723 (1991); E. J. Heller, Reply to Comment on: Semiclassical time evolution without root searches: Comments and perspective, J. Chem. Phys. 95, 9431 (1991).

29. S. Levit and U. Smilansky, The Hamiltonian path integrals and the uniform semiclassical approximations for the propagator, Ann. Phys. 108, 165 (1977).

30. G. Campoleti and P. Brumer, Semiclassical collision theory in the initial value representation: Efficient numerics and reactive formalism, J. Chem. Phys. 96, 5969 (1992); Phys. Rev. A 50, 997 (1994).

31. S. Keshavamurthy and W. H. Miller, Semi-classical correction for quantum-mechanical scattering, Chem. Phys. Lett. 218, 189 (1994); B. W. Spath and W. H. Miller, Semiclassical calculation of Franck-Condon intensities for reactive systems, Chem. Phys. Lett. 262, 486 (1996).

32. M. A. Sepulveda and F. Grossmann, Time-dependent semiclassical mechanics, Adv. Chem. Phys. 96, 191 (1996).

33. K. G. Kay, Integral expressions for the semiclassical time-dependent propagator, J. Chem. Phys. 100, 4377 (1994).

34. K. G. Kay, Numerical study of semiclassical initial value methods for dynamics, J. Chem. Phys. 100, 4432 (1994).

35. K. G. Kay, Semiclassical propagation for multidimensional systems by an initial value method, J. Chem. Phys. 101, 2250 (1994).

36. D. Zor and K. G. Kay, Globally uniform semiclassical expressions for time-independent wave functions, Phys. Rev. Lett. 76, 1990 (1996).

37. K. G. Kay, IVR formulation of Miller's correspondence relations, J. Phys. Chem. A 105, 2535 (2001).

Classical S-Matrix Theory

38. W. H. Miller, Classical-limit quantum mechanics and the theory of molecular collisions, Adv. Chem. Phys. 25, 69 (1974).

39. W. H. Miller, Semiclassical theory of atom-diatom collisions: Path integrals and the classical S matrix, J. Chem. Phys. 53, 1949 (1970).

40. W. H. Miller, Classical S matrix: Numerical application to inelastic collisions, J. Chem. Phys. 53, 3578 (1970).

41. R. A. Marcus, Theory of semiclassical transition probabilities (*S* matrix) for inelastic and reactive collisions, J. Chem. Phys. 54, 3965 (1971).

42. Y. Elran and K. G. Kay, Uniform semiclassical IVR treatment of the S-matrix, J. Chem. Phys. 114, 4362 (2001); Y. Elran and K. G. Kay, Semiclassical IVR calculations of reactive collisions, J. Chem. Phys. 116, 10577 (2002).

Coherent States

43. J. R. Klauder and B.-S. Skagerstam, *Coherent States* (World Scientific, Singapore, 1985).

44. W. H. Louisell, *Quantum Statistical Properties of Radiation* (Wiley, New York, 1973).

45. L. Mandel and E. Wolf, *Optical Coherence and Quantum Optics* (Cambridge University Press, New York, 1994).

46. J. von Neumann, Die Eindeutigkeit der Schrödingerschen Operatoren, Math. Ann. 104, 570 (1931).

47. A. M. Perelomov, On the completeness of a system of coherent states, Teor. Mat. Fiz. 6, 213 (1971). (English translation: Theor. Math. Phys. 11, 156 (1971).)

48. A. M. Perelomov, Generalized coherent states and some of their applications, Sov. Phys. Usp. 20, 703 (1977).

49. V. Bargmann, P. Butera, L. Girardello and J. R. Klauder, On the completeness of the coherent states, Rep. Math. Phys. 2, 221 (1971).

50. M. Boon and J. Zak, Discrete coherent states on the von Neumann lattice, Phys. Rev. B 18, 6744 (1978).

51. M. J. Davis and E. J. Heller, Semiclassical Gaussian basis set method for molecular vibrational wave functions, J. Chem. Phys. 71, 3383 (1979).

Coherent State Path Integration

52. J. R. Klauder, Continuous representations and path integrals, revisited, in *Path Integrals and their Applications in Quantum, Statistical and Solid State Physics, Proceedings of the NATO Advanced Study Institute*, eds. G. J. Papadopoulos and J. T. Devreese (Plenum, New York, 1978).

53. J. R. Klauder, Path integrals and stationary-phase approximations, Phys. Rev. D19, 2349 (1979).

54. E. J. Heller, Wavepacket path integral formulation of semiclassical dynamics, Chem. Phys. Lett. 34, 321 (1975).

55. E. J. Heller, Generalized theory of semiclassical amplitudes, J. Chem. Phys. 66, 5777 (1977).

56. Y. Weissman, Semiclassical approximation in the coherent states representation, J. Chem. Phys. 76, 4067 (1982).

57. J. R. Klauder, Some recent results on wave equations, path integrals, and semiclassical approximations, in *Random Media,* the IMA Volume in Mathematics and its Applications, ed. G. Papanicolauou, (Springer, New York, 1987).

58. F. Grossmann, Semiclassical coherent state path integrals for scattering, Phys. Rev. A, 57, 3256 (1998).

59. M. Baranger, M. A. M. de Aguiar, F. Keck, H. J. Korsch and B. Schellhaas, Semiclassical approximations in phase space with coherent states, J. Phys. A: Math. Gen. 34, 7227 (2001).

Wavepackets and Semiclassical Mechanics

60. E. J. Heller, Time-dependent approach to semiclassical dynamics, J. Chem. Phys. 62, 1544 (1975).

61. E. J. Heller, Classical S-matrix limit of wavepacket dynamics, J. Chem. Phys. 65, 4979 (1976).

62. R. G. Littlejohn, The semiclassical evolution of wavepackets, Phys. Rep. 138, 193 (1986).

63. D. Huber, E. J. Heller and R. G. Littlejohn, Generalized Gaussian wavepacket dynamics, Schrödinger equation, and stationary phase approximation, J. Chem. Phys. 89, 2003 (1988).

64. M. Boiron and M. Lombardi, Complex trajectory method in semiclassical propagation of wavepackets, J. Chem. Phys. 108, 3431 (1998).

65. F. Grossmann, A hierarchy of semiclassical approximations based on Gaussian wavepackets, Comments in Atomic and Molecular Physics 34, 141 (1999).

The Heller–Herman–Kluk–Kay Method

66. E. J. Heller, Frozen Gaussians: A very simple semiclassical approximation, J. Chem. Phys. 75, 2923 (1981).

67. M. F. Herman and E. Kluk, A semiclassical justification for the use of non-spreading wavepackets in dynamics calculations, Chem. Phys. 91, 27 (1984).

68. M. F. Herman, Time reversal and unitarity in the frozen Gaussian approximation for semiclassical scattering, J. Chem. Phys. 85, 2069 (1986).

69. E. Kluk, M. F. Herman and H. L. Davis, Comparison of the propagation of semiclassical frozen Gaussian wave functions with quantum propagation for a highly excited anharmonic oscillator, J. Chem. Phys. 84, 326 (1986).

70. K. G. Kay, Integral expressions for the semiclassical time-dependent propagator, J. Chem. Phys. 100, 4377 (1994).

71. K. G. Kay, Numerical study of semiclassical initial value methods for dynamics, J. Chem. Phys. 100, 4432 (1994).

72. K. G. Kay, Semiclassical propagation for multidimensional systems by an initial value method, J. Chem. Phys. 101, 2250 (1994).

73. A. Walton and D. Manolopoulos, A new semiclassical initial value method for Franck-Condon spectra, Molec. Phys. 87, 961 (1996).

74. A. Walton and D. Manolopoulos, Application of the frozen Gaussian approximation to the photodissociation of CO_2, Chem. Phys. Lett. 244, 448 (1995).

75. F. Grossmann, Time-dependent semiclassical calculation of resonance lifetimes, Chem. Phys. Lett. 262, 470 (1996).

76. S. Garashchuk and D. J. Tannor, Wavepacket correlation function approach to $H_2(v) + H \rightarrow H + H_2(v')$: Semiclassical implementation, Chem. Phys. Lett. 262, 477 (1996).

77. F. Grossmann and A. L. Xavier, Jr., From the coherent state path integral to a semiclassical initial value representation of the quantum mechanical propagator, Phys. Lett. A 243, 243 (1998).

78. J. Shao and N. Makri, Forward-backward semiclassical dynamics in the interaction representation, J. Chem. Phys. 113, 3681 (2000).

79. W. H. Miller, The semiclassical initial value representation: A potentially practical way for adding quantum effects to classical molecular dynamics simulations, J. Phys. Chem. A 105, 2942 (2001).

80. M. Ovchinnikov, V. A. Apkarian and G. A. Voth, Semiclassical molecular dynamics computation of spontaneous light emission in the condensed phase: Resonance Raman spectra, J. Chem. Phys. 114, 7130 (2001).

81. W. H. Miller, On the relation between the semiclassical initial value representation and an exact quantum expansion in time-dependent coherent states, J. Phys. Chem. B 106, 8132 (2002).

82. J. Burant and V. S. Batista, Real time path integrals using the Herman-Kluk propagator, J. Chem. Phys. 116, 2748 (2002).

83. M. S. Child and D. V. Shalashilin, Locally coupled coherent states and Herman-Kluk dynamics, J. Chem. Phys. 118, 2061 (2003).

Chapter 11

Numerical Methods for Solving the Time-Dependent Schrödinger Equation

The formal framework for quantum mechanics is an infinite dimensional Hilbert space. In any numerical calculation, however, it is necessary to truncate the basis at some finite dimension, N. An elegant way of thinking about the dynamics on this reduced Hilbert space is that the reduced space *could*, in principle, be the complete space for a different quantum mechanical problem. Therefore, why not demand that all the usual formalism of quantum mechanics apply on this reduced Hilbert space, for example, orthogonality relations, completeness relations, unitary transformations and so forth. In favorable cases we may even expect that commutator relations and the uncertainty principle may take the same form as on the full Hilbert space. The key to providing a rigorous formulation of the quantum dynamics on the reduced Hilbert space is to realize that the truncation to N basis functions can be described by a projection operator, P_N, projecting onto the space spanned by the basis. The Hamiltonian viewed as a mapping on this subspace (see Section 11.1) becomes

$$H_N = P_N H P_N = P_N T P_N + P_N V P_N \tag{11.1}$$

and the Time-Dependent Schrödinger Equation becomes

$$i\hbar \frac{\partial \Psi_N}{\partial t} = H_N \Psi_N, \tag{11.2}$$

where $\Psi_N(t=0) = P_N \Psi(t=0)$.

However, the most prevalent methods for solving the TDSE today do not use conventional orthogonal bases, but rather grid representations: a continuous wavefunction is represented in terms of a discrete set of time-evolving complex amplitudes at a set of grid points. What relationship, if any, is there between the representation in terms of complex amplitudes at grid points and a representation in terms of a conventional basis of orthogonal functions? As will be shown in this chapter, the amplitudes at the grid points can be interpreted as the coefficients of localized basis functions. The conventional basis of orthogonal functions is referred to as a "spectral" basis, while the basis of localized functions is called a "pseudospectral" basis. If these two bases are related by a unitary transformation, the projection operators associated with the two bases are identical. As a result, completeness relations on the subspace that can be expressed in terms of the spectral basis can be expressed in terms of the pseudospectral basis as well. It is usually convenient to evaluate the kinetic energy, T_N, in a spectral basis, and the potential energy, V_N, in a pseudospectral basis. Ultimately, a single representation has to be chosen for $H_N = T_N + V_N$, but this is easily accomplished provided the unitary transformation between the two representations is known. The equivalence of the projection operators in the two representations guarantees that the reduced Hilbert space on which H is represented is defined consistently. Two important examples of this combination of spectral and pseudospectral representation are the Fourier method and the DVR (discrete variable representation) method, both of which will be described in detail in this chapter.

In addition to the projection onto a subspace of the Hilbert space associated with P_N, there is a second source of numerical error, coming from the approximate evaluation of the operators within this reduced Hilbert space. For the systems of interest in this chapter, T_N can be calculated *exactly* in the spectral basis, using derivative relations for the orthogonal functions; however, the calculation of V_N is generally approximate. Consider the evaluation of V_N in the spectral representation. Each element involves integrals of the form $\int \phi_m^*(x) V(x) \phi_n(x)\, dx$, with $V(x)$, ϕ_m and ϕ_n normally quite complicated, and hence in general the integral cannot be evaluated analytically. However, if V_N is represented in a pseudospectral basis, the matrix elements now involve integrals of products of the complicated $V(x)$ with *localized* basis functions. The simple procedure of essentially treating the pseudospectral basis functions as if they were δ-functions turns out to be remarkably accurate, leading to a diagonal representation of V_N whose elements are simply the value of $V(x)$ at the position of the associated δ-function. As we shall show in this chapter, this approximation can be described formally by the replacement

$$P_N V(\hat{x}) P_N = V_N(\hat{x}) \rightarrow V(\hat{x}_N) = V(P_N \hat{x} P_N). \tag{11.3}$$

The representation of T_N, while not diagonal or banded in the pseudospectral basis, is still straightforward to calculate by applying an appropriate unitary transformation to the T_N in the spectral representation. Combining the two approximations to the Hamiltonian—the projection onto a subspace and the evaluation of the matrix elements—we have

$$H_N \approx T_N(\hat{p}) + V_N(\hat{x}_N). \tag{11.4}$$

Finally, there is a third source of numerical error. This arises because the evaluation of the evolution operator, $U(t)\psi = e^{-iHt/\hbar}\psi = e^{-i(\hat{p}^2/2m+V(\hat{x}))t/\hbar}\psi$, is more complicated than evaluation of $H\psi$, since in the former the x and p operators are entangled through the exponentiation. There are several major approaches to evaluating the evolution operator. One approach is to construct the H matrix and diagonalize it, that is, $e^{-iHt/\hbar}\psi = U e^{-i\lambda t/\hbar} U^{-1}\psi$. Then propagation is simple, but the price is that diagonalization is expensive. Alternatively, the H matrix can be used to integrate the Time-Dependent Schrödinger Equation via repeated evaluation of $H\psi$ to find $\dot{\psi}$. This can be accomplished via several methods, for example, by rewriting the Time-Dependent Schrödinger Equation in a form that displays the symplectic (i.e., canonical) structure of classical mechanics and employing well-developed integrators for the classical problem. A second approach, called the Split Operator method, approximates $e^{-i(\hat{p}^2/2m+V(\hat{x}))t/\hbar}\psi \approx e^{-i(\hat{p}^2 t/2m\hbar)} e^{-iV(\hat{x})t/\hbar}\psi$. A third class of approaches evaluates $H\psi$, and then represents $e^{-iHt/\hbar}\psi = \sum_n a_n P_n(H)\psi$, avoiding diagonalization; $P_n(H)$ is some polynomial of the Hamiltonian operator, whose operation on ψ can be evaluated by iterative operations of H on ψ. Many choices of $P_n(H)$

are possible; the different possibilities can be divided into two types, those that specify the polynomials in advance (uniform methods) and those that do not (nonuniform methods).

In addition to the formal treatment of numerical methods, and the error incurred, one of the goals of this chapter is to provide a practical guide to the most commonly used algorithms. The reader will find the central equations, for example, for the DVR method, the Fourier Grid Hamiltonian method, the FFT method, the Split Operator method, and the Chebyshev method in Sections 11.5–11.7.

11.1   Spectral Projection and Collocation

11.1.1   Spectral Projection

Preliminaries

A spectral representation refers to the representation of a wavefunction and the operators that act on it in terms of a basis set of orthogonal functions. Because any numerical representation is inherently finite, the infinite set of orthogonal functions must necessarily be truncated at a finite value, N. This truncation can be expressed in terms of a projection operator,

$$P_N = \sum_{n=1}^{N} |\phi_n\rangle\langle\phi_n|, \tag{11.5}$$

which projects the dynamics onto an N-dimensional subspace of the infinite Hilbert space.

Consider, for example, constructing a matrix representation of the Hamiltonian,

$$H = \hat{p}^2/2m + V(\hat{x}), \tag{11.6}$$

using a set of N orthogonal basis functions. Assume also that the matrix elements $\langle\phi_m|H|\phi_n\rangle$ are known exactly. Truncation of the matrix H at $N \times N$ can be expressed in terms of the projection operator P_N as

$$H_N = P_N H P_N = T_N + V_N. \tag{11.7}$$

Defining $Q_N = 1 - P_N$, the TDSE may be written as a set of two coupled differential equations:

$$i\hbar\frac{\partial\Psi_N}{\partial t} = P_N H P_N \Psi_N + P_N H Q_N \Psi_\perp \tag{11.8}$$

$$i\hbar\frac{\partial\Psi_\perp}{\partial t} = Q_N H P_N \Psi_N + Q_N H Q_N \Psi_\perp, \tag{11.9}$$

where $\Psi_N = P_N\Psi$ and $\Psi_\perp = Q_N\Psi = \Psi - \Psi_N$. We will consider in this chapter only numerical methods that neglect the contribution from $\Psi_\perp$. In the numerical analysis literature this is known as the Galerkin approximation (see, e.g., Gottlieb, 1977) and corresponds to the solution of the time-dependent variational formulation of the Schrödinger equation in a finite basis described in Section 9.6. If the contribution of $\Psi_\perp$ is neglected, the TDSE takes the form

$$i\hbar\frac{\partial\Psi_N}{\partial t} = P_N H P_N \Psi_N, \tag{11.10}$$

and in integral form becomes

$$\Psi_N(t) = e^{-iH_N t/\hbar}\Psi_N(0) = e^{-iP_N H P_N t/\hbar}\Psi_N(0) \approx P_N e^{-iHt/\hbar}P_N\Psi(0). \quad (11.11)$$

The approximation in Eq. 11.11 is exact if $[P_N, H] = 0$. If the basis functions ϕ_n were the exact eigenfunctions of H, the commutator $[P_N, H]$ would indeed be 0; to the extent that the basis functions are close to the exact eigenfunctions the approximation is a reasonable one. Diagonalizing H_N gives energies E_n^N and eigenstates ψ_n^N, which approximate the true energies E_n and eigenstates ψ_n. Propagation of the wavepacket using these approximate energies gives

$$\Psi_N(t) = \sum_{n=1}^{N} a_n^N \psi_n^N e^{-iE_n^N t/\hbar}. \quad (11.12)$$

Equation 11.12 is an example of a "spectral" propagation method, "spectral" referring to a basis of orthogonal eigenstates.

▶ **Exercise 11.1**   Derive Eq. 11.12.

Phase Space Interpretation of the Reduced Hilbert Space Projection Operator

Every spectral method, and hence every pseudospectral method (Section 11.2 and Eq. 11.49), can be given a phase space interpretation. Once a choice of basis functions is made this defines a projection operator for a subspace of the Hilbert space,

$$P_N = \sum_{n=1}^{N} |\phi_n\rangle\langle\phi_n|. \quad (11.13)$$

The Wigner transform of the projection operator gives a representation of the region of phase space spanned by the basis

$$P_N^W(p, q) = \sum_{n=1}^{N} \frac{1}{2\pi\hbar} \int_{-\infty}^{\infty} \phi_n^* \left(q - \frac{s}{2}\right) \phi_n \left(q + \frac{s}{2}\right) e^{ips/\hbar}\, ds. \quad (11.14)$$

If the basis spans the lowest N eigenstates of a Hamiltonian, one expects intuitively that the Wigner function of the projection operator will be a function of uniform density 1 in regions below $H(p, q) \approx E_n$ and of uniform density 0 in regions above $H(p, q) \approx E_n$, with a smooth transition region in between, where $H(p, q)$ is the *classical* Hamiltonian function. Numerical tests show that this is approximately true, although oscillations persist in the interior region of phase space as may be seen in Figure 11.1 (Poirier, 2000).

The Spectral Projector Acting on a Wavefunction

We now consider the action of a spectral projection operator on a wavefunction. Although these results are simple and straightforward we present them here in order to contrast them with the action of the collocation projection operator in Section 11.1.2, which is most easily understood in terms of its action on a wavefunction.

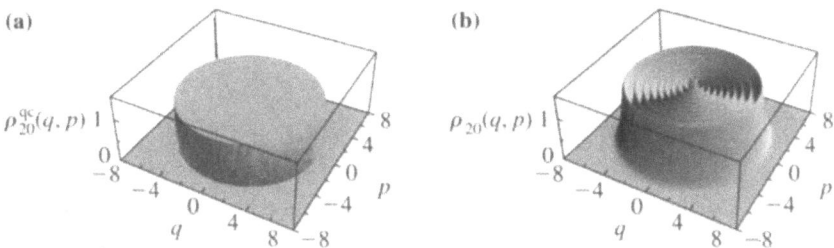

Figure 11.1 Phase space distributions $\rho(p, q)$ corresponding to the projection operator $P_N = \sum_{n=1}^{N} |\phi_n\rangle\langle\phi_n|$ for the one-dimensional harmonic oscillator $H(p, q) = (q^2 + p^2)/2$ with $N = 20$. (a) quasiclassical distribution; (b) exact quantum distribution based on the Wigner representation of the projection operator, Eq. 11.14. Adapted from Poirier (2000).

Consider an arbitrary wavefunction $\psi(x)$. We may decompose $\psi(x)$ in terms of an infinite, orthonormal basis $\{\phi_n\}$ as follows:

$$\psi(x) = \sum_{n=1}^{\infty} a_n \phi_n(x), \tag{11.15}$$

where

$$\int \phi_m^*(x)\phi_n(x)dx = \delta_{mn}, \qquad a_n = \int \phi_n^*(x)\psi(x)\,dx, \qquad m, n = 1, \ldots, \infty. \tag{11.16}$$

We define the spectral projection operator P_N such that

$$\tilde{\psi}(x) = P_N \psi(x) = \sum_{n=1}^{N} a_n \phi_n(x), \tag{11.17}$$

with a_n given by Eq. 11.16 for $n = 1, \ldots, N$. It is easy to show that

$$P_N \phi_n(x) = \begin{cases} \phi_n(x) & n = 1, \ldots, N \\ 0 & n = N + 1, \ldots. \end{cases} \tag{11.18}$$

It follows that $P_N^2 = P_N$:

$$P_N^2 \psi(x) = P_N \sum_{n=1}^{N} a_n \phi_n(x) = \sum_{n=1}^{N} a_n P_N \phi_n(x) = \sum_{n=1}^{N} a_n \phi_n(x) = P_N \psi(x). \tag{11.19}$$

▶ **Exercise 11.2**   Show that $P_N = \sum_{n=1}^{N} |\phi_n\rangle\langle\phi_n|$ satisfies Eq. 11.17 with a_n given by Eq. 11.16. We have used a more cumbersome notation in Eqs. 11.16 and 11.17 in anticipation of our discussion of the collocation projector.

11.1.2   Collocation

Definition of the Collocation Projector

Collocation refers to the numerical solution of a set of equations such that the result is exact at a discrete set of points. One can associate with collocation a projection operator defined as follows:

$$\tilde{\psi}(x) = P_N \psi(x) = \sum_{n=1}^{N} b_n \phi_n(x), \tag{11.20}$$

where the b_n are determined by the condition that $\tilde{\psi}(x) = \psi(x)$ at the N collocation points $\{x_i\}$, $i = 1, \ldots, N$, that is,

$$\tilde{\psi}(x_i) = P_N \psi(x_i) = \sum_{n=1}^{N} b_n \phi_n(x_i) = \psi(x_i) \quad i = 1, \ldots, N. \tag{11.21}$$

Equation 11.21 defines an interpolation scheme at points other than the collocation points via the relation

$$\tilde{\psi}(x) = \sum_{n=1}^{N} b_n \phi_n(x) \approx \psi(x). \tag{11.22}$$

Consider now the action of P_N^2:

$$P_N^2 \psi(x_i) = P_N \tilde{\psi}(x_i) = \sum_{m=1}^{N} c_m \phi_m(x_i), \tag{11.23}$$

where Eq. 11.23 defines the $\{c_n\}$. But from Eq. 11.21 $P_N \tilde{\psi}(x_i) = \tilde{\psi}(x_i)$ and hence

$$P_N^2 \psi(x_i) = P_N \tilde{\psi}(x_i) = \tilde{\psi}(x_i) = \sum_{n=1}^{N} b_n \phi_n(x_i). \tag{11.24}$$

For Eqs. 11.23 and 11.24 to be consistent we must have $c_n = b_n$, $n = 1, \ldots, N$ and hence $P_N^2 = P_N$.

Equation 11.21 can be written in matrix form as

$$\bar{\Psi} = \Phi \mathbf{b} \tag{11.25}$$

where $\bar{\Psi}_j = \psi(x_j)$ and the matrix elements are $\Phi_{jn} = \phi_n(x_j)$. Provided that the $\phi_n(x_j)$ are linearly independent, the solution is

$$\mathbf{b} = \Phi^{-1} \bar{\Psi}. \tag{11.26}$$

One of the principal properties of the collocation projection is that it obeys a composition property

$$\tilde{\psi}_1(x) \tilde{\psi}_2(x) = \widetilde{\psi_1 \psi_2}(x). \tag{11.27}$$

▶ **Exercise 11.3**   Verify Eq. 11.27.

Contrasting the Spectral and Collocation Projectors: Orthogonal versus Nonorthogonal Projection

We now observe that P_N for both spectral projection and collocation project onto the same subspace of the Hilbert space, which we may call $\mathcal{H}_N \equiv \mathrm{span}\{\phi_1, \ldots, \phi_N\}$. The two projectors differ not in the space onto which they project but in the state that they produce in the reduced subspace, which manifests itself in terms of the difference in the coefficients $\{a_n\}$

versus $\{b_n\}$. The difference in the state produced can be given a geometrical interpretation as follows. Define the quantity

$$\chi = \psi - \tilde{\psi}. \tag{11.28}$$

For spectral projection we have

$$\chi(x) = \sum_{n=1}^{\infty} a_n \phi_n(x) - \sum_{n=1}^{N} a_n \phi_n(x) = \sum_{n=N+1}^{\infty} a_n \phi_n(x). \tag{11.29}$$

It is easily seen that $\langle \chi | \tilde{\psi} \rangle = 0$:

$$\langle \chi | \tilde{\psi} \rangle = \int \left(\sum_{n=N+1}^{\infty} a_n^* \phi_n^*(x) \right) \left(\sum_{m=1}^{N} a_n \phi_m(x) \right) dx = 0. \tag{11.30}$$

The orthogonality of the residue χ to the projected part of the state, $\tilde{\psi}$, shows that the spectral projector is an *orthogonal projector*. Now consider the collocation projector. In this case the residue becomes

$$\chi(x) = \sum_{n=1}^{\infty} a_n \phi_n(x) - \sum_{n=1}^{N} b_n \phi_n(x). \tag{11.31}$$

Now $\langle \chi | \tilde{\psi} \rangle \neq 0$:

$$\langle \chi | \tilde{\psi} \rangle = \int \left(\sum_{n=1}^{N} (a_n - b_n)^* \phi_n^*(x) \right) \left(\sum_{m=1}^{N} b_n \phi_m(x) \right) dx \neq 0. \tag{11.32}$$

The collocation projector is a *nonorthogonal projector*. The fact that two projectors can project a vector onto the same subspace but produce two different vectors is shown geometrically in Figure 11.2.

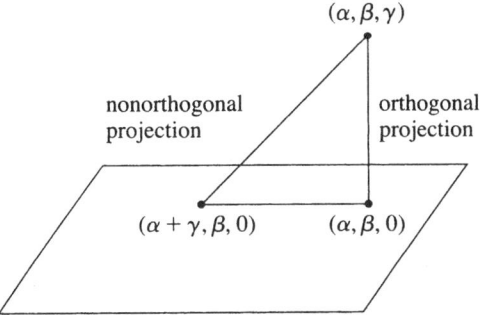

Figure 11.2 Geometrical interpretation of orthogonal and nonorthogonal projection from a three-dimensional to a two-dimensional Euclidean space. Note that both the orthogonal and nonorthogonal projectors have no z-component in the final vector. In the orthogonal projection, the z-component in no way influences the value of x or y in the final vector, while in the nonorthogonal projection it does.

► Exercise 11.4

a. Show that both

$$\mathbf{P}_1 = \begin{pmatrix} 1 & 0 & 0 \\ 0 & 1 & 0 \\ 0 & 0 & 0 \end{pmatrix} \quad \text{and} \quad \mathbf{P}_2 = \begin{pmatrix} 1 & 0 & 1 \\ 0 & 1 & 0 \\ 0 & 0 & 0 \end{pmatrix} \tag{11.33}$$

satisfy the relationship $\mathbf{P}^2 = \mathbf{P}$, but only the first is Hermitian. The Hermitian character of the matrix is the signature of an orthogonal projection.

b. Calculate the action of $\mathbf{P}_1$ and $\mathbf{P}_2$ on the column vector $(\alpha, \beta, \gamma)^T$. What happens to the third component of the *final* vector under orthogonal projection? Under nonorthogonal projection? What happens to the third component of the *initial* vector under orthogonal projection? Under nonorthogonal projection? Interpret your result geometrically using Figure 11.2.

► Exercise 11.5

a. Show that if $\psi(x)$ lies within the projected Hilbert space, that is, $\psi(x) = \sum_{n=1}^{N} a_n \phi_n(x)$, then $\tilde{\psi}(x)$ for spectral projection and for collocation give the same result.

b. Show that the $\{a_n\}$ in both spectral projection and collocation satisfy the relationship

$$a_n = \int \phi_n^*(x) \tilde{\psi}(x) \, dx \tag{11.34}$$

(note that $\tilde{\psi}(x)$ and not $\psi(x)$ appears on the RHS).

Orthogonal Collocation

Is it possible for a projector that satisfies the collocation conditions exactly to also be an orthogonal projector? Consider the case that the expansion functions $\phi_n(x)$ obey a set of discrete orthogonality relations at the collocation points:

$$\sum_{j=1}^{N} \phi_m^*(x_j) \phi_n(x_j) \Delta_j = \delta_{mn}, \quad m, n = 1, \ldots, N. \tag{11.35}$$

Equation 11.35 represents a set of orthogonality relations that are analogous to those of Eq. 11.16, with the integral replaced by a sum over values at the collocation points, x_j, and dx replaced by a generally j-dependent weight factor Δ_j. Despite the flexibility in choosing the points $\{x_j\}$ and the weights $\{\Delta_j\}$, Eq. 11.35 is nontrivial to satisfy: the same points and weights must simultaneously satisfy all the orthogonality relations for $1 \leq m, n \leq N$.

Equation 11.35 allows for direct inversion of the coefficients b_n in Eq. 11.21 by left multiplying by $\phi_m^*(x_j)\Delta_j$ and summing over j:

$$b_n = \sum_{j=1}^{N} \psi(x_j) \phi_n^*(x_j) \Delta_j. \tag{11.36}$$

With the $\{b_n\}$ chosen according to Eq. 11.36 the conditions for collocation are satisfied precisely. Yet clearly Eq. 11.36 is also a discrete approximation to the relation $a_n = \int \phi_n^*(x)\psi(x)\,dx$, which according to Eq. 11.16 is the condition for an orthogonal projection! Because Eq. 11.36 is only an approximation to the integral Eq. 11.16, it does not exactly define an orthogonal projection; nevertheless, because it provides a close approximation we will refer to schemes satisfying the property 11.35 (and hence 11.36) as orthogonal collocation schemes.

Orthogonal collocation schemes can be recast in a simple and powerful way. We begin by defining

$$\Phi_n(x_j) \equiv \sqrt{\Delta_j} \phi_n(x_j). \tag{11.37}$$

With this definition Eq. 11.35 can be written as

$$\sum_{j=1}^{N} \Phi_m^*(x_j) \Phi_n(x_j) = \delta_{mn} \qquad 1 \le m, n \le N. \tag{11.38}$$

Defining $\Phi_n(x_j) \equiv \Phi_{jn}$ we may rewrite Eq. 11.38 in matrix notation as

$$\Phi^\dagger \Phi = 1. \tag{11.39}$$

Equation 11.39 indicates that the transformation matrix Φ is unitary. The unitarity of the transformation implies a second relation:

$$\Phi \Phi^\dagger = 1, \tag{11.40}$$

or in component form

$$\sum_{n=1}^{N} \Phi_n(x_i) \Phi_n^*(x_j) = \delta_{ij} \qquad 1 \le i, j \le N. \tag{11.41}$$

Despite the apparent similarity between Eq. 11.41 and Eq. 11.38, the physical significance is completely different. Equation 11.41 is effectively an orthogonality relation for the different grid points, the analog of the relation

$$\langle x'|x \rangle = \sum_{n=1}^{\infty} \langle x'|n \rangle \langle n|x \rangle = \delta(x - x') \tag{11.42}$$

on the infinite Hilbert space. Below, we will refer to Φ as the orthogonal collocation matrix. Relations of the type in Eq. 11.39 will be referred to as basis orthogonality relations and relations of the type in Eq. 11.40 as grid orthogonality relations. The grid orthogonality relation, Eq. 11.41, is the starting point for our discussion of the pseudospectral basis in the next section.

11.2 The Pseudospectral Basis

11.2.1 Definition of the Pseudospectral Basis

Although we have defined the projection operator P_N in Eq. 11.5 in terms of the basis $\{\phi_n\}$, P_N is an operator and therefore formally basis independent. This hints that P_N may be represented in a basis other than the spectral basis without any loss of accuracy, provided that the new basis is related to the spectral basis by a unitary transformation on the reduced Hilbert space.

An important special case is the representation of P_N in a pseudospectral basis—a basis of spatially *localized* functions, each one concentrated around a different spatial center. Collectively, the N pseudospectral basis functions $\{\theta_j\}$ span *exactly* the same Hilbert space as the N spectral basis functions from which they derive via a unitary (or orthogonal) transformation. Hence the projector in the pseudospectral basis is fundamentally identical to the projector in the spectral basis,

$$P_N = \sum_{n=1}^{N} |\phi_n\rangle\langle\phi_n| = \sum_{j=1}^{N} |\theta_j\rangle\langle\theta_j|.$$

The major properties of the pseudospectral representation will be derived in this section.

It is interesting to see the pseudospectral basis emerge from the theory of orthogonal collocation. If we replace $\Phi_n(x_i) \to \phi_n(x)$ in only the first factor in Eq. 11.41 we obtain

$$\sum_{n=1}^{N} \phi_n(x)\Phi_n^*(x_j) \equiv \theta_j(x), \tag{11.43}$$

where we have defined the pseudospectral basis functions, $\{\theta_j(x)\}$. Because the LHS of Eq. 11.43 is in some sense close to the LHS of Eq. 11.41, we expect the functions $\{\theta_j(x)\}$ to be close to δ-functions. In fact, the functions $\{\theta_j\}$ are localized, each one around a different value of x_j. Moreover, they satisfy the modified Kronecker δ-function property

$$\theta_j(x_i) = \Delta_j^{-1/2}\delta_{ij}, \tag{11.44}$$

as follows immediately from Eq. 11.43 together with Eqs. 11.37 and 11.41. Illustrations of the functions $\theta_j(x)$ may be seen later in the chapter in Figures 11.4 and 11.6.

Equation 11.43 can be written in matrix–vector notation as

$$\boldsymbol{\Phi}^\dagger\boldsymbol{\phi}(x) = \boldsymbol{\theta}(x), \tag{11.45}$$

showing that the functions $\{\theta_j(x)\}$ can be viewed as an alternative basis to the original basis $\{\phi_n\}$. Because the basis $\{\theta_j\}$ and the original basis $\{\phi_n\}$ are related by a unitary transformation $\boldsymbol{\Phi}^\dagger$, they project onto the same subspace of the Hilbert space. Thus, every unitary (or in the case of real basis functions, orthogonal) collocation relation of the form in Eq. 11.38 implies the existence of a set of localized basis functions that span the same space as the original orthogonal functions; the form of the localized basis is determined completely by the points and weights that enter into Eq. 11.38, and the primary basis of orthogonal functions $\{\phi_n\}$.

11.2.2 Completeness and Orthogonality of Pseudospectral Basis Functions

The localized basis functions θ_n have properties of completeness and orthogonality entirely analogous to those of the original basis, ϕ. To show this, we first invert Eq. 11.43 by multiplying by $\phi_m^*(x)$ and integrating over the domain. This gives

$$\Phi_n^*(x_j) = \langle\phi_n|\theta_j\rangle. \tag{11.46}$$

The grid orthogonality relation, Eq. 11.41, then becomes

$$\sum_{n=1}^{N} \Phi_n(x_i)\Phi_n^*(x_j) = \delta_{ij} = \sum_{n=1}^{N} \langle\theta_i|\phi_n\rangle\langle\phi_n|\theta_j\rangle = \langle\theta_i|\theta_j\rangle. \tag{11.47}$$

Similarly, the basis orthogonality relation, Eq. 11.38, becomes

$$\sum_{j=1}^{N} \Phi_m^*(x_j)\Phi_n(x_j) = \delta_{mn} = \sum_{j=1}^{N} \langle\phi_m|\theta_j\rangle\langle\theta_j|\phi_n\rangle. \tag{11.48}$$

Equation 11.47 is a statement of the orthonormality of the different θ basis functions. This orthogonality may be understood qualitatively, given that each θ function is peaked around a different value of x_i and vanishes at all other values x_j, but the precise orthogonality includes the cancellation of the oscillatory portions of the θ functions away from the peak as well. Equation 11.48 is a completeness (or closure) relationship with respect to summation over all the grid points j. The property of completeness says in essence that representation onto the localized θ functions represents projection onto the exact same subspace of Hilbert space as the projection defined by the original basis of orthogonal functions. Both the completeness and orthogonality of the θ functions follow immediately from the fact that the transformation matrix with elements $\Phi_n(x_i)$ is unitary. Thus, the formal statement of the closure relation is

$$P_N = \sum_{n=1}^{N} |\phi_n\rangle\langle\phi_n| = \sum_{j=1}^{N} |\theta_j\rangle\langle\theta_j|. \tag{11.49}$$

Consider applying the projection operator P_N to a Dirac delta function:

$$P_N\delta(x - x_i) = \sum_{j=1}^{N} \langle x|\theta_j\rangle\langle\theta_j|x_i\rangle = \Delta_i^{-1/2}\theta_i(x), \tag{11.50}$$

where we have used Eq. 11.44. Equation 11.50 indicates that the functions $\{\theta_i\}$ can be viewed as "smeared" delta functions that approach actual delta functions in the limit that the size of the basis goes to infinity.

11.2.3 The Collocation Projector in the Pseudospectral Basis

We now return to the collocation relation, Eq. 11.20. Note that Eq. 11.43 can be inverted to give

$$\phi_n(x) = \sum_{i=1}^{N} \Phi_n(x_i)\theta_i(x), \tag{11.51}$$

or in matrix–vector notation

$$\mathbf{\Phi}\boldsymbol{\theta}(x) = \boldsymbol{\phi}(x). \tag{11.52}$$

Using this relation we can rewrite the collocation relation as

$$\tilde{\psi}(x) = \sum_{n=1}^{N} b_n \phi_n(x) = \sum_{i=1}^{N} \left\{ \sum_{n=1}^{N} b_n \Phi_n(x_i) \right\} \theta_i(x) \tag{11.53}$$

$$= \sum_{i=1}^{N} \psi(x_i) \theta_i(x) \Delta_i^{1/2} \approx \psi(x), \tag{11.54}$$

where we have used Eqs. 11.21 and 11.37. At the collocation points $\{x_i\}$ we have

$$\tilde{\psi}(x_i) = \sum_{i=1}^{N} \psi(x_i) \theta_i(x_i) \Delta_i^{1/2} = \psi(x_i). \tag{11.55}$$

Comparing Eq. 11.55 with Eq. 11.21 we see that it is a collocation formula with the basis functions $\{\theta_i\}$ replacing $\{\phi_n\}$. The coefficients of the basis functions are the values of the exact function $\psi(x)$ at the collocation points $\{x_i\}$—the same points as in the collocation expression using $\{\phi_n\}$—with a weight factor of $\Delta_i^{1/2}$.

Multiplying both sides of Eq. 11.54 by $\theta_j(x)$ and integrating we obtain

$$\langle \theta_j | \tilde{\psi} \rangle = \sum_{i=1}^{N} \psi(x_i) \langle \theta_j | \theta_i \rangle \Delta_i^{1/2} = \psi(x_j) \Delta_j^{1/2}. \tag{11.56}$$

The significance of Eq. 11.56 is that for any state $\tilde{\psi}$ that is in the space $\mathcal{H}_N$, its inner product with the θ function centered at point x_i is just given by the value of $\tilde{\psi}$ at the point x_i multiplied by a number that depends on the point but not on the function $\tilde{\psi}$.

With the aid of Eq. 11.54 we can derive a simple formula for the collocation projector in Dirac notation. We have

$$\langle x | \tilde{\psi} \rangle = \sum_{i=1}^{N} \psi(x_i) \theta_i(x) \Delta_i^{1/2} \tag{11.57}$$

$$= \sum_{i=1}^{N} \langle x | \theta_i \rangle \langle x_i | \psi \rangle \Delta_i^{1/2}. \tag{11.58}$$

But $\langle x | \tilde{\psi} \rangle = \langle x | P_N | \psi \rangle$ so we see from Eq. 11.58 that the collocation projector may be written formally as

$$P_N = \sum_{i=1}^{N} |\theta_i\rangle \langle x_i | \Delta_i^{1/2}. \tag{11.59}$$

▶ Exercise 11.6

a. Show that P_N given by Eq. 11.59 satisfies the properties $P_N^2 = P_N$ but $P_N^\dagger \neq P_N$, consistent with P_N for collocation being a nonorthogonal projector.

b. Use Eq. 11.59 to show the composition property of collocation projection $\tilde{\psi}_1 \tilde{\psi}_2 = \widetilde{\psi_1 \psi_2}$ (cf. Exercise 11.3).

11.2.4   Evaluation of Integrals by Sampling at Collocation Points

Now consider the inner product $\langle \tilde{\psi}_1 | \tilde{\psi}_2 \rangle$. Using the form for the collocation projector, Eq. 11.59, we may obtain a simple result about the exactness of evaluating this integral by evaluating the integrand at the collocation points. In particular,

$$\langle \tilde{\psi}_1 | \tilde{\psi}_2 \rangle = \langle \tilde{\psi}_1 | P_N^\dagger P_N | \tilde{\psi}_2 \rangle = \sum_{i,j=1}^{N} \langle \tilde{\psi}_1 | x_j \rangle \langle \theta_j | \theta_i \rangle \langle x_i | \tilde{\psi}_2 \rangle \Delta_i^{1/2} \Delta_j^{1/2} \tag{11.60}$$

$$= \sum_{i} \langle \tilde{\psi}_1 | x_i \rangle \langle x_i | \tilde{\psi}_2 \rangle \Delta_i = \sum_{i=1}^{N} \tilde{\psi}_1^{*}(x_i) \tilde{\psi}_2(x_i) \Delta_i, \tag{11.61}$$

where we have used Eqs. 11.59 and 11.47. Equation 11.61 says that the evaluation of the inner product of two functions that lie within the Hilbert space $\mathcal{H}_N$ is exact if the integrand is evaluated at the collocation points and summed with the weighting factor Δ_i. The key step in the derivation is inserting $P_N^\dagger$ and P_N in the first step; this is permissible since both $\tilde{\psi}_1$ and $\tilde{\psi}_2$ lie within the space $\mathcal{H}_N$ and for functions within this space the collocation projector is transparent. One may show that Eq. 11.61 implies that

$$\int f(x)\,dx = \sum_{i=1}^{N} f(x_i) \Delta_i \tag{11.62}$$

for $f(x)$ in $\mathcal{H}_{2N-1}$. Equation 11.62 is a remarkable result. It indicates that with just N quadrature points and weights we can evaluate exactly the integral of any function that exists within the associated $2N - 1$ dimensional Hilbert space.

11.3   Gaussian Quadrature

A major component of our exposition of pseudospectral methods and their accuracy was the assumption of an orthogonal collocation relation of the form of Eq. 11.35. But we have not discussed how one might go about finding the points $\{x_i\}$ and weights $\{\Delta_i\}$ that will simultaneously make the set of N^2 orthonormality relations hold. For a Fourier basis the points and weights can be guessed intuitively: the points are evenly spaced and the weights at every point are equal. However, for a general basis of orthogonal functions it is not obvious how to find the points and weights. We will present here a method to find the points and weights for a basis of orthogonal functions based on the classical polynomials. It turns out that, in addition to allowing the points and weights for orthogonal collocation to be found in a fully automatic way, there is an additional bonus: the evaluation of inner products using the collocation points and weights is exact provided that the integrand lies within $\mathcal{H}_{2N}$, rather than just $\mathcal{H}_{2N-1}$.

The property that an integral can be evaluated exactly for an integrand lying within a $2N$-dimensional Hilbert space using only N points and weights is the characteristic feature of integration by Gaussian quadrature integration. In what follows, we will show explicitly the connection between the method presented for finding orthogonal collocation points and weights and the theory of Gaussian quadrature. In Section 11.3.1 we will begin by assuming the existence of Gaussian quadrature points and weights. In Section 11.3.3 we will show that the theory of Gaussian quadrature can actually be derived from first principles from the properties of the pseudospectral functions.

11.3.1   Quadrature Points and Weights by Matrix Diagonalization

We begin with a brief statement of the main results of the theory of Gaussian quadrature. Consider approximating the integral of a function $f(x)$ by a finite sum:

$$\int_a^b w(x)f(x)\,dx \approx \sum_{i=1}^N W_i f(x_i), \tag{11.63}$$

where $w(x)$ is a positive weight function and the N values $\{W_i\}$ are the weights to be given to the N function values $f(x_i)$. If the points are not fixed there are $2N$ undetermined parameters (the $\{W_i\}$ and the $\{x_i\}$), precisely the number of parameters needed to determine a polynomial of degree $2N - 1$ (the extra parameter is needed to determine the constant term in the polynomial). In Gaussian quadrature, the $\{x_i\}$ and the $\{W_i\}$ are chosen such that the sum in Eq. 11.63 yields the integral exactly when $f(x)$ is a polynomial of degree $2N - 1$ or less. As we shall see below, the $\{x_i\}$ are zeros of a polynomial p_N of degree N orthogonal (with respect to the weight function $w(x)$) to the space of all lower-degree polynomials (see Eq. 11.96). Tables of points and weights for the classical orthogonal polynomials for different numbers of quadrature points are given by Abramowitz (1964), for example.

▶ **Exercise 11.7**   The Lagrange interpolating polynomials are defined as

$$L_i(x) = \prod_{\substack{j=1 \\ j \neq i}}^N \frac{x - x_j}{x_i - x_j}. \tag{11.64}$$

a.   Show that $L_i(x)$ is a polynomial of degree $N - 1$.

b.   Show that $L_i(x)$ is equal to 1 at $x = x_i$ and equal to 0 at the $N - 1$ nodes $x = x_j$.

c.   Let $f(x) = L_i(x)$ in Eq. 11.63. Show that

$$W_i = \int_a^b w(x)L_i(x)\,dx. \tag{11.65}$$

It turns out that there is a simple way to calculate the points and weights for integration by Gaussian quadrature for any common set of classical polynomials. What makes the procedure remarkable is not only the ease and accuracy of the algorithm, but also the conceptual bridge it establishes between integration by Gaussian quadrature and pseudospectral theory.

Consider a set of polynomials, $\{p_n, n = 0, \dots, N - 1\}$, satisfying the orthogonality relationship $\int_a^b w(x)p_m(x)p_n(x)\,dx = \delta_{mn}$. This is a set of orthogonality relations reminiscent of Eq. 11.35 but with the inner product defined with respect to the positive weight function $w(x)$. If the points and weights are chosen according to an N-point Gaussian quadrature, Eq. 11.63 becomes an equality for polynomials up to degree $2N - 1$, and therefore

$$\int_a^b w(x) p_m(x) p_n(x)\, dx = \sum_{i=0}^{N-1} W_i p_m(x_i) p_n(x_i) = \delta_{mn}. \tag{11.66}$$

Defining the function

$$\phi_n(x) = \sqrt{w(x)}\, p_n(x), \tag{11.67}$$

and the corresponding matrix Φ with elements

$$\Phi_{in} = \sqrt{W_i}\, p_n(x_i), \tag{11.68}$$

we can rewrite Eq. 11.66 as

$$\Phi^\dagger \Phi = 1, \tag{11.69}$$

indicating that Φ is a unitary matrix. Note the complete correspondence between Eqs. 11.68 and 11.69 and Eqs. 11.37 and 11.39 if we identify $\Delta_j^{1/2} \phi_n(x_j) = \sqrt{W_j}\, p_n(x_j)$ or

$$W_i = \Delta_i w(x_i). \tag{11.70}$$

Consider now the integral $\int_a^b w(x) p_m(x) x p_n(x)\, dx$. This is a polynomial of degree no higher than $2N - 1$, and so Gaussian quadrature must be exact:

$$\int_a^b w(x) p_m(x) x p_n(x)\, dx = \sum_{i=0}^{N-1} W_i p_m(x_i) x_i p_n(x_i) = \sum_{i,j=0}^{N-1} \Phi^\dagger_{mi} x_{ij} \Phi_{jn} = X_{mn} \tag{11.71}$$

or in matrix notation

$$\mathbf{X} = \Phi^\dagger \mathbf{x} \Phi, \tag{11.72}$$

where $\mathbf{X}$ is an $N \times N$ matrix with elements X_{mn} defined by the RHS of Eq. 11.71 and $\mathbf{x}$ is a diagonal matrix with elements x_{ij}. The decomposition in Eq. 11.72 is unique (for a given ordering of the $\{x_i\}$) and hence x_{ii} is the ith Gaussian quadrature node and Φ_{in} must satisfy Eq. 11.68. Eq. 11.72 suggests the following simple procedure for finding Gaussian quadrature points and weights:

1. Construct the $N \times N$ matrix representation of the $\hat{x}$ operator, $\mathbf{X}$, using the first N basis functions:

$$X_{mn} = \int_a^b \phi_m(x) x \phi_n(x)\, dx = \int_a^b w(x) p_m(x) x p_n(x)\, dx. \tag{11.73}$$

The matrix $\mathbf{X}$ is tridiagonal, a consequence of the three-term recursion relation satisfied by classical orthogonal polynomials. Examples of $\mathbf{X}$ for several of the more common classical polynomials are given in Table 11.1.

2. Diagonalize $\mathbf{X}$ by a unitary transformation:

$$\mathbf{U}^{-1} \mathbf{X} \mathbf{U} = \Phi \mathbf{X} \Phi^\dagger = \mathbf{x}. \tag{11.74}$$

Table 11.1 Properties of the classical orthogonal polynomials and the matrix representation of $\hat{x}$ in their basis

	Chebyshev (2nd kind)	Legendre
	$U_n(x); x \in [-1, 1]$	$P_n(x); x \in [-1, 1]$
$w(x)$	$(1 - x^2)^{1/2}$	1
X	$\begin{pmatrix} \cdot & \frac{1}{2} & \cdot & \cdot \\ \frac{1}{2} & \cdot & \frac{1}{2} & \cdot \\ \cdot & \frac{1}{2} & \cdot & \ddots \\ \cdot & \cdot & \ddots & \ddots \end{pmatrix}$	$\begin{pmatrix} \cdot & \frac{1}{(1\cdot 3)^{1/2}} & \cdot & \cdot \\ \frac{1}{(1\cdot 3)^{1/2}} & \cdot & \frac{2}{(3\cdot 5)^{1/2}} & \cdot \\ \cdot & \frac{2}{(3\cdot 5)^{1/2}} & \cdot & \frac{3}{(5\cdot 7)^{1/2}} \\ \cdot & \cdot & \frac{3}{(5\cdot 7)^{1/2}} & \ddots \\ \cdot & \cdot & \cdot & \ddots \end{pmatrix}$

	Laguerre	Hermite
	$L_n; x \in [0, \infty)$	$H_n(x); x \in (-\infty, \infty)$
$w(x)$	e^{-x}	e^{-x^2}
X	$\begin{pmatrix} 1 & -1 & \cdot & \cdot & \cdot \\ -1 & 3 & -2 & \cdot & \cdot \\ \cdot & -2 & 5 & -3 & \cdot \\ \cdot & \cdot & -3 & 7 & \ddots \\ \cdot & \cdot & \cdot & \ddots & \ddots \end{pmatrix}$	$\begin{pmatrix} \cdot & \sqrt{\frac{1}{2}} & \cdot & \cdot & \cdot \\ \sqrt{\frac{1}{2}} & \cdot & \sqrt{\frac{2}{2}} & \cdot & \cdot \\ \cdot & \sqrt{\frac{2}{2}} & \cdot & \sqrt{\frac{3}{2}} & \cdot \\ \cdot & \cdot & \sqrt{\frac{3}{2}} & \cdot & \ddots \\ \cdot & \cdot & \cdot & \ddots & \ddots \end{pmatrix}$

The eigenvalues $\{x_i\}$ are the integration points of Eq. 11.71 and hence the Gaussian quadrature points of Eq. 11.63. The weights can be easily extracted from the first row of the transformation matrix, $\mathbf{U} = \mathbf{\Phi}^\dagger$ (first column of $\mathbf{\Phi}$). Since the lowest member of every set of classical polynomials is a constant, we have from Eq. 11.66 that $p_0(x_i) = (\int_a^b w(x)\, dx)^{-1/2}$. Substituting into Eq. 11.68 we obtain an explicit equation for the weights:

$$W_i = \Phi_{i0}^2 \int_a^b w(x)\, dx. \qquad (11.75)$$

Quadrature points and weights for several of the more common classical polynomials are shown in Figure 11.3 for the case $N = 12$.

Historically, the procedure of constructing $\mathbf{X}$ and diagonalizing was introduced by Harris, Engerholm and Gwinn (Harris, 1965; henceforth HEG) in the context of a novel procedure for calculating matrix elements of the form $\int \phi_n(x) V(x) \phi_m(x)\, dx$ (Section 11.4). The relationship between their method and Gaussian quadrature was established later by Dickinson and Certain (Dickinson, 1968), who showed that the characteristic equation for the points and weights obtained by matrix diagonalization agrees with known expressions from Gaussian quadrature theory. Our discussion above represents an equally rigorous proof, bypassing the lengthy algebra of Dickinson and Certain by invoking the uniqueness of the decomposition Eq. 11.72. The relationship between diagonalizing $\mathbf{X}$ and Gaussian

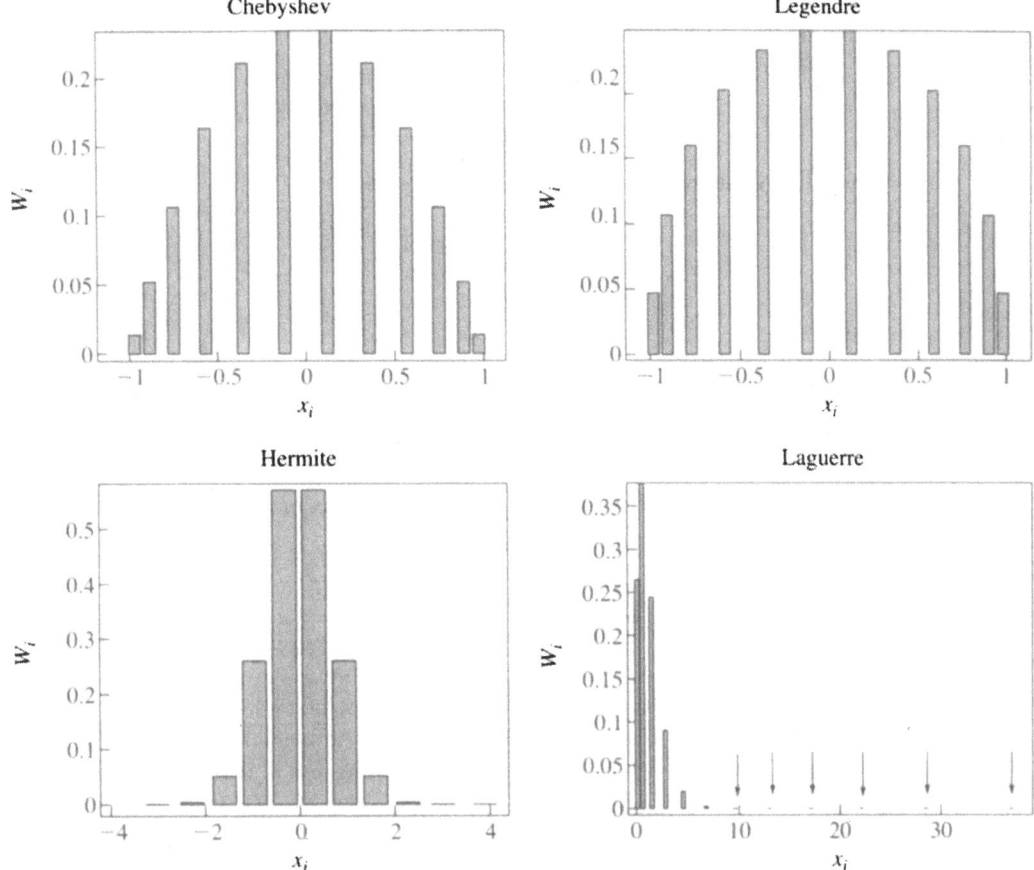

Figure 11.3 Quadrature points and weights for several of the more common orthogonal polynomials. In all cases $N = 12$, although the smallest weights are not always visible.

quadrature was discovered independently in the applied mathematics community by Golub and Welsch (1969).

11.3.2 Properties of the Gaussian Quadrature Pseudospectral Functions

Having obtained an orthogonal collocation matrix in the previous section whose points and weights are based on Gaussian quadrature, in this section we explore the properties of the corresponding pseudospectral functions.

We begin by noting that from Eq. 11.74, the columns of $\Phi^\dagger$ are the eigenvectors of $\mathbf{X}$, hence the *eigenfunctions* of $\hat{x}_N$ are given by

$$\theta_i(x) = \sum_{n=1}^{N} \Phi_{ni}^\dagger \phi_n(x), \quad \text{with } \Phi_{ni}^\dagger = \langle \phi_n | \theta_i \rangle. \tag{11.76}$$

This is exactly Eq. 11.43 (and 11.46), and the $\{\theta_i\}$ are the localized pseudospectral functions that are linear combinations of the delocalized spectral basis functions.

The explicit form of the Gaussian quadrature pseudospectral functions may be obtained by substituting Eqs. 11.67 and 11.68 into Eq. 11.41:

$$\theta_j(x) = \sum_{n=1}^{N} \phi_n(x)\Phi_n^*(x_j) = \sum_{n=1}^{N} w^{1/2}(x)p_n(x)W_j^{1/2}p_n(x_j). \tag{11.77}$$

Comparing this equation with the grid orthogonality relation

$$\sum_{n=1}^{N} W_i^{1/2}p_n(x_i)W_j^{1/2}p_n(x_j) = \delta_{ij} \tag{11.78}$$

we find that

$$\left(\frac{W_i}{w(x_i)}\right)^{1/2} \theta_j(x_i) = \delta_{ij}. \tag{11.79}$$

Using Eq. 11.77, we may conclude that the Gaussian quadrature pseudospectral functions are related in a simple way to the Lagrange interpolating polynomials (see Exercise 11.7) with nodes at the Gaussian quadrature points:

$$L_j(x) = \prod_{i\neq j=1}^{N} \frac{x-x_i}{x_j-x_i} = \left(\frac{W_j}{w(x)}\right)^{1/2} \theta_j(x). \tag{11.80}$$

This follows from the following observations. First, from Eq. 11.77 we see that $\theta_j(x)/w^{1/2}(x)$ is a polynomial of degree $N-1$. The nodes of this polynomial determine the polynomial uniquely up to an overall multiplicative factor. Substituting $x=x_i$ in Eq. 11.80 we see that the LHS and the RHS have nodes at the same location. Finally, Eq. 11.79 guarantees that the overall scaling factor in Eq. 11.80 is the correct one. Figure 11.4 shows the (scaled) Legendre polynomials and the (scaled) associated pseudospectral functions, for the case $N=12$.

It is useful to rewrite Eq. 11.73 in Dirac notation:

$$X_{mn} = \langle\phi_m|\hat{x}|\phi_n\rangle = \langle\phi_m|P_N\hat{x}P_N|\phi_n\rangle, \tag{11.81}$$

where we are allowed to introduce the orthogonal projection operator P_N on both sides of $\hat{x}$ since both ϕ_n and ϕ_m are unaffected by this projection operator. For future use we define the operator

$$\hat{x}_N = P_N\hat{x}P_N. \tag{11.82}$$

The ijth element of Eq. 11.74 can now be rewritten as

$$\left(\mathbf{\Phi X \Phi}^\dagger\right)_{ij} = \sum_{m,n=1}^{N} \langle\theta_i|\phi_m\rangle\langle\phi_m|\hat{x}_N|\phi_n\rangle\langle\phi_n|\theta_j\rangle \tag{11.83}$$

$$= \langle\theta_i|\hat{x}_N|\theta_j\rangle = x_i\delta_{ij}, \tag{11.84}$$

showing in a basis independent representation that the localized pseudospectral functions are eigenfunctions of the projected $\hat{x}$ operator, $\hat{x}_N$. In the coordinate representation the eigenvalue equation takes the form

$$\hat{x}_N\theta_i(x) = x_i\theta_i(x). \tag{11.85}$$

(a)

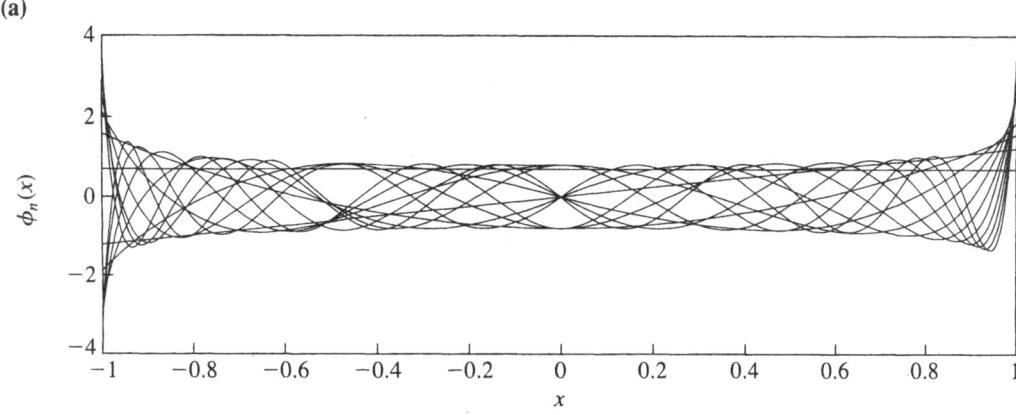

(b)

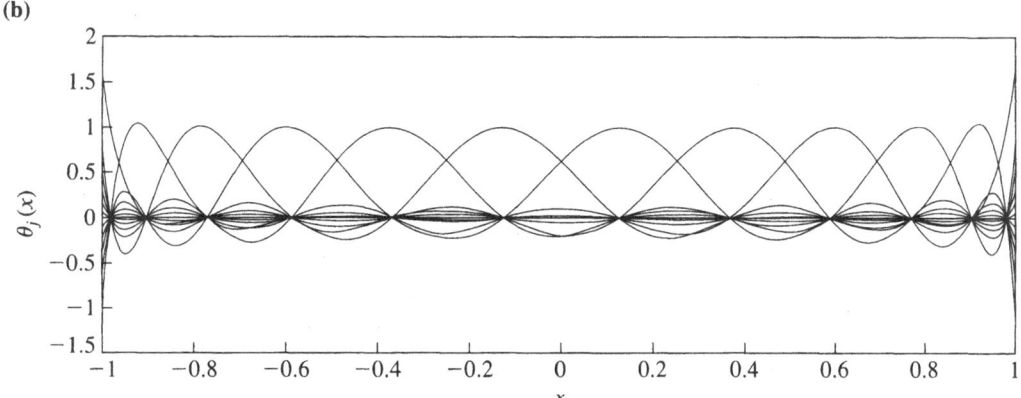

Figure 11.4 (a) The first 12 (scaled) Legendre polynomials, $\phi_n(x) = w^{1/2}(x)p_n(x)$. (b) The 12 (scaled) DVR functions, $L_j(x) = (W_j^{1/2}/w^{1/2}(x))\theta_j(x)$ (see text). The scaled DVR functions are identical to the Lagrange interpolating functions, and thus have the property that they are 1 at their own grid point and vanish at each other grid point.

11.3.3   Derivation of the Theory of Gaussian Quadrature from the Properties of the Pseudospectral Functions

In Section 11.3.1 we assumed the existence of Gaussian quadrature points and weights that give exact values for integrals up to degree $2N - 1$. We now show that we do not need to assume this; the theory of Gaussian quadrature can be derived from first principles from the properties of the pseudospectral basis functions.

We begin by returning to Eq. 11.73. Inserting two complete sets of the $\{\theta_i\}$ we obtain

$$X_{mn} = \sum_{i,j=1}^{N} \langle \phi_m | \theta_i \rangle \langle \theta_i | \hat{x}_N | \theta_j \rangle \langle \theta_j | \phi_n \rangle = \sum_{i=1}^{N} \phi_m^*(x_i) x_i \phi_n(x_i) \Delta_i, \qquad (11.86)$$

where we have used Eqs. 11.49, 11.37 and 11.84. Comparing with Eq. 11.73 and using Eq. 11.70 we see that

$$\int_a^b w(x) p_m(x) x p_n(x)\, dx = \sum_{i=0}^{N-1} W_i p_m(x_i) x_i p_n(x_i). \qquad (11.87)$$

This is exactly the Gaussian quadrature relation that was *assumed* in Eq. 11.71. Here the exact equivalence between the integral and a weighted sum of specific values of the integrand is not assumed, but emerges automatically from the properties of the $\{\theta_i\}$.

We can use Eq. 11.84 to show that the Gaussian quadrature points and weights give an exact result for the integral of the form of Eq. 11.63 with $f(x)$ *any* polynomial of degree $2N - 1$ or lower; that is, we can use the properties of the $\{\theta_i\}$ to prove the usual statement of Gaussian quadrature in Eq. 11.63 (Degani, 2005). Consider the integral

$$\int_a^b w(x) x^{n_1} x x^{n_2} \, dx \qquad 0 \le n_1, n_2 \le N - 1. \tag{11.88}$$

We define $\bar{x}^n = w^{1/2}(x) x^n$. Since both $\bar{x}^{n_1}$ and $\bar{x}^{n_2}$ lie within the space $\mathcal{H}_N$ we can introduce the projection operator P_N before and after the central factor of x to obtain

$$\int_a^b \bar{x}^{n_1} x \bar{x}^{n_2} \, dx = \int_a^b w(x) x^{n_1} \hat{x}_N x^{n_2} \, dx \tag{11.89}$$

$$= \sum_{i,j=1}^N \langle \bar{x}^{n_1} | \theta_i \rangle \langle \theta_i | \hat{x}_N | \theta_j \rangle \langle \theta_j | \bar{x}^{n_2} \rangle \tag{11.90}$$

$$= \sum_i \Delta_i^{1/2} \bar{x}_i^{n_1} x_i \Delta_i^{1/2} \bar{x}_i^{n_2} = \sum_i W_i x_i^{n_1 + n_2 + 1}, \tag{11.91}$$

where we have used Eqs. 11.56, 11.85 and 11.70. Equation 11.91 establishes that for any monomial $f(x) = x^j$, $j = 0, \ldots, 2N - 1$,

$$\int_a^b w(x) f(x) \, dx = \sum_i W_i f(x_i). \tag{11.92}$$

By linearity, Eq. 11.92 holds for any *polynomial* up to degree $2N - 1$.

A well-known property of Gaussian quadrature points is that they fall on the zeros of the polynomial of degree N that is orthogonal to the space of the N lower polynomials of degree $0, \ldots, N - 1$. This may be derived simply as follows (Degani, 2005). Consider

$$\langle \theta_i | \bar{x}^{N+1} \rangle = \langle \theta_i | \hat{x} | \bar{x}^N \rangle = \langle \theta_i | \hat{x}_N | \bar{x}^N \rangle = x_i \langle \theta_i | \bar{x}^N \rangle. \tag{11.93}$$

Continuing this procedure recursively we find that

$$\langle \theta_i | \bar{x}^{N+1} \rangle = W_i^{1/2} x_i^{N+1}, \tag{11.94}$$

where the factor $W_i^{1/2}$ comes from the last factor

$$\langle \theta_i | \bar{1} \rangle = \int_a^b w^{1/2}(x) \theta_i(x) dx = W_i^{1/2} \tag{11.95}$$

(cf. Eqs. 11.56, 11.70). By linearity we have that $\langle \theta_i | \bar{p} \rangle = W_i^{1/2} p(x_i)$ for $w^{1/2} p(x)$ in $\mathcal{H}_{N+1}$ (note that the result applies to $N + 1$, not just to N). Now consider a polynomial $p_\perp$ of degree $N + 1$ that is orthogonal to the space $\mathcal{H}_N$. Then

$$0 = \langle \theta_i | \bar{p}_\perp \rangle = W_i^{1/2} p(x_i), \qquad i = 1, \ldots, N. \tag{11.96}$$

Thus, the polynomial of degree $N + 1$ that is orthogonal to $\mathcal{H}_N$ must vanish at the points $\{x_i\}$, $i = 1, \ldots, N$, and conversely, the N quadrature points must fall at the zeros of the polynomial $p_\perp$ that is of degree $N + 1$ and orthogonal to $\mathcal{H}_N$.

▶ Exercise 11.8   Show that Eq. 11.95

$$\langle \theta_i | \bar{1} \rangle = \int_a^b w^{1/2}(x)\theta_i(x)dx = W_i^{1/2} \tag{11.97}$$

is consistent with Eqs. 11.80 and 11.65.

11.4   Representation of the Hamiltonian in the Reduced Space

In the previous sections we introduced the notions of spectral projection, collocation and the pseudospectral basis. The key results were as follows:

1. Spectral projection and collocation are generally defined by two distinct projection operators.

2. Spectral projection can be represented in either a delocalized spectral basis or a localized pseudospectral basis. These two representations of the spectral projection operator are completely equivalent as long as the transformation between the spectral and pseudospectral basis is unitary (orthogonal):

$$P_N = \sum_{n=1}^{N} |\phi_n\rangle\langle\phi_n| = \sum_{i=1}^{N} |\theta_i\rangle\langle\theta_i|. \tag{11.98}$$

3. The projection operator for collocation is conveniently represented in the pseudospectral basis as

$$P_N = \sum_{i=1}^{N} |\theta_i\rangle\langle x_i|. \tag{11.99}$$

Until now we have focused primarily on the action of these projectors on wavefunctions. We now turn to the construction of the Hamiltonian operator $H = T(\hat{p}) + V(\hat{x})$ using these projectors. In this section we will discuss construction of the Hamiltonian in the spectral basis, $P_N = \sum_{n=1}^{N} |\phi_n\rangle\langle\phi_n|$. In the following two sections we will discuss construction of the Hamiltonian under orthogonal projection in a pseudospectral basis, $P_N = \sum_{i=1}^{N} |\theta_i\rangle\langle\theta_i|$. When this is done using pseudospectral functions derived from classical polynomials the method is known as the Discrete Variable Representation (Section 11.5). When the pseudospectral functions are based on complex exponentials it is known as the Fourier method (Section 11.6). The collocation projector $P_N = \sum_{i=1}^{N} |\theta_i\rangle\langle x_i|$ will appear at various points in our discussion.

Our treatment here will be cast in terms of one-dimensional systems. One can construct a multidimensional extension in a straightforward way using the direct product. Consider a Hamiltonian in two dimensions,

$$H = T_x + T_y + V(x, y), \tag{11.100}$$

and a finite basis representation that is of direct product form,

$$|\Phi_{mn}\rangle = |\phi_m^x\rangle \otimes |\phi_n^y\rangle, \tag{11.101}$$

where $\otimes$ is the tensor product. The Hamiltonian matrix in multidimensions takes the form

$$\langle \Phi_{m'n'}|H|\Phi_{mn}\rangle = \langle \phi_{m'}^x| \otimes \langle \phi_{n'}^y|T_x + T_y + V(x, y)|\phi_m^x\rangle \otimes \phi_n^y\rangle \tag{11.102}$$

$$= \langle \phi_{m'}^x|T_x|\phi_m^x\rangle \otimes \delta_{nn'}^y + \delta_{mm'}^x \otimes \langle \phi_{n'}^y|T_y|\phi_n^y\rangle$$

$$+ \langle \phi_{m'}^x| \otimes \langle \phi_{n'}^y|V(x, y)|\phi_m^x\rangle \otimes |\phi_n^y\rangle. \tag{11.103}$$

The size of the direct product basis grows exponentially with the number of degrees of freedom, and becomes prohibitively large at 5–10 degrees of freedom. In the pseudospectral basis, one can similarly build a representation of the Hamiltonian in direct product form. Defining

$$|\Theta_{ij}\rangle = |\theta_i^x\rangle \otimes |\theta_j^y\rangle \tag{11.104}$$

we can write the Hamiltonian matrix in multidimensions as

$$\langle \Theta_{i'j'}|H|\Theta_{ij}\rangle = \langle \theta_{i'}^x|T_x|\theta_i^x\rangle \otimes \delta_{jj'}^y + \delta_{ii'}^x \otimes \langle \theta_{j'}^y|T_y|\theta_j^y\rangle + \langle \theta_{i'}^x| \otimes \langle \theta_{j'}^y|V(x, y)|\theta_i^x\rangle \otimes |\theta_j^y\rangle. \tag{11.105}$$

However, one of the great attractions of the use of a pseudospectral basis is the possibility of avoiding the direct product form of Eq. 11.105. This is a subject of considerable current interest (Littlejohn, 2002; Dawes, 2004; Degani, 2005).

11.4.1 Spectral Projection of the Hamiltonian

Consider a Hamiltonian that consists of a sum of kinetic and potential energy operators:

$$H = \hat{p}^2/2m + V(\hat{x}) = T(\hat{p}) + V(\hat{x}). \tag{11.106}$$

If one applies spectral projection down to an N-dimensional Hilbert space the truncated operators can be written as

$$H_N = P_N T(\hat{p}) P_N + P_N V(\hat{x}) P_N, \tag{11.107}$$

where

$$P_N = \sum_{n=1}^{N} |\phi_n\rangle\langle\phi_n|. \tag{11.108}$$

Inserting Eq. 11.108 into Eq. 11.107 we see that the necessary matrix elements are of the form

$$\langle \phi_n|T(\hat{p})|\phi_m\rangle \quad \text{and} \quad \langle \phi_n|V(\hat{x})|\phi_m\rangle. \tag{11.109}$$

(The index m should not be confused with the mass, the latter appearing only in the combination $\frac{1}{2m}$.) While the kinetic energy operator can actually become very complicated in multidimensions, in one dimension the T operator is generally quite simple. As a result, in one dimension the matrix elements in the spectral representation, $T_{mn} = \langle \phi_m | T | \phi_n \rangle$, can generally be written down analytically. For example, if $T = \hat{p}^2/2m$, then

$$T_{mn} = \langle \phi_m | T | \phi_n \rangle = -\frac{\hbar^2}{2m} \int_a^b \phi_m^*(x) \frac{\partial^2}{\partial x^2} \phi_n(x)\, dx \qquad (11.110)$$

for which there are generally analytic expressions. The potential energy matrix elements

$$V_{mn} = \langle \phi_m | V(\hat{x}) | \phi_n \rangle = \int_a^b \phi_m^*(x) V(x) \phi_n(x)\, dx \qquad (11.111)$$

generally are not known analytically in the spectral representation. However, one may employ Gaussian quadrature to approximate the integral of a function $f(x)$ by a finite sum:

$$\int_a^b w(x) f(x)\, dx \approx \sum_{i=1}^N W_i f(x_i). \qquad (11.112)$$

Using the expression $\phi_l(x) = w^{1/2}(x) p_l(x)$, $l = n, m$, we make the identification $f(x) = p_n(x) V(x) p_m(x)$ in the quadrature formula and obtain the following approximate expression for the individual matrix elements:

$$V_{mn} \approx \sum_{i=1}^n W_i p_m(x_i) V(x_i) p_n(x_i), \qquad (11.113)$$

with the points and weights taken from the theory of Gaussian quadrature.

11.4.2   The HEG Algorithm

In constructing the different matrix elements of V, one would a priori use different numbers and locations of quadrature points to calculate each of the different matrix elements. However, the procedure developed by HEG (Section 11.3.1) allows one to calculate all the $N \times N$ potential energy matrix elements at once, using the same quadrature points for all integrals. The HEG procedure for calculating potential energy matrix elements is as follows (Harris, 1965):

1. Construct the $N \times N$ matrix representation of the $\hat{x}$ operator, $\mathbf{X}_N$, using the first N basis functions. This matrix has elements

$$X_{mn} = \int_a^b \phi_m(x) x \phi_n(x)\, dx = \int_a^b w(x) p_m(x) x p_n(x)\, dx. \qquad (11.114)$$

   The matrix $\mathbf{X}_N$ is tridiagonal, a consequence of the form of the three-term recursion relation satisfied by most classical orthogonal polynomials.

2. Diagonalize $\mathbf{X}_N$ by a unitary transformation:

$$\mathbf{U}_N^\dagger \mathbf{X}_N \mathbf{U}_N = \mathbf{x}, \qquad (11.115)$$

   where $\mathbf{x}_{ij} = x_i \delta_{ij}$. These first two steps in the HEG procedure are identical to the procedure described in Section 11.3.1 to calculate Gaussian quadrature points and weights by matrix diagonalization.

3. Compute $V(\mathbf{x})$. Since $\mathbf{x}$ is diagonal, $(V(\mathbf{x}))_{ij} = V(x_i)\delta_{ij}$.

4. Transform back to the basis of orthogonal polynomials:

$$\mathbf{V}_N^{\text{HEG}} = \mathbf{U}_N V(\mathbf{x}) \mathbf{U}_N^\dagger. \tag{11.116}$$

Multiplying out the matrices in Eq. 11.116 and using Eq. 11.68 we obtain

$$(\mathbf{V}_N^{\text{HEG}})_{mn} = \sum_{i=1}^n \mathbf{U}_{mi} V(\mathbf{x}_{ii})(\mathbf{U}^\dagger)_{in} = \sum_{i=1}^n W_i p_m(x_i) V(x_i) p_n(x_i). \tag{11.117}$$

Equation 11.117 is identical with Eq. 11.113, the expression obtained with an N-point Gaussian quadrature approximation to the integral $(\mathbf{V}_N(\hat{x}))_{mn}$, if the $\{x_i\}$ are identified with the Gaussian quadrature points and the $\{W_i\}$ are identified with the Gaussian weights (see Section 11.3.1).

An alternative perspective on the approximation inherent in the HEG method is obtained by noting that

$$\mathbf{V}_N^{\text{HEG}} = \mathbf{U}_N V(\mathbf{x}) \mathbf{U}_N^\dagger = V(\mathbf{X}_N). \tag{11.118}$$

The proof is straightforward. From Eq. 11.115 we have $\mathbf{U}_N^\dagger \mathbf{X}_N \mathbf{U}_N = \mathbf{x}$, and therefore

$$\mathbf{U}_N^\dagger \mathbf{X}_N^n \mathbf{U}_N = (\mathbf{U}_N^\dagger \mathbf{X}_N \mathbf{U}_N)^n = \mathbf{x}^n. \tag{11.119}$$

Since Eq. 11.119 holds for all powers n, it is easily seen to hold for polynomial $V(x)$:

$$\mathbf{U}_N^\dagger V(\mathbf{X}_N) \mathbf{U}_N = V(\mathbf{x}), \tag{11.120}$$

which is equivalent to Eq. 11.118. Thus, the key approximation in the HEG method can be expressed as the replacement

$$\mathbf{V}_N \rightarrow V(\mathbf{X}_N). \tag{11.121}$$

The matrices with subscript N are all defined here in a particular basis, namely that of the orthogonal functions. However, the approximation inherent in the HEG method can also be expressed in a basis-independent fashion. Formally, the matrix $\mathbf{X}_N$ corresponds to the operator

$$\hat{x}_N = P_N \hat{x} P_N \tag{11.122}$$

(cf. Eq. 11.82). Comparing Eq. 11.124 with Eq. 11.107, and using Eqs. 11.118 and 11.122, we see that the essential approximation in the HEG method may be expressed more abstractly as the replacement

$$V_N(\hat{x}) = P_N V(\hat{x}) P_N \rightarrow V(\hat{x}_N) = V(P_N \hat{x} P_N), \tag{11.123}$$

where we now have a basis-independent representation for the approximation. The advantage to this basis-independent representation is that it indicates that the approximation is independent of the basis, and hence the numerical results should be identical whether calculated in a spectral basis as in the HEG method or in a pseudospectral basis as in the DVR method described below. Moreover, if the order of approximation can be quantified in one basis we know what it must be in the other basis as well.

This completes the calculation of the matrix elements V_{nm}. These matrix elements can then be used to compute eigenvalues and eigenfunctions of the full Hamiltonian as follows (Harris, 1965):

5. Construct the approximate Hamiltonian matrix by adding the kinetic and potential contributions:

$$\mathbf{H}_N^{\mathrm{HEG}} \equiv \mathbf{T}_N + \mathbf{V}_N^{\mathrm{HEG}}. \tag{11.124}$$

6. Diagonalize the matrix $\mathbf{H}_N^{\mathrm{HEG}}$ to solve the TISE, or to propagate the TDSE according to Eq. 11.10 or 11.11.

11.5   The Discrete Variable Representation

The Discrete Variable Representation or DVR method is a pseudospectral method for calculating matrix elements of the Hamiltonian (Light, 1985). What is specific to the DVR method is that the pseudospectral basis is related to the original spectral basis by an orthogonal collocation matrix based on Gaussian quadrature points and weights. Since many of the properties of the DVR method are common to all pseudospectral methods we begin by presenting the elements of the DVR method within the more general context of the evaluation of the Hamiltonian in a pseudospectral basis. In Section 11.5.2 we present the specifics of the DVR algorithm, with some further discussion of its inherent assumptions and approximations.

11.5.1   Evaluation of the Hamiltonian in a Pseudospectral Basis

As in Section 11.4.1 we consider a Hamiltonian of the form $H = T(\hat{p}) + V(\hat{x})$. We seek an orthogonal projection onto $\mathcal{H}_N$, however now, instead of expressing the projector in terms of the spectral basis we use the pseudospectral form

$$P_N = \sum_{i=1}^{N} |\theta_i\rangle \langle \theta_i|. \tag{11.125}$$

Considering $H_N = P_N T(\hat{p}) P_N + P_N V(\hat{x}) P_N$ in the pseudospectral basis it is clear that the necessary matrix elements are of the form

$$\langle \theta_i | T(\hat{p}) | \theta_j \rangle \quad \text{and} \quad \langle \theta_i | V(\hat{x}) | \theta_j \rangle. \tag{11.126}$$

In Section 11.4.1 we dismissed the kinetic energy terms very quickly but gave a lengthy discussion of the evaluation of the potential energy by Gaussian quadrature. Here the situation is to some extent reversed. In the pseudospectral basis, the evaluation of the potential energy matrix elements is given by a simple prescription. Since the basis functions $\{\theta_i\}$ are localized and satisfy the orthonormality relation $\langle \theta_i | \theta_j \rangle = \delta_{ij}$ it seems plausible to make the replacement

$$V_{ij} = \langle \theta_i | V(\hat{x}) | \theta_j \rangle \rightarrow V(x_i) \delta_{ij}. \tag{11.127}$$

The replacement in Eq. 11.127 is the key step in all pseudospectral methods. In particular, it is the central approximation of the DVR method described in this section as well as in the Fourier method, described in Section 11.6.

For the DVR method it turns out that the replacement in Eq. 11.127 represents an approximation identical to the Gaussian quadrature approximation in the HEG method. We will express this approximation explicitly in terms of Gaussian quadrature at the end of Section 11.5.2, after we have presented the DVR algorithm. However, we can already see that the approximation is identical to that of HEG by the following consideration. Since in the DVR method the orthogonal collocation matrix is based on Gaussian quadrature points and weights, the pseudospectral basis functions satisfy the equation $\hat{x}_N |\theta_i\rangle = x_i |\theta_i\rangle$ (cf. Eq. 11.85). Combining this relation with our earlier result that the central approximation in the HEG method can be written as $P_N V(\hat{x}) P_N \to V(P_N \hat{x} P_N) = V(\hat{x}_N)$ (Eq. 11.123) and using the pseudospectral form for the projector (Eq. 11.125), the DVR pseudospectral matrix elements take the form

$$V(\hat{x}_N)_{ij} = \sum_{k,l=1}^{N} \langle \theta_i | V \left(|\theta_k\rangle \langle \theta_k | \hat{x} | \theta_l\rangle \langle \theta_l| \right) |\theta_j\rangle \tag{11.128}$$

$$= \sum_{k=1}^{N} \langle \theta_i | V \left(|\theta_k\rangle x_k \langle \theta_k| \right) |\theta_j\rangle \tag{11.129}$$

$$= \sum_{k=1}^{N} \langle \theta_i | \theta_k \rangle V(x_k) \langle \theta_k | \theta_j \rangle = V(x_i) \delta_{ij}. \tag{11.130}$$

It is interesting to note that one may give a different interpretation to the replacement in Eq. 11.127. If P_N is taken to be the nonorthogonal projector associated with collocation, $P_N = \sum_{i=1}^{N} |\theta_i\rangle \langle x_i|$, then

$$P_N \hat{x} P_N^{\dagger} = \sum_{i,j=1}^{N} |\theta_i\rangle \langle x_i | V(\hat{x}) | x_j\rangle \langle \theta_j| = \sum_{i=1}^{N} |\theta_i\rangle V(x_i) \langle \theta_i| \tag{11.131}$$

with no approximation. Does it make any difference which of these two interpretations is adopted? Possibly yes, in that it could lead to different prescriptions for how the projector for the kinetic energy should be chosen. Suppose that the kinetic energy operator is different depending on whether it is evaluated under spectral projection or under collocation projection, and suppose that its representation can be evaluated exactly for both projectors. It could potentially be preferable to use collocation projection, which is nonorthogonal but according to which the potential energy matrix elements have been evaluated exactly, rather than spectral projection, which is orthogonal, but according to which the potential energy matrix elements are only approximate. These remarks are highly speculative at this point, but the concept is interesting and may be worth pursuing.

We now turn to the calculation of the kinetic energy matrix in a pseudospectral basis. The calculation of these matrix elements is more complicated in the pseudospectral than in the spectral basis, but is not particularly difficult. It requires the calculation of second derivative matrix elements T_{ij} between two localized basis functions associated with different grid points i and j. Although at first glance this seems complicated, there is a quite simple procedure:

$$(\mathbf{T}^\theta)_{ij} = \langle \theta_i | T | \theta_j \rangle \tag{11.132}$$

$$= \sum_{n,m=1}^{N} \langle \theta_i | \phi_n \rangle \langle \phi_n | T | \phi_m \rangle \langle \phi_m | \theta_j \rangle \tag{11.133}$$

$$= \sum_{n,m=1}^{N} \Phi_{in} (\mathbf{T}^\phi)_{nm} \Phi_{mj}^\dagger \tag{11.134}$$

$$= (\mathbf{\Phi} \mathbf{T}^\phi \mathbf{\Phi}^\dagger)_{ij}, \tag{11.135}$$

where we have used Eqs. 11.46 and 11.47. Equation 11.135 expresses the second derivative matrix between pseudospectral basis functions, $\mathbf{T}^\theta$, in terms of a unitary transformation on the second derivative matrix in the spectral basis $\mathbf{T}^\phi$. Once the transformation $\mathbf{\Phi}$ between the spectral and the pseudospectral basis is known, $\mathbf{T}^\theta$ is readily calculated.

A different perspective is obtained by rewriting Eq. 11.132 in the following form:

$$(\mathbf{T}^\theta)_{ij} = \sum_{n=1}^{N} \langle \theta_i | \phi_n \rangle \langle \phi_n | T | \theta_j \rangle \tag{11.136}$$

$$\approx -\frac{\hbar^2}{2m} \sum_{n=1}^{N} \phi_n(x_i) \left. \frac{\partial^2 \phi_n}{\partial x^2} \right|_{x_j} . \tag{11.137}$$

Equation 11.137 gives an approximate route to calculating the second derivative matrix between two pseudospectral functions. The second derivative of the function θ_j is evaluated at grid point j by representing the value at the grid point as a sum over a basis of orthogonal functions, and calculating (analytically) the second derivative of the basis functions at the grid point. The approximation comes from the fact that

$$\langle \theta_j | T | \phi_n \rangle \approx \langle x_j | T | \phi_n \rangle. \tag{11.138}$$

Equation 11.137, which calculates derivatives by means of a spectral method, has "global" accuracy, as opposed to the "local" accuracy of derivatives calculated by finite differencing methods. Global methods converge exponentially with the number of basis functions, while local methods converge polynomially. In particular, local methods for calculating derivatives can make significant errors in the high k components of the wavefunction, as shown in Figure 11.5. The nonlocal evaluation of the derivative operators is consistent with the uncertainty principle: since the evaluation of the coordinate operation is local it is natural that the evaluation of the momentum operation be nonlocal, when represented on a coordinate space grid.

11.5.2 The DVR Algorithm

To implement the DVR method the first two steps are exactly as in the method of HEG. However, in the third step the unitary transformation is performed on the *kinetic* energy

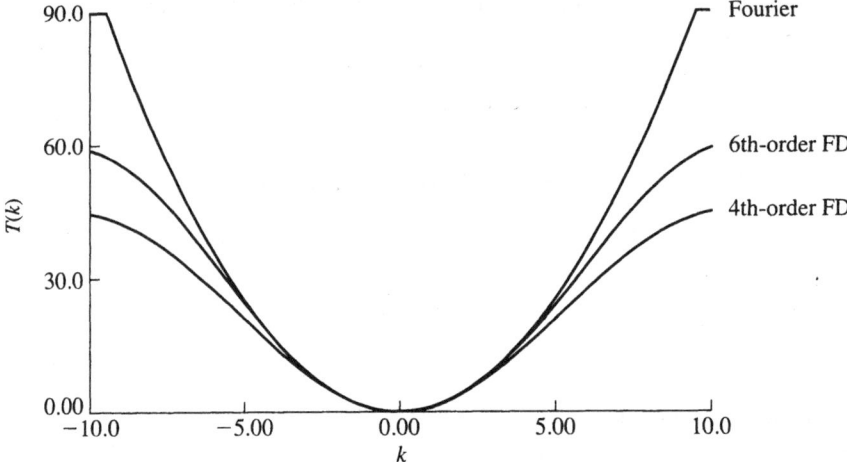

Figure 11.5 Comparison of the kinetic energy operator spectrum for a spectral (Fourier) method with the fourth- and sixth-order finite difference methods. The spectral method, which is global, converges exponentially while finite difference methods, which are local, converge polynomially. Adapted from Kosloff (1993).

matrix instead of the potential energy matrix to leave the Hamiltonian in the pointwise representation. The procedure is as follows:

1. Construct the matrix representation of $\hat{x}$ in a truncated basis of N orthogonal polynomials, $\mathbf{x}_N$. This matrix is tridiagonal; the form of the matrix elements for some common orthogonal polynomials are given in Table 11.1.

2. Diagonalize $\mathbf{X}_N$ by a unitary transformation

$$\mathbf{U}^\dagger \mathbf{X}_N \mathbf{U} = \mathbf{x}, \tag{11.139}$$

where $(\mathbf{x})_{ij} = x_i \delta_{ij}$.

3. Compute $V(\mathbf{x})$. Since $\mathbf{x}$ is diagonal, $(V(\mathbf{x}))_{ij} = V(x_i)\delta_{ij}$. It is this diagonal form of the potential energy matrix, and its construction by simple evaluation of $V(x)$ at the grid points, that leads to both the attraction of the DVR method and its close kinship to the Fourier method described in the next section.

4. Calculate the kinetic energy matrix in the basis of orthogonal polynomials, which we denote $\mathbf{T}_N^\phi$. The matrix elements are normally known analytically and this matrix is typically diagonal or tridiagonal. In order that the Hamiltonian $\mathbf{H}_N = \mathbf{T}_N + \mathbf{V}_N$ be expressed completely in the pseudospectral basis, $\mathbf{T}_N^\phi$ must now be transformed by the same transformation $\mathbf{U}$ that was used to transform the matrix $\mathbf{V}_N$ to the pseudospectral basis:

$$\mathbf{T}_{ij}^{\text{DVR}} = \left(\mathbf{U}^\dagger \mathbf{T}^\phi \mathbf{U}\right)_{ij}. \tag{11.140}$$

Note that this expression matches the general structure of Eq. 11.135.

5. Construct the DVR Hamiltonian by adding the kinetic and potential energies:

$$\mathbf{H}_N^{DVR} = \mathbf{T}_N^{DVR} + \mathbf{V}_N^{DVR} \tag{11.141}$$

6. Diagonalize the DVR Hamiltonian, $\mathbf{H}_N^{DVR}$.

From the DVR algorithm it is clear that the pseudospectral matrix $\mathbf{H}_N^{DVR}$ differs from the spectral matrix $\mathbf{H}_N^{HEG}$ of the previous section only in the choice of representation. Whereas the HEG constructs the orthogonal projection of the Hamiltonian onto $\mathcal{H}_N$ using the spectral basis $\{\phi_i\}$, the DVR method constructs the Hamiltonian in the pseudospectral basis $\{\theta_i\}$. As discussed above, the two bases are related by an orthogonal (or unitary) transformation and thus

$$\mathbf{U}\mathbf{H}_N^{DVR}\mathbf{U}^{\dagger} = \mathbf{H}_N^{HEG}. \tag{11.142}$$

Since the Hamiltonians differ only by a unitary transformation, they must have identical eigenvalues and if used to propagate an initial wavefunction will give identical dynamics. This is consistent with what we have already seen at Eq. 11.123, that in a basis-independent formulation both algorithms make the replacement $P_N V(\hat{x}) P_N \rightarrow V(P_N \hat{x} P_N)$.

It follows that the DVR method contains the same approximation of N-point Gaussian quadrature integration as does the HEG method. From the discussion at Eq. 11.123 we can conclude that the pseudospectral matrix elements of the potential in the DVR method will be evaluated exactly, that is,

$$\langle\theta_i|V(\hat{x})|\theta_j\rangle = \int \theta_i(x)V(x)\theta_j(x)\,dx = \sum_{k=1}^{N} W_k\theta_i(x_k)V(x_k)\theta_j(x_k) = V(x_i)\delta_{ij}, \tag{11.143}$$

provided that $\theta_i(x)V(x)\theta_j(x) = w(x)f(x)$ with $f(x)$ a polynomial of degree $2N - 1$ or less. In summary, the replacement in Eq. 11.127 is an approximation to the evaluation of $P_N H P_N$ at an accuracy equivalent to an N-point Gaussian quadrature approximation to the integral.

11.6   The Fourier Method

The Fourier method (Kosloff, 1983, 1988, 1993; Feit, 1982; Colbert, 1992; Willner, 2001) is a special case of a pseudospectral method on a grid of evenly spaced points. As in any pseudospectral method, the potential energy matrix is diagonal and is just given by the value of the potential at the grid point. The Fourier method has two implementations: the Fourier Grid Hamiltonian method, which involves construction of the Hamiltonian matrix in a pseudospectral Fourier basis (Section 11.6.4), and the Dynamic Fourier Method, in which the construction of matrices is avoided by calculating their action on the wavefunction directly via Fast Fourier Transform procedure (Section 11.6.5).

11.6.1   The Spectral Projector in the Fourier Method

In the Fourier method, the orthogonal basis functions are of the general form

$$\phi_k(x) \propto e^{ikx}. \tag{11.144}$$

The projector in the Fourier method depends on the choice of the range of x and k, but there is some residual freedom in the projector that depends on the exact choice of functions $\{k\}$. We will discuss here two choices, a continuous basis in k and a discrete basis in k.

Continuous k-Basis

It is ubiquitous in the literature on the Fourier method to restrict the Hilbert space to $-K \leq k \leq K$. Functions that have no components of $|k|$ (or whose k components decay exponentially) beyond K are called "band-limited." If the coordinate space range is infinite, then the (k-normalized) basis functions of the form in Eq. 11.144 are

$$\phi_k(x) = \frac{e^{ikx}}{\sqrt{2\pi}}, \qquad -K \leq k \leq K, \qquad -\infty < x < \infty. \tag{11.145}$$

The basis set corresponding to continuous values of k, Eq. 11.145, defines a projector operator that we denote $P_\sqcap$:

$$P_\sqcap(\hat{p}) = \int_{-K}^{K} |\phi_k\rangle\langle\phi_k| \, dk, \tag{11.146}$$

where the symbol $\sqcap$ is meant to suggest visually the band-limited range of k.

Discrete k-Basis

If one assumes that the Hilbert space spans only a finite range of coordinate space, $0 \leq x \leq L$, the Hilbert space is said to have "finite support." We may define a projection operator associated with this finite range,

$$P_\sqcap(\hat{x}) = \int_{0}^{L} |x\rangle\langle x| \, dx. \tag{11.147}$$

The normalized basis functions of the form in Eq. 11.144 are then determined by the condition

$$\int_{0}^{L} \phi_k^*(x)\phi_{k'}(x) \, dx = \delta_{kk'} = \frac{1}{L} \int_{0}^{L} e^{-ikx} e^{ik'x} \, dx. \tag{11.148}$$

Note that Eq. 11.148 can be satisfied only for discrete values of k, satisfying $k - k' = \frac{n2\pi}{L}$ or

$$\Delta k = \frac{2\pi}{L}, \tag{11.149}$$

where Δk is the difference in k between two neighboring basis functions. Thus, the normalized basis functions are given by

$$\phi_k(x) = \frac{e^{ikx}}{\sqrt{L}} \qquad 0 \leq x \leq L, \quad k = \kappa \, \Delta k, \quad -\infty < \kappa < \infty. \tag{11.150}$$

The restriction to discrete values of k corresponds to a projection operator $P_\sqcup(\hat{p})$ defined by the equation

$$P_\sqcup(p) = \sum_{\kappa=-\infty}^{\infty} |p_\kappa\rangle\langle p_\kappa|, \tag{11.151}$$

where $p_\kappa = \hbar\kappa\,\Delta k$. The symbol $\sqcup$ is meant to suggest visually the discrete, picket-fence-like structure of the projector in k. From the theory of Fourier series we know that the basis defined by Eq. 11.150 is complete with respect to the expansion of any periodic function with period $L = 2\pi/\Delta k$ (see Section 6.2.3). Note that $P_\sqcup(p)$ is close to but not identical to $P_\sqcap(x)$. Applying the former to a function $\Psi(x)$ produces the same function as the latter on the domain $0 \le x \le L$, but produces in addition an infinite, periodically repeating series of replicas. Despite the fact that the two projectors are not identical, the functions they return have the same information content: from the periodic function one can construct the function on the interval $[0, L]$, and from the function on the interval $[0, L]$ one can construct the periodic function. This is essentially the content of the Nyquist theorem, which states that for a function of finite support on $[0, L]$, there is no loss of information by sampling it at discrete values of k with $\Delta k = 2\pi/L$.

The Combined k-Basis

Can we combine the restriction of finite support on $0 \le x \le L$ with the restriction of band limit, $-K \le k \le K$? Strictly speaking, no: since $P_\sqcap(\hat{x})$ and $P_\sqcap(\hat{p})$ do not commute (because $\hat{x}$ and $\hat{p}$ do not), the product of the two is not Hermitian and therefore cannot be a projector. Note, however, that the close relative of $P_\sqcap(\hat{x})$, $P_\sqcup(\hat{p})$, does commute with $P_\sqcap(\hat{p})$ since they are both functions of $\hat{p}$. This combined projector can be written as

$$P = P_\sqcap(\hat{p})P_\sqcup(\hat{p}), \tag{11.152}$$

where here and henceforth we assume that $2K/\Delta k = KL/\pi = N$, where N is an integer (if this condition is not satisfied it is always possible to increase the value of K slightly). The combined projector has a simple geometrical interpretation in k-space: the operator $P_\sqcap(\hat{p})$ corresponds to multiplication of a function in k-space by a rectangular envelope, while the operator $P_\sqcup(\hat{p})$ corresponds to multiplication of a function in k-space by a picket fence. Clearly the order of these multiplications is immaterial and hence the operators commute.

The combined projection operator, Eq. 11.152, corresponds to a normalized basis $\{\phi_\kappa\}$,

$$P = P_\sqcap(\hat{p})P_\sqcup(\hat{p}) = \sum_{\kappa=1}^{N} |\phi_\kappa\rangle\langle\phi_\kappa|, \tag{11.153}$$

where

$$\phi_\kappa(x) = \frac{e^{i\kappa\Delta kx}}{\sqrt{L}} = \frac{e^{i2\pi\kappa x/L}}{\sqrt{L}}, \qquad -\frac{N}{2}+1 \le \kappa \le \frac{N}{2}, \quad -\infty < x < \infty. \tag{11.154}$$

Although strictly speaking the basis functions in Eq. 11.154 are defined over the domain $-\infty < x < \infty$, from the periodic function on the infinite domain one can trivially recover the function on the unit domain $[0, L]$.

11.6.2   The Orthogonal Collocation Matrix in the Fourier Method

There are two Fourier pseudospectral schemes, corresponding to the two different choices of k-basis (discrete and continuous) described in the previous section. We begin with the discrete basis defined by Eq. 11.154, and then turn to the continuous basis defined by Eq. 11.145.

Discrete k-Basis

In the discrete basis, the pseudospectral matrix is chosen to make the orthogonality relation exact:

$$\int_0^L \phi_{\kappa'}^*(x)\phi_\kappa(x)\,dx = \delta_{\kappa\kappa'} = \sum_{j=1}^N \phi_{\kappa'}^*(x_j)\phi_\kappa(x_j)\Delta x = \sum_{j=1}^N \Phi_{\kappa'}^*(x_j)\Phi_\kappa(x_j). \quad (11.155)$$

Taking

$$\Delta x = \frac{L}{N}, \qquad x_j = j\Delta x = \frac{jL}{N}, \quad (11.156)$$

the basis orthogonality relation is solved for

$$\Phi_\kappa(x_j) = \frac{e^{\sqrt{-1}2\pi\kappa j/N}}{\sqrt{N}} \quad (11.157)$$

(cf. Eqs. 11.149–11.150). Although $\Phi_\kappa(x_j)$ differs from $\phi_k(x_j)$ only in the factor $\frac{1}{\sqrt{N}}$ versus $\frac{1}{\sqrt{L}}$, this difference is important for normalization, and illustrates the general method for calculating the elements of the pseudospectral matrix. The corresponding grid orthogonality relation takes the form (see Exercise 11.19)

$$\sum_{\kappa=-\frac{N}{2}+1}^{\frac{N}{2}} \Phi_\kappa(x_i)\Phi_\kappa^*(x_j) = \sum_{\kappa=-\frac{N}{2}+1}^{\frac{N}{2}} \frac{e^{\sqrt{-1}2\pi\kappa i/N}}{\sqrt{N}} \frac{e^{-\sqrt{-1}2\pi\kappa j/N}}{\sqrt{N}} = \delta_{ij}. \quad (11.158)$$

Continuous k-Basis

We now turn to the continuous basis defined by Eq. 11.145. Again, the orthogonal collocation matrix is chosen to make the basis orthogonality relation exact:

$$\int_{-\infty}^\infty \phi_{k'}^*\phi_k(x)\,dx = \int_{-\infty}^\infty \frac{e^{i(k-k')x}}{2\pi}\,dx = \delta(k-k')$$

$$= \sum_{j=-\infty}^\infty \phi_{k'}^*(x_j)\phi_k(x_j)\Delta x = \sum_{j=-\infty}^\infty \Phi_{k'}^*(x_j)\Phi_k(x_j). \quad (11.159)$$

Taking $\Delta x = \frac{2\pi}{2K}$, $x_j = j\Delta x = \frac{j\pi}{K}$, the basis orthogonality relation is satisfied for

$$\Phi_k(x_j) = \frac{e^{ikx_j}}{\sqrt{2K}} \quad (11.160)$$

(see Exercise 11.20). Note that the "matrix" is continuous in the index k and discrete but infinite in the index x_j. The corresponding grid orthogonality relation takes the form (see Exercise 11.20)

$$\int_{-K}^K \Phi_k(x_i)\Phi_k^*(x_j)\,dk = \int_{-K}^K \frac{e^{ik(x_i-x_j)}}{2K}\,dk = \delta_{ij}. \quad (11.161)$$

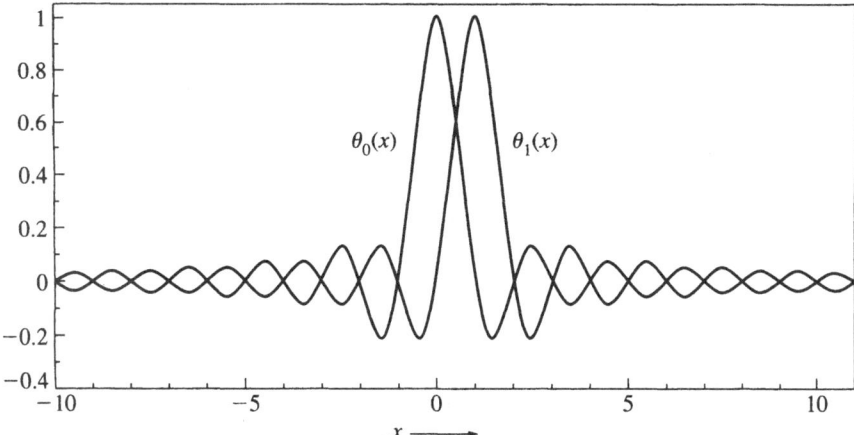

Figure 11.6 Depiction of the pseudospectral basis functions, $\theta_j(x) = \text{sinc}[K(x - x_j)]$ in the second variant of the Fourier method (see text). These functions display the characteristic properties of all pseudospectral bases; the basis functions are orthonormal and each basis function vanishes at all grid points except the one where it is centered.

11.6.3   Pseudospectral Basis Functions in the Fourier Method

The pseudospectral interpretation of the Fourier method has two variants, corresponding to the two choices of discrete or continuous basis described above. The continuous variant is obtained by substituting Eqs. 11.145 and 11.160 into Eq. 11.43:

$$\int_{-K}^{K} \phi_k(x)\Phi_k^*(x_j)\, dk = \theta_j(x) = \int_{-K}^{K} \frac{e^{ikx}}{\sqrt{2\pi}} \frac{e^{-ikx_j}}{\sqrt{2K}}\, dk$$

$$= \frac{\sin(K(x - x_j))}{\sqrt{\pi K}(x - x_j)} = \sqrt{\frac{K}{\pi}}\,\text{sinc}[K(x - x_j)] \qquad (11.162)$$

Each sinc function is centered on a different grid point, the space between grid points being $\Delta x = \frac{\pi}{K}$. Moreover, the width of each sinc function is approximately $\frac{\pi}{K}$.

The behavior of the sinc function is depicted in Figure 11.6. Note that the sinc functions have properties that correspond to those of pseudospectral basis functions in general: they are 1 at their own grid point, vanish at all other grid points, are orthogonal to each other, and form a complete, orthogonal set in a subspace of $L^2(-\infty, \infty)$ that consists of band-limited funtions (functions whose Fourier transforms vanish outside the interval $[-K, K]$). The latter statement will be formalized in the next section, using projection operators. The basis of sinc functions is sometimes called the Hardy basis.

The discrete pseudospectral functions are obtained by substituting Eqs. 11.154 and 11.157 into Eq. 11.158:

$$\sum_{\kappa=-(N/2-1)}^{N/2} \phi_\kappa(x)\Phi_\kappa^*(x_j) = \theta_j(x) = \sum_{\kappa=-(N/2-1)}^{N/2} \frac{e^{i2\pi\kappa x/L}}{\sqrt{L}} \frac{e^{-i2\pi\kappa x_j/L}}{\sqrt{N}}. \qquad (11.163)$$

To appreciate the form of the resulting functions, note that the discreteness of the representation in k implies periodicity in the x-representation (see Section 6.2); specifically, it implies periodic boundary conditions in x with period $L = \frac{\pi}{\Delta k}$. Thus, the pseudospectral basis functions corresponding to Eq. 11.154 are periodic trains of sinc functions.

11.6.4 The Fourier Grid Hamiltonian

The Fourier Grid Hamiltonian (FGH) method (Marston, 1989) refers to the evaluation of the matrix elements of the Hamiltonian using the Fourier pseudospectral scheme. Normally, this matrix is diagonalized to obtain eigenvalues and eigenvectors. The starting point of the FGH is the exact expression for the Hamiltonian operator in the coordinate representation:

$$\langle x|H|x'\rangle = \langle x|T(\hat{p}) + V(\hat{x})|x'\rangle \tag{11.164}$$

$$= \int_{-\infty}^{\infty} \langle x|k\rangle \frac{\hbar^2 k^2}{2m} \langle k|x'\rangle dk + V(x)\delta(x - x')$$

$$= \frac{\hbar^2}{2m} \int_{-\infty}^{\infty} \frac{e^{\sqrt{-1}kx}}{\sqrt{2\pi}} k^2 \frac{e^{-\sqrt{-1}kx'}}{\sqrt{2\pi}} dk + V(x)\delta(x - x'). \tag{11.165}$$

In the Fourier pseudospectral scheme, the coordinates x, x' are discretized at N evenly spaced points,

$$x \to i\Delta x \qquad x' \to j\Delta x, \tag{11.166}$$

and thus the Hamiltonian matrix in the FGH method is an $N \times N$ matrix with H_{ij} corresponding to position $i\Delta x$, $j\Delta x$. (In this section the letter i is used as an index and we use $\sqrt{-1}$ for the imaginary number.) The potential energy matrix elements are given by the usual DVR prescription, that the off-diagonal elements vanish and that the diagonal elements are given by the local evaluation of $V(x)$:

$$\langle x_i|V(\hat{x})|x_j\rangle \to V(x_j)\delta_{ij}. \tag{11.167}$$

The kinetic energy matrix elements take the following form:

$$\langle x_i|T(\hat{p})|x_j\rangle = \int_{-\infty}^{\infty} \langle x_i|k\rangle \langle k|T(\hat{p})|k\rangle \langle k|x_j\rangle \, dk = \frac{\hbar^2}{2m} \int_{-\infty}^{\infty} \frac{e^{\sqrt{-1}ki\Delta x}}{\sqrt{2\pi}} k^2 \frac{e^{-\sqrt{-1}kj\Delta x}}{\sqrt{2\pi}} \, dk. \tag{11.168}$$

In the FGH method one truncates the range of integration to a region $[-K, K]$, where $K = \pi/\Delta x$, and discretizes the integral over k, with the spacing in k given by $\Delta k = \frac{2\pi}{N\Delta x}$, corresponding to the basis implicit in the Fourier pseudospectral scheme. Making the replacement $k = \kappa \, \Delta k$ and using the relation

$$\Delta k \, \Delta x = \frac{2\pi}{N}, \tag{11.169}$$

we obtain

$$\langle x_i|T(\hat{p})|x_j\rangle \approx \frac{\hbar^2}{2m} \sum_{\kappa=-N/2+1}^{N/2} \frac{e^{\sqrt{-1}2\pi\kappa i/N}}{\sqrt{2K}} (\kappa \Delta k)^2 \frac{e^{-\sqrt{-1}2\pi\kappa j/N}}{\sqrt{2K}} \Delta k \tag{11.170}$$

$$= \frac{\hbar^2}{2m} \left(\frac{2K}{N}\right)^2 \sum_{\kappa=-N/2+1}^{N/2} \frac{e^{\sqrt{-1}2\pi\kappa i/N}}{\sqrt{N}} \kappa^2 \frac{e^{-\sqrt{-1}2\pi\kappa j/N}}{\sqrt{N}} \tag{11.171}$$

(cf. Eq. 11.158 and Exercise 11.19). The summation in Eq. 11.171 can be performed analytically. For N even, this takes the form

$$\langle x_i | T(\hat{p}) | x_j \rangle = \frac{\hbar^2}{2m} \begin{cases} \frac{K^2}{3} \left(1 + \frac{2}{N^2} \right) & (i = j) \\ \frac{2K^2}{N^2} \frac{(-1)^{j-i}}{\sin^2(\pi \frac{j-i}{N})} & (i \neq j). \end{cases} \tag{11.172}$$

(For N odd replace the 2 by 1 in the equation for $i = j$ and add a factor $\cos(\pi \frac{j-i}{N})$ in the numerator of the equation for $i \neq j$.) Note that Eq. 11.171 has the form

$$\langle x_i | T(\hat{p}) | x_j \rangle = \sum_{\kappa = -(N/2-1)}^{N/2} \phi_\kappa(x_j) \frac{\hbar^2 \kappa^2}{2m} \phi_\kappa^*(x_i) = (\mathbf{U}^\dagger \mathbf{T} \mathbf{U})_{ji} \tag{11.173}$$

exactly as in any pseudospectral scheme (cf. Eq. 11.137).

As described above, there is a variation on the original Fourier method in which the underlying basis is taken as $\frac{e^{\sqrt{-1}kx}}{\sqrt{2\pi}}$, where k is a continuous variable, $k \in [-K, K]$. This suggests a variation in the FGH method in which the region of integration is truncated *without discretizing* k (Willner, 2001). Again, the matrix elements can be evaluated analytically, with the resulting T matrix now taking a somewhat simpler form (Colbert, 1992):

$$\langle x_i | T(\hat{p}) | x_j \rangle \rightarrow \frac{\hbar^2}{2m} \int_{-K}^{K} \frac{e^{\sqrt{-1}ki\Delta x}}{\sqrt{2K}} k^2 \frac{e^{-\sqrt{-1}kj\Delta x}}{\sqrt{2K}} dk \tag{11.174}$$

$$= \frac{\hbar^2}{2m} \begin{cases} \frac{K^2}{3} & (i = j) \\ \frac{2K^2}{\pi^2} \frac{(-1)^{j-i}}{(j-i)^2} & (i \neq j). \end{cases} \tag{11.175}$$

In the limit $N \rightarrow \infty$, Eq. 11.172 is equivalent to Eq. 11.175. Since in the second procedure k is not discretized, this procedure avoids introducing periodic boundary conditions in x.

Adding the kinetic and potential energies (Eqs. 11.172 (or 11.175) and 11.167) we obtain

$$\mathbf{H}_{ij}^{\text{FGH}} = \mathbf{T}_{ij}^{\text{FGH}} + \mathbf{V}_{ij}^{\text{FGH}}. \tag{11.176}$$

In principle, the approximation of the FGH matrix is the replacement

$$H \rightarrow H^{\text{FGH}} = P_\sqcup(\hat{p}) P_\sqcap(\hat{p}) H P_\sqcap(\hat{p}) P_\sqcup(\hat{p}) \tag{11.177}$$

or

$$H \rightarrow H^{\text{FGH}} = P_\sqcap(\hat{p}) H P_\sqcap(\hat{p}), \tag{11.178}$$

again depending on the choice of underlying basis. This is in addition to the approximation inherent in the evaluation of the potential energy matrix elements.

11.6.5 The Dynamic Fourier Method

The Dynamic Fourier Method (traditionally known as the Fast Fourier Transform, or FFT method), was pioneered by Feit and Fleck (1982) and Kosloff and Kosloff (1983). Although similar in spirit to the Fourier Grid Hamiltonian method, the Dynamic Fourier Method avoids ever constructing the Hamiltonian matrix. Historically, the Dynamic Fourier Method was the first widely used pseudospectral method in quantum mechanics. It is simple to use, very accurate and very efficient.

In the Dynamic Fourier Method, the operation of $H\psi$ is calculated as follows:

1. The Hamiltonian operator is partitioned in the usual fashion, $H = T + V$. The strategy is to calculate each operator locally. The potential operator is already local in coordinate space, and therefore its operation is simply a multiplication of $V(x_j)$ by $\psi(x_j)$.

2. A local operation of the kinetic energy operator is possible in momentum space where it becomes a multiplication by the kinetic energy discrete spectrum: $T(k) = \hbar^2 k^2/2m$. Specifically, the kinetic energy operator is calculated by transforming ψ to momentum space by a backward Discrete Fourier Transform (DFT), multiplying by $T(k)$ and performing a forward DFT back to coordinate space. (Note that the use of Fourier transforms to calculate derivatives was described in a more general context in Exercise 6.10.)

The explicit matrix–vector representation of $H\psi = (T + V)\psi$ is as follows:

$$
\left\{ \frac{1}{N} \begin{pmatrix} 1 & 1 & 1 & \cdot & 1 \\ 1 & w & w^2 & \cdot & w^{N-1} \\ 1 & w^2 & w^4 & \cdot & w^{2(N-1)} \\ \cdot & \cdot & \cdot & \cdot & \cdot \\ 1 & w^{N-1} & w^{2(N-1)} & \cdot & w^{(N-1)^2} \end{pmatrix} \begin{pmatrix} T_0 & 0 & 0 & \cdot & 0 \\ 0 & T_1 & 0 & \cdot & 0 \\ 0 & 0 & T_2 & \cdot & 0 \\ \cdot & \cdot & \cdot & \cdot & \cdot \\ 0 & 0 & 0 & \cdot & T_{-1} \end{pmatrix} \times \right.
$$

$$
\begin{pmatrix} 1 & 1 & 1 & \cdot & 1 \\ 1 & w^{-1} & w^{-2} & \cdot & w^{-(N-1)} \\ 1 & w^{-2} & w^{-4} & \cdot & w^{-2(N-1)} \\ \cdot & \cdot & \cdot & \cdot & \cdot \\ 1 & w^{-(N-1)} & w^{-2(N-1)} & \cdot & w^{-(N-1)^2} \end{pmatrix} + \left. \begin{pmatrix} V(x_0) & 0 & 0 & \cdot & 0 \\ 0 & V(x_1) & 0 & \cdot & 0 \\ 0 & 0 & V(x_2) & \cdot & 0 \\ \cdot & \cdot & \cdot & \cdot & \cdot \\ 0 & 0 & 0 & 0 & V(x_{N-1}) \end{pmatrix} \right\}
$$

$$
\times \begin{pmatrix} \psi(x_0) \\ \psi(x_1) \\ \psi(x_2) \\ \cdot \\ \psi(x_{N-1}) \end{pmatrix}, \qquad (11.179)
$$

where $w = e^{2\pi i/N}$, $T_\kappa = \frac{\hbar^2 (\kappa \Delta k)^2}{2m}$, and we have factored out $\frac{1}{\sqrt{N}}$ from the orthogonal collocation matrices. (Note that T_κ runs over $[-N/2 + 1, \ldots, 0, \ldots, N/2]$ but to match the indexing of the FFT matrix the index of T is shifted to run from $[0, \ldots, N/2, -N/2 + 1, \ldots, -1]$.)

3. Given $H\psi(x, t_0)$, the advancement in time may be performed by any of a variety of methods to obtain $\psi(x, t_1)$, discussed below. Steps 1–3 are then iterated to obtain $\psi(t_2)$, and so on. The use of the DFT, because of its discrete sampling in k-space, is inextricably linked to periodic boundary conditions in x, that is, to basis Eq. 11.154, not to basis Eq. 11.145.

In practice, the DFT is implemented by the Fast Fourier Transform (FFT) algorithm, in which the number of multiplications is $\frac{1}{2}N \ln N$ rather than N^2 for straightforward implementation of the DFT. The use of the FFT gives a significant computational savings for large vectors. The FFT is based on exploiting the symmetry of the factors w^i. The basic idea of the FFT involves three steps (see Brigham (1974), Strang (1988); we present the steps here in terms of $\tilde{\psi} \rightarrow \psi$, since that is the forward FFT):

1. Split $\tilde{\psi}$ into $\tilde{\psi}'$ and $\tilde{\psi}''$ by separating its even- and odd-numbered components.
2. Transform $\tilde{\psi}' \rightarrow \psi'$, $\tilde{\psi}'' \rightarrow \psi''$ using the DFT matrix of size $N/2$.
3. Construct ψ from ψ' and ψ'' using the formulas

$$\psi_j = \psi'_j + w^j \psi''_j, \quad j = 0, \ldots, N/2 - 1 \tag{11.180}$$

$$\psi_{j+N/2} = \psi'_j - w^j \psi''_j, \quad j = 0, \ldots, N/2 - 1. \tag{11.181}$$

▶ **Exercise 11.9**   Consider the FFT for the case $N = 4$. In matrix form, the three steps above can be written as

$$\begin{pmatrix} 1 & 1 & 1 & 1 \\ 1 & i & i^2 & i^3 \\ 1 & i^2 & i^4 & i^6 \\ 1 & i^3 & i^6 & i^9 \end{pmatrix} = \begin{pmatrix} 1 & 0 & 1 & 0 \\ 0 & 1 & 0 & i \\ 1 & 0 & -1 & 0 \\ 0 & 1 & 0 & -i \end{pmatrix} \begin{pmatrix} 1 & 1 & 0 & 0 \\ 1 & -1 & 0 & 0 \\ 0 & 0 & 1 & 1 \\ 0 & 0 & 1 & -1 \end{pmatrix} \begin{pmatrix} 1 & 0 & 0 & 0 \\ 0 & 0 & 1 & 0 \\ 0 & 1 & 0 & 0 \\ 0 & 0 & 0 & 1 \end{pmatrix}$$

$$\tag{11.182}$$

a. Verify that the product of these matrices gives the DFT for $N = 4$. Explain why this decomposition corresponds to the three steps described above.

b. Verify that each of the 2×2 blocks in the middle matrix of the RHS is itself a DFT for $N = 2$. Work out the three-step decomposition for each of these 2×2 blocks.

The example in Exercise 11.9 illustrates the following general principle: a DFT of size $N = 2^l$ can be decomposed into a product of three matrices, the middle matrix consisting of two blocks, each of which is a DFT of size $\frac{N}{2} = 2^{l-1}$. As a result, the number of multiplications required for a DFT of size N, instead of being N^2, is just twice the number required for the DFT of size $\frac{N}{2}$ (second step) plus $\frac{N}{2}$ additional multiplications by w^j, $j = 1, \ldots N/2$ (third step). Representing by O_N the number of operations for the DFT of dimension N, we have

$$O_N = 2O_{N/2} + \frac{N}{2}. \tag{11.183}$$

Starting with $O_1 = 0$ and iterating, we find that

$$O_N = \frac{1}{2}Nl = \frac{1}{2}N \ln N. \tag{11.184}$$

▶ **Exercise 11.10**   Verify Eq. 11.184 by induction.

Solution   Assume that $O_N = \frac{1}{2}Nl$. Then $O_{2N} = \frac{1}{2}(2N)(l+1) = 2(\frac{1}{2}Nl) + N = 2O_N + N$, consistent with Eq. 11.183. Since $O_1 = 0$ the assertion is proved.

The Dynamic Fourier Method has two apparent advantages over both the FGH and other methods:

1. By exploiting the FFT algorithm to perform the DFT, the method scales semilinearly with the number of points ($N \ln N$), rather than as N^2 as with matrix multiplication or N^3 as with matrix diagonalization.

2. No matrices are ever constructed or stored. The FFT algorithm works simply by rearranging the elements of the input vector (i.e., the wavefunction).

These advantages of the FFT method should be regarded with some caution, however, since there are ways of obtaining comparable advantages with other methods as well.

11.6.6   Phase Space Analysis and the Scaled Fourier Method

In Section 11.1.1 we described a general approach to phase space analysis of spectral and pseudospectral methods. In the case of the Fourier method there is an interesting alternative way to understand the phase space structure (Kosloff, 1993).

In the Fourier method, the maximum wave number, K, is related to the sampling spacing Δx by $K = \pi / \Delta x$. Defining $k_{\text{range}} = k_{\text{max}} - k_{\text{min}} = 2K$ we have

$$k_{\text{range}} = 2\pi / \Delta x. \qquad (11.185)$$

Similarly,

$$x_{\text{range}} = 2\pi / \Delta k. \qquad (11.186)$$

Thus, grid *spacing* in coordinate space corresponds to grid *range* in momentum space, and vice versa: grid range in coordinate space corresponds to grid spacing in momentum space. The volume in phase space covered by the Fourier representation is calculated as the product of the range of the grid in coordinate and momentum. Using the relations $x_{\text{range}} = L$, $\Delta x = L/N$ and $p_{\text{range}} = \hbar k_{\text{range}}$, where N is the number of grid points, we find the phase space volume as

$$\text{volume} = x_{\text{range}} \times p_{\text{range}} = Lh/\Delta x = Nh \qquad (11.187)$$

(see Figure 11.7a). This simple result has the following appealing interpretation: the volume in phase space is proportional to the number of grid points, where the phase space volume per grid point is Planck's constant. A phase space representation is also useful for analyzing the efficiency of the Fourier method (see Figure 11.7b).

The Fourier method is restricted to evenly spaced grid points, and therefore converges very slowly with respect to the number of grid points for certain problems, such as the Coulomb problem. However, it is possible to perform a nonlinear coordinate transformation that effectively distorts the potential energy function and makes an evenly spaced grid suitable (Fattal, 1996). In one dimension, the original coordinate, q, is mapped to a new coordinate, $Q = Q(q, \alpha)$, where α is a parameter for the mapping (see Figure 11.8). Wavefunctions are transformed as follows: $\Psi(Q) = \psi(q(Q))$. This results in a new scalar product, $\langle \Psi | \Phi \rangle = \int \Psi^*(Q) \Phi(Q) J \, dQ = \int \psi^*(q)\phi(q) \, dq$, and in $T = -\frac{\hbar^2}{2m}(J^{-1}(Q)\frac{\partial}{\partial Q})^2$, where $J = \frac{\partial q}{\partial Q}$ in one dimension. Clearly, the mapping mixes the coordinates with the ki-

(a)

(b)

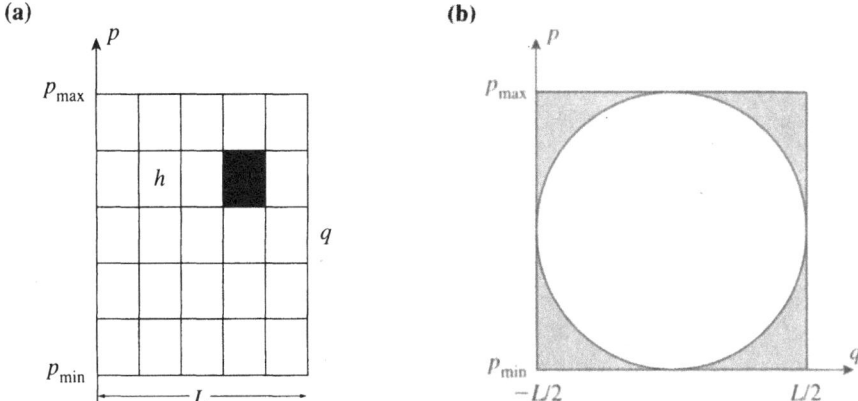

Figure 11.7 (a) The phase space structure underlying the Fourier method, $2Lp_{max} = Nh$. (b) The Fourier method is based on a square (or rectangular) phase space while for the harmonic oscillator the maximum energy contour is a circle. This implies a maximum efficiency for the Fourier method of $\pi/4$. This efficiency limit can be overcome using the mapped Fourier method (see text). Adapted from Kosloff (1993).

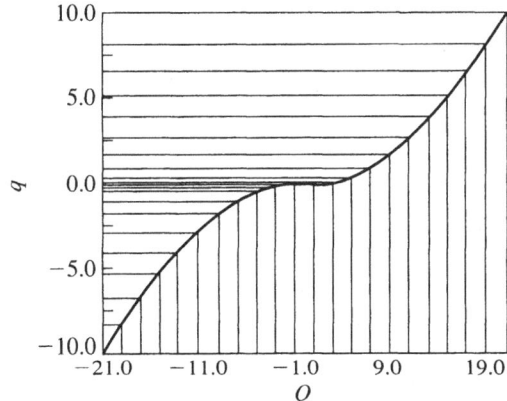

Figure 11.8 The relation between grids in the Cartesian coordinate system q and the mapped coordinate Q, given by the mapping function in the text with $\beta = 0.1$ and $A = 9.9$. Notice the high density of sampling points near the origin of the q-coordinate, as appropriate for the Coulomb problem. Adapted from Fattal (1996).

netic energy operator, inducing a correlation between the kinetic and the potential energy operators.

For the Coulomb problem the mapping $q = Q - A \arctan(\beta Q)$ is appropriate. Figure 11.9a shows the phase space included in the energy shells $n = 1, \ldots, 20$ contained within the unscaled grid, while Figure 11.9b shows the portion of the phase space contained within the scaled grid. Here p is the momentum conjugate to q, while P is conjugate to Q.

(a)   **(b)**

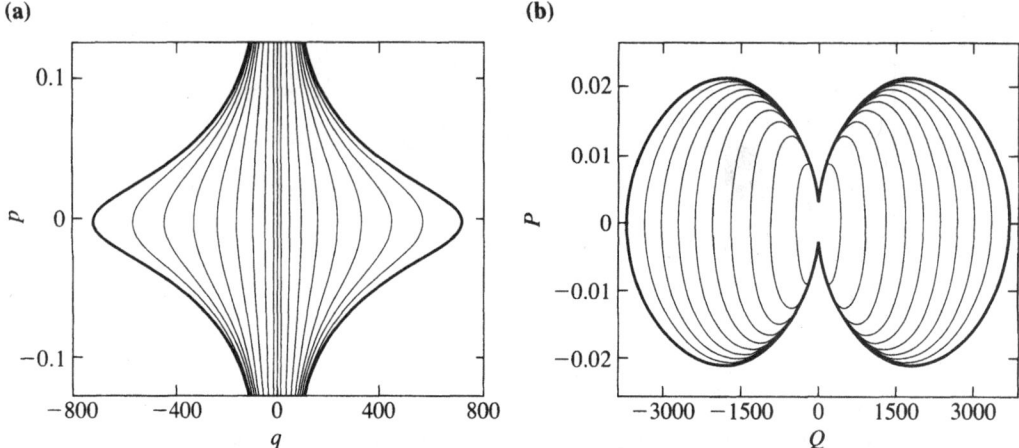

Figure 11.9 (a) The part of the energy shells $n = 1, 3, \ldots, 20$, which is represented by the grid rectangle in the unscaled phase space. It is seen that the grid cannot represent the momentum portion of the phase space. (b) The part of the energy shells $n = 1, 3, \ldots, 20$, which is represented by the grid rectangle in the scaled phase space. The same number of grid points are used as in (a), but now the grid covers all the shells up to $n = 20$. Adapted from Fattal (1996).

The mapped Fourier method is an active area of research, with recent ideas exploiting semiclassical concepts of the local de Broglie wavelength to optimize the grid spacing in one and multidimensions (Kokoouline, 1999; Willner, 2002; Goldfarb, 2004).

11.7   Time Propagation

So far we have discussed numerical methods for performing the operation $H\psi$ on a grid of points. We now turn to the question of the numerical implementation of $e^{-iHt/\hbar}\psi$. There are three basic strategies for doing this. The first is the straightforward, almost trivial, method of using the approximate eigenvalues and eigenvectors of H to propagate. The main drawbacks of this approach are that it is limited to time-independent Hamiltonians and it requires the construction and diagonalization of the Hamiltonian matrix, the latter scaling as N^3. The second approach is to retain the exponent structure of the propagator, but to approximate the propagator as a product of a kinetic and a potential factor (Feit, 1982):

$$e^{-iH\Delta t/\hbar} = e^{-i(\hat{p}^2/2m + V(\hat{x}))\Delta t/\hbar} \tag{11.188}$$

$$\approx e^{-i(T\Delta t/\hbar)}e^{-i(V\Delta t/\hbar)}, \tag{11.189}$$

where $T = \hat{p}^2/2m$. This approach is called the "Split Operator" method, and explicitly preserves unitarity. The third major strategy is to consider $e^{-iHt/\hbar}$ as a function of the Hamiltonian, that is, $f(H)$, and look for a suitable polynomial approximation to this function (Leforestier, 1991; Kosloff, 1994):

$$e^{-iHt/\hbar} \approx \sum_{n=0}^{N} c_n P_n(H). \tag{11.190}$$

Because we know how to calculate $H\psi$, for example by pseudospectral methods, we can calculate $H(H\psi)$ and ultimately $H^n\psi$ for any n by iteration; therefore, $P_n(H)\psi$ can be calculated for any polynomial of H. There are many variations on the third approach (some of which do not even look superficially to be in this form), which differ in the choice of the polynomial used. Below, we describe in more detail all of these propagation strategies.

11.7.1   The Split Operator Method

The Split Operator method is one of the simplest and most popular methods for time propagation of wavepackets (Feit, 1982). The method begins by representing the propagator over the global time interval $[0, \ t]$ as a product of propagators over short time intervals, Δt, where $N\Delta t = t$. Thus,

$$U(t, 0) = e^{-iHt/\hbar} = \underbrace{e^{-iH\Delta t/\hbar}e^{-iH\Delta t/\hbar}\cdots e^{-iH\Delta t/\hbar}}_{N\text{times}}. \tag{11.191}$$

The strategy is then to approximate each short time propagator as a product of a kinetic factor and a potential factor:

$$e^{-iH\Delta t/\hbar} = e^{-i(\hat{p}^2/2m+V(\hat{x}))\Delta t/\hbar}$$
$$\approx e^{-i(T\Delta t/\hbar)}e^{-i(V\Delta t/\hbar)} + O(\Delta t^2), \tag{11.192}$$

where $T = \hat{p}^2/2m$. Note that this product representation would be exact if T and V commuted; we expect the error, therefore, to be proportional to the commutator $[T, V]$.

▶ **Exercise 11.11**   Find the leading order corrections and show that they are indeed $O(\Delta t^2)$ and proportional to $[T, V]$.

Solution   Expand $e^{-i(T+V)\Delta t/\hbar}$:

$$e^{-i(T+V)\Delta t/\hbar} = 1 - i(T + V)\frac{\Delta t}{\hbar} + \frac{(-i)^2(T + V)^2 \, \Delta t^2}{2\hbar^2} + \cdots$$
$$= 1 - i(T + V)\frac{\Delta t}{\hbar} - \frac{(T^2 + V^2 + TV + VT) \, \Delta t^2}{2\hbar^2} + \cdots \tag{11.193}$$

Now expand $e^{-i(T\Delta t/\hbar)}e^{-i(V\Delta t/\hbar)}$:

$$e^{-i(T\Delta t/\hbar)}e^{-i(V\Delta t/\hbar)}$$
$$= \left(1 - iT\frac{\Delta t}{\hbar} + \frac{(-iT\Delta t)^2}{2\hbar^2} + \cdots\right)\left(1 - iV\frac{\Delta t}{\hbar} + \frac{(-iV\Delta t)^2}{2\hbar^2} + \cdots\right)$$
$$= 1 - iT\frac{\Delta t}{\hbar} - iV\frac{\Delta t}{\hbar} - \frac{T^2\Delta t^2}{2\hbar^2} - \frac{V^2\Delta t^2}{2\hbar^2} - TV\frac{\Delta t^2}{\hbar^2} + \cdots \tag{11.194}$$

Comparing the approximation, Eq. 11.194, with the exact expression, Eq. 11.193, we see that the error is

$$\text{Error} = \frac{TV - VT}{2}\frac{\Delta t^2}{\hbar^2} + \cdots = \frac{[T, V]}{2}\frac{\Delta t^2}{\hbar^2} + \cdots \tag{11.195}$$

It is easily verified that choosing the opposite order of the products,

$$e^{-iH\Delta t/\hbar} \approx e^{-i(V\Delta t/\hbar)}e^{-i(T\Delta t/\hbar)} \tag{11.196}$$

gives the same order of error. However, the leading order error can be eliminated by forming a symmetrized product of the kinetic and potential factors:

$$e^{-iH\Delta t/\hbar} \approx \{e^{-i(V\Delta t/2\hbar)}e^{-i(T\Delta t/2\hbar)}\}\{e^{-i(T\Delta t/2\hbar)}e^{-i(V\Delta t/2\hbar)}\} + O(\Delta t^3)$$
$$= e^{-i(V\Delta t/2\hbar)}e^{-i(T\Delta t/\hbar)}e^{-i(V\Delta t/2\hbar)} + O(\Delta t^3) \tag{11.197}$$

▶ **Exercise 11.12**   Find the leading order corrections to the symmetrized product and show that they are indeed $O(\Delta t^3)$.

Solution   Expand $e^{-i(T+V)\Delta t/\hbar}$:

$$e^{-i(T+V)\Delta t/\hbar} = 1 - i(T+V)\frac{\Delta t}{\hbar} + \frac{(-i)^2(T+V)^2\Delta t^2}{2\hbar^2} + \frac{(-i)^3(T+V)^3\,\Delta t^3}{3!\hbar^3} + \cdots$$

$$= 1 - i(T+V)\frac{\Delta t}{\hbar} - \frac{(T^2+V^2+TV+VT)\,\Delta t^2}{2\hbar^2}$$

$$+ i\frac{(T^3+TVT+T^2V+TV^2+VT^2+V^2T+VTV+V^3)}{3!\hbar^3} + \cdots \tag{11.198}$$

Now expand $e^{-iV\Delta t/2\hbar}e^{-iT\Delta t/\hbar}e^{-iV\Delta t/2\hbar}$:

$$e^{-iV\Delta t/2\hbar}e^{-iT\Delta t/\hbar}e^{-iV\Delta t/2\hbar}$$

$$= \left[1 - iV\frac{\Delta t}{2\hbar} + \frac{(-iV\Delta t)^2}{8\hbar^2} + \cdots\right]\left[1 - iT\frac{\Delta t}{\hbar} + \frac{(-iT\,\Delta t)^2}{2\hbar^2} + \cdots\right]$$

$$\times \left[1 - iV\frac{\Delta t}{2\hbar} + \frac{(-iV\Delta t)^2}{8\hbar^2} + \cdots\right]$$

$$= 1 - i(T+V)\frac{\Delta t}{\hbar} - \frac{(T^2+V^2+TV+VT)\Delta t^2}{2\hbar^2}$$

$$+ i\left(\frac{V^3}{6} + \frac{T^3}{6} + \frac{VTV}{4} + \frac{V^2T}{8} + \frac{VT^2}{4} + \frac{T^2V}{4} + \frac{TV^2}{8}\right)\frac{\Delta t^3}{\hbar^3} + \cdots \tag{11.199}$$

Comparing the approximation, Eq. 11.199, with the exact expression, Eq. 11.198, we see that the error of $O(\Delta t^2)$ vanishes and the leading error term is

$$\text{Error} = i\frac{\Delta t^3}{\hbar^3}\left(\frac{T[V,T]}{12} + \frac{[T,V]T}{12} + \frac{[T,V],V}{24} + \frac{V[V,T]}{24}\right)$$

$$= i\frac{\Delta t^3}{\hbar^3}\left(\frac{[T,[V,T]]}{12} + \frac{[V,[V,T]]}{24}\right). \tag{11.200}$$

When multiple time steps are concatenated, the half-time steps with evolution under V coalesce, and one obtains

$$e^{-iHt/\hbar} =$$
$$\{e^{-i(V\Delta t/2\hbar)}e^{-i(T\Delta t/\hbar)}e^{-i(V\Delta t/2\hbar)}\}\{e^{-i(V\Delta t/2\hbar)}e^{-i(T\Delta t/\hbar)}e^{-i(V\Delta t/2\hbar)}\}\cdots$$
$$\underbrace{\{e^{-i(V\Delta t/2\hbar)}e^{-i(T\Delta t/\hbar)}e^{-i(V\Delta t/2\hbar)}\}}_{N \text{ times}}$$
$$= e^{-i(V\Delta t/2\hbar)}e^{-i(T\Delta t/\hbar)}e^{-i(V\Delta t/\hbar)}e^{-i(T\Delta t/\hbar)}e^{-i(V\Delta t/\hbar)}\cdots$$
$$\underbrace{e^{-i(T\Delta t/\hbar)}e^{-i(V\Delta t/\hbar)}}_{N-1 \text{ times}}\ e^{-i(T\Delta t/\hbar)}e^{-i(V\Delta t/2\hbar)}.$$

$$(11.201)$$

The operation $e^{-iV(\hat{x})\Delta t/\hbar}\psi$ is calculated by multiplication, $e^{-iV(x)\Delta t/\hbar}\psi(x)$; and $e^{-iT(\hat{p})\Delta t/\hbar}\psi = e^{-i\hat{p}^2\Delta t/2m\hbar}\psi$ is calculated by $Ze^{-ip^2\Delta t/2m\hbar}Z^\dagger\psi(x)$, where $Z^\dagger$ is the transformation between the coordinate and momentum representations. In most applications the Split Operator method is implemented in the Fourier basis. In the Fourier basis $Z^\dagger$ is just a Discrete Fourier Transform (matrix elements of the form $Z^\dagger_{ij} = \frac{1}{\sqrt{N}}e^{ip_i x_j/\hbar}$), which may be calculated very efficiently using an FFT. However, it is worth noting that the Split Operator method is not limited to the Fourier basis; in fact it can be used with other spectral bases provided that the Discrete Fourier Transform is replaced by the corresponding pseudospectral $\leftrightarrow$ spectral transformation matrix. The Split Operator method is manifestly unitary, and is recommended whenever the Hamiltonian can be written as a sum of operators that depend on coordinates and operators that depend on momenta. However, the method cannot handle operators that mix coordinates and momenta, such as an operator of the form $e^{i\hat{p}\hat{x}}$.

The Split Operator method may still be used, but requires slightly more work if the dynamics is evolving on multiple electronic states. At every time step, in addition to transforming between the coordinate and momentum representations, the coordinate space wavefunction has to be transformed into a representation in which the potential is diagonal. For two electronic states this transformation can be performed analytically. Since this is a common case we present the algebra explicitly (Schwendner, 1997). Consider the two-electronic-state evolution operator:

$$\exp\left[-\frac{i}{\hbar}\begin{pmatrix} T+V_1 & V_{12} \\ V_{21} & T+V_2 \end{pmatrix}\Delta t\right] =$$
$$\exp\left[-\frac{i}{\hbar}\begin{pmatrix} T & 0 \\ 0 & T \end{pmatrix}\frac{\Delta t}{2}\right]\exp\left[-\frac{i}{\hbar}\begin{pmatrix} V_1 & V_{12} \\ V_{21} & V_2 \end{pmatrix}\Delta t\right]\exp\left[-\frac{i}{\hbar}\begin{pmatrix} T & 0 \\ 0 & T \end{pmatrix}\frac{\Delta t}{2}\right]$$
$$+ O(\Delta t^3). \qquad (11.202)$$

The complication arises since the potential energy, V, is now a matrix, and is no longer diagonal. This difficulty can be overcome by transforming the 2×2 potential matrix into a representation in which it is diagonal, and then transforming back after the multiplication operation on the wavefunction is performed. For a 2×2 Hermitian matrix this transformation can be done analytically (see Exercise 8.74). Note that since the V matrix is coordinate dependent, so is the transformation matrix, U, that diagonalizes it. Denoting this matrix by **U**, since **U** is unitary we can write

$$\exp\left[-\frac{i}{\hbar}\begin{pmatrix} V_1 & V_{12} \\ V_{21} & V_2 \end{pmatrix}\Delta t\right]$$

$$= \mathbf{U}\exp\left[-\frac{i}{\hbar}\mathbf{U}^\dagger\begin{pmatrix} V_1 & V_{12} \\ V_{21} & V_2 \end{pmatrix}\mathbf{U}\,\Delta t\right]\mathbf{U}^\dagger = \mathbf{U}\begin{pmatrix} e^{-\frac{i}{\hbar}\lambda_1\Delta t} & 0 \\ 0 & e^{-\frac{i}{\hbar}\lambda_2\Delta t} \end{pmatrix}\mathbf{U}^\dagger \qquad (11.203)$$

$$= \exp\left[-\frac{i}{\hbar}(V_1 + V_2)\frac{\Delta t}{2}\right]\left[\cos\left(\sqrt{D}\frac{\Delta t}{2\hbar}\right)\begin{pmatrix} 1 & 0 \\ 0 & 1 \end{pmatrix} + i\frac{\sin(\sqrt{D}\frac{\Delta t}{2\hbar})}{\sqrt{D}}\begin{pmatrix} V_2 - V_1 & -2V_{12} \\ -2V_{21} & V_1 - V_2 \end{pmatrix}\right],$$

$$(11.204)$$

where we have used the result of Exercise 8.74.

► **Exercise 11.13**   Verify Eq. 11.204.

Equation 11.204 is in a convenient form for numerical work, requiring only straightforward matrix multiplication. Combined with Eq. 11.202, it gives a full prescription for the Split Operator propagator for the two-surface case. For three or more potentials, however, there is no such simple analytical transformation, and a numerical matrix diagonalization must be done for every coordinate. For time-independent potentials, this numerical diagonalization does not need to be repeated at every time step, but can be performed once at the beginning of the propagation and the transformation matrices stored. However, for time-dependent potentials— for example, evolution in the presence of pulsed laser fields—the diagonalization must be performed at every time step.

11.7.2   Symplectic Integrators

Symplectic integrators are integrators that respect the symplectic symmetry properties of a dynamical system; they preserve the canonical relationship between p and q. Such integrators have been widely used in classical mechanics, and less so in quantum mechanics. Recently, some encouraging preliminary applications of these integrators to quantum mechanics have been demonstrated (Gray, 1996).

The starting point is that the Time-Dependent Schrödinger Equation can be written in discrete form as

$$i\frac{d}{dt}c(t) = H \cdot c(t). \qquad (11.205)$$

Take H to be a real symmetric, time-independent matrix. Define

$$q(t) = \sqrt{2}\mathrm{Re}(c(t)), \qquad p(t) = \sqrt{2}\mathrm{Im}(c(t)), \qquad (11.206)$$

and the Hamiltonian function

$$h(q, p) = \frac{1}{2}\sum_{ij} H_{ij}(p_i p_j + q_i q_j). \qquad (11.207)$$

Then Eq. 11.205 is equivalent to the equations

$$\frac{d}{dt}q(t) = \frac{\partial}{\partial p}h(q, p) = H \cdot p, \tag{11.208}$$

$$\frac{d}{dt}p(t) = -\frac{\partial}{\partial q}h(q, p) = -H \cdot q, \tag{11.209}$$

which are just Hamilton's equations for a large set of quadratically coupled harmonic oscillators. $h(q, p)$ has the physical interpretation of the mean wavepacket energy. A discrete procedure for integrating these equations is

$$p_j = p_{j-1} - b_j \tau H \cdot q_{j-1}, \tag{11.210}$$

$$q_j = q_{j-1} + a_j \tau H \cdot p_j, \quad j = 1, 2, \ldots, m. \tag{11.211}$$

Equation 11.211 involves numerical work equivalent to $2m$ evaluations of H on a real vector, but the evolution matrix contains terms up to $(H\tau)^{2m}$. In essence, what makes Eq. 11.211 symplectic is the use of p_j and not p_{j-1} in the determination of q_j at each iteration.

11.7.3   Polynomial Methods

Polynomial methods represent the propagator as $e^{-iHt/\hbar}\psi = \sum_n a_n P_n(H)\psi$, where $P_n(H)$ is some polynomial of the Hamiltonian operator. Once a polynomial sequence is chosen, the a_n are uniquely determined. Polynomial methods can be divided into two categories: those that choose the polynomial in advance, and those that do not (Leforestier, 1991; Kosloff, 1994). An example of a method that chooses the polynomial in advance is the Chebyshev method (Tal-Ezer, 1984). An example of the second type of polynomial expansion is the Short Iterative Lanczos propagator (Park, 1986). We now discuss both of these methods in some detail.

Chebyshev Propagation

For functions of scalar arguments, $f(x)$, x in the interval $[-1, 1]$, $f(x) = \sum a_n P_n(x)$, the representation in terms of N Chebyshev polynomials minimizes the maximum error in the representation of the function over the interval. This minimization of the maximum error allows the use of Chebyshev polynomials to reach the level of machine accuracy in numerical computations.

Since we are interested in expanding the propagator in terms of Chebyshev polynomials, we are interested in the representation of exponential functions in terms of Chebyshev polynomials. For exponential functions this expansion takes the specific form

$$e^{-i\alpha x} = \sum a_n(\alpha)\Phi_n(-ix), \tag{11.212}$$

where

$$a_n(\alpha) = \int_{-i}^{i} \frac{e^{i\alpha x}\Phi_n(x)\, dx}{(1 - x^2)^{1/2}} = 2J_n(\alpha) \tag{11.213}$$

with

$$a_0(\alpha) = J_0(\alpha), \tag{11.214}$$

where the $J_n(\alpha)$ are Bessel functions. The Chebyshev recurrence relation is

$$\Phi_{n+1} = -2ix\Phi_n + \Phi_{n-1}. \tag{11.215}$$

In implementing this expansion for the propagator, the argument $-iHt/\hbar$ must first be mapped onto the domain $[-i, i]$:

$$H_{\text{norm}} = 2\frac{H - I(\Delta E_{\text{grid}}/2 + E_{\text{min}})}{\Delta E_{\text{grid}}}, \tag{11.216}$$

where $\Delta E_{\text{grid}} = E_{\text{max}} - E_{\text{min}}$ is the range of energy supported by the grid. Here, $E_{\text{max}} = V_{\text{max}} + T_{\text{max}}$ and $E_{\text{min}} = V_{\text{min}}$ are the maximum and minimum energy supported by the grid, and $T_{\text{max}} = \hbar^2 k_{\text{range}}^2/2m$ is the maximum kinetic energy supported by the grid, with $k_{\text{range}} = \pi/\Delta x$. With this mapping, the wavefunction is given by

$$\psi(t) \approx e^{-i(E_{\text{min}}t/\hbar + \alpha)} \sum_n a_n(\alpha)\Phi_n(-iH_{\text{norm}})\psi(0), \tag{11.217}$$

where the Φ_n are the complex Chebyshev polynomials and $\alpha = \frac{\Delta E_{\text{grid}}t}{2\hbar}$. The polynomials are generated by the recurrence relation,

$$\phi_{n+1} = -2iH_{\text{norm}}\phi_n + \phi_{n-1}, \tag{11.218}$$

where

$$\phi_n = \Phi_n(-iH_{\text{norm}})\phi(0); \tag{11.219}$$

the recurrence is started with

$$\phi_0 = \psi(0) \tag{11.220}$$

and

$$\phi_1 = -iH_{\text{norm}}\psi(0). \tag{11.221}$$

The Chebyshev propagator effectively does the propagation in a single time step, and intermediate time results are not automatically obtained. However, since all the time dependence and none of the spatial dependence is in the Bessel function coefficients, once the calculation of the Chebyshev polynomials is done, information at any intermediate time can be obtained cheaply. Similarly, time-correlation functions can be obtained by calculating the overlaps of the Chebyshev polynomials with the wavefunction of interest, and then weighting these time-independent overlaps by Bessel functions of time-dependent arguments.

Short Iterative Lanczos Propagator

As an example of the second type of polynomial expansion, one for which the polynomial is not determined in advance, we consider the Short Iterative Lanczos propagator. We first introduce the concept of a Krylov space. This is a subspace of the full Hilbert space, obtained by acting with a linear operator on an initial state N times. For example, if the linear operator is the Hamiltonian, the Krylov space is spanned by the vectors $u_j = H^j\psi(0)$. The Lanczos method constructs a matrix representation of the Hamiltonian in a basis spanned by the Krylov space; the Hamiltonian matrix is then diagonalized and the diagonal representation is used to propagate the initial state within the Krylov space. Since the operator H is used to build the Krylov space, the Krylov space is in fact tailored to include that portion of the Hilbert space that the wavefunction explores in the near future.

The functions that define the Krylov space, $u_j = H^j \psi(0)$, are generally not orthogonal. However, in practice, it is usually more convenient to work with orthogonal functions. In the Lanczos method, as each new Krylov vector is constructed it is also orthogonalized to the previous Krylov vectors. Specifically, the first basis function, q_0, is just the initial state itself:

$$q_0 = \psi(0). \tag{11.222}$$

The second basis function, q_1, is determined by operating with H on q_0, and subtracting off the component of q_0 in the result:

$$H q_0 = \alpha_0 q_0 + \beta_0 q_1, \tag{11.223}$$

where $\alpha_0 = \langle q_0 | H | q_0 \rangle$ and $\beta_0 = \langle q_1 | H | q_0 \rangle$. The third member of the basis, q_2, is determined by operating with H on q_1, and subtracting off the component of both q_1 and q_0 in the result:

$$H q_1 = \beta_0 q_0 + \alpha_1 q_1 + \beta_1 q_2, \tag{11.224}$$

where $\alpha_1 = \langle q_1 | H | q_1 \rangle$ and $\beta_1 = \langle q_2 | H | q_1 \rangle$. The general expression takes the form

$$H q_j = \beta_{j-1} q_{j-1} + \alpha_j q_j + \beta_j q_{j+1}, \tag{11.225}$$

where the coefficients are given by

$$\alpha_j = \langle q_j | H q_j \rangle \tag{11.226}$$

and

$$\beta_{j-1} = \langle q_j | H q_{j-1} \rangle. \tag{11.227}$$

The striking feature of the general expression is that it never has more than three terms; for example, $H q_2$ has no q_0 term. This is a direct result of Hermiticity: since $\langle q_2 | H | q_0 \rangle = 0$ (which can be seen by taking the inner product of Eq. 11.223 with q_2), it follows that $\langle q_0 | H | q_2 \rangle$ must also be zero. We now note that the various coefficients in Eqs. 11.223, 11.224, and 11.225 are matrix elements of the Hamiltonian operator in the Lanczos basis. This matrix is evidently tridiagonal, and takes the following form:

$$H_N = \begin{pmatrix} \alpha_0 & \beta_0 & 0 & \cdots & \cdots & \cdots & 0 \\ \beta_0 & \alpha_1 & \beta_1 & 0 & \cdots & \cdots & 0 \\ 0 & \beta_1 & \alpha_2 & \beta_2 & 0 & \cdots & 0 \\ \vdots & \vdots & \ddots & \ddots & \vdots & \vdots & 0 \\ 0 & \cdots & \cdots & \cdots & \beta_{N-3} & \alpha_{N-2} & \beta_{N-2} \\ 0 & \cdots & \cdots & \cdots & \cdots & \beta_{N-2} & \alpha_{N-1} \end{pmatrix}, \tag{11.228}$$

where we have assumed the Hermiticity of the matrix to equate the lower and upper half of the H matrix.

▶ **Exercise 11.14**   According to Eq. 11.225, $Hq_2 = \beta_1 q_1 + \alpha_2 q_2 + \beta_2 q_3$. Show explicitly that the coefficient of q_1 is indeed the same as β_1 defined by Eq. 11.224.

Solution   Define the quantity $\gamma_1 \equiv \langle q_1 | H q_2 \rangle$. The task then is to show that $\gamma_1 = \beta_1$ as defined by Eq. 11.224. We have

$$\gamma_1 = \langle q_1 | H q_2 \rangle$$

$$= \left\langle q_1 \left| H \frac{H q_1 - \alpha_1 q_1}{\beta_1} \right. \right\rangle$$

$$= \frac{\langle q_1 | H^2 | q_1 \rangle - \alpha_1 \langle q_1 | H | q_1 \rangle}{\beta_1}, \tag{11.229}$$

where we have used the definition of q_2 from Eq. 11.224. Since q_2 is assumed to be normalized, we have from Eq. 11.224

$$\beta_1^2 = \langle H q_1 - \alpha_1 q_1 | H q_1 - \alpha_1 q_1 \rangle$$

$$= \langle q_1 | H^2 | q_1 \rangle - 2\alpha_1 \langle q_1 | H | q_1 \rangle + \alpha_1^2$$

$$= \langle q_1 | H^2 | q_1 \rangle - \alpha_1 \langle q_1 | H | q_1 \rangle, \tag{11.230}$$

where we have used the relation $\alpha_1 = \langle q_1 | H | q_1 \rangle$. Dividing both sides of Eq. 11.230 by β_1 and comparing with Eq. 11.229 completes the proof.

▶ **Exercise 11.15**   Use a similar approach to show explicitly that $\langle q_1 | H q_3 \rangle = 0$.

Diagonalization of the matrix representation of the Hamiltonian gives

$$Z^\dagger H_N Z = D_N, \tag{11.231}$$

where D_N is a diagonal matrix of eigenvalues of H_N. Using the transformation matrix Z, one obtains the following expression for the propagator:

$$U(\Delta t) = e^{-i H_N \Delta t / \hbar} = Z e^{-i Z^\dagger D_N Z \Delta t / \hbar} Z^\dagger = Z e^{-i D_N \Delta t / \hbar} Z^\dagger. \tag{11.232}$$

The propagated wavefunction is then given by

$$\psi(\Delta t) = Z e^{-i D_N \Delta t / \hbar} Z^\dagger \psi(0). \tag{11.233}$$

Note that in the Lanczos basis, the initial state $\psi(0)$ is just the column vector $(1, 0, \ldots 0)^T$. Hence, if we view $Z e^{-i D_N \Delta t / \hbar} Z^\dagger$ as an $N \times N$ matrix, only the first column of this matrix affects the calculation of $\psi(\Delta t)$.

We have commented above that the Krylov space spanned by $u_j = H^j \psi(0)$ spans that subspace of the Hilbert space where $\psi(0)$ will evolve in short times. The reasoning is that this is the same subspace spanned by the Taylor series expansion of the propagator, which is valid at short times. The size of the Krylov space determines the time for which the evolving wavepacket will be confined within the Krylov space; the larger the Krylov space generated, the longer it should be able to describe the packet dynamics. Indeed, one can show that for a finite Hilbert space—an N-level system—a Krylov space of size N generates the exact dynamics, within numerical error. In practice, there is a trade-off between generating a large

Krylov space and using it for long time step propagation, and generating a small Krylov space but using it for short time steps and updating it frequently, generating a new Krylov space with the current initial state. The latter is called the Short Iterative Lanczos method. There is some empirical evidence that the optimal numerical efficiency is obtained with a Krylov space whose dimension is between six and twenty, depending on the system. The Short Iterative Lanczos method can be used with a variable time step. A convenient way to estimate the time step is to calculate the quantity $||\frac{\partial}{\partial t} - \frac{H}{i\hbar}\psi(0)||$, where $\frac{\partial \psi}{\partial t}$ is obtained as the finite difference between $\psi(\Delta t)$ calculated by the Lanczos method and $\psi(t)$. The difference between these two quantities is an estimate of the numerical error; by calculating this quantity and maintaining it below a desired error tolerance, one can propagate with an adaptive step size.

11.7.4   Spectral Analysis of Time-Correlation Functions

The goal of wavepacket propagation is often to calculate some form of energy-resolved transition probability. This transition probability may be, for example, an absorption spectrum, $\sigma(E)$, or a scattering matrix element, $S_{\beta\alpha}(E)$. These transition probabilities, as well as many others, can be formulated as Fourier transforms of wavepacket time-correlation functions, that is, $\sigma(E) = \int_{-\infty}^{\infty} C(t)e^{iEt/\hbar}\, dt$. Thus, one is often in a situation of having a time signal of some form and wanting to know its Fourier transform.

Conventional wisdom is that the maximum resolution of the frequency spectrum is determined by the length of the time signal according to the time–energy uncertainty principle, $\Delta E \approx \hbar/T$. This is true if the frequency points are evenly spaced. However, if the actual spectrum has detailed structure in some regions and is flat in other regions a set of evenly spaced frequencies will not be optimal and it is possible to do much better than the uncertainty principle limit. We describe in this section a remarkable method for obtaining high-resolution spectra from short time signals. The method does not actually violate the uncertainty principle, but simply circumvents it, essentially by exploiting prior information on the functional form of the time signal. The method can be applied not only to calculations but also to experimental data, providing high-resolution spectral information from a short time signal.

The usual way to construct a frequency spectrum from a time signal is via a Discrete Fourier Transform (DFT):

$$\sigma(\omega_m) = \sum_{n=1}^{N} C(t_n)e^{i\omega_m t_n}, \tag{11.234}$$

where the t_n are the times at which the signal is known. In the DFT, the frequencies ω_m are specified a priori; since we assume no a priori knowledge of the spectrum, the ω_m are usually taken to be evenly spaced. The DFT amplitudes can then be viewed as the optimal solution (in a least squares sense) to the following fitting problem: find the set of $\sigma(\omega_m)$, $m = 1, \ldots, N$, such that

$$C(t_n) = \frac{1}{N}\sum_{m=1}^{N} \sigma(\omega_m)e^{-i\omega_m t_n}. \tag{11.235}$$

The equivalence of Eqs. 11.234 and 11.235 follows from the properties of the inverse DFT (see Eq. 6.54). A consequence of the use of evenly spaced points is indeed that the highest frequency resolution possible in the DFT is given by the uncertainty principle, $\Delta E \approx \hbar / T$, where T is the length of the time signal.

Consider now allowing both the frequencies and the amplitudes to be optimized. Writing the signal as

$$C(t_n) = \sum_{k=1}^{K} d_k e^{-i \omega_k t_n}, \quad n = 1, 2, \ldots, N, \tag{11.236}$$

if the number N of sampling points in time is larger than the number $2K$ of unknowns contributing to the signal, which consists of K frequencies ω_k with their amplitudes d_k, one should be able in principle to extract all the unknowns. This problem, sometimes called "harmonic inversion," is a fundamental problem in physics, electrical engineering and many other diverse fields. Unfortunately, as opposed to the DFT, the harmonic inversion problem with variable frequencies is a nonlinear fitting problem.

In this section we describe a new and efficient approach to this fitting problem that has recently emerged from the literature on wavepacket correlation functions. The approach, invented by Wall and Neuhauser (WN) in terms of continuous time signals (Wall, 1995) and later reformulated by Mandelshtam and Taylor (MT) in terms of discrete time signals (Mandelshtam, 1997), involves two crucial ideas. The first is to associate the signal $C(t)$ with an autocorrelation function of a dynamical system:

$$C(t) = \langle \Phi_0 | e^{-it\hat{\Omega}} | \Phi_0 \rangle. \tag{11.237}$$

This establishes an equivalence between the problem of extracting spectral information from the signal with the one of diagonalizing the evolution operator $e^{-it\hat{\Omega}}$ of the underlying dynamical system. Because the operator $e^{-it\hat{\Omega}}$ has restrictions on the form of its eigenvalues and eigenfunctions, Eq. 11.237 is equivalent to assuming something about the functional form of the time signal.

The second crucial idea is that neither the identity of the operator $e^{-it\hat{\Omega}}$ nor the basis functions need be known explicitly: the matrix elements of $e^{-it\hat{\Omega}}$ can be expressed in terms of $C(t)$ alone. Essentially, the one-dimensional sequence $C(t)$ is converted to a two-dimensional array; the latter can be diagonalized, so that the original nonlinear fitting problem has been transformed into a standard problem in linear algebra. Note that from the point of view of information content the matrix contains no more information than the original sequence $C(t)$, but it doesn't need to: the final goal is a set of $2K$ parameters (the K eigenvalues and K overlaps of the eigenvectors), which is a one-dimensional sequence.

To develop the first idea, we express the evolution operator, $e^{-it\hat{\Omega}}$, in terms of its eigenvalues and eigenvectors:

$$e^{-it\hat{\Omega}} = \sum_{k=1}^{K} e^{i\omega_k t} |u_k\rangle \langle u_k|, \tag{11.238}$$

where the set of eigenvectors $\{u_k\}$ form an orthonormal basis set. Inserting Eq. 11.238 into Eq. 11.237 we obtain

$$C(t) = \sum_{k=1}^{K} d_k e^{-i\omega_k t}, \tag{11.239}$$

with

$$d_k \equiv |\langle u_k | \Phi_0 \rangle|^2, \tag{11.240}$$

which is identical with Eq. 11.236. In keeping with conventional quantum mechanics, we assume here that the Hamiltonian $\hat{\Omega}$ is Hermitian, implying that both ω_k and d_k are real with $d_k > 0$. In this special case the signal has time reversal symmetry: $C(-t) = C^*(t)$. Note though that in a more general formalism (see below) these assumptions are not necessary.

Equations 11.239–11.240 show the equivalence of the harmonic inversion problem to one of matrix diagonalization, reducing this centuries-old problem to a standard linear algebra calculation. For the case of a discrete time signal, $C(t_n) = C(n\tau)$, it is convenient to define the evolution operator for a single time step, τ:

$$\hat{U} = e^{-i\tau\hat{\Omega}} = \sum_{k=1}^{K} e^{i\omega_k \tau} |u_k\rangle \langle u_k|. \tag{11.241}$$

Then $e^{-it\hat{\Omega}} = \hat{U}^n$ and the discrete analog of Eq. 11.239 is

$$C(t_n) = C(n\tau) = \sum_{k=1}^{K} d_k e^{-i\omega_k n\tau}. \tag{11.242}$$

The remaining step is how to construct a two-dimensional array from the one-dimensional time signal $c_n = C(n\tau)$. Consider the Krylov-type basis

$$\Psi_j \equiv \Psi(z_j) = \sum_{n=0}^{M} \left(\hat{U}/z_j\right)^n \Phi_0, \tag{11.243}$$

for a set of complex values $z_j \equiv e^{i\phi_j}$, $j = 1, 2, \ldots, N$ taken on the unit circle (real ϕ_j). The function $\Psi(z_j)$ will be dominated by the eigenvectors $|u_k\rangle$ whose eigenvalues $u_k = e^{i\omega_k \tau}$ are close to z_j. This allows for control of the energy window of eigenvalues by the choice of ϕ_j. The decomposition of the problem into separate energy bands by the choice of the ϕ_j is known as filter diagonalization, which is an optional, but normally a desirable, component of the harmonic inversion scheme.

The matrix elements of the evolution operator in this Krylov-type basis take the form

$$U_{jj'} = \langle \Psi_j | \hat{U} | \Psi_{j'} \rangle. \tag{11.244}$$

Since the basis is not orthogonal we also need the overlap matrix

$$S_{jj'} = \langle \Psi_j | \Psi_{j'} \rangle. \tag{11.245}$$

Inserting Eq. 11.243 into Eqs. 11.244 and 11.245 we obtain

$$U_{jj'} = \sum_{n'=0}^{M} \sum_{n=0}^{M} c_{n-n'+1} z_j^{-n} z_{j'}^{n'} \tag{11.246}$$

$$S_{jj'} = \sum_{n'=0}^{M} \sum_{n=0}^{M} c_{n-n'} z_j^{-n} z_{j'}^{n'}. \tag{11.247}$$

The double sum in Eqs. 11.246–11.247 can be reduced to a single sum (see Eqs. 11.256–11.257 below for a similar but more general result). Although this simplifies the construction of the matrix U, it should not obscure the main point: Eq. 11.246 shows that the matrices $U_{jj'}$ and $S_{jj'}$ can be constructed from the one-dimensional sequence, c_n, allowing the use of matrix diagonalization techniques to solve the harmonic inversion problem.

Since the basis functions $\{\Psi_j\}$ are generally not orthogonal, we must solve the generalized eigenvalue problem (see Exercises 8.79–8.80):

$$\mathbf{U}\,\mathbf{B}_k = u_k \mathbf{S}\,\mathbf{B}_k, \tag{11.248}$$

where the bold characters $\mathbf{U}$ and $\mathbf{S}$ define the corresponding matrices of size $N \times N$. The eigenvectors $\mathbf{B}_k$ are then column vectors with elements B_{jk} satisfying

$$|u_k\rangle = \sum_{j=1}^{K} B_{jk} |\Psi_j\rangle. \tag{11.249}$$

Once we have the eigenvectors $\mathbf{B}_k$ the amplitudes d_k can be obtained as

$$d_k \equiv |\langle u_k|\Phi_0\rangle|^2 = \left| \sum_{j=1}^{K} B_{jk} \langle \Psi_j|\Phi_0\rangle \right|^2. \tag{11.250}$$

To generalize the formalism to complex $\{\omega_k\}$, it is convenient to use a different convention for the inner product on the Hilbert space than we have used elsewhere in this book. Thus, instead of $\langle u|\hat{A}|v\rangle \equiv \int_{-\infty}^{\infty} u^*(x)(\hat{A}v(x))\,dx$ we use $(u|\hat{A}|v) \equiv \int_{-\infty}^{\infty} u(x)(\hat{A}v(x))\,dx$. Then

$$C(t) = \left(\Phi_0 | e^{-it\hat{\Omega}} | \Phi_0 \right) \tag{11.251}$$

$$c_n = \left(\Phi_0 | \hat{U}^n | \Phi_0 \right) = \left(\Phi_0 | e^{-in\tau\hat{\Omega}} | \Phi_0 \right) \tag{11.252}$$

$$\hat{U} = \sum_{k=1}^{K} u_k \, |u_k) \, (u_k| \,. \tag{11.253}$$

Using the alternate inner product definition, the matrix elements of $\hat{U}$ take the form

$$U_{jj'} = \sum_{n'=0}^{M} \sum_{n=0}^{M} c_{n+n'+1} z_j^{-n} z_{j'}^{-n'} = \sum_{n'=0}^{M} \sum_{n=0}^{M} (z_{j'}/z_j)^n c_{n+n'+1} z_{j'}^{-(n+n')} \tag{11.254}$$

and the overlap matrix has elements

$$S_{jj'} = \sum_{n'=0}^{M} \sum_{n=0}^{M} c_{n+n'} z_j^{-n} z_{j'}^{-n'} = \sum_{n'=0}^{M} \sum_{n=0}^{M} (z_{j'}/z_j)^n c_{n+n'} z_{j'}^{-(n+n')}. \tag{11.255}$$

Making the substitution $l = n + n'$, and summing over n one obtains

$$U_{jj'} = \frac{1}{z_j - z'_j} \left[z_j \sum_{l=0}^{M} c_{l+p} z_{j'}^{-l} - z - j^{-M} \sum_{l=M+1}^{2M} c_{l+p} z_{j'}^{M-l+1} \right.$$
$$\left. - z_{j'} \sum_{l=0}^{M} c_{l+p} z_j^{-l} + z_{j'}^{-M} \sum_{l=M+1}^{2M} c_{l+p} z_j^{M-l+1} \right] \tag{11.256}$$

$$U_{jj} = \sum_{l=0}^{2M} (M - |M - l| + 1) c_{l+p} z_j^{-l}, \tag{11.257}$$

with similar expressions for $S_{jj'}$.

To avoid spurious results, the time grid must be dense enough to adequately represent the maximum frequency contributing to the signal:

$$-\frac{\pi}{\tau} < \text{Re}\, \omega_k < \frac{\pi}{\tau}, \tag{11.258}$$

Another condition, albeit not a very strict one, is on the imaginary part of ω_k:

$$\text{Im}\, \omega_k \leq O(N\tau). \tag{11.259}$$

This is also quite natural since it makes the signal $C(t)$ bounded on the segment $[0, N\tau]$. We note, however, that in most physically interesting situations the dynamics are strictly dissipative, corresponding to all nonpositive $\text{Im}\, \omega_k$. This is equivalent to having all the eigenvalues of $\hat{U}$ inside the unit circle in the complex plane, $|u_k| \leq 1$ (see Figure 11.10).

In a slightly different form, the harmonic inversion problem was already formulated two hundred years ago as a fit of a signal $C(t)$ to the sum of purely exponentially decaying terms,

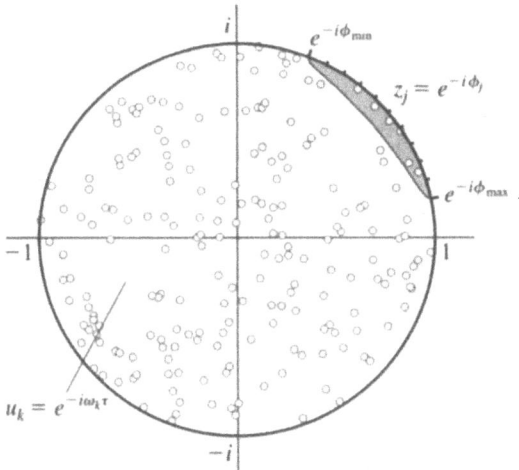

Figure 11.10 A schematic plot of the eigenvalues of the operator $\hat{U}$ for the case of strictly dissipative dynamics. Only a small portion (in the shadowed region) of the eigenvalues u_k are extracted by the filter-diagonalization procedure.

and an algorithm for its solution was discovered by Baron de Prony. There are several approaches related to the Prony method currently in use for extracting a high-resolution spectrum from short time signals, such as the modern versions of the Prony method, MUSIC, ESPRIT, and so forth. (see, e.g., Marple (1987); Roy (1991)). All these methods differ from the Fourier transform in that they are not based on a linear transformation of the signal and are in fact highly nonlinear by their nature. However, the common feature present in most versions of these methods is converting the nonlinear fitting problem to a linear algebra problem.

Problems

▶ **Exercise 11.16** Show that the projection operator for the discrete k-basis can be written in the alternative, more symmetric forms:

$$P = P_{\sqcap}(\hat{p}) P_{\sqcup}(\hat{p}) P_{\sqcap}(\hat{p}) = P_{\sqcup}(\hat{p}) P_{\sqcap}(\hat{p}) P_{\sqcup}(\hat{p}). \tag{11.260}$$

Solution These identities follow immediately from the fact that $P_{\sqcap}(\hat{p})$ and $P_{\sqcup}(\hat{p})$ commute and the projection operator property $P_{\sqcap}(\hat{p}) P_{\sqcap}(\hat{p}) = P_{\sqcap}(\hat{p})$ and $P_{\sqcup}(\hat{p}) P_{\sqcup}(\hat{p}) = P_{\sqcup}(\hat{p})$.

▶ **Exercise 11.17** Use the Nyquist sampling theorem to show that

$$P_{\sqcup}(\hat{p}) P_{\sqcap}(\hat{x}) P_{\sqcup}(\hat{p}) = P_{\sqcup}(\hat{p}) \qquad P_{\sqcup}(\hat{x}) P_{\sqcap}(\hat{p}) P_{\sqcup}(\hat{x}) = P_{\sqcup}(\hat{x}) \tag{11.261}$$

$$P_{\sqcap}(\hat{p}) P_{\sqcup}(\hat{x}) P_{\sqcap}(\hat{p}) = P_{\sqcap}(\hat{p}) \qquad P_{\sqcap}(\hat{x}) P_{\sqcup}(\hat{p}) P_{\sqcap}(\hat{x}) = P_{\sqcap}(\hat{x}). \tag{11.262}$$

Solution The first identity follows by noting that $P_{\sqcup}(\hat{p})$, by sampling the wavefunction discretely in p-space, renders the function periodic in x-space. Making the function periodic in x, truncating it at the unit interval, and then making it again periodic in x is identical to making it periodic the first time around. The other identities can be arrived at by similar reasoning.

▶ **Exercise 11.18** Consider the operators $P_{\sqcap}(\hat{x}) P_{\sqcap}(\hat{p})$ and $P_{\sqcap}(\hat{x}) P_{\sqcup}(\hat{p})$. Do these operators satisfy the equation $P^2 = P$? Are they Hermitian? Are they projection operators?

▶ **Exercise 11.19**

a. Noting that the basis orthogonality relation, Eq. 11.155, can be rewritten as $\Phi^{\dagger}\Phi = 1$, find the corresponding grid orthogonality relation.

b. Find the continuous analog of this discrete grid orthogonality relation.

Solution

a. The grid orthogonality relation, $\Phi\Phi^{\dagger} = 1$, takes the form

$$\sum_{\kappa=-\frac{N}{2}+1}^{\frac{N}{2}} \Phi_{\kappa}(x_i)\Phi_{\kappa}^*(x_j) = \sum_{\kappa=-\frac{N}{2}+1}^{\frac{N}{2}} \frac{e^{\sqrt{-1}2\pi\kappa i/N}}{\sqrt{N}} \frac{e^{-\sqrt{-1}2\pi\kappa j/N}}{\sqrt{N}} = \delta_{ij}. \tag{11.263}$$

Equation 11.263 may be rewritten as

$$\sum_{\kappa=-\frac{N}{2}+1}^{\frac{N}{2}} \bar{\phi}_\kappa(x_i)\bar{\phi}_\kappa^*(x_j)\,\Delta k = \sum_{\kappa=-\frac{N}{2}+1}^{\frac{N}{2}} \frac{e^{\sqrt{-1}2\pi\kappa\Delta ki/2K}}{\sqrt{2K}}\frac{e^{-\sqrt{-1}2\pi\kappa\Delta kj/2K}}{\sqrt{2K}}\Delta k = \delta_{ij}.$$

$$(11.264)$$

where

$$\bar{\phi}_\kappa(x_j) = \frac{e^{-i2\pi\kappa\Delta kj/2K}}{\sqrt{2K}} \quad \text{and} \quad \Delta k = \frac{2K}{N}. \qquad (11.265)$$

b. Equation 11.264 is the discrete version of the orthogonality relation

$$\int_{-K}^{K} \bar{\phi}_k(x_i)\bar{\phi}_k^*(x_j)\,dk = \int_{-K}^{K} \frac{e^{ik(x_i-x_j)}}{2K}\,dk = \delta_{ij}. \qquad (11.266)$$

To see this, note that if $i = j$ the integral gives 1. If $i \neq j$ the integral gives

$$\int_{-K}^{K} \frac{e^{-ik(x_i-x_j)}}{2K}\,dk = \frac{e^{-ik(x_i-x_j)}}{-i(x_i-x_j)}\Bigg|_{-K}^{K} = \frac{e^{-i\pi(i-j)} - e^{i\pi(i-j)}}{-i(x_i-x_j)} = 0, \quad (11.267)$$

where we have used the relation $K\,\Delta x = \pi$. Equation 11.264 plays an important role in the Fourier Grid Hamiltonian method.

▶ **Exercise 11.20**

a. Verify that Eq. 11.160 satisfies the basis orthogonality relation in Eq. 11.159.

b. Find the corresponding grid orthogonality relation.

Solution

a. Substitution of Eq. 11.160 into Eq. 11.159 yields

$$\sum_{j=-\infty}^{\infty} \frac{e^{i(k-k')x_j}}{2K} = \sum_{n=-\infty}^{\infty} \delta(k-k'-n2K) = \delta(k-k'), \qquad (11.268)$$

where the first equality follows from Eq. 6.83 and the second equality follows from the restrictions $-K \leq k, k' \leq K$. Note that the pseudospectral "matrix," $\Phi_k(x_j)$, has an infinite number of rows since the $\{x_j\}$ are discrete but infinite, and an infinite number of columns since k is a continuous variable.

b. To find the corresponding grid orthogonality relation we note that Eq. 11.159 can be written in matrix form as $\mathbf{\Phi}^\dagger\mathbf{\Phi} = \mathbf{1}$. The grid orthogonality relation, $\mathbf{\Phi}\mathbf{\Phi}^\dagger = \mathbf{1}$, takes the form

$$\int_{-K}^{K} \Phi_k(x_i)\Phi_k^*(x_j)\,dk = \int_{-K}^{K} \frac{e^{ik(x_i-x_j)}}{2K}\,dk = \delta_{ij}. \qquad (11.269)$$

▶ **Exercise 11.21**   Show that the complete set of sinc functions, Eq. 11.162, gives the same projector as that of Eq. 11.146.

Solution The sinc functions can be written in Dirac notation as

$$|\theta_j\rangle = P_\sqcap(\hat{p}) \frac{e^{-i\hat{p}x_j/\hbar}}{\sqrt{2K}} |0\rangle, \quad j = -\infty, ..., \infty, \qquad \Delta x = \frac{\pi}{K}, \qquad (11.270)$$

where

$$P_\sqcap(p)\tilde{\Psi}(p) = \begin{cases} \tilde{\Psi}(p) & \text{if } p \in [-K, K] \\ 0 & \text{otherwise} \end{cases}$$

and

$$\langle x|0\rangle = \delta(0).$$

Forming the projection operator from Eq. 11.270 we obtain

$$P = \sum_{j=-\infty}^{\infty} |\theta_j\rangle\langle\theta_j| \qquad (11.271)$$

$$= P_\sqcap(\hat{p}) \left(\sum_{j=-\infty}^{\infty} \frac{e^{-i\hat{p}x_j/\hbar}}{\sqrt{2K}} \middle|0\rangle\langle 0\middle| \frac{e^{i\hat{p}x_j/\hbar}}{\sqrt{2K}} \right) P_\sqcap(\hat{p}) \qquad (11.272)$$

$$= P_\sqcap(\hat{p}) \sum_{j=-\infty}^{\infty} |x_j\rangle\langle x_j| P_\sqcap(\hat{p}) \qquad (11.273)$$

$$\equiv P_\sqcap(\hat{p}) P_\sqcup(\hat{x}) P_\sqcap(\hat{p}) \qquad (11.274)$$

$$= P_\sqcap(\hat{p}), \qquad (11.275)$$

where in the last line we have used Eq. 11.262.

▶ **Exercise 11.22** In the original formulation of filter diagonalization by Wall and Neuhauser, the basis functions are

$$\Psi_j = \int_{-\infty}^{\infty} e^{-iHt/\hbar} e^{i\omega_j t} e^{-\alpha t^2} \Phi_0 \, dt. \qquad (11.276)$$

a. Compare Eq. 11.276 with the spectral method for constructing eigenfunctions (Section 6.3). Discuss why Ψ_j will have significant components of only these eigenstates with energy in the range near $\hbar\omega_j$.

b. Show that Eq. 11.243 is essentially the discrete counterpart of Eq. 11.276.

c. Discuss why Eq. 11.243 represents a Krylov-type basis. In what way is it similar and in what way is it different from the standard Krylov basis discussed in Section 11.7.3?

References

General Numerical Methods, Gaussian Quadrature, Fast Fourier Transform

1. B. Carnahan, H. A. Luther and J. O. Wilkes, *Applied Numerical Methods* (Wiley, New York, 1969).

2. W. H. Press, B. P. Flannery, S. A. Teukolsky and W. T. Vetterling, *Numerical Recipes: The Art of Scientific Computing* (Cambridge University Press, 1986).

3. G. Strang, *Linear Algebra and Its Applications*, 3rd ed. (Harcourt Brace Jovanovich, San Diego, 1988).

4. E. Oran Brigham, *The Fast Fourier Transform* (Prentice Hall, Englewood Cliffs, NJ, 1974).

5. P. J. Davis and P. Rabinowitz, *Methods of Numerical Integration,* 2nd ed. (Academic Press, Orlando, 1984).

6. *Handbook of Mathematical Functions with Formulas, Graphs, and Mathematical Tables*, eds. M. Abramowitz and I. A. Stegun (National Bureau of Standards Applied Mathematics Series 55, 1964).

Spectral and Pseudospectral Representation

7. D. Gottlieb and S. A. Orszag, *Numerical Analysis of Spectral Methods: Theory and Application* (SIAM, Philadelphia, 1977).

Gaussian Quadrature via Matrix Diagonalization

8. D. O. Harris, G. G. Engerholm and W. D. Gwinn, Calculation of matrix elements for one-dimensional quantum-mechanical problems and the application to anharmonic oscillators, J. Chem. Phys. 43, 151 (1965).

9. A. S. Dickinson and P. R. Certain, Calculation of matrix elements for one-dimensional quantum-mechanical problems, J. Chem. Phys. 49, 4209 (1968).

10. G. H. Golub and J. H. Welsch, Calculation of Gauss quadrature rules, Math. Comp. 23, 221 (1969).

11. I. Degani, J. Schiff and D. J. Tannor, Commuting extensions and cubature formulae, Numerische Mathematik (2005).

Discrete Variable Representation

12. J. C. Light, I. P. Hamilton and J. V. Lill, Generalized discrete variable approximation in quantum mechanics, J. Chem. Phys. 82, 1400 (1985).

13. J. C. Light, Discrete variable representations in quantum dynamics, in *Time Dependent Quantum Molecular Dynamics*, eds. J. Broeckhove and L. Lathouwers (Plenum, New York, 1992), p. 185.

14. R. G. Littlejohn, M. Cargo, T. Carrington, Jr., K. A. Mitchell and B. Poirier, A general framework for discrete variable representation basis sets, J. Chem. Phys. 116, 8691 (2002).

15. J. C. Light and T. Carrington, Jr., Discrete variable representations and their utilization, Adv. Chem. Phys. 114, 263 (2000).

16. R. Dawes and T. Carrington, Jr., A multidimensional discrete variable representation basis obtained by simultaneous diagonalization, J. Chem. Phys. 121, 726 (2004).

Phase Space Representation

17. M. J. Davis and E. J. Heller, Semiclassical Gaussian basis set method for molecular vibrational wave functions, J. Chem. Phys. 71, 3383 (1979).

18. R. Kosloff, Time-dependent quantum-mechanical methods for molecular dynamics, J. Phys. Chem. 92, 2087 (1988).

19. E. Fattal, R. Baer and R. Kosloff, Phase space approach for optimizing grid representations: The mapped Fourier method, Phys. Rev. E 53, 1217 (1996).

20. B. Poirier, Algebraically self-consistent quasiclassical approximation on phase space, Found. Phys. 30, 1191 (2000); B. Poirier and J. C. Light, Phase space optimization of quantum representations: Direct product basis sets, J. Chem. Phys. 111, 4869 (1999); B. Poirier and J. C. Light, Phase space optimization of quantum representations: Three-body systems and the bound states of HCO, J. Chem. Phys. 113, 211 (2000).

Fourier Method

21. D. Kosloff and R. Kosloff, A Fourier method solution for the time dependent Schrödinger equation as a tool in molecular dynamics, J. Comput. Phys. 52, 35 (1983); R. Kosloff and D. Kosloff, A Fourier method solution for the time dependent Schrödinger equation: A study of the reaction $H^+ + H_2$, $D^+ + HD$, and $D^+ + H_2$, J. Chem. Phys. 79, 1823 (1983).

22. M. D. Feit, J. A. Fleck, Jr. and A. Steiger, Solution of the Schrödinger equation by a spectral method, J. Comput. Phys. 47, 412 (1982); M. D. Feit and J.A. Fleck, Jr., Solution of the Schrödinger equation by a spectral method II: Vibrational energy levels of triatomic molecules, J. Chem. Phys. 78, 301 (1983); M. D. Feit and J. A. Fleck, Jr., Wavepacket dynamics and chaos in the Hènon-Heiles system, J. Chem. Phys. 80, 2578 (1984).

23. R. Kosloff, Time dependent methods in molecular dynamics, J. Phys. Chem. 92, 2087 (1988).

24. R. Kosloff, The Fourier method, in *Numerical Grid Methods and Their Application to Schrödinger's Equation*, ed. C. Cerjan (Kluwer, Boston, 1993).

25. C. C. Marston and G. G. Balint-Kurti, The Fourier grid Hamiltonian method for bound state eigenvalues and eigenfunctions, J. Chem. Phys. 91, 3571 (1989).

26. D. T. Colbert and W. H. Miller, A novel discrete variable representation for quantum mechanical reactive scattering via the S-matrix Kohn method, J. Chem. Phys. 96, 1982 (1992).

Mapped Fourier Method

27. F. Gygi, Adaptive Riemannian metric for plane-wave electronic-structure calculations, Europhys. Lett. 19, 617 (1992).

28. E. Fattal, R. Baer and R. Kosloff, Phase space approach for optimizing grid representations: The mapped Fourier method, Phys. Rev. E 53, 1217 (1996).

29. V. Kokoouline, O. Dulieu, R. Kosloff and F. Masnou-Seeuws, Mapped Fourier methods for long-range molecules: Application to perturbations in the Rb_2 (O_u^+) photoassociation spectrum, J. Chem. Phys. 110, 9865 (1999).

30. Y. Goldfarb and D. J. Tannor, in preparation.

31. K. Willner, O. Dulieu and F. Masnou-Seeuws, Mapped grid methods for long-range molecules and cold collisions, J. Chem. Phys. 120, 548 (2004).

Time Propagation

32. *Time Dependent Methods for Quantum Dynamics*, ed. K. C. Kulander, Comput. Phys. Commun. 63 (1991).

33. *Time Dependent Quantum Molecular Dynamics*, eds. J. Broeckhove and L. Lathouwers (Plenum, New York, 1992).

34. E. A. McCullough, Jr. and R. E. Wyatt, Dynamics of the collinear $H + H_2$ reaction. I. Probability density and flux, J. Chem. Phys. 54, 3578 (1971).

35. A. Askar and A. S. Cakmak, Explicit integration method for the time-dependent Schrödinger equation for collision problems, J. Chem. Phys. 68, 2794 (1978).

36. M. D. Feit, J. A. Fleck, Jr. and A. Steiger, Solution of the Schrödinger equation by a spectral method, J. Comput. Phys. 47, 412 (1982).

37. H. Tal-Ezer and R. Kosloff, An accurate and efficient scheme for propagating the time dependent Schrödinger equation, J. Chem. Phys. 81, 3967 (1984).

38. T. J. Park and J. C. Light, Unitary quantum time evolution by iterative Lanczos reduction, J. Chem. Phys. 85, 5870 (1986).

39. C. Leforestier et al., A comparison of different propagation schemes for the time dependent Schrödinger equation, J. Comput. Phys. 94, 59 (1991).

40. P. Schwendner, F. Seyl and R. Schinke, Photodissociation of Ar_2^+ in strong laser fields, Chem. Phys. 217, 233 (1997).

41. R. Kosloff, Propagation methods for quantum molecular dynamics, Ann. Rev. Phys. Chem. 45, 145 (1994).

42. S. K. Gray and D. E. Manolopoulos, Symplectic integrators tailored to the time-dependent Schrödinger equation, J. Chem. Phys. 104, 7099 (1996).

Spectral Analysis of Time Signals

43. S. Marple, Jr., *Digital Spectral Analysis with Applications*, Prentice Hall, Englewood Cliffs, NJ, 1987.

44. R. Roy, B. G. Sumpter, G. A. Pfeffer, S. K. Gray and D. W. Noid, Novel methods for spectral analysis, Comput. Phys. Rep. 205, 109 (1991).

45. M. R. Wall and D. Neuhauser, Extraction, through filter diagonalization, of general quantum eigenvalues or classical normal mode frequencies from a small number of residues or a short-time segment of a signal. I. Theory and application to a quantum-dynamics model, J. Chem. Phys. 102, 8011 (1995).

46. D. Neuhauser, Bound state eigenfunctions from wavepackets: Time → energy resolution, J. Chem. Phys. 93, 2611 (1990).

47. V. A. Mandelshtam and H. S. Taylor, A low-storage filter diagonalization method for quantum eigenenergy calculation or for spectral analysis of time signals, J. Chem. Phys. 106, 5085 (1997).

48. V. A. Mandelshtam and H. S. Taylor, Spectral analysis of time correlation function for a dissipative dynamical system using filter diagonalization: Application to calculation of unimolecular decay rates, Phys. Rev. Lett. 78, 3274 (1997).

Part III

Applications

Chapter 12

Introduction to Molecular Dynamics

Before starting on the Applications per se, we give a brief review of some basic concepts in molecular reaction dynamics and spectroscopy. In particular, we will introduce the Born–Oppenheimer approximation, leading to adiabatic molecular potential energy surfaces; the concept of asymptotic arrangement channels and a transition state that divides them; and the coupling between potential energy surfaces, either due to nonadiabatic interactions or interaction with light via a transition dipole moment. These concepts will be central to all the applications of wavepackets to molecular dynamics that we present in subsequent chapters. To keep things as simple as possible, and to keep the focus on the dynamics, we focus on the internal degrees of freedom of the molecule. The external degrees of freedom, translation and rotation, would take us off our main line. Moreover, the calculus of angular momentum is a complex subject to which we could not do justice here, and we simply refer the reader to several excellent texts on the subject.

12.1   The Born–Oppenheimer Approximation

The full molecular Hamiltonian may be written as

$$H = \sum_i -\frac{\hbar^2 \nabla_{e,i}^2}{2m} + \sum_{j>i} \frac{e^2}{|r_i - r_j|} + \sum_i -\frac{\hbar^2 \nabla_{N,i}^2}{2M_i} + \sum_{j>i} \frac{Z_i Z_j e^2}{|R_i - R_j|} - \sum_{ij} \frac{Z_j e^2}{|r_i - R_j|}$$

$$(12.1)$$

$$\equiv T_e + V_e + T_N + V_N + V_{eN} \tag{12.2}$$

where we use $\{r, \nabla_e\}$ to refer to the electron coordinates and momenta and $\{R, \nabla_N\}$ to refer to the nuclear coordinates and momenta. Z_i refers to the nuclear charge on nucleus i. Equation 12.2 defines a shorthand notation for each of the five terms in Eq. 12.1, namely, electron kinetic energy, electron–electron potential energy, nuclear kinetic energy, nuclear–nuclear potential energy, and electron–nuclear potential energy. The TISE in the full space of electronic and nuclear coordinates is

$$H(r, R)\Psi(r, R) = E\Psi(r, R), \tag{12.3}$$

where $\Psi(r, R)$ is an energy eigenfunction in the full coordinate space. To find an approximate solution of Eq. 12.3, we consider the TISE for the electrons only, at a fixed internuclear geometry, R:

$$H_e \psi(r; R) = E(R)\psi(r; R) \tag{12.4}$$

where $H_e = T_e + V_e + V_{eN}$. We *assume* that the full electronic–nuclear adiabatic wavefunction can be written as

$$\Psi(r, R) = \psi(r; R)\chi(R). \tag{12.5}$$

Substituting $\Psi(r, R) = \psi(r; R)\chi(R)$ back into the TISE for the *full* molecular Hamiltonian, and using Eq. 12.4 we obtain

$$H(r, R)\{\psi(r; R)\chi(R)\} = (E(R) + V_N)\{\psi(r; R)\chi(R)\}$$

$$+ \sum_i -\frac{\hbar^2}{2M_i}\left(\psi(r; R)\nabla^2\chi(R) + 2\nabla\psi(r; R)\cdot\nabla\chi(R) + \nabla^2\psi(r; R)\chi(R)\right) \tag{12.6}$$

$$= E\psi(r; R)\chi(R), \tag{12.7}$$

where here and henceforth, ∇ will refer to ∇_N. The last two terms in Eq. 12.7 involve derivatives of the electronic wavefunction with respect to the nuclear coordinates. The first of these terms is more important than the second, since it involves only the first derivative with respect to nuclear coordinates. However, both these terms are proportional to the mass ratio of electrons to nuclei to some power, and they are typically orders of magnitude smaller than the size of the other terms. If these terms are neglected we obtain

$$H(r, R)\chi(R) = \left(T_N + E(R) + V_N(R)\right)\chi(R) = E\chi(R). \tag{12.8}$$

Equation 12.8 is a TISE for the nuclei under an *adiabatic Hamiltonian* whose effective potential is given by the averaged electron–electron and electron–nuclear forces, contained in $E(R)$, and the instantaneous nuclear–nuclear repulsion, contained in $V_N(R)$. The neglect of the last two terms in Eq. 12.6 is known as the Born–Oppenheimer approximation.

The physical picture that emerges from Eq. 12.8 is shown in Figure 12.1a. For every value of R we need to solve a TISE for the electrons. The spectrum of eigenvalues is a function of R; connecting the values while preserving the ordering of levels gives rise to a continuous set of adiabatic electronic eigenvalue curves. However, these same curves are seen (when V_N is added to them) to give the effective potential under which the nuclei move in Eq. 12.8. Thus, these curves, including V_N, are called potential energy curves, or adiabatic potentials, or Born–Oppenheimer (B.O.) potentials. The B.O. approximation has been called the most important approximation in quantum chemistry.

All these curves have several features in common. As the internuclear distance goes to zero the potential energy goes to infinity; this is due to the infinite nuclear–nuclear electrostatic repulsion at zero separation (we are neglecting the so-called strong and weak nuclear forces in this discussion, which are responsible, for example, for fusion). At large internuclear separation, all these potentials go to an asymptotic value: if the atoms are sufficiently far apart the internuclear forces vanish. At intermediate internuclear separations there are two classes of curves: those with a minimum and those without. Those with a minimum

(a) **(b)**

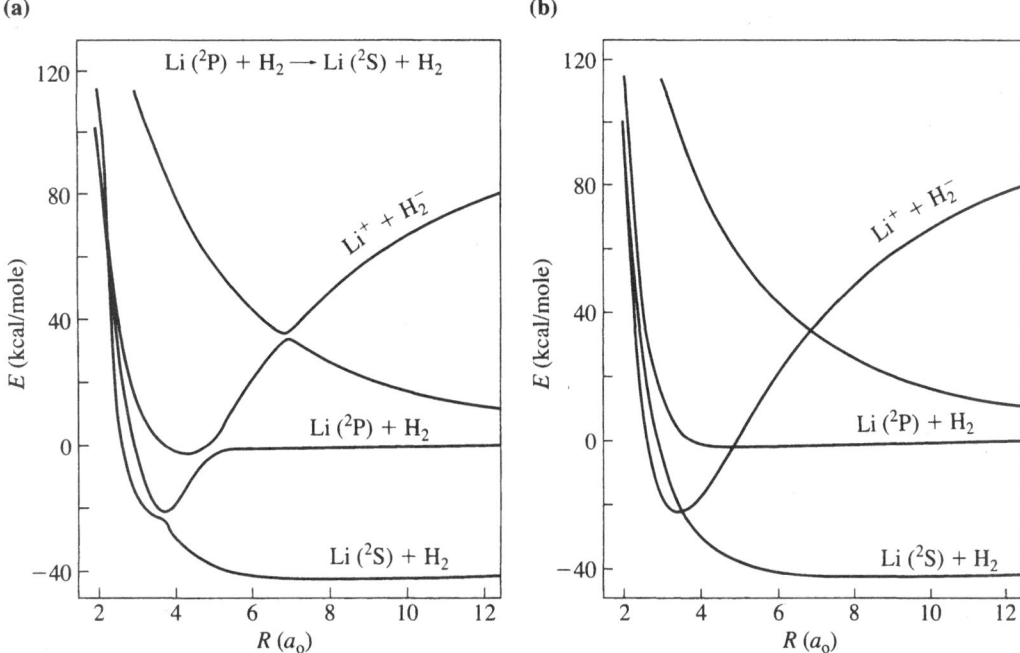

Figure 12.1 Slices of the four lowest potential energy surfaces of a particular symmetry of LiH_2 as a function of Li–H_2 separation. (The orientation of the H_2 internuclear axis is 60° relative to the Li–H_2 direction, and the H_2 internuclear distance is 1.4 Å.) (a) Adiabatic representation. Note that the adiabatic potentials, being of the same symmetry, satisfy a "no-crossing" rule. At the avoided crossings, the physical character of the adiabatic electronic state generally changes, for example, from ionic to covalent. (b) Diabatic representation. The diabatic potentials are continuous, and maintain a consistent physical character. Adapted from Tully (1976).

imply that the molecule is stable in that electronic state (provided that the vibrational zero point energy is lower than the asymptotic value of the potential; otherwise the molecule has no true bound states); the equilibrium bond length is the value of the internuclear coordinate where the potential is minimum. The difference between the asymptotic potential energy at large separations and the potential energy at the minimum is the bond dissociation energy. Those potential energy functions with no minimum correspond to unstable molecular species.

Although the terms involving $\nabla \psi(r; R)$ in Eq. 12.6 are ordinarily small, near an avoided crossing of the type shown in Figure 12.1a, they can become quite significant. This is because the adiabatic electronic states are defined to preserve the ordering of the energy levels, but in general not their physical character. Thus, in the region of an avoided crossing, the adiabatic states may change their character dramatically (e.g., from ionic to covalent), and hence derivatives of the type in Eq. 12.6 can be quite large. In this case, it may be more convenient to consider *diabatic* electronic states, as illustrated in Figure 12.1b. These states do actually cross, and preserve the physical character of the electronic eigenfunction. Pioneering work by Landau (1932), Zener (1932), and Stückelberg (1932) showed that the collision equations using the diabatic electronic states lend themselves to a treatment in which the interaction between diabatic states can be treated in the high-energy limit as a small perturbation causing inelastic transitions. As a general rule of thumb, when the nuclear motion is slow, the adiabatic electronic states are more appropriate; when the nuclear motion is fast, the diabatic electronic states are more appropriate (cf. the discussion in Section 9.4.2). In the next section, we elaborate on the use of adiabatic and diabatic electronic states as a complete basis for solution of the Schrödinger equation.

12.2   Adiabatic versus Diabatic Representations

In this section we go beyond the treatment in the previous section to discuss the use of the adiabatic electronic states, as well as the alternative diabatic electronic states, as a complete basis. This section also includes a discussion of the possibility of intersection of potential energy curves in one dimension, or potential energy surfaces in multidimensions. These topics are not required for understanding the rest of the chapter, and can be skipped on a first reading.

12.2.1   Adiabatic Representation

It is possible to go beyond the Born–Oppenheimer approximation by writing a formally exact solution in the adiabatic basis:

$$\Psi(r; R) = \sum_{n=0}^{\infty} \psi_n(r; R) \chi_n(R). \tag{12.9}$$

Strictly speaking, the discrete sum over n spans only the bound electronic states. For a complete representation one needs to include also an integral over the unbound electronic states; for simplicity, we neglect this subtlety in what follows. Substituting Eq. 12.9 into Eq. 12.6 and projecting from the left with $\psi_m(r; R)$ leads to the set of coupled equations:

$$\sum_n H_{mn}(R) \chi_n(R) = \sum_n \{(T_N + E_n(R) + V_N(R))\delta_{mn} + 2T_{mn}^{(1)}(R) \cdot \nabla + T_{mn}^{(2)}(R)\} \chi_n(R)$$

$$= E \chi_m(R), \tag{12.10}$$

where

$$T_{mn}^{(1)}(R) = \langle \psi_m | \nabla \psi_n \rangle \qquad T_{mn}^{(2)}(R) = \langle \psi_m | \nabla^2 \psi_n \rangle; \tag{12.11}$$

here and in the remainder of this section the Dirac notation indicates integration over electronic coordinates only. Here and henceforth, the operator ∇ is a multidimensional derivative with respect to mass-weighted nuclear coordinates, described more fully later in this chapter. Note that $T_{mn}^{(1)}$ is a vector, the number of whose components is equal to the number of vibrational degrees of freedom (strictly speaking, there can be contributions from the rotational degrees of freedom as well), and $T_{mn}^{(2)}$ is a scalar.

Taking the electronic wavefunctions as normalized to the same value at all nuclear geometries,

$$\langle \psi_m | \psi_n \rangle = \delta_{nm}, \tag{12.12}$$

it follows that the matrix $T_{mn}^{(1)}$ is anti-Hermitian, that is,

$$T_{mn}^{(1)} = -T_{nm}^{(1)*}. \tag{12.13}$$

▶ **Exercise 12.1**   Derive Eq. 12.13 by considering the expression $\nabla \langle \psi_m | \psi_n \rangle$.

It is always possible to choose the electronic wavefunctions to be real-valued, in which case it follows from Eq. 12.13 that the diagonal elements of the matrix $T_{mn}^{(1)}$ vanish:

$$T_{nn}^{(1)}(R) = 0. \tag{12.14}$$

Therefore, if we keep just the diagonal terms in Eq. 12.10 we obtain the adiabatic Schrödinger equation for the nuclei, Eq. 12.8, with an additional contribution from $T_{nn}^{(2)}$, which is usually small but nonvanishing:

$$(T_N + E_n(R) + V_N(R) + T_{nn}^{(2)}(R))\chi_n(R) = E\chi_n(R). \tag{12.15}$$

As discussed above, in the region of an avoided crossing the adiabatic states may change their character dramatically (e.g., from ionic to covalent), and hence derivatives of the type in Eq. 12.7 can be quite large. Moreover, the nonadiabatic matrix elements of the form in Eq. 12.10 are normally quite inconvenient to calculate. As a result, in regions of an avoided crossing, other choices of basis may prove more convenient than the adiabatic basis. This is the subject of the next subsection.

12.2.2   Diabatic Representation

An alternative to the adiabatic basis is to seek a *diabatic* basis, for which all electronic matrix elements of the derivatives of the form $\frac{\partial}{\partial R}$ and $\frac{\partial^2}{\partial R^2}$ are absent, or as small as possible. It turns out that these derivatives can be made to vanish in a trivial way, and in fact in an infinite number of ways, simply by choosing the electronic basis functions to be independent of the nuclear coordinates: $\psi(r, R) \rightarrow \psi(r, R_0)$, where R_0 is any suitably chosen nuclear geometry, as in the asymptotic region of the potential. A choice of electronic wavefunctions that do not change with nuclear geometry is sometimes called a "crude adiabatic" basis.

In the crude adiabatic basis, the wavefunction has the form

$$\Psi(r; R) = \sum_n \psi_n(r; R_0)\chi_n^{(0)}(R). \tag{12.16}$$

We take $\psi_n(r; R_0)$ as the solutions of the electronic Schrödinger equation at the reference geometry, R_0:

$$H(r; R_0)\psi_n(r; R_0) = \{T_e + V(r; R_0)\}\psi_n(r; R_0) = E_n(R_0)\psi_n(r; R_0), \tag{12.17}$$

where $V(r; R_0)$ includes V_e, V_N and V_{eN} (cf. Eq. 12.2). Substituting Eq. 12.16 into the TISE, Eq. 12.3, and projecting on the left with $\psi_m(r, R_0)$ we obtain the matrix equation

$$\sum_n \{T_{nn}\delta_{nm} + U_{mn}\}\chi_n^{(0)} = E\chi_m^{(0)}, \tag{12.18}$$

where

$$U_{mn} = \langle \psi_m(R_0)|V(r, R) - V(r, R_0)|\psi_n(R_0)\rangle; \tag{12.19}$$

again the bra-ket notation refers to integration over electronic coordinates only. Note the simplicity of Eq. 12.18 relative to Eq. 12.10! By changing from the adiabatic to the diabatic representation, the coupling between the electronic potentials has moved from derivative coupling, $(d/dR)_{mn}$, to coordinate coupling, $(U(R))_{mn}$. The diagonal elements of U are the diabatic potentials, which, within an energy shift, follow the dashed curves in Figure 12.1b. Transitions between diabatic surfaces are caused by the off-diagonal elements of U_{mn}.

In principle, the set of coupled equations in the crude adiabatic representation, Eq. 12.18, can be solved by applying standard numerical methods known as close-coupling methods. However, the crude adiabatic basis required is generally much larger than the normal adiabatic basis; the large basis required is simply a reflection of the fact that the electronic wavefunctions *do* in fact change significantly with nuclear geometry. In practice, the crude adiabatic basis is convenient only when the off-diagonal elements U_{mn} are sufficiently small, that is, when

$$|U_{mn}| \ll |U_m - U_n|. \tag{12.20}$$

Ideally, what one would like is the best of both worlds—the compactness of the adiabatic basis along with the convenience of calculation of the diabatic basis. For many problems of chemical interest only one or two adiabatic states contribute. One would hope that by transforming from the adiabatic basis to an optimal diabatic basis one would be able to significantly reduce the nonadiabatic coupling without increasing the size of the basis. It turns out that the complete elimination of the nonadiabatic coupling is possible only for diatomics. For polyatomics the story is more complicated—in general, only part of the nonadiabatic coupling can be transformed away. In order to explore when a rigorous diabatic representation is possible, we have to first develop the form of the vibronic Schrödinger equation under a change of basis. This is done in the next few exercises.

▶ Exercise 12.2

    a. Define an orthogonal (unitary) transformation matrix, $A^\dagger$, that transforms the electronic basis from the adiabatic representation, $\psi(r, R)$, to the diabatic representation, $\psi'(r, R)$:

$$A^\dagger_{mn}(R) = \langle \psi'_m(R)|\psi_n(R)\rangle \qquad \psi(r, R) = A^\dagger(R)\psi'(r, R). \tag{12.21}$$

    Show that the transformation that transforms the nuclear basis from the adiabatic to the diabatic representation is the adjoint matrix, A, so that

$$\chi(R) = A(R)\chi'(R). \tag{12.22}$$

    b. Use the definitions from part a to show that the nuclear Schrödinger equation in the transformed representation is

$$A\nabla^2\chi'(R) + 2(\nabla A + T^{(1)}A) \cdot \nabla\chi'(R)$$

$$+ \left[(T^{(2)} + \nabla^2 + 2T^{(1)} \cdot \nabla)A - \frac{2}{\hbar^2}(V - E)A\right]\chi'(R) = 0. \tag{12.23}$$

▶ **Exercise 12.3**   Show that if the transformation matrix A satisfies the equation

$$\nabla A + T^{(1)} A = 0, \tag{12.24}$$

then

$$\nabla^2 A + 2 T^{(1)} \cdot \nabla A + T^{(2)} A = 0 \tag{12.25}$$

and the resulting Schrödinger equation simplifies to

$$\nabla^2 \chi' - \frac{2\mu}{\hbar^2} (W - E) \chi' = 0, \tag{12.26}$$

where

$$W \equiv A^{\dagger} V A \tag{12.27}$$

is the potential in the diabatic representation (Smith, 1969; Baer, 1985). Note the consistency of Eqs. 12.26 and 12.18, with the role of W being played by U in the crude adiabatic basis. It follows from Eq. 12.27 that

$$V = A W A^{\dagger}. \tag{12.28}$$

This shows that diagonalizing the potential matrix in the diabatic representation gives the adiabatic potentials as the eigenvalues and the diabatic–adiabatic transformation matrix, A, as the eigenfunctions.

Solution   Calculate the divergence of Eq. 12.24 to obtain

$$\nabla^2 A + \nabla \cdot T^{(1)} A + T^{(1)} \cdot \nabla A = 0. \tag{12.29}$$

Since $T^{(1)}_{mn} = \langle \psi_m(R) | \nabla \psi_n(R) \rangle$ and $T^{(2)}_{mn} = \langle \psi_m(R) | \nabla^2 \psi_n(R) \rangle$, it follows that

$$\nabla \cdot T^{(1)} = -T^{(1)} \cdot T^{(1)} + T^{(2)}. \tag{12.30}$$

Substituting Eq. 12.30 into Eq. 12.29 and using Eq. 12.24 we obtain Eq. 12.25. Equation 12.26 is now the straightforward result.

Does Eq. 12.24 always have a solution? Taking the curl of Eq. 12.24, and noting that $\nabla \times \nabla \Phi = 0$ for any Φ, we obtain the following necessary condition for Eq. 12.24 to be satisfied:

$$\nabla \times (T^{(1)} A) = 0. \tag{12.31}$$

In diatomics, $T^{(1)}$ has only one component; $\nabla \times T^{(1)} = 0$, and Eq. 12.31 is trivially satisfied. In polyatomics, however, where $T^{(1)}$ has multiple components, Eq. 12.31 can be satisfied in general only if the matrix A is as large as the full space of adiabatic electronic states (Mead, 1982). But this defeats the purpose of the diabatic basis, which was intended to provide a compact representation in the presence of an avoided crossing. In fact, it can be shown that the diabatic states obtained from solving Eq. 12.31 for A are independent of nuclear coordinates in the limit that the adiabatic space is complete; they are just the crude adiabatic states (Mead, 1982). The matrix $T^{(1)}(R)$ can, however, be decomposed into a longitudinal and a tranverse part,

$$T^{(1)}(R) = T_l^{(1)}(R) + T_t^{(1)}(R), \tag{12.32}$$

where

$$\nabla \times T_t^{(1)}(R) = \nabla \times T^{(1)}(R), \qquad \nabla \cdot T_t^{(1)}(R) = 0, \tag{12.33}$$

and

$$T_l^{(1)}(R) = \nabla \Phi^{(1)}(R), \tag{12.34}$$

and where we have defined $\Phi^{(1)}(R)$ as the scalar potential associated with $T^{(1)}(R)$. Note that Eq. 12.33 as written is valid only for triatomics, since $\nabla \times$ is defined only in three degrees of freedom. For a general number of degrees of freedom, $\nabla \times$ should be replaced by curl. Equation 12.24 does have a solution in general if $T^{(1)}$ is replaced by $T_l^{(1)}(R) = \nabla \Phi^{(1)}(R)$:

$$\nabla A + \nabla \Phi^{(1)} A = 0 \tag{12.35}$$

(Mead, 1982; Kuppermann, 1996). Defining A by Eq. 12.35 and left-multiplying Eq. 12.23 by $A^\dagger$ we obtain

$$\nabla^2 \chi'(R) + 2(T_t^{(1)}(R)) \cdot \nabla \chi'(R) + \left[(T^{(2)d}(R) - \frac{2\mu}{\hbar^2}(W - E) \right] \chi'(R) = 0, \tag{12.36}$$

where $T^{(2)d} = A^\dagger T^{(2)} A$ and the superscript d stands for diabatic. In practice, the two additional terms in Eq. 12.36 relative to Eq. 12.26 are generally neglected, which is a good approximation in the neighborhood of a conical intersection (see below), where the longitudinal component of $T^{(1)}$ dominates.

12.2.3   Intersection of Adiabatic Potential Energy Surfaces

The discussion of adiabatic and diabatic representation leads naturally to a more general consideration of the nature of the intersection of molecular potential energy curves in one dimension and potential energy surfaces in multidimensions. Specifically, we examine the possibility of intersection of *adiabatic* potentials, since the diabatic potentials have a residual potential coupling between the surfaces; nevertheless, the diabatic representation facilitates the derivation.

Consider first a diatomic. As a function of internuclear separation, two adiabatic curves belonging to electronic states of the same symmetry quite generally do not cross, but exhibit avoided crossings as in Figure 12.1a (von Neumann, 1927). The proof of this noncrossing rule involves consideration of the potential matrix in the diabatic representation:

$$U = \begin{pmatrix} U_{11} & U_{12} \\ U_{21} & U_{22} \end{pmatrix}. \tag{12.37}$$

Diagonalizing the matrix U gives the adiabatic potentials

$$V_\pm = \frac{U_{11} + U_{22}}{2} \pm \frac{1}{2}\sqrt{(U_{11} - U_{22})^2 + 4|U_{12}|^2}. \tag{12.38}$$

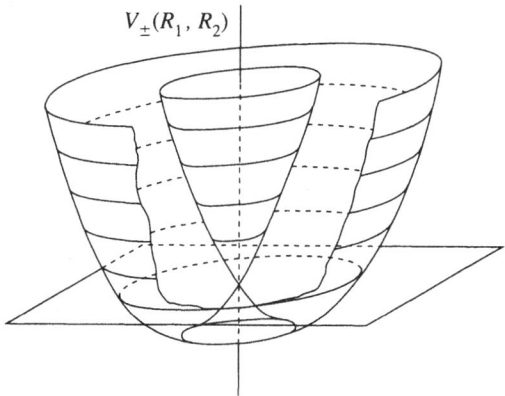

$V_{\pm}(R_1, R_2)$

Figure 12.2 Geometry of a conical intersection in the potential energy functions $V_{\pm}$ as a function of two nonsymmetric nuclear coordinates. Adapted from Herzberg (1966).

The separation between the two surfaces is given by

$$\Delta V = V_+ - V_- = \sqrt{(U_{11} - U_{22})^2 + 4|U_{12}|^2}. \qquad (12.39)$$

For the two curves to cross, the two terms under the square root in Eq. 12.39 must vanish independently, and thus

$$U_{11}(R) = U_{22}(R) \qquad (12.40)$$

$$U_{12}(R) = 0. \qquad (12.41)$$

Thus, two equations must be satisfied with only one independent parameter, the internuclear separation, R; hence, in one dimension the curves generally do not cross.

In two dimensions, there is generally a single (or a discrete set) of isolated points where the surfaces touch; these points are called "conical" intersections, because of the geometrical structure of the intersection (see Figure 12.2) (Jahn, 1937, Teller, 1937). In $N > 2$ dimensions, the intersection is normally a manifold of dimension $N - 2$, due to the two constraints in Eq. 12.41. Conical intersections can have a profound effect on the molecular dynamics, even if the dynamics never comes close to the intersection. For the conventional choice of real, adiabatic electronic wavefunctions, the electronic wavefunction has the bizarre property that it develops a phase of π for a circuit by 2π around a conical intersection (Herzberg, 1963; Longuet-Higgins, 1958); this was one of the first known examples of the geometric phase (Section 9.4). Since the total wavefunction (electronic times nuclear) must be single-valued, the nuclear wavefunction must also develop a phase of π for a circuit around the conical intersection, even if the nuclei never approach the intersection region! There are two approaches that avoid this discontinuity. The first is to transform to a diabatic basis. Although, as discussed above, a strict diabatic basis cannot generally be found for $N \geq 2$, it turns out that in the neighborhood of a conical intersection the first derivative adiabatic coupling is dominated by the longitudinal part, which can be transformed away. The second derivative adiabatic coupling diverges at the conical intersection, but the divergence is also transformed away by the diabatization process. Although the diabatic basis is well defined for this case, it requires the use of two potential energy surfaces where one is sufficient from the point of view of energetics. The second approach to avoiding the discontinuity is to use complex electronic and nuclear wavefunctions. By multiplying the electronic and nuclear wavefunctions by suitable phases that are equal and opposite, the wavefunctions can be made to be individually single valued (Mead, 1979). The price for this choice of electronic and vibrational wavefunctions is that the *diagonal* matrix element of the first derivative coupling no longer vanishes in general (cf. Eq. 12.14) (Mead, 1979). Thus, there are at least *three* alternative approaches to describing the electronic–vibrational wavefunction near a conical intersection, each with its own numerical disadvantage. In the first approach, using real, adiabatic electronic wavefunctions, the nuclear wavefunction changes sign for a circuit around the conical intersection, and hence is somewhere discontinuous. In the second approach, using real diabatic electronic wavefunctions, the dynamics must be followed on two potential energy surfaces. In the third approach, using complex adiabatic electronic wavefunctions, there is a diagonal matrix element of the first derivative term that is difficult to calculate.

The next exercise shows the origin of conical intersections of potential energy surfaces in a simple two-dimensional model, as well as the origin of a geometric phase for adiabatic circuits around the point of degeneracy.

▶ **Exercise 12.4**   Consider a molecule with two independent coordinates, x and y. If we take the origin as a point of degeneracy, the simplest form for the 2×2 potential energy matrix is

$$\begin{pmatrix} U_{11} + hR_1 & lR_2 \\ lR_2 & U_{11} - hR_1 \end{pmatrix}. \tag{12.42}$$

a.   Show that the two eigenvalues of this matrix are

$$V_{\pm}(R_1, R_2) = U_{11} \pm \sqrt{(hR_1)^2 + (lR_2)^2},$$

that is, they form a double cone with its vertex at the origin.

b.   Show that if the electronic wavefunction is taken to be real, the normalized eigenvector associated with the lower branch of the double cone has the form $(c_1, c_2) = (\sin(\frac{1}{2}\theta), -\cos(\frac{1}{2}\theta))$. Note that as θ increases from 0 to 2π, c_1 and c_2 both change sign.

c.   Show that multiplying c_1, c_2 by the phase factor $e^{i\theta/2}$ makes the electronic wavefunction single valued as θ goes from 0 to 2π, but at the expense of making ψ complex.

Solution   We will confine ourselves here to the solution of part b. Define polar coordinates, (R, θ), such that $hR_1 = R \cos \theta$, $lR_2 = R \sin \theta$. The eigenvalue for the lower branch of the cone becomes $U_{11} - R$, and hence the eigenvector equation for the coefficients of the wavefunction on the lower branch is

$$\begin{pmatrix} R + R \cos \theta & R \sin \theta \\ R \sin \theta & R - R \cos \theta \end{pmatrix} \begin{pmatrix} c_1 \\ c_2 \end{pmatrix}. \tag{12.43}$$

The coefficients c_1, c_2 therefore satisfy the equation

$$\frac{c_1}{c_2} = \frac{-\sin\theta}{1+\cos\theta} = \frac{\cos\theta - 1}{\sin\theta} = -\tan\frac{1}{2}\theta. \tag{12.44}$$

If the coefficients c_1, c_2 are to be real, the unique normalized eigenvector within an overall sign is

$$c_1 = \sin\frac{1}{2}\theta, \qquad c_2 = -\cos\frac{1}{2}\theta. \tag{12.45}$$

Writing the overall electronic wavefunction, ψ, as $\psi = c_1\phi_1 + c_2\phi_2$, we see that ψ changes sign as θ goes from 0 to 2π.

12.3   Potential Energy Surfaces

12.3.1   Qualitative Features

Figure 12.1 showed the potential energy as a function of internuclear separation for a diatomic molecule. In that case, there is only one coordinate of interest, and the potential energy function is a one-dimensional curve. In the case of a polyatomic molecule there are many more internuclear coordinates. The generalization of the potential energy curve in a diatomic is the potential energy surface, or *hypersurface*, in a polyatomic.

Consider, for example, the carbon dioxide molecule, $O'CO''$, where we use the primes to distinguish the two oxygens. For simplicity, we will temporarily freeze the bend angle and consider the potential energy as a function of the two stretch coordinates. Figure 12.3 shows both a stereoscopic projection of the potential energy function, as well as a contour map of curves of equal energy. Note that there is now a two-dimensional potential energy minimum, at the value of R_1 and R_2 corresponding to the equilibrium bond lengths. Note, in addition, that there are now two asymptotic regions of the potential, one corresponding to $R_1 \to \infty$, and the other corresponding to $R_2 \to \infty$. These regions correspond to $O' + C$—O'' and O'—$C + O''$, respectively. As in the diatomic case, at large distances the interaction of the fragments (now an atom with the diatom) vanishes. Still, at these distances, the diatomic has its own internal potential energy function, as seen in the figure. Finally, there is a region corresponding to O'—C—O'', for three-body breakup. These asymptotic regions, which correspond to distinct fragmentations ($O' + CO''$, $O'C + O''$, $O' + C + O''$) are called *chemical arrangement channels*.

Note the symmetry of the potential energy surface about the line $R_1 = R_2$. Since the two O atoms are equivalent, there is no reason that the potential energy for stretching one of them should be different from that for stretching the other. This is the simplest case of symmetry considerations in potential energy functions, which will be put into a more complete framework below. In fact, the ultimate origin of the symmetry of the potential is the symmetry of the Hamiltonian with respect to permutation of identical nuclei.

At a fixed bend angle, $O'CO''$ has two asymptotic two-body arrangement channels, one corresponding to $O' + CO''$ and one to $O'C + O''$. However, triatomic molecules generally have three two-body arrangement channels; for example, in $O'CO''$, there is not only $O'C + O''$ and $O' + CO''$, but also $C + O'O''$. The third arrangement channel cannot be

(a)   **(b)**

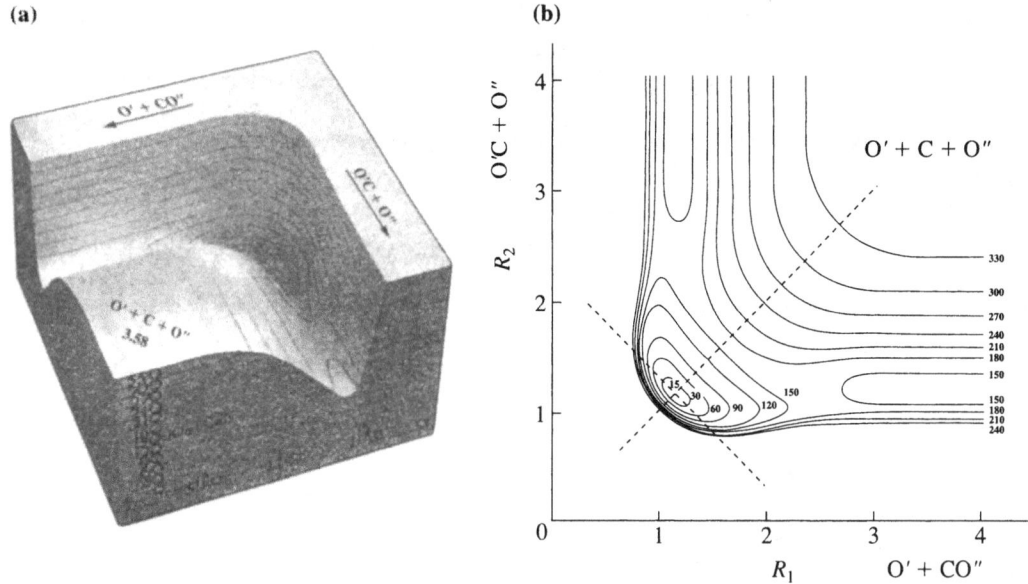

Figure 12.3 Potential energy surface for collinear OCO in the ground electronic state. (a) Stereoscopic projection; (b) equipotential contour plot. Adapted from Herzberg (1966).

visualized in Figure 12.3 because it corresponds to a large excursion in the bend coordinate. The full range of triatomic arrangement channels is nicely visualized in hyperspherical coordinates, as will be shown in the last section of this chapter. Moreover, for molecules like H_3 and O_3, in which all three atoms are identical, hyperspherical coordinates show the full permutational symmetry of the potential and the three two-body arrangement channels.

12.3.2   The Problem of Nondiagonal Kinetic Energy

We now turn to the form of the kinetic energy operator. Keeping with the example of $O'CO''$, the Hamiltonian for collinear $O'CO''$, written in terms of the bond coordinates R_1 and R_2, takes the following form:

$$H = T + V = -\frac{\hbar^2}{2\mu_1}\frac{\partial^2}{\partial R_1^2} - \frac{\hbar^2}{2\mu_2}\frac{\partial^2}{\partial R_2^2} + \frac{\hbar^2}{\mu_3}\frac{\partial^2}{\partial R_1 \partial R_2} + V(R_1, R_2), \quad (12.46)$$

where $\mu_1 \equiv (m_A^{-1} + m_B^{-1})^{-1}$, $\mu_2 \equiv (m_B^{-1} + m_C^{-1})^{-1}$, $\mu_3 = m_B$, and m_A, m_B, m_C are the masses of atoms A, B and C respectively. Henceforth we will use the generic labels ABC for the atoms $O'CO''$. Note the cross-term in the kinetic energy, that is, the term of the form $\frac{\hbar^2}{\mu_3}\frac{\partial^2}{\partial R_1 R_2}$.

The cross-term in the kinetic energy, and ways of rewriting the kinetic energy operator to remove the cross-term, are issues that go back to the very early days of molecular dynamics. In the early 1930s, after the concept of adiabatic potential energy surfaces had just been introduced, researchers wanted to learn qualititative relationships between potential surface properties and reaction probabilities. This was well before computers were invented, and the only avenue of exploration was analog calculation. Michael Polanyi and coworkers built potential energy surfaces for collinear H_3 as a function of the two bond lengths out of papier-mâché, and rolled balls from one asymptotic region to the other. The position of the ball represented the coordinates of the atoms in the two-dimensional space of R_1, R_2, and gravity played the role of the internuclear potential. It was soon realized that for this form of analog computation to be meaningful the kinetic energy terms of the ball had to match the kinetic energy terms of the molecule; alas, the kinetic energy of the ball had only a single mass and was diagonal in each coordinate, while the molecular kinetic energy in general had several masses and cross-terms between them. The ingenious solution to this problem is discussed in the next section.

In Exercise 12.5 we derive Eq. 12.46. This is the simplest example of a general machinery developed by Wilson and coworkers (Wilson, 1955) for finding the form of the kinetic energy in internal coordinates. The central object in Wilson's approach is the **G** matrix, a matrix of inverse masses. Off-diagonal elements of the **G** matrix correspond to nondiagonal kinetic energy. The key idea in the derivation is to use the theory of canonical transformation (Section 10.4.1) to find the momentum variables conjugate to the bond coordinates, and then to express the kinetic energy in terms of these momentum variables.

▶ **Exercise 12.5**   Show that the kinetic energy for a collinear triatomic in bond coordinates takes the form written in Eq. 12.46.

Solution   Consider the classical mechanical problem of the coordinate transformation from the Cartesian coordinates x_A, x_B, x_C at the atomic centers to the bond coordinates $R_1 = x_B - x_A$, $R_2 = x_C - x_B$, $R_3 = (m_A x_A + m_B x_B + m_C x_C)/m_{ABC}$, where R_3 is the center-of-mass coordinate and $m_{ABC} = m_A + m_B + m_C$. Although the coordinate R_3 is necessary to completely define the coordinate transformation it is not necessary for the problem at hand and we omit it in what follows. This coordinate transformation can then be written in matrix form as

$$\begin{pmatrix} R_1 \\ R_2 \end{pmatrix} = \begin{pmatrix} -1 & 1 & 0 \\ 0 & -1 & 1 \end{pmatrix} \begin{pmatrix} x_A \\ x_B \\ x_C \end{pmatrix} \tag{12.47}$$

or simply $\bar{R} = \mathbf{B}\bar{x}$. The new Hamiltonian is derived from the old Hamiltonian via a canonical transformation. The F_2 generator (see Section 10.4.1) is

$$F_2(P, x) = \bar{P}^{\mathrm{T}} \mathbf{B} \bar{x}, \tag{12.48}$$

where T indicates transpose. The old momenta are obtained from the relationship $p_i = \frac{\partial F_2}{\partial x_i}$. In vector form this becomes

$$\bar{p}^{\mathrm{T}} = \bar{P}^{\mathrm{T}} \mathbf{B} \quad \text{or} \quad \bar{p} = \mathbf{B}^{\mathrm{T}} \bar{P}. \tag{12.49}$$

The kinetic energy can now be written as

$$T = \frac{p_A^2}{2m_A} + \frac{p_B^2}{2m_B} + \frac{p_C^2}{2m_C} = \frac{1}{2} \bar{p}^{\mathrm{T}} \mathbf{g} \bar{p} \tag{12.50}$$

$$= \frac{1}{2} \bar{P}^{\mathrm{T}} \mathbf{B} \mathbf{g} \mathbf{B}^{\mathrm{T}} \bar{P} = \frac{1}{2} \bar{P}^{\mathrm{T}} \mathbf{G} \bar{P}, \tag{12.51}$$

where

$$\mathbf{g} \equiv \begin{pmatrix} \frac{1}{m_A} & 0 & 0 \\ 0 & \frac{1}{m_B} & 0 \\ 0 & 0 & \frac{1}{m_C} \end{pmatrix} \tag{12.52}$$

and

$$\mathbf{G} \equiv \mathbf{BgB}^T = \begin{pmatrix} \frac{1}{m_A} + \frac{1}{m_B} & -\frac{1}{m_B} \\ -\frac{1}{m_B} & \frac{1}{m_B} + \frac{1}{m_C} \end{pmatrix} . \tag{12.53}$$

Replacing the classical variable P_i by the operator $-i\hbar \frac{\partial}{\partial R_i}$ gives Eq. 12.46.

12.3.3 Skewed Coordinates

In the previous section we saw how nondiagonal kinetic energies can arise, and how this imposes an obstacle to analog calculation of molecular dynamics. In this section we will see that nondiagonal kinetic energies can be eliminated by a scaling of the coordinates R_1 and R_2 such that the new kinetic energy is expressed in terms of a single mass and has no cross-terms (Eyring, 1931; Hirschfelder, 1936; Smith, 1959; Hirschfelder, 1969). It turns out that there are many choices of linear combinations of the R_1 and R_2 that diagonalize the kinetic energy. One specific choice is

$$Q_1 = aR_1 + bR_2 \cos \beta, \qquad Q_2 = bR_2 \sin \beta, \tag{12.54}$$

where a, b and $\cos \beta$ depend only on the masses:

$$a = \left[\frac{m_A m_{BC}}{m_{ABC}} \right]^{1/2} = \mu_{A,BC}{}^{1/2} \qquad b = \left[\frac{m_C m_{AB}}{m_{ABC}} \right]^{1/2} = \mu_{C,AB}{}^{1/2} \tag{12.55}$$

$$\cos^2 \beta = \frac{m_A m_C}{m_{BC} m_{AB}} \tag{12.56}$$

and $m_{AB} = m_A + m_B$, $m_{BC} = m_B + m_C$, $m_{ABC} = m_A + m_B + m_C$. The geometrical interpretation of this change of variables is illustrated in Figure 12.4a. If Q_1 and Q_2 are taken as two Cartesian axes, the effect of the transformation is to skew the two bond distances at the angle β to each other. Although there are many choices of coordinates that diagonalize the

(a)   **(b)**

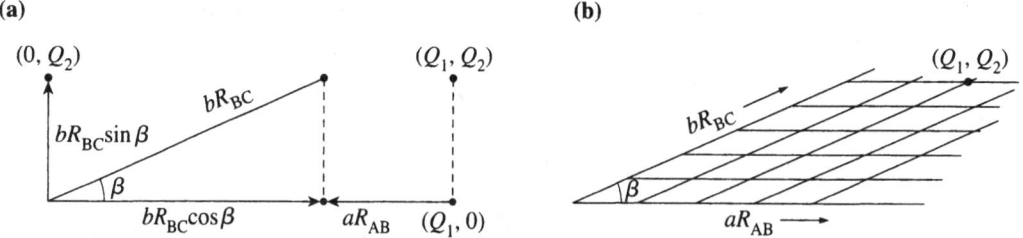

Figure 12.4 (a) The construction of the skewed coordinates Q_1 and Q_2 in terms of the internuclear bond coordinates R_{AB} and R_{BC}. (b) The grid corresponding to the skewed coordinate system. The point (Q_1, Q_2) on the grid is the same as the point (Q_1, Q_2) in (a). Adapted from Hirschfelder (1969).

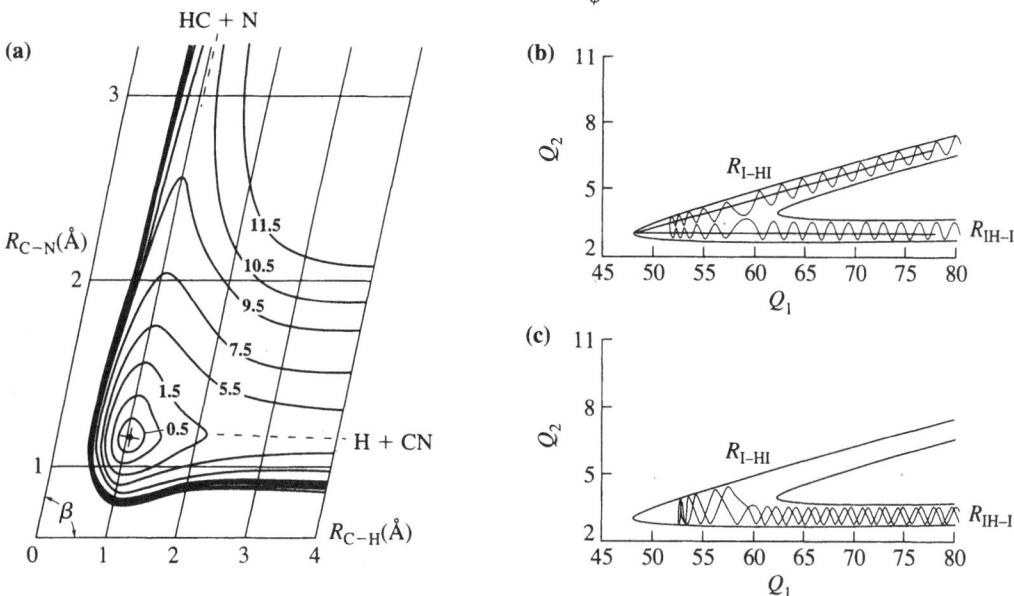

Figure 12.5 (a) Ground electronic potential energy surface for HCN, as a function of the H–C and C–N stretch coordinates. The skewed coordinates diagonalize the kinetic energy (see text). Within an overall rotation, the skewed coordinates are equivalent to the normal mode coordinates, which point along the major and minor axes of the elliptical contours at the minimum of the potential. Adapted from Herzberg (1966). (b)–(c) Ground potential energy surface for IHI as a function of the I–H and H–I stretch coordinates. The figure shows an example of a reactive trajectory (b) and a non-reactive trajectory (c). A system with a small skew angle such as this one shows extreme sensitivity of reactivity to initial conditions. Adapted from Skodje (1988).

kinetic energy, the skew angle β is always the same, since it depends only on the masses. Below, in Exercise 12.8, we give a simple derivation of Eq. 12.56.

It is instructive to examine the relationship between skew angle and the mass combination. For a collinear ABC molecule that is light-heavy-light, there is little or no skewing (the skew angle is 90°); physically, this is because the heavy middle atom almost completely prevents coupling of vibrational energy between the stretches of the two end atoms, just as the 90° angle minimizes energy transfer between coordinates. For three equal masses the skew angle is 60°. Finally, for a heavy-light-heavy mass combination, the skew angle can approach zero arbitrarily closely. This leads generally to many bounces in the interaction region, and great sensitivity of the fragmentation pattern to the initial conditions. The potential energy surface corresponding to the two bond stretching coordinates of HCN is depicted in skewed coordinates in Figure 12.5a, while the corresponding potential energy surface for the unstable molecule IHI is shown in Figure 12.5b.

▶ **Exercise 12.6**   Use Eq. 12.56 to verify the properties of the skew angle described in the previous paragraph, in (a) the limit of a light-heavy-light mass combination, (b) the limit of a heavy-light-heavy mass combination and (c) three equal masses.

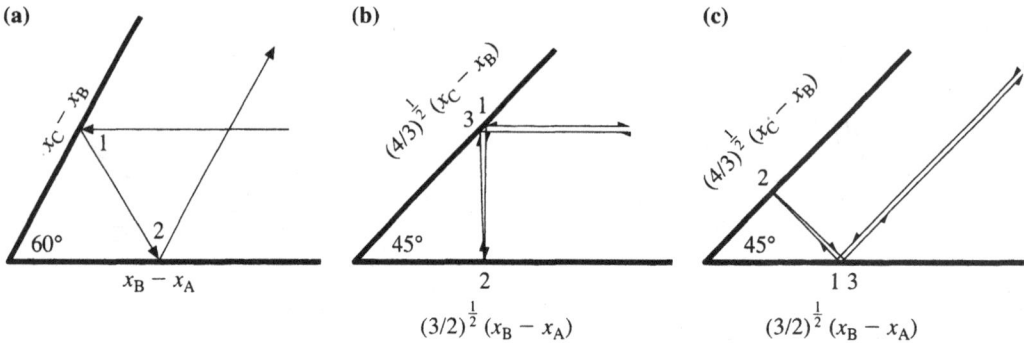

Figure 12.6 Billiard ball collision of three particles in a line. (a) $m_A = m_B = m_C$, b) $m_A = 3m_B$ and $m_C = 2m_B$, (c) same as (b) but for a different initial condition. The figure shows that a change in the skew angle changes the trajectory from reactive to nonreactive. After Hirschfelder (1969).

One may rightly ask, in the age of digital computers, is the transformation to skewed coordinates still of interest? The answer is yes; the skewed coordinates that arise from the transformation have an enormous impact of reaction kinematics, and ultimately give qualitative insight into reaction dynamics. A simple but instructive set of examples involves the collision of three hard spheres, described in Exercise 12.7.

▶ **Exercise 12.7**   Figure 12.6 shows some special trajectories for three hard spheres on a line.

   a.   For each of the three figures, describe in words what each of the three particles is doing initially, and after the events in circles 1, 2 and 3, respectively.

   b.   For each of the three figures, calculate the final momentum of all the particles, taking the initial momentum of the incoming particle as having magnitude u. Verify that momentum and energy are conserved in the collision.

In the next exercise, the skew angle is derived from the **G** matrix expression for the kinetic energy, Eq. 12.51. The key idea is to define a transformation on the bond momenta that diagonalizes the kinetic energy and then find the corresponding coordinates. Finally, the angle between the bond coordinates is expressed in terms of these new coordinates that diagonalize the kinetic energy. An interesting intermediate result of this derivation is a simple formula expressing the skew angle solely in terms of the **G** matrix elements.

▶ **Exercise 12.8**   Derive the formula for the skew angle, Eq. 12.56, from the **G** matrix expression for the kinetic energy, Eq. 12.51, using the method of canonical transformations.

Solution   Defining a new matrix **M** such that

$$\mathbf{G} = \mathbf{M}^T \mathbf{M}, \tag{12.57}$$

we may define a new momentum vector, $\bar{P}' = \mathbf{M}\bar{P}$. It is easily seen that $\bar{P}'$ diagonalizes the kinetic energy:

$$T = \frac{1}{2}\bar{P}^{\mathrm{T}}\mathbf{G}\bar{P} = \frac{1}{2}\bar{P}^{\mathrm{T}}\left(\mathbf{M}^{\mathrm{T}}\mathbf{M}\right)\bar{P} = \frac{1}{2}\bar{P}'^{\mathrm{T}}\bar{P}'. \tag{12.58}$$

The coordinates $\{Q_1, Q_2\}$ that are canonically conjugate to the new momenta, $\bar{P}'$, may be derived from the F_3 generator (see Section 10.4.1),

$$F_3(\bar{P}, \bar{Q}) = -\bar{Q}^{\mathrm{T}}\mathbf{M}\bar{P}, \tag{12.59}$$

using the relationship $R_i = -\frac{\partial F_3}{\partial P_i}$. In vector form this becomes

$$-\bar{R}^{\mathrm{T}} = -\bar{Q}^{\mathrm{T}}\mathbf{M} \qquad \text{or} \qquad \bar{R} = \mathbf{M}^{\mathrm{T}}\bar{Q}. \tag{12.60}$$

(Alternatively, the new coordinates may be obtained by making appropriate substitutions in Eqs. 12.47 ($\bar{R} = \mathbf{B}\bar{x}$) and 12.49.) In the basis of the new coordinates we may write

$$\bar{R}_1 = (M_{11}Q_1, M_{21}Q_2) \qquad \bar{R}_2 = (M_{12}Q_1, M_{22}Q_2). \tag{12.61}$$

By definition, the new coordinates Q_1 and Q_2 form an orthogonal basis since in this coordinate system T is diagonal, and without loss of generality can be taken to be orthonormal. The skew angle, β, is determined from the equation

$$\bar{R}_1^{\mathrm{T}} \cdot \bar{R}_2 = |\bar{R}_1||\bar{R}_2| \cos\beta. \tag{12.62}$$

Substituting Eq. 12.61 into Eq. 12.62 we obtain

$$\cos(\beta) = \frac{\bar{R}_1^{\mathrm{T}} \cdot \bar{R}_2}{|\bar{R}_1^{\mathrm{T}} \cdot \bar{R}_1|^{1/2}|\bar{R}_2^{\mathrm{T}} \cdot \bar{R}_2|^{1/2}} = \frac{M_{11}^{\mathrm{T}}M_{12} + M_{12}^{\mathrm{T}}M_{22}}{\left(M_{11}^{\mathrm{T}}M_{11} + M_{12}^{\mathrm{T}}M_{21}\right)^{1/2}\left(M_{21}^{\mathrm{T}}M_{12} + M_{22}^{\mathrm{T}}M_{22}\right)^{1/2}}$$

$$= \frac{(\mathbf{M}^{\mathrm{T}}\mathbf{M})_{12}}{(\mathbf{M}^{\mathrm{T}}\mathbf{M})_{11}^{1/2}(\mathbf{M}^{\mathrm{T}}\mathbf{M})_{22}^{1/2}} = \frac{\mathbf{G}_{12}}{\mathbf{G}_{11}^{1/2}\mathbf{G}_{22}^{1/2}} = \left(\frac{m_{\mathrm{A}}m_{\mathrm{C}}}{(m_{\mathrm{A}} + m_{\mathrm{B}})(m_{\mathrm{B}} + m_{\mathrm{C}})}\right)^{1/2}, \tag{12.63}$$

in agreement with Eq. 12.56. In passing to the fourth expression we have used the fact that each of the factors in the third expression can be viewed as the inner product of a row of $\mathbf{M}^{\mathrm{T}}$ with a column of $\mathbf{M}$; in passing to the final expression we have used the definition of the $\mathbf{G}$-matrix, Eq. 12.53. Note the intuitively appealing form of Eq. 12.63, which provides an explicit connection between the skew angle and the off-diagonal elements of the kinetic energy matrix. A related procedure can be used to derive the formulas for the mass scaling factors a and b defined by Eq. 12.55 (see Exercises 12.20–12.21).

The quantitative formulas for the mass-skewed coordinates described in this section are relevant only to a pair of stretching coordinates. As discussed in the next section, the more standard approach of finding normal modes of vibration also diagonalizes the kinetic energy, and is applicable to any type and number of internal molecular coordinates, not restricted to a pair of stretches. Nevertheless, the dynamics of two coupled stretching coordinates remains an important reduced model for both internal energy transfer and reactive scattering, and the *qualitative* dependence of the skew angle on mass combinations remains useful for understanding kinematic effects in these processes.

12.4   Normal Modes of Vibration

In the previous section we discussed the mass-skewed coordinates that diagonalize the kinetic energy operator. It turns out that if the potential contains no terms higher than quadratic it is possible to find coordinates that diagonalize not only the kinetic, but also the potential energy; these are the normal modes, a relatively standard topic in mechanics courses. The physical significance of the normal modes is that they define collective coordinates of the N particles, such that if a single mode is displaced from equilibrium it oscillates about equilibrium periodically for all time with a characteristic frequency, completely decoupled from the other normal modes of the system. The characteristic frequency of vibration is experimentally observable, manifesting itself as the position of a line in the infrared absorption spectrum, for example.

The normal coordinates are thus collective coordinates composed of linear combinations of the Cartesian displacements of each atom. In a polyatomic molecule with N atoms, there are $3N$ independent Cartesian coordinates, x, y and z, for each atom. The number of independent coordinates cannot change in the transformation from Cartesian to normal coordinates, and hence there are $3N$ normal coordinates. Quite generally, three of the normal coordinates involve overall translation of the molecule in the x, y and z directions, respectively, and hence no internal motion of the atoms relative to one another. Moreover, three of the normal coordinates involve overall rotation of the molecule around the x, y and z axes, respectively (for a linear molecule, there are only two overall rotations, since rotation along the molecular axis is ill defined). Thus, the number of normal modes of vibration—involving internal motion of the atoms relative to one another—is $3N - 6$ ($3N - 5$ for linear molecules). Diagrams showing the form of these collective coordinates for a bent and for a linear triatomic molecule are shown in Figure 12.7a and b, respectively. The vibrational modes are referred to as (i) symmetric stretch, (ii) bend and (iii) asymmetric stretch. Note that for the linear molecule the bending vibration becomes doubly degenerate, involving motion in each of the two perpendicular planes that include the molecular axis. The existence of an additional bending vibration in the linear molecule is closely related to the absence of a rotation along the molecular axis: one can form a linear combination of the in-plane and out-of-plane bending motion, which is a rotation around the linear axis with frequency zero, in this way recovering the third free rotation.

No true molecular system is harmonic, and thus the normal mode treatment in molecular dynamics is an approximation to the true dynamics of anharmonic systems, valid if the displacement from equilibrium is small. This may be seen geometrically in Figure 12.5. The normal coordinate axes are oriented along the the major and minor axes of the ellipse near the potential minimum in Figure 12.5. For small displacements, motion along one of these axes will continue for all time, without coupling to the other coordinate. This is consistent with our earlier statement, that the normal coordinates diagonalize the potential as well as the kinetic energy, that is, they eliminate cross-terms that would lead to energy transfer among the coordinates. The distortion of the equipotential contours from elliptical at higher energies is the signature of anharmonicity, and it is intuitively clear that at these higher energies simple harmonic motion is no longer possible.

Since the normal coordinates diagonalize the multidimensional kinetic energy operator, and are uniquely defined, there must be some relationship between the normal coordinates and the mass-skewed coordinates. In fact, these two coordinate systems introduce precisely

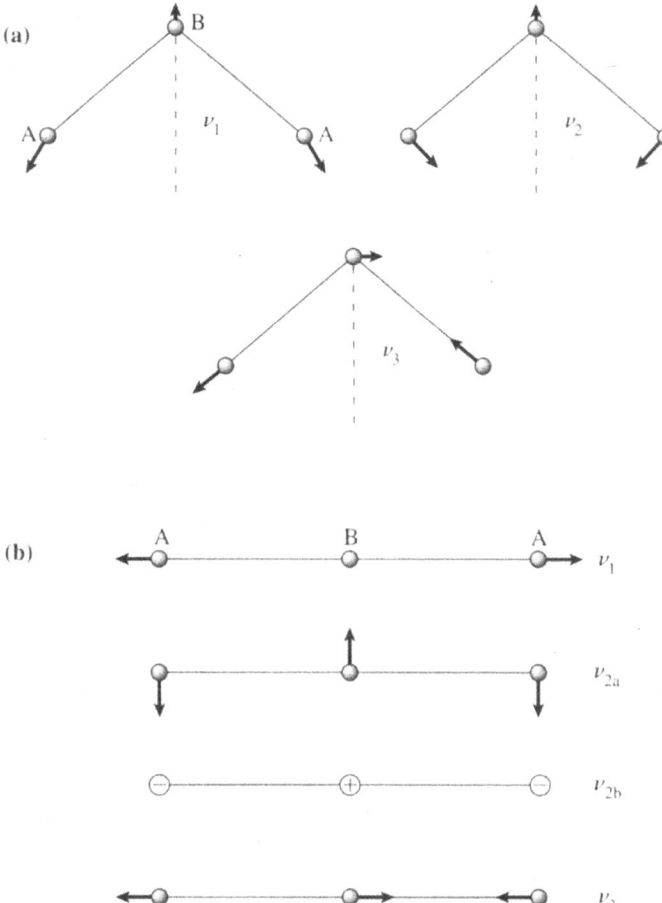

Figure 12.7 (a) Normal modes of vibration for a bent symmetric triatomic molecule, ABA. The modes are referred to as symmetric stretch (ν_1), bend (ν_2), and asymmetric stretch (ν_3). (b) Normal modes of vibration for a linear triatomic molecule, ABA. There are still the symmetric and asymmetric stretch modes, but now there are two degenerate bending vibrations involving bending motion in the two perpendicular planes. Adapted from Herzberg (1966).

the same mass scaling, but are rotated with respect to one another: the mass-skewed coordinates define Q_2 to point along R_2, thus treating R_1 and R_2 asymmetrically, while the normal coordinates, Q_1 and Q_2, involve linear combinations of R_1 and R_2 that diagonalize the potential energy, which do not point along either R_1 or R_2.

Procedures for finding normal modes of vibrations for an N-atom system with any geometry are well established, starting from both Cartesian coordinates and internal coordinates for the component atoms. We will now review these procedures briefly.

Consider the set of $3N$ Cartesian coordinates, $\{\xi_k\}$, corresponding to the displacement in x, y and z, respectively, of each of the N atoms in the molecule from its equilibrium geometry. The normal mode coordinates, Q_K, are defined as the linear combinations of the

ξ_k, which bring the kinetic and potential energies, T and V, simultaneously to the diagonal forms

$$T = \frac{1}{2} \sum_{K=1}^{3N} (\dot{Q}_K)^2 \qquad V = \frac{1}{2} \sum_{K=1}^{3N} \omega_K^2 Q_K^2, \tag{12.64}$$

where $\dot{Q}_K = dQ_K/dt$. (The lack of restoring forces for translation and rotation makes six of the ω_K equal zero.) If the canonical momentum is designated as

$$P_K \equiv \frac{\partial T}{\partial \dot{Q}_K}, \tag{12.65}$$

the classical Hamiltonian consists of a sum of terms:

$$H = T + V = \sum_{K=1}^{3N} H_K = \sum_{K=1}^{3N} \frac{1}{2}(P_K^2 + \omega_K^2 Q_K^2). \tag{12.66}$$

To find the transformation that relates Q_K to the ξ_k, it is convenient to transform from the ξ_k to "mass-weighted" Cartesian coordinates:

$$q_k = \sqrt{m_k}\,\xi_k, \tag{12.67}$$

where m_k is the mass of the atom having displacement ξ_k. In terms of the q_k, the Hamiltonian is transformed from

$$H = \sum_k \frac{1}{2} m_k \dot{\xi}_k^2 + \sum_{k,l} \frac{1}{2} \frac{\partial^2 V}{\partial \xi_k \partial \xi_l} \xi_k \xi_l \tag{12.68}$$

to

$$H = \sum_k \frac{1}{2} \dot{q}_k^2 + \sum_{k,l} \frac{1}{2} \frac{\partial^2 V}{\partial q_k \partial q_l} q_k q_l \tag{12.69}$$

$$= \sum_k \frac{1}{2} p_k^2 + \sum_{k,l} \frac{1}{2} f_{kl} q_k q_l, \tag{12.70}$$

where the p_k are the momenta conjugate to the q_k. Since the kinetic energy term is now expressed as a simple sum of squares, any orthogonal transformation of coordinates will preserve this form. Hence, one is free to choose the transformation so as to reduce the potential energy to a sum of squared terms with no cross-terms. Defining the coordinates

$$q_k = \sum_K a_{kK} Q_K \tag{12.71}$$

we obtain

$$V = \frac{1}{2} \sum_{k,l,K,L} f_{kl} a_{kK} a_{lL} Q_K Q_L \tag{12.72}$$

$$= \frac{1}{2} \sum_{KL} [\mathbf{a}^{-1}\mathbf{fa}]_{KL} Q_K Q_L. \tag{12.73}$$

We want this to equal a sum of terms of the form $\frac{1}{2}\omega_K^2 Q_K^2$. This will be true if

$$[\mathbf{a}^{-1}\mathbf{fa}]_{KL} = \omega_K^2 \delta_{KL}. \tag{12.74}$$

After some rearrangement, this can be written as

$$\sum_l [f_{kl} - \omega_K^2 \delta_{kl}] a_{lK} = 0 \qquad k = 1, 2, \ldots, 3N. \tag{12.75}$$

For a nontrivial solution to these equations, ω_K must satisfy the secular equation

$$|f_{kl} - \omega_K^2 \delta_{kl}| = 0. \tag{12.76}$$

Solution of Eq. 12.75 yields $3N$ characteristic frequencies, ω_K, and associated with each, a set of $3N$ coefficients a_{lK} that define the corresponding normal coordinate. As discussed above, for a nonlinear molecule, three of these normal coordinates correspond to overall translation along the axes x, y or z, three correspond to overall rotations around the axes x, y or z, and $3N - 6$ correspond to vibrations. Note that the ω_K associated with each of the three translational and three rotational coordinates vanish, since these motions are unhindered.

▶ **Exercise 12.9** For HOH, there are three vibrational normal coordinates, whose form is shown in Figure 12.7. Draw the corresponding pictures of the three translational and three rotational normal coordinates.

▶ **Exercise 12.10** Consider finding the normal modes of the unstable species H_3 in the region of the barrier top. What differences would one expect relative to the normal modes of a stable species at its minimum?

Solution Generally, we would expect to get six eigenvalues equal to 0 (three translations and three rotations), $3N - 7$ positive eigenvalues corresponding to bound motion (ω^2 positive implies that ω is real) and one negative eigenvalue corresponding to unbound motion in the coordinate along the reaction path (ω^2 negative implies that ω is imaginary).

Although mass-weighted Cartesian coordinates simplify the general theory, they are often not the most practical for calculation. With only a slight change in the formulation of the problem one can work with nonweighted Cartesian coordinates:

$$\sum_l [f_{kl} - \omega_K^2 m_{kk}] a_{lK} = 0, \qquad k = 1, 2, \ldots, 3N, \tag{12.77}$$

where m_{kk} are elements of a diagonal mass matrix $\mathbf{m}$ and ω_K must satisfy the secular equation

$$|f_{kl} - \omega_K^2 m_{kk}| = 0; \tag{12.78}$$

however, the coordinate transformation is no longer orthogonal. More generally, one can use non-Cartesian coordinates η_k in which there are cross-terms in the kinetic energy as well as in the potential energy. In this case one again obtains a generalized eigenvalue equation and secular equation of the form of Eqs. 12.77 and 12.78, respectively, but with a nondiagonal mass matrix; m_{kk} is replaced by m_{kl}.

A particularly useful approach is the use of *internal coordinates*. In this method one eliminates the six redundant coordinates at the very beginning by using coordinates internal to the molecule, such as bond lengths and angles. This reduces the size of the secular equation from $3N \times 3N$ to $(3N - 6) \times (3N - 6)$, which greatly simplifies the solution. In this approach, one defines a force constant matrix, $\mathbf{F}$, and a matrix of inverse mass combinations and geometrical factors, $\mathbf{G}$, expressed in internal coordinates of bond lengths and angles. The simple case of the $\mathbf{G}$ matrix for collinear triatomics was discussed in Section 12.3.2. Wilson gave the explicit form for these matrices for molecules of essentially arbitrary complexity. The simultaneous diagonalization of both the $\mathbf{F}$ and $\mathbf{G}$ matrices, which is a generalized eigenvalue problem of the form

$$\mathbf{Fv} = \lambda \mathbf{G}^{-1}\mathbf{v}, \tag{12.79}$$

gives again the normal modes of vibration. This leads to a secular equation for the eigenvalues of the form

$$|\mathbf{F} - \lambda \mathbf{G}^{-1}| = 0 \tag{12.80}$$

or

$$|\mathbf{GF} - \mathbf{I}\lambda| = 0. \tag{12.81}$$

This approach is treated in detail in the excellent book by Wilson, Decius and Cross (1955).

▶ Exercise 12.11

    a. Solve the vibrational eigenvalue problem for a symmetric, collinear triatomic molecule with masses m, M and m, using Cartesian coordinates. Use the form of the secular equation

$$|\mathbf{f} - \mathbf{g}^{-1}\lambda| = 0, \tag{12.82}$$

where $\mathbf{f}$ and $\mathbf{g}^{-1}$ are 3×3 matrices. Interpret your eigenvalues and eigenvectors.

    b. Solve the same problem as in a, but using the two bond coordinates R_1 and R_2. The secular equation can be written in the form

$$|\mathbf{GF} - \mathbf{I}\lambda| = 0 \tag{12.83}$$

where $\mathbf{G}$ and $\mathbf{F}$ are 2×2 matrices and $\mathbf{I}$ is the 2×2 unit matrix. The two eigenvalues and eigenvectors should agree with the result of part a. Which eigenvalue/eigenvector of part a is missing and why?

12.5   Chemical Reactions and Transition State Theory

12.5.1   Potential Energy Surfaces for Bimolecular Reactions

In our study of potential energy surfaces (Section 12.3), we were primarily concerned with potential energy surfaces corresponding to stable chemical species such as OCO. The potential energy surfaces associated with such species have minima at the coordinates corresponding to the molecule's equilibrium geometry. The evolution of a trajectory from the region of the potential minimum to one of the fragment arrangement channels, such

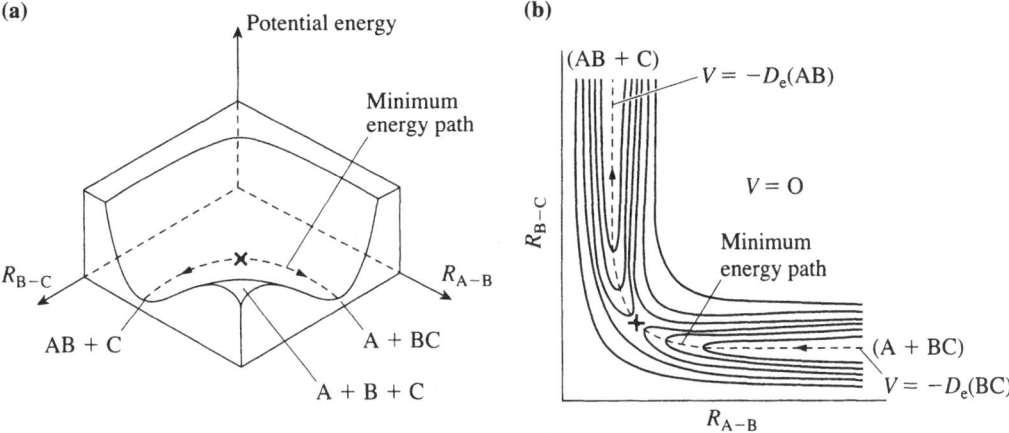

Figure 12.8 Potential energy surface for collinear $A + BC \rightarrow AB + C$. (a) stereoscopic projection; (b) equipotential contours. Adapted from Levine (1974). Note that the "transition state" between reactants and products has a saddle-like structure: a maximum along the reaction coordinate and a minimum in the perpendicular coordinate(s).

as $O + CO$, is called a unimolecular reaction. Unimolecular reactions can be induced by collision (collision-induced dissociation) as well as by light (photodissociation). The latter process will be discussed in some detail in Chapters 16 and 17.

In this section we will be concerned with chemical reactions of the form $A + BC \rightarrow ABC^{\ddagger} \rightarrow AB + C$, in which $ABC^{\ddagger}$ is not a stable species. Such reactions are inherently *bimolecular*, and require overcoming an energy, or "activation" barrier in order to proceed. Despite the fact that $ABC^{\ddagger}$ is not stable, one can still construct the potential energy surface for the three atoms as a function of internuclear distance. Now the potential will not have a minimum in the region of $ABC^{\ddagger}$, but a *saddle point*. Consider, for instance, the three-atom system H—H—H, which in its ground electronic state is an unstable species. Figure 12.8 shows the potential energy surface for collinear geometries. If we follow the minimum energy path from reactants to products we see that it reaches a maximum somewhere in the middle, the so-called "transition state" region. Perpendicular to the reaction path at the transition state, the potential is generally bound. The combination of a maximum in one direction and a minimum in the other gives rise to a characteristic saddle shape, hence the name saddle point. The region around the saddle point is characterized by hyperbolic equipotentials, as opposed to the elliptical equipotentials that characterize potential minima and maxima.

This minimum energy path referred to above, also called the "reaction path," defines a coordinate system consisting of the local reaction path coordinate and the perpendicular coordinates. In particular, one may diagonalize the matrix of mass-weighted second derivatives of the potential at each point along the reaction path; for simple reaction coordinates as in Figure 12.8, one will find one unstable mode of imaginary (or zero) frequency parallel to the reaction path and $3N - 7$ normal modes perpendicular to the reaction path, which are generally stable. At the transition state the unstable normal mode has asymmetric stretch character, while the perpendicular normal modes have symmetric stretch and bend character. In the asymptotic regions of the potential, the form of the normal coordinates is completely different: the unstable normal mode is now translational motion, while the perpendicular coordinates are vibration and rotation.

12.5.2   Jacobi Coordinates and the Skew Angle Revisited

Another way to define the coordinates that is primarily relevant to chemical reactions of the type $A + BC \rightarrow AB + C$ is in terms of so-called Jacobi coordinates (Child, 1991; Manolopoulos, 1999). In this system, R is defined as the distance from A to the center of mass of BC, and r is the distance between B and C (see Figure 12.9):

$$R = \frac{m_B x_B + m_C x_C}{m_{BC}} - x_A = \lambda R_{BC} + R_{AB} \tag{12.84}$$

$$r = x_C - x_B = R_{BC}, \tag{12.85}$$

where $\lambda = \frac{m_C}{m_{BC}}$. This choice of coordinates is suitable when $R \gg r$, that is, when A is well separated from BC. One may similarly define an alternative set of Jacobi coordinates in which r' is the distance between A and B, and R' is the distance between B and C:

$$R' = x_C - \frac{m_A x_A + m_B x_B}{m_{AB}} = \lambda' R_{AB} + R_{BC} \tag{12.86}$$

$$r' = x_B - x_A = R_{AB} \tag{12.87}$$

where $\lambda' = \frac{m_A}{m_{AB}}$. This choice of coordinates is suitable when $R' \gg r'$, that is, when C is well separated from AB. For the reaction $A + BC \rightarrow AB + C$, the first choice of Jacobi coordinates is suitable for describing the reactants, while the second choice is suitable for describing the products. The Hamiltonian can be expressed in either system of Jacobi coordinates:

$$H = \frac{1}{2m} P_R^2 + \frac{1}{2\mu} p_r^2 + V(R, r) \qquad \text{(reactants)} \tag{12.88}$$

$$H = \frac{1}{2m'} P_R'^2 + \frac{1}{2\mu'} p_r'^2 + V(R', r') \qquad \text{(products),} \tag{12.89}$$

where

$$m = m_A m_{BC}/m_{ABC}, \qquad \mu = m_B m_C/m_{BC} \tag{12.90}$$

$$m' = m_C m_{AB}/m_{ABC}, \qquad \mu' = m_A m_B/m_{AB}. \tag{12.91}$$

▶ **Exercise 12.12**   Use canonical transformations to derive these Hamiltonians starting from the Hamiltonian in the three Cartesian coordinates, $H(x_1, x_2, x_3, p_1, p_2, p_3)$. [Hint: adapt the procedure from Exercise 12.8.]

In the next exercise, the skew angle discussed in Section 12.3.3 is derived from a new perspective: the skew angle emerges as a kinematic angle of rotation between mass-scaled Jacobi coordinates of reactants and products. Pictorially, the skew angle is the angle between the translational asymptotes of reactants and products, when the potential is plotted in terms of either set of mass-scaled Jacobi coordinates (see Figure 12.9).

(a)

(b)

(c)

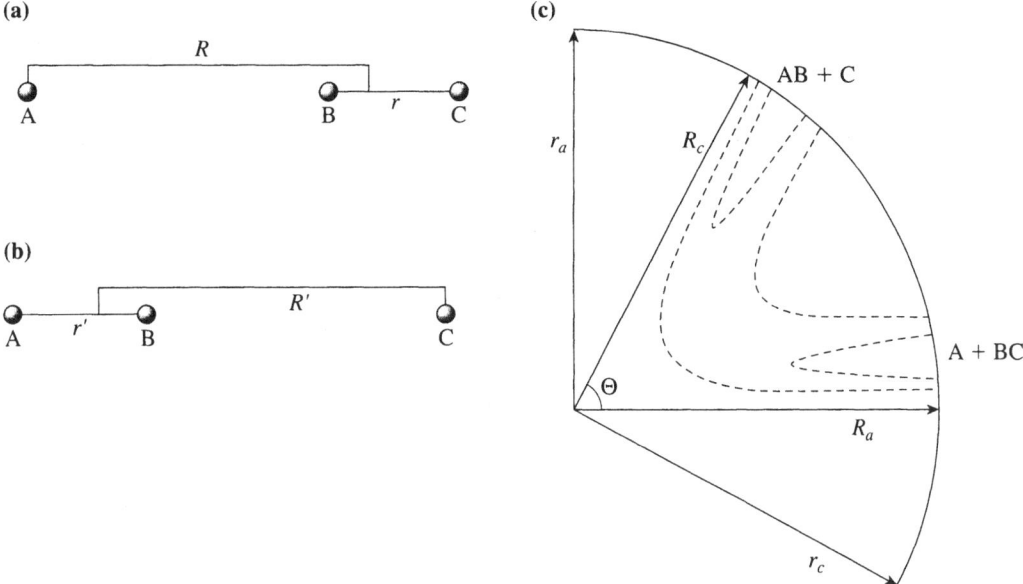

Figure 12.9 The definition of the Jacobi coordinates for (a) reactants and (b) products. (c) Representation of the *mass-scaled* Jacobi coordinates for reactants (R_a, r_a) and for products (R_c and r_c). Note that the transformation angle Θ between the two sets of mass-scaled Jacobi coordinate systems (Eq. 12.98) takes on a geometrical significance as the skew angle β. Adapted from Manolopoulos (1999).

▶ Exercise 12.13

a. Use canonical transformations to further transform to mass-scaled reactant and product Jacobi coordinates in the following way:

$$R_a = \lambda_a R \qquad r_a = \lambda_a^{-1} r \qquad \text{(reactants)} \qquad (12.92)$$

$$R_c = \lambda_c R' \qquad r_c = \lambda_c^{-1} r' \qquad \text{(products)}, \qquad (12.93)$$

where

$$\lambda_a = \left(\frac{m_A m_{BC}^2}{m_B m_C m_{ABC}} \right)^{1/4} \qquad \lambda_c = \left(\frac{m_C m_{AB}^2}{m_A m_B m_{ABC}} \right)^{1/4}. \qquad (12.94)$$

b. Show that the new Hamiltonian can be written as

$$H = \frac{1}{2\mu}(P_{R_a}^2 + p_{r_a}^2) + V(R_a, r_a) \qquad \text{(reactants)} \qquad (12.95)$$

or

$$H = \frac{1}{2\mu}(P_{R_c}^2 + p_{r_c}^2) + V(R_c, r_c) \qquad \text{(products)}, \qquad (12.96)$$

where

$$\mu = \left(\frac{m_A m_B m_C}{m_{ABC}} \right)^{1/2}. \qquad (12.97)$$

The appearance of a single reduced mass in both the reactant and product expressions for H indicates that these mass-scaled coordinates provide a kinematically democratic description of the reaction.

c.   Show that

$$\begin{pmatrix} R_c \\ r_c \end{pmatrix} = \begin{pmatrix} \cos\Theta & \sin\Theta \\ \sin\Theta & -\cos\Theta \end{pmatrix} \begin{pmatrix} R_a \\ r_a \end{pmatrix} \qquad (12.98)$$

and

$$\begin{pmatrix} P_{R_c} \\ p_{r_c} \end{pmatrix} = \begin{pmatrix} \cos\Theta & \sin\Theta \\ \sin\Theta & -\cos\Theta \end{pmatrix} \begin{pmatrix} P_{R_a} \\ p_{r_a} \end{pmatrix}, \qquad (12.99)$$

where

$$\Theta = \arctan\left(\frac{m_B}{\mu}\right). \qquad (12.100)$$

[Hint: express both (R_c, r_c) and (R_a, r_a) in terms of a transformation of the bond coordinates (R_{AB}, R_{BC}).] Θ has the physical interpretation of the angle between the asymptotic mass-scaled translational coordinate for reactants and products (see Figure 12.9c). Verify that Eq. 12.100 for Θ agrees with the formula for the skew angle β in Eq. 12.56.

12.5.3   Transition State Theory

The previous subsection dealt with some of the general features of the potential energy surface for bimolecular reactions. In this subsection we describe some features of the classical dynamics on these surfaces. The study of both unimolecular and bimolecular dynamics using classical mechanics is a rich and fascinating field, with many close ties to recent developments in classical mechanics, including periodic orbit theory, resonances, cantori and classical chaos. These developments deserve a book on their own. Here I limit the discussion to one or two topics that give a flavor for the interplay of translational and internal motion in reaction dynamics, an effect that will be present in many of the quantum wavepacket applications in subsequent chapters. This interplay can be approximated in a number of useful ways, resulting in formulations of chemical reaction rates and probabilities in which the dynamics no longer appears. The most important example of such a formulation is classical transition state theory, which is now described.

One of the key concepts in the theory of rates of chemical reactions is that of the transition state, which corresponds to the unstable species at the maximum of the reaction path between reactants and products. For example, in the reaction $A + BC \rightarrow AB + C$, if there is an energy of activation to the reaction (see Figure 12.10) the intermediate complex ABC is known as the transition state. The celebrated Arrhenius formula expresses the thermal rate constant for a chemical reaction as an exponential that decays with increasing barrier height:

$$k(T) = A(T)e^{-E^{\ddagger}/k_B T}, \qquad (12.101)$$

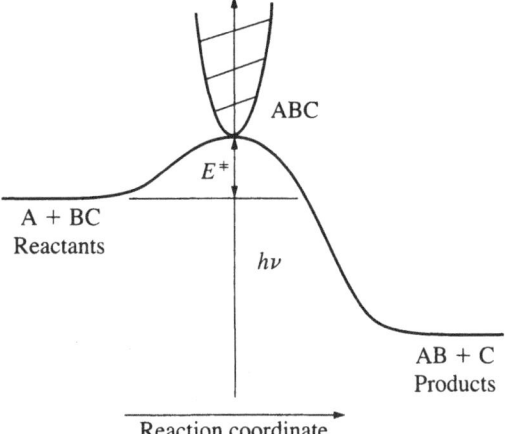

Figure 12.10 Schematic representation of the structure of the transition state. The single most important factor that determines the rate constant for reaction is $E^{\ddagger}$, the height of the barrier relative to asymptotic reactants. However, in addition to the energetics of the reaction path, the rate constant is also determined by the frequencies in the $3N - 1$ molecular motions perpendicular to the reaction path, relative to their asymptotic values for reactants. Simply put, high frequencies mean a narrow transition state, which is hard to pass, while low frequencies mean a broad transition state, which, statistically speaking, is easier to pass. The available phase space perpendicular to the reaction path may be thought of as defining an entropy factor, which, together with the energy factor for the barrier height, defines a free energy of activation.

where $E^{\ddagger}$ is the energy of the transition state barrier relative to that of the reactants, k_B is the Boltzmann constant and T is the temperature. To gain insight into the prefactor $A(T)$ in Eq. 12.101, we note that at constant temperature the Helmholtz free energy, A, is the more natural thermodynamic variable than E (do not confuse the Helmholtz free energy, $A^{\ddagger}$, with the Arrhenius prefactor, $A(T)$). Denoting the Helmholtz free energy at the transition state as $A^{\ddagger} = E^{\ddagger} - T S^{\ddagger}$ we have

$$k(T) = A'(T)e^{-A^{\ddagger}/k_B T}. \tag{12.102}$$

This leads to the identification

$$A(T) = A'(T)e^{T S^{\ddagger}}. \tag{12.103}$$

Hence, the prefactor $A(T)$ includes an entropy factor. If the entropy associated with these perpendicular coordinates is larger at the transition state than for reactants, this increases the rate constant; if the entropy factor for these coordinates is smaller it decreases the reaction rate.

At the next level of sophistication one seeks a microscopic interpretation for this entropy prefactor. From statistical mechanics we know the entropy is a measure of the number of available states, or the size of the available phase space of the system. Thus this prefactor measures the size of the available phase space perpendicular to the reaction coordinate, in the transition state region relative to the reactant region. If there is no change from reactants to transition state this factor drops out; but if there is a narrowing in the perpendicular degrees of freedom this presents a bottleneck and decreases the reaction rate. Conversely, if there is a widening in the perpendicular directions this opens more states and increases the reaction rate. This idea is embodied in the so-called "absolute rate theory" expression of Pelzer and Wigner (Pelzer, 1932) and Eyring (1935) for the rate constant:

$$k(T) = \frac{k_B T}{h} \frac{Q^{\ddagger}}{Q} e^{-E^{\ddagger}/k_B T}. \tag{12.104}$$

In this expression, $Q^{\ddagger}$ is the partition function at the barrier in the $3N - 1$ coordinates perpendicular to the transition state; Q is the partition function in the $3N$ degrees of freedom of the reactants. Thus, the absolute rate theory formula expresses the entropy factor as $k_B T / h$ times the ratio of the partition function at the transition state divided by the partition function of the reactants. Both these partition functions require knowledge of the vibrational frequencies, the first at the transition state and the second in the reactant region. The latter is probed by standard spectroscopies; the former is much more elusive, since the transition state is not a stable species (see Figure 12.10). Thus, it has been a goal of chemical physics for many years to probe the frequencies of the transition state directly, spectroscopically, rather than learning about the transition state only indirectly via highly averaged rate constant measurements.

The rigorous formulation of the rate constant is in terms of trajectories that start in the reactant region of the potential and may or may not end up in the product region. The transition state theory expression for the rate constant (see below) can then be cast as a well-defined approximation to this formulation. The continuity equation for particle flow is

$$\frac{\partial \rho}{\partial t} + \nabla \cdot (\rho \mathbf{v}) = 0. \tag{12.105}$$

We define a dividing surface S that separates reactants from products and use Gauss's theorem,

$$\int_V \frac{\partial \rho}{\partial t} dV = -\int_S (\rho \mathbf{v} \cdot \mathbf{n}) \, dS, \tag{12.106}$$

where $\mathbf{n}$ is the unit outward normal to dS. Equation 12.106 represents the rate of loss of reactant in terms of the *flux* through a dividing surface that separates reactants and products. Both the volume, V, and the surface, S, are properly defined not in configuration space but in phase space (Keck, 1967). For simplicity, we assume that the phase space dividing surface is actually a coordinate space dividing surface of dimension $3N - 1$ times a momentum dividing surface of dimension $3N$, although more general definitions are possible. We therefore write

$$\int_V \rho \, dV = \int_V \rho(\mathbf{p}, \mathbf{q}) \prod_{i=1}^{3N} dp_i \prod_{i=1}^{3N} dq_i \equiv h^{3N} Q \tag{12.107}$$

$$\int_S (\rho \mathbf{v} \cdot \mathbf{n}) \, dS = \int_S (\rho(\mathbf{p}, \mathbf{q}) \mathbf{v} \cdot \mathbf{n}) \prod_{i=1}^{3N} dp_i \prod_{i=1}^{3N-1} dq_i, \qquad (12.108)$$

where $\mathbf{p}$ and $\mathbf{q}$ are $3N$ dimensional vectors in momentum and coordinate, respectively. The factor h^{3N} appears as the usual normalization factor of classical phase space integrals in order to approach the quantum mechanical result at high temperatures. Note that the product in the surface integral is over only the $3N - 1$ coordinates perpendicular to the dividing surface. In what follows we will use the notation $d\mathbf{p} = \prod_{i=1}^{3N} dp_i, d\mathbf{q} = \prod_{i=1}^{3N} dq_i$, for the elements of the multidimensional momentum and coordinate volume elements, $d\mathbf{q}_s = \prod_{i=1}^{3N-1} dq_i$ for the multidimensional coordinate space dividing surface element, and q_r and p_r for the coordinate and momentum perpendicular to the coordinate space dividing surface. Identifying the rate of loss of reactant with $\frac{\partial Q}{\partial t}$, and noting that this loss rate is given as the product of the rate constant, k, and the reactant concentration, Q, we have

$$\frac{\partial Q}{\partial t} = -kQ = -\frac{1}{h^{3N}} \int_S (\rho \mathbf{v} \cdot \mathbf{n}) \, dS = -\frac{1}{h^{3N}} \int \rho \, p_r \, d\mathbf{p} d\mathbf{q}_s, \qquad (12.109)$$

where we have used Eqs. 12.106–12.107. Here and henceforth we assume mass-weighted coordinates.

Equation 12.109 is a nonequilibrium ensemble average; it is possible, however, to replace the nonequilibrium distribution, $\rho(\mathbf{p}, \mathbf{q}, t)$, with an equilibrium distribution, $e^{-H/k_B T}$. At equilibrium, however, the total flux through this surface is zero, since the amount of formation of new product is exactly balanced by the amount of formation of new reactant. Therefore, to extract a rate constant from an equilibrium distribution one must use just the *reactive* flux through this dividing surface—the flux corresponding to particles that begin as reactants and end as products (Pechukas, 1976).

This leads to the reactive flux expression for the rate constant:

$$\frac{\partial Q}{\partial t} = -kQ = -\frac{1}{h^{3N}} \int e^{-H(\mathbf{p}, \mathbf{q}_s)/k_B T} p_r \chi(\mathbf{p}, \mathbf{q}) d\mathbf{p} d\mathbf{q}_s. \qquad (12.110)$$

We have introduced the factor $\chi(\mathbf{p}, \mathbf{q})$, which takes the value 1 if $(\mathbf{p}, \mathbf{q})$ lies on a reactive trajectory, and is zero otherwise; $\chi(\mathbf{p}, \mathbf{q})$ thus ensures that only the reactive flux is counted in the formula. Here and below we set the mass to unity, which is appropriate if the calculation is done in mass-scaled Cartesian coordinates.

The central assumption of transition state theory (TST) is that all trajectories passing through the dividing surface with positive momentum will end up as products, and conversely, trajectories passing through with negative momentum wlll end up as reactants. This is equivalent to replacing $\chi(\mathbf{p}, \mathbf{q})$ by a function $\chi_+(\mathbf{p}, \mathbf{q})$,

$$\chi_+(\mathbf{p}, \mathbf{q}) = \begin{cases} 1, & p_r > 0 \\ 0, & \text{otherwise} \end{cases} \qquad (12.111)$$

(see Figure 12.11). This replacement allows for an analytical integration in Eq. 12.110 over the momentum variable along the reaction coordinate. To see this, note that the Hamiltonian can be written as

$$H(\mathbf{p}, \mathbf{q}_s) = \mathbf{p}^2/2 + V(\mathbf{q}_s) = p_r^2/2 + \mathbf{p}_s^2/2 + V(\mathbf{q}_s). \qquad (12.112)$$

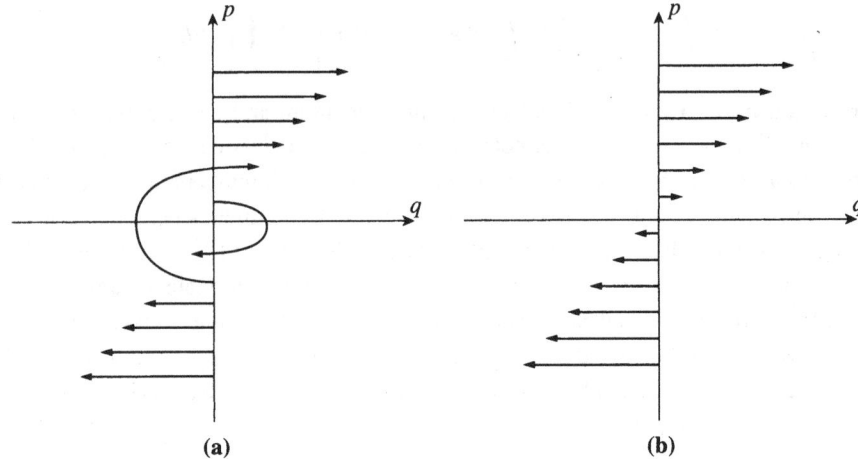

Figure 12.11 (a) Schematic picture of the phase space orbits associated with the barrier shown in Figure 12.10. Note the existence of trajectories that recross the transition state (i.e., the barrier top). (b) Phase space structure assumed by transition state theory: lacking any prior knowledge of the phase space structure, transition state theory assumes there are no recrossing trajectories.

If we substitute Eqs. 12.111–12.112 into Eq. 12.110, the integral over the momentum in the reaction coordinate gives $k_B T$, leading to the transition state theory rate constant, k_{TST}:

$$k_{TST} = \frac{k_B T}{h^{3N} Q} \int d\mathbf{p}_s d\mathbf{q}_s e^{-(\mathbf{p}_s^2/2 + V(\mathbf{q}_s))/k_B T} = \frac{k_B T}{h} \frac{Q^\ddagger}{Q} e^{-E^\ddagger/k_B T}, \qquad (12.113)$$

where we have defined the partition function at the transition state, $Q^\ddagger$, as the reduced dimensional phase space integral in the $N - 1$ coordinates perpendicular to the transition state:

$$h^{3N-1} Q^\ddagger e^{-E^\ddagger/k_B T} = \int d\mathbf{p}_s d\mathbf{q}_s e^{-(\mathbf{p}_s^2/2 + V(\mathbf{q}_s))/k_B T}. \qquad (12.114)$$

This reduced partition function is referenced to a zero of energy for asymptotic reactants at the potential minimum along the $N - 1$ coordinates perpendicular to the reaction path; this accounts for the separate factor of $e^{-E^\ddagger/k_B T}$. Note the agreement between Eq. 12.113 and the Eyring formula, Eq. 12.104.

An alternative way to think about the χ function, which is perhaps more transparent, is to notice that since the integral in Eq. 12.110 is at the dividing surface we need only the value of χ at the surface. This function may be written as a Heaviside function in the following way:

$$\chi(\mathbf{p}, \mathbf{q}_s) = \lim_{t \to \infty} \theta[q_r(t); \mathbf{p}, \mathbf{q}_s], \qquad (12.115)$$

where q_r is the position of the trajectory along the reaction coordinate at time t given that it started at time zero at the dividing surface with initial conditions $\mathbf{p}, \mathbf{q}_s$, and

$$\theta[q_r(t)] = \begin{cases} 1 & \text{if } q_r(t) > 0 \\ 0 & \text{if } q_r(t) < 0. \end{cases} \qquad (12.116)$$

The central assumption of transition state theory in this notation is the replacement of

$$\chi(\mathbf{p}, \mathbf{q}_s) = \lim_{t \to \infty} \theta[q_r(t); \mathbf{p}, \mathbf{q}_s] \tag{12.117}$$

by

$$\chi_+(\mathbf{p}, \mathbf{q}_s) = \theta[p_r(0); \mathbf{p}, \mathbf{q}_s], \tag{12.118}$$

that is, removal of all dynamics from the χ function, and assigning its value as 1 if the component of the velocity of a trajectory along the reaction coordinate is positive at the dividing surface and otherwise assigning its value as 0:

$$\theta[p_r(0)] = \begin{cases} 1 & \text{if } p_r(0) > 0 \\ 0 & \text{if } p_r(0) < 0. \end{cases} \tag{12.119}$$

What the transition state theory rate constant, Eq. 12.113, neglects is the possibility of barrier recrossing: that trajectories passing through the dividing surface with positive momentum might not end up as products, and conversely, that trajectories passing through with negative momentum might not end up as reactants (see Figure 12.11). Substituting Eq. 12.111, the TST approximation, into Eq. 12.110 shows that both these types of recrossing decrease the true rate constant with respect to the TST approximation: trajectories with *positive* momentum at the top of the barrier that end as *reactants* are weighted by $\chi = 0$ and hence do not contribute to the exact rate constant; and trajectories with *negative* momentum at the top of the barrier that end as *products* are weighted by $\chi = 1$ and hence their negative flux contributes to the exact rate constant. Since TST neglects both types of recrossings it provides an upper bound to the true classical rate:

$$k_{\text{TST}} \geq k. \tag{12.120}$$

One may write quite generally

$$k = \kappa k_{\text{TST}} \quad 0 \leq \kappa \leq 1, \tag{12.121}$$

where the "transmission coefficient," κ, is an average measure of recrossing.

▶ **Exercise 12.14**   Show that the neglect of recrossings can be expressed succinctly in terms of the following inequality:

$$\mathbf{p}_s \chi_+(\mathbf{p}, \mathbf{q}_s) \geq \mathbf{p}_s \chi(\mathbf{p}, \mathbf{q}_s). \tag{12.122}$$

If the reactive flux is calculated exactly, the rate constant k is independent of the choice of dividing surface between reactants and products. This is no longer true once the assumption of no recrossing trajectories is invoked: k_{TST} will in general depend on the choice of dividing surface. But since k_{TST} is always an upper bound to the true rate constant, k, the best dividing surface is the one that minimizes the TST rate constant, since this provides a lowest upper bound to the full classical rate constant. This is the key idea behind the variational TST method, also due to Wigner, in which one searches the space between reactants and products for a dividing surface that minimizes k_{TST}.

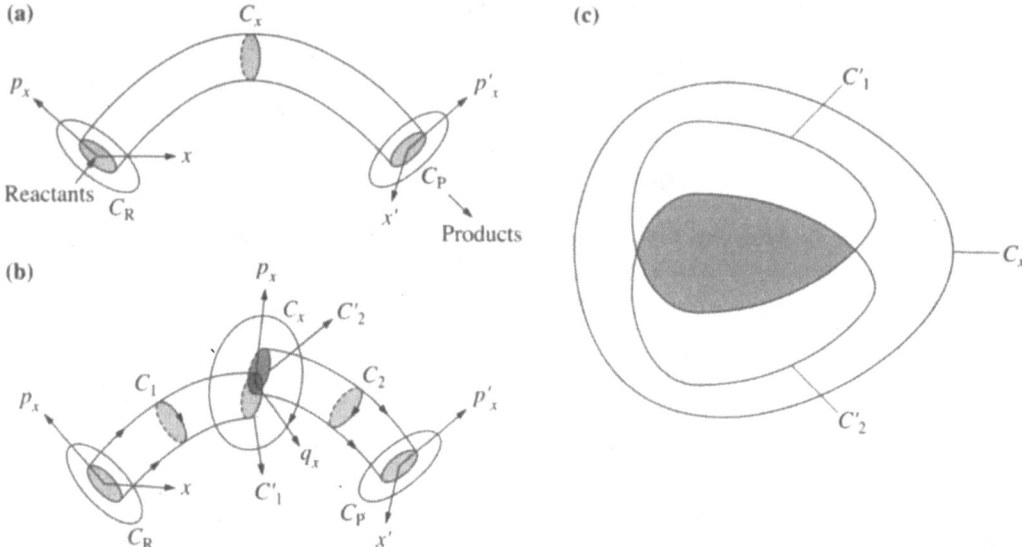

Figure 12.12 (a) Schematic representation of the dynamical flow of phase space area from reactants (C_R) to products (C_P) where there is no barrier recrossing. (b) Same as (a) but where there is barrier recrossing. (c) Slice of the phase space in (b) denoted C_x. The area C_1' evolves from C_R, and the area C_2' evolves from C_P. Their overlap, given by the shaded area, corresponds to direct reactive flux. Adapted from Pollak (1980).

A somewhat more sophisticated approach to understanding recrossing trajectories is based on the phase space representation of the incoming and outgoing flux, shown in Figure 12.12 (Pollak, 1980). The overlap of these fluxes gives the direct reactive flux: the portion of amplitude that passes from reactants to products without recrossing the barrier. In Figure 12.12a all the reaction is direct; in Figure 12.12b there is a direct and an indirect component. A magnification of the dividing surface of Figure 12.12b is shown in Figure 12.12c. The outer boundary, C_x, is the energetically accessible phase space. The area C_1' evolves from C_R, and the area C_2' evolves from C_P. Both C_1' and C_2' provide upper bounds to the rate constant. Their overlap, given by the shaded area, corresponds to direct reactive flux. This intersection provides a *lower* bound on the rate constant, since it discards trajectories that may cross on future passes. Of the sections of the manifold that do not overlap, part will pass from reactants to products after recrossing the transition state dividing surface an even number of times and part will return to reactants after recrossing the dividing surface an odd number of times. Associated with each number of recrossing is a manifold of trajectories. Formally, one can bound the exact rate constant from above as well as from below ever more precisely by summing the contributions from manifolds representing successive recrossings.

There is a fascinating connection between the structure of the phase space diagrams in Figure 12.12 and the existence of periodic trajectories in the vicinity of the transition state, shown in Figure 12.13. At energies sufficiently close to the energy at the saddle point, the classical trajectory perpendicular to the reaction path that passes through the saddle point is an unstable periodic orbit (central dark line in Figure 12.13a) (Pechukas, 1977). If the orbit of this periodic trajectory is plotted in phase space, it defines a surface

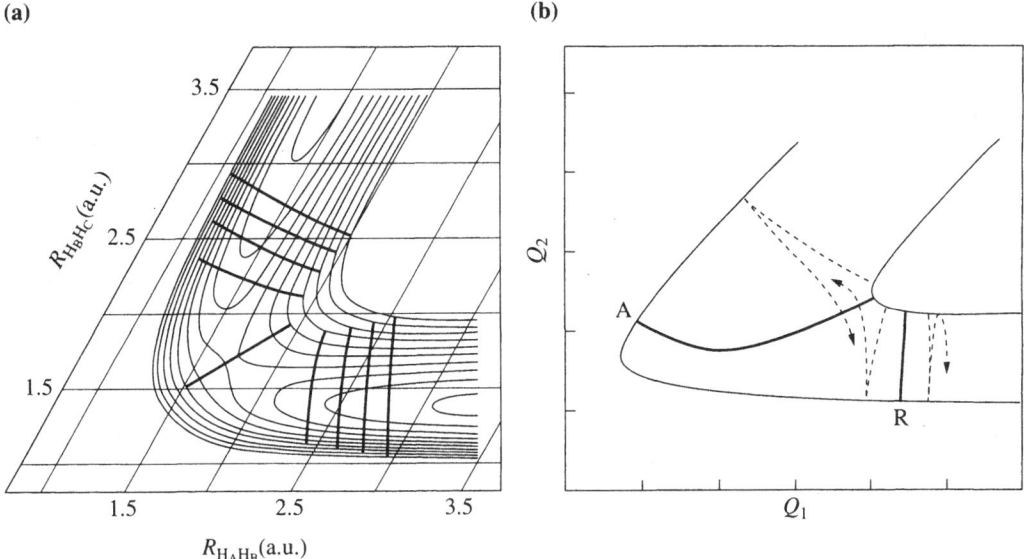

Figure 12.13 (a) Locations of periodic orbit dividing surfaces (PODS) at various energies for the $H + H_2$ reaction. The existence of secondary PODS away from the barrier top signals a flux manifold as in Figure 12.12b. Adapted from Pollak, 1978. (b) Paths of trajectories initiated close to repulsive (R) and attractive (A) PODS. Adapted from Child, 1991.

known as a periodic orbit dividing surface, or PODS, corresponding to the area C_x in Figure 12.12a. This dividing surface is further termed a "repulsive" PODS, since any trajectory that crosses it never comes back. As the energy is increased, typically a pair of new unstable periodic trajectories appears, one on either side of the transition state. These new PODS are repulsive—any trajectory crossing them in the direction away from the saddle point never comes back, while the original PODS has become attractive—any trajectory that crosses one of the repulsive PODS in the direction of the saddle point must cross the attractive PODS before crossing another repulsive PODS. The energy at which the original PODS changes from repulsive to attractive correlates precisely with the change in phase space structure between Figure 12.12a and 12.12b. It can be shown that as long as there is a single PODS, transition state theory is exact, while at energies at which there are multiple PODS, transition state theory is no longer exact. The properties of the PODS can be used to prove refined upper and lower bounds on the rate constant (Pollak, 1980).

▶ **Exercise 12.15** Use the fact that C_x is an attractive PODS to prove the following upper bound on the area in Figure 12.12c corresponding to the microcanonical reactive flux:

$$C_x - (C_1' \cup C_2') + (C_1' \cap C_2'), \tag{12.123}$$

where $\cup$ is the union of two sets and $\cap$ is the intersection. [Hint: Consider what happens to a trajectory in the region $C_1' - (C_1' \cap C_2')$. Explain why such a trajectory, if it is ultimately reactive, cannot pass to products in the region $C_2' - (C_1' \cap C_2')$.]

▶ **Exercise 12.16**   In the discussion above, the periodic orbit dividing surfaces clearly represent a dynamical divider between reactive and unreactive trajectories. There is an alternative perspective on the PODS, as surfaces that minimize the equilibrium reactive flux in the TST approximation. Show that the PODS have this property.

Solution   We specialize to the case of a single coordinate only perpendicular to the reaction path. In this case the microcanonical (constant E) reactive flux integral can be written in the TST approximation as

$$F(E) = \int \int \int dp_s dp_r dq_s \, p_r \theta(p_r) \delta \left(E - \left(\frac{p_s^2}{2} + \frac{p_r^2}{2} + V(q_s) \right) \right). \quad (12.124)$$

Take p_r as fixed and integrate over p_s:

$$F(E) = 2 \int_0^{p_{r_{max}}} \int_{q_{s_{min}}}^{q_{s_{max}}} dp_r dq_s \, p_r \left[2 \left(E - V(q_s) - \frac{p_r^2}{2} \right) \right]^{-1/2}, \quad (12.125)$$

where $q_{s_{max}}$ means $(q_s)_{max}$, and so on, we have used the property of δ-functions

$$\delta(f(x) - f_0) = \sum \frac{1}{|f'(x_0)|} \delta(x - x_0) \quad (12.126)$$

(see Appendix A), and used the Heaviside function to set the lower bound of the integral on p_r to zero. The overall factor of 2 comes from the two choices of p_s consistent with a given p_r and q_s at energy E. Using the definite integral

$$\int_0^{a^{1/2}} \frac{x}{(a - x^2)^{1/2}} dx = a^{1/2} \quad (12.127)$$

we find that

$$F(E) = 2^{3/2} \int_{q_{s_{min}}}^{q_{s_{max}}} dq_s \left(E - V(q_s) \right)^{1/2}. \quad (12.128)$$

Equation 12.128 is an action integral, and according to the principle of least action in classical mechanics q_s should be the configuration space path of a classical trajectory of energy E.

The case of multiple coordinates perpendicular to the reaction path is left for the reader.

12.5.4   Final Product Distributions

The previous subsection dealt with the overall rate constant for a chemical reaction. In this section we look briefly at the factors that determine the final product distributions in the reaction. In particular, we will focus on the factors affecting the vibrational final product distribution.

Trajectories crossing the barrier generally have components of motion both along and perpendicular to the reaction path. To what extent does motion parallel to the reaction path at the transition state stay parallel until the asymptotic region, and similarly, to what extent does motion perpendicular to the reaction coordinate stay perpendicular? The answer to this question has implications for what seems at first glance to be an unrelated question: to facilitate a chemical reaction at a given total energy, is it more efficient to put the energy in translation motion of the reagents or in the internal vibration of one of the reagents? The two questions are flip sides of the same coin; by time reversal symmetry, the energy disposition of exiting products indicates the correct conditions for incoming reactants if the reaction is run in reverse.

A simple set of rules was discovered by Evans and Polanyi in the early days of molecular dynamics: early barriers (barriers that are closer to the reactant portion of the reaction path) favor translational activation, while late barriers (barriers closer to the product portion of the reaction path) favor vibrational activation. Some sample trajectories illustrating these ideas are shown in Figure 12.14. These trends can be interpreted as indicating that the asymptotic partitioning of energy is approximately preserved right up to the transition state, and the

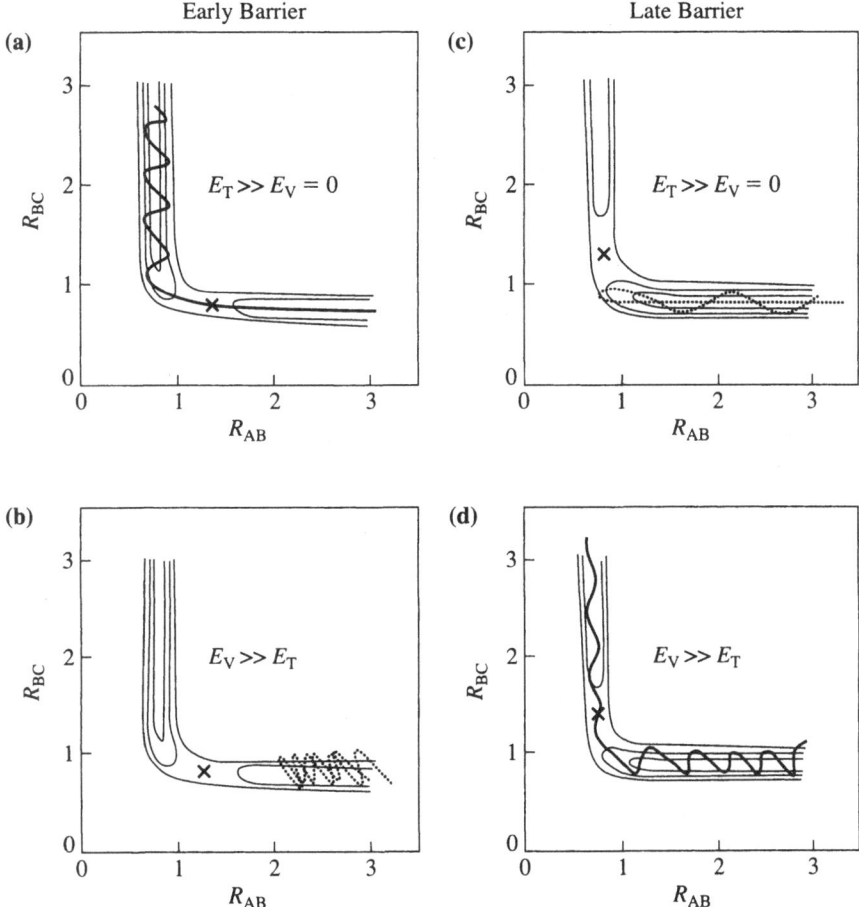

Figure 12.14 Illustration of the Evans and Polanyi rules. E_V and E_T refer to the incident energy in vibration and translation, respectively. Early barriers—barriers located in the reactant channel—are more easily crossed if the energy of reactants is deposited in translation (a) rather than in vibration (b). Moreover, the outgoing products tend to have a disproportionate amount of their energy in internal vibration (also seen in (a)). For late barriers the situation is reversed: depositing reactant energy in translation is unfavorable for reaction (c) while depositing it in vibration is favorable (d). Adapted from Levine (1974).

differences in reactivity in different systems arise from the orientation of the transition state coordinates relative to those of reactants. Moreover, by time reversal symmetry, we see that if products are formed vibrationally excited, then vibrational excitation would be effective in promoting the reverse reaction; similarly if products are formed vibrationally cold, translationally hot, then this distribution should be effective in promoting the reverse reaction. Note that early barriers in one direction of reaction imply late barriers for the other direction of reaction; thus, the conditions favoring reactivity in one direction can be quite different from those favoring reactivity in the reverse direction. Note, however, that these pictures do not take into account the skew angle (ironic, since the concept of the skew angle was introduced by Eyring and Polanyi), which can have a profound effect on the outcome of the reaction (see Smith (1959) for an illuminating discussion).

A somewhat more sophisticated approach to final product distributions uses the phase space representation of the incoming and outgoing flux shown in Figure 12.12 (Pollak, 1980). Since the PODS represents the boundary between the phase space manifolds that correlate with reactants and those that correlate with products, a small perturbation of the PODS in one direction or the other will propagate exclusively to either reactants or products. Once reaching the asymptotic region, the deformed boundary defines the boundary of the internal (i.e., vibrational) distribution in the asymptotic region that is reactive. These boundaries tend to be very simple for the direct component of the flux, but can become very complicated for the indirect components. It is this author's impression that there is still interesting research to be done on the relationship between the mass-skewing angle, the Evans–Polanyi rules and the propagation of the perturbed PODS into the asymptotic regions.

12.6   Symmetry and Permutations

12.6.1   General Considerations

Even a casual familiarity with molecular structures indicates that often there is a high degree of geometrical symmetry in molecules, for example, CH_4, C_6H_6, or C_{60} (!). These are symmetries in molecular *structure*, and our main interest in this chapter is in molecular *dynamics*. Yet it is striking that the symmetries in molecular structure, the so-called *point group* symmetries, reflect deeper underlying symmetries in molecular Hamiltonians, which trace to the symmetry with respect to permutation of identical nuclei. The symmetries in the molecular Hamiltonian manifest themselves both in the form of the potential energy surfaces we have seen above, as well as in the electronic wavefunctions of the molecule. These symmetries in turn dictate symmetries in the molecular dynamics, as manifested both in the form of the normal modes of vibration and more generally in the form of the quantum eigenstates and wavepackets.

The mathematical significance of the symmetry operations derives from the fact that they commute with the molecular Hamiltonian. As a result, it is possible to find simultaneous eigenfunctions of both the symmetry operation and the molecular Hamiltonian, and hence label molecular eigenstates by quantum numbers associated with their transformation properties under different symmetry operations. These symmetry labels are important for several reasons. First, they are helpful for understanding the anatomy of the energy eigenfunctions, which manifest the symmetry of the molecular Hamiltonian in different ways, depending on the specific transformation properties of the wavefunction. Second, they are important for discovering "selection rules" for transitions among molecular eigenstates by perturbations having a specific symmetry. Third, it is generally easier to find eigenfunctions of the symmetry operations than of the Hamiltonian; this leads to a useful numerical procedure in which one first constructs eigenfunctions of the symmetry operations and then uses these eigenfunctions as a basis for representing the Hamiltonian. The Hamiltonian takes a block-diagonal form, each block associated with a different symmetry label.

12.6.2   Point Symmetry versus Permutation–Inversion Symmetry

The conventional approach to the systematic enumeration of the symmetries of molecular Hamiltonians uses the group of point transformation on the molecule, the so-called *point group*. The point group identifies the structural symmetries of the molecule at its equilibrium geometry. For example, the point group treats HOH as a rigid, isosceles triangle, and enumerates the full set of finite symmetry operations that map the triangle into an indistinguishable configuration. These operations consist of the identity (E), reflection in the plane containing the molecule (σ'_v), reflection in the perpendicular plane (σ_v), and rotation by 180° about an axis perpendicular to the plane of the molecule (C_2). The set of point symmetry operations of HOH form a group, called C_{2v}.

▶ **Exercise 12.17**   Show that these operations indeed form a *group*. The requirements are that there is an identity, that every element has an inverse, and that the composition of any two elements gives another element in the group.

An important tool in the application of group theory to quantum mechanics is the concept of a "representation" of a group. A group representation has the property that the multiplication of the representation of two elements obeys the same product rules as do the elements themselves. Figure 12.15b shows the full set of the so-called "irreducible" representations (or "irreps") of the group C_{2v}, irreducible in the sense that they cannot be

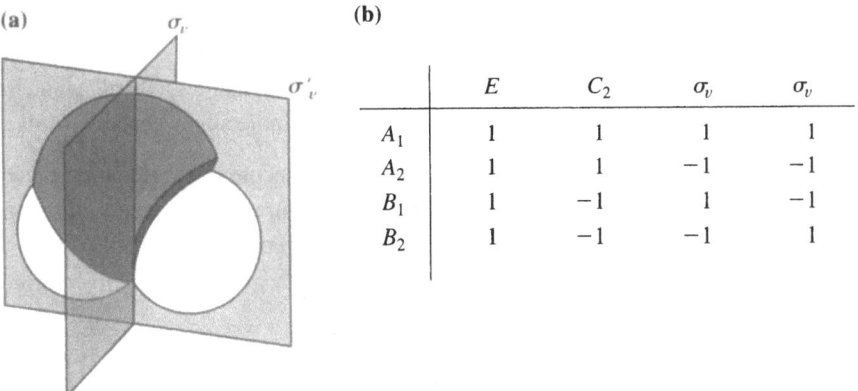

(b)	E	C_2	σ_v	σ_v
A_1	1	1	1	1
A_2	1	1	-1	-1
B_1	1	-1	1	-1
B_2	1	-1	-1	1

Figure 12.15 (a) Definition of the reflection symmetry elements in a molecule belonging to the C_{2v} point group. (b) Irreducible representations of the C_{2v} point group.

decomposed into simpler matrices (an observation that is trivial in the case of C_{2v}, since the matrices are 1×1). Note that the irreps are orthogonal to each other, and are complete with respect to the vector space whose order, h, is the number of elements of the group.

▶ **Exercise 12.18** Each of the normal coordinates of HOH must correspond to one of the irreps of C_{2v}. Use Figure 12.15b to assign the symmetry of the vibrations (Figure 12.7) as well as the translations and rotations (Exercise 12.9).

▶ **Exercise 12.19** In the previous exercise, the symmetry of each of the normal modes of HOH was assigned *a posteriori*. However, group theory can also be used to predict *a priori* the number of normal modes of each symmetry type. The procedure has three parts. 1) One builds a basis of $3N$ Cartesian coordinates, from the x, y and z coordinate of each of the N atoms. 2) One applies each of the symmetry operations of the group to the full set of basis vectors to obtain a $3N \times 3N$ matrix representation in the Cartesian basis. The entries in these matrices may be calculated by inspection, by applying the symmetry operation in question to the jth Cartesian basis vector and representing the result in terms of its projection on the full set of Cartesian basis vectors. These matrices form a *reducible* representation of the group. 3) Finally, one decomposes the reducible representation obtained in this way into a sum of irreps. The multiplicity of each irrep in the decomposition is the number of normal modes having the symmetry of that irrep (with the degenerate irreps weighted by their degeneracy). The details can be found in any of the standard texts on group theory (e.g., Cotton, 1971; Atkins, 1983). Use the procedure above to find the number of normal modes of HOH of each symmetry type.

Clearly, for molecules such as CH_4, C_6H_6, or C_{60} the structural symmetries are much higher, and the number of symmetry operations in the molecular point group therefore is much higher as well. As a result, the number of irreps is not only larger, but more complicated, involving multidimensional irreps. There are many excellent texts that develop the point group approach fully, for molecules of any symmetry. Our goal here, in including a section on molecular symmetry, is not to repeat this relatively standard—and beautiful— theory, but to raise conceptual difficulties with the point group approach, and answer these difficulties by showing how the point group symmetries have their ultimate origin in the symmetry of permutation of identical particles.

Here are some of the conceptual difficulties and paradoxes that arise when trying to apply point group theory to analyze the symmetry of nuclear dynamics, as well as electronic wavefunctions, where the high structural symmetry of the molecule is generally broken.

1. The point group symmetry implicitly treats the molecule as a rigid structure with a finite symmetry. It is reasonable that this rigid structure would be relevant so long as the displacements from the symmetric equilibrium geometry are small. Yet the point group operations of reflection, inversion, and so forth, as conventionally interpreted, generate large displacements of the atoms in space.

2. The point group operations include rotations; yet rotational symmetry is external to the molecule and does not require identical nuclei. What is the relationship of the point group operation and the external rotation?

3. Is the point group symmetry useful for large displacements, far from the original, symmetrical geometry? The equivalence of identical nuclei does not break down at large displacements, so the underlying symmetry considerations cannot disappear; yet, at sufficiently large geometrical distortions the equilibrium structure becomes irrelevant, if not incorrect, such as CH_4 distorted from tetrahedral to coplanar geometry.

4. The point group treats the molecule OOO, which at equilibrium is an isosceles triangle with a 120° bend angle, as having the same symmetry as OCO. Shouldn't there be a higher symmetry because the middle atom is identical to the two end atoms?

5. Is the point group symmetry relevant in the case of a chemical reaction, for example, $H + H_2 \rightarrow H_2 + H$? From a comparison of Figures 12.8 and 12.3, there would seem to be much in common in terms of symmetry between the unstable species, HHH, and the stable species, OCO; yet HHH has no equilibrium geometry, let alone a rigid one around which one can consider small displacements. Furthermore, even if one assigned a point group symmetry of C_{2v} or D_{3h} to the HHH complex, how would one take into account that when the collision partners are at a great distance, that is, when they form $H_2 + H$, the symmetry of the fragments is lower than that of the complex?

6. Does the point group operation apply to nuclear coordinates, electronic coordinates or both? This is important in considering the symmetry of transition dipole moments and/or nonadiabatic coupling functions, as seen below.

All these questions can be answered at once, by realizing that fundamentally the point group symmetry has its origin in the permutational symmetry of identical particles. This is a profound insight, first clearly enunciated and explored by Longuet-Higgins (1963), which clears up many conceptual issues with point group symmetry in one sweep. Viewed from the perspective of the permutation–inversion (PI) group, each point group operation is *isomorphic* with a permutation of identical nuclei. For molecules of high symmetry, there are generally many permutations of identical nuclei that are not *feasible*, and then the point group is isomorphic only to the subgroup of feasible permutations in the PI group.

For example, consider the set of PI operations in HOH, $S_2^* = \{E, (12), E^*, (12)^*\}$. The operation (12) signifies permutation of the labels of nuclei 1 and 2, while * signifies inversion of all particles (i.e., electrons as well as nuclei) through the origin. It is readily checked that the four elements in S_2^* form a group. Note that the operations in S_2^* are not identical to the operations in C_{2v}; the two sets of operations are contrasted in Figures 12.16–12.18. For example, as seen in Figure 12.16, the point group operation C_{2x} involves a small displacement of the nuclei, while the PI operation (12) involves a large displacement; the point group operation C_{2x} operates on the electrons as well as the nuclei, while the PI operation (12) operates only on the nuclei. Nevertheless, it is clear from Figures 12.16–12.18 that there is a one-to-one correspondence between the elements of S_2^* and those of C_{2v}; moreover, as illustrated in Figures 12.16–12.18, the elements of S_2^* can be composed with a

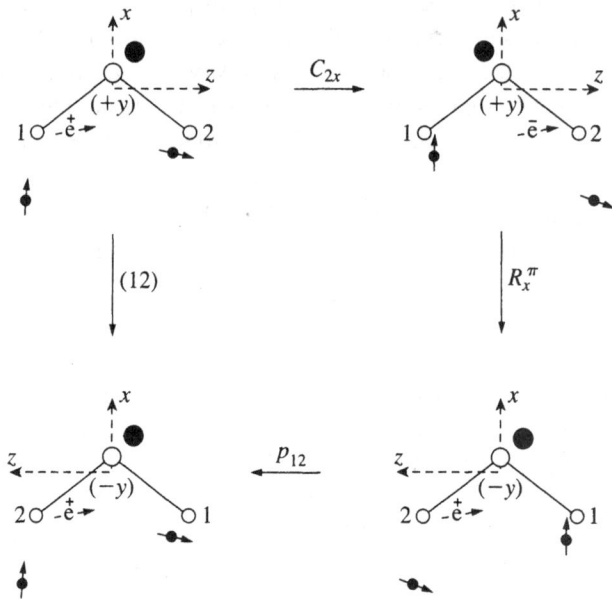

Figure 12.16 The effect of the successive operations C_{2x}, R_x^π and p_{12} on a distorted water molecule and the equivalence of this to the permutation (12). The molecule-fixed (x, y, z) axes are right-handed in all figures. The open circles are the nuclei at the symmetric reference geometry, while the filled circles are the actual positions of the nuclei at the distorted geometry. A representative electron is designated by "e," and $\pm$ represents out of or into the page, respectively. The arrows are a schematic representation of electron or nuclear spin. Adapted from Bunker (1979).

supplementary spatial rotation and an exchange of nuclear spin, to yield the corresponding point group operation precisely. Specifically, the composition relations take the form

$$E = E R^0 p_0 \tag{12.129}$$

$$C_{2x} = (12) R_x^\pi p_{12} \tag{12.130}$$

$$\sigma_{xz} = E^* R_y^\pi p_0 \tag{12.131}$$

$$\sigma_{xy} = (12)^* R_z^\pi p_{12}, \tag{12.132}$$

where the operations R_x^π, R_y^π and R_z^π represent rotations by π of all particles in the system (i.e., electrons and nuclei) around the x, y or z axis, respectively, and the operation p_{12} is an exchange of nuclear spin variables. Note the profound difference between the rotation operator R_x^π and the point group operation C_{2x} (Figure 12.16): although both operations operate on both the electrons and the nuclei, R_x^π is a true spatial rotation, reflecting the isotropy of space and not any molecular symmetry, and hence exchanges the positions of the nuclei, while the point group operation C_{2x} characterizes only small-amplitude operations about the nuclear frame, and hence does not exchange the labels of the nuclei.

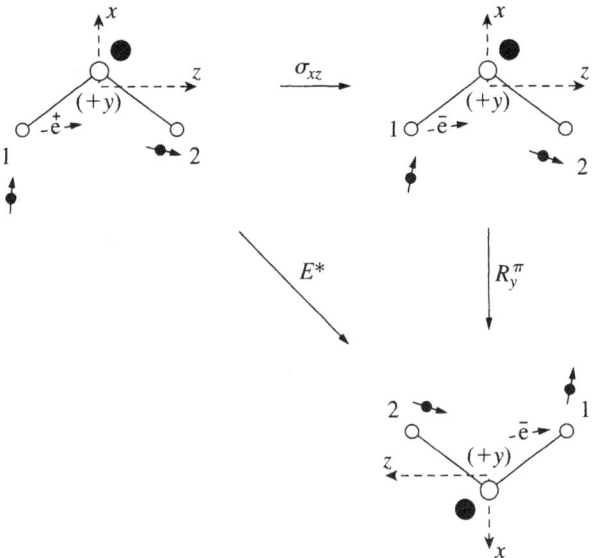

Figure 12.17 The effect of the successive operations σ_{xz} and R_y^π on a distorted water molecule and the equivalence of this to the inversion E^*. The molecule-fixed (x, y, z) axes are right-handed in all figures. Adapted from Bunker (1979).

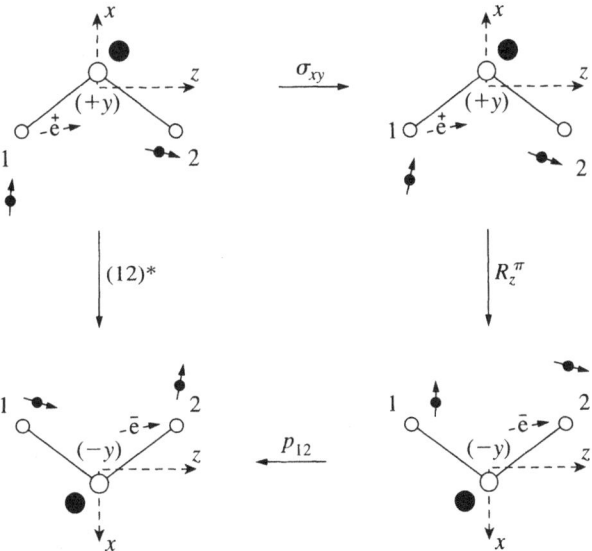

Figure 12.18 The effect of the successive operations σ_{xy}, R_z^π and p_{12} on a distorted water molecule and the equivalence of this to the permutation–inversion operation $(12)^*$. The molecule-fixed (x, y, z) axes are right-handed in all figures. Adapted from Bunker (1979).

With this understanding we have the answer to the first three questions we posed. The point group operations are, when properly interpreted, indeed small-amplitude motions. The rotations and reflections within point group symmetry are the composition of two fundamental symmetry operations—permutation and spatial rotation–inversion—each of which involves large-amplitude displacement in the laboratory frame, but when combined give a small displacement! The point group rotation operation differs fundamentally from a spatial rotation in that a spatial rotation keeps the labels on the atoms unchanged as they rotate rigidly, while the point group operation leaves the atoms near their initial position.

Moreover, the PI group is in no way limited to small-amplitude motion. Since the fundamental symmetries exploited in the PI group are in no way tied to a rigid geometry, the PI symmetries apply equally well to stable (isomerizing) as well as unstable systems (e.g., dissociative and reactive systems). In these cases, the composition of the PI operation and the spatial rotation gives a large distortion between symmetrically related geometries; the use of point group symmetry for these large displacements is questionable, if not impossible. The application of the PI group to unstable (dissociating or reacting) systems has an extra twist, which we will discuss below after first introducing the concept of *feasible* permutations.

We turn to the question of when the permutational symmetry is higher than the point group symmetry, as in ozone, O_3. At equilibrium, this molecule has a bend angle of $120°$, and thus point group symmetry would label this as C_{2v}, the same point group as HOH. Yet clearly the permutation group is higher, S_3^* versus S_2^*. The point group does not recognize that the middle atom is identical to the two end atoms, while the PI group does!

For most practical purposes, the fact that the middle atom is indistinguishable from the two end atoms is irrelevant. Although in principle the middle atom can exchange position with an end atom, at room temperature the time scale for this process is so long that it can be neglected. Under these circumstances the PI group consisting of all *feasible* operations is isomorphic with C_{2v}. Nevertheless, we can imagine situations, perhaps at higher energy, perhaps induced by collison or excitation by light, in which these atoms exchange and the higher symmetry becomes revealed. Under these circumstances, the PI group consisting of all feasible operations, called S_3^*, is isomorphic with D_{3h}.

The concept of feasible permutations can be further clarified by considering the global potential energy surface of ozone as a function of its three vibrational coordinates; symmetry dictates that there must be three equivalent minima, one with O_1 in the middle, one with O_2 in the middle and one with O_3 in the middle. The potential energy hypersurface is a function of the three internal coordinates, which is difficult to visualize; however, with the right choice of just two of these internal coordinates the three minima may be seen. Figure 12.19a shows equipotentials for the ground electronic state of ozone as a function of the antisymmetric stretch and the bend, and indeed the three equivalent minima are seen. If we consider just the region around the upper of these three minima the potential locally is seen to have only reflection symmetry, similar to the reflection symmetry in HOH. For vibrational spectroscopy and low-energy processes there is effectively no communication with the other wells and hence no indication that there is a larger global symmetry. However, for higher-energy processes or in electronic transitions to highly displaced or dissociative excited electronic states, amplitude may travel from the region of one well to that of another. Figure 12.19b shows the excited electronic state of ozone; in place of the three equivalent minima there are now three equivalent saddle points.

(a)

(b)

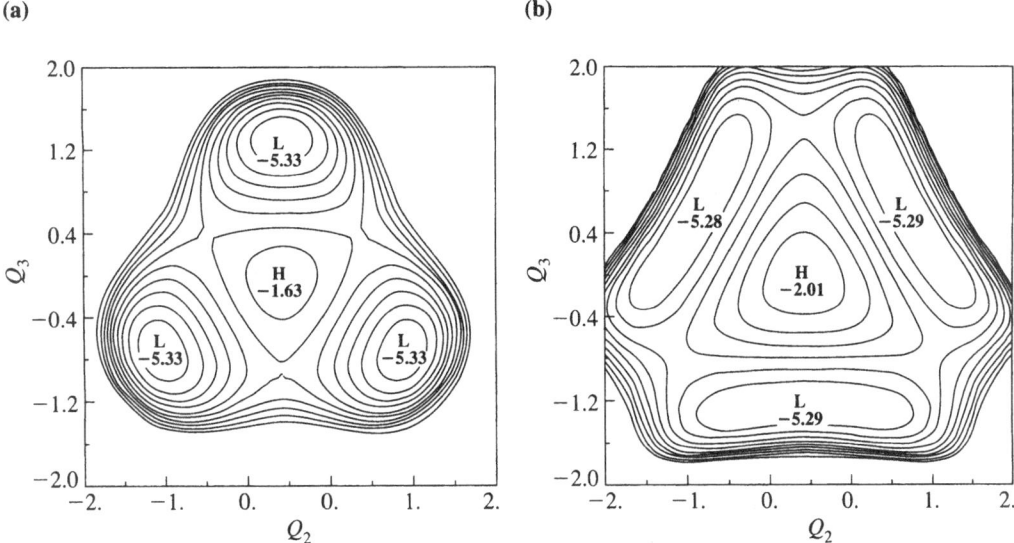

Figure 12.19 Potential energy surfaces for ozone, as a function of the asymmetric stretch and the bend, Q_2 and Q_3, respectively. This choice of coordinates shows the latent three-fold symmetric in the ozone potential. (a) Ground electronic state potential. (b) Excited electronic state potential.

We now return to reactive systems. We said before that the PI group applies equally well to stable systems (OOO) and unstable (reactive) systems (HHH). Yet in the case of unstable systems, should one use the PI group of reactants, products, or the full set of particles? A beautiful paper by Quack (1977) shows that there can be symmetry selection rules for chemical reactions, when one takes into account the permutation group for the complex and how that group correlates with the permutation group of reactants and products. Quack suggests that if not all symmetry-allowed transitions are observed between reactants and products, this gives valuable information on the effective permutation group of the complex, and hence the subset of feasible permutations possible during the rearrangement of reactants and products.

12.6.3   Symmetry of Electronic Wavefunctions and Transition Dipole Moments

Many of the applications of wavepacket dynamics in Part III involve electronic transitions mediated by absorption of light. The key molecular property that mediates its interaction with the light is the *transition dipole moment*; we will therefore devote a little space here to the definition of the transition dipole moment and discuss its coordinate dependence and its symmetry properties.

The dipole moment operator, $d(r, R)$, is defined as

$$d(r, R) = \sum_i -er_i + \sum_i Z_i e R_i. \qquad (12.133)$$

It is a product of charge times distance, summed over all charged particles; in particular, the dipole operator contains contributions from both the electrons and the nuclei. The transition dipole moment, $\mu(R)$, is the matrix element of the dipole moment operator sandwiched between two different electronic wavefunctions, ψ_a for the electronic ground state and ψ_b for the electronic excited state.

$$\mu_{ba}(R) \equiv \int dr\, \psi_b(r; R)\{d(r, R)\}\psi_a(r; R). \tag{12.134}$$

Note that in constructing the transition dipole moment the electronic coordinates are integrated out, leaving behind the nuclear coordinate dependence. In analyzing the structure of $\mu_{ba}(R)$, the first observation is that since the electronic wavefunctions viewed as functions of r are orthogonal to each other at each value of R, the nuclear dipole operator doesn't couple them; therefore only the first term in Eq. 12.133 contributes:

$$\mu_{ba}(R) \equiv \int dr\, \psi_b(r; R) \left\{ \sum_i -er_i \right\} \psi_a(r; R). \tag{12.135}$$

For a molecule with a symmetrical equilibrium geometry, it is not uncommon for the overlap at the equilibrium geometry to vanish. In this case the transition is called "electronically forbidden." Away from the symmetrical equilibrium geometry the overlap may become nonvanishing; in such a case the transition is called "electronically forbidden, vibronically allowed." These ideas are illustrated in Figure 12.20, for the lowest-lying electronic transition in HOH. In each of the six figures, the electronic wavefunction for the totally symmetric (nodeless) electronic ground state, $\psi_a(r; R)$, is shown on the left, while the electronic wavefunction of the antisymmetric excited electronic state wavefunction, $\psi_b(r; R)$, is shown at the top. The transition dipole overlap integral involves first multiplying the ground state wavefunction, $\psi_a(r; R)$, by the dipole operator r_i, which can be visualized as putting a node along the direction r_i (Figure 12.20). This one-node wavefunction is then multiplied by the excited state wavefunction $\psi_b(r; R)$ and integrated over all space. It is clear from Figure 12.20 that this transition is allowed at the equilibrium geometry for x-polarized light (Figure 12.20a), and forbidden for y- and z-polarized light (Figure 12.20d and e, respectively).

If the molecule is distorted from the symmetrical geometry, the simple reflection symmetry of the electronic wavefunction is broken (cf. the wavefunctions on the left and right in Figure 12.20c and f). However, closer inspection of these figures reveals that there remains a global symmetry to the electronic wavefunction. As discussed in Section 12.6.2, the key is to "interpret" the point group operations as operating on both the nuclei and the electrons, specifically small-amplitude operations on the nuclei and large amplitude operations on the electron (cf. Figures 12.16–12.18). With this proviso, one may use the conventional symmetry labels for these electronic states, according to the point group C_{2v}, even at distorted geometries. Comparison of the transformation properties of the electronic wavefunctions under the point group operations with the irreps in Figure 12.15b leads to the assignment of the symmetry labels A_1 and B_1, respectively.

Combining this global symmetry of the electronic wavefunctions with the symmetry of the dipole moment operator leads to selection rules that determine whether the transition is electronically and/or vibronically allowed: if the direct product of the irreducible representations (irreps) corresponding to the two electronic wavefunctions and the polarization of the light contains a component of A_1 symmetry, the transition is electronically allowed. If the direct product transforms as some non-A_1 irrep, the transition can become vibronically

(a)

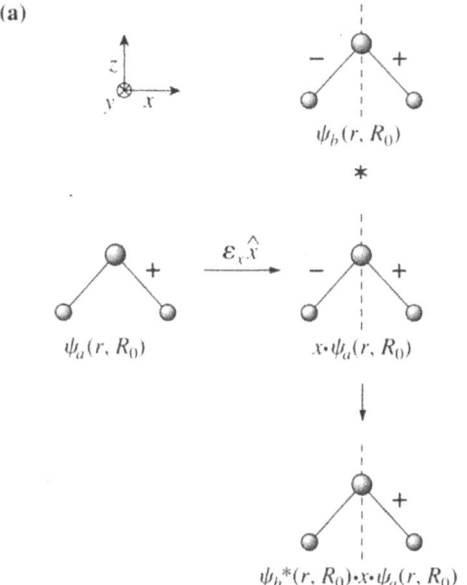

(b)

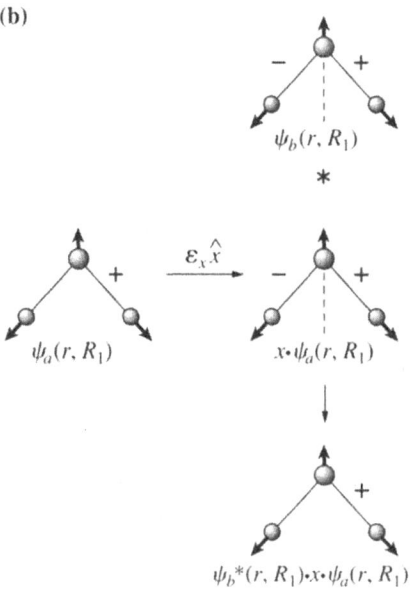

Figure 12.20 (a) The x-component of the dipole operator puts a node in the ground state wavefunction along the x-axis. This leads to overlap with an electronic state of the same symmetry, illustrating an electronically allowed transition. (b) For symmetrical geometry distortions, the overlap may change but symmetry says nothing about the magnitude or sign of the change. (Figure continued on page 380.)

(c)

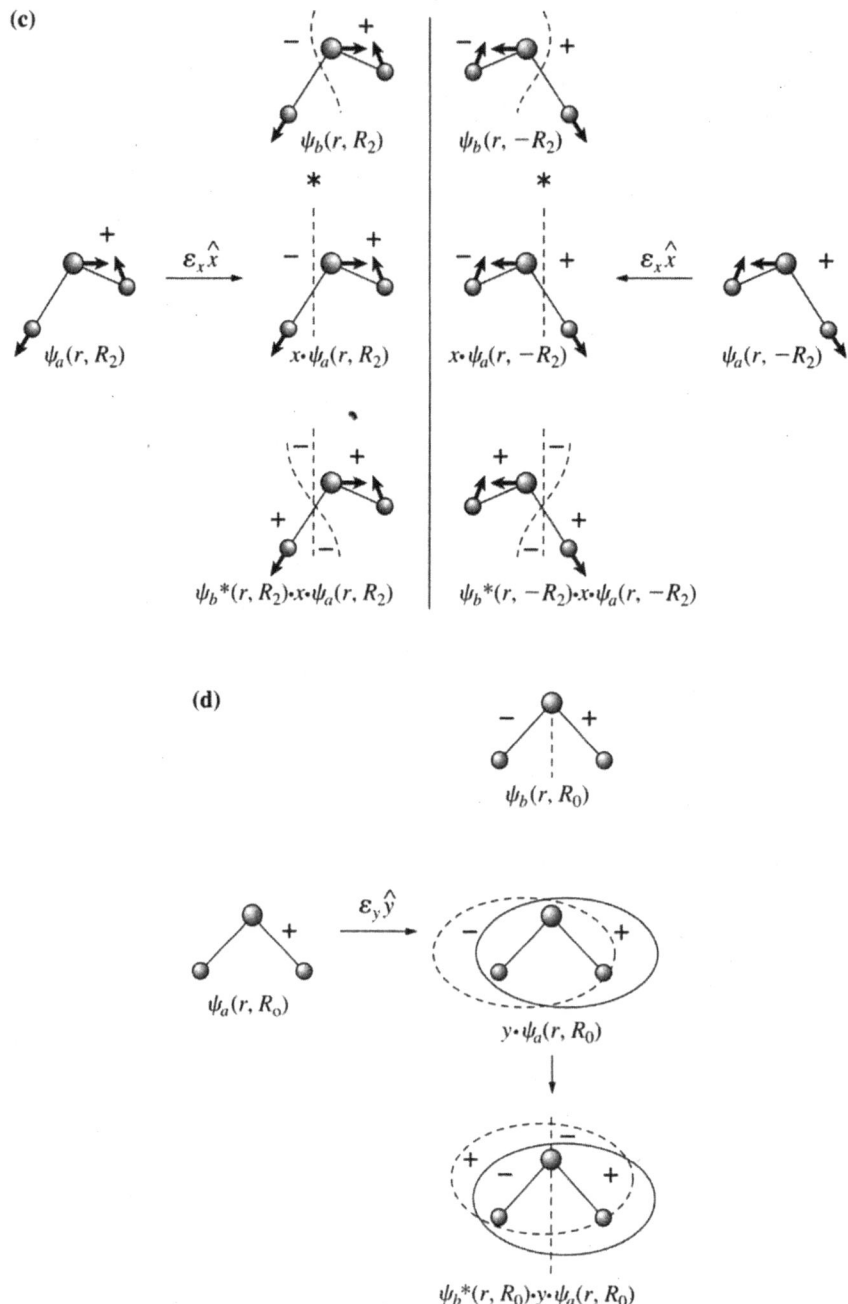

(d)

Figure 12.20 *(continued)* (c) The x-component of the transition dipole operator for a geometry distortion with the same symmetry as the dipole operator. Careful inspection of the figure shows that the change in the dipole must be a symmetric function of the geometry distortion. (d) The y-component of the dipole. The transition is not only electronically forbidden, but forbidden for any geometry distortion. (Figure continued on page 381.)

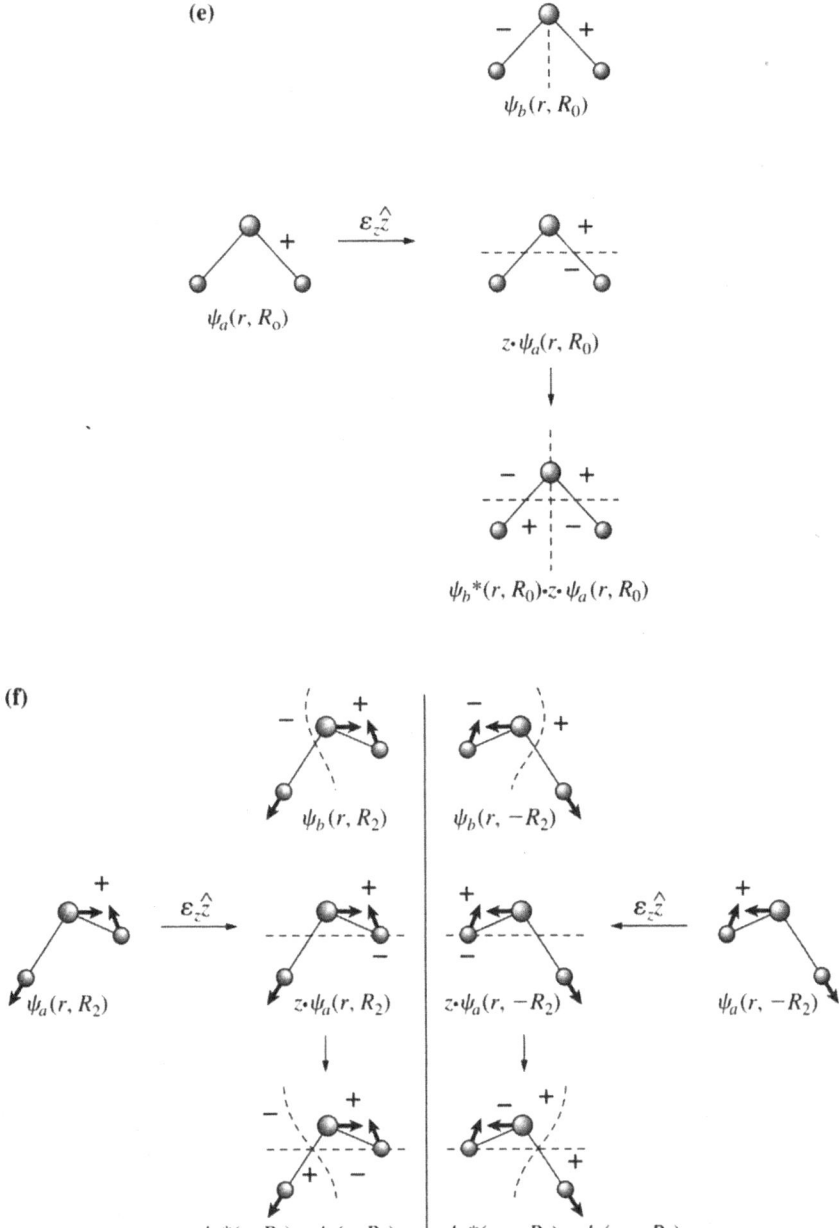

Figure 12.20 *(continued)* (e) The z-component of the transition dipole operator puts a node in the ground state wavefunction along the z-axis. At the equilibrium geometry the transition to the excited electronic state is forbidden by symmetry. (f) Geometry distortions along the asymmetric stretch, R_3, remove the symmetry in (e), making the z-component of the transition "electronically forbidden, vibronically allowed." Adapted from Tannor (1984).

allowed if there is a vibration that transforms as the same non-A_1 irrep, or more generally, if the direct product with the vibration contains a component of A_1. For the proof of this statement, see any of the standard references on group theory in quantum mechanics, cited in the references.

12.7 Hyperspherical Coordinates

We have dealt in this chapter primarily with geometry changes involving only stretching motion. However, in general a triatomic molecule, ABC, undergoes bending motion as well as stretching, and the full dynamics is three-dimensional. As discussed above, this bending motion is not necessarily small, and in fact, if allowed to become sufficiently large can lead to an entirely different chemical arrangement channel, as in the case of HOH $\rightarrow$ H$_2$ + O. In fact, even the division of coordinates into two stretches and bend is artificial, since two of the interatomic distances are being treated differently from the third. These considerations suggest that there may be value in defining coordinates in which all three interatomic distances are treated symmetrically. A convenient way to do this is via hyperspherical coordinates.

In hyperspherical coordinates, there is a single coordinate, the "hyperradius," which measures the overall perimeter of the molecule; the remaining $3N - 7$ internal coordinates are hyperspherical angles. We will begin our discussion with the definition of hyperspherical coordinates for the collinear triatomic molecule, in which there is a single hyperspherical angle, and then proceed to the nonlinear triatomic, where there are two hyperspherical angles.

For the collinear triatomic, the hyperspherical coordinates are simply the polar coordinates:

$$\rho = \left(R_\alpha^2 + r_\alpha^2 \right)^{1/2} \tag{12.136}$$

and

$$\theta_\alpha = \arctan\left(\frac{r_\alpha}{R_\alpha} \right), \tag{12.137}$$

where α refers to one of the chemical arrangement channels (a or c) and R_α and r_α are the mass-weighted Jacobi coordinates associated with channel α. The hyperradius is independent of the arrangement channel label α, as may be seen from Eq. 12.136. It is easily verified that the hyperangles θ_a and θ_c corresponding to different arrangement channels must sum to the skewing angle, Θ:

$$\theta_a + \theta_c = \Theta. \tag{12.138}$$

At a sufficiently large hyperspherical radius, ρ, the hyperspherical radius becomes parallel to the translational coordinate, R_α, and the hyperspherical angle, θ_α, becomes parallel to the vibrational coordinate, r_α, in each arrangement channel.

The kinetic energy operator in hyperspherical coordinates takes a relatively simple form. The collinear Hamiltonian can be written as

$$H = -\frac{\hbar^2}{2\mu} \left(\frac{\partial^2}{\partial \rho^2} + \frac{1}{\rho} \frac{\partial}{\partial \rho} + \frac{1}{\rho^2} \frac{\partial}{\partial \theta_a^2} \right) + V(\rho, \theta_a), \tag{12.139}$$

Unscaled Jacobi coordinates (R, r)

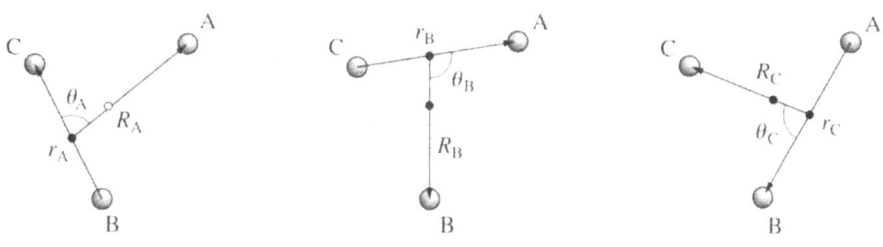

Mass-scaled Jacobi coordinates (S, s)

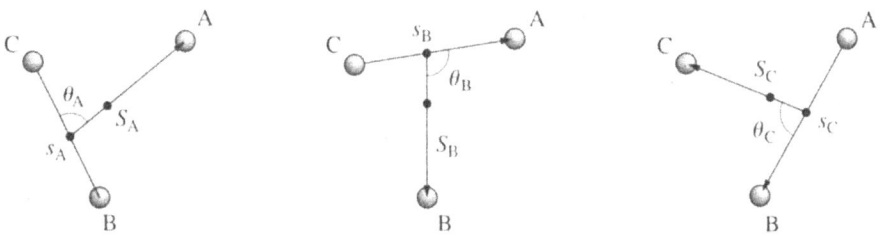

Figure 12.21 Definition of the unscaled and scaled Jacobi coordinates for the three different arrangement channels in a triatomic molecule, ABC.

where $\mu = (m_A m_B m_C / m_{ABC})^{1/2}$. The volume integral of an arbitrary function $f(\rho, \theta_a)$ takes the form

$$\int_0^\infty \int_0^\Theta f(\rho, \theta_a) \rho \, d\rho \, d\theta_a. \tag{12.140}$$

Generalizing to three-dimensional space, we have six coordinates once the center-of-mass coordinates are removed. These six coordinates may be defined in terms of two mass-scaled Jacobi vectors:

$$\begin{pmatrix} \mathbf{S}_\tau \\ \mathbf{s}_\tau \end{pmatrix} = \begin{pmatrix} X_\tau \\ Y_\tau \\ Z_\tau \\ x_\tau \\ y_\tau \\ z_\tau \end{pmatrix}, \tag{12.141}$$

where τ is an index for the arrangement channel. As shown in Figure 12.21, there are three such choices of Jacobi vectors, corresponding to the three choices of arrangement channel (A + BC, AB + C and AC + B). We can define a matrix $\mathbf{T}(\chi_{\zeta\tau})$ that transforms between arrangement channel τ and arrangement channel ζ:

$$\begin{pmatrix} \mathbf{S}_\zeta \\ \mathbf{s}_\zeta \end{pmatrix} = \mathbf{T}(\chi_{\zeta\tau}) \begin{pmatrix} \mathbf{S}_\tau \\ \mathbf{s}_\tau \end{pmatrix}, \tag{12.142}$$

where $\mathbf{T}$ is a 6×6 matrix of the form

$$\mathbf{T}(\chi_{\zeta\tau}) = \begin{pmatrix} \cos(\chi_{\zeta\tau})\mathbf{1} & \sin(\chi_{\zeta\tau})\mathbf{1} \\ -\sin(\chi_{\zeta\tau})\mathbf{1} & \cos(\chi_{\zeta\tau})\mathbf{1} \end{pmatrix}. \tag{12.143}$$

(a)   **(b)**

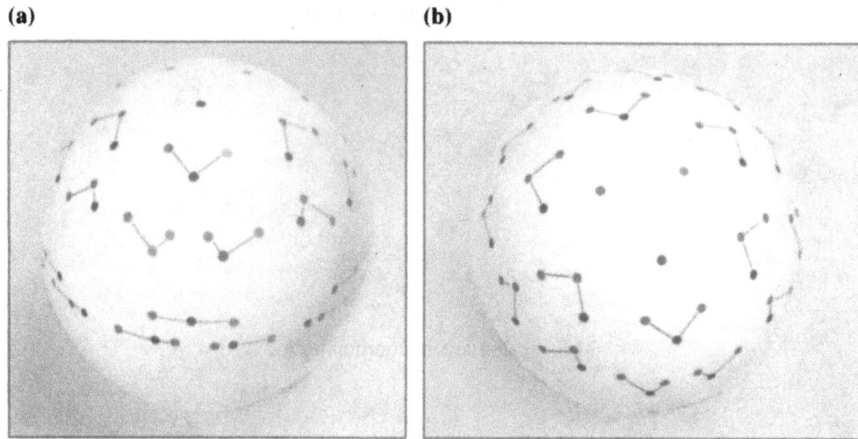

Figure 12.22 Depiction of static triatomic geometries on the hypersphere. The north pole corresponds to the equilateral triangle geometry, while the equator corresponds to all possible arrangement channels for linear configurations. (a) View from approximately 45° latitude. (b) View from near the north pole.

The six coordinates defined by the Jacobi vectors can be transformed to three Euler angles and three internal coordinates. In doing so, it is convenient to generalize the transformation angle χ connecting the mass-weighted Jacobi coordinates associated with different arrangement channels to be a continuous variable. We refer the reader to the original paper by Pack and Parker (1987) for the details, and present here just the final formulas for the three internal hyperspherical coordinates, ρ, θ and χ in terms of the original mass-weighted Jacobis:

$$\rho = (\mathbf{S}_\tau^2 + \mathbf{s}_\tau^2)^{1/2} \tag{12.144}$$

$$\tan \theta = \frac{\left[(\mathbf{S}_\tau^2 - \mathbf{s}_\tau^2)^2 + (2\mathbf{S}_\tau \cdot \mathbf{s}_\tau)^2\right]^{1/2}}{2|\mathbf{S}_\tau||\mathbf{s}_\tau| \sin \Theta_\tau} \tag{12.145}$$

$$\sin(2\chi_\tau) = \frac{2\mathbf{S}_\tau \cdot \mathbf{s}_\tau}{\left[(\mathbf{S}_\tau^2 - \mathbf{s}_\tau^2)^2 + (2\mathbf{S}_\tau \cdot \mathbf{s}_\tau)^2\right]^{1/2}} \qquad \cos(2\chi_\tau) = \frac{(\mathbf{S}_\tau^2 - \mathbf{s}_\tau^2)}{\left[(\mathbf{S}_\tau^2 - \mathbf{s}_\tau^2)^2 + (2\mathbf{S}_\tau \cdot \mathbf{s}_\tau)^2\right]^{1/2}}.$$
$$\tag{12.146}$$

Although the arrangement channel index τ appears in the final expressions, ρ and θ are actually independent of τ, and χ is simply shifted by $\pi/3$. As χ goes from 0 to 2π each arrangement channel actually appears twice in this system of hypersphericals, with the second appearance related to the first by inversion. There is an alternative definition of hyperspherical coordinates in which the counterpart of the hyperspherical angle χ is doubled, so that as χ goes from 0 to 2π each arrangement channel appears only once.

The range of θ is $0 \leq \theta \leq \pi/2$, while $0 \leq \chi \leq 2\pi$ and $0 \leq \rho < \infty$. Therefore θ and χ can be visualized as the polar and azimuthal angles spanning the upper half of a sphere (see Figure 12.22), while ρ determines the radius of this sphere. The north pole corresponds to equilateral triangle geometry while the equator corresponds to all linear geometries; thus, the angle θ, which measures latitude, is a measure of the departure from equilateral

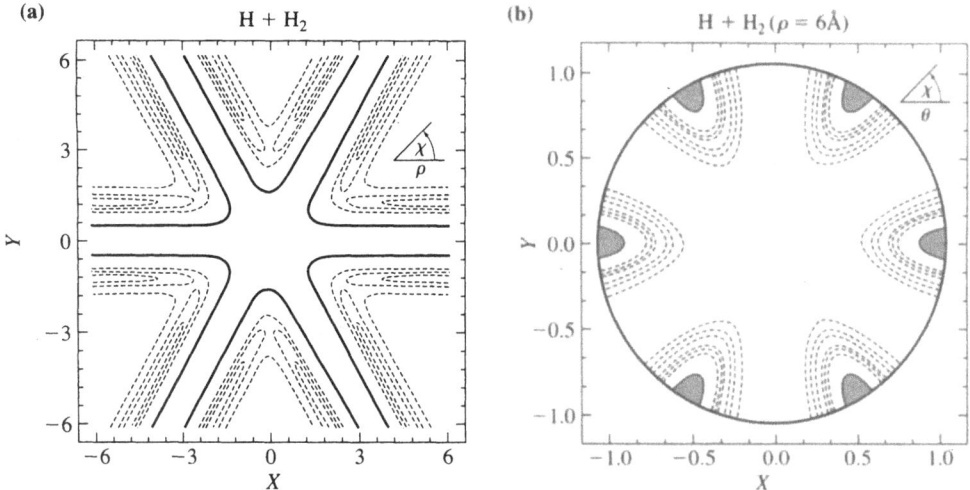

Figure 12.23 (a) Collinear plane of the PK2 H_3 potential energy surface. X is the z-component of S_A and Y is the z-component of s_A in a.u. This plot corresponds to the $\rho-\chi$ plane, with $\theta = \pi/2$. Arrangement channel A is near $\chi = 0$, B is near $\chi = -2\pi/3$ and C is near $\chi = 2\pi/3$. The other channels result from inversion. (b) Stereographic projection of the H_3 potential energy surface as a function of θ and χ on the surface of a sphere with $\rho = 6.0$ Å. The shaded regions lie above 10.0 eV. Adapted from Pack (1987).

geometry. The longitudinal angle, χ, determines the arrangement of the three atoms; for example, which of the three atoms occupies the middle position. As the angle χ is changed continuously the identity of the middle atom varies in a cyclic fashion, from $ABC \rightarrow BAC \rightarrow BCA \rightarrow \dots$.

The following simple geometrical construction can be used to find the geometry corresponding to every point on the hypersphere (cf. Figure 12.22). Place an equilateral triangle with three differently colored balls at the center of the sphere, in the plane defined by the equator. When viewed from above, this triangle indeed looks equilateral. However, when viewed from the equator this triangle looks linear; whatever the visual projection of the object from the infinite vantage point, that is the geometry drawn on the sphere at the corresponding finite radius. Clearly, according to this prescription the geometry changes from equilateral to linear as one moves one eye in the latitudinal direction. By the same token, if one fixes one's vantage point at the equator but moves one's eye in the longitudinal direction, one will be viewing the equilateral triangle sidelong, from 0 to 2π. Clearly, the atom of the triangle that occupies the central position will cycle continuously and return to its original value after rotation of 2π. This interchange of identity of atoms is called pseudorotation. It leads to the accumulation of an adiabatic phase if, as is often the case, the electronic energy is degenerate at the equilateral triangle geometry. Furthermore, all permutational symmetry of the Hamiltonian manifests itself in the symmetry properties of the potential energy hypersurface with respect to the χ angle. For example, consider the potential energy hypersurface corresponding to H_3 (Figure 12.23; note that the hypersurface for O_3, for example, has the same permutational symmetry). When viewed in the coordinates R_1, R_2, one observes only the symmetry associated with the permutation of the two end atoms (H_1—H_3 or O_1—O_3). When viewed in the hyperspherical angle, χ, one sees a six-fold symmetry, corresponding to the previous left-right symmetry but duplicated three times, for the three

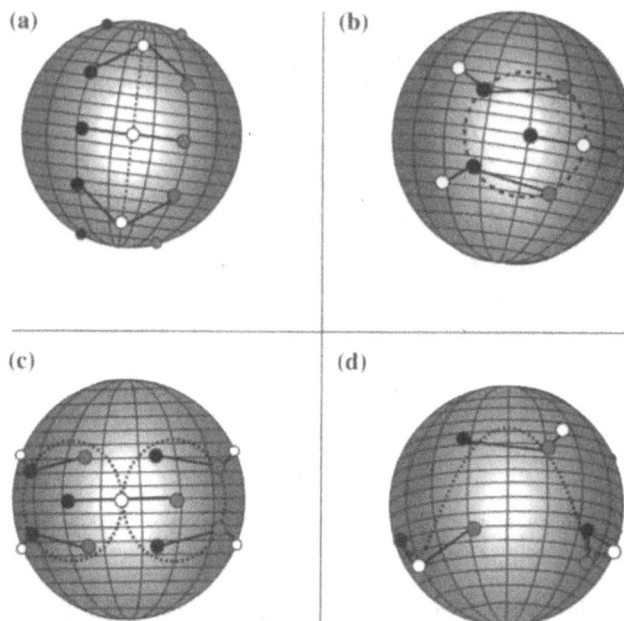

Figure 12.24 Depicting classes of triatomic motion on the hypersphere. Although the hyperspherical coordinates in the text are defined only for the upper hemisphere, in the figure for continuity the lower hemisphere is treated as the mirror image of the upper hemisphere. (a) Bent → linear → bent motion; (b) free diatomic rotation, for example, at large separations of A + BC; (c) a concerted motion in which rotating AB+C goes to A+ rotating BC and back to rotating AB + C; (d) pseudorotation, a motion in which there is a cyclical permutation of distorted arrangements of the form AB + C → B + AC → BC + A → · · · . In the figure, the pseudorotation is accompanied by a bending motion as well.

possible choices of H (or O) in the middle. There is a final doubling due to parity, leading to a 12-fold symmetry in the present, but not in all hyperspherical coordinate systems.

The hyperspherical coordinates are very useful for visualization. Figure 12.24 shows how different classes of triatomic motion can be visualized on the hypersphere. Hyperspherical coordinates have the advantage of treating all the arrangement channels (at least for triatomics) on an equal footing, and they provide a natural framework for taking advantage of permutational symmetry in numerical calculations. However, they are not always optimal in numerical calculations: in a fragmentation process, as the fragments separate there is a numerical simplification in that only one chemical arrangement channel needs to be followed, whereas hyperspherical coordinates are even-handed in their treatment of the chemical arrangement channels at all points in the process of the reaction. Hence, a hybrid approach is usually used, in which Jacobi coordinates, which are tailored to a specific arrangement channel, are used for both the early and late stages of the reactive collision, while hyperspherical coordinates are used during the intermediate dynamics when chemical rearrangement is going on. For further details about the hyperspherical coordinates the reader is referred to the References.

History and Further Reading

The adiabatic separation of electronic and nuclear motion has a subtle history. The original paper of Born and Oppenheimer (1927) is very different from modern treatments: it treats the nuclear kinetic energy as a perturbation on the full rovibronic wavefunction and obtains the adiabatic separation of electronic and nuclear motion only by a combination of tedious algebra and imaginative extrapolation. The awkwardness of the original derivation is aptly expressed by Pauling and Wilson (1935, p. 261): "Even in its simplest form the argument of Born and Oppenheimer is very long and complicated. On the other hand, the results of their treatment can be very simply and briefly described." The text by Born and Huang (1954) devotes ten pages to reviewing the method of Born–Oppenheimer, and then in a short appendix (Appendix VIII) gives an alternative derivation based on Born (1951). The alternative derivation, in addition to being enormously simpler than the original derivation, includes two improvements: a correction to the adiabatic energies and a rigorous expression for the coupling between adiabatic states.

The first discussion of a rigorous *diabatic* representation for diatomics is given by Felix Smith (1969). Baer extended Smith's formalism to polyatomics and developed the curl condition, which must be satisfied for Eq. 12.24 to have a solution (Baer, 1975; Baer, 1985, and references therein). Mead and Truhlar showed that in polyatomics this curl condition cannot generally be satisfied unless the matrix A (see Eq. 12.31) is as large as the full space of adiabatic electronic states (Mead, 1982). But this defeats the purpose of the diabatic basis, which was intended to provide a compact representation in the presence of an avoided crossing. In fact, the diabatic states obtained from solving Eq. 12.31 for A become independent of nuclear coordinates in the limit that the adiabatic space is complete: they are just the crude adiabatic states (Mead, 1982). The no-crossing rule for adiabatic states of diatomics of the same symmetry was given by von Neumann and Wigner (1927). That the no-crossing rule does not hold for two potential energy surfaces of a polyatomic was realized in a general context by Teller (1937) and in a more specialized context by Jahn (1937). The non–single valuedness of the electronic wavefunction for circuits around a degeneracy, the precursor of the geometric phase, was noted in passing by Longuet-Higgins et al. in a study of Jahn–Teller molecules (Longuet-Higgins, 1958, p. 12; Herzberg, 1966, vol. 3, p. 63). Those authors also noted the requirement for a compensatory phase in the nuclear wavefunction such that the overall wavefunction is single valued. Herzberg and Longuet-Higgins (1963) generalized this result, realizing that the non–single valuedness of the electronic wavefunction resulted simply from the conical geometry of the intersection, and was therefore more general than the Jahn–Teller systems with their high permutational symmetry. Mead and Truhlar (1979) showed that the adiabatic electronic wavefunction can be single valued and continuous if it is allowed to be complex, but this introduces a new term, with a vector potential character, into the nuclear Schrödinger equation. They also developed a new expression for the phase change in a closed circuit that later figured prominently in Berry's general analysis of the geometric phase. A survey of the theory of the adiabatic–diabatic transformation and the geometrical phase effect, with application to the $H + H_2$ reaction, is given by Kuppermann (1996).

London (1928) was the first to realize that a chemical reaction could take place on a single adiabatic potential energy surface that governs reactants, transition state and products. Eyring and Polanyi (1931) introduced skewed coordinates, but they made an arithmetic mistake in calculating the skew angle, which was corrected only after some years by Hirschfelder (1936). An amusing account is given in Hirschfelder (1984). Transition state theory has its roots before quantum mechanics; however, the beginning of modern transition state theory is associated with Polanyi and Wigner (1928), who used quantum mechanical partition functions for the activated complex but did not have the concept yet of a potential energy surface. The thermodynamic formulation of transition state theory is due to Pelzer and Wigner (1932), Eyring (1935) and Evans and Polanyi (1935). Wigner (1938) gives an insightful review of the thermodynamic approach; the same paper by Wigner introduces the concept of a dividing surface between reactants and products and formulates the no-crossing approximation inherent in the use of transition state theory. Wigner, in a separate work submitted at almost the same time (1937), formulates the rate constant in terms of the time evolution of the dividing surface. In both these works, Wigner makes a point of the transition state theory rate constant as being an upper bound. The *Trans. Faraday Soc.* 34, 1938, is a wonderful historical document for the early history of transition state theory; this volume is summarized engagingly by Miller (1998). Transition state theory underwent a renaissance in the 1970s, beginning with the reformulation of the no-crossing dynamical assumption by Pechukas and McLafferty (1973). Pechukas and Pollak showed the relationship between variationally optimal transition state theory and unstable periodic trajectories (Pechukas, 1976; Pollak, 1978), and showed that the existence of a single such unstable periodic orbit was a necessary and sufficient condition for transition state theory to be exact (Pechukas, 1979). Pollak, Child and Pechukas (1980a, 1980b) showed that the stable and unstable periodic orbit dividing surfaces of Pollak (1978) could be used to obtain a *lower* bound for classical transition state theory, and in fact to obtain a succession of increasingly more accurate upper and lower bounds.

The beginnings of hyperspherical coordinates were developed independently by Jepsen and Hirschfelder (1959), and by Delves (1959). Jepsen and Hirschfelder showed how, using mass-scaled Jacobi coordinates, the kinetic energy could be put into diagonal form in any number of ways, by drawing "mobile" diagrams clustering the particles into successively larger groups. It was Smith (1959) who recognized that in the mass-scaled Jacobi coordinates of Jepsen and Hirschfelder, the transformations between the coordinates of the three arrangement channels become simple kinematic rotations. Smith also gave a physical interpretation to the angular coordinate and corresponding quantum number appearing in Delves's work (1959). Additional properties of symmetrized hyperspherical coordinates were derived by Kuppermann (1993). An illuminating historical review is contained in the introduction of the comprehensive article by Pack and Parker (1987).

Problems

▶ **Exercise 12.20**   Derive the formulas for the mass scaling factors a and b defined by Eq. 12.55.

Solution   The expression $\bar{R} = \mathbf{M}^{\mathrm{T}} \bar{Q}$ implies that

$$\bar{Q} = \left(\mathbf{M}^{\mathrm{T}}\right)^{-1} \bar{R} \equiv \mathbf{M}'' \bar{R}. \tag{12.147}$$

To compare with Eq. 12.54 we write the latter in matrix form:

$$\begin{pmatrix} Q_1 \\ Q_2 \end{pmatrix} = \begin{pmatrix} a & b\cos\beta \\ 0 & b\sin\beta \end{pmatrix} \begin{pmatrix} R_1 \\ R_2 \end{pmatrix} \qquad (12.148)$$

or $\bar{Q} = \mathbf{M}'\bar{R}$. To check consistency of Eq. 12.147 with Eq. 12.148 we compare $\mathbf{M}'^{\mathrm{T}}\mathbf{M}'$ with $\mathbf{M}''^{\mathrm{T}}\mathbf{M}''$:

$$\mathbf{M}'^{\mathrm{T}}\mathbf{M}' = \begin{pmatrix} a & 0 \\ b\cos\beta & b\sin\beta \end{pmatrix} \begin{pmatrix} a & b\cos\beta \\ 0 & b\sin\beta \end{pmatrix} = \begin{pmatrix} a^2 & ab\cos\beta \\ ab\cos\beta & b^2 \end{pmatrix}, \qquad (12.149)$$

which is equal to

$$\mathbf{M}''^{\mathrm{T}}\mathbf{M}'' = \mathbf{M}^{-1}\left(\mathbf{M}^{\mathrm{T}}\right)^{-1} = \left(\mathbf{M}^{\mathrm{T}}\mathbf{M}\right)^{-1} = \mathbf{G}^{-1} = \frac{1}{M}\begin{pmatrix} m_{\mathrm{A}}(m_{\mathrm{B}}+m_{\mathrm{C}}) & m_{\mathrm{A}}m_{\mathrm{C}} \\ m_{\mathrm{A}}m_{\mathrm{C}} & m_{\mathrm{C}}(m_{\mathrm{A}}+m_{\mathrm{B}}) \end{pmatrix}, \qquad (12.150)$$

where $M = m_{\mathrm{A}} + m_{\mathrm{B}} + m_{\mathrm{C}}$ is the total mass, and we have used the property that the inverse of the transpose is the transpose of the inverse.

▶ **Exercise 12.21** Show that replacing the matrix $\mathbf{M}$ in Eq. 12.57 by $\mathbf{M}' = \mathbf{UM}$, where $\mathbf{U}$ is an arbitrary orthogonal matrix, still diagonalizes the kinetic energy and leaves β, a and b invariant. Give a physical interpretation to this arbitrariness in the definition of (Q_1, Q_2).

Solution We define a new momentum vector, $\bar{P}'' = \mathbf{M}'\bar{P} = \mathbf{UM}\bar{P}$. Then

$$T = \frac{1}{2}\bar{P}^{\mathrm{T}}\mathbf{G}\bar{P} = \frac{1}{2}\bar{P}^{\mathrm{T}}\mathbf{M}^{\mathrm{T}}\mathbf{M}\bar{P} = \frac{1}{2}\bar{P}^{\mathrm{T}}\mathbf{M}^{\mathrm{T}}\left(\mathbf{U}^{\mathrm{T}}\mathbf{U}\right)\mathbf{M}\bar{P} = \frac{1}{2}\bar{P}^{\mathrm{T}}\mathbf{M}'^{\mathrm{T}}\mathbf{M}'\bar{P} = \frac{1}{2}\bar{P}''^{\mathrm{T}}\bar{P}'', \qquad (12.151)$$

leaving the kinetic energy diagonal. Moreover, replacing $\mathbf{M}$ by $\mathbf{M}'$ and $\mathbf{M}^{\mathrm{T}}$ by $\mathbf{M}'^{\mathrm{T}}$ in Eq. 12.63 gives

$$\mathbf{M}'^{\mathrm{T}}\mathbf{M}' = \mathbf{M}^{\mathrm{T}}\mathbf{U}^{\mathrm{T}}\mathbf{UM} = \mathbf{M}^{\mathrm{T}}\mathbf{M}, \qquad (12.152)$$

leaving β invariant. Finally, replacing $\mathbf{M}$ by $\mathbf{M}'$ in Eq. 12.147 gives $\mathbf{M}'' = \left(\mathbf{M}'^{\mathrm{T}}\right)^{-1} = \left((\mathbf{UM})^{\mathrm{T}}\right)^{-1} = (\mathbf{M}^{\mathrm{T}}\mathbf{U}^{\mathrm{T}})^{-1}$ and $\mathbf{M}''^{\mathrm{T}} = \left(\mathbf{M}'\right)^{-1} = (\mathbf{UM})^{-1}$. As a result,

$$\mathbf{M}''^{\mathrm{T}}\mathbf{M}'' = (\mathbf{UM})^{-1}\left(\mathbf{M}^{\mathrm{T}}\mathbf{U}^{\mathrm{T}}\right)^{-1} = \left(\left(\mathbf{M}^{\mathrm{T}}\mathbf{U}^{\mathrm{T}}\right)(\mathbf{UM})\right)^{-1} = \left(\mathbf{M}^{\mathrm{T}}\mathbf{M}\right)^{-1} = \mathbf{G}^{-1}, \qquad (12.153)$$

leaving a and b invariant.

The general expression for (Q_1, Q_2) can be written as

$$\begin{pmatrix} Q_1 \\ Q_2 \end{pmatrix} = \begin{pmatrix} \cos\alpha & -\sin\alpha \\ \sin\alpha & \cos\alpha \end{pmatrix} \begin{pmatrix} a & b\cos\beta \\ 0 & b\sin\beta \end{pmatrix} \begin{pmatrix} R_1 \\ R_2 \end{pmatrix}$$

$$= \begin{pmatrix} a\cos(\alpha) & b\cos(\alpha+\beta) \\ a\sin(\alpha) & b\sin(\alpha+\beta) \end{pmatrix} \begin{pmatrix} R_1 \\ R_2 \end{pmatrix}, \qquad (12.154)$$

where α is an arbitrary clockwise rotation angle of the new coordinates. If α is chosen so that the axes (Q_1, Q_2) line up with the major and minor axes of the elliptical potential energy contours near the minimum of the potential (cf. Figure 12.5), then Q_1 and Q_2 are the normal coordinates.

▶ Exercise 12.22

    a. Show that

$$\sin^2 \beta = \frac{m_B m_{ABC}}{m_{BC} m_{AB}} = \left(\frac{m_B m_C}{m_{BC}} \right) \left(\frac{m_{ABC}}{m_C m_{AB}} \right) = \frac{\mu_{BC}}{\mu_{C,AB}}.$$

    b. Show the Q_1 and Q_2 defined by Eq. (12.54) can be written as

$$Q_1 = a\left(R_1 + \frac{m_C}{m_{BC}} R_2\right) = \mu_{A,BC}^{1/2} R$$

$$Q_2 = b \sin \beta R_2 = \mu_{BC}^{1/2} r,$$

where (R, r) are Jacobi coordinates (Eqs. (12.84)–(12.85)). Since the kinetic energy remains diagonal if Q_1 and Q_2 are scaled by a common factor, one finds several forms of mass-scaled Jacobi coordinates in common use: $(\mu_{A,BC}^{1/2} R, \mu_{BC}^{1/2} r)$; $((\frac{\mu_{A,BC}}{\mu_{BC}})^{1/2} R, r)$ (Figures 14.12, 14.13a); $(R, (\frac{\mu_{BC}}{\mu_{A,BC}})^{1/2} r)$ (Figure 18.8); $((\frac{\mu_{ABC}}{\mu_{BC}})^{1/4} R, (\frac{\mu_{BC}}{\mu_{A,BC}})^{1/4} r)$ (Figures 12.9, 12.23).

References

General

1. L. Pauling and E. B. Wilson, *Introduction to Quantum Mechanics with Applications to Chemistry* (McGraw-Hill, New York, 1935).

2. M. Karplus and R. N. Porter, *Atoms and Molecules* (Benjamin, Menlo Park, 1970).

3. P. W. Atkins, *Molecular Quantum Mechanics* (Oxford, New York, 1983).

4. D. A. McQuarrie, *Quantum Chemistry* (University Science Books, 1983).

5. G. C. Schatz and M. A. Ratner, *Quantum Mechanics in Chemistry* (Prentice Hall, Englewood Cliffs, NJ, 1993).

6. G. Herzberg, *Molecular Spectra and Molecular Structure. I. Spectra of Diatomic Molecules; II. Infrared and Raman Spectra; III. Electronic Spectra of Polyatomic Molecules* (van Nostrand, New York, 1950; 1945; 1966).

7. D. R. Herschbach, Molecular dynamics of elementary chemical reactions, Nobel lecture (1986); Y. T. Lee, Molecular beam studies of elementary chemical processes, Nobel lecture (1986); J. C. Polanyi, Some concepts in reaction dynamics, Nobel lecture (1986). These lectures may be found in *Nobel Lectures, Chemistry 1981–1990*, editor-in-charge T. Frängsmyr, ed. B. G. Malmström (World Scientific, Singapore, 1992), or at http://nobelprize.org/chemistry/laureates/1986.

8. R. D. Levine and R. B. Bernstein, *Molecular Reaction Dynamics and Chemical Reactivity* (Oxford, New York, 1974); revised 1985.

9. *Dynamics of Molecular Collisions*, ed. W. H. Miller (Plenum, New York, 1976), 2 vols.

10. M. S. Child, *Molecular Collision Dynamics* (Academic, London, 1974).

11. M. S. Child, *Semiclassical Mechanics with Molecular Applications* (Clarendon, Oxford, 1991).

12. *Theory of Chemical Reaction Dynamics*, M. Baer, ed., 4 vols. (CRC Press, Boca Raton, FL 1985).

13. *Molecular Collision Dynamics*, ed. J. M. Bowman (Springer, Berlin, 1983).

14. *Dynamics of Molecules and Chemical Reactions*, ed. R. E. Wyatt and J. Z. H. Zhang (Dekker, New York, 1996).

15. J. Z. H. Zhang, *Theory and Application of Quantum Molecular Dynamics* (World Scientific, Singapore, 1999).

16. J. N. Murrell and S. D. Bosanac, *Introduction to the Theory of Atomic and Molecular Collisions* (Wiley, New York, 1989).

17. R. N. Zare, *Angular Momentum: Understanding Spatial Aspects in Chemistry and Physics* (Wiley, New York, 1988).

18. J. N. Murrell, S. Carter, S. C. Farantos, P. Huxley and A. J. C. Varandas, *Molecular Potential Energy Functions* (Wiley, New York, 1984).

Adiabatic and Diabatic Representations

19. M. Born and R. Oppenheimer, Zur Quantentheorie der Molekeln, Annalen der Physik 84, 457 (1927); M. Born, Nachr. Akad. Wiss. Göttingen. II. Math-Physik. Kl. (1951), Nr. 6; M. Born and K. Huang, *Dynamical Theory of Crystal Lattices* (Clarendon Press, Oxford, 1954), Appendix VIII.

20. L. D. Landau, Zur Theorie der Energieübertragung II, Physik. Z. Sowjetunion 2, 46 (1932); English translation: A theory of energy transfer II, in *Collected Papers of L. D. Landau*, ed. D. ter Haar (Gordon and Breach, New York, 1965).

21. C. Zener, Non-adiabatic crossing of energy levels, Proc. Roy. Soc. London, A137, 696 (1932).

22. E. C. G. Stückelberg, Theorie der unelastischen Stösse zwischen Atomen, Helv. Phys. Acta 5, 369 (1932).

23. F. T. Smith, Diabatic and adiabatic representations for atomic collision problems, Phys. Rev. 179, 111 (1969).

24. M. Baer, Adiabatic and diabatic representations for atom-molecule collisions: Treatment of the collinear arrangement, Chem. Phys. Lett. 35, 112 (1975).

25. M. Baer, The theory of electronic nonadiabatic transitions in chemical reactions, in *Theory of Chemical Reaction Dynamics*, ed. M. Baer, vol. II (CRC Press, Boca Raton, FL 1985).

26. C. A. Mead and D. G. Truhlar, Conditions for the definition of a strictly diabatic electronic basis for molecular systems, J. Chem. Phys. 77, 6090 (1982).

27. J. von Neumann and E. Wigner, Über das Verhalten von Eigenwerten bei adiabatischen Prozessen, Physik Z. 30, 467 (1927).

28. E. Teller, The crossing of potential energy surfaces, J. Phys. Chem. 41, 109 (1937).

29. H. A. Jahn and E. Teller, Stability of polyatomic molecules in degenerate electronic states. I: Orbital degeneracy, Proc. Roy. Soc. A, 161, 220 (1937).

30. Longuet-Higgins, Öpik, Pryce and Sack, Studies of the Jahn-Teller effect II. The dynamical problem, Proc. Roy. Soc. A, 244, 1 (1958).

31. G. Herzberg and H. C. Longuet-Higgins, Intersection of potential energy surfaces in polyatomic molecules, Discussions Faraday Soc. 35, 77 (1963).

32. C. A. Mead and D. G. Truhlar, On the determination of Born-Oppenheimer nuclear motion wave functions including complications due to conical intersections and identical nuclei, J. Chem. Phys. 70, 2282 (1979).

33. A. Kuppermann, The geometric phase in reaction dynamics, in *Dynamics of Molecules and Chemical Reactions*, ed. R. E. Wyatt and J. Z. H. Zhang (Dekker, New York, 1996).

34. M. Baer, On the Longuet-Higgins phase and its relation to the electronic adiabatic-diabatic transformation angle, J. Chem. Phys. 107, 2694 (1997).

35. J. Tully, Nonadiabatic processes in molecular collisions, in *Dynamics of Molecular Collisions*, ed. W. H. Miller, (Plenum, New York, 1976), part B.

36. L. Butler, Chemical reaction dynamics beyond the Born-Oppenheimer approximation, Ann. Rev. Phys. Chem. 49, 125 (1998).

Potential Energy Surfaces, Skewed Coordinates and Coordinate Transformations

37. F. London, *Sommerfeld Festschrift* (S. Hirzel, 1928), p. 104.

38. H. Eyring and M. Polanyi, Über einfache Gasreaktionen, Z. Physikal. Chem. B12, 279 (1931).

39. J. Hirschfelder, H. Eyring and N. Rosen, Calculation of the energy of H_3 and H_3^+, J. Chem. Phys. 4, 121 (1936).

40. F. T. Smith, Participation of vibration in exchange reactions, J. Chem. Phys. 31, 1352 (1959).

41. J. O. Hirschfelder, Coordinates which diagonalize the kinetic energy of relative motion, J. Quant. Chem. 3S, 17 (1969).

42. J. O. Hirschfelder, My adventures in theoretical chemistry, Annual Rev. Phys. Chem. 34, 1 (1983).

43. B. Podolsky, Quantum-mechanically correct form of Hamiltonian function for conservative systems, Phys. Rev. 32, 812 (1928).

Normal Modes of Vibration

44. E. B. Wilson, Jr., J. C. Decius and P. C. Cross, *Molecular Vibrations* (Dover, New York, 1955).

45. M. Tinkham, *Group Theory and Quantum Mechanics* (McGraw-Hill, New York, 1964).

46. H. Goldstein, *Classical Mechanics* (Addison-Wesley, Reading, MA 1950).

47. G. Strang, *Linear Algebra and Its Applications* (Harcourt Brace Jovanovich, San Diego, 1986), pp. 343–345.

Transition State Theory

48. M. Polanyi and E. Wigner, Zeits. f. Physik. Chemie A (Haber Band), 439 (1928).

49. H. Pelzer and E. Wigner, Über die Geschwindigkeitskonstante von Austauschreaktionen, Zeits. f. physik. Chemie B15, 445 (1932).

50. H. Eyring, The activated complex in chemical reactions, J. Chem. Phys. 3, 107 (1935).

51. M. C. Evans and M. Polanyi, Some applications of the transition state method to the calculation of reaction velocities, especially in solution, Trans. Faraday Soc. 31, 875 (1935).

52. E. Wigner, The transition state method, Trans. Faraday Soc. 34, 29, (1938).

53. E. Wigner, Calculation of the rate of elementary association reactions, J. Chem. Phys. 5, 720 (1937).

54. J. C. Keck, Variational theory of reaction rates, Adv. Chem. Phys. 13, 85 (1967).

55. P. Pechukas and F. J. McLafferty, J. Chem. Phys. 58, 1622 (1973); F. J. McLafferty and P. Pechukas, Chem. Phys. Lett. 27, 511 (1974).

56. P. Pechukas, Statistical approximations in collision theory, in *Dynamics of Molecular Collisions*, ed. W. H. Miller (Plenum, New York, 1976), part B.

57. P. Pechukas and E. Pollak, Trapped trajectories at the boundary of reactivity bands in molecular collisions, J. Chem. Phys. 67, 5976 (1977); E. Pollak and P. Pechukas, Transition states, trapped trajectories, and classical bound states embedded in the continuum, J. Chem. Phys. 69, 1218 (1978).

58. P. Pechukas and E. Pollak, Classical transition state theory is exact if the transition state is unique, J. Chem. Phys. 71, 2062 (1979).

59. E. Pollak, M. S. Child and P. Pechukas, Classical transition state theory: A lower bound to the reaction probability, J. Chem. Phys. 72, 1669 (1980a); E. Pollak and M. S. Child, Classical mechanics of a collinear exchange reaction: A direct evaluation of the reaction probability and product distribution, J. Chem. Phys. 73, 4373 (1980b).

60. D. G. Truhlar, W. L. Hase and J. T. Hynes, Current Status of Transition State Theory, J. Phys. Chem. 87, 2664 (1983).

61. D. Chandler, Statistical mechanics of isomerization dynamics in liquids and the transition state approximation, J. Chem. Phys. 68, 2959 (1978); J. A. Montgomery, D. Chandler and B. J. Berne, Trajectory analysis of a kinetic theory for isomerization dynamics in condensed phase, J. Chem. Phys. 70, 4056 (1979).

62. W. H. Miller, Quantum and semiclassical theory of chemical reaction rates, Faraday Disc. 110, 1 (1998).

Product Distributions

63. M. G. Evans and M. Polanyi, Notes on the luminescence of sodium vapour in highly dilute flames, Trans. Faraday Soc. 35, 178 (1939).

64. F. T. Smith, Participation of vibration in exchange reactions, J. Chem. Phys. 31, 1352 (1959).

65. E. Pollak and M. S. Child, Classical mechanics of a collinear exchange reaction: A direct evaluation of the reaction probability and product distribution, J. Chem. Phys. 73, 4373 (1980).

66. R. T. Skodje and M. J. Davis, A phase space analysis of the collinear I+HI reaction, J. Chem. Phys. 88, 2429 (1988).

Symmetry

67. F. A. Cotton, *Chemical Applications of Group Theory* (Wiley, New York, 1971), 2nd ed.

68. M. Tinkham, *Group Theory and Quantum Mechanics* (McGraw-Hill, New York, 1964).

69. E. P. Wigner, *Group Theory* (Academic Press, New York, 1959).

70. W. G. Harter, *Principles of Symmetry, Dynamics and Spectroscopy* (Wiley, New York, 1993).

71. P. R. Bunker, *Molecular Symmetry and Spectroscopy* (Academic, New York, 1979).

72. G. S. Ezra, *Symmetry Properties of Molecules* (Springer, New York, 1982).

73. H. C. Longuet-Higgins, The symmetry groups of non-rigid molecules, Mol. Phys. 6, 445 (1963).

74. M. Quack, Detailed symmetry selection rules for reactive collisions, Mol. Phys. 34, 477 (1977).

Transition Dipole Moment Symmetry

75. D. J. Tannor, Time dependent approach to resonance Raman scattering, Ph.D. thesis, UCLA (1984); D. J. Tannor, Anomalously polarized emission from molecules undergoing large-amplitude motion, J. Phys. Chem. 92, 3341 (1988).

76. D. S. Talaga and J. I. Zink, Choosing a model and appropriate transition dipole moments for time-dependent calculations of intervalence electronic transitions, J. Phys. Chem. 100, 8712 (1996).

Hyperspherical Coordinates

77. D. W. Jepsen and J. O. Hirschfelder, Set of co-ordinate systems which diagonalize the kinetic energy of relative motion, Proc. Natl. Acad. Sci. 45, 249 (1959); J. O. Hirschfelder, Coordinates which diagonalize the kinetic energy of relative motion, Int. J. Quantum Chem. Symp. 3, 17 (1969).

78. L. M. Delves, Tertiary and general-order collisions, Nucl. Phys. 9, 391 (1959); L. M. Delves, Tertiary and general-order collisions (II), Nucl. Phys. 20, 275 (1960).

79. F. T. Smith, Participation of vibration in exchange reactions, J. Chem. Phys. 31, 1352 (1959); F. T. Smith, Generalized angular momentum in many-body collisions, Phys. Rev. 120, 1058 (1960).

80. A. Kuppermann, A useful mapping of triatomic potential energy surfaces, Chem. Phys. Lett. 32, 374 (1975); A. Kuppermann, A new look at hyperspherical coordinates, *Advances in Molecular Vibrations and Collision Dynamics*, vol. 2B (JAI Press, 1993).

81. R. T Pack and G. A. Parker, Quantum reactive scattering in three dimensions using hyperspherical (APH) coordinates. Theory. J. Chem. Phys. 87, 3888 (1987).

82. D. Manolopoulos, Reactive scattering, in *Encyclopedia of Computational Chemistry*, ed. P. Von R. Schleyer, (Wiley, 1999).

Chapter 13

Femtosecond Pulse Pair Excitation

We are now positioned to bring together the principles of molecular dynamics from the previous chapter with the first- and second-order time-dependent perturbation theory formulas derived in Chapter 9. The combination of these two leads immediately to a physically appealing treatment of the optical excitation of molecules with femtosecond laser pulses (10^{-15} s). The beauty of this approach is that the molecular dynamics on the Born–Oppenheimer potential energy surfaces is treated together with the time dependence of the laser pulse or pulses in a single unified framework. In this chapter we will concentrate on the effect on a molecule of one or several *unshaped* femtosecond pulses. The applications of this treatment include the fascinating field of *femtochemistry*, including the real-time probing and clocking of chemical reactions, as well as nonlinear femtosecond spectroscopy. We will defer to Chapter 16 the discussion of many additional interesting effects that occur when one or several optical pulses are *shaped* in amplitude or phase using sophisticated techniques of pulse shaping.

In this chapter we will be dealing with first-, second- and even third-order perturbation theory. The Hamiltonian will be written $H = H_0 + H_1$, where H_0 is the bare molecular Hamiltonian and H_1 is the perturbation, taken to be $-\mu \cdot \varepsilon(t)$, where μ is the molecular dipole moment vector and $\varepsilon(t)$ is the instantaneous electric field associated with the laser pulse. In the first part of the chapter we consider the excited-state amplitudes created by the pulse sequences; the absolute value squared of these amplitudes is the excited-state population, which is assumed to be proportional to a fluorescence or ionization signal. In the second part of the chapter we consider the *coherences* formed between different electronic states. The coherences can be measured via nonlinear spectroscopy, which can provide a directional, zero-background signal.

13.1 First-Order Processes: Wavepacket Interferometry

Adopting the Born–Oppenheimer approximation, and specializing to two B.O. states, H_0 can be written as

$$H_0 = \begin{pmatrix} H_a & 0 \\ 0 & H_b \end{pmatrix} \tag{13.1}$$

and H_1 as

$$H_1 = \begin{pmatrix} 0 & -\mu_{ab}\varepsilon^*(t) \\ -\mu_{ba}\varepsilon(t) & 0 \end{pmatrix}, \tag{13.2}$$

where a and b refer to the ground and excited electronic states, respectively, and $\mu_{ab/ba}$ is the transition dipole moment (we are neglecting the permanent dipole moment and higher-order multipoles; moreover, for much of our treatment below we assume there is only one component to the transition dipole vector and we neglect its vector character). We will generally assume that before the excitation there is no amplitude in the excited electronic state, so the initial state can be written

$$\psi^{(0)}(0) = \begin{pmatrix} \psi_a \\ 0 \end{pmatrix}. \tag{13.3}$$

Substituting Eqs. 13.1, 13.2 and 13.3 into the generic first-order perturbation theory formula, Eq. 9.37, we obtain the following formula for first-order optical processes:

$$\psi^{(1)}(x, t) = \frac{1}{i\hbar} \int_0^t e^{-\frac{i}{\hbar} H_0(t-t')} H_1(t') e^{-\frac{i}{\hbar} H_0 t'} \psi^{(0)}(0) \, dt' \tag{13.4}$$

$$= \frac{1}{i\hbar} \int_0^t \begin{pmatrix} e^{-\frac{i}{\hbar} H_a(t-t')} & 0 \\ 0 & e^{-\frac{i}{\hbar} H_b(t-t')} \end{pmatrix} \begin{pmatrix} 0 & -\mu_{ab}\varepsilon^*(t') \\ -\mu_{ba}\varepsilon(t') & 0 \end{pmatrix} \tag{13.5}$$

$$\times \begin{pmatrix} e^{-\frac{i}{\hbar} H_a t'} & 0 \\ 0 & e^{-\frac{i}{\hbar} H_b t'} \end{pmatrix} \begin{pmatrix} \psi_a(x, 0) \\ 0 \end{pmatrix} \, dt'$$

$$= \begin{pmatrix} \psi_a(x, t) \\ \psi_b(x, t) \end{pmatrix} \tag{13.6}$$

$$= \begin{pmatrix} 0 \\ \frac{1}{i\hbar} \int_0^t e^{-\frac{i}{\hbar} H_b(t-t')} \{-\mu_{ba}\varepsilon(t')\} e^{-\frac{i}{\hbar} H_a t'} \psi_a(x, 0) \, dt' \end{pmatrix}. \tag{13.7}$$

The vector notation for $\psi^{(1)}$ is pedantic, since the first element is zero and we simply write

$$\psi^{(1)}(x, t) = \frac{1}{i\hbar} \int_0^t e^{-\frac{i}{\hbar} H_b(t-t')} \{-\mu_{ba}\varepsilon(t')\} e^{-\frac{i}{\hbar} H_a t'} \psi_a(x, 0) \, dt'. \tag{13.8}$$

Equation 13.8 has a very appealing physical interpretation, reading this equation from right to left. The initial wavefunction, ψ_a, evolves from $t = 0$ until time t' under the ground electronic state Hamiltonian, H_a. At time t' the electric field of amplitude $\varepsilon(t')$ interacts with the transition dipole moment, promoting amplitude to the excited electronic state. This amplitude evolves under the influence of H_b from time t' until time t. The integral dt' indicates that one must take into account all instants in time t' at which the interaction with the field could have taken place. In general, if the field has some envelope of finite duration in time, the promotion to the excited state can take place at any instant under this envelope, and there will be interference from portions of the amplitude that are excited at one instant and portions that are excited at another. The various steps in the process may be visualized schematically with the use of Feynman diagrams. The Feynman diagram for the process just described is shown in Figure 13.1. The time t' on the Feynman diagram should be understood as being any instant between 0 and t; what the Feynman diagram emphasizes is that there is a correlation between the time at which the photon impinges and the subsequent molecular evolution, as if the photon and the molecule synchronize their clocks. This synchronization will become clearer in our discussion of second-order processes, particularly in the next chapter.

If we assume that the initial state, ψ_a, is the vth vibrational state of H_a ($H_a \psi_v = E_v \psi_v$), there is no spatial evolution on the ground electronic state potential energy surface, just the

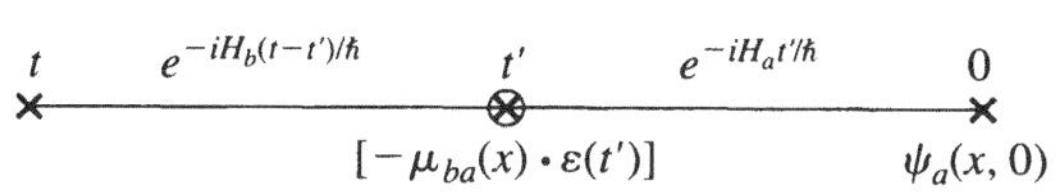

$$t \qquad\quad e^{-iH_b(t-t')/\hbar} \qquad\quad t' \qquad\quad e^{-iH_at'/\hbar} \qquad\quad 0$$

$$[-\mu_{ba}(x) \cdot \varepsilon(t')] \qquad\qquad \psi_a(x, 0)$$

Figure 13.1 Feynman diagram for the first-order process described in the text. Time goes from right to left.

accumulation of an overall phase factor. Then the quantity $e^{-iH_at'/\hbar}\psi_a(x, 0)$ can be replaced by $e^{-iE_vt'/\hbar}\psi_v(x, 0)$. The most common case of interest is $v = 0$, but to retain generality we will call the initial state ψ_i, where $H_a\psi_i = E_i\psi_i$. This allows us to rewrite Eq. 13.8 as

$$\psi^{(1)}(x, t) = \frac{1}{i\hbar} \int_0^t \{\varepsilon(t')e^{-\frac{i}{\hbar}E_it'}\}e^{-\frac{i}{\hbar}H_b(t-t')}\{-\mu_{ba}(x)\psi_i(x, 0)\}\, dt' \qquad (13.9)$$

$$= -\frac{1}{i\hbar} \int_0^t \{\varepsilon(t')e^{-\frac{i}{\hbar}E_it'}\}e^{-\frac{i}{\hbar}H_b(t-t')}\phi_i(x, 0)\, dt', \qquad (13.10)$$

where we have defined the state

$$\phi_i(x, 0) = \mu_{ba}(x)\psi_i(x, 0). \qquad (13.11)$$

According to Eq. 13.10, $\phi_i(x, 0)$ is the effective initial state that evolves on the excited-state potential. Normally, the dependence of μ_{ba} on x is weak and to a good first approximation can be taken as independent of x. This is called the Condon approximation. In this case the initial state on the excited-state potential is an exact replica of the initial vibrational state ψ_i. One speaks of the initial vibrational state ψ_i as making a vertical, or Franck–Condon, transition to the excited-electronic state, and the region of the excited state potential where the initial state is born is called the Franck–Condon region.

Although the Condon approximation is of very broad applicability, there are cases where it fails completely. For example, as we saw in Section 12.6.3, there are times when by symmetry the transition dipole moment vanishes at the Franck–Condon geometry but is nonvanishing at asymmetrical geometries. In this case the nuclear coordinate dependence of the transition dipole moment is crucial; otherwise one obtains no electronic transition amplitude.

We will now proceed to work through some applications of Eq. 13.10 to different pulse sequences. Perhaps the simplest is to consider the case of a δ-function excitation by light:

$$\varepsilon(t') = \delta(t' - t_1). \qquad (13.12)$$

In this case, the first-order amplitude reduces to

$$\psi^{(1)}(x, t) = \frac{1}{i\hbar}e^{-\frac{i}{\hbar}H_b(t-t_1)}\{-\mu_{ba}\}e^{-\frac{i}{\hbar}E_it_1}\psi_i(x, 0) \qquad (13.13)$$

$$= -\frac{1}{i\hbar}e^{-\frac{i}{\hbar}E_it_1}e^{-\frac{i}{\hbar}H_b(t-t_1)}\phi_i(x, 0). \qquad (13.14)$$

The first-order amplitude is simply a constant times the effective initial state, $\phi_i(x, 0) = \mu_{ba}(x)\psi_i(x, 0)$, propagated on the excited-state potential energy surface! This process can be visualized by drawing the initial vibrational wavefunction, multiplied by the transition

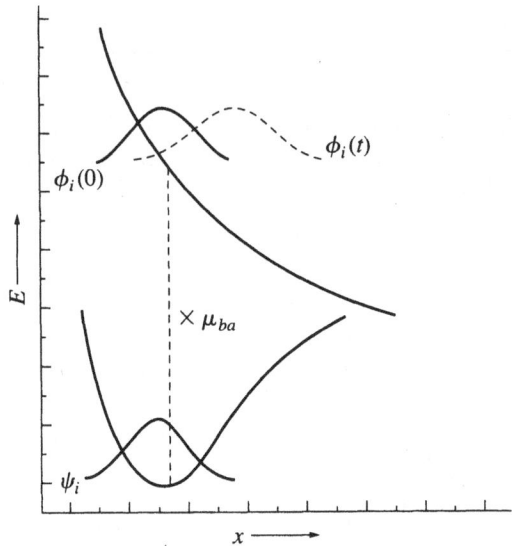

Figure 13.2 Wavepacket picture corresponding to the first-order process described in the text. The initial vibrational state, ψ_i, propagates on the ground state surface until time t_1, but because it is an eigenstate of this surface it only develops a phase factor. At time t_1 a photon impinges and promotes the initial vibrational state to an excited electronic state, for which it is not an eigenstate. The effective state, $\phi_i = \mu_{ba}\psi_i$, is now a *wavepacket* and begins to move according to the Time-Dependent Schrödinger Equation. At short times, the ensuing motion is very classical-like, the initial motion being along the gradient of the excited-state potential. Shown here is a case that the excited-state potential is dissociative and the wavepacket leaves the Franck–Condon region permanently. If the excited-state potential is bound, the wavepacket returns to the Franck–Condon region, roughly at multiples of the excited-state vibrational period.

dipole moment, displaced vertically to the excited-state potential (see Figure 13.2). Although the initial vibrational state was an eigenstate of H_a, in general ϕ_i is not an eigenstate of H_b, and starts to evolve as a coherent wavepacket. For example, if the excited-state potential energy surface is repulsive as in Figure 13.2, the wavepacket will evolve away from the Franck–Condon region toward the asymptotic region of the potential, corresponding to separated atomic or molecular fragments. If the excited-state potential is bound, the wavepacket will leave the Franck–Condon region, but after half a period will reach a classical turning point and return to the Franck–Condon region for a complete or partial recurrence.

An alternative perspective is as follows. A δ-function pulse in time has an infinitely broad frequency range. Thus, the pulse promotes transitions to all the excited-state vibrational eigenstates having good overlap (Franck–Condon factors) with the initial vibrational state. The pulse, by virtue of its coherence, in fact prepares a coherent superposition of all these excited-state vibrational eigenstates. From the earlier chapters, we know that each of these eigenstates evolves with a different time-dependent phase factor, leading to coherent spatial translation of the wavepacket.

The δ-function excitation is not only the simplest case to consider; it is the fundamental building block, in the sense that the more complicated pulse sequences can be interpreted as superpositions of δ functions, giving rise to *superpositions of wavepackets* that can in principle interfere.

The simplest case of this interference is the case of two δ-function pulses:

$$\varepsilon(t') = \delta(t' - t_1)e^{-i\omega_L t_1} + \delta(t' - t_2)e^{-i\omega_L t_2}e^{i\phi}. \qquad (13.15)$$

We will explore the effect of three parameters: $t_2 - t_1$, ω_L and ϕ, which are the time delay between the pulses, the carrier frequency of the laser (relative to resonance with a transition to a particular excited-state vibrational level) and the relative phase of the two pulses. We will assume that the initial vibrational state is $\psi_i(x, 0) = \psi_0(x, 0)$, namely, the ground vibrational state of the ground electronic state. Using Eq. 13.7,

$$\psi^{(1)}(x, t) = \frac{1}{i\hbar} e^{-\frac{i}{\hbar} H_b(t-t_1)} \{-\mu_{ba} e^{-i\omega_L t_1}\} e^{-\frac{i}{\hbar} E_0 t_1} \psi_0(x, 0)$$

$$+ \frac{1}{i\hbar} e^{-\frac{i}{\hbar} H_b(t-t_2)} \{-\mu_{ba} e^{-i\omega_L t_2}e^{i\phi}\} e^{-\frac{i}{\hbar} E_0 t_2} \psi_0(x, 0) \qquad (13.16)$$

$$= \frac{1}{i\hbar} \left(e^{\frac{i}{\hbar} H_b(t_2-t_1)} e^{-i\omega_L(t_2-t_1)} e^{-i\omega_0(t_2-t_1)} e^{i\phi} + 1 \right) \psi^{(1,1)}(x, t), \quad (13.17)$$

where we have defined

$$\psi^{(1,1)}(x, t) \equiv e^{-\frac{i}{\hbar} H_b(t-t_1)} \{-\mu_{ba} e^{-i\omega_L t_1}\} e^{-\frac{i}{\hbar} E_0 t_1} \psi_0(x, 0); \qquad (13.18)$$

$\psi^{(1,1)}$ has the physical interpretation of the contribution to the amplitude coming from the first pulse.

To simplify the notation, we define $\tilde{H}_b = H_b - \tilde{E}_{00}$ and $\tilde{\omega}_L = (\omega_L + \omega_0) - \tilde{E}_{00}/\hbar$, where $\tilde{E}_{00} = E_{00} + E_{0b}$, E_{00} is the separation of the minimum of the two potentials and E_{0b} is the vibrational zero point energy in the excited state. This shift of the energy scale references everything to the lowest vibrational level of the excited electronic state, simplifying the form of Eq. 13.20 below (see the energy-level diagram in Figure 13.3). Equation 13.17 can now be rewritten as

$$\psi^{(1)}(x, t) = \frac{1}{i\hbar} (e^{\frac{i}{\hbar} \tilde{H}_b \tau} e^{-i\tilde{\omega}_L \tau} e^{i\phi} + 1) \psi^{(1,1)}(x, t), \qquad (13.19)$$

where we have defined $\tau = t_2 - t_1$.

To proceed further, we specialize to an excited-state harmonic oscillator potential with frequency ω. In this case,

$$e^{-\frac{i}{\hbar} \tilde{H}_b \tau_0} \psi(x, 0) = \psi(x, 0), \qquad (13.20)$$

where $\tau_0 = \frac{2\pi}{\omega}$ is the excited-state vibrational period; any wavefunction in the harmonic oscillator returns exactly to its original spatial distribution after one period. (Note that the zero of energy was chosen to eliminate an additional phase factor in Eq. 13.20.) Moreover, for the harmonic oscillator level spacing we may write

$$\tilde{\omega}_L = n\omega + \Delta, \qquad (13.21)$$

where Δ is the detuning from an excited-state vibrational eigenstate (another convenient consequence of the zero of energy). We now assume that the time delay between pulses is equal to the natural period of motion, that is, $\tau = \tau_0$. Substituting Eqs. 13.20 and 13.21 into Eq. 13.19 gives

$$\psi^{(1)}(x, t) = \frac{1}{i\hbar}(e^{\frac{i}{\hbar}\tilde{H}_b\tau}e^{-i(n\omega+\Delta)\tau}e^{i\phi} + 1)\psi^{(1,1)}(x, t) \qquad (13.22)$$

$$= \frac{1}{i\hbar}\left(e^{-i(\Delta\tau-\phi)} + 1\right)\psi^{(1,1)}(x, t). \qquad (13.23)$$

To illustrate the dependence on detuning, Δ, time delay, τ, and phase difference, ϕ, it is worth considering some special cases.

1. If $\Delta = \phi = 0$ then the first term in braces gives $1 + 1 = 2$. In this case, the two pulses create two wavepackets that add constructively, giving two units of amplitude or four units of excited-state population.

2. If $\Delta = 0$ and $\phi = \pm\pi$ then the first term in braces gives $-1 + 1 = 0$. In this case, the two pulses create two wavepackets that add destructively, giving no excited-state population! Emission versus additional absorption is therefore controlled by the phase of the second pulse relative to the first.

3. If $\Delta = 0$ and $\phi = \pm\frac{\pi}{2}$ then the first term in braces gives $\pm i + 1$. In this case, the excited-state population $\langle\psi^{(1)}|\psi^{(1)}\rangle$ is governed by the factor $(-i + 1)(i + 1) = 2$. The amplitudes created by the two pulses overlap, but have no net interference contribution. This result prefigures the phenomenon of "photon locking," which will be discussed in Chapter 15.

4. If $\Delta = \frac{\omega}{2}$ and $\phi = 0$ then the first term in braces gives $-1 + 1 = 0$. This is the ultra-short excitation counterpart of tuning the excitation frequency between vibrational resonances in a single frequency excitation: no net excited-state population is produced. As in the case above, of the two pulses π out of phase, the two wavepackets destructively interfere. In this case, the destructive interference comes from the off-set of the carrier frequency from resonance, leading to a phase factor of $\frac{\omega}{2}\tau = \pi$. For time delays τ that are significantly different from τ_0, the first wavepacket is not in the Franck–Condon region when the second packet is promoted to the excited state, and the packets do not interfere; two units of population are prepared on the excited state, as in the case of a $\pm\frac{\pi}{2}$ phase shift. These different cases are summarized in Figure 13.3.

▶ **Exercise 13.1**   Calculate the Fourier transform of the pulse sequence in Eq. 13.15. Examine the dependence of the spectrum on the same three parameters just examined in the time domain, namely, Δ, τ and ϕ. From the point of view of the frequency domain, discuss why the total absorption depends only on the combination of parameters $\Delta\tau - \phi$, and not their individual values.

Figure 13.4 shows the experimental results of Scherer et al. on excitation of I_2 using pairs of phase-locked pulses. By the use of heterodyne detection, those authors were able to measure just the interference contribution to the total excited-state fluorescence (i.e., the difference in excited-state population from the two units of population that would be

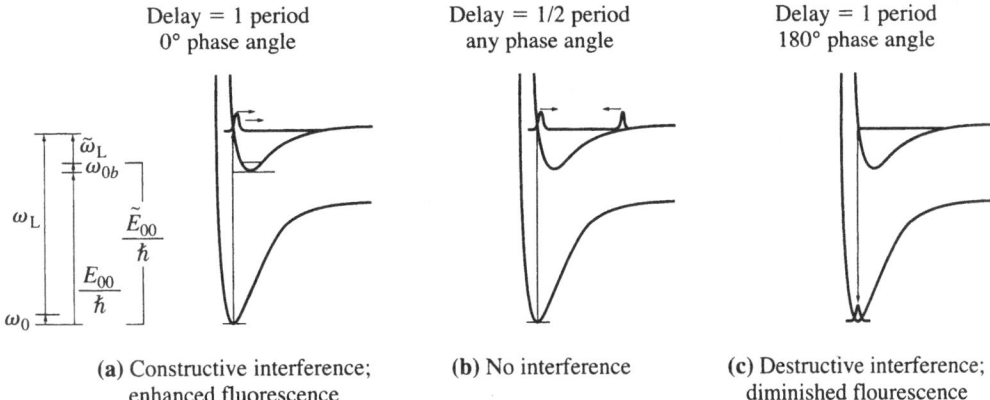

(a) Constructive interference; enhanced fluorescence

(b) No interference

(c) Destructive interference; diminished flourescence

Figure 13.3 Schematic diagram of wavepacket interferometry. (a) Delay of one period, 0° phase angle. The two wavepackets interfere constructively. (b) Delay of 1/2 period. The two wavepackets do not interfere because they do not overlap in phase space. (c) Delay of one period, 180° phase angle. The wavepackets interfere destructively. Reprinted from Scherer et al. (1991).

prepared if there were no interference). The basic qualitative dependence on time delay and phase is the same as that predicted by the harmonic model: significant interference is observed only at multiples of the excited-state vibrational frequency, and the relative phase of the two pulses determines whether that interference is constructive or destructive.

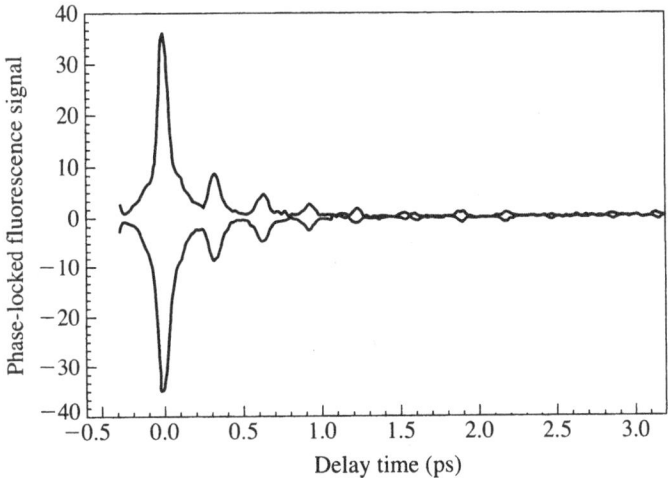

Figure 13.4 Wavepacket interferometry. The *interference contribution* to the excited-state fluorescence of I_2 is measured by heterodyne detection, as a function of the time delay between a pair of ultrashort pulses. The upper trace shows the case of 0° phase angle between the pulses, corresponding to Figure 13.3a, while the lower trace shows the case of 180° phase angle between the pulses, corresponding to Figure 13.3c. Note that the structure in the interferogram occurs only at multiples of 300 fs, the excited-state vibrational period of I_2; at other times the situation is as in Figure 13.3b. Adapted from Scherer et al. (1991).

Consider now the case of continuous wave (CW) excitation, that is, excitation with a field of the form $Ae^{-i\omega_L t'}$, where the envelope function A is a constant in time. In this case, conventional wisdom is that if ω_L is tuned to a vibrational energy of the excited state the corresponding vibrational eigenstate is prepared, while if ω_L is between resonances nothing is prepared. How can one understand this result using the concepts in the previous paragraphs? One can think of the CW excitation as a train of δ-function pulses with a precise phase relationship. Each δ-function pulse excites an additional piece of wavepacket amplitude from the ground to the excited state, and that portion starts to evolve. At the same time, new amplitude is arriving on the excited state from later portions of the pulse. The new amplitude will interfere with portions of the excited-state wavepacket that are returning to the Franck–Condon region, and this interference will be either constructive, if ω_L is on resonance with a vibrational eigenstate, or destructive, if ω_L is between resonances, exactly as in the two-pulse case discussed above. That the spatial distribution of the state prepared is proportional to the excited-state eigenstate at that energy may be seen from the formula for $\psi^{(1)}(x, t)$:

$$\psi^{(1)}(x, t) = \frac{1}{i\hbar} \int_0^t e^{-\frac{i}{\hbar} H_b(t-t')} \{-\mu_{ba} A e^{-i\omega_L t'}\} e^{-\frac{i}{\hbar} E_0 t'} \psi(x, 0) \, dt' \qquad (13.24)$$

$$= \frac{1}{i\hbar} \int_{t_0}^t e^{-\frac{i}{\hbar} \tilde{H}_b(t-t')} (-\mu_{ba} A e^{-i\tilde{\omega}_L t'}) \psi(x, t_0) \, dt' \qquad (13.25)$$

Note that $t = 0$ has been replaced by $t = t_0$, and we are interested in the limit that $t_0 \to -\infty$. This is because for true CW excitation the field must have already been on in the infinite past. Changing integration variables to $s = t - t'$ yields

$$\psi^{(1)}(x, t) = \frac{1}{i\hbar} \int_0^\infty e^{-\frac{i}{\hbar} \tilde{H}_b s} \{-\mu_{ba} A e^{i\tilde{\omega}_L s}\} e^{-i\tilde{\omega}_L t} \psi(x, 0) \, ds. \qquad (13.26)$$

This expression for $\psi^{(1)}(x, t)$ is quite close to that of the spectral representation of eigenstates discussed in Chapter 6. The only difference, aside from an overall multiplicative factor, is that the lower limit of the integral is 0, as opposed to $-\infty$. This is not just a mathematical artifact. The lower limit of integration, $s = 0$, essentially applies a step function to the time integration. The physical interpretation of this step function can be seen by noting that $s = t - t'$ is the *time difference* between the excitation and the current time. Wavepackets that were prepared in the distant past interfere with wavepackets that were excited both earlier and later; however, the wavepacket that is currently being excited can interfere only with past wavepackets; the future packets have not been promoted yet. For bound systems, excited on resonance, this contribution from $s = 0$ is insignificant compared with the amplitude that has accumulated by constructive interference since $t' = -\infty$. However, for dissociative systems, or off-resonance excitation, the $s = 0$ contribution is important, and in fact can play a leading role in second-order processes, for example, two-photon absorption or Raman scattering. We will return to this point in Chapter 14.

For pulse durations that are finite, but not infinite, there is an interplay between pulse duration, pulse shape, carrier frequency, and the time scale of the excited-state dynamics. Assuming a smooth envelope function for the excitation, to a first approximation $\psi^{(1)}$ begins to resemble the excited-state eigenstate at the excitation frequency (or to destructively interfere, if the carrier frequency is between resonances) on time scales of the order of one vibrational period.

There is a good analogy between the effects of pulse pairs and pulse shapes, and Fresnel and Fraunhofer diffraction in optics. Fresnel diffraction refers to the interference pattern obtained by light passing through two slits; interference from the wavefronts passing through the two slits is the spatial analog of the interference from the two pulses in time discussed above. Fraunhofer diffraction refers to interference arising from the finite width of a single slit. The different subportions of a single slit can be thought of as independent slits that happen to adjoin; wavefronts passing through each of these subslits will interfere. This is the analog of a single pulse with finite duration: there is interference from excitation coming from different subportions of the pulse, which may be insignificant if the pulse is short but can be important for longer pulse durations.

There is an alternative, and equally instructive, way of viewing the effect of different pulse sequences, by Fourier transforming the pulse train to the frequency domain. In the time domain, the wavefunction produced is the convolution of the pulse sequence with the excited-state dynamics; in frequency it is simply the product of the frequency envelope with the Franck–Condon spectrum (the latter is simply the spectrum of overlap factors between the initial vibrational state and each of the excited vibrational states). The Fourier transform of δ-function excitation is simply a constant excitation in frequency, which excites the entire Franck–Condon spectrum. The Fourier transform of a sequence of two δ functions in time with spacing τ is a spectrum having peaks with a spacing of $\frac{2\pi}{\tau}$. If the carrier frequency of the pulses is resonant and the relative phase between the pulses is zero, the frequency spectrum of the pulses will lie on top of the Franck–Condon spectrum and the product will be nonzero; if, on the other hand, the carrier frequency is between resonances, or the relative phase is π, the frequency spectrum of the pulses will lie in between the features of the Franck–Condon spectrum, signifying zero net absorption. Similarly, a single pulse of finite duration may have a frequency envelope that is smaller than that of the entire Franck–Condon spectrum. The absorption process will depend on the overlap of the frequency spectrum with the Franck–Condon spectrum, and hence on both pulse shape and carrier frequency.

13.2   Second-Order Processes: Clocking Chemical Reactions

We now turn to second-order processes. The second-order amplitude is given by

$$\psi^{(2)}(x, t)$$
$$= \left(\frac{1}{i\hbar}\right)^2 \int_0^t \int_0^{t'} e^{-\frac{i}{\hbar}H_c(t-t')}\{-\mu_{cb}\varepsilon(t')\}e^{-\frac{i}{\hbar}H_b(t'-t'')}\{-\mu_{ba}\varepsilon(t'')\}e^{-\frac{i}{\hbar}H_a t''}\psi(x, 0)\, dt' dt''.$$

$$(13.27)$$

This expression may be interpreted in a very similar spirit to that given above for one-photon processes. Now there is a second interaction with the electric field and the subsequent evolution is taken to be on a third surface, with Hamiltonian H_c. In general, there is also a second-order interaction with the electric field through μ_{ab} that returns a portion of the excited-state amplitude to surface a, with subsequent evolution on surface a. The Feynman diagram for this second-order interaction is shown in Figure 13.5.

Second-order effects include Ahmed Zewail's experiments on "clocking chemical reactions" (Zewail, 1988). The experiments are shown schematically in Figures 13.6–13.7.

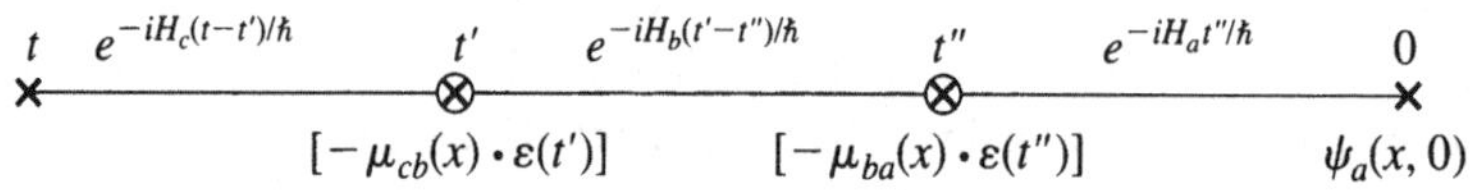

$$t \quad e^{-iH_c(t-t')/\hbar} \quad t' \quad e^{-iH_b(t'-t'')/\hbar} \quad t'' \quad e^{-iH_a t''/\hbar} \quad 0$$

$$\times \underset{[-\mu_{cb}(x) \cdot \varepsilon(t')]}{\overset{}{\longrightarrow}} \otimes \underset{[-\mu_{ba}(x) \cdot \varepsilon(t'')]}{\overset{}{\longrightarrow}} \otimes \underset{\psi_a(x,0)}{\overset{}{\longrightarrow}} \times$$

Figure 13.5 Feynman diagram for the second-order process described in the text. Time goes from right to left.

An initial 100–150 femtosecond pulse moves amplitude from the bound ground state to the dissociative first excited state in ICN. A second pulse, time delayed from the first then moves amplitude from the first excited state to the second excited state, which is also dissociative. By noting the frequency of light absorbed from the second pulse, the energy difference between the two excited-state surfaces can be estimated. By correlating this energy difference with the time delay between pulses, one can infer the motion of the wavepacket prepared on the first excited state and the time it takes to reach the asymptotic region of the potential (Figure 13.6).

A dramatic set of experiments by Zewail (1988) involves the use of femtosecond pulse pairs to probe the wavepacket dynamics at the crossing between covalent and ionic states of NaI. A first pulse promotes wavepacket amplitude from the ionic to the covalent potential curve. The packet begins to move out, but most of the amplitude is reflected back from the crossing between the covalent and ionic curves, that is, the adiabatic potential changes character to ionic at large distances, and this curve is bound, leading to wavepacket reflection back to the FC region. The result is a long progression of wavepacket recurrences, with a slow overall decay coming from amplitude that dissociates on the diabatic curve every period (Figure 13.7). The process of impeded exit on the covalent state is referred to picturesquely as the "harpooning" mechanism. The I is the whale and the Na is the whaler: as the Na and I neutrals separate, an electron—the harpoon—leaves the Na and binds to the I, forming Na^+ and I^-. The attractive Coulomb force between the ions draws them back together to form NaI.

Another interesting application of ultrashort two-pulse interactions, by the Gerber group, is to photoelectron spectroscopy, where the first photon prepares the wavepacket on an intermediate electronic state and the second pulse ionizes the molecule (Baumert (1992)). The total number of molecular ions or electrons is detected as a function of time delay between the two femtosecond laser pulses.

The difference between ionization in the second step versus excitation to a neutral potential energy surface is that the electron can come off with any kinetic energy, E_k, between 0 and $E_{max} = E_{ex} - E_I$, where E_{ex} is the sum of the initial energy of the neutral plus the energy deposited by the field, and E_I is the lowest energy state of the ion. To understand the effect of E_k on the second-order amplitude, note that the larger is E_k, the smaller is the residual energy in the neutral. Thus, the role of E_k is equivalent to an effective reduction in the excitation frequency of the second field, as expressed in the following formula (Meier, 1993):

$$\psi_I^{(2)}(x,t) = \frac{1}{i\hbar} \int_0^t e^{-\frac{i}{\hbar}H_I(t-t')}\{-\mu_{Ib}(E_k)f(t'-\tau)e^{-i(\omega - E_k/\hbar)t'}\}\psi^{(1)}(t')\,dt' \quad (13.28)$$

$$= \left(\frac{1}{i\hbar}\right)^2 \int_0^t \int_0^{t'} e^{-\frac{i}{\hbar}H_I(t-t')}\{-\mu_{Ib}(E_k)f(t'-\tau)\}e^{-i(\omega - E_k/\hbar)t'}e^{-\frac{i}{\hbar}H_b(t-t'')}$$

$$\{-\mu_{ba}f(t'')e^{-i\omega t'}\}e^{-\frac{i}{\hbar}H_a t'}\psi(x,0)\,dt'dt''. \quad (13.29)$$

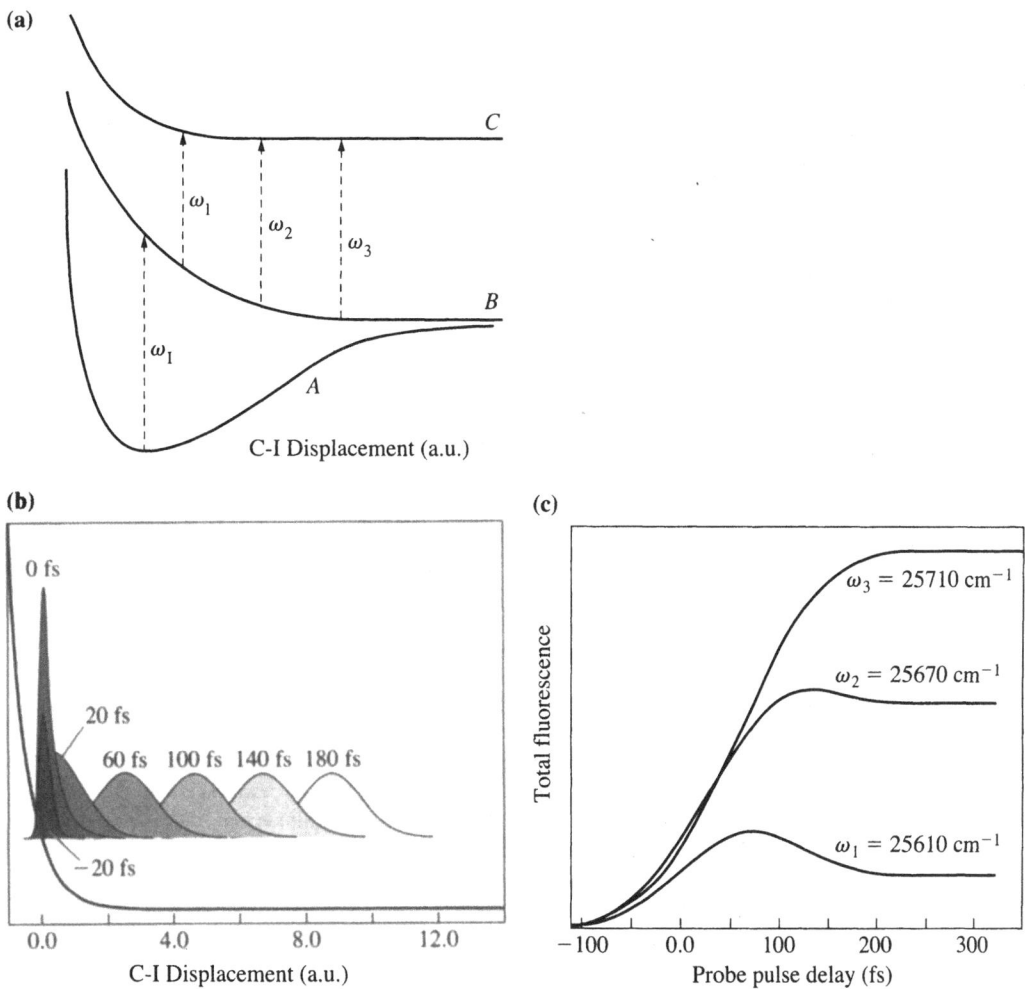

Figure 13.6 (a) A schematic representation of the three potential energy surfaces of ICN participating in the clocking of the dissociation time. (b) Theoretical quantum mechanical simulations for the reaction ICN → ICN* → [I—CN]$^{\dagger *}$ → I + CN. The wavepacket moves and spreads in time, with its center evolving about 5 Å in 200 fs. The wavepacket dynamics refers to motion on the intermediate potential energy surface B. (c) The calculated femtosecond transient absorption signal (FTS), corresponding to total fluorescence from state C, as a function of the time delay between the first excitation pulse ($A \to B$) and the second excitation pulse ($B \to C$). In accordance with spectroscopic units, ω is given in units of 1/wavelength. Adapted from S. O. Williams and D. G. Imre (1988).

Here $\psi^{(1)}$ is the first-order amplitude on the neutral state prepared by the first excitation and τ is the center of the ionization pulse. Note that we have written $\varepsilon(t') = f(t')e^{-i(\omega - E_k/\hbar)t'}$, and similarly for $\varepsilon(t'')$, that is, we have separated off the envelope function $f(t)$ from the carrier wave $e^{-i(\omega - E_k/\hbar)t}$ to highlight the fact that E_k serves effectively just to shift the carrier frequency ω. Note that in principle the transition dipole moment to the ionic state, $\mu_{Ib}(E_k)$, is a function of the electron kinetic energy, although this is generally a relatively minor effect. Clearly, a second-order amplitude with different spatial dependence is prepared for each different value of E_k, due to the convolution of the wavefunction with fields of different effective frequency $\hbar\omega - E_k$.

(a)   **(b)**

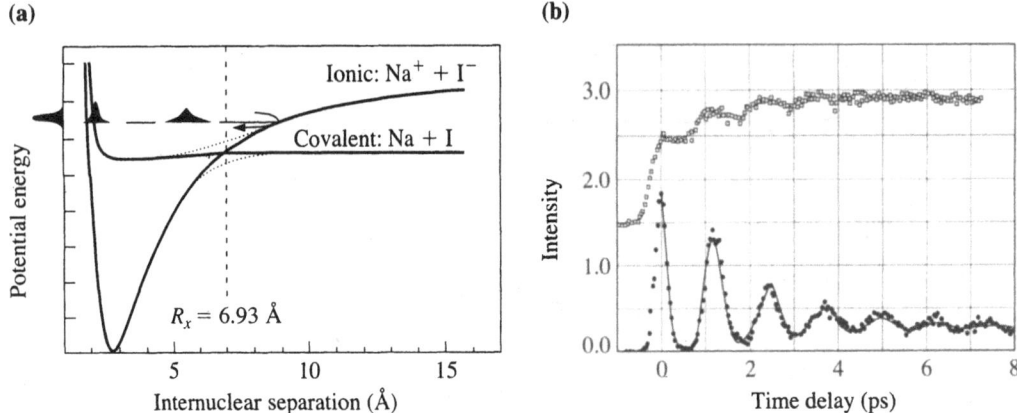

Figure 13.7 (a) Potential energy curves for the reactions of the alkali halides, displaying the ionic and covalent states of the reaction. The pump pulse (λ_1), indicated by the bell-shaped curve on the vertical axis, takes the initial Na^+I^- state to the covalent surface NaI^*. As the bond stretches, $[Na—I]^{†*}$, the probe (λ_2) monitors the evolution of the wavepacket both in the transition state region (λ_2^*) and at "infinite" separation (λ_2^∞). Note the avoided crossing between the ionic and covalent curves and the possible trapping of the packet that may result from this electron harpooning at the crossing point, R_x. (b) Experimental FTS results for the reaction of the alkali halides $NaI^* \rightarrow [Na—I]^{†*} \rightarrow Na + I$. The signal at the bottom shows the oscillation of the wavepacket in the well. The recurrence time is the time for round trip motion in the well; the decay of the oscillations is due to the finite ($\approx .1$) probability for escape each time the wavepacket encounters the crossing region. The experiment was performed by exciting at 310 nm and detecting off-resonance to the transition of free Na atoms at 589 nm. The signal at the top was obtained using detection on resonance with the asymptotic transition (i.e., $\lambda_2 = \lambda_2^\infty$). This signal probes the wavepacket amplitude that has escaped the well region, integrated from the crossing point to ∞. The steps come from new amplitude that leaks out of the well region every time the wavepacket encounters the crossing, while the plateaus indicate the constancy, between encounters with the crossing, of the integrated amplitude that has escaped. Reprinted from A. H. Zewail (1988).

The total ionization probability is

$$P_{\mathrm{I}} = \int_0^{E_{\max}} dE_{\mathrm{k}} \lim_{t \to \infty} \int dx |\psi_{\mathrm{I}}^{(2)}(x, t, E_{\mathrm{k}})|^2. \tag{13.30}$$

Note that P_{I} depends on the time delay, τ, between the two pulses. A plot of the low-lying potential energy curves for Na_2 is shown in Figure 13.8a. Figure 13.8b–d show plots of the time-dependent wavepacket dynamics on the $|s\rangle$ state (left side of each panel) and the dependence on time delay of the photoelectron energy (right side of each panel). Note how the time-resolved photoelectron spectrum mirrors the wavepacket dynamics on the intermediate $|s\rangle$ state. When the wavepacket is near the outer turning point, the photoelectron energy is a minimum, while when the wavepacket is near the inner turning point the photoelectron energy is a maximum. This can be understood from Figure 13.8a, by noting that at the inner turning point the energy separation of the $|s\rangle$ and $|I\rangle$ states is smaller than at the outer turning point, and therefore, for fixed excitation energy E_{ex}, more energy is available for electron kinetic energy if ionization takes place at the inner turning point. In addition to the striking correspondence of wavepacket dynamics and photoelectron signal is the dramatic sequence of wavepacket spreading, followed by fractional revival, followed by full revival,

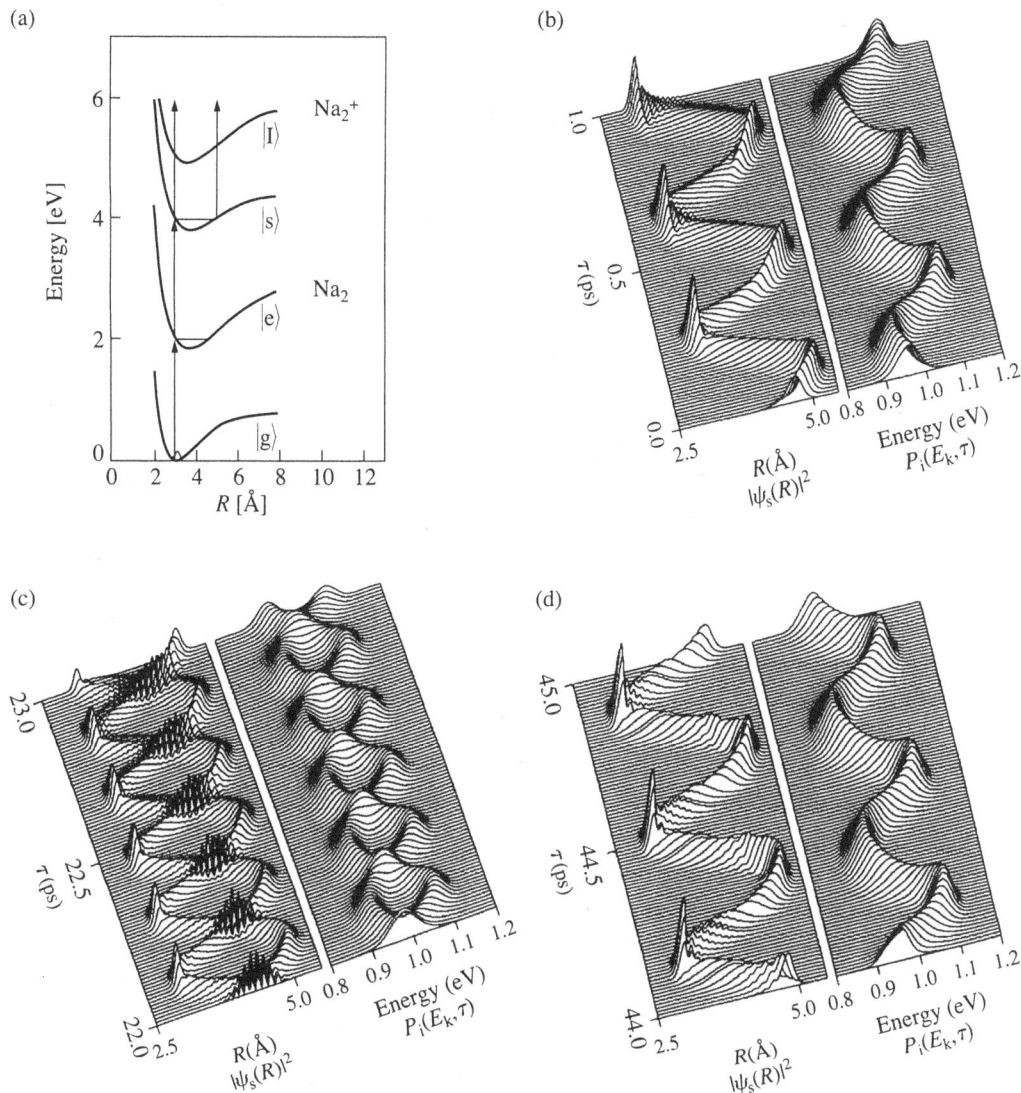

Figure 13.8 (a) Three-photon ionization of Na_2: the lower three potentials correspond to different electronic states of the neutral molecule which can be coupled by an electronic dipole transition. The upper potential energy curve is the ground state of the molecular ion. (b) Wavepacket dynamics (left side of panel) and time-resolved photoelectron spectra (right side of panel) in the state $|s\rangle$ for short delay times (up to 1 ps); (c) same as (b) but for time delays of around 22 ps. Note the wavepacket fractional revival: the two replicas of the initial, compact wavepacket and the signature of these compact replicas on the spectrum (see Section 4.3); (d) same as (b) but for time delays of about 44 ps. In (d), the full revival of the vibrational wavepacket results in electron kinetic energy distributions that resemble those for short times. Adapted from Meier (1995).

as one traces through Figure 13.8b–d. As discussed in Chapter 4, the phenomenon of wave-packet revivals is a ubiquitous property of wavepacket propagation in one-dimensional anharmonic systems. These revivals in Na_2 were first predicted theoretically and have since been observed experimentally, in data almost identical to that in Figure 13.8.

13.3   Coherent Nonlinear Spectroscopy

13.3.1   Preliminaries

In the previous two sections we have examined the quantities $\psi^{(1)}(t)$ and $\psi^{(2)}(t)$. Implicitly, we constructed the populations $\langle\psi^{(1)}(t)|\psi^{(1)}(t)\rangle$ and $\langle\psi^{(2)}(t)|\psi^{(2)}(t)\rangle$. These populations are assumed to be proportional to the fluorescence signal, as in the case of the Zewail experiments, or the ion count, as in the case of the Gerber experiments. This assumption holds provided that the final state is bound and there are no radiationless transitions. There is an additional class of experiments that measures not populations, but coherences, such as $\langle\psi^{(0)}(t)|\psi^{(1)}(t)\rangle$ and $\langle\psi^{(1)}(t)|\psi^{(2)}(t)\rangle$, or more precisely, transition dipole moments of the form $\langle\psi^{(0)}(t)|\mu|\psi^{(1)}(t)\rangle$ and $\langle\psi^{(1)}(t)|\mu|\psi^{(2)}(t)\rangle$. These coherences manifest themselves as a macroscopic *polarization* of the sample, which enters as a new source term in Maxwell's equations and in general leads to emission in a direction other than that of any of the incident beams (see Figure 13.9). This leads to a coherent optical signal that is directional and background-free, features that are very attractive from the experimental point of view. We will reserve the term *coherent* nonlinear spectroscopy for the cases where such a directional signal is produced and measured.

What information is contained in coherent nonlinear spectroscopy? For gas phase experiments—experiments in which the state of the system undergoes little or no dissipation—the goal of coherent nonlinear spectroscopy is generally as in incoherent linear or nonlinear spectroscopy: revealing the quantum energy-level structure of the molecule, both in the ground and the excited electronic states. For example, nonlinear spectroscopy allows transitions that are symmetry forbidden in linear spectroscopy, and hence nonlinear spectroscopy allows the spectroscopic study of many systems that are otherwise dark. Moreover, nonlinear spectroscopy allows access to highly excited vibrational levels that cannot be accessed in these linear spectroscopies, as in the example of time-dependent CARS spectroscopy below. In these cases, coherent nonlinear spectroscopy may give a sensitive and elegant method of detection of quantum energy levels, but is not being used specifically to probe the molecular coherences.

Consider now a molecule in solution. The decay of the coherences is an indicator of molecule–solvent interactions. One normally distinguishes two sources of decay of the

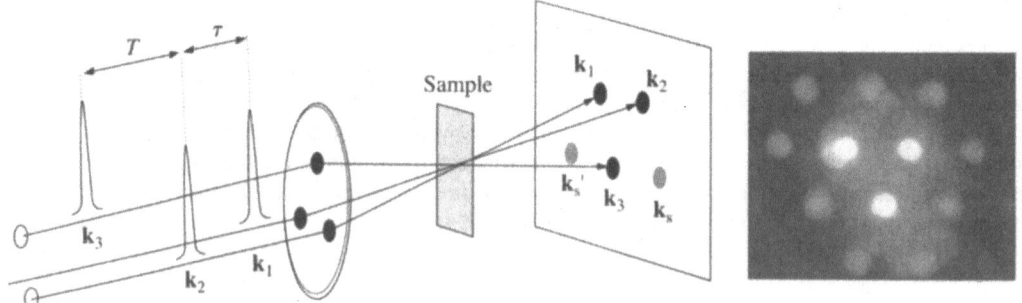

Figure 13.9 Diagram showing the directionality of the signal in coherent spectroscopy. Associated with the carrier frequency of each interaction with the light is a wave vector, **k**. The output signal in coherent spectroscopies is determined from the direction of each of the input signals via momentum conservation (courtesy of Graham Fleming.)

coherence: inhomogeneous decay, which arises from static differences in the environment of different molecules; and homogeneous decay, which arises from the dynamic interaction with the surroundings and is the same for all molecules. Both these sources of decay contribute to the *linewidth* of spectral lines; in many cases the inhomogeneous decay is faster than the homogeneous decay, masking the latter. In echo spectroscopies, which are related to a particular subset of nonlinear spectroscopies, one can at least partially discriminate between homogeneous and inhomogeneous decay. We will not be able to do justice to this fascinating application of nonlinear spectroscopy. For details on this subject the reader is referred to the texts by Shen (1984) and Mukamel (1995).

Finally, nonlinear spectroscopy has emerged as a powerful probe of molecules in anisotropic environments. For molecules at interfaces between a bulk solid and a liquid or gas, there is a nonlinear signal that is absent for molecules in an isotropic environment. Detection of this signal becomes a sensitive probe of structure and dynamics of the interface molecules without background from the bulk. For further reading on these and other topics in nonlinear spectroscopy the reader is referred to the excellent general introductions in the references (Boyd, 2003; Bloembergen, 1965; Shen, 1984; Mukamel, 1995).

13.3.2   Polarization as the Rate of Change of Population and Energy

We begin our study of coherent spectroscopy with some relations that show the centrality of the polarization in determining the rate of change of population and energy. The derivations in this section are nonperturbative, that is, they are applicable to fields of arbitrary strength (Kosloff, 1992, 1984; Lee, 1995). Later in this chapter we will apply the general relations to the perturbative regime.

Change of Population

Consider the instantaneous rate of change of the ground state population, $N_a = \langle \psi_a | \psi_a \rangle$:

$$\frac{dN_a}{dt} = \langle \dot{\psi}_a | \psi_a \rangle + \langle \psi_a | \dot{\psi}_a \rangle \tag{13.31}$$

$$= 2\text{Re}\left\{ \left\langle \frac{H_a}{i\hbar}\psi_a - \frac{\mu_{ab}}{i\hbar}\varepsilon^*(t)\psi_b \,\middle|\, \psi_a \right\rangle \right\} \tag{13.32}$$

$$= 2\text{Re}\left\{ \left\langle \psi_a \,\middle|\, \frac{H_a}{-i\hbar} \,\middle|\, \psi_a \right\rangle + \left\langle \psi_b \,\middle|\, \frac{\mu_{ba}}{i\hbar} \,\middle|\, \psi_a \right\rangle \varepsilon(t) \right\}, \tag{13.33}$$

where in the second line we have used the TDSE for $\dot{\psi}_a$ with H given by Eqs. 13.1–13.2. The first term in braces is purely imaginary, so its real part vanishes. Writing the time dependence of ψ_a and ψ_b explicitly we have

$$\frac{dN_a}{dt} = \frac{2}{\hbar}\text{Im}\{\langle \psi_b(t)|\mu_{ba}|\psi_a(t)\rangle \varepsilon(t)\} \tag{13.34}$$

$$= \frac{2}{\hbar}\text{Im}\{P_{ba}(t)\varepsilon(t)\}, \tag{13.35}$$

where

$$P_{ba}(t) \equiv \langle \psi_b(t)|\mu_{ba}|\psi_a(t)\rangle \tag{13.36}$$

is the ba component of the polarization. From Eq. 13.35, the integrated excited-state population growth (the negative of the change of ground state population) is

$$\Delta N_a = \int \frac{dN_a}{dt}\, dt = \frac{2}{\hbar}\text{Im}\int P_{ba}(t)\varepsilon(t)\, dt$$

$$= \frac{4\pi}{\hbar}\text{Im}\int_{-\infty}^{\infty} \tilde{P}_{ab}^*(\omega)\tilde{\varepsilon}(\omega)\, d\omega \equiv \int_{-\infty}^{\infty} \Delta N_a(\omega)\, d\omega, \tag{13.37}$$

where

$$\tilde{P}_{ab}(\omega) \equiv \frac{1}{2\pi}\int_{-\infty}^{\infty} P_{ab}(t)e^{i\omega t}dt \qquad \tilde{\varepsilon}(\omega) \equiv \frac{1}{2\pi}\int_{-\infty}^{\infty} \varepsilon(t)e^{i\omega t}dt, \tag{13.38}$$

and we have used the Fourier inner product relation, Eq. 6.17, and Exercise 6.8.

Note that in Eq. 13.38, and indeed in virtually all time–frequency Fourier transforms in quantum theory, the phase $e^{i\omega t}$ rather than $e^{-i\omega t}$ enters into the time integral. It is convenient, and general convention, to use the tilde as we have used it in Eq. 13.38 although strictly speaking this corresponds to an inverse Fourier transform (see Eq. 6.12 and the discussion following Exercise 6.3). All the results of Fourier theory in Chapter 6 can still be applied with appropriate modification; for example, in the Fourier inner product relation, the Fourier convolution theorem and Parseval's relation the factor of 2π gets replaced by its reciprocal.

Change of Energy

It is interesting to consider not only the rate of population change but also the power, or rate of energy absorption. The latter is also determined by the polarization, $P(t)$:

$$\frac{dE}{dt} = \frac{d\langle\psi|H|\psi\rangle}{dt} \tag{13.39}$$

$$= \langle\dot{\psi}|H|\psi\rangle + \langle\psi|H|\dot{\psi}\rangle + \left\langle\psi\left|\frac{\partial H}{\partial t}\right|\psi\right\rangle \tag{13.40}$$

$$= \left\langle\psi_a, \psi_b\left|\begin{pmatrix} 0 & -\mu_{ab}\dot{\varepsilon}^*(t) \\ -\mu_{ba}\dot{\varepsilon}(t) & 0 \end{pmatrix}\right|\begin{matrix}\psi_a \\ \psi_b\end{matrix}\right\rangle \tag{13.41}$$

$$= -2\text{Re}[\langle\psi_b|\mu_{ba}|\psi_a\rangle\dot{\varepsilon}(t)] \tag{13.42}$$

$$= -2\text{Re}[P_{ba}(t)\dot{\varepsilon}(t)], \tag{13.43}$$

where E is the energy.

The total energy absorbed, ΔE, is given by

$$\Delta E = \int_{-\infty}^{\infty} \frac{dE}{dt}dt = -2\text{Re}\int_{-\infty}^{\infty} P_{ba}(t)\dot{\varepsilon}(t)\, dt$$

$$= -4\pi\,\text{Im}\int_{-\infty}^{\infty} \omega\tilde{P}_{ab}^*(\omega)\tilde{\varepsilon}(\omega)d\omega \equiv \int_{-\infty}^{\infty} \Delta E(\omega)\, d\omega. \tag{13.44}$$

In obtaining the last relation we have used the Fourier inner product relation (Eq. 6.17) and the Fourier derivative relation (Eq. 6.19 with a sign change coming from the different definition of the tilde).

Comparing Change of Population and Change of Energy

Comparing Eq. 13.44 with Eq. 13.37, we see that

$$\Delta E(\omega) = -\hbar\omega\, \Delta N_a(\omega), \tag{13.45}$$

that is, the *integrated* energy absorption at frequency ω is just the *integrated* excited-state population growth at frequency ω (the negative of the change of ground state population) times the energy of each photon, $\hbar\omega$. For the special case of monochromatic excitation, $\varepsilon(t) = \varepsilon e^{-i\omega t}$, there is also a relation between change of population and change of energy on the differential level. Equation 13.43 gives

$$\frac{dE}{dt} = 2\text{Re}[P_{ba}(t)i\omega\varepsilon e^{-i\omega t}] \tag{13.46}$$

$$= -2\omega\text{Im}[P_{ba}(t)\varepsilon(t)]. \tag{13.47}$$

Comparing Eq. 13.47 with Eq. 13.35 we see that

$$\frac{dE}{dt} = -\hbar\omega\frac{dN_a}{dt} = \hbar\omega\frac{dN_b}{dt}, \tag{13.48}$$

where $N_b = \langle\psi_b|\psi_b\rangle$ is the population in electronic state b. The last relation follows from the fact that in a two-electronic state system, $\frac{dN_a}{dt} + \frac{dN_b}{dt} = 0$. Equation 13.48 states that for monochromatic excitation, the rate of energy absorption is just the rate of photon absorption (the negative of the change in ground state population) times the energy of each photon, $\hbar\omega$. We will use this result below to argue that the absorption spectrum may be defined equivalently in terms of the rate of energy absorption or the rate of photon absorption.

In summary, Eq. 13.45 for the integrated change is valid for arbitrary time-dependent pulses while Eq. 13.48, relating to the instantaneous rate of change, is valid only for monochromatic excitation. Both results are valid for arbitrary field strengths.

Definition of the Nonperturbative Spectrum

We now proceed to the spectrum, or frequency-dependent response. The absorption spectrum may be defined as the ratio of the total number of photons absorbed at each frequency to the number of incident photons per unit area at that frequency. Hence the spectrum has units of area, corresponding to an effective cross section for collision with a photon. We identify the integrated population transfer on the RHS of Eq. 13.37 with the total number of photons absorbed at frequency ω. The number of incident photons per unit area at a given frequency is given by

$$\frac{N}{A} = \frac{|\tilde{\varepsilon}(\omega)|^2 c}{\hbar\omega}. \tag{13.49}$$

Constructing the ratio of Eq. 13.37 to Eq. 13.49 at a given frequency we obtain the nonperturbative absorption spectrum:

$$\sigma(\omega) = -\frac{4\pi\omega}{c}\frac{\text{Im}(\tilde{P}_{ab}^*(\omega)\tilde{\varepsilon}(\omega))}{|\tilde{\varepsilon}(\omega)|^2}, \tag{13.50}$$

where the negative sign reflects the fact that absorption corresponds to negative ΔN_a.

▶ **Exercise 13.2**   Show Eq. 13.49.

Solution   According to classical electromagnetism, the energy E in the field is given by

$$E = \int dV \frac{\mathbf{E}^2 + \mathbf{B}^2}{8\pi}. \tag{13.51}$$

It is convenient to rewrite the volume element as $dV = Ac\,dt$, where A is the area of the beam and $c\,dt$ is the distance traversed by the beam in time dt. This gives

$$E = \int \frac{\mathbf{E}^2(t) + \mathbf{B}^2(t)}{8\pi} Ac\,dt. \tag{13.52}$$

Taking $\mathbf{E}^2 = (\varepsilon + \varepsilon^*)^2 = \varepsilon^2 + \varepsilon^{*2} + 2|\varepsilon|^2$, only the last term on the RHS survives the time integral since the other terms are rapidly varying. Noting that the average energies of the electric and magnetic fields are equal, we obtain

$$E = \int \frac{|\varepsilon(t)|^2}{2\pi} Ac\,dt = \int |\tilde{\varepsilon}(\omega)|^2 Ac\,d\omega, \tag{13.53}$$

where in the second relation we have used Parseval's theorem (Eq. 6.26). But by the Einstein relation we know that the energy of a single photon of frequency ω is given by $\hbar\omega$, and hence the total energy in the field at frequency ω is

$$E = \int N(\omega)\hbar\omega\,d\omega, \tag{13.54}$$

where $N(\omega)$ is the number of photons per unit frequency at frequency ω. Combining Eqs. 13.53 and 13.54 we find that the total number of photons impinging per unit area at frequency ω is given by

$$\frac{N(\omega)}{A} = \frac{|\tilde{\varepsilon}(\omega)|^2 c}{\hbar\omega}. \tag{13.55}$$

▶ **Exercise 13.3**   Because the absorption spectrum is a ratio, it is amenable to other interpretations. Show that the absorption spectrum can also be interpreted as the ratio of energy absorbed at frequency ω to incident energy per unit area at frequency ω.

13.3.3   Series Expansion of the Polarization

In the previous sections we have described the interaction of the electromagnetic field with matter, noting the way the material is affected by the presence of the field. But there is a second, reciprocal perspective: the excitation of the material by the electromagnetic field generates a dipole (polarization) where none existed previously. Over a sample of finite size this dipole is macroscopic, and serves as a new source term in Maxwell's equations. For weak fields, the source term, $\mathbf{P}$, is linear in the field strength,

$$\mathbf{P} = \chi\varepsilon, \tag{13.56}$$

where the proportionality constant χ, called the (linear) susceptibility, is generally frequency dependent and complex. The imaginary part of the linear susceptibility determines the absorption spectrum, as we shall see below, while the real part determines the dispersion, or refractive index, of the material. There is a universal relationship between the real part and the imaginary part of the linear susceptibility, known as the Kramers–Kronig relation, which establishes a relationship between the absorption spectrum and the frequency-dependent refractive index (see, for example, Mukamel 1995).

For stronger fields the relationship between the macroscopic polarization and the incident field is nonlinear. The general relation between $\mathbf{P}$ and ε is written as

$$\mathbf{P} = \chi^{(1)}\varepsilon + \chi^{(2)}\varepsilon : \varepsilon + \chi^{(3)}\varepsilon : \varepsilon : \varepsilon + \cdots \equiv \mathbf{P}^{(1)} + \mathbf{P}^{(2)} + \mathbf{P}^{(3)} + \cdots \quad (13.57)$$

The microscopic origin of χ and hence of $\mathbf{P}$ is the nonuniformity of the charge distribution in the medium. To lowest order this is given by the dipole moment, which in turn can be related to the dipole moments of the component molecules in the sample. Thus, on a microscopic quantum mechanical level we have the relation

$$\mathbf{P} = \rho^{(0)}\langle\psi|\mu|\psi\rangle = \rho^{(0)}\boldsymbol{P}. \quad (13.58)$$

where $\mathbf{P}$ is the polarization per unit volume, $\rho^{(0)}$ is the number of molecules per unit volume and $\boldsymbol{P}$ is the polarization per molecule. Assuming that the system has no permanent dipole moment, the existence of $\boldsymbol{P}$ depends on a nonstationary ψ induced by an external electric field. For weak fields, we may expand the polarization in orders of the perturbation. For simplicity, in what follows we will neglect the vectorial aspects of $\boldsymbol{P}$ and write

$$P(t) \equiv \langle\psi|\mu|\psi\rangle = P^{(0)}(t) + P^{(1)}(t) + P^{(2)}(t) + P^{(3)}(t) + \cdots. \quad (13.59)$$

We can then identify each term in the expansion with one or more terms in the perturbative expansion of $\psi(t)$:

$$P^{(0)}(t) \equiv \langle\psi^{(0)}(t)|\mu|\psi^{(0)}(t)\rangle \quad (13.60)$$

$$P^{(1)}(t) \equiv \langle\psi^{(0)}(t)|\mu|\psi^{(1)}(t)\rangle + \text{c.c.} \quad (13.61)$$

$$P^{(2)}(t) \equiv \langle\psi^{(0)}(t)|\mu|\psi^{(2)}(t)\rangle + \text{c.c.} + \langle\psi^{(1)}(t)|\mu|\psi^{(1)}(t)\rangle \quad (13.62)$$

and

$$P^{(3)}(t) \equiv \langle\psi^{(0)}(t)|\mu|\psi^{(3)}(t)\rangle + \text{c.c.} + \langle\psi^{(1)}(t)|\mu|\psi^{(2)}(t)\rangle + \text{c.c.}, \quad (13.63)$$

and so forth. Note that for an isotropic medium, terms of the form $P^{(2n)}(t)$ $(P^{(0)}(t), P^{(2)}(t),$ etc.) do not survive orientational averaging. For example, the first term, $\langle\psi^{(0)}|\mu|\psi^{(0)}\rangle$, is the permanent dipole moment, which gives zero when averaged over an isotropic medium. At an interface, however, such as between air and water, these even orders of $P(t)$ do not vanish, and in fact are sensitive probes of interface structure and dynamics.

The central dynamical objects that enter into the polarization are the coherences of the form $\langle\psi^{(0)}(t)|\mu|\psi^{(1)}(t)\rangle$ and $\langle\psi^{(1)}(t)|\mu|\psi^{(2)}(t)\rangle$, and so on. These quantities are overlaps between wavepackets moving on different potential energy surfaces (Tannor, 1985, 1987; Pollard, 1990; Lee, 1995): the instantaneous overlap of the wavepackets creates a non-vanishing transition dipole moment that interacts with the light. This view is appropriate both in the regime of weak fields, where perturbation theory is valid, and for strong fields, where perturbation theory is no longer valid. In the next section we will focus on the linear term $P^{(1)}$. After that we will explore the nonlinear term $P^{(3)}$. Since the $P^{(2)}$ vanishes in an isotropic medium, the $P^{(3)}$ term tends to be the central object of study in nonlinear optical spectroscopy.

13.3.4 Linear Response

In the perturbative regime one may decompose the coherences described in the previous section into a contribution from the field and a part that is intrinsic to the matter; the latter is referred to as the *response* function. For example, note that the expression $P_{01}^{(1)}(t) = \langle \psi^{(0)}(t)|\mu|\psi^{(1)}(t)\rangle$ is not simply an intrinsic function of the molecule: it depends on the functional form of the field, since $\psi^{(1)}(t)$ does. However, since the dependence on the field is linear it is possible to write $P_{01}^{(1)}$ as a *convolution* of the field with a response function that depends on the material. Using the definition of $\psi^{(1)}$,

$$\psi^{(1)} = \frac{1}{i\hbar} \int_{-\infty}^{t} e^{-iH_b(t-t')/\hbar}\{-\mu\varepsilon(t')\}e^{-iE_0t'/\hbar}\psi^{(0)}\, dt', \tag{13.64}$$

we find that

$$P_{01}^{(1)}(t) = \frac{1}{i\hbar} \int_{-\infty}^{t} \langle \psi^{(0)}(t)|\mu e^{-iH_b(t-t')/\hbar}\{-\mu\varepsilon(t')\}e^{-iE_0t'/\hbar}|\psi^{(0)}\rangle dt' \tag{13.65}$$

$$= \frac{i}{\hbar} \int_{0}^{\infty} \langle \psi^{(0)}(t)|\mu e^{-iH_b\tau/\hbar}\mu|\psi^{(0)}\rangle \varepsilon(t-\tau)\, d\tau. \tag{13.66}$$

Defining the *wavepacket correlation function*, $C_{00}(t)$, as

$$C_{00}(t) \equiv \langle \psi^{(0)}|\mu e^{-iH_b t/\hbar}\mu|\psi^{(0)}\rangle, \tag{13.67}$$

we find that $P_{01}^{(1)}(t)$ can be written as

$$P_{01}^{(1)}(t) = \frac{i}{\hbar}\{\varepsilon(t) \otimes S_{00}(t)\}, \tag{13.68}$$

where $S_{00}(t)$ is the half, or *causal*, form of the autocorrelation function:

$$S_{00}(t) = C_{00}(t) \quad t > 0 \tag{13.69}$$

$$= 0 \qquad t < 0 \tag{13.70}$$

and $\otimes$ signifies convolution.

By the Fourier convolution theorem (cf. Exercise 6.12),

$$\tilde{P}_{01}^{(1)}(\omega) \equiv \frac{1}{2\pi} \int_{-\infty}^{\infty} P_{01}^{(1)}(t)e^{i\omega t}\, dt = \left\{\frac{2\pi i}{\hbar}\tilde{\varepsilon}(\omega)\tilde{S}_{00}(\omega)\right\}, \tag{13.71}$$

where

$$\tilde{S}_{00}(\omega) = \frac{1}{2\pi} \int_{-\infty}^{\infty} S_{00}(t)e^{i\omega t}\, dt = \frac{1}{2\pi} \int_{0}^{\infty} S_{00}(t)e^{i\omega t}\, dt. \tag{13.72}$$

Substituting $\tilde{P}_{01}^{(1)}(\omega)$ into Eq. 13.50, we find that the linear absorption spectrum is given by

$$\sigma(\omega) = -\frac{4\pi\omega}{c}\frac{\text{Im}(\tilde{P}_{01}^{(1)*}(\omega)\tilde{\varepsilon}(\omega))}{|\tilde{\varepsilon}(\omega)|^2} \tag{13.73}$$

$$= -\frac{4\pi\omega}{c}\frac{\text{Im}(-\frac{2\pi i}{\hbar}\tilde{\varepsilon}^*(\omega)\tilde{S}_{00}^*(\omega)\tilde{\varepsilon}(\omega))}{\tilde{\varepsilon}^*(\omega)\tilde{\varepsilon}(\omega)} \tag{13.74}$$

$$= \frac{8\pi^2\omega}{\hbar c}\text{Re}\{\tilde{S}_{00}(\omega)\} = \frac{2\pi\omega}{\hbar c}\int_{-\infty}^{\infty} C_{00}(t)e^{i\omega t}\,dt \tag{13.75}$$

$$= \frac{2\pi\omega}{\hbar c}\int_{-\infty}^{\infty}\langle\psi^{(0)}|\mu e^{-iH_b t/\hbar}\mu|\psi^{(0)}\rangle e^{i\omega t}\,dt, \tag{13.76}$$

where we have used the property that $C_{00}(t) = C_{00}^*(-t)$. Equation 13.76 indicates that a first-order spectrum may be interpreted as the Fourier transform of a wavepacket autocorrelation function. This interpretation will play a central role in the next chapter on spectroscopy. Equation 13.76 will be derived there from a completely different viewpoint, as the rate of absorption of monochromatic radiation.

Equation 13.75 may be given an alternative interpretation in terms of the complex susceptibility. Comparing Eq. 13.71 with the definition of the susceptibility, χ (Eq. 13.56), and using Eq. 13.58, we see that

$$\chi^{(1)}(\omega) = \frac{2\pi i\rho^{(0)}}{\hbar}\tilde{S}_{00}(\omega). \tag{13.77}$$

Note that $\tilde{S}_{00}(\omega)/\hbar$ has dimensions of volume, while $\chi^{(1)}(\omega)$ is dimensionless. Using Eq. 13.77, Eq. 13.75 can be written as

$$\sigma(\omega) = \frac{4\pi\omega}{c\rho^{(0)}}\text{Im}\{\chi^{(1)}(\omega)\}, \tag{13.78}$$

establishing the statement above (at Eq. 13.56) that the absorption spectrum is related to the imaginary part of the susceptibility χ at frequency ω.

13.3.5   Coherent Anti-Stokes Raman Spectroscopy (CARS)

We now turn to the nonlinear material response, $P^{(3)}(t)$. Since $P^{(2)}$ vanishes in an isotropic medium, this is the most commonly encountered nonlinear term. The terms that contribute to this order are

$$P^{(3)}(t) = \langle\psi^{(0)}(t)|\mu|\psi^{(3)}(t)\rangle + \langle\psi^{(1)}(t)|\mu|\psi^{(2)}(t)\rangle + \text{c.c.} \tag{13.79}$$

The terms of the form

$$\langle\psi^{(0)}(t)|\mu|\psi^{(3)}(t)\rangle + \text{c.c.} \tag{13.80}$$

are responsible for the process of Coherent Anti-Stokes Raman Spectroscopy (CARS). Similarly, the terms of the form

$$\langle\psi^{(1)}(t)|\mu|\psi^{(2)}(t)\rangle + \text{c.c.} \tag{13.81}$$

are responsible for the process of stimulated emission or excited-state absorption spectroscopy.

We begin by considering the CARS process. To arrive at a dynamical interpretation of this process it is instructive to write the formula for the associated term in $P^{(3)}$ explictly (Tannor, 1985):

$$P^{(3)}(t) = \langle \psi^{(0)}(t)|\mu|\psi^{(3)}(t)\rangle + \text{c.c.} \tag{13.82}$$

$$= \frac{(-)^3}{(i\hbar)^3} \int_{-\infty}^{t_4} dt_3 \int_{-\infty}^{t_3} dt_2 \int_{-\infty}^{t_2} dt_1 \langle \psi^{(0)}(t')|\{\mu\}e^{-iH_b(t-t_3)/\hbar}\{\mu\varepsilon_3(t_3)\}e^{-iH_a(t_3-t_2)/\hbar}$$

$$\times \{\mu\varepsilon_2(t_2)\}e^{-iH_b(t_2-t_1)/\hbar}\{\mu\varepsilon_1(t_1)\}e^{-iH_a t_1}|\psi^{(0)}\rangle + \text{c.c.}, \tag{13.83}$$

where in the second line we have substituted explicitly for the third-order wavefunction, $\psi^{(3)}(t)$. Equation 13.83, although slightly longer than the expressions for the first- and second-order amplitude discussed in Sections 13.1–13.2, has the same type of simple dynamical interpretation. Reading Eq. 13.83 from right to left, we see that the initial wavepacket $\psi^{(0)}$ propagates on the ground surface, a, up until time t_1. At time t_1, the molecule interacts with the field and propagates on surface b for time $t_2 - t_1$; at time t_2 it interacts a second time with the field and propagates on the ground surface a for time $t_3 - t_2$; at time t_3 it interacts a third time with the field and propagates on surface b until variable time t. The third-order wavepacket on surface b is projected onto the initial wavepacket on the ground state; this overlap is a measure of the coherence that determines both the magnitude and phase of the CARS signal. Formally, the expression involves an integral over three time variables, reflecting the coherent contribution of all possible instants at which the interaction with the light took place, for each of the three interactions. However, if the interaction is with pulses that are short compared with a vibrational period, as we saw in Eq. 13.14, one can approximate the pulses by δ functions in time, and eliminate the three integrals, and the simple dynamical interpretation above becomes precise.

Qualitatively, the delay between interactions 1 and 2 is a probe of excited-state dynamics, while the delay between interactions 2 and 3 reflects ground state dynamics. If pulses 1 and 2 are coincident, the combination of the first two pulses prepares a vibrationally excited wavepacket on the ground state potential energy surface; the time delay between pulses 2 and 3 then determines the time interval for which the wavepacket evolves on the ground state potential, and is thus a probe of ground state dynamics (Tannor, 1985; Meyer, 2000; Pausch, 2000). If a second delay—the delay between pulses 1 and 2—is introduced, this allows large wavepacket excursions on the excited state before coming back to the ground state. The delay between pulses 1 and 2 can be used in a very precise way to tune the level of ground state vibrational excitation, and can prepare ground vibrational wavepackets with extremely high energy content (Knopp, 2000). The sequence of pulses is shown in Figure 13.10a. The wavepacket dynamics that results from the series of three pulses is illustrated in Figure 13.10b. According to Eq. 13.83, the amplitude of the time-dependent CARS signal is proportional to the overlap of the wavepacket on the excited electronic state potential after the third pulse with the initial vibrational state on the ground state surface.

13.3.6 Dynamic Absorption Spectrum, $\sigma^{(3)}(\omega)$

We now turn to excited-state absorption, arising from the terms in $P^{(3)}(t)$ of the form $\langle \psi^{(1)}(t)|\mu|\psi^{(2)}(t)\rangle + \text{c.c.}$ By analogy to the case of $P^{(1)}$, this term in $P^{(3)}$ can be written as (Pollard, 1990; Lee, 1995)

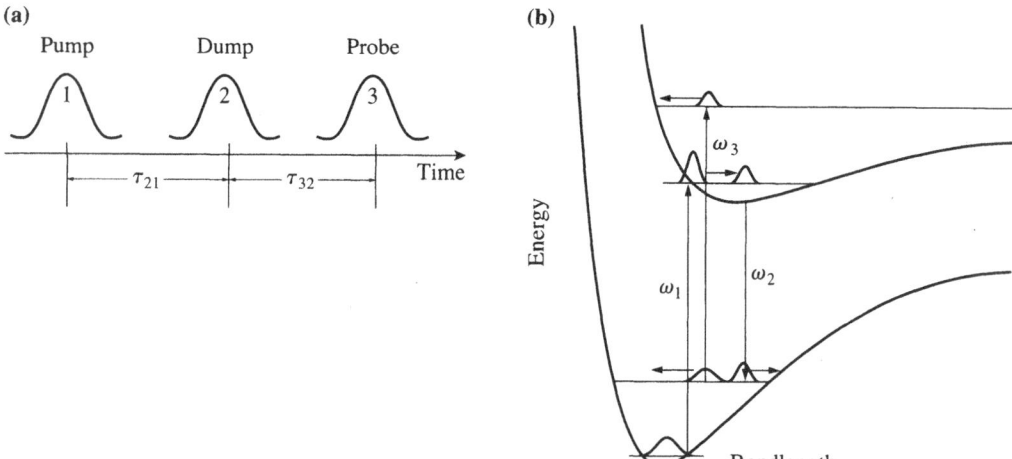

Figure 13.10 (a) Pulse sequence for two–time delay CARS spectroscopy ((TD)^{2}CARS). (b) Wave-packet creation and evolution generated by the three-pulse sequence in (a). The amplitude of the time-dependent CARS signal is proportional to the overlap, after the third pulse, of the wavepacket on the excited electronic state potential with the initial vibrational state on the ground state surface.

$$P^{(3)}(t) = \int_{-\infty}^{\infty} \langle \psi^{(1)}(t') | \mu e^{-i H_c(t-t')/\hbar} \mu | \psi^{(1)}(t') \rangle \varepsilon(t') \, dt'. \tag{13.84}$$

However, because of the t' dependence in $\psi^{(1)}(t')$ one cannot write that $P^{(3)} = \varepsilon(t) \otimes S_{11}(t)$. For the latter to hold, it is necessary to go to the limit of a probe pulse that is short compared with the dynamics on surface b. In this case, $\psi^{(1)}(t')$ is essentially frozen and we can write $\psi^{(1)} \approx \psi_\tau^{(1)}$, where we have indicated explicitly the parametric dependence on the pump–probe delay time, τ. Then

$$P^{(3)} = \varepsilon(t) \otimes S_{11}(t), \tag{13.85}$$

where

$$S_{11}(t) = C_{11}(t) \quad t > 0 \tag{13.86}$$
$$= 0 \quad t < 0 \tag{13.87}$$
$$C_{11}(t) = \langle \phi_\tau^{(1)} | e^{-i H_c t/\hbar} | \phi_\tau^{(1)} \rangle, \tag{13.88}$$

and

$$|\phi_\tau^{(1)}\rangle \equiv \mu | \psi_\tau^{(1)} \rangle. \tag{13.89}$$

Therefore,

$$\sigma^{(3)}(\omega) = 2\omega \mathrm{Re} \int_0^{\infty} dt \, e^{i\omega t} C_{11}(t) = \omega \int_{-\infty}^{\infty} dt \, e^{i\omega t} C_{11}(t). \tag{13.90}$$

The correlation function $C_{11}(t)$ represents the overlap of the state $\phi_\tau^{(1)}$, propagated on surface c, with the unpropagated state $\phi_\tau^{(1)}$ (see Figure 13.11). This is just first-order spectroscopy on the frozen state, $\phi_\tau^{(1)}$, on surface c. Note the residual dependence of the frozen state on τ, the pump–probe delay: different values of τ define different states, $\phi_\tau^{(1)}$.

(a) Stimulated emission

(b) Excited state absorption

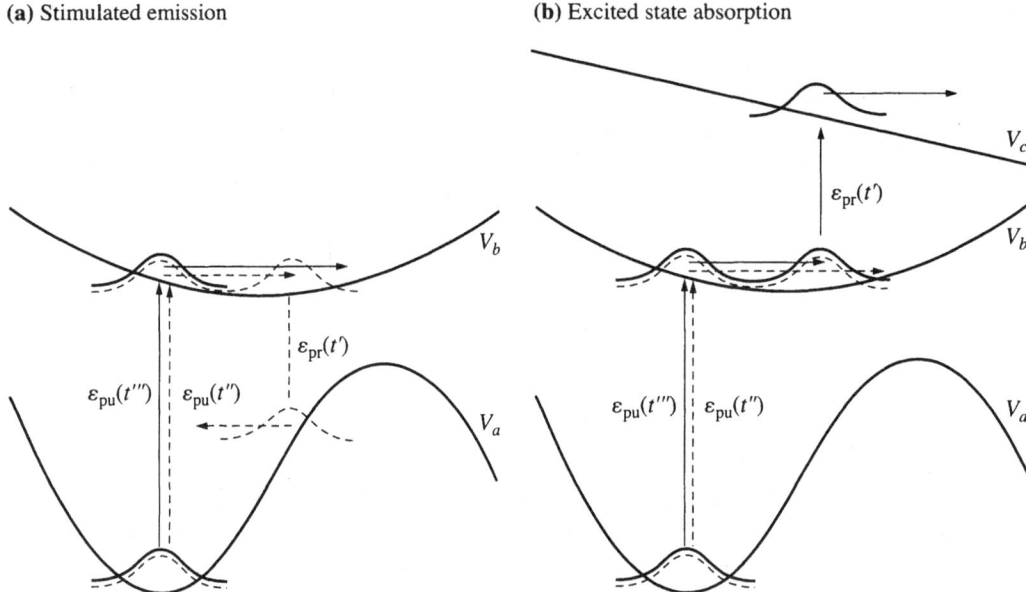

Figure 13.11 Bra and ket wavepacket dynamics that determine the coherence overlap, $\langle \phi^{(1)} | \phi^{(2)} \rangle$. Horizontal arrows indicate free propagation on the potential surface. Solid lines are used for the ket wavepacket, while dashed lines indicate the bra wavepacket. (a) Stimulated emission. (b) Excited-state (transient) absorption. Adapted from Lee (1995).

Thus, variation of the variables (ω, τ) generates a two-dimensional dynamic absorption spectrum. It is important to note that the pair of variables (ω, τ) is not limited by some form of time–energy uncertainty principle. This is because, although the absorption is finished when the probe pulse is finished, the spectral analysis of which frequency components were absorbed depends on the full time evolution of the system, beyond its interaction with the probe pulse. Thus, the dynamic absorption signal can give high resolution both in time (i.e., time delay between pump and probe pulses) and frequency, simultaneously.

As with the first-order spectrum, one may integrate the dynamic absorption spectrum over frequency to obtain the integrated spectrum:

$$\Delta I^{(3)} = \int_{-\infty}^{\infty} \omega \mathrm{Im}\{\tilde{\varepsilon}_{\mathrm{pr}}(\omega) \tilde{P}^{(3)}(\omega)\} d\omega, \tag{13.91}$$

where $\tilde{\varepsilon}_{\mathrm{pr}}(\omega)$ is the probe pulse in the frequency domain.

If one first removes the ω prefactor from the absorption and then integrates, one obtains the total photon loss from pump + probe, minus the photon loss from probe alone:

$$\Delta N^{(3)} = \int_{-\infty}^{\infty} d\omega \, \mathrm{Im}\{\tilde{\varepsilon}_{\mathrm{pr}}(\omega) \tilde{P}^{(3)}(\omega)\}. \tag{13.92}$$

▶ **Exercise 13.4** Show that, analogous to the first-order case, $\Delta N^{(3)}$ is proportional to $\langle \psi^{(2)} | \psi^{(2)} \rangle$.

Femtosecond pump–probe dynamic absorption spectroscopy has recently been applied to systems of biological significance. A primary example has been the application of pump–

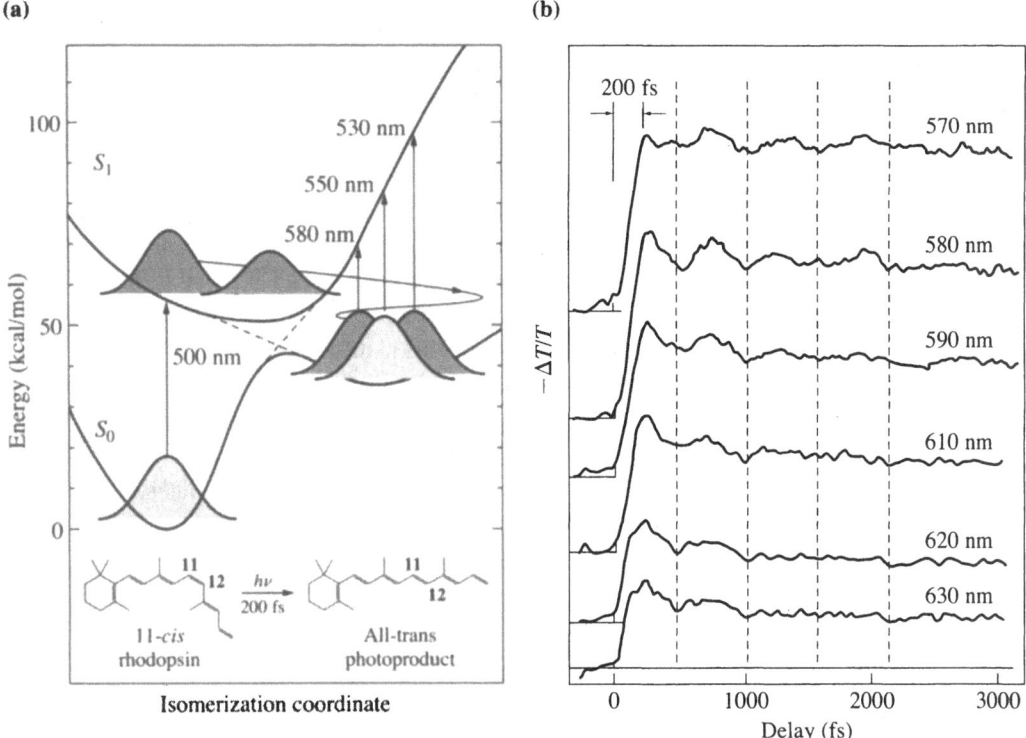

Figure 13.12 (a) Wavepacket dynamics on the first excited singlet state of the visual pigment, rhodopsin. (b) Dynamic absorption spectrum of rhodopsin. Reprinted from Wang (1994).

probe techniques to study the first step in the vision process, the isomerization of the visual pigment rhodopsin from 11-*cis* to 11-*trans* (Wang, 1994). The dynamic absorption spectrum shows a pronounced spectral shift, and coherent oscillations at times less than 200 fs following excitation (Figure 13.12). The authors infer from the ultrashort dynamic absorption spectrum that the isomerization takes place in under 200 fs.

13.4   Density Operator Formulation of Optical Perturbations

There is an alternative formulation of pump–probe spectroscopy that is based on a perturbative expansion for the density operator, instead of for the wavefunction (Bloembergen, 1965; Shen, 1984). We note that the populations and coherences described in the previous sections may be expressed conveniently using the density operator. For example, the populations may be written as

$$\langle \psi^{(1)} | \psi^{(1)} \rangle = \mathrm{Tr}(\rho_{bb}^{(2)}) \tag{13.93}$$

and the coherences as

$$\langle \psi^{(0)} | \mu | \psi^{(1)} \rangle = \mathrm{Tr}(\mu \rho_{ba}^{(1)}). \tag{13.94}$$

The main reason for giving up the simplicity of the wavefunction formulation described in previous sections is to incorporate the effects of dissipative processes; since dissipation destroys the purity of the state, these effects are included much more naturally in the density operator formulation. The density operator approach distinguishes between certain processes involving an overlap of wavepacket bra and ket, depending on whether the interaction with the transition dipole took place first with the ket and then with the bra, or vice versa. The distinction between these processes is highlighted by a change in integration variables, to the time *differences* between bra and ket interactions, or, more generally, left/right interactions of the field with the density operator. Depending on whether the time difference between these interactions is greater or less than zero the process is given a different name and a different physical interpretation. Specifically, in the presence of relaxation processes (dephasing and population relaxation) different time orderings of left/right interactions lead to different propagation intervals for diagonal versus off-diagonal elements of the density operator, which in turn lead to different durations for which population versus coherence relaxation processes are in effect. In order to treat consistently the time intervals for which these processes act, one must synchronize the time variables in bra and ket, that is, have a single time variable for the density operator (Mukamel, 1995).

As an example, consider the third-order density matrix element

$$\rho_{ab}^{(3)} \equiv |\psi_a^{(2)}\rangle\langle\psi_b^{(1)}| + |\psi_a^{(0)}\rangle\langle\psi_b^{(3)}|. \tag{13.95}$$

The first term may be written explicitly as

$$|\psi_a^{(2)}\rangle\langle\psi_b^{(1)}|$$

$$= \frac{1}{(i\hbar)^2} \int_{-\infty}^t dt' \int_{-\infty}^{t'} dt'' e^{-iH_a(t-t')}\{-\mu_{ab}\varepsilon(t')\}e^{-iH_b(t'-t'')}\{-\mu_{ba}\varepsilon(t'')\}e^{-iH_a t''}|\psi_a^{(0)}\rangle$$

$$\times \frac{1}{i\hbar} \int_{-\infty}^t dt''' \langle\psi_a^{(0)}|e^{iH_a t'''}\{-\mu_{ab}\varepsilon(t''')\}e^{iH_b(t-t''')} \tag{13.96}$$

$$= -\frac{1}{(i\hbar)^3} \int_{-\infty}^t dt' \int_{-\infty}^{t'} dt'' \int_{-\infty}^t dt''' \varepsilon(t')\varepsilon(t'')\varepsilon(t''')$$

$$\times e^{-iH_a(t-t')}\mu_{ab}e^{-iH_b(t'-t'')}\mu_{ba}e^{-iH_a t''}|\psi_a^{(0)}\rangle\langle\psi_a^{(0)}|e^{iH_a t'''}\mu_{ab}e^{iH_b(t-t''')}, \tag{13.97}$$

where here and for the remainder of this section, we take the field, $\varepsilon(t)$, to be real. This expression for $|\psi_a^{(2)}\rangle\langle\psi_b^{(1)}|$ is a triple integral over interaction times t', t'' and t'''. In the wavefunction formulation that we have used to construct this density matrix element, t'' is restricted to come before t' (these are the two interaction times for the ket), but the interaction time for the bra, t''', is unrestricted relative to these ket times, as long as it comes before the final time, t. As a result, one may distinguish three different contributions to this triple integral, depending on whether the interaction time with the bra, t''', comes before, in between or after the two interactions with the ket. The second term, $|\psi_a^{(0)}\rangle\langle\psi_b^{(3)}|$, may also be written as a triple integral over time; however, since all the interaction times are with the ket no separate contributions of the type above may be distinguished.

These three different contributions are shown schematically in Figure 13.13. The first two of these processes may be classified as stimulated emission: since the bra and ket are

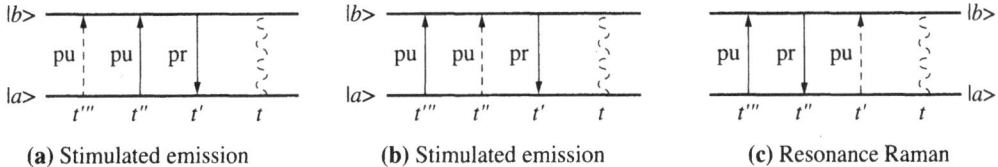

Figure 13.13 Time-ordered energy-level ladder diagrams representing the three terms in the density matrix $\rho_{ab}^{(3)}$. The solid and dashed vertical arrows mark the transitions of the ket and bra wavefunctions, respectively, between the electronic surfaces. Time moves forward from left to right. The first two diagrams correspond to stimulated emission while the third corresponds to resonance Raman. Adapted from D. Lee and A. C. Albrecht (1985).

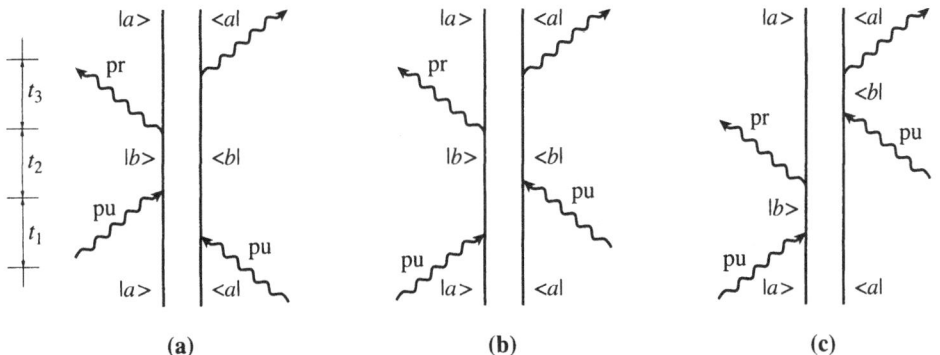

Figure 13.14 A representation equivalent to that in Figure 13.13, using double-sided Feynman diagrams to show the interaction time of the ket (left) and the bra (right). Time moves forward from down to up. Note that the three diagrams here are in a one-to-one correspondence with those in Figure 13.13.

both excited before the third interaction takes place, there is a first-order *population* on the excited state when the third interaction impinges. The third of these processes may be classified as resonance Raman. It is different from the previous terms in the sense that the first two interactions with the field are both with the ket; since the third interaction occurs when there is no population on the excited state, it is a virtual process of the Raman type. The physical significance of the distinction between these processes will be really meaningful only when we introduce decay processes below. There we will see that the Raman type processes are not subject to population decay factors, since no intermediate population is created, while the stimulated emission (or resonance fluorescence type processes, if the emission is spontaneous) are subject to this form of decay. An alternative depiction of the different processes is shown Figure 13.14. This is a generalization of the Feynman diagrams we saw earlier; such diagrams are referred to as double-sided, or *dual*, Feynman diagrams.

Mathematically, the separate contributions to the triple integral, Eq. 13.97, emerge automatically by developing the perturbation theory expansion to third order explicitly in terms of the density matrix:

$$\rho_{ba}^{(1)}(t) = \frac{i}{\hbar} \int_{-\infty}^{t} dt' e^{-\gamma_{ba}(t-t')} e^{-iH_b(t-t')/\hbar} \mu_{ba} \varepsilon(t') |\psi_a\rangle\langle\psi_a| e^{iH_a(t-t')/\hbar} \tag{13.98}$$

$$\rho_{bb}^{(2)}(t) = \frac{i}{\hbar} \int_{-\infty}^{t} dt' e^{-\gamma_{bb}(t-t')} e^{-iH_b(t-t')/\hbar} \left(\mu_{ba}\varepsilon(t')\rho_{ab}^{(1)}(t') - \rho_{ba}^{(1)}(t')\mu_{ab}\varepsilon(t') \right) e^{iH_b(t-t')/\hbar}$$

$$\tag{13.99}$$

$$\rho_{aa}^{(2)}(t) = \frac{i}{\hbar} \int_{-\infty}^{t} dt' e^{-\gamma_{aa}(t-t')} e^{-iH_a(t-t')/\hbar} \left(\mu_{ab}\varepsilon(t')\rho_{ba}^{(1)}(t') - \rho_{ab}^{(1)}(t')\mu_{ba}\varepsilon(t') \right) e^{iH_a(t-t')/\hbar}$$

$$\tag{13.100}$$

$$\rho_{ab}^{(3)}(t) = \frac{i}{\hbar} \int_{-\infty}^{t} e^{-\gamma_{ba}(t-t')} e^{-iH_a(t-t')/\hbar} \left(\mu_{ab}\varepsilon(t')\rho_{bb}^{(2)}(t') - \rho_{aa}^{(2)}(t')\mu_{ab}\varepsilon(t') \right) e^{iH_b(t-t')/\hbar}.$$

$$\tag{13.101}$$

We have introduced the electronic dephasing rate, γ_{ba}, and the electronic population decay rate, γ_{bb}. The electronic population decay rate, γ_{aa}, is included for didactic purposes, but should be thought of as having magnitude zero. Aside from these decay rates, the expression for $\rho_{ab}^{(3)}(t)$ is identical to the expression above, Eq. 13.97. To see the correspondence explicitly, note that since $\rho_{aa}^{(2)}(t)$ and $\rho_{bb}^{(2)}(t)$ each consist of two terms, $\rho_{ab}^{(3)}(t)$ consists of four terms. One of these terms, which can be described as having all three interactions with $\mu\varepsilon$ on the left, corresponds to the term $|\psi_a^{(0)}\rangle\langle\psi_b^{(3)}|$. The other three terms, which may be described as having interaction sequences of the form right-left-left, left-right-left and left-left-right, correspond to the partitioning of the term $|\psi_a^{(2)}\rangle\langle\psi_b^{(1)}|$ into the three contributions described above.

Why does this partitioning emerge naturally in the density matrix formulation? Since the density matrix treats bra and ket as a single, combined entity, there is only a single time line for the interaction of the field with the density matrix. The price for having this single, consistent time for bra/ket evolution is a proliferation of terms, reflecting different sequences of left/right interactions. While this proliferation is cumbersome, it is precisely the partitioning we sought, each term corresponding to a different sequence of interactions with bra and ket (or more generally, left and right applications of the field to the density operator).

Writing all terms explicitly, and using the time difference variables, $t_1 = t - t'$, $t_2 = t' - t''$ and $t_3 = t'' - t'''$, we can write $\rho_{ab}^{(3)}$ as

$$\rho_{ab}^{(3)} = |\psi_a^{(2)}\rangle\langle\psi_b^{(1)}| + |\psi_a^{(0)}\rangle\langle\psi_b^{(3)}| \tag{13.102}$$

$$= -\frac{1}{(i\hbar)^3} \int_0^{\infty} dt_3 \int_0^{\infty} dt_2 \int_0^{\infty} dt_1 \, \varepsilon(t-t_3)\varepsilon(t-t_2-t_3)\varepsilon(t-t_1-t_2-t_3)$$

$$\times \left\{ e^{-\gamma_{ba}t_3} e^{-\gamma_{bb}t_2} e^{-\gamma_{ba}t_1} \left\{ e^{-iH_a t_3/\hbar} \mu_{ab} e^{-iH_b t_2/\hbar} \mu_{ba} e^{-iH_a t_1/\hbar} \rho^{(0)} \mu_{ab} e^{iH_b(t_1+t_2+t_3)/\hbar} \right. \right.$$

$$\left. + e^{-iH_a t_3/\hbar} \mu_{ab} e^{-iH_b(t_1+t_2)/\hbar} \mu_{ba} \rho^{(0)} e^{iH_a t_1/\hbar} \mu_{ab} e^{iH_b(t_2+t_3)/\hbar} \right\}$$

$$+ e^{-\gamma_{ba}t_3} e^{-\gamma_{aa}t_2} e^{-\gamma_{ba}t_1} \left\{ e^{-iH_a(t_2+t_3)/\hbar} \mu_{ab} e^{-iH_b t_1/\hbar} \mu_{ba} \rho^{(0)} e^{iH_a(t_1+t_2)/\hbar} \mu_{ab} e^{iH_b t_3/\hbar} \right.$$

$$\left. \left. + e^{-iH_a(t_1+t_2+t_3)/\hbar} \rho^{(0)} \mu_{ab} e^{iH_b t_1/\hbar} \mu_{ba} e^{iH_a t_2/\hbar} \mu_{ab} e^{iH_b t_3/\hbar} \right\} \right\}$$

$$\tag{13.103}$$

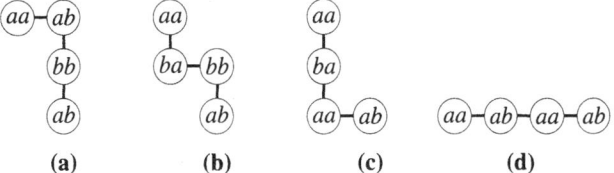

Figure 13.15 The four pathways in Liouville space that correspond to the processes of stimulated emission ((a) and (b)), resonance Raman (c), and CARS (d). Interactions of the density matrix with the field from the right (left) is signified by a horizontal (vertical) step. Note the one-to-one correspondence of (a)–(c) with the diagrams in Figures 13.13 and 13.14. The advantage to the Liouville lattice representation is that populations are clearly identified as diagonal lattice points, while coherences are off-diagonal points. This allows immediate identification of the processes that are subject to population decay.

Three comments about Eq. 13.103:

1. Note that t_1, t_2 and t_3 run over positive times only. This constraint is equivalent to the requirement that $t''' < t'' < t' < t$. This provides another perspective on the emergence of three terms, where previously there had been only one: the triple integral must be subdivided so that each term respects this definition of time ordering.

2. Notice that the inclusion of the electronic population relaxation and dephasing factors is expressed very naturally in terms of the time difference variables, reflecting the fact that physically it is the time *interval* between interactions that determines the duration for which a population relaxation or a dephasing process acts.

3. Finally, note that we have abandoned the wavefunction bra/ket notation completely. Indeed, under relaxation processes of a general nature, one cannot even speak of bras and kets any longer, and the only time variables of any significance are those that refer to the time line of the density operator. The generalization of sequences of interactions with the bra and ket is now sequences of left versus right interactions with the density operator: different time sequences of left/right interactions determine whether coherences or populations are formed, and are referred to as different pathways in "Liouville space."

The different pathways in Liouville space are conveniently diagrammed on a lattice, where right interactions are horizontal steps and left interactions are vertical (Mukamel, 1995). The diagonal vertices represent populations and the off-diagonal vertices are coherences. The three different time orderings for contributions to $|\psi_a^{(2)}\rangle\langle\psi_b^{(1)}|$, Eq. 13.96, correspond to the three Liouville pathways shown in Figure 13.15. From such a diagram one sees at a glance which pathways pass through intermediate populations (diagonal vertices) and hence are governed by population decay processes, and which pathways do not.

As an example, we return to Eq. 13.103. With the lattice representation of Liouville pathways in hand, we are in a better position to appreciate the distinction between the labels for the different terms in Figure 13.13. We will consider the important case of *spontaneous* light emission, where the processes refer to resonance fluorescence (instead of stimulated emission) and resonance Raman. As discussed above, the former process is associated with some intermediate time electronic population relaxation process, while the latter is subject only to electronic dephasing processes. This is clear from the diagrams for the different Liouville pathways in Figure 13.15. Resonance fluorescence follows pathways (a) and (b); that is, each of these pathways includes an intermediate vertex that is diagonal (a population) after the second interaction, and hence is subject to population decay processes. Resonance Raman scattering, which follows pathway (c), includes only off-diagonal intermediate vertices (coherences), and hence is subject only to dephasing processes. (Note that the converse is not true: resonance fluorescence pathways pass through off-diagonal intermediate vertices and hence are subject to dephasing during a portion of their intermediate time evolution.)

▶ **Exercise 13.5**   Use the bra–ket time line, the double-sided Feynman diagram and the Liouville lattice representation to diagram all the pathways that contribute to excited-state absorption. Discuss how an energy separation between surfaces *b* and *c* with a different transition frequency than between *a* and *b* would discriminate against some of these diagrams, if the carrier frequency of the pump and probe pulses were identical.

It is worth pointing out that there is a growing literature involving numerically accurate propagation of the density operator that is nonperturbative in the field interaction. In these methods the consistency of the order of interaction with the field and the appropriate relaxation process is achieved automatically (Seidner, 1995).

References

General

1. *Femtochemistry*, ed. A. H. Zewail, vols. 1 and 2 (World Scientific, Singapore, 1994).
2. *Femtosecond Chemistry*, eds. J. Manz and L. Wöste (VCH, Heidelberg, 1995).
3. *Nobel Symposium on Femtochemistry*, Stockholm, 1996, ed. V. Sundström (Imperial College Press).
4. G. R. Fleming, *Chemical Applications of Ultrafast Spectroscopy* (Oxford University Press, New York, 1986).
5. D. J. Tannor, Interaction of light with matter: A coherent perspective, in *Encyclopedia of Chemical Physics and Physical Chemistry*, eds. J. H. Moore and N. D. Spencer (Institute of Physics, Philadelphia, 2000).

Wavepacket Interferometry

6. N. F. Scherer, R. J. Carlson, A. Matro, M. Du, A. J. Ruggiero, V. Romero-Rochin, J. A. Cina, G. R. Fleming and S. A. Rice, Fluorescence-detected wavepacket interferometry: Time resolved molecular spectroscopy with sequences of femtosecond phase-locked pulses, J. Chem. Phys. 95, 1487 (1991).
7. N. F. Scherer, A. Matro, L. D. Ziegler, M. Du, J. A. Cina and G. R. Fleming, Fluorescence-detected wavepacket interferometry. II. Role of rotations and determination of the susceptibility, J. Chem. Phys. 96 4180 (1992).

8. V. Engel and H. Metiu, Two-photon wavepacket interferometry, J. Chem. Phys. 100, 5448 (1994).

9. V. Blanchet, M. A. Bouchene and B. Girard, Temporal coherent control in Cs_2: Theory and experiment, J. Chem. Phys. 108, 4862 (1997).

Pump and Probe Spectroscopy

10. A. H. Zewail, Laser femtochemistry, Science 242, 1645 (1988).

11. D. J. Tannor and S. A. Rice, Coherent pulse sequence control of product formation in chemical reactions, Adv. Chem. Phys. 70, 441 (1988).

12. V. Engel, H. Metiu, R. Almeida, R. A. Marcus and A. H. Zewail, Molecular state evolution after excitation with an ultrashort laser pulse: A quantum analysis of NaI and NaBr dissociation, Chem. Phys. Lett. 152, 1 (1988).

13. T. Baumert, V. Engel, C. Röttgermann, W. T. Strunz and G. Gerber, Femtosecond pump–probe study of the spreading and recurrence of a vibrational wavepacket in Na_2, Chem. Phys. Lett. 191, 639 (1992).

14. Ch. Meier and V. Engel, Electron kinetic energy distributions from multiphoton ionization of Na_2 with femtosecond laser pulses, Chem. Phys. Lett. 212, 691 (1993).

15. Q. Wang, R. W. Schoenlein, L. A. Peteanu, R. A. Mathies and C. V. Shank, Vibrationally coherent photochemistry in the femtosecond primary event of vision, Science 266, 422 (1994).

16. S. O. Williams and D. G. Imre, Determination of real-time dynamics in molecules by femtosecond laser excitation, J. Phys. Chem. 92, 6648 (1988).

17. M. Shapiro, Real-time dependence of photodissociation and continuum Raman experiments, J. Phys. Chem. 97, 7396 (1993).

18. Ch. Meier and V. Engel, Pump-probe ionization spectroscopy of a diatomic molecule, in *Femtosecond Chemistry*, eds. J. Manz and L. Wöste (VCH, Heidelberg, 1995), Chapter 11.

19. G. Haran, E. A. Morlino, J. Matthes, R. H. Callender and R. M. Hochstrasser, Femtosecond polarized pump-probe and stimulated emission spectroscopy of the isomerization reaction of rhodopsin, J. Phys. Chem. A 103, 2202 (1999).

20. B. Hou, N. Friedman, S. Ruhman, M. Sheves and M. Ottolenghi, Ultrafast spectroscopy of the protonated Schiff bases of free and $C_{13}=C_{14}$ locked retinals, J. Phys. Chem. B 105, 7042 (2001).

Nonlinear Spectroscopy

21. N. Bloembergen, *Nonlinear Optics* (Benjamin-Cummings, Reading, MA 1965).

22. Y. R. Shen, *The Principles of Nonlinear Optics* (Wiley, New York, 1984).

23. R. W. Boyd, *Nonlinear Optics* (Academic Press, San Diego, 2003), 2nd edition.

24. S. Mukamel, *Principles in Nonlinear Optical Spectroscopy* (Oxford University Press, New York, 1995).

25. D. Lee and A. C. Albrecht, A unified view of Raman, resonance Raman, and fluorescence spectroscopy (and their analogues in two-photon absorption) in *Advances in Infrared and Raman Spectroscopy*, vol. 12, p. 179, eds. R. J. H. Clark and R. E. Hester (Wiley Heyden, London, 1985).

26. D. J. Tannor, S. A. Rice and P. M. Weber, Picosecond CARS as a probe of ground electronic state intramolecular vibrational redistribution, J. Chem. Phys. 83, 6158 (1985).

27. D. J. Tannor and S. A. Rice, Photon echoes in multilevel systems, in *Understanding Molecular Properties*, eds. J. Avery et al. (Reidel, 1987), p. 205.

28. W. T. Pollard, S.-Y. Lee and R. A. Mathies, Wavepacket theory of dynamic absorption spectra in femtosecond pump-probe experiments, J. Chem. Phys. 92, 4012 (1990).

29. S.-Y. Lee, Wavepacket model of dynamic dispersed and integrated pump-probe signals in femtosecond transition-state spectroscopy, in *Femtosecond Chemistry*, eds. J. Manz and L. Wöste (VCH, Heidelberg, 1995).

30. S. Meyer and V. Engel, Femtosecond time-resolved CARS and DFWM spectroscopy on gas-phase I_2: A wavepacket description, J. Raman Spectrosc. 31, 33 (2000).

31. G. Knopp, I. Pinkas and Y. Prior, Two-dimensional time-delayed coherent anti-Stokes Raman spectroscopy and wavepacket dynamics of high ground-state vibrations, J. Raman Spectrosc. 31, 51 (2000).

32. R. Pausch, M. Heid, T. Chen, H. Schwoerer and W. Kiefer, Quantum control by stimulated Raman scattering, J. Raman Spectrosc. 31, 7 (2000).

33. I. Pastirk, E. J. Brown, B. I. Grimberg, V. V. Lozovoy and M. Dantus, Sequences for controlling laser excitation with femtosecond three-pulse four-wave mixing, Faraday Discuss. 113, 401 (1999).

34. J. Faeder, I. Pinkas, G. Knopp, Y. Prior and D. J. Tannor, Vibrational polarization beats in femtosecond CARS: A signature of dissociative pump-dump-pump wavepacket dynamics, J. Chem. Phys. 115, 8440 (2001).

35. T. Joo, Y. Jia, J.-Y. Yu, M. J. Lang and G. R. Fleming, Third-order nonlinear time domain probes of solvation dynamics, J. Chem. Phys. 104, 6089 (1996).

36. S. M. Gallagher Faeder and D. Jonas, Two-dimensional electronic correlation and relaxation spectra: Theory and model calculations, J. Phys. Chem. A 102, 10489 (1999).

37. M. Ovchinnikov, V. A. Apkarian and G. A. Voth, Semiclassical molecular dynamics computations of spontaneous light emission in the condensed phase: Resonance Raman spectra, J. Chem. Phys. 114, 7130 (2001).

38. M. T. Zanni, N. H. Ge, Y. S. Kim and R. M. Hochstrasser, Two-dimensional IR spectroscopy can be designed to eliminate the diagonal peaks and expose only the cross peaks needed for structure determination, Proc. Natl. Acad. Sci. USA 98, 11265 (2001).

39. L. Seidner, G. Stock and W. Domcke, Nonperturbative approach to femtosecond spectroscopy—General theory and application to multidimensional nonadiabatic photoisomerization processes, J. Chem. Phys. 103, 4002 (1995).

40. R. Kosloff and M. A. Ratner, Beyond linear response: Line shapes for coupled spins or oscillators via direct calculation of dissipated power, J. Chem. Phys. 80, 2352 (1984).

41. R. Kosloff, A. D. Hammerich and D. Tannor, Excitation without demolition: Radiative excitation of ground-surface vibration by impulsive stimulated Raman scattering with damage control, Phys. Rev. Lett. 69, 2172 (1992).

42. G. Ashkenazi, U. Banin, A. Bartana, S. Ruhman and R. Kosloff, Quantum description of the impulsive photodissociation dynamics of I_3^- in solution, Adv. Chem. Phys. 100, 229 (1997).

Chapter 14

One- and Two-Photon Electronic Spectroscopy

We now possess the tools to discuss one- and two-photon spectroscopies such as electronic absorption spectroscopy and resonance Raman spectroscopy. The basic formulas can be applied, with only minor modification, to a wide range of spectroscopies. For example, the basic one-photon formula can be used to describe electronic absorption, emission, photoionization. and photodetachment/transition state spectroscopy. Similarly, the same basic two-photon formula describes resonance Raman scattering, two-photon absorption and dispersed fluorescence spectroscopy. The focus of this chapter will be on the formulation and the interpretation of this wide range of spectroscopies in terms of the Fourier transform of wavepacket time-correlation functions. However, before proceeding with the time-domain formulation, a brief orientation will be given to these different spectroscopies.

The conventional approach to electronic absorption spectroscopy is cast in the energy, rather than in the time, domain. The conventional formula for the vibronic (vibrational + electronic) absorption spectrum is

$$\sigma(\omega_I) = \frac{4\pi^2 \omega_I}{3\hbar c} \sum_n |\langle \psi_n | \mu | \psi_i \rangle|^2 \delta(\omega_I + \omega_i - \omega_n), \tag{14.1}$$

where ω_I is the incident light frequency and ω_i is the frequency of the initial state—the initial vibrational level in the initial electronic state. The coefficients of the δ function in the sum are called *Franck–Condon* factors, and reflect the overlap of the initial state with the excited state ψ_n at energy $E_n = \hbar\omega_n$ (see Figure 14.1). Formally, Eq. 14.1 gives a "stick" spectrum of the type shown in Figure 14.1b.

If the excited electronic state is unbound, the final states are called predissociative (Figure 14.2a). A typical electronic absorption spectrum for a predissociative state is shown in Figure 14.2b. Note the diffuse vibronic spectrum. The smoothing out of the vibronic spectrum is the result of the coupling of the quasibound states with the continuum.

Electronic absorption can also take place from vibrationally excited initial states. In this case, the vibrational progression generally no longer consists of a single mode; it is multimodal, with the number of modes in the spectrum equal to the number of *nodes* in the initial vibrational states plus one.

The above formula for the absorption spectrum can be applied, with minor modifications, to other one-photon spectroscopies, such as photoionization and photodetachment spectroscopy, as well as emission spectroscopy. Photoionization spectroscopy refers to electronic absorption, generally of a neutral molecule to produce a positive ion: $A + h\nu \rightarrow A^+ + e^-$. In photoionization spectroscopy, normally a single excitation frequency is used

(a)   **(b)**

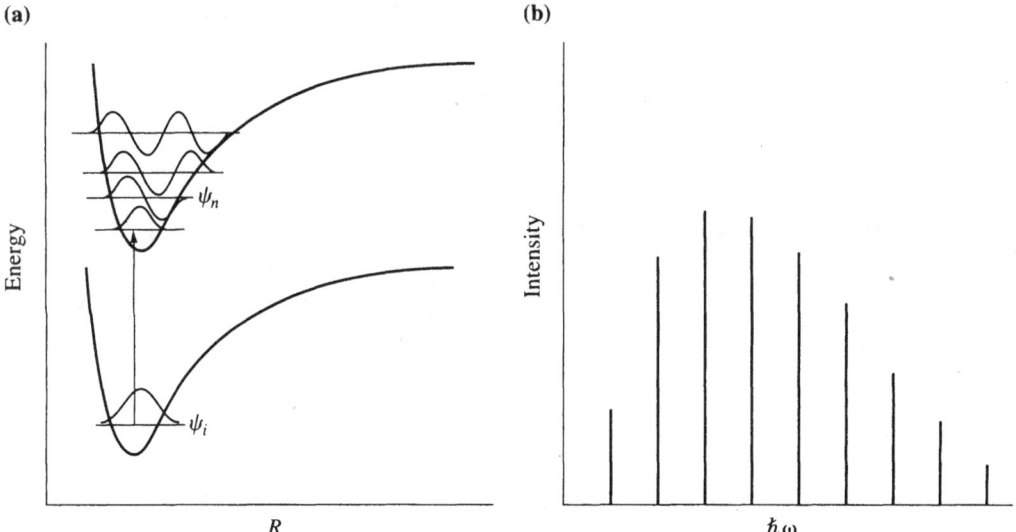

Figure 14.1 (a) Potential energy curves for two electronic states. The vibrational wavefunctions of the excited electronic state and for the lowest level of the ground electronic state are shown superimposed. (b) Stick spectrum representing the Franck–Condon factors (the squares of the overlap integrals) between the vibrational wavefunction of the ground electronic state and the vibrational wavefunctions of the excited electronic state.

and the distribution of kinetic energies of the exiting electrons is measured. By conservation of energy, the energy of the positive ion can be deduced, and thus the spectrum of energies of the exiting electrons can be interpreted as the spectrum of the positive ion. As in the case with the electronic absorption spectrum, the photoionization spectrum generally shows pronounced vibrational features corresponding to the vibrational levels in the ground electronic state of the positive ion. Photodetachment spectroscopy is very similar to photoionization spectroscopy, except that the initial species corresponds to a negative ion and the final vibrational states are associated with the neutral molecule: $A^- + h\nu \rightarrow A + e^-$. Since much of conventional chemistry involves reactions among neutral species, photodetachment spectroscopy is a particularly valuable probe, allowing access to highly excited vibrational levels of the neutral that is ordinarily not possible in one-photon experiments. As will be discussed below, in some of the most interesting examples the neutral molecule is actually unstable, corresponding to a transient intermediate in a reaction of other species.

Electronic emission spectroscopy refers to either spontaneous or stimulated emission from an excited electronic state back to the ground electronic state. For spontaneous emission, the prefactor of ω_I in Eq. 14.1 is replaced by ω_S^3, where ω_S is the frequency of the spontaneous emission, whereas for stimulated emission, ω_I is replaced by ω_S. The extra factor of ω_S^2 in spontaneous emission is due to the density of states of vacuum field states that induce the spontaneous emission, which increases quadratically with frequency. Note that in emission spectroscopy the roles of the ground and excited-state potential energy surfaces are reversed: the initial wavepacket starts from the vibrational state $i = v'$ of the *excited* electronic state and makes transitions to $n = v''$ on the *ground* electronic state. (We use "i" to represent the initial state and "n" to represent the final state, to highlight the anal-

(a)

(b)

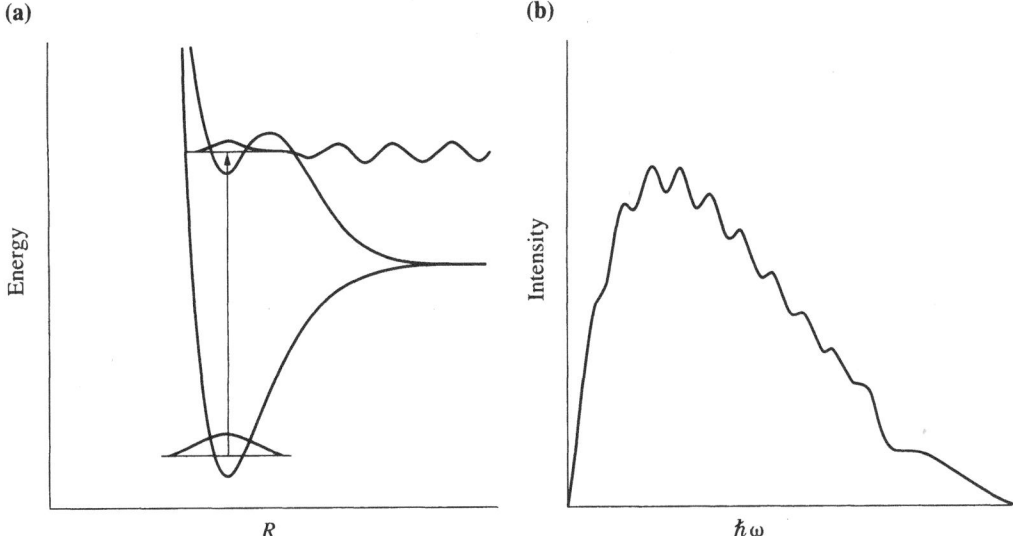

Figure 14.2 (a) Potential energy curves for two electronic states. The electronically excited state supports no true bound states. Electronic excitation produces "predissociation"—quasibound motion that leads ultimately to decay into the dissociative continuum. (b) Diffuse vibronic absorption spectrum, corresponding to the electronically excited state in (a). The smoothing out of the vibronic spectrum is the result of the coupling of the quasibound states with the continuum.

ogy with absorption spectroscopy. In giving labels to these initial and final states, we follow the conventional spectroscopic notation of v' for the vibrational level of the excited electronic state and v'' for the vibrational level of the ground electronic state.) As a result, the emission spectrum is to a first approximation the mirror image of the absorption spectrum, since the higher the energy of the final vibrational state on the ground electronic state, the lower the frequency of the transition.

The most common type of emission spectrum is from $v' = 0$—the ground vibrational state of the excited electronic state; this is because in the presence of collisions, an electronically excited molecule generally undergoes vibrational relaxation on a time scale that is fast compared with the radiative lifetime. However, it is also possible to measure emission spectra from specific, vibrationally excited states in the electronically excited state. As in the case of absorption, the emission spectrum tends to reflect the number of nodes in the initial state, with a single mode spectrum if $v' = 0$ and a multimode spectrum if $v' > 0$.

So far we have discussed one-photon spectroscopies; we now turn to two-photon spectroscopies. Arguably, the most important two-photon spectroscopy is Raman scattering, where the final electronic state is the same as the initial electronic state, but there is a change in vibrational quantum number. In Raman scattering, absorption of an incident photon of frequency ω_I carries the intial wavefunction, ψ_i, from the lower potential to the upper. The emission of a photon of frequency ω_S returns the system to the lower potential, to state ψ_f, where f is the vibrational quantum number of the final level in the ground electronic state. If $\omega_S = \omega_I$ then the scattering is elastic and the process is called Rayleigh scattering. Raman scattering occurs when $\omega_S \neq \omega_I$ and in that case $\psi_f \neq \psi_i$. The measured quantity is the Raman intensity, $I(\omega_I; \omega_S)$. The amplitude of the incident and emitted fields are taken as ε_I

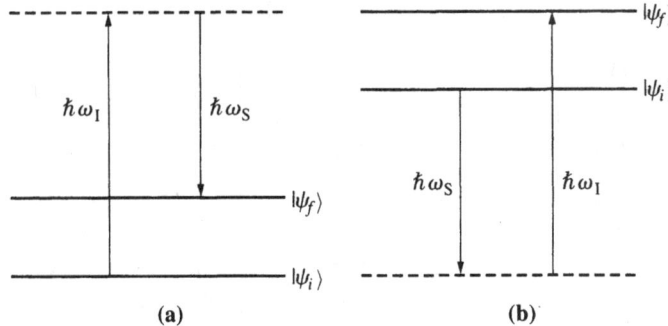

Figure 14.3 Stokes Raman scattering, in which the final vibrational state is higher in energy than the initial vibrational state. There are two contributions, differing in the order of the interaction with the two photons: (a) ω_I precedes ω_S; (b) ω_S precedes ω_I. If the one-photon process with ω_I is resonant, diagram (a) greatly dominates that in (b), the latter corresponding to the improbable event of emission before absorption.

and ε_S. Typically in Raman scattering, the initial vibrational state is the ground vibrational state of the ground electronic state, that is, $i = 0$, while the final vibrational state is excited, that is, $f > 0$. If the final vibrational state is $f = 1$, the line in the Raman spectrum is called a "fundamental"; if $f > 1$ the line is called an overtone. Note that in a polyatomic molecule there are many vibrational coordinates, and hence many fundamentals and at times many overtones as well. Moreover, in polyatomics there may be combination final states, $f = (v_1'', v_2'')$, that is, scattering to final states in which more than one vibrational mode is excited. If the initial state is vibrationally excited, $i > 0$, it is possible to observe Raman scattering to final states with a *lower* vibrational quantum number than the initial state. This is called anti-Stokes Raman scattering, as opposed to the usual case where the final vibrational state has a higher quantum number than the initial state, called Stokes Raman scattering (see Figures 14.3 and 14.4).

Raman scattering, as is the case with all two-photon spectroscopies, is characterized by two terms, which differ in the sequence of the interactions with the two different frequencies of light. Under conditions where one of these frequencies is near-resonant with an intermediate electronic state, one of these terms (the resonant term) will tend to dominate the two-photon spectrum. However, a distinguishing feature of two-photon spectroscopies is that there is no need for an intermediate (i.e., one-photon) resonance; similarly, in an N-photon process, only the sum of N-photon frequencies needs to be resonant with a molecular transition; intermediate resonances, although they enhance the signal, are not required. These two terms are shown in Figures 14.3 and 14.4, for Stokes and anti-Stokes Raman scattering, respectively.

We now proceed to the quantitative description of Raman scattering, for which the conventional, energy-domain formula is the Kramers–Heisenberg–Dirac (KHD) expression. The Raman scattering cross section, integrated over all directions and polarizations, is (Albrecht, 1961; Myers, 1987)

$$\sigma_{fi}(\omega_I) = \frac{8\pi\omega_I\omega_S^3}{9c^4}\sum_{\rho\lambda}\left|\left(\alpha_{fi}(\omega_I)\right)_{\rho\lambda}\right|^2, \tag{14.2}$$

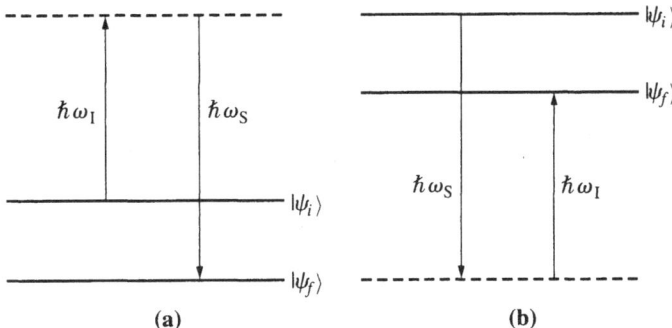

Figure 14.4 Anti-Stokes Raman scattering, in which the final vibrational state is lower in energy than the initial vibrational state. Anti-Stokes Raman scattering necessarily originates with an excited vibrational level. Again, there are two contributions, differing in the order of the interaction with the two photons: (a) ω_I precedes ω_S; (b) ω_S precedes ω_I. If the one-photon process is resonant, diagram (a) greatly dominates that of (b).

where the polarizability, α, is given by

$$\left(\alpha_{fi}(\omega_I)\right)_{\rho\lambda} = -\sum_k \sum_n \left[\frac{\langle \psi_f | \mu_{1k,\rho} | \psi_n \rangle \langle \psi_n | \mu_{k1,\lambda} | \psi_i \rangle}{E_i + \hbar\omega_I - E_{kn} + i\hbar\gamma} + \frac{\langle \psi_f | \mu_{1k,\lambda} | \psi_n \rangle \langle \psi_n | \mu_{k1,\rho} | \psi_i \rangle}{E_i - \hbar\omega_S - E_{kn} + i\hbar\gamma} \right].$$

(14.3)

The subscripts ρ and λ refer to x, y and z components of the vector μ. The summation index k runs over all electronic states (excluding the ground electronic state), while the index n runs over all vibrational states within a particular electronic state. We write $\alpha_{fi}(\omega_I)$ despite the appearance of ω_S on the RHS, since by energy conservation $\hbar\omega_S = E_i + \hbar\omega_I - E_f$.

Equation 14.3 is the KHD expression for the polarizability α. Inspection of the denominators indicates that the first term grows large on resonance, where $E_i + \hbar\omega_I \approx E_n$, while the second term does not; hence the first term is called the "resonant term" and the second the "nonresonant term." Note the product of Franck–Condon factors in the numerator—one corresponding to the amplitude for excitation and the other to the amplitude for emission. The KHD formula is sometimes called the "sum-over-states" formula, since it requires a sum over all intermediate states labelled by k, n, each intermediate state participating according to how far it is from resonance and the size of the matrix elements that connect it to the states ψ_i and ψ_f. Although formally there is a sum over all excited electronic states, if the incident photon is resonant with a particular excited electronic state, then to a good approximation the contribution of the other electronic states can be neglected and the sum over k reduces to a single term.

▶ **Exercise 14.1**   Consider the case of excitation far off resonance. Let $E_{kn} \to E_k(\mathbf{x}_0)$, $E_i \to E_1(\mathbf{x}_0)$ and $\omega_I + \omega_i$, $\omega_S - \omega_i \to \omega_I$. Show that one may write

$$\left(\alpha_{fi}(\omega_I)\right)_{\rho\lambda} = -\sum_k \left[\frac{\langle \psi_f | \mu_{1k,\rho}\mu_{k1,\lambda} | \psi_i \rangle}{E_i(\mathbf{x}_0) + \hbar\omega_I - E_k(\mathbf{x}_0) + i\hbar\gamma} + \frac{\langle \psi_f | \mu_{1k,\lambda}\mu_{k1,\rho} | \psi_i \rangle}{E_i(\mathbf{x}_0) - \hbar\omega_I - E_k(\mathbf{x}_0) + i\hbar\gamma} \right].$$

(14.4)

Equation 14.4 can be written as

$$\left(\alpha_{fi}(\omega_I)\right)_{\rho\lambda} = \langle\psi_f|P_{\rho\lambda}|\psi_i\rangle, \tag{14.5}$$

where $P_{\rho\lambda}$ define the elements of the Placzek polarizability tensor. Note that because the states ψ_i and ψ_f are orthogonal, Raman transitions are made possible only through the coordinate dependence of $P_{\rho\lambda}$. This coordinate dependence can be traced to the coordinate dependence of the $\{\mu_{k1}\}$. This indicates that off-resonance Raman transitions arise from static properties of the transition dipole moment and not vibrational wavepacket dynamics.

Because of the two frequencies, ω_I and ω_S, that enter into the Raman spectrum, Raman spectroscopy may be thought of as a "two-dimensional" form of spectroscopy. Normally, one fixes ω_I and looks at the intensity as a function of ω_S; however, one may vary ω_I and probe the intensity as a function of $\omega_I - \omega_S$. This is called a Raman excitation profile. Particularly interesting is the effect on the Raman spectrum of detuning ω_I from resonance with the excited electronic state. This will be discussed is some detail in Section 14.3.4 below.

In the literature, the distinction is sometimes drawn between resonance Raman scattering and resonance fluorescence, or dispersed fluorescence spectroscopy. In the absence of relaxation processes, these two spectra are the same. In the presence of relaxation processes one normally observes a set of sharp lines in the fluorescence spectrum, which shift with the excitation frequency, and a broad background that does not depend on excitation frequency. The former is normally called resonance Raman scattering and the latter resonance fluorescence. These two contributions can be correlated with different histories of interaction with the environment. To a good approximation, resonance Raman scattering is associated with the early time emission from the electronically excited molecule, while resonance fluorescence is associated with the late time emission.

Two-photon processes also include two-photon absorption, illustrated schematically in Figure 14.5. Again, note the presence of two contributions, differing in the order of the interaction of the system with the photons of different frequencies. Equation 14.3 can be used equally well for the description of two-photon absorption, with the modification $-\omega_S \rightarrow \omega_I'$. There are a wide variety of other two-photon processes possible, including all pairwise combinations of the one-photon processes we discussed above; the quantitative description of all these processes is given by Eq. 14.3, once the appropriate replacement is made for ω_I and ω_S.

With this introduction, we now turn to the time-domain formulation of one-photon and two-photon spectroscopies. The general approach will be as in the previous chapter, to treat the interaction with the light using first- or second-order time-dependent perturbation theory, with the Born–Oppenheimer molecular Hamiltonian providing the zeroth-order description. However, now instead of the perturbation from the field being a short pulse or pulse pair, the field is taken as continuous wave (CW), characterized by a single frequency. The spectrum is defined in terms of the rate of absorption or emission of energy, which is proportional to the rate of change of population induced by the light. Equations 14.1 and 14.3, the central equations of one- and two-photon spectroscopy in the energy frame formulation, will emerge automatically from the time-domain derivation.

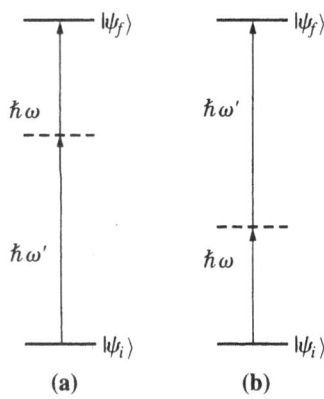

Figure 14.5 Energy-level diagram for two-photon absorption. There are two contributions to the two-photon absorption process, shown in (a) and (b), differing in the sequence of the interactions with the two photons. Note that there is no need for an intermediate resonance at the one-photon energy, although such a resonance would enhance the two-photon absorption.

14.1   Electronic Absorption and Emission Spectroscopy

For a one-photon process, we apply first-order perturbation theory, where the perturbing Hamilitonian is the matter–radiation interaction. As in the previous chapter we employ the dipole approximation and we treat the light field classically. Now, however, we take the field as continuous wave, given by $\varepsilon(t) = \varepsilon_0 \cos(\omega t) = \frac{1}{2}\varepsilon_0(e^{i\omega t} + e^{-i\omega t})$. As we shall see below, one of these exponentials is resonant and the other is nonresonant, depending on the process, and to a good approximation the nonresonant term can be neglected. This is known as the rotating wave approximation or RWA. Therefore, the matter–radiation interaction can be written

$$H_1(t) = \begin{cases} -\frac{1}{2}\mu \cdot \varepsilon_I e^{-i\omega_I t}, & \text{absorption} \\ -\frac{1}{2}\mu \cdot \varepsilon_S e^{+i\omega_S t}, & \text{emission,} \end{cases} \tag{14.6}$$

where the incident (scattered) light has frequency $\omega_I(\omega_S)$ and μ is the (possibly coordinate-dependent) transition dipole moment for going from the lower state to the upper state. Formally, Eq. 14.6 calls for the vector dot product of μ and ε_I or ε_S. We will neglect this vector aspect in what follows, using instead $\mu\varepsilon$, and reintroduce the vector aspect when necessary.

Equation 14.6 for the matter–radiation interaction Hamiltonian represents a light field that is "on" all the time from $-\infty$ to ∞. This interaction will continuously move population from the lower state to the upper state. The propagating packets on the upper states will interfere with one another—constructively if the incident light is resonant with a transition from an eigenstate of the lower surface to an eigenstate of the upper surface, destructively if not. Since for a one-photon process we have two potential energy surfaces, and therefore we have, in effect, two different H_0's—one for before excitation (call it H_a) and one for after (call it H_b). With this in mind, we can use the results of the previous chapter (Eq. 13.8) to write down the first-order correction to the unperturbed wavefunction:

$$|\psi^{(1)}(t)\rangle = -\frac{1}{2i\hbar} \int_{-\infty}^{t} e^{-\frac{i}{\hbar}H_b(t-t')} \{\mu\varepsilon_I e^{-i\omega_I t'}\} e^{-\frac{i}{\hbar}H_a t'} |\psi_i(-\infty)\rangle \, dt'. \tag{14.7}$$

If $|\psi_i(-\infty)\rangle$ is an eigenstate of the ground state Hamiltonian, H_a, then

$$|\psi^{(1)}(t)\rangle = -\frac{1}{2i\hbar} \int_{-\infty}^{t} e^{-\frac{i}{\hbar}H_b(t-t')} \{\mu\varepsilon_1 e^{-i\omega_1 t'}\} \, e^{-\frac{i}{\hbar}E_i t'} |\psi_i(-\infty)\rangle \, dt'. \quad (14.8)$$

Note that we have neglected a phase factor that, loosely speaking, is of the form $e^{-i\omega_i(-\infty)}$, associated with the propagation of ψ_i from $-\infty$ to 0. This phase factor, if it were included, would cancel below against an identical phase factor in the bra. By the same token, we will simplify the notation for the initial state below and write ψ_i instead of $\psi_i(-\infty)$.

In the previous chapter we defined the absorption spectrum as the ratio of number of photons absorbed at each frequency to the number of incident photons per unit area at that frequency (see discussion preceding Eq. 13.49). In this chapter we will define the absorption spectrum as the ratio of the number of photons absorbed per unit *time* to the number of incident photons per unit area per unit *time*. We shall see that the final expression emerging from these two different definitions is identical.

The number of photons absorbed per unit time is equal to the rate of change of the excited-state population. The latter is given in the perturbative limit by the expression $\langle\psi^{(1)}(t)|\psi^{(1)}(t)\rangle$. Defining $\tilde{\omega} = E_i/\hbar + \omega_1$, we can write

$$\langle\psi^{(1)}(t)|\psi^{(1)}(t)\rangle = \frac{1}{4\hbar^2} \int_{-\infty}^{t} dt' \int_{-\infty}^{t} dt'' \langle\psi_i|\mu\varepsilon_1 e^{-\frac{i}{\hbar}H_b(t''-t')}\mu\varepsilon_1|\psi_i\rangle e^{i\tilde{\omega}(t''-t')}. \quad (14.9)$$

We may differentiate Eq. 14.9 with respect to t and, after making a judicious change of variables, obtain an expression for the time derivative of the upper state population:

$$\frac{d}{dt}\langle\psi^{(1)}(t)|\psi^{(1)}(t)\rangle = \frac{\varepsilon_1^2}{4\hbar^2} \int_{-\infty}^{\infty} \langle\psi_i|\mu e^{-\frac{i}{\hbar}H_b t}\mu|\psi_i\rangle e^{i\tilde{\omega}t} \, dt. \quad (14.10)$$

▶ **Exercise 14.2** Show Eq. 14.10.

Solution   Differentiating Eq. 14.9 with respect to t we obtain two terms:

$$\frac{d}{dt}\langle\psi^{(1)}(t)|\psi^{(1)}(t)\rangle = \frac{\varepsilon_1^2}{4\hbar^2} \left\{ \int_{-\infty}^{t} dt' \langle\psi_i|\mu e^{-\frac{i}{\hbar}H_b(t-t')}\mu|\psi_i\rangle e^{i\tilde{\omega}(t-t')} \right.$$
$$\left. + \int_{-\infty}^{t} dt'' \langle\psi_i|\mu e^{-\frac{i}{\hbar}H_b(t''-t)}\mu|\psi_i\rangle e^{i\tilde{\omega}(t''-t)} \right\}. \quad (14.11)$$

Changing variables to $\tau = t - t'$ ($\int_{-\infty}^{t} dt' \to \int_{0}^{\infty} d\tau$) in the first integral and to $\tau = t'' - t$ ($\int_{-\infty}^{t} dt'' \to \int_{-\infty}^{0} d\tau$) in the second integral we obtain

$$\frac{d}{dt}\langle\psi^{(1)}(t)|\psi^{(1)}(t)\rangle$$
$$= \frac{\varepsilon_1^2}{4\hbar^2} \left\{ \int_{0}^{\infty} d\tau \langle\psi_i|\mu e^{-\frac{i}{\hbar}H_b\tau}\mu|\psi_i\rangle e^{i\tilde{\omega}\tau} + \int_{-\infty}^{0} d\tau \langle\psi_i|\mu e^{-\frac{i}{\hbar}H_b\tau}\mu|\psi_i\rangle e^{i\tilde{\omega}\tau} \right\} \quad (14.12)$$

$$= \frac{\varepsilon_1^2}{4\hbar^2} \int_{-\infty}^{\infty} \langle\psi_i|\mu e^{-\frac{i}{\hbar}H_b t}\mu|\psi_i\rangle e^{i\tilde{\omega}t} dt, \quad (14.13)$$

where in the last line we have combined the two integrals and renamed the integration variable t.

To obtain the absorption spectrum we need to divide Eq. 14.10 by the number of incident photons per unit area per unit time (incident photon flux). The latter is given by

$$\frac{Nc}{V} = \frac{\varepsilon_I^2 c}{8\pi \hbar \omega_I}. \tag{14.14}$$

▶ Exercise 14.3   Derive Eq. 14.14.

Solution   According to classical electromagnetism, the energy E in the field is given by

$$E = \int dV \frac{\mathbf{E}^2 + \mathbf{B}^2}{8\pi}. \tag{14.15}$$

Taking $\mathbf{E} = \varepsilon_I \left(\frac{e^{i\omega_I t} + e^{-i\omega_I t}}{2} \right)$ we have $\mathbf{E}^2 = \frac{\varepsilon_I^2}{4} \left(e^{2i\omega_I t} + e^{-2i\omega_I t} + 2 \right)$. Only the last term survives the volume integral since the other terms are rapidly varying. Noting that the average energies of the electric and magnetic fields are equal we obtain

$$E = \int dV \frac{\varepsilon_I^2}{8\pi} \tag{14.16}$$

(compare with Exercise 13.2 with the identification $\varepsilon \rightarrow \frac{\varepsilon_I}{2} e^{-i\omega_I t}, \varepsilon^* \rightarrow \frac{\varepsilon_I}{2} e^{i\omega_I t}$). But by the Einstein relation we know that the energy of a single photon of frequency ω is given by $\hbar \omega_I$, and hence the total energy in the field is

$$E = N\hbar\omega_I, \tag{14.17}$$

where N is the number of photons. Combining Eqs. 14.16 and 14.17 we obtain

$$\varepsilon_I = \left(\frac{8\pi N \hbar \omega_I}{V} \right)^{1/2} \tag{14.18}$$

and hence $\frac{Nc}{V} = \frac{\varepsilon_I^2 c}{8\pi \hbar \omega_I}$.

Dividing Eq. 14.10 by Eq. 14.14 we obtain the formula for the absorption cross section (Kulander, 1978):

$$\sigma(\omega_I) = \frac{8\pi \hbar \omega_I}{\varepsilon_I^2 c} \frac{d}{dt} \langle \psi^{(1)}(t) | \psi^{(1)}(t) \rangle \tag{14.19}$$

$$= \frac{2\pi \omega_I}{\hbar c} \int_{-\infty}^{\infty} \langle \psi_i | \mu e^{-i H_b t / \hbar} \mu | \psi_i \rangle e^{i(\omega_i + \omega_I)t} \, dt. \tag{14.20}$$

At this point we recall the vectorial aspect of the interaction, $\mu \cdot \varepsilon_I$. Replacing μ in Eq. 14.20 by $\mu \cdot \mathbf{e}$, where $\mathbf{e}$ is the unit vector in the direction of the field polarization, gives a factor of $\cos^2 \theta$ in Eq. 14.20. Rotational averaging of this factor gives $1/3$ (Exercise 14.4) and we obtain

$$\sigma(\omega_I) = \frac{2\pi\omega_I}{3\hbar c} \int_{-\infty}^{\infty} \langle \phi_i | \phi_i(t) \rangle e^{i\tilde{\omega}t} \, dt, \tag{14.21}$$

where we have defined $|\phi_i\rangle \equiv \mu|\psi_i\rangle$ and $\tilde{\omega} \equiv \omega_i + \omega_I$, and where $|\phi_i(t)\rangle = e^{-iH_b t/\hbar}|\phi_i\rangle$. Equation 14.21 (or Eq. 14.20) will play a central role in the later parts of this chapter. Note that these expressions are identical to Eq. 13.76, derived in the previous chapter from linear response theory.

Since the absorption spectrum is a ratio it is amenable to other interpretations. One such interpretation is that the absorption spectrum is the ratio of energy absorbed to energy incident. From this perspective, the quantity $\hbar\omega_I \frac{d}{dt}\langle\psi^{(1)}|\psi^{(1)}\rangle$ is interpreted as the rate of *energy* absorption (per unit volume) since according to Eq. 13.48 $\frac{dE}{dt} = \hbar\omega_I \frac{dN_b}{dt}$, while the quantity $\frac{\varepsilon_I^2 c}{8\pi}$ is interpreted as the incident *energy* flux, which depends only on the field intensity and is independent of frequency.

▶ **Exercise 14.4**   To calculate the rotational average of $\cos^2 \theta$ one evaluates the ratio

$$\frac{\int_0^{2\pi} \int_0^{\pi} \cos^2(\theta) \sin(\theta) \, d\theta \, d\phi}{\int_0^{2\pi} \int_0^{\pi} \sin(\theta) \, d\theta \, d\phi}. \tag{14.22}$$

Verify that this gives a factor of $1/3$.

▶ **Exercise 14.5**   Show that the absorption cross section has dimensions of area, both in its basic definition, as well as in the final expression, Eq. 14.20.

Solution   That the definition of the absorption cross section has dimensions of area can be seen as follows:

$$\sigma(\omega_I) = \frac{\text{transition rate}}{\text{incident photon flux}} \propto \frac{\frac{dN}{dt}}{\frac{Nc}{V}} \propto \frac{(\text{time}^{-1})}{(\text{length})(\text{time}^{-1})/(\text{length}^3)} \propto \text{area}. \tag{14.23}$$

To see that Eq. 14.20 also has dimensions of area, note that the dipole μ has dimensions of $(e \cdot r)$, where e is charge. Therefore, the quantity

$$I(\omega) = \int_{-\infty}^{\infty} \langle\psi|\mu e^{-iHt/\hbar}\mu|\psi\rangle e^{i\omega t} dt \tag{14.24}$$

has dimensions of $(e \cdot r)(e \cdot r)t = \frac{e^2}{r}r^3 t = (\text{energy})(\text{volume})(\text{time})$. Noting that $\hbar$ has dimensions of (energy)(time), we find that

$$\sigma(\omega) = \frac{2\pi\omega}{3\hbar c} I(\omega) \propto \left(\frac{(\text{time}^{-1})}{(\text{energy})(\text{time})(\text{length})(\text{time}^{-1})}\right)\left((\text{energy})(\text{volume})(\text{time})\right) = \text{area}. \tag{14.25}$$

A convenient way to think about this area is to remove the factor e^2 from $I(\omega)$ and regroup it with the prefactor. Noting that $\frac{e^2}{r}$ has dimensions of energy we find that

$$\sigma(\omega) \propto \left(\frac{\text{(energy)}}{\text{(energy)(time)}} \right) (\text{length}^2)(\text{time}) = \text{area}. \qquad (14.26)$$

Thus, the scale of area of the cross section can be viewed as arising from the square of the length scale of the dipole multiplied by other dimensionless factors.

Equation 14.21 expresses the absorption spectrum as the Fourier transform of a wave-packet correlation function. The identical result was derived in Chapter 13 from linear response theory (compare Eq. 13.76). This is a result of central importance. In Chapter 6 we derived essentially the same formula, as an intrinsic property of the spectrum of a wavepacket, without any need to invoke light absorption or first-order perturbation theory. However, in the context of one-photon processes the Fourier transform relationship between the wavepacket autocorrelation function and the absorption spectrum provides a powerful tool for interpreting absorption spectra in terms of the underlying nuclear wave-packet dynamics that follows the optically induced transition. For example for absorption spectroscopy, the initial wavepacket is $\phi_i = \mu_{ba}\psi_i$, and the relevant correlation function is that of the moving wavepacket on the excited-state potential energy surface, $\phi_i(t)$, with the initial wavepacket $\phi_i(0) = \phi_i$. In this case, the spectrum is a probe of excited-state wave-packet dynamics, particularly in the Franck–Condon region (i.e., the region accessed by a vertical transition at $t = 0$). This is illustrated in Figure 14.6.

Since often only short or intermediate dynamics enter in the spectrum (for example, because of photodissociation or radiationless transitions to other electronic states) computation of the time-correlation function can be much simpler than evaluation of the spectrum in terms of Franck–Condon overlaps, which formally can involve millions of eigenstates for an intermediate-sized molecule. As mentioned in the introduction to this chapter, the same formula with minor modifications applies to emission spectra, photoionization spectra and photodetachment spectra (photoionization of a negative ion). Because of the widely different physics of these processes we now show illustrations of each of these different types of spectroscopies. In Chapter 17 we will derive a very similar formula for the photodissociation spectrum, and again illustrate it by example.

▶ **Exercise 14.6**   Following the derivation in Chapter 6, insert a complete set of excited-state eigenstates into Eq. 14.21 and derive the usual energy frame formula for the absorption spectrum, involving Franck–Condon overlaps of stationary states:

$$\sigma(\omega_{\mathrm{I}}) = \frac{4\pi^2 \omega_{\mathrm{I}}}{3\hbar c} \sum_n |\langle \psi_n | \mu | \psi_i \rangle|^2 \delta(\tilde{\omega}_{\mathrm{I}} - \omega_n). \qquad (14.27)$$

We may apply the mathematical manipulations of Exercise 14.6 directly to the intermediate result, Eq. 14.10. This gives

$$\frac{d}{dt} \langle \psi^{(1)}(t) | \psi^{(1)}(t) \rangle = 2\pi \hbar \frac{\varepsilon_{\mathrm{I}}^2}{4\hbar^2} \sum_n |\langle \psi_n | \mu | \psi_i \rangle|^2 \delta(\tilde{E}_{\mathrm{I}} - E_n), \qquad (14.28)$$

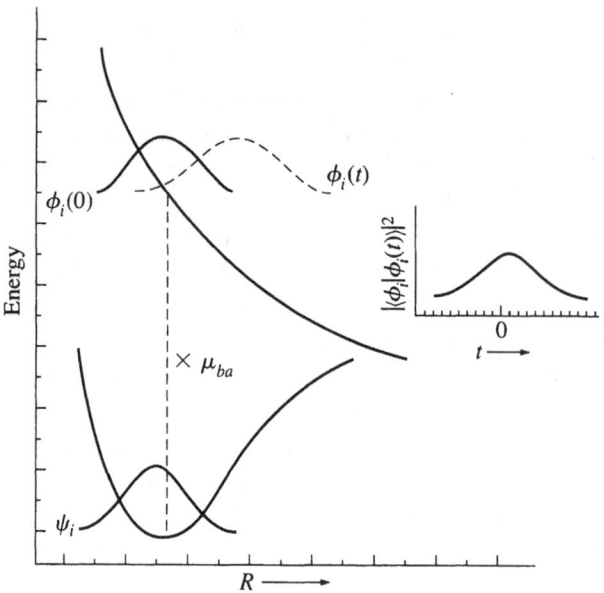

Figure 14.6 Wavepacket interpretation of the formula for the one-photon absorption spectrum, $\sigma(\omega_I)$ (Eq. 14.21). The state $\phi_i = \mu_{ba}\psi_i$ is a *wavepacket* on the excited electronic state potential and begins to move according to the Time-Dependent Schrödinger Equation. The absolute value squared of the autocorrelation function constructed from the overlap of $\phi_i(t)$ with $\phi_i(0)$ is shown in the inset. The Fourier transform of the autocorrelation function at frequency $\tilde\omega_I$ determines the absorption spectrum (see Eq. 14.21).

where $\tilde E_I = \hbar\tilde\omega_I$. Changing the summation to an integration over energy gives

$$\frac{d}{dt}\langle\psi^{(1)}(t)|\psi^{(1)}(t)\rangle = \frac{2\pi}{\hbar}\frac{\varepsilon_I^2}{4}\int_0^\infty \overline{|\langle\psi_n|\mu|\psi_i\rangle|^2}\rho(E)\delta(\tilde E_I - E)dE \qquad (14.29)$$

$$= \frac{2\pi}{\hbar}\frac{\varepsilon_I^2}{4}\overline{|\langle\psi_n|\mu|\psi_i\rangle|^2}\rho(\tilde E_I), \qquad (14.30)$$

where $\overline{|\langle\psi_n|\mu|\psi_i\rangle|^2}$ is the average square matrix-element and $\rho(\tilde E_I)$ is the density of states in an energy band between $\tilde E_I$ and $\tilde E_I + d\tilde E_I$. The LHS is the rate of population transfer to the excited state. Equation 14.30 is of the form of Fermi's Golden Rule for rate k:

$$k = \frac{2\pi}{\hbar}\overline{|V|^2}\rho(E). \qquad (14.31)$$

Thus, the rate is proportional to the product of the (average) coupling squared times the density of states.

We now proceed to some examples of this Fourier transform view of optical spectroscopy.

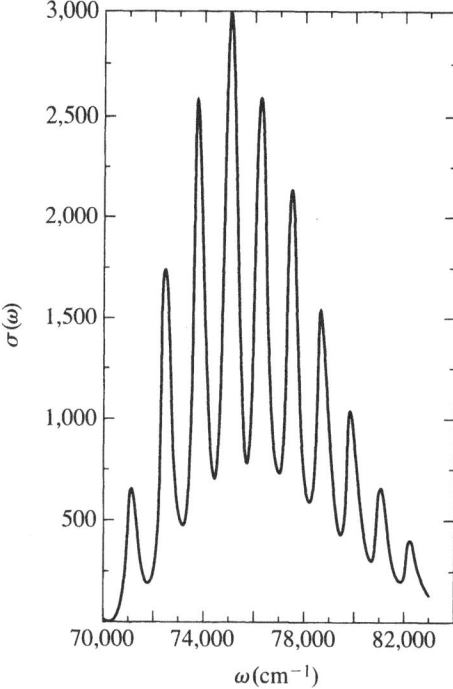

Figure 14.7 Idealized UV absorption spectrum of CO_2. Note the regular progression of intermediate resolution vibrational progression. In the frequency regime this structure is interpreted as a Franck–Condon progression in the symmetric stretch, with broadening of each of the lines due to predissociation. Adapted from Pack (1976).

14.1.1   Predissociation

Although our expression for the vibrational absorption spectrum seems to imply that the spectrum is a series of sharp lines, in fact experimental absorption spectra range from narrow lines to diffuse but separated features to broad, featureless spectra. For example, the UV absorption spectrum of CO_2, shown in Figure 14.7, is seen to have a long progression of vibrational features of fairly uniform shape and width. What is the physical interpretation of this vibrational progression and what is the origin of the width of the features? If we can come up with a dynamical model that leads to a wavepacket autocorrelation function whose Fourier transform agrees with the spectrum in Figure 14.7, we shall have supplied the sought-after physical interpretation (Heller, 1978).

Figure 14.8a gives a plausible dynamical model leading to such an autocorrelation function. In (1), equipotential contours of the excited-state potential energy surface of CO_2 are shown as a function of the two bond lengths, R_1 and R_2, or equivalently, as a function of the symmetric and antisymmetric stretch coordinates, v and u (the latter are linear combinations of the former). Along the axis $u = 0$ the potential has a minimum; along the axis $v = 0$ (the local "reaction path") the potential has a maximum. Thus, the potential in the region $u = 0$, $v = 0$ has a "saddle point." There are two symmetrically related exit

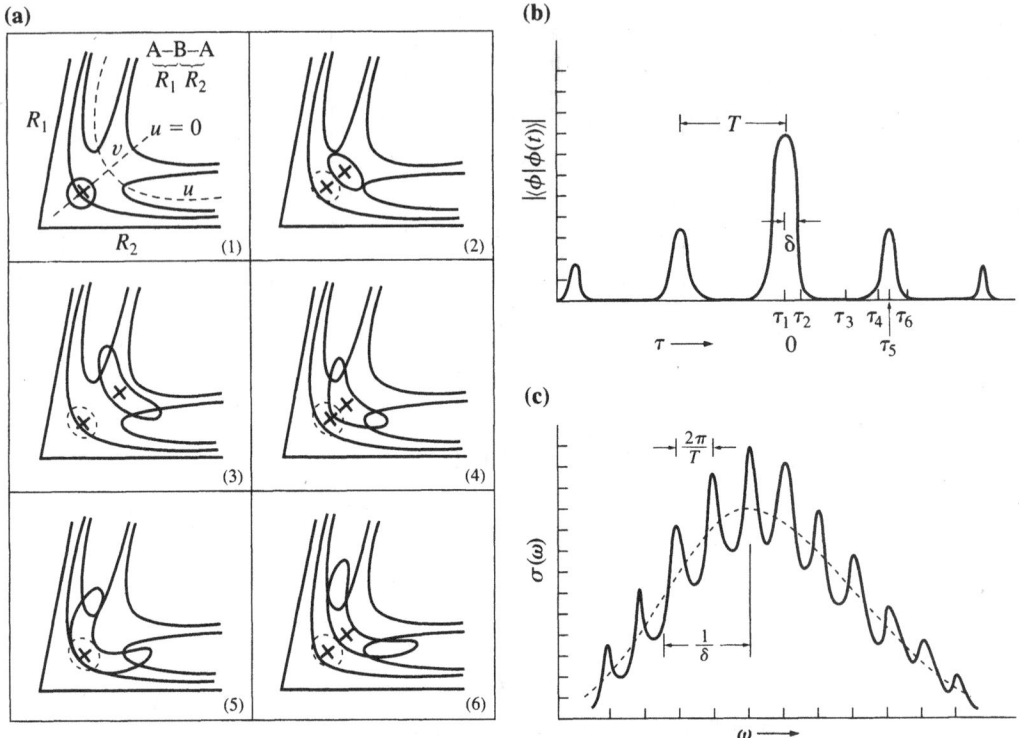

Figure 14.8 (a) Qualitative diagram showing evolution of $\phi(t)$ on the upper potential surface of CO_2. Note the oscillation along the v (symmetric stretch) coordinate, and the spreading along the u (antisymmetric stretch) coordinate. (b) The absolute value of the correlation function, $|\langle\phi|\phi(t)\rangle|$ versus t for the dynamical situation shown in (a). τ_1 corresponds to the first frame in (a), etc. (c) The Fourier transform of $\langle\phi|\phi(t)\rangle$, giving the absorption spectrum. Note that the central lobe in the correlation function, with decay constant δ, gives rise to the overall width of the absorption spectrum, on the order of $\frac{2\pi}{\delta}$. Furthermore, the recurrences in the correlation on the time scale T give rise to the oscillations in the spectrum on the time scale $\frac{2\pi}{T}$. Adapted from Heller (1978).

channels, for large values of R_1 and R_2, respectively, corresponding to the formation of $O'C + O''$ versus $O' + CO''$. Figure 14.8a (1) also shows the initial wavepacket, which is approximately a two-dimensional Gaussian. Its center is displaced from the minimum in the symmetric stretch coordinate. Figure 14.8a (2)–(6) shows the subsequent dynamics of the wavepacket. It moves downhill along the v coordinate, while at the same time spreading. After one vibrational period in the v coordinate the center of the wavepacket comes back to its starting point in v, but has spread in u (Figure 14.8a (5)). The resulting wavepacket autocorrelation function is shown in Figure 14.8b. At $t = 0$ the autocorrelation function is 1. On a time scale τ_2 the correlation function has decayed to nearly 0, reflecting the fact that the wavepacket has moved away from its initial Franck–Condon location (Figure 14.8a (2). At time τ_5 the wavepacket has come back to the Franck–Condon region in the v coordinate, and the autocorrelation function has a recurrence. However, the magnitude of the recurrence is much smaller than the initial value, since there is irreversible spreading of the wavepacket in the u coordinate. Note that there are further, smaller recurrences at multiples of τ_5.

The spectrum obtained by Fourier transform of Figure 14.8b is shown in Figure 14.8c. Qualitatively, it has all the features of the spectrum in Figure 14.7: a broad envelope with resolved vibrational structure underneath, but with an ultimate, unresolvable linewidth. The correspondence between features in the time and frequency domain follows exactly along the lines of the discussion in Chapter 6: the shortest time decay, δ, determines the overall envelope in frequency, $1/\delta$; the recurrence time, T, determines the vibrational frequency spacing, $2\pi/T$; the overall decay time determines the width of the vibrational features. As in Chapter 6 we note that decays in time correspond to widths in frequency, while recurrences in time correspond to spacings in frequency. Moreover, the shortest time scales correspond to the broadest frequency features and vice versa.

The photodissociation dynamics in Figure 14.8 is indirect and is called "predissociation." The recognition that predissociation is responsible for intrinsically unresolvable vibrational features in vibrational spectra is as old as vibrational spectroscopy itself. The precise formulation of the spectrum as the Fourier transform of a wavepacket autocorrelation function, and the systematic interpretation of features in this autocorrelation function in terms of the underlying wavepacket dynamics, began with Heller in the late 1970s.

14.1.2   Photoionization and Photoelectron Spectroscopy

Photoionization generally refers to the excitation of a neutral molecule to an excited electronic state corresponding to a positively charged ion + electron. The formula for the photoionization spectrum is essentially identical to that for photoabsorption, with the role of the excited-state potential energy surface being played by the potential energy surface of the cation. Photoelectron spectroscopy has essentially the same information content, but instead of measuring the absorption as a function of incident light frequency one measures the kinetic energy spectrum of the exiting electron for a fixed incident light frequency. Since the total energy is fixed by the incident light frequency, the different electron kinetic energies leave the positive ion with different vibrational energy contents, and thus the electron kinetic energy spectrum reflects the ion's vibrational spectrum. Figure 14.9a shows the photoelectron spectrum of HCN. The potential energy surface as a function of the two bond stretching coordinates is shown in Figure 14.9c. The center of the wavepacket translates in the symmetric stretch coordinate, while spreading in the asymmetric stretch. At longer times some fragmentation to H + CN is observed. The general behavior is not unlike that of CO_2 discussed above, but the asymmetry here and the bound character of the ion potential along the CN coordinate lead to a much more complicated-looking autocorrelation function (Figure 14.9b), which is reflected in the variegated widths and features in the photoelectron spectrum (Figure 14.9a).

14.1.3   Emission Spectroscopy

Emission spectroscopy refers to the process whereby a molecule initially in a vibrational state of the excited electronic state spontaneously emits a photon, returning to the ground electronic state. Since the process is not triggered directly by an incident photon it is more natural to speak of the emission *rate*, rather than the cross section. The emission rate $k_\downarrow$ is given by Eq. 14.10 with the following modifications (see Figure 14.10):

1. the replacement $\varepsilon_I \rightarrow \varepsilon_S$;
2. the identification of ψ_i with the initial vibrational state of the excited electronic state;

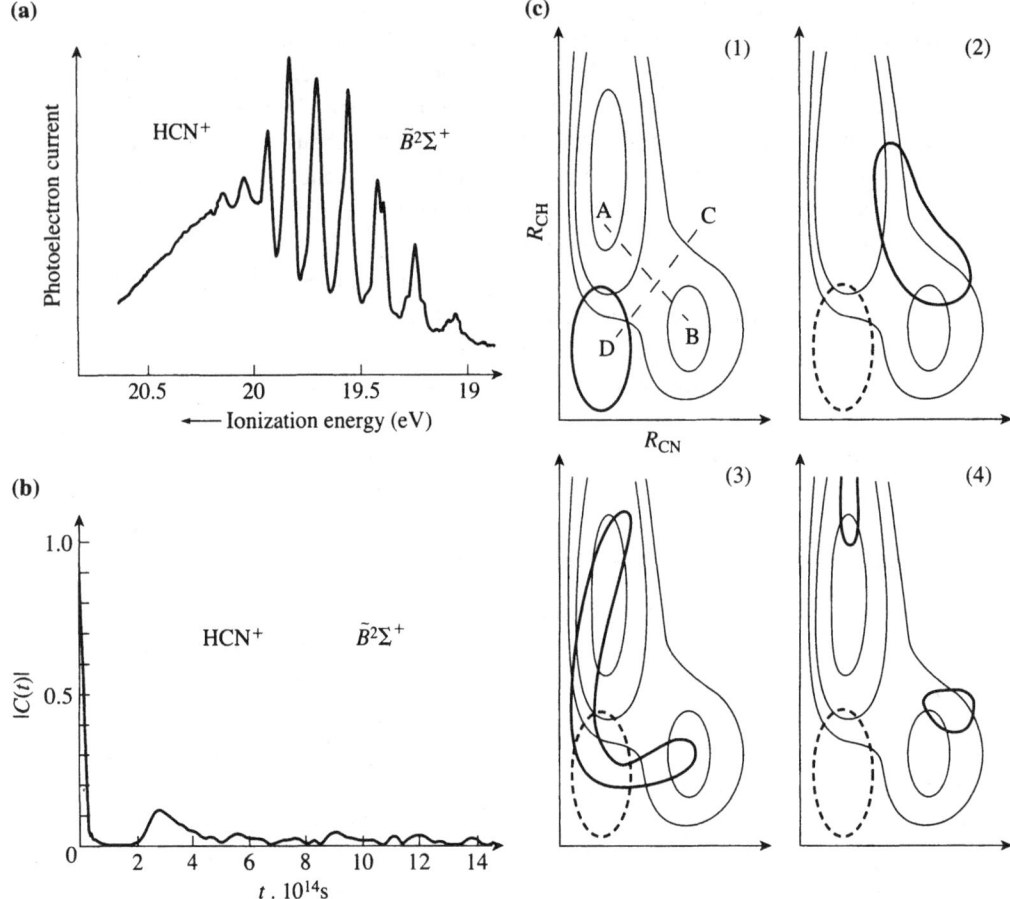

Figure 14.9 (a) Absolute value of the correlation function of the wavepacket as it evolves on the $\tilde{B}^2\Sigma^+$ state of HCN$^+$. (b) Photoionization spectrum corresponding to the autocorrelation function in (a). (c) Potential energy surface of the $\tilde{B}^2\Sigma^+$ state of HCN$^+$. The moving wavepacket (represented by one thick contour line) is shown superimposed. (1) The initial wavepacket, $\phi(0)$, created by ionization. (2)–(4) Subsequent motion of the wavepacket, $\phi(t)$, on the ionic potential energy surface. The wavepacket oscillates along direction CD while spreading along coordinate AB. Adapted from (Lorquet, 1982).

3. the replacement $\tilde{\omega}_I = \omega_i + \omega_I \to \tilde{\omega}_S = \omega_i - \omega_S$;

4. the inclusion of a factor accounting for the density of states of the radiation field for spontaneous emission.

Thus,

$$k_\downarrow = \frac{d}{dt}\langle \psi^{(1)}|\psi^{(1)}\rangle = \int \frac{\varepsilon_S^2}{4\hbar^2}\rho(E_S)\int_{-\infty}^{\infty}\langle\psi_i(0)|\mu e^{-\frac{i}{\hbar}H_a t}\mu|\psi_i(0)\rangle e^{i\tilde{\omega}_S t}\,dt\,dE_S, \quad (14.32)$$

where $\tilde{\omega}_S = \omega_i - \omega_S$. The effective value of ε_S is found by noting that the photon density is given by (Eq. 14.14) $\frac{\varepsilon_S^2}{8\pi\hbar\omega_S} = \frac{N}{V}$. Setting $N = 1$ for spontaneous emission gives

$$\varepsilon_S^2 = \frac{8\pi\hbar\omega_S}{V}. \quad (14.33)$$

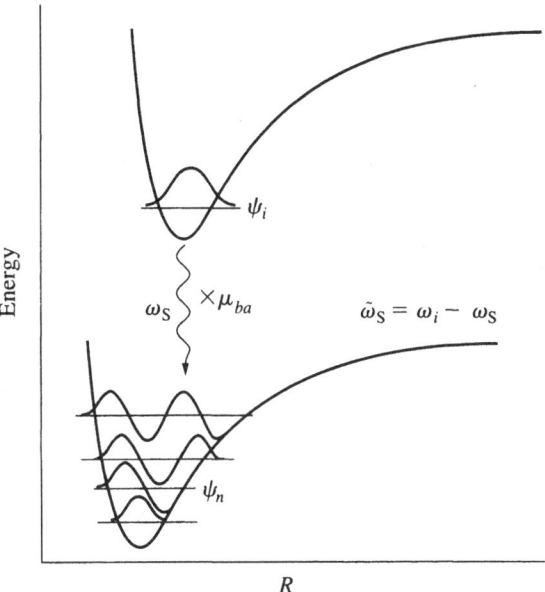

Figure 14.10 Typical arrangement of the electronic potential energy surfaces in emission spectroscopy. The most common situation is that the initial vibrational state, ψ_i, is the ground vibrational state of the *excited* electronic state, and the emission takes place to the vibrational states, ψ_n, of the ground electronic state.

The factor $\rho(E_S)$ is given by (Exercise 14.7)

$$\rho(E_S) = \frac{V}{(2\pi c)^3} \frac{\omega_S^2}{\hbar} d\Omega. \tag{14.34}$$

Combining Eqs. 14.32–14.34 we may write

$$\frac{d^2 k_\downarrow}{dE_S \, d\Omega} = \frac{1}{(2\pi\hbar)^2} \frac{\omega_S^3}{c^3} \int_{-\infty}^{\infty} \langle \psi_i | \mu e^{-iH_a t/\hbar} \mu | \psi_i \rangle e^{i\tilde{\omega}_S t} dt. \tag{14.35}$$

Recall that ψ_i is the initial vibrational state on the *excited* electronic state. Equation 14.35 indicates that the emission rate can be calculated by propagating ψ_i under the ground state Hamiltonian, H_a, and Fourier transforming the autocorrelation function at frequency $\tilde{\omega}_S = \omega_i - \omega_S$. Equation 14.35 is analogous to Eq. 14.20 for the absorption cross section with the replacements $H_b \to H_a$, $\tilde{\omega}_I \to \tilde{\omega}_S$. Note that the factor ω_I for absorption is replaced by ω_S^3 for emission.

▶ **Exercise 14.7** Show that the density of photon states is

$$\rho(E) = \frac{V}{(2\pi c)^3} \frac{\omega^2}{\hbar} d\Omega. \tag{14.36}$$

Solution Consider a volume $V = L^3$. The energy levels of the photon states are

$$E_\mathbf{n} = \frac{\hbar^2 \pi^2}{L^2}(n_x^2 + n_y^2 + n_z^2). \tag{14.37}$$

The total number of states at energy E is

$$N(E) = \sum_\mathbf{n} \Delta n_x \Delta n_y \Delta n_z, \tag{14.38}$$

where Δn_x, Δn_y and Δn_z are just integer steps, and hence each of these quantities is equal to 1; this artificial notation for writing 1 is convenient for the change of variables that follows. The sum over $\mathbf{n} \equiv (n_x, n_y, n_z)$ is over $n_x^2 + n_y^2 + n_x^2 \le E$. Noting that we can express Eq. 14.37 in terms of the wave vector $\mathbf{k}$,

$$E_n = \hbar^2(k_x^2 + k_y^2 + k_z^2), \tag{14.39}$$

we have the relation $\Delta k_x = \frac{\pi}{L}\Delta n_x$ and so forth, and hence

$$N(E) = \sum_\mathbf{k} \left(\frac{L}{\pi}\right)^3 \Delta k_x \Delta k_y \Delta k_z. \tag{14.40}$$

Passing from the sum to an integral and using the relationship $dk_x\, dk_y\, dk_z = k^2\, dk\, d\Omega$ between Cartesian and polar volume elements we obtain

$$N(E) = \frac{V}{\pi^3} \int \frac{k^2}{8} dk\Omega = \frac{V}{\pi^3} \int \frac{\omega^2 d\omega}{8c^3} d\Omega. \tag{14.41}$$

In the last step we have used the relation $k = \omega/c$ for photons. The factor of 1/8 comes from the fact that the sum over n (and hence k) is over positive integers only and hence the integral in three dimensions is over one octant only. Finally, noting that $\rho(E) = \frac{dN}{dE} = \frac{1}{\hbar}\frac{dN}{d\omega}$ we obtain

$$\rho(E) = \frac{V}{(2\pi c)^3} \frac{\omega^2}{\hbar} d\Omega. \tag{14.42}$$

An example of the use of the wavepacket autocorrelation function formulation to interpret an emission spectrum is shown in Figure 14.11. As an additional twist, this example shows a case in which the emission spectrum has a single characteristic frequency of 550 cm^{-1}, but this frequency does not correspond to any of the normal mode frequencies of the ground electronic state. This phenomenon was discussed in Chapter 6, as an example of the Fourier convolution theorem. The key observation is that a recurrence of a wavepacket autocorrelation function requires a simultaneous return to the initial state in all degrees of freedom. In the example in Chapter 6, the correlation function was factorizable into the product of a correlation function in two separable coordinates, but the phenomenon is much more general: the wavepacket motion takes place in a space of high dimensionality, and the recurrence time may not bear any simple relation to a simple one-dimensional coordinate. This has been called the missing mode effect, or MIME, and many examples have been identified.

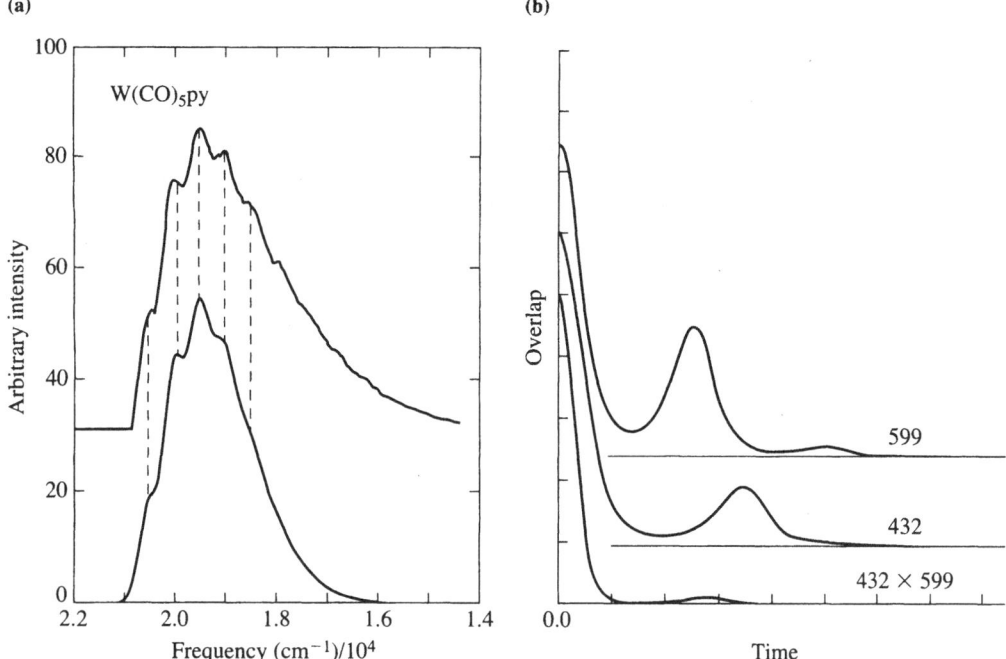

(a)

(b)

Figure 14.11 (a) Emission spectrum illustrating the missing mode effect (MIME). The upper curve is the experimental emission spectrum of $W(CO)_5py$, while the lower curve is the calculated spectrum. Note the low-resolution vibrational progression at 550 cm^{-1}, a frequency that does not correspond to any individual ground state vibration. (b) Upper curve: Overlap $|\langle\phi_{599}|\phi_{599}(t)\rangle|e^{-\Gamma^2t^2/2}$ versus t. Middle curve: $|\langle\phi_{432}|\phi_{432}(t)\rangle|e^{-\Gamma^2t^2/2}$ versus t. Bottom curve: Absolute value of the product $\langle\phi|\phi(t)\rangle$ of the two correlation functions. Note that the recurrence of the product is at a time in between those of the individual modes. The Fourier transform of the correlation function gives the frequency spectrum shown in (a), containing the 550 cm^{-1} MIME frequency. Adapted from Tutt (1982).

14.2   Transition State Spectroscopy

In Chapter 12 we learned the central concepts of transition state theory for rate constants for chemical reactions. As discussed in Chapter 12, the Arrhenius expression gives the rate constant as

$$k(T) = Ae^{-E^{\ddagger}/k_BT}. \tag{14.43}$$

We mentioned there that the prefactor may be viewed as an entropy factor. The "absolute rate theory" expression introduced by Eyring, Eq. 12.104, gives an explicit form for the prefactor:

$$k(T) = \frac{kT}{h}\frac{Q^{\ddagger}}{Q}e^{-E^{\ddagger}/kT}. \tag{14.44}$$

In this expression, $Q^{\ddagger}$ is the partition function in the coordinates perpendicular to the transition state; Q is the partition function for the reactants. In other words, the entropy factor is given in terms of the ratio of the internal partition function at the transition state to the internal partition function in the reactant region, multiplied by a thermal flux factor, k_BT/h.

Both the partition functions require knowledge of the vibrational frequencies, the first at the transition state and the second in the reactant region. The latter is probed by standard spectroscopies; the former is much more elusive, since the transition state is not a stable species (see Figure 12.10). Thus, it has been a goal of chemical physics for many years to probe the frequencies of the transition state directly, spectroscopically, rather than learning about the transition state only indirectly via highly averaged rate constant measurements.

In the last fifteen years, a beautiful method, initiated by Neumark, has emerged for probing the frequencies at the transition state spectroscopically. The key is the realization that although the transition state itself is unstable and therefore short-lived, the negative ion corresponding to the geometry of the unstable species is often, in fact, stable. This suggests starting with the stable negative ion species and photodetaching the extra electron, thereby preparing a Franck–Condon transition to the neutral potential energy surface. In favorable cases, the Franck–Condon region on the neutral potential will be close to the transition state between reactants and products, and the vibrational structure in the photodetachment spectrum gives detailed information on the vibrational frequencies at the transition state in the coordinates perpendicular to the reaction path—precisely the information needed for input into a rate constant calculation.

Figure 14.12a shows the potential energy surface of the unstable BrHBr molecule, as a function of the two bond lengths (or equivalently, as a function of the symmetric and antisymmetric stretch coordinates). Superimposed is the time-evolving wavepacket, which originated at the equilibrium geometry of the $BrHBr^-$ ion geometry. Note that, by symmetry, the $BrHBr^-$ equilibrium geometry is at 0 in the antisymmetric stretch coordinate, while it is displaced along the symmetric stretch coordinate. The subsequent motion of the wavepacket is along the symmetric stretch, while at the same time it bifurcates symmetrically in the antisymmetric stretch. Figure 14.12b shows the modulus of the time autocorrelation function on the neutral surface, $|C(t)| = |\langle \phi | \phi(t) \rangle|$, obtained from the wavepacket dynamics in Figure 14.12a. The Fourier transform of the autocorrelation function gives the $BrHBr^-$ photodetachment spectrum (dashed), shown in Figure 14.12c superimposed on the experimental spectrum (solid).

The photodetachment spectrum in this and similar examples gives detailed, direct information on the frequencies and motion in the coordinates perpendicular to the transition state on the potential energy surface of the neutral. As described above, it is exactly these perpendicular frequencies that enter into the entropy factor in transition state theory, since they determine the width of the bottleneck to reaction. It is only in recent years that direct spectroscopic observation of these frequencies has become available, through photodetachment of negative ions, to obtain the spectra of the resulting unstable neutrals. The modelling of the photodetachment using the wavepacket autocorrelation function formulation is in some sense only incidental; the key is in the information content of the spectrum itself and its connection to rate theory. Nevertheless, the wavepacket formulation is very convenient for these systems, both because the spectrum is intrinsically diffuse and hence the dynamics is relatively short time, and because of the conceptual connection it presents between the spectrum and the transition state dynamics.

We turn now to one more example of transition state spectroscopy, involving the $F + H_2$ reaction (Manolopoulos, 1993). Next to the $H + H_2$ reaction, this is perhaps the most studied reaction in the field of molecular reaction dynamics. Various aspects of the asymptotic product energy and angular distributions of the $F + H_2$ reaction have been measured in chemical laser, infrared chemiluminescence, and crossed molecular beam experiments.

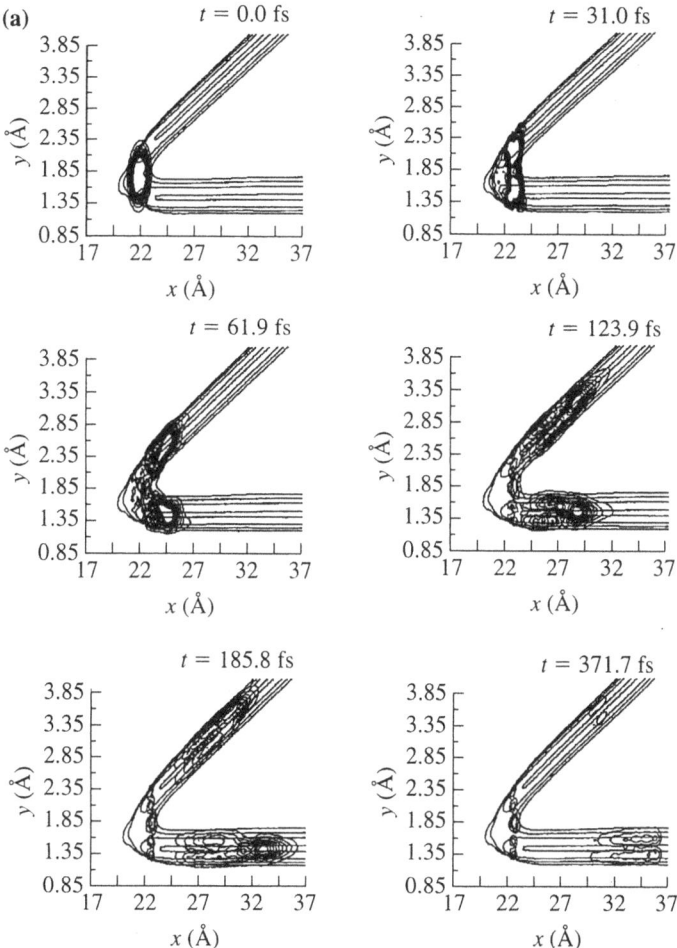

Figure 14.12 Transition state spectroscopy: photodetachment of $BrHBr^- \to BrHBr + e^-$. (a) Time evolution of the initially prepared wavepacket $\phi(0)$ on $Br + HBr$ potential surface. $x = (\mu_{Br, HBr}/\mu_{HBr})^{1/2} R_{Br, HBr}$, $y = R_{HBr}$. *(Continued on next page)*

This reaction turns out to be an ideal case for photodetachment spectroscopy. Ab initio calculations show that the precursor anion, FH_2^-, is essentially F^- weakly bound to H_2, so that the photodetachment should access the reactant side of the $F + H_2$ potential energy surface (see Figure 14.13a). A two-dimensional cut through the transition state region of the $F + H_2$ potential, as a function of the $F—H_2$ distance and angle, is shown in Figure 14.13b. The contours of the para FH_2^- vibrational ground state wavefunction are superimposed (the corresponding ortho wavefunction may be obtained by reversing the sign of one of the two lobes). Notice that the anion wavefunction has a good overlap with the neutral transition state. The FH_2^- equilibrium geometry is believed to be linear with $F—H$ and $H—H$ distances of 1.690 and 0.770 Å, and the corresponding distances at the two degenerate saddle points on the $F + H_2$ surface are similar, at 1.541 and 0.772 Å. The fact that the $H—H$ bond length changes so little from the anion to the neutral implies that Figure 14.13 contains virtually all the relevant information. Notice further that, in contrast to the linear equilibrium

(b)

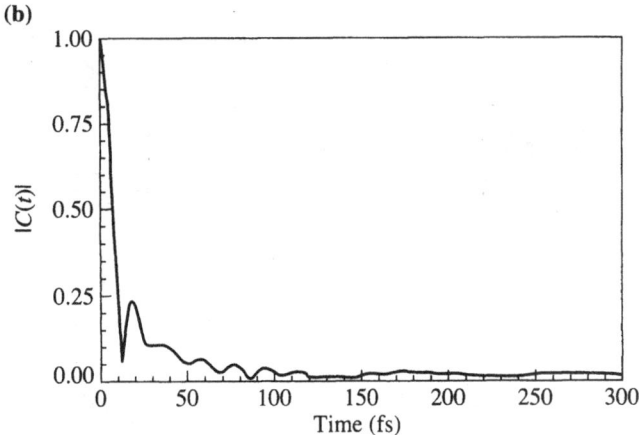

(c)

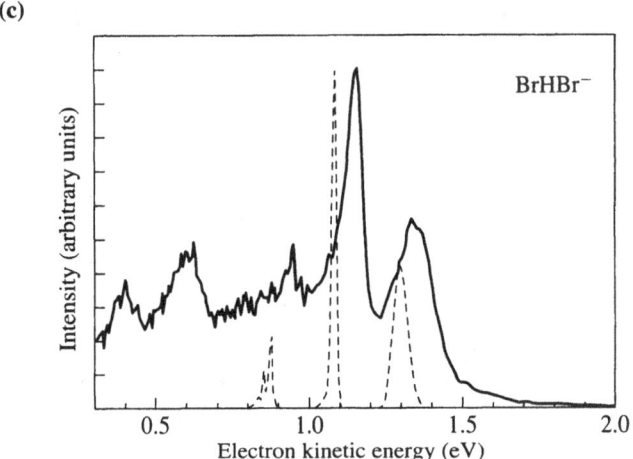

Figure 14.12 *(continued)* (b) Modulus of the time autocorrelation function, $|C(t)|$, obtained from the wavepacket dynamics. (c) The calculated BrHBr⁻ photodetachment spectrum (dashed) superimposed on the experimental spectrum (solid). Adapted from R. Metz et al. (1992).

geometry of the anion, the saddle points on the $F + H_2$ surface are bent at an F—H—H angle of 119°. The coordinate that experiences the largest change in potential energy on photodetachment is therefore the bending coordinate, which explains why the observed FH_2^- photodetachment spectra are dominated by bending progressions. The comparison of the experimental and simulated photodetachment spectra, for both para and normal FH_2^-, is shown in Figure 14.14. The former contains only transitions to bending states that correlate asymptotically with $F + H_2$ (even j) while the latter contains transitions to states that correlate with both $F + H_2$ (odd j) and $F + H_2$ (even j) in the usual 3:1 ortho:para ratio. The fact that the largest geometry change is in the bending coordinate explains the large difference between the photodetachment spectra of the para and the normal FH_2^- species.

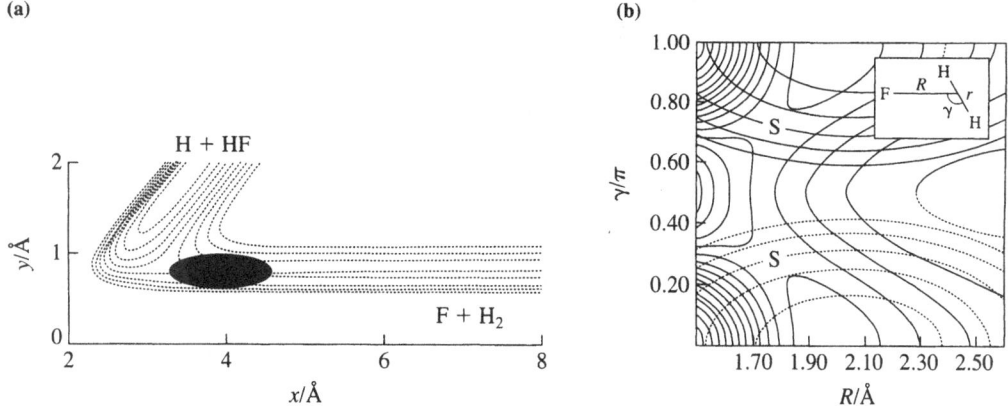

Figure 14.13 (a) Collinear section of the T5a potential energy surface for the $F + H_2$ reaction, with the Franck–Condon region shaded. $x = (\mu_{F, HH}/\mu_{HH})^{1/2} R_{F, HH}$, $y = R_{HH}$. Adapted from Neumark (1993). (b) The transition state region of the $F + H_2$ PES and its overlap with the FH_2^- anion wave function. The solid lines are contours of the $F + H_2$ PES in steps of 0.4 kcal/mol from 0.4 to 2.8 kcal/mol, relative to the bottom of the asymptotic $F + H_2$ valley. The long dashed lines are similar contours from -2.8 to 0.0 kcal/mol. The short dashed lines are contours of the para FH_2^- anion wave function, with its maximum at $\gamma/\pi = 0$ and 1, and each successive contour representing a decrease in the square modulus of the wave function by a factor of 10. The coordinates of the plot are the F—H_2 distance R, and the Jacobi angle γ (see inset); the H—H bond length r is fixed at 0.771 Å throughout. The $F + H_2$ transition state saddle points are signified by S. Adapted from Manolopoulos (1993).

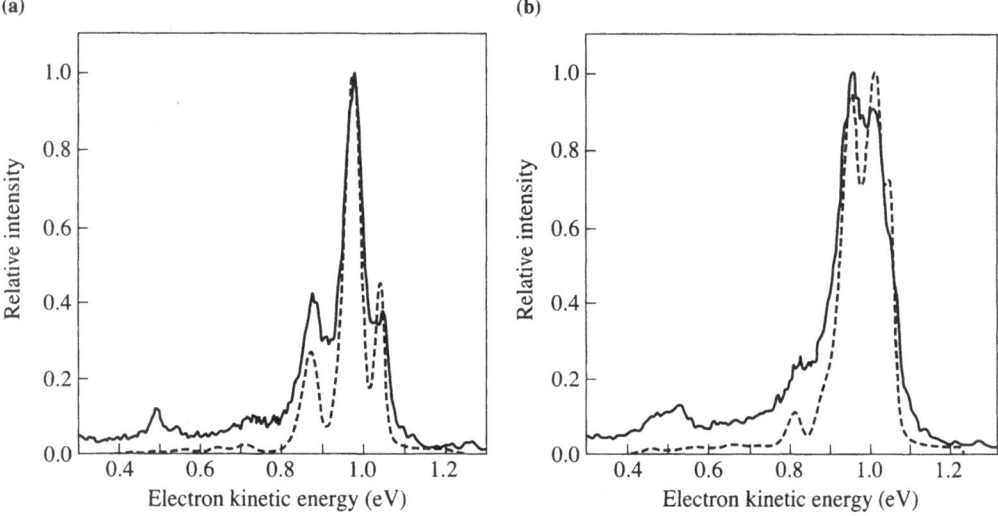

Figure 14.14 Comparison between experimental (solid line) and theoretical (dashed line) FH_2^- photodetachment spectra. (a) para FH_2^- and (b) ortho FH_2^-. Adapted from Manolopoulos (1993).

14.3 Resonance Raman Spectroscopy

We will now look at two-photon processes. We will concentrate on Raman scattering, although two-photon absorption can be handled using the same approach. In Raman scattering, an incident photon of frequency ω_I promotes the initial wavefunction, ψ_i, from the lower to the upper electronic state. The scattering of a photon of frequency ω_S returns the system to the lower electronic state and vibrational state ψ_f. If $\omega_S = \omega_I$ then the scattering is elastic and the process is called Rayleigh scattering. Raman scattering occurs when $\omega_S \neq \omega_I$ and in that case $\psi_f \neq \psi_i$. The measured quantity is the Raman intensity, $I(\omega_I; \omega_S)$.

If the frequency of the incident photon is off resonance with any of the excited electronic states the process is called ordinary, or off-resonance, Raman scattering. In this case, all the excited electronic states participate approximately equally in the scattering process. The scattering reflects little or no excited-state wavepacket dynamics and arises from the coordinate-dependent transition dipole moment. In the traditional treatments of ordinary Raman scattering the central quantity is the "polarizability," which is, essentially, a weighted sum of products of the transition dipole moment to all intermediate electronic states (cf. Eq. (14.4)).

We will be interested primarily in resonance Raman scattering, where the incident photon is on resonance or near resonance with a particular excited electronic state. In this case we can ignore the contribution to the scattering coming from the other excited electronic states and focus just on the contribution coming from the resonant electronic state. We will derive the following central expression for the resonance Raman cross section:

$$\sigma^{f \leftarrow i} = \frac{8\pi \omega_I \omega_S^3}{9c^4} \sum_{\rho \lambda} | \left(\alpha_{fi}(\omega_I) \right)_{\rho \lambda} |^2, \tag{14.45}$$

where $\left(\alpha_{fi}(\omega_I) \right)_{\rho \lambda}$ is the polarizability, which we will show is equal to

$$\left(\alpha_{fi}(\omega_I) \right)_{\rho \lambda} = \frac{-1}{i\hbar} \int_0^\infty \langle \phi_{f\rho} | \phi_{i\lambda}(t) \rangle e^{i\tilde{\omega}t} \, dt \; + \; \text{(nonresonant term)}. \tag{14.46}$$

Equation 14.46 indicates that the wavepacket dynamics on the excited electronic state can play a very important role. In fact, as we shall see, the resonance Raman spectrum is a sensitive probe of wavepacket dynamics on the excited electronic state, and hence of geometry changes between the ground and excited electronic state.

14.3.1   Derivation of the Resonance Raman Intensity Formula

Since Raman scattering involves two interactions with light, it is described by second-order perturbation theory. We start from the expression for the second-order wavefunction (Eq. 13.27):

$$|\psi^{(2)}(t)\rangle =$$

$$-\frac{1}{4\hbar^2} \int_{-\infty}^{t} dt' \int_{-\infty}^{t'} dt'' \, e^{-\frac{i}{\hbar}H_a(t-t')} \{ \mu \cdot \varepsilon_S e^{+i\omega_S t'} \} e^{-\frac{i}{\hbar}H_b(t'-t'')} \{ \mu \cdot \varepsilon_I e^{-i\tilde{\omega}_I t''} \} |\psi_i(-\infty)\rangle,$$

$$\tag{14.47}$$

where H_a (H_b) is the Hamiltonian for the lower (upper) potential energy surface and, as before, $\tilde{\omega}_I = \omega_I + \omega_i$. In words, Eq. 14.47 says that the second-order wavefunction is obtained by propagating the initial wavefunction on the ground state surface until time t'', at which time it is excited up to the excited state upon which it evolves until time t', when it is returned to the ground state where it propagates until time t. Note that as in the first-order treatment we have neglected a phase factor that, loosely speaking, is of the form $e^{-i\tilde{\omega}_I(-\infty)}$, corresponding to propagation from $-\infty$ to 0. As in the first-order treatment, we will simplify the notation by writing ψ_i instead of $\psi_i(-\infty)$. Moreover, we will write $\mu \cdot \varepsilon_S = \mu_\rho \varepsilon_S$ and $\mu \cdot \varepsilon_I = \mu_\lambda \varepsilon_I$.

According to Eq. 9.35, the second-order wavefunction can also be expressed in the following form:

$$|\psi^{(2)}(t)\rangle = \frac{-\varepsilon_S}{2i\hbar} \int_{-\infty}^{t} dt' \, e^{-\frac{i}{\hbar}H_a(t-t')} \mu_\rho e^{+i\omega_S t'} |\psi^{(1)}(t')\rangle. \tag{14.48}$$

Consider now the first-order wavefunction, $\psi^{(1)}(t')$, that enters into Eq. 14.48:

$$|\psi^{(1)}(t')\rangle = \frac{-\varepsilon_I}{2i\hbar} \int_{-\infty}^{t'} e^{-\frac{i}{\hbar}H_b(t'-t'')} \mu_\lambda e^{-i\tilde{\omega}_I t''} |\psi_i\rangle \, dt''. \tag{14.49}$$

We can rewrite this expression by multiplying and dividing by a phase factor and changing variables:

$$|\psi^{(1)}(t')\rangle = -e^{-i\tilde{\omega}_I t'} \frac{\varepsilon_I}{2i\hbar} \int_{-\infty}^{t'} e^{-\frac{i}{\hbar}H_b(t'-t'')} \mu_\lambda e^{+i\tilde{\omega}_I(t'-t'')} |\psi_i\rangle \, dt''$$

$$= -e^{-i\tilde{\omega}_I t'} \frac{\varepsilon_I}{2i\hbar} \int_{0}^{\infty} e^{-\frac{i}{\hbar}H_b \tau} \mu_\lambda e^{+i\tilde{\omega}_I \tau} |\psi_i\rangle \, d\tau \tag{14.50}$$

$$= -e^{-i\tilde{\omega}_I t'} \frac{\varepsilon_I}{2i\hbar} |R_{\omega_I}\rangle, \tag{14.51}$$

where we have defined the "Raman wavefunction," R_{ω_I} (Heller, 1982),

$$|R_{\omega_I}\rangle \equiv \int_{0}^{\infty} e^{-\frac{i}{\hbar}H_b \tau} \mu_\lambda e^{+i\tilde{\omega}_I \tau} |\psi_i\rangle \, d\tau. \tag{14.52}$$

The Raman wavefunction has an interesting physical interpretation that is discussed in Section 14.3.4. For now, just note that the Raman wavefunction is independent of time and as a result $\psi^{(1)}$ has a trivial time dependence—just an overall time-dependent phase.

Substituting Eq. 14.51 back into Eq. 14.47 we obtain

$$|\psi^{(2)}(t)\rangle = -\frac{\varepsilon_I \varepsilon_S}{4\hbar^2} \int_{-\infty}^{t} dt' \, e^{-\frac{i}{\hbar}H_a(t-t')} \mu_\rho e^{+i\omega_S t'} e^{-i\tilde{\omega}_I t'} |R_{\omega_I}\rangle \tag{14.53}$$

$$= -\frac{\varepsilon_I \varepsilon_S}{4\hbar^2} \int_{-\infty}^{t} dt' \, e^{-\frac{i}{\hbar}H_a(t-t')} \mu_\rho e^{+i\tilde{\omega}_S t'} |R_{\omega_I}\rangle, \tag{14.54}$$

where we have defined $\tilde{\omega}_S = \omega_S - \tilde{\omega}_I$. Note that, due to the trivial time dependence of $\psi^{(1)}(t)$, $\psi^{(2)}$ has the structure of a *first-order* amplitude in which the role of the initial state is played by R_{ω_I}. We can therefore proceed exactly as in the one-photon case, with the Raman wavefunction playing the role of the initial state and the propagation being on the ground state instead of the excited state. Taking the time derivative of $\langle \psi^{(2)}(t)|\psi^{(2)}(t)\rangle$ yields (Heller, 1982)

$$\frac{d}{dt} \langle \psi^{(2)}(t)|\psi^{(2)}(t)\rangle = \frac{\varepsilon_I^2 \varepsilon_S^2}{16\hbar^4} \int_{-\infty}^{\infty} \langle \tilde{R}_{\omega_I}|e^{-\frac{i}{\hbar}H_a t}|\tilde{R}_{\omega_I}\rangle e^{-i\tilde{\omega}_S t} \, dt, \tag{14.55}$$

where $|\tilde{R}_{\omega_I}\rangle = \mu_\rho |R_{\omega_I}\rangle$.

In Eq. 14.47, the interaction $\mu \cdot \varepsilon_I$ preceded the interaction $\mu \cdot \varepsilon_S$. However, at the level of second-order perturbation theory there is another term that enters, in which the interactions are in the opposite order, that is, $\mu \cdot \varepsilon_S$ precedes $\mu \cdot \varepsilon_I$. This gives rise to the so-called nonresonant term, corresponding to the counterintuitive (and relatively unlikely) process in which emission precedes absorption. Following through the algebra for this form of $\psi^{(2)}$ gives the nonresonant contribution to the Raman wavefunction:

$$|R_{\omega_S}\rangle = \int_0^\infty e^{-\frac{i}{\hbar}H_b\tau} \mu_\rho e^{-i\tilde{\omega}_S\tau}|\psi_i\rangle\, d\tau \qquad \text{(nonresonant contribution)} \quad (14.56)$$

Formally speaking, $\tilde{R}_{\omega_I}$ in Eq. 14.55 should be replaced by $\tilde{R}_{\omega_I} + \tilde{R}_{\omega_S}$, where $\tilde{R}_{\omega_S} = \mu_\lambda R_{\omega_S}$.

Note the similarity of Eq. 14.55 with Eq. 14.10 for the one-photon process. Equation 14.55 indicates that the rate of change of the second-order probability is proportional to the autocorrelation of the *Raman wavefunction* propagated on the *ground* state potential surface, whereas Eq. 14.10 indicates that the rate of change of the first-order probability is proportional to the autocorrelation function of the *initial* wavefunction propagated on the *excited* state potential surface. In particular, the Raman wavefunction plays the role of an effective *initial* state from which the emission labelled Raman scattering can be thought to emanate.

If we now insert a complete set of final ground state wavefunctions in Eq. 14.55, we obtain

$$\frac{d}{dt}\langle\psi^{(2)}(t)|\psi^{(2)}(t)\rangle = \frac{\varepsilon_I^2\varepsilon_S^2}{16\hbar^4}\sum_f \int_{-\infty}^\infty dt\, \langle\tilde{R}_{\omega_I}|\psi_f\rangle\langle\psi_f|\tilde{R}_{\omega_I}\rangle e^{-i(\omega_f + \tilde{\omega}_S)t}. \quad (14.57)$$

Performing the integral over time yields a delta function in frequency. Thus (Heller, 1982)

$$\frac{d}{dt}\langle\psi^{(2)}(t)|\psi^{(2)}(t)\rangle = \frac{2\pi\varepsilon_I^2\varepsilon_S^2}{16\hbar^4}\sum_f |\langle\psi_f|\tilde{R}_{\omega_I}\rangle|^2\delta(\omega_f + \omega_S - (\omega_I + \omega_i)) \quad (14.58)$$

$$= \frac{2\pi\varepsilon_I^2\varepsilon_S^2}{16\hbar^2}\sum_f |\alpha_{fi}(\omega_I)|^2\delta(\omega_f + \omega_S - (\omega_I + \omega_i)), \quad (14.59)$$

where we have defined the frequency-dependent polarizability, $\alpha_{fi}(\omega_I)$:

$$\alpha_{fi}(\omega_I) = \frac{-1}{i\hbar}\langle\psi_f|\tilde{R}_{\omega_I}\rangle. \quad (14.60)$$

Equation 14.58 expresses the Raman scattering in terms of a Franck–Condon overlap between the Raman wavefunction and the final vibrational state on the ground electronic state. This further supports the viewpoint that Raman scattering may be viewed as emission from an effective initial state, R_{ω_I}. In a very precise sense, R_{ω_I} corresponds with what was loosely called the "virtual state" in the older resonance Raman literature.

So far we have ignored the fact that Raman scattering involves sponanteous, rather than stimulated, emission. As discussed in Section 14.1.3, the explicit inclusion of spontaneous emission has two effects that must be included in the formula for $\frac{d}{dt}\langle\psi^{(2)}(t)|\psi^{(2)}(t)\rangle$: 1) the density of states of the radiation field must be taken into account, and 2) a suitable choice of ε_S, the amplitude of the scattered radiation, must be found. Switching variables from ω to E ($\delta(\omega) = \hbar\delta(E)$), the density of states of the radiation field enters by writing

$$\frac{d}{dt}\langle\psi^{(2)}(t)|\psi^{(2)}(t)\rangle = \frac{2\pi\varepsilon_I^2\varepsilon_S^2}{16\hbar}\sum_f |\alpha_{fi}|^2\delta(E - E_f) \tag{14.61}$$

$$\rightarrow \frac{2\pi\varepsilon_I^2\varepsilon_S^2}{16\hbar}\int_0^\infty |\alpha_{fi}|^2\rho(E)\delta(E - E_f)\,dE \tag{14.62}$$

$$= \frac{2\pi\varepsilon_I^2\varepsilon_S^2}{16\hbar}|\alpha_{fi}|^2\rho(E). \tag{14.63}$$

Using $\varepsilon_S^2 = \frac{8\pi\hbar\omega_S}{V}$ (Eq. 14.33) and $\rho(E) = \frac{V}{(2\pi c)^3}\frac{\omega_S^2}{\hbar}d\Omega$ (Eq. 14.42) and dividing by the incident photon flux gives the spontaneous Raman cross section:

$$d\sigma_{\rho\lambda}^{f\leftarrow i} = \frac{\frac{d}{dt}\langle\psi^{(2)}(t)|\psi^{(2)}\rangle_{\rho\lambda}}{\frac{\varepsilon_I^2 c}{8\pi\hbar\omega_I}} \tag{14.64}$$

$$= \frac{\omega_I\omega_S^3}{c^4}|(\alpha_{fi}(\omega_I))_{\rho\lambda}|^2 d\Omega, \tag{14.65}$$

where the subscripts ρ, λ refer to the polarization of the scattered and incident light, respectively. Averaging over all molecular orientations, summing over all incident and scattered polarizations, and integrating over 4π solid angle gives

$$\sigma^{f\leftarrow i} = \frac{8\pi\omega_I\omega_S^3}{9c^4}\sum_{\rho\lambda}|(\alpha_{fi}(\omega_I))_{\rho\lambda}|^2, \tag{14.66}$$

where the summation indices ρ, λ are taken here to run over the values of the three Cartesian axes x, y and z. As a check on the dimensions notice that the Raman cross section, $\sigma^{f\leftarrow i}$, should have units of area. From Eq. 14.66, we have

$$\sigma^{f\leftarrow i} \propto \frac{(\text{time}^{-1})(\text{time}^{-3})}{(\text{length}^4)(\text{time}^{-4})}(\text{length}^6) = \text{area}. \tag{14.67}$$

14.3.2   Resonance Raman Amplitude as the Fourier Transform of Wavepacket Cross-Correlation Functions

Equation 14.60 expresses α_{fi} as a stationary Franck–Condon overlap. However, there is an alternative perspective on α_{fi} that is very appealing. Substituting the expressions for the resonant and nonresonant terms in R_{ω_I}, Eqs. 14.52 and 14.56, respectively, into Eq. 14.58, we find

$$\alpha_{fi}(\omega_I) = \frac{-1}{i\hbar}\langle\psi_f|\tilde{R}_{\omega_I}\rangle \tag{14.68}$$

$$= \frac{-1}{i\hbar}\left\{\int_0^\infty \langle\psi_f|\mu_\rho e^{-\frac{i}{\hbar}H_b t}\mu_\lambda|\psi_i\rangle e^{+i\omega_I t}\,dt + \int_0^\infty \langle\psi_f|\mu_\lambda e^{-\frac{i}{\hbar}H_b t}\mu_\rho|\psi_i\rangle e^{-i\omega_S t}\,dt\right\}. \tag{14.69}$$

(a)   **(b)**

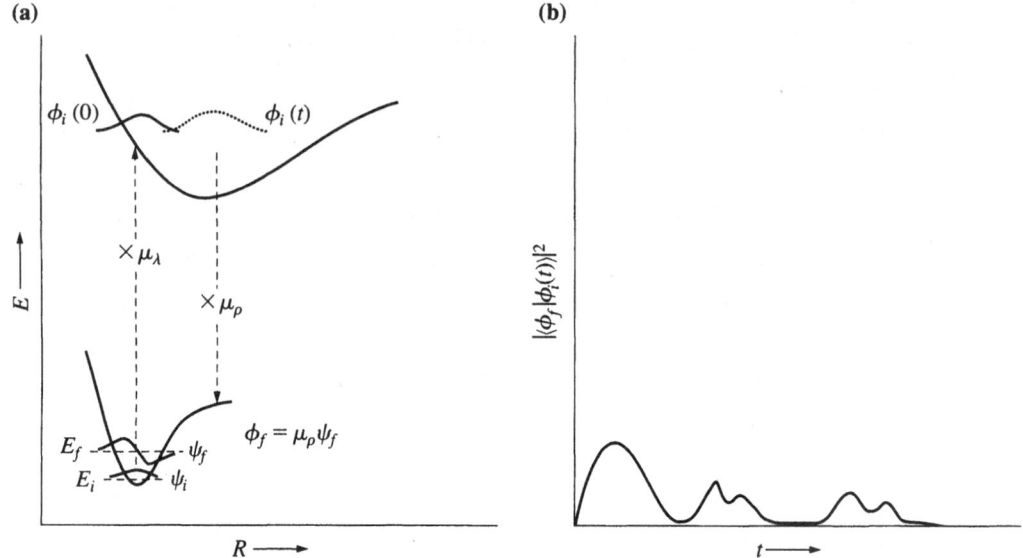

Figure 14.15 (a) Schematic diagram showing the promotion of the initial wavepacket, followed by free evolution on the excited electronic state. The diagram corresponds to the first term in Eq. 14.7. (b) The cross-correlation functions with the excited vibrational states of the ground state surface determine the resonance Raman amplitude to those final states (Lee, 1979). The subscripts λ and ρ on ϕ are omitted for simplicity.

Defining $|\phi_{i\lambda,\rho}\rangle = \mu_{\lambda,\rho}|\psi_i\rangle$, $|\phi_{f\lambda,\rho}\rangle = \mu_{\lambda,\rho}|\psi_f\rangle$, and $|\phi_i(t)\rangle = e^{-\frac{i}{\hbar}H_b t}|\phi_i\rangle$, $\alpha_{fi}(\omega_{\mathrm{I}})$ can be written as (Lee, 1979)

$$\alpha_{fi}(\omega_{\mathrm{I}}) = \frac{-1}{i\hbar}\left\{\int_0^\infty \langle\phi_{f\rho}|\phi_{i\lambda}(t)\rangle\, e^{i\bar{\omega}_{\mathrm{I}} t}\, dt + \int_0^\infty \langle\phi_{f\lambda}|\phi_{i\rho}(t)\rangle\, e^{-i\bar{\omega}_{\mathrm{S}} t}\, dt\right\}, \qquad (14.70)$$

matching the form of Eq. 14.46. The second term in α_{fi} is called the nonresonant term, a designation that will become clearer in Exercise 14.10.

Equation 14.70 has a simple physical interpretation. We will neglect the nonresonant term in what follows, although it may be given an analogous physical interpretation. At $t = 0$ the initial state, ψ_i, is multiplied by μ. This product, denoted ϕ_i, constitutes an initial wavepacket that begins to propagate on the excited-state potential energy surface (Figure 14.15). If μ is independent of R (Condon approximation), initially the wavepacket will have overlap only with ψ_i, and will be orthogonal to all other ψ_f on the ground state surface. As a wavepacket begins to move on the excited state, however, it will develop overlap with electronically ground, vibrationally excited states of ever increasing number. Eventually, if the excited potential is bound, the wavepacket will reach a turning point and begin moving back toward the Franck–Condon region of the excited-state surface, now overlapping electronically ground, vibrationally excited states in decreasing order of their quantum number. These time-dependent overlaps determine the Raman intensities via Eqs. 14.70 and 14.66. If the excited state is dissociative, then the wavepacket never returns to the Franck–Condon region and the Raman spectrum has a monotonically decaying envelope (see Figures 14.22–14.23). If the wavepacket bifurcates on the excited state due to a bistable potential, then it will have nonzero overlaps only with ground vibrational states that are of

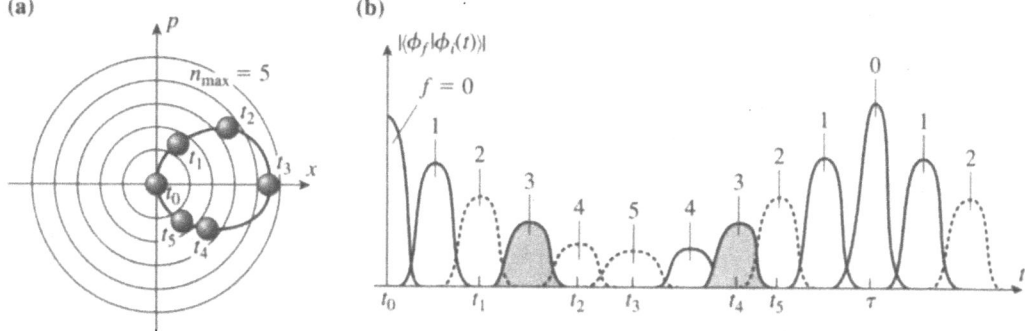

Figure 14.16 (a) Phase space evolution of the excited-state wavepacket (elliptical orbit) superimposed on the contours of the vibrationally excited eigenstates of the ground electronic state (concentric rings). (b) The absolute value of the overlap, $|\langle \phi_f | \phi_i(t) \rangle|$, for increasing values of f. The timing of the overlaps is easily understood from the phase space picture in (a) (see text).

even parity; the Raman spectrum will then have "missing" lines (see Figure 14.24). In multidimensional systems, there are ground vibrational states corresponding to each mode of vibration. The Raman intensities then contain information about the extent to which different coordinates participate in the wavepacket motion, and even in what sequence (see discussion at Figure 14.21). Clearly, resonance Raman intensities can be a very sensitive probe of wavepacket dynamics on the excited-state potential.

Figure 14.15a shows the promotion of the initial wavepacket to the excited electronic state, followed by free evolution on the excited-state potential. The cross-correlation function with the first excited vibrational state of the ground state surface is shown in Figure 14.15b. Note that the cross-correlation function starts at 0 at $t = 0$, since the moving wavepacket at $t = 0$ is just the ground vibrational state and hence is orthogonal to all excited vibrational states of the ground potential surface.

▶ **Exercise 14.8**   The previous statement is within the Condon approximation (the assumption of a coordinate-independent transition dipole). If the transition dipole moment is coordinate dependent, it can lead to Raman transitions even without any excited-state wavepacket dynamics. Show this both algebraically and pictorially.

The wavepacket cross-correlation functions, $C_{fi}(t) = \langle \phi_f | e^{-iHt/\hbar} | \phi_i \rangle$, have an interesting interpretation in terms of phase space dynamics. Figure 14.16a shows the elliptical orbit in phase space of the $\phi_i(t)$ evolving on the excited electronic state potential energy surface. Superimposed are the schematic phase space contours of the excited vibrational states of the *ground* electronic state. Figure 14.16b shows the approximate $|\langle \phi_f | \phi_i(t) \rangle|$ versus t for all ϕ_f having appreciable overlap. The magnitudes of the different overlaps are readily understood from Figure 14.16a. In particular, the following features of $|\langle \phi_f | \phi_i(t) \rangle|$ are evident from the phase space dynamics:

1. The higher eigenstates develop overlap later and with lower amplitude than the lower eigenstates.

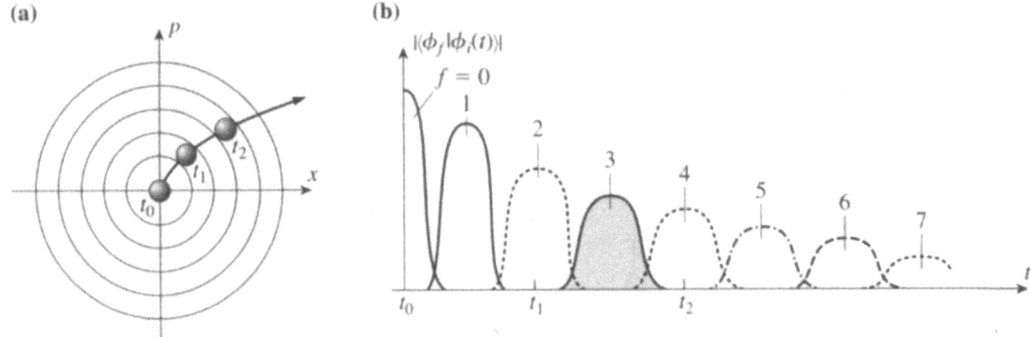

Figure 14.17 Same as in Figure 14.16 except that the excited state is dissociative. Again, the timing of the overlaps in (b) is easily understood from the phase space picture in (a) (see text).

2. The overlap is typically double-humped, centered at multiples of the period τ. This is because for bound motion, the wavepacket overlaps a given excited vibrational level both on its way out and on its way back in to the Franck–Condon region.

3. The double-humped structure separates for higher vibrational eigenstates, eventually coalescing for f_{max} around $\tau/2$, where f_{max} is the highest vibrational eigenstate that can be accessed given the energy content of $\phi_i(t)$ as determined by the size of the Franck–Condon displacement of the ground and the excited potential energy surfaces.

Figure 14.17a shows the analogous phase space dynamics if the excited-state potential energy surface is dissociative. The dynamics is reflected in the form of the $|\langle\phi_f|\phi_i(t)\rangle|$ shown in Figure 14.17b. For the first half-period, the behavior is almost identical to that obtained with bound motion, Figure 14.16; however, at longer times the recurrences that were due to return to the Franck–Condon region are missing.

Note that if the transition dipole moments are taken to be unity, the time-dependent probabilities $P_t(f|i)$ obey the sum rule

$$\sum_f P_t(f|i) = \sum_f |\langle\phi_f|\phi_i(t)\rangle|^2 = |\langle\phi_i(t)|\phi_i(t)\rangle|^2 = 1. \qquad (14.71)$$

This is a statement of conservation of probability, using the vibrational eigenstates $\{\psi_f\}$ as the complete basis for representing the moving wavepacket $\phi_i(t)$.

▶ **Exercise 14.9**   Show that the polarizability α has dimensions of volume.

Solution

$$\alpha = \frac{-1}{i\hbar}\int_0^\infty \langle\psi|\mu e^{-iHt/\hbar}\mu|\psi\rangle e^{i\omega t}\,dt \propto \frac{(e\cdot r)(e\cdot r)(\text{time})}{(\text{energy})(\text{time})}$$

$$= \frac{e^2}{\text{length}}(\text{length}^3)\frac{(\text{time})}{(\text{energy})(\text{time})} = \text{volume.} \qquad (14.72)$$

▶ **Exercise 14.10** For excitation that is resonant with only a single excited-electronic state k, the Kramers–Heisenberg–Dirac (KHD) expression for the frequency-dependent polarizability takes the form (cf. Eq. 14.3 with $\mu = \mu_{k1}$ (or μ_{1k}):

$$\alpha_{fi}(\omega_I) = -\sum_n \left[\frac{\langle \psi_f | \mu_\rho | \psi_n \rangle \langle \psi_n | \mu_\lambda | \psi_i \rangle}{E_i + \hbar\omega_I - E_n + i\hbar\gamma} + \frac{\langle \psi_f | \mu_\lambda | \psi_n \rangle \langle \psi_n | \mu_\rho | \psi_i \rangle}{E_i - \hbar\omega_S - E_n + i\hbar\gamma} \right]. \quad (14.73)$$

The quantity γ is included to prevent the formula from diverging at exact resonance, and may be interpreted as corresponding to the rate of spontaneous emission.

    a. Explain why the first term is resonant and the second term is nonresonant.

    b. Show that the time-domain formula for α, Eq. 14.69, is equivalent to Eq. 14.73, by inserting a complete set of vibrational states $\{n\}$ of electronic state k and integrating over time.

▶ **Exercise 14.11** Note that in Eq. 14.70 we have retained the dependence of α on the direction of polarization of the incident and scattered light. It is customary to speak of the 3×3 polarizability tensor $(\alpha_{fi}(\omega_I))_{\rho\lambda}$, where λ refers to the direction of polarization of the incident light (x, y or z) and ρ refers to the direction of polarization of the scattered light (x, y or z). The differences between excitation/scattering by/into x, y and z polarized light result from differences in the coordinate dependence of the transition dipole along x, y and z molecular axes. Substitute the qualitative form for $\mu_{x/y/z}$ from Figure 12.20 into Eq. 14.70, and analyze each of the nine elements of the polarizability tensor. Specifically, analyze the initial state that is prepared, its evolution on the excited state and its subsequent emission (Tannor, 1988). Explain why the polarizability tensor is symmetric ($\alpha_{\lambda\rho} = \alpha_{\rho\lambda}$) off resonance but not on resonance. The asymmetry of the tensor on resonance leads to an effect called "anomalous polarization."

14.3.3 Cross-Correlation Functions as a Probe of Dynamics

As discussed above, Raman spectroscopy is really a "two-dimensional" form of spectroscopy since there are two frequencies involved—ω_I and ω_S. So-called "Raman excitation profiles" are obtained when the intensity of scattering at a fixed ω_S is monitored as a function of ω_I. The area under a Raman excitation profile may be interpreted as the amount of time the moving wavepacket spends in the neighborhood of the final ground state wavefunction (Heller, 1982). We may define a function $P_\Gamma(f|i)$ as

$$P_\Gamma(f|i) = 2\Gamma \int_0^\infty |\alpha_{fi}(\omega_I)|^2 \, d\omega_I, \quad (14.74)$$

where Γ is a phenomenological lifetime of the excited state. By inserting the definition of the Raman polarizability, setting the lower limit of the ω_I integral to $-\infty$ and performing the ω_I integral first, one sees that

$$P_\Gamma(f|i) = 2\Gamma \int_0^\infty |\langle \phi_f | \phi_i(t) \rangle|^2 \, e^{-2\Gamma t} \, dt. \quad (14.75)$$

This integral is simply the total time the wavepacket $|\phi_i(t)\rangle$ has a nonzero overlap with $|\phi_f\rangle$ and is called the "survival probability." If we define a time-dependent probability, $P_t(f|i)$, as $P_t(f|i) = |\langle \phi_f | \phi_i(t) \rangle|^2$ then the survival probability is simply

$$P_\Gamma(f|i) = 2\Gamma \int_0^\infty P_t(f|i) \, e^{-2\Gamma t} \, dt. \tag{14.76}$$

As indicated in Eq. 14.71, for all transition dipole moments set to unity the time-dependent probability $P_t(f|i)$ obeys the sum rule

$$\sum_f P_t(f|i) = 1 \tag{14.77}$$

and so the survival probability must obey the nontrivial sum rule:

$$\sum_f P_\Gamma(f|i) = 1. \tag{14.78}$$

This sum rule states that the sum of the areas under all the excitation profiles must be one. So, the integrated excitation profiles are a probe of the amount of time the system stays in the region of the final state.

If we now fix ω_I and look at the Raman intensity as a function of ω_S, we can learn something about the virtual state represented by the Raman wavefunction. Recall that

$$I(\omega_I; \omega_S) \propto \int_{-\infty}^\infty \langle R_{\omega_I} | e^{-\frac{i}{\hbar} H_a t} | R_{\omega_I} \rangle \, e^{-i\tilde{\omega}_S t} \, dt. \tag{14.79}$$

To obtain the total dispersed fluorescence, we integrate the spectrum over ω_S. Exchanging the order of the time and frequency integrals yields

$$I_{TDF} \equiv \int I(\omega_I; \omega_S) \, d\omega_S = \langle R_{\omega_I} | R_{\omega_I} \rangle. \tag{14.80}$$

The total dispersed fluoresence is seen to be equal to the norm of the Raman wavefunction, a result that makes intuitive sense, since the latter is simply the amplitude in the virtual state created by the first photon (see Section 14.3.4). If we assume that there are no radiationless transitions and that the transition dipole moment is a constant (Condon approximation), then within a proportionality constant the total dispersed fluoresence is equal to the absorption spectrum, as we now show. Since

$$|\tilde{R}_{\omega_I}\rangle = \mu \int_0^\infty e^{i\tilde{\omega}_I t - \Gamma t} |\phi_i(t)\rangle \, dt, \tag{14.81}$$

the total dispersed fluoresence may be written as

$$I_{TDF}(\omega_I) = \frac{1}{2} \int_{-\infty}^\infty du \int_{|u|}^\infty ds \, e^{i\tilde{\omega}_I u} e^{-\Gamma s} \langle \phi_i | e^{\frac{i}{2\hbar} H_b(s-u)} \mu^2 e^{-\frac{i}{2\hbar} H_b(s+u)} | \phi_i \rangle, \tag{14.82}$$

where the variables of integration have been changed from t and t' to $u = t - t'$ and $s = t + t'$. Since we are assuming that the dipole moment is a constant, it may be removed from the integrals. Carrying out the s integral we are left with (Heller, 1982)

$$I_{TDF}(\omega_I) = \frac{\mu^2}{2\Gamma} \int_{-\infty}^\infty du \, e^{i\tilde{\omega}_I u} e^{-\Gamma |u|} \langle \phi_i | \phi_i(u) \rangle \propto I_{abs}(\omega_I), \tag{14.83}$$

which is the formula for a damped absorption spectrum.

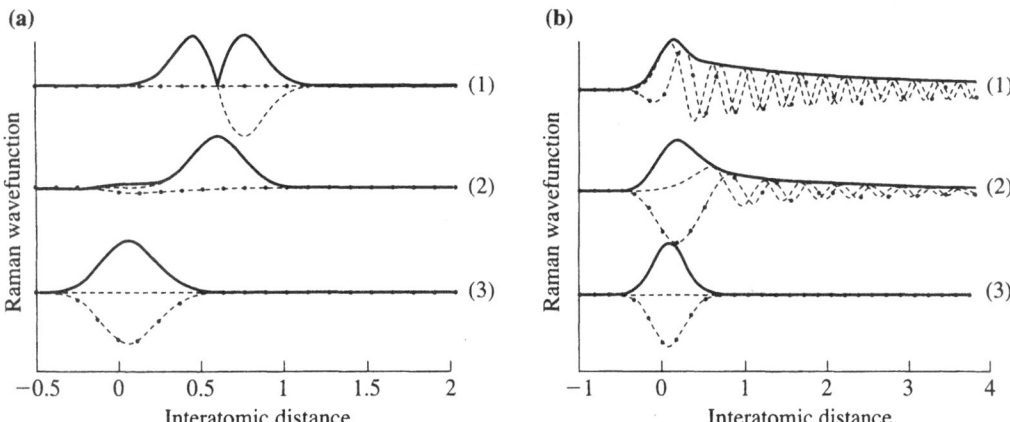

Figure 14.18 (a) Raman wavefunctions for a bound excited state, for laser excitation (1) resonant to $v' = 1$, (2) resonant to $v' = 0$ and (3) $10\ \hbar\omega$ below the bottom of the excited-state potential well. The thick solid line is the absolute value, the dashed line is the real part and the chain-linked line is the imaginary part. (b) Raman wavefunctions for a repulsive excited state, for laser excitation (1) on the center of the absorption band, (2) off the center of the absorption band and (3) below the threshold of the absorption band. Adapted from Williams (1988).

14.3.4   The Raman Wavefunction and the Role of Detuning

In this section we return to the Raman wavefunction, which was introduced in Section 14.3.1 as a convenient mathematical entity. But the Raman wavefunction has an interesting physical interpretation that, once understood, gives a new perspective on the Raman process. According to Eq. 14.58, the amplitude of Raman scattering is given by the overlap of the Raman wavefunction with the various final vibrational states of the ground state surface, indicating that the Raman wavefunction is the state from which emission takes place, the "virtual state" in the older Raman literature. Note the close resemblance between Eq. 14.52 for $|R_{\omega_{\mathrm{I}}}\rangle$ and Eq. 6.64, the formula for the spectral representation of an eigenstate. The only difference, aside from an overall multiplicative factor, is that the lower limit of the integral over τ is 0, rather than $-\infty$. This causes the Raman wavefunction to be complex, with a real part that is proportional to the molecular eigenfunction at the energy corresponding to the excitation but with an additional imaginary component. For bound systems on resonance, the imaginary part is dwarfed by the real part, and the Raman wavefunction is essentially just the molecular eigenstate; however, for unbound systems the real and imaginary parts are of comparable magnitude. Moreover, for off-resonance excitation, the real part vanishes and the only contribution is from the imaginary part. These features are illustrated in Figure 14.18.

▶ **Exercise 14.12**   Show that the real part of the Raman wavefunction is proportional to the molecular eigenstate at energy $\hbar\tilde{\omega}_{\mathrm{I}}$.

To understand these features of the Raman wavefunction we return to the definition, Eq. 14.52. Following the approach introduced in Chapter 6, we interpret the integral as a superposition of the propagating wavepacket at different times, added with the phase factor

$e^{i\tilde{\omega}t}$. In Chapter 6, we noted that at frequencies ω corresponding to an eigenvalue the spatial distribution of the eigenfunction is built up over time through constructive interference of the wavepackets, while at frequencies away from an eigenvalue the wavepackets will interfere destructively, and the resulting superposition will vanish. In the case of the Raman wavefunction this situation is modified, since the lower bound of the time integral is 0 and not $-\infty$. The real part of the Raman wavefunction is still identical to the eigenfunction produced by integrating $(-\infty, \infty)$, but now the imaginary part does not cancel completely because of the absence of negative time contributions.

This imaginary part of the Raman wavefunction, although always present, is sometimes swamped by the real part. In particular, for bound systems with resonant excitation the wavepacket returns periodically or quasiperiodically to the Franck–Condon region, interfering constructively with previous wavepackets. The real part of the Raman wavefunction is spatially just the molecular eigenfunction, but the amplitude continues to grow with each return pass of the wavepacket. In contrast, the imaginary part of the Raman wavefunction suffers destructive interference at subsequent passes, and only a small transient around $t = 0$ survives. However, there are other cases where the imaginary part of the Raman wavefunction plays a significant, or even crucial, role. If the system is unbound, there is no quasiperiodic return of the wavepacket to the Franck–Condon region: the real and the imaginary parts of the Raman wavefunction record the track of the dissociating wavepacket, and display very similar amplitude and spatial behavior, as seen in Figure 14.18b.

The imaginary part of the Raman wavefunction plays a crucial role if the excitation is detuned from resonance. In this case, the real part of the Raman wavefunction vanishes and the imaginary part, although it decreases with detuning, is the only contribution. Inspection of the Raman wavefunction in Figure 14.18a (3) and b (3) indicates that it doesn't resemble any excited-state eigenfunction, but rather reflects the short time excited-state dynamics. Indeed, for a detuning of $\Delta\omega$, the time interval that contributes to the integral for the Raman wavefunction is effectively $[0, \tau]$, where $\tau \approx 1/\Delta\omega$ is the "effective" lifetime on the excited-state surface (see Figure 14.19). This reciprocal relationship between time and detuning suggests that detuning may serve as a convenient experimental control over the time interval that contributes to the dynamics, and hence as a kind of molecular clock.

An additional perspective on this detuning clock can be obtained by returning to the second-order perturbation theory formula, Eq. 14.52. Equation 14.52 shows that the physical

Effective lifetime on upper surface

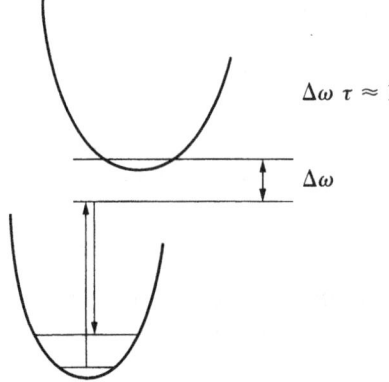

$\Delta\omega \, \tau \approx 1$

$\Delta\omega$

Figure 14.19 Schematic diagram showing the time–energy uncertainty principle operative in resonance Raman scattering. If the incident light is detuned from resonance by an amount $\Delta\omega$, the effective lifetime on the excited state is $\tau \approx 1/\Delta\omega$.

significance of the integration variable τ in the Raman wavefunction is the time *interval* between the incidence of the first photon at t'' and the scattering of the second photon at t'. If the time interval between incidence and scattering is short, the process is "scattering"; if it is long, the process is essentially a sequence of independent processes of absorption and emission. Furthermore, Eq. 14.52 indicates that there is a synchronization between the light and the matter: the time interval $t' - t''$ between incident and scattered photons is precisely the duration of the excited-state wavepacket evolution. Thus, the longer this interval, the longer the effective propagation time on the excited-state potential. When ω_I is on resonance with the excited electronic state, large values of $t' - t''$ contribute to the integral, implying both a long time interval between incidence and scattering and correspondingly long-time excited-state wavepacket dynamics; this manifests itself in a spatially extended Raman wavefunction. However, as ω_I is detuned from resonance only small values of $t' - t''$ contribute to the integral, corresponding to a short time interval between incidence and scattering, and a Raman wavefunction that is localized to the Franck–Condon region.

Finally, note that according to Eq. 14.58 the overlap between the Raman wavefunction and the vibrational eigenstates of the ground electronic state determines the Raman spectrum. This indicates that the Raman spectrum may be considered as an emission spectrum in which the Raman wavefunction, rather than an eigenstate of the excited-state potential, plays the role of the initial state. But this is just the defining property of the mysterious "virtual state" in the older Raman literature—the effective intermediate state from which the scattering can be viewed as originating, a state that exists at energies for which there are no eigenstates and that goes smoothly into an eigenstate on resonance with a bound state. Indeed, the Raman wavefunction may aptly be viewed as a rigorous definition of the Raman virtual state.

Many of the properties of the Raman wavefunction can be understood by comparing the integral $\int_0^\infty e^{i\omega t - \gamma t}\, dt$ with the integral $\int_{-\infty}^\infty e^{i\omega t - \gamma |t|}\, dt$ in the limit $\gamma \to 0$. One may show that (Appendix A)

1. the real parts of the two integrals are equivalent within a factor of 2;
2. the imaginary part of the first integral does not vanish but the imaginary part of the second integral does;
3. the imaginary part of the first integral goes as $\frac{1}{\omega}$; however, at $\omega = 0$ the imaginary part vanishes;
4. the real part of the first integral is proportional to $\delta(\omega)$.

The properties of the integral $\lim_{\gamma \to 0} \int_0^\infty e^{i\omega t - \gamma t}\, dt$ map onto those of the Raman wavefunction. On resonance (corresponding to $\omega = 0$), the Raman wavefunction is real. Away from resonance ($\omega \neq 0$), the Raman wavefunction is nonvanishing but is imaginary, and decays with detuning as $1/\omega$. From the time-domain perspective, the nonvanishing imaginary part of the Raman wavefunction is a result of the lower bound of 0 in the time integral, corresponding to a sharp onset of the integrand at positive times. This sharp onset destroys the balance between positive and negative time contributions that leads to destructive interference in the imaginary part of the integral $(-\infty, \infty)$. After a transient time, $\tau \approx 1/\omega$, the positive time imaginary contributions begin to interfere destructively *among themselves*, but the contributions from short positive times, $\tau < 1/\omega$, are never cancelled unless and until the wavepacket returns to the $t = 0$ region of phase space.

(a)   **(b)**

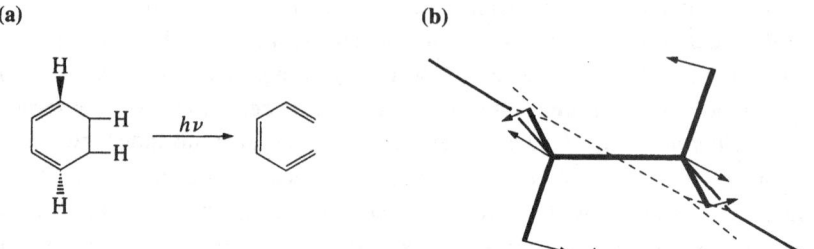

Figure 14.20 (a) Ring opening of 1,3-cyclohexadiene. (b) Geometry changes in 1,3-cyclohexadiene 10 fs after photoexcitation. The arrows represent the superposition of the motion along the 507, 948, 1321, and 1578 cm^{-1} normal coordinates on the $^1B_2$ potential surface. The displacements have been multiplied by 20 and the motions of the diene have been suppressed for clarity. Adapted from Trulson (1989).

14.3.5 Resonance Raman Scattering from Isomerizing and Dissociating Molecules

Resonance Raman scattering is thus a frequency-domain probe of the excited-state wave-packet within the first few femtoseconds after optical excitation. Therefore, it can serve as a powerful tool for elucidating mechanisms of optically induced reactions. A good example is the ring opening of 1,3-cyclohexadiene. The Woodward–Hoffmann rules, based on a symmetry analysis of the molecular orbitals in the excited electronic state, predict that the ring opening will be conrotatory, that is, the two branches after cleavage will rotate in the same direction (see Figure 14.20a). This prediction is borne out by the resonance Raman spectrum.

Figure 14.21a shows the proposed arrangement of the $^2A_1$ and $^1B_2$ surfaces in 1,3-cyclohexadiene. The abscissa represents C=C stretching and other components of the conrotatory ring opening coordinate. Figure 14.21b shows the resonance Raman spectrum of 1,3-cyclohexadiene excited at 257.3 nm. The peaks at 393 and 1623 cm^{-1} are assigned as

(a)   **(b)**

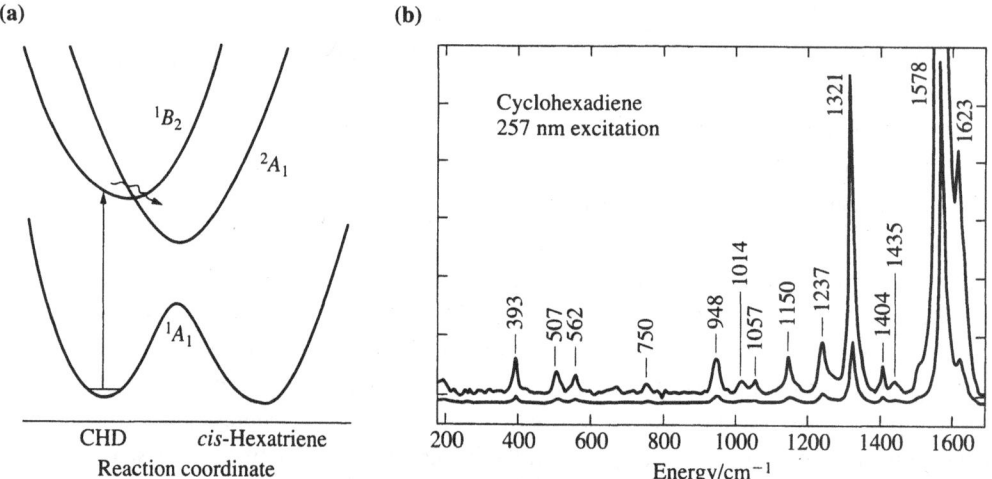

Figure 14.21 (a) Proposed arrangement of the $^2A_1$ and $^1B_2$ surfaces in 1,3-cyclohexadiene. The abscissa represents C=C stretching and other components of the conrotatory ring opening coordinate. (b) Resonance Raman spectrum of 1,3-cyclohexadiene excited at 257.3 nm. The peaks at 393 and 1623 cm^{-1} are the ethylenic torsion and stretch, respectively, of cis-1,3,5-hexatriene. Adapted from Trulson (1989).

the ethylenic torsion and stretch, respectively, of *cis*-1,3,5-hexatriene. Figure 14.20b shows a simulation of the geometry changes in 1,3-cyclohexadiene 10 fs after photoexcitation. The arrows represent the superposition of the motion along the 507, 948, 1321, and 1578 cm^{-1} normal coordinates on the $^1B_2$ potential surface. The picture that emerges from the resonance Raman spectrum is thus entirely consistent with short time conrotatory dynamics.

Although resonance Raman scattering is a rich probe of wavepacket dynamics on the excited-state potential, it can be difficult to unravel the underlying dynamics. If, however, the spectrum is dominated by short time dynamics, the Raman spectrum provides a relatively unambiguous probe. A fascinating example is resonance Raman scattering from dissociating molecules. In what follows, we will adopt the correlation function formulation of the Raman spectrum, Eq. 14.70, although one could equally well focus on the Raman spectrum as the overlap of the dissociative Raman wavefunction with the ground vibrational states, as described at the end of the previous section.

As the molecule dissociates in the excited electronic state, it emits light; if the dissociation is direct the total number of photons emitted is small, but the frequencies of emission give a rich profile of the trail of the wavepacket on the route to dissociation. Figure 14.22 shows this concept schematically in one dimension. As the wavepacket exits on the

(a)   **(b)**

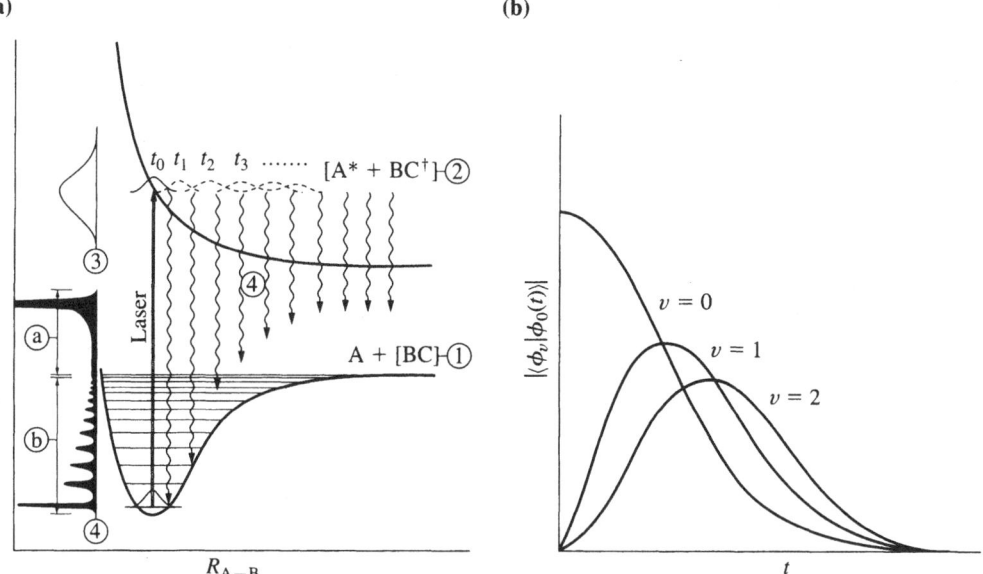

Figure 14.22 (a) Schematic illustration of a radiative recombination process (A* + BC → ABC + $h\nu$). The initial energy spread (ΔE) implies a broad and featureless absorption spectrum, while the resonance Raman spectrum is quite structured. The figure also shows schematically the relationship between different probes of photodissociation dynamics. The laser transfers the ground state wavefunction to the repulsive excited state where it evolves (dashed wavepackets t_1, t_2, t_3, ..., etc.) into A* + BC†. Indicated in the figure by numbers are accessible experimental probes: (1) equilibrium geometry and spectroscopic constants of the final BC product, (2) internal state, angular, and velocity distributions of the final products, (3) absorption (photodissociation) spectrum, (4) emission spectrum, (a – wing emission, b – discrete emission). (b) Schematic representation of the magnitude of the cross-correlation functions $\langle\phi_0|\phi_0(t)\rangle$, $\langle\phi_1|\phi_0(t)\rangle$, $\langle\phi_2|\phi_0(t)\rangle$. Note the later onset for higher overtones. Moreover, note the short time behavior: $\nu = 0$ decays as $1 - t^2$, $\nu = 1$ grows as t and $\nu = 2$ grows as t^2. For a direct dissociation process these correlation functions have no recurrences. Reprinted from Imre et al. (1984).

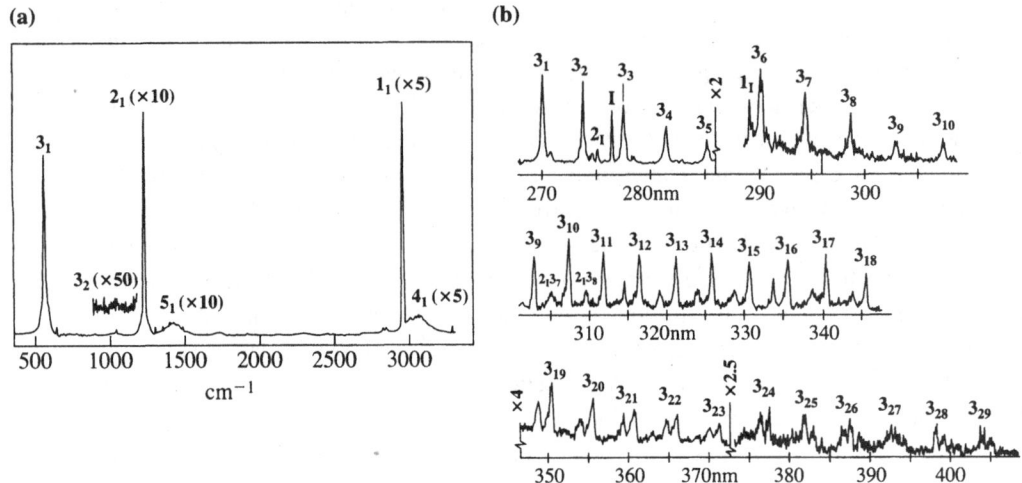

Figure 14.23 (a) CH_3I Raman spectrum obtained by exciting at 355 nm (off resonance). Note the weak overtone in the CH_3—I stretch coordinate (3_2). Because of the large detuning from resonance, the effective lifetime in the excited state is very short, and hence there is very little contribution to the Raman spectrum from excited-state dynamics. The Raman activity that is seen can be traced to the coordinate-dependent polarizability (which arises from the coordinate dependence of the transition dipole moment to all excited electronic states). (b) CH_3I Raman spectrum obtained by exciting at 266 nm (on resonance). Note the long progression in the CH_3—I stretch coordinate (3_n). On resonance, the effective lifetime in the excited electronic state is very long, leading to extensive wavepacket dynamics and an extended overtone progression. Reprinted from Imre et al. (1984).

excited electronic state, it overlaps progressively higher and higher overtones of the ground state. This is reflected in the cross-correlation functions with $v = 0$, $v = 1$ and $v = 2$: these correlation functions rise in strict sequence and have no future recurrences.

Figure 14.23 shows this concept in action in the experimental RRS spectrum from photodissociating CH_3I. The spectrum is dominated by a long progression in the CH_3—I stretching coordinate, which is the dissociation coordinate (b). If the incident frequency is detuned from resonance, the effective lifetime on the excited state is shortened and only the first few lines appear in the Raman spectrum (a). The main contribution to the Raman spectrum in this case is from the coordinate dependence of $\mu(R)$.

A second interesting example is the RRS spectrum from photodissociating ozone. The potential surface for O_3 in the two bond-length coordinates is shown in Figure 14.24a. The Franck–Condon region is along the line of symmetry, and is marked with a black dot. The photodissociation dynamics is shown schematically: at short times the wavepacket begins to move along the symmetric stretch, but soon the unstable dynamics in the antisymmetric stretch comes into play. The antisymmetric stretch motion essentially correlates with the bond-breaking coordinate (O—OO or OO—O) while the symmetric stretch motion correlates to some extent with vibrational motion in the asymptotic region. (Note that the true wavepacket bifurcates and thus retains symmetry during the dissociation process, while the trajectory in the figure shows only one-half of this process.) The RRS spectrum from ozone is shown in Figure 14.24b. There is a long progression of overtones and combinations in the symmetric and antisymmetric stretches (the bending motion is not observed). Note

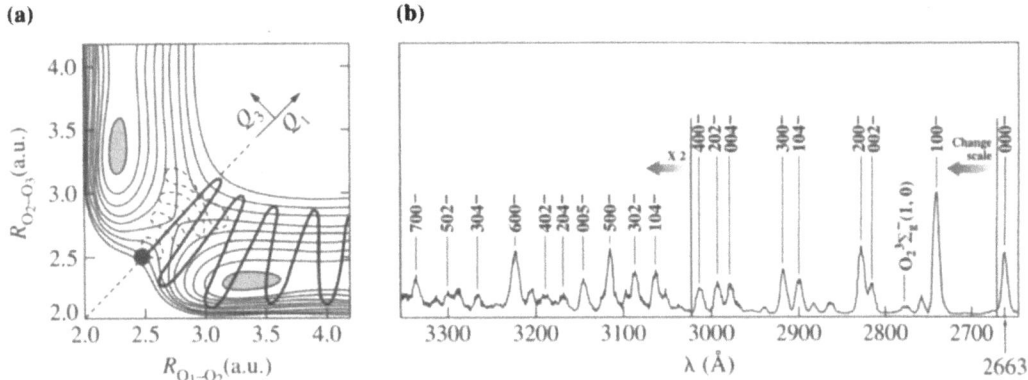

Figure 14.24 (a) Ozone B_2 excited-state surface along the two stretch coordinates Q_1 and Q_3, with the bending coordinate fixed at the ground state value. A photodissociating trajectory is shown in heavy lines. The spreading wavepacket is shown in dashed lines. Note the symmetry of the potential and the time-evolving wavepacket with respect to the two end oxygen atoms. Note further that there are quasibound regions in the exit channels, leading to long-lived trajectories. (b) Ozone resonance Raman spectrum obtained by exciting at 266 nm. Bands are labelled by three numbers: ν_1 (symmetric stretch), ν_2 (bend) and ν_3 (antisymmetric stretch). Reprinted from Imre et al. (1984).

that the RRS spectrum has both even and odd overtones in the symmetric stretch, but only even overtones and combinations in the antisymmetric stretch. This reflects the symmetry of the potential in the antisymmetric stretch and the resulting symmetry of the evolution of the quantum wavepacket in the excited state with respect to the antisymmetric stretch coordinate.

14.4   Dispersed Fluorescence Spectroscopy

As a final example of the power of the correlation function approach to problems in one- and two-photon spectroscopy, consider the dispersed fluorescence spectra of a series of alkyl-substituted benzene rings. Figure 14.25 shows a series of experimental absorption spectra for the alkylbenzenes with increasing length of alkyl tail. The absorption spectra show no apparent dependence on tail length ((a)–(c)). However, the dispersed fluorescence spectra (DFS) show a significant dependence on tail length ((e)–(f)), increasing in complexity with increasing tail length. We may partition the complex DFS in (e) into sharp lines and broad background; (f) shows a blow-up of the DFS of pentylbenzene, with the sharp lines (marked with asterisks) clearly distinguished from the broad background. To understand the origin of these features, we note that the electronic excitation is localized to the benzene ring. The sharp lines in the DFS are due to emission from the initial ring excited state, while the broad background is due to emission from states in which the energy has redistributed from the ring to the alkyl tail. A short tail has only a small number of available modes, but a long tail has many low-frequency modes into which the initial quantum of ring energy can be converted; thus, if the energy has a chance to fully redistribute in the long tail the DFS will be quite complicated, reflecting emission from many electronically excited low-frequency tail states.

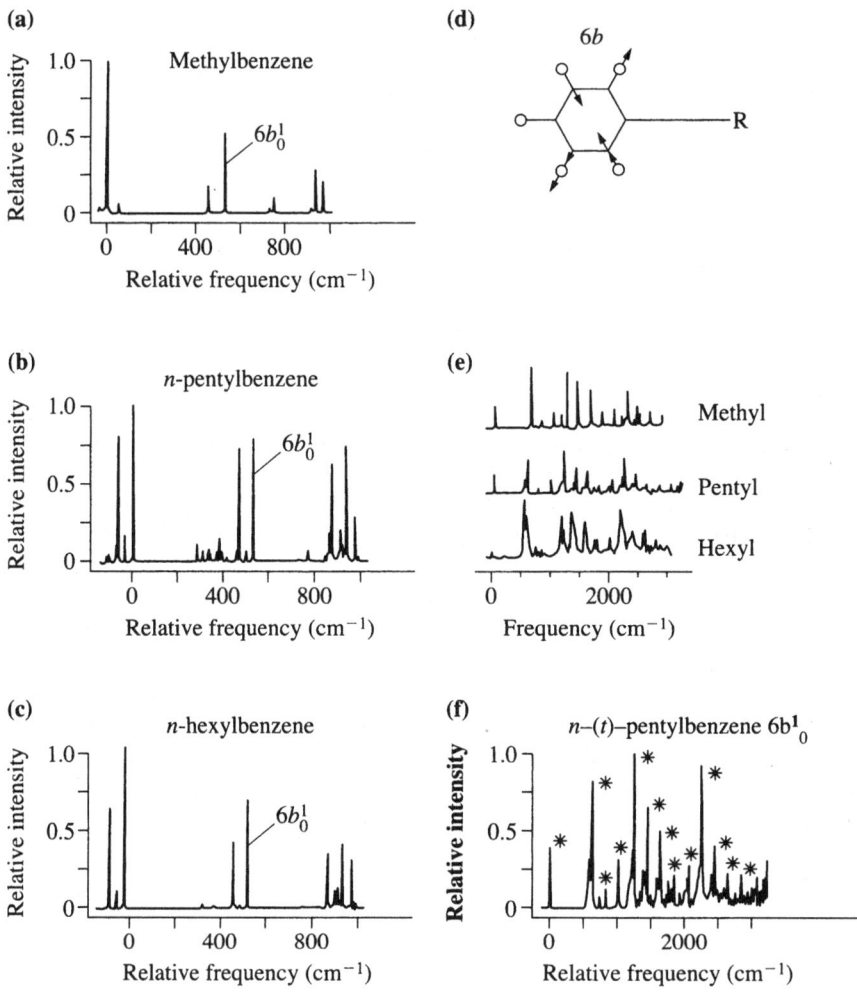

Figure 14.25 (a) Absorption spectrum for jet-cooled methylbenzene. The arrow indicates the $6b_0^1$ vibrational feature. (b) Same as (a) but for pentylbenzene. (c) Same as (a) but for hexylbenzene. (d) Form of the $6b_0^1$ normal mode, which involves no tail motion. The position of the absorption feature associated with this vibration does not shift position as the length of the tail is increased. (e) Dispersed fluorescence spectrum (DFS) of jet-cooled methylbenzene, pentylbenzene and hexylbenzene. The incident light was tuned to the $6b_0^1$ vibrational feature. (f) Expanded trace of DFS for jet-cooled pentylbenzene. Asterisks indicate the sharp lines or "tense" fluorescence. Broad features represent relaxed fluorescence. Reprinted from J. B. Hopkins et al., I and II (1980).

The vibrational energy transfer from the ring to the tail can be described in terms of wavepacket evolution, although there are some conceptual differences from the wavepackets we have been dealing with so far in this chapter. We may define a Hamiltonian as

$$H = H_S + H_B + H_{SB}, \qquad (14.84)$$

where H_S is the Hamiltonian of the system (benzene ring), H_B is the Hamiltonian of the bath (alkyl tail) and H_{SB} is the system–bath coupling. In thinking about the energy transfer from ring to tail it is convenient to define a zeroth-order Hamiltonian

$$H_0 = H_S + H_B \tag{14.85}$$

with zeroth-order eigenstates, $|l_n\rangle$

$$H_0|l_n\rangle = E_n|l_n\rangle. \tag{14.86}$$

Note that the states $|l_n\rangle$ are not eigenstates of the full ring + tail Hamiltonian; thus if one were to prepare an initial state $|l_i\rangle$ (i is one of the n), it would could correspond to a superposition of eigenstates of the full H. Thus, as time evolves this initial state will have changing phase factors for each of its component eigenstates, leading to evolution, in this case from the ring excitation to tail excitation. Note that the time scale here for the wavepacket dynamics, which is the time scale for energy transfer between the ring and the tail, is on the order of picoseconds for the short time dynamics and on the order of nanoseconds for the long time dynamics. To avoid confusion, we emphasize that the experimental plots in Figure 14.25 correspond to CW excitation, and not short pulse (picosecond) excitation, which would literally prepare the initial state $|l_i\rangle$. Nevertheless, in the same way that wavepacket dynamics is a useful construct in interpreting the RRS spectrum in the previous section—despite the CW conditions of the experiment—here too, the wavepacket dynamics provides a useful picture and the relevant wavepacket is the zeroth-order state $|l_i\rangle$.

Having established the basic analogy with the early parts of this chapter we now elaborate on two further aspects of this analogy. The first is that the sharp emission, corresponding to emission from the vibrationally excited benzene ring with a cold tail, is the *analog* of Rayleigh scattering. Although the sharp emission consists of many lines, collectively all these lines have the character of no tail excitation and hence correspond to emission from the initial vibrational *mode* without any time evolution into other vibrations (i.e., the tail). Rayleigh scattering is the emission to the original vibrational *state* without evolution to other states, even in the same degree of freedom. The sharp emission may therefore be thought of as playing the role, on the picosecond time scale, of the Rayleigh line on the femtosecond time scale, its depletion relative to the other lines measuring energy transfer from the initial state/mode. Similarly, all the broad lines reflect emission from at least a partially hot tail; they are inelastic scattering in the sense that the excitation has left the initial portion of the molecule in which it was deposited and hence is the analog of Raman scattering, in which the emission is to a state other than the initial one. In what follows, this analogy will emerge precisely from the equations.

The second extension is the idea of detuning. Although the time scales are now picosecond instead of femtosecond, the time–energy uncertainty relation $\Delta\omega\,\tau \simeq 1$ still applies. Tuning the exciting laser directly to one of the lines in the ring electronic absorption spectrum, $\Delta\omega = 0$, results in the infinite-time (in this case, nanosecond) limit of the emission spectrum. If the exciting laser is detuned from resonance, then the emission will probe shorter (picosecond) time scales. The dispersed fluoresence spectrum of the long "tail" alkylbenzenes should simplify upon detuning, resembling that of shorter and shorter "tail" alkylbenzenes and eventually benzene itself. Moreover, the order of disappearance of lines in the DFS as the exciting laser is detuned contains a great deal of information on the "pathway" of energy transfer as it migrates from the ring to the tail and within the tail: the lines that disappear first correspond to modes that received the energy later, and conversely,

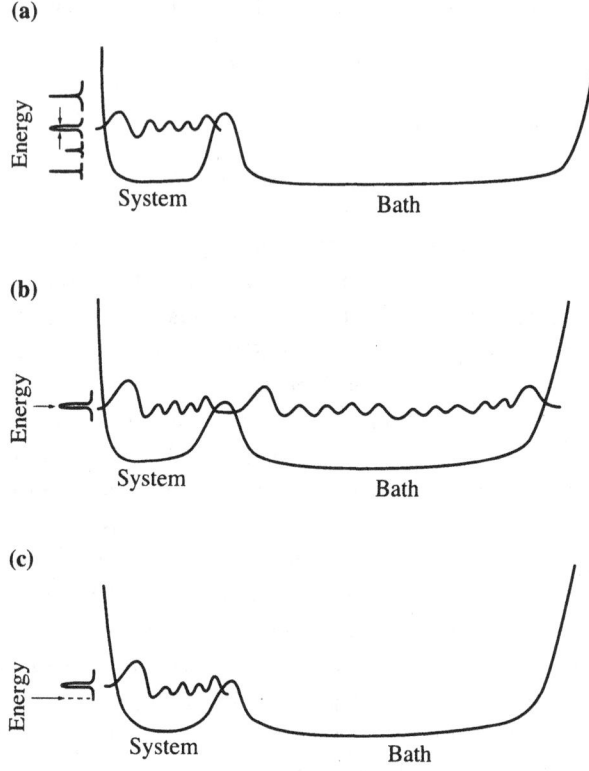

Figure 14.26 (a) Schematic zeroth-order vibrational feature (indicated by arrows) and corresponding zeroth-order state for a system with a large bath. (b) Schematic Raman wavefunction for the same potential as in (a), ·for $\omega_i + \omega_I$ tuned directly to the zeroth-order vibrational feature. (c) Schematic Raman wavefunction for the same potential as in (a) but $\omega_i + \omega_I$ is detuned from the vibrational feature. Reprinted from D. J. Tannor et al. (1984).

the lines that disappear last correspond to modes that received the energy first. One way of understanding this simplification is in terms of the Raman wavefunction. On resonance, the Raman wavefunction is excited in both the ring and the tail; if the incident light is detuned from the vibrational absorption feature, the Raman wavefunction will reflect only short time dynamics: vibrational excitation only in the ring (see Figure 14.26). The projection of the Raman wavefunction onto the eigenstates of the ground state potential surface gives the Raman spectrum, and hence detuning should lead to a Raman spectrum with only sharp lines and no broad background (a cold tail).

The mathematical description is facilitated by defining projection operators, $\mathcal{P}_S$ and $\mathcal{P}_B$, where $\mathcal{P}_S$ projects onto states in the joint system-bath space that are cold (unexcited) in the bath and $\mathcal{P}_B = 1 - \mathcal{P}_S$ projects onto everything else. In addition, we define projection operators for disjoint regions of the energy spectrum. The well-defined vibrational structure of the electronic absorption spectrum implies that one can replace the initial state, ϕ_i, with one of the zeroth-order system states, l_i; the other zeroth-order states, l_n, $n \neq i$, are removed in energy and to a good approximation do not contribute to either the absorption or the scattering, provided the excitation is sufficiently close in energy to l_i. One can exploit this observation by defining a projector, $\mathcal{P}_n$, onto the energy range surrounding each of the zeroth-order states, l_n, where $\sum_n \mathcal{P}_n = 1$; we will be interested in the projector for $n = i$, $\mathcal{P}_i$. With the aid of these two projectors, $\mathcal{P}_S$ and $\mathcal{P}_n$, we now proceed to a quantitative analysis.

The ratio of "sharp" emission to the total emission is defined by

$$\frac{S(\omega_I)}{I_{\text{TDF}}(\omega_I)} = \frac{\sum_{f'} I^{f'}(\omega_I)}{\sum_f I^f(\omega_I)}, \tag{14.87}$$

where the prime on the index of the sum in the numerator indicates that the sum is restricted to final states that are cold in the tail. Recall that the intensity of Raman scattering into a final state, $I^f(\omega_I)$, is proportional to the square of the magnitude of the polarizability of the transition, $\alpha_{fi}(\omega_I)$. Since $\alpha_{fi} = \langle \phi_f | R_{\omega_I} \rangle$, then

$$\sum_{f'} I^{f'}(\omega_I) = \underbrace{\sum_{f'} \langle R_{\omega_I} | \phi_{f'} \rangle \langle \phi_{f'} | R_{\omega_I} \rangle}_{\substack{\text{complete set of states} \\ \text{for } cold\ tail\ only}} = \langle R_{\omega_I} | \mathcal{P}_S | R_{\omega_I} \rangle. \tag{14.88}$$

Also recall that the total dispersed fluoresence is given by Eq. 14.80:

$$I_{\text{TDF}}(\omega_I) = \sum_f I^f(\omega_I) = \langle R_{\omega_I} | R_{\omega_I} \rangle. \tag{14.89}$$

Thus,

$$\frac{S(\omega_I)}{I_{\text{TDF}}(\omega_I)} = \frac{\langle R_{\omega_I} | \mathcal{P}_S | R_{\omega_I} \rangle}{\langle R_{\omega_I} | R_{\omega_I} \rangle}. \tag{14.90}$$

Equation 14.90 has the interesting physical interpretation that the sharp emission is the emission from the system portion of the Raman wavefunction. To proceed further, we make three observations. First, the zeroth-order vibrational states $\{l_n\}$ can be used as a complete basis to expand the *system* Hilbert space, and therefore we can write

$$\mathcal{P}_S = \sum_n |l_n\rangle \langle l_n| \otimes |0\rangle \langle 0|. \tag{14.91}$$

Second, for excitation at an energy near the absorption feature l_i, the scattering is dominated by this single resonant feature. This leads to the conclusion (see Exercise 14.13) that the energy spread of the Raman wavefunction, R_{ω_I}, is localized within the energy range defined by $\mathcal{P}_i$, and therefore

$$\mathcal{P}_i R_{\omega_I} = R_{\omega_I}. \tag{14.92}$$

Third, applying the projection operator $\mathcal{P}_i$ to both sides of Eq. 14.91 gives

$$\mathcal{P}_i \mathcal{P}_S = \mathcal{P}_i \sum_n |l_n\rangle \langle l_n| \otimes |0\rangle \langle 0| = |l_i\rangle \langle l_i| \otimes |0\rangle \langle 0|, \tag{14.93}$$

where the last equation follows from the definition of $\mathcal{P}_i$. Inserting Eqs. 14.92 and 14.93 into Eq. 14.90 gives (Tannor, 1984b)

$$\frac{S(\omega_I)}{I_{\text{TDF}}(\omega_I)} = \frac{\langle R_{\omega_I} | \mathcal{P}_i \mathcal{P}_S \mathcal{P}_i | R_{\omega_I} \rangle}{\langle R_{\omega_I} | R_{\omega_I} \rangle} = \frac{|\langle l_i | R_{\omega_I} \rangle|^2}{\langle R_{\omega_I} | R_{\omega_I} \rangle}. \tag{14.94}$$

▶ **Exercise 14.13** Verify the statement above, that if the scattering is dominated by a single resonant absorption feature, l_i, then the energy spread of the Raman wavefunction, R_{ω_I}, is localized within the energy range defined by $\mathcal{P}_i$, and therefore $\mathcal{P}_i R_{\omega_I} = R_{\omega_I}$.

Solution   Note that since the initial state, ϕ_i, is localized to the system, the system zeroth-order states $\{l_n\}$ form a complete basis for the expansion $\phi_i = \sum_n c_n l_n$. Therefore, the Raman wavefunction can be written as

$$R_{\omega_I} = \int_0^\infty \phi_i(t) e^{i(\omega_i + \omega_I)t - \Gamma t}\, dt$$

$$= \sum_n c_n \int_0^\infty l_n(t) e^{i(\omega_i + \omega_I)t - \Gamma t}\, dt$$

$$\approx c_n \int_0^\infty l_n(t) e^{i(\omega_n + \Delta\omega)t - \Gamma t}\, dt, \tag{14.95}$$

where the Raman detuning has been implicitly defined by requiring that $\omega_i + \omega_I = \omega_n + \Delta\omega$, where n is the index of the group of lines whose energy is closest to the excitation energy. In Eq. 14.95 we have deleted the summation, since the detuning factor strongly discriminates against zeroth-order states other than the one closest to the laser excitation frequency. Since $c_n l_n = \mathcal{P}_n \phi_i$, we find that

$$R_{\omega_I} = \int_0^\infty \mathcal{P}_n \phi_i(t) e^{i(\omega_i + \omega_I)t - \Gamma t}\, dt = \mathcal{P}_n R_{\omega_I}, \tag{14.96}$$

since $\mathcal{P}_n$ can be taken outside the integral.

Substituting Eq. 14.95 into both the numerator and denominator in Eq. 14.94, and using Eq. 14.83, we obtain the central result of this section (Tannor, 1984b):

$$\frac{S(\omega_I)}{I_{\text{TDF}}(\omega_I)} = \frac{\left| \int_0^\infty \langle l_i | l_i(t) \rangle e^{i(\tilde{\omega}_0 + \Delta\omega)t - \Gamma|t|}\, dt \right|^2}{\frac{1}{2\Gamma} \int_{-\infty}^\infty \langle l_i | l_i(t) \rangle e^{i(\tilde{\omega}_0 + \Delta\omega)t - \Gamma|t|}\, dt}, \tag{14.97}$$

where we have defined $\tilde{\omega}_0 = \omega_n$, $n = i$ (note that i here is the index for the zeroth-order state that is in resonance). Equation 14.97 expresses the ratio of "sharp" to total emission as a ratio of the square of the half-Fourier transform of the wavefunction autocorrelation function to the full Fourier transform. Since $\langle l_i | l_i(t) \rangle$ can be determined from the high-resolution absorption spectrum, all the quantities needed to apply this theory can be experimentally determined. Note that the numerator, which is a measure of the total amount of sharp emission, has the structure of an equation for Rayleigh scattering, except that $|\phi_i\rangle$ is replaced by $|l_i\rangle$ in both the bra and the ket. In physical terms, the replacement in the ket is allowed because the frequency of excitation is nearly resonant with one of the l_n, allowing the neglect of all the others. The replacement in the bra is allowed because the sum over all cold-tail states in the ground electronic state spans the same subspace as the sum over all cold-tail states in the excited state. The latter subspace is spanned by the $\{l_n\}$, that is, all sharp emission must come from one of the l_n. However, since all the l_n are, by assumption, in disjoint regions of the energy spectrum, the time evolution of l_i cannot access any other of the l_n; thus all the sharp emission may be viewed as coming from the short time dynamics of a single zeroth-order state, l_i. This beautifully confirms our qualitative observation at the beginning of this section, that the sharp emission, collectively, is the analog of Rayleigh scattering. We see that l_i plays the dual role as the effective initially prepared state (the rest of ϕ_i is filtered by the excitation process) and as the probe for the collective emission to any of the sharp lines.

Using Eq. 14.97 one can determine the behavior of the ratio of "sharp" emission to total emission in some limits. For short times, corresponding to large detunings, $\langle l_i | l_i(t) \rangle = 1 + \cdots$ and so

$$\frac{S(\omega_{\mathrm{I}})}{I_{\mathrm{TDF}}(\omega_{\mathrm{I}})} = 1. \tag{14.98}$$

At large detunings, the sharp spectrum corresponding to emission from the benzene ring is recovered. On the other hand, for resonant excitation, $\Delta\omega = 0$. If we take the band center, $\tilde{\omega}_0$, as zero, and assume that $\langle l_i | l_i(t) \rangle$ is real, then Eq. 14.97 may be written as

$$\frac{S(\omega_{\mathrm{I}})}{I_{\mathrm{TDF}}(\omega_{\mathrm{I}})} = \frac{\int_0^\infty \langle l_i | l_i(t) \rangle e^{-\Gamma t}\, dt}{\int_0^\infty e^{-\Gamma t}\, dt}. \tag{14.99}$$

This is simply the ratio of the amount of time the wavefunction is in the system to the total time of the energy transfer (set by the lifetime Γ). If one makes the common assumption that $\langle l_i | l_i(t) \rangle \sim e^{-\alpha|t|}$, then the ratio is

$$\frac{S(\omega_{\mathrm{I}})}{I_{\mathrm{TDF}}(\omega_{\mathrm{I}})} = \frac{\Gamma}{\Gamma + \alpha}. \tag{14.100}$$

Surprisingly, this result has no dependence on the detuning. Since the Fourier transform of an exponential is a Lorentzian, and since a Lorentzian has very slowly decaying wings, one can never truly detune. However, $e^{-\alpha|t|}$ has an unphysical cusp at $t = 0$. A more physically reasonable form for the correlation function is the so-called Kubo form,

$$\langle l_i | l_i(t) \rangle = e^{\frac{\Delta^2}{\gamma^2}(\gamma|t| - 1 + e^{-\gamma|t|})}. \tag{14.101}$$

By varying γ and Δ the Kubo form may be "tuned" between a Gaussian at short times to an exponential at long times. For this form of the correlation function, the ratio of "sharp" to total emission does, indeed, vary from 0 to almost 1 as the exciting laser is detuned from resonance (see Figure 14.27).

Figure 14.28 shows a calculation on a three-dimensional model system (x, y, z). The ground state potential is three harmonic oscillators ($\omega_x = 1.0$, $\omega_y = 1.1$, $\omega_z = 1.0$), while the excited-state potential is three harmonic oscillators (again $\omega_x = 1.0$, $\omega_y = 1.1$, $\omega_z = 1.0$) coupled together by cubic terms and displaced in the x coordinate only. The initial state, $v = 0$ of the ground state, consists of a superposition of eigenstates of the excited-state *zeroth-order Hamiltonian*, which is the three-dimensional harmonic oscillator. This is a superposition of the zeroth-order states, l_n; conceptually we can think of a single, global wavepacket corresponding to the $v = 0$ state evolving on the excited state, or of the l_n states, each of which has a much narrower energy range (and correspondingly, a much longer time scale for evolution) evolving independently. Sharp emission corresponds to RRS to any of

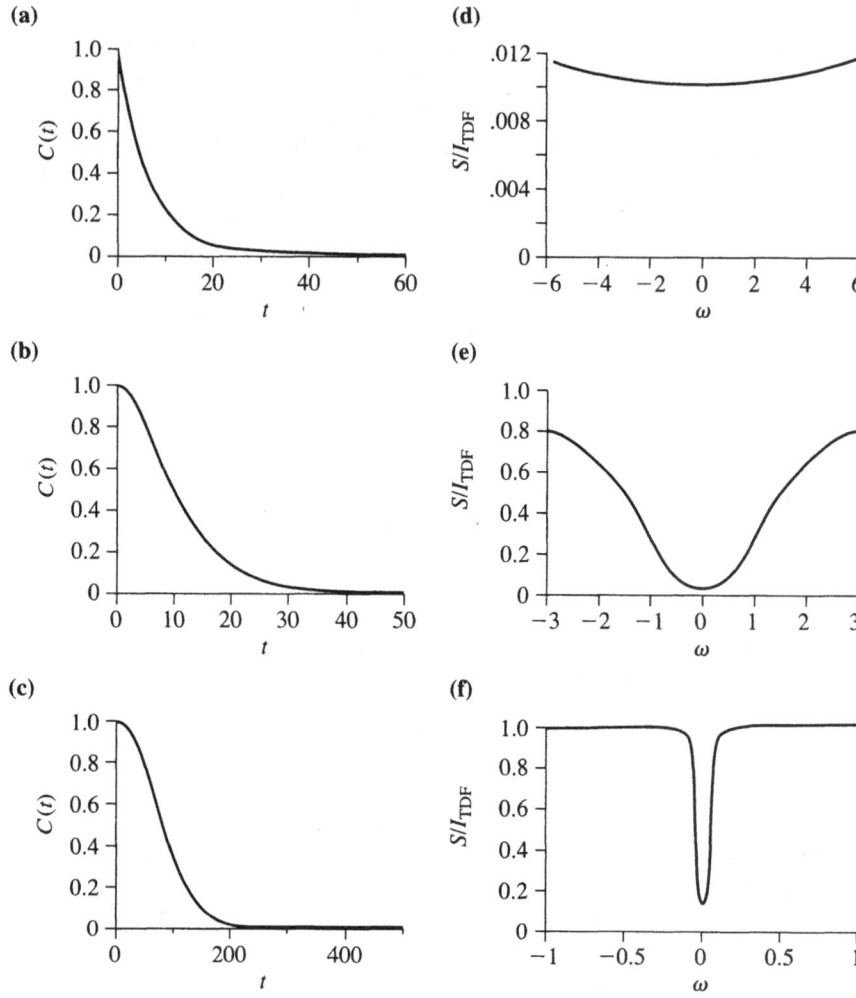

Figure 14.27 Correlation functions and S/I_{TDF} using the Kubo model, $C(t) = \langle l_i | l_i(t) \rangle = e^{-\frac{\Delta^2}{\gamma^2}(\gamma t - 1 + e^{-\gamma t})}$. Δ, γ and Γ (see Eq. (14.99)) all have dimensions of inverse time. (a) $C(t)$ for $\Delta = 0.15$, $\gamma = 15$, $\Gamma = 0.0015$. (b) Same as (a) but $\Delta = 0.15$, $\gamma = 0.15$. (c) Same as (a) but $\Delta = 0.015$, $\gamma = 0.0015$. (d) $\frac{S(\omega)}{I_{TDF}(\omega)}$ versus ω for the parameters in (a) ($\hbar$). (e) $\frac{S(\omega)}{I_{TDF}(\omega)}$ versus ω for the parameters in (b). (f) $\frac{S(\omega)}{I_{TDF}(\omega)}$ versus ω for the parameters in (c). Reprinted from D. J. Tannor et al. (1984).

the ground surface eigenstates with no tail (y, z) excitation, while broad emission is defined as RRS to a ground surface eigenstate with any tail excitation whatsoever. The ratio of the total sharp emission to the total dispersed fluorescence is shown in Figure 14.28. Clearly, around each of the zeroth-order absorption features is a dip in this ratio, with the magnitude of dip increasing at higher energies where the density of bath states is higher. Thus, each absorption feature, corresponding to a different l_n, functions independently, evolving from ring to tail when the excitation is resonant and not having a chance to evolve completely when the excitation is detuned from resonance.

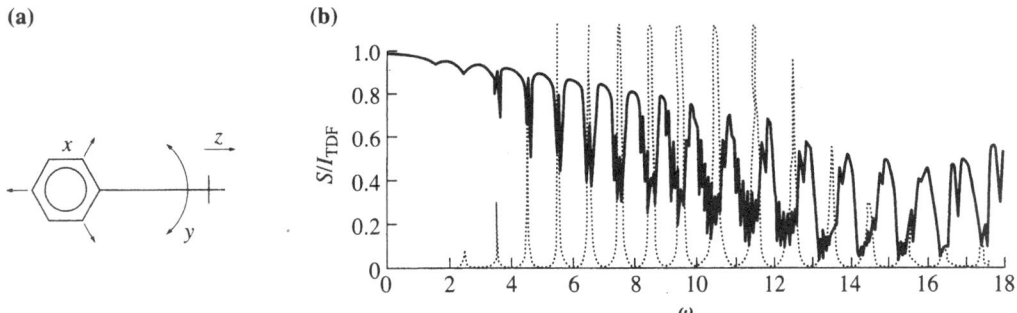

(a)   **(b)**

Figure 14.28 (a) Form of the normal mode coordinates x, y, z for the three-dimensional anharmonic system discussed in the text. (b) $\frac{S(\omega)}{I_{TDF}(\omega)}$ versus ω for a three-dimensional anharmonic system. Dashed line is the absorption spectrum. Reprinted from Tannor (1984).

Problems

▶ **Exercise 14.14** Using the generating function for the Hermite polynomials, show that for the harmonic oscillator, with $H = \frac{1}{2}m\omega^2 x^2$, the cross-correlation function $C_{n0}(t)$ is given by

$$C_{n0}(t) \equiv \langle \psi_n | e^{-\frac{i}{\hbar}Ht} | \psi_0 \rangle$$
$$= e^{-\frac{\Delta^2}{2}\left(1 - e^{-i\omega t}\right)}\left(1 - e^{-i\omega t}\right)^n \frac{(-1)^n \Delta^n}{(2^n n!)^{1/2}}, \tag{14.102}$$

where ψ_0 is the displaced ground state

$$\psi_0 = \left(\frac{m\omega}{\pi\hbar}\right)^{1/4} e^{-\frac{m\omega}{2\hbar}(x-x_0)^2} \tag{14.103}$$

and $\Delta = \sqrt{m\omega/\hbar}\, x_0$.

▶ **Exercise 14.15** This exercise and those that follow are based on a short time expansion of $C_{n0}(t)$, and hence the resulting formulas are intended to be used for moderate to large detunings of ω_I from the absorption maximum, as well as for dissociative excited states (Heller, 1982).

a. Show that a short time expansion of $C_{n0}(t)$ gives

$$C_{n0}(t) = e^{-\frac{\sigma^2 t^2}{2} - i\frac{Et}{\hbar}}(i\omega t)^n \frac{\Delta^n(-1)^n}{(2^n n!)^{1/2}}, \tag{14.104}$$

where we have defined $\sigma^2 \equiv \frac{\omega^2 \Delta^2}{2}$, and $E \equiv \frac{\Delta^2 \hbar\omega}{2} + \frac{\hbar\omega}{2} = \langle H \rangle$ is the expectation value of the excited-state energy in the Franck–Condon region.

b. Show that for an excited-state harmonic oscillator we have $\omega^2 \Delta^2 = \frac{(V_x)^2}{\hbar\omega}$, where V_x is the slope of the excited-state potential in mass-weighted (but not dimensionless) coordinates. (The expression in terms of Franck–Condon slopes, although derived from the harmonic expression, is in fact a more general form at short times.)

c. Using the result of part b, show that the Raman intensity for scattering into a fundamental vibration is given by

$$|\alpha_{10}(\omega_I)|^2 = \frac{V_x^2}{2\omega_0\sigma^4} F_1\left(\frac{\tilde{\omega}_I - \Omega}{\sigma}\right) \tag{14.105}$$

and that the Raman intensity for scattering into an overtone vibration is given by

$$|\alpha_{20}(\omega_I)|^2 = \frac{V_x^4}{8\omega_0^2\sigma^6} F_2\left(\frac{\tilde{\omega}_I - \Omega}{\sigma}\right), \tag{14.106}$$

where we have defined the function

$$F_n(\omega) = \left| \int_0^\infty e^{-t^2/2 + i\omega t} t^n \, dt \right|^2 \tag{14.107}$$

and $\Omega = E/\hbar$.

▶ Exercise 14.16

a. Use the formula from part c of Exercise 14.15 to show that the ratio of Raman intensities into two different fundamentals, in coordinates k and k', goes as (Heller, 1982)

$$\frac{I_{10}^k}{I_{10}^{k'}} = \frac{\omega_{0k'}}{\omega_{0k}} \left(\frac{V_k}{V_{k'}}\right)^2. \tag{14.108}$$

b. Show that the ratio of the Raman intensity into the overtone relative to the fundamental is

$$\frac{I_{20}^k}{I_{10}^k} = \frac{V_k^2}{4\omega_{0k}\sigma^2} \frac{F_2(\frac{\tilde{\omega}_I - \Omega}{\sigma})}{F_1(\frac{\tilde{\omega}_I - \Omega}{\sigma})}. \tag{14.109}$$

c. Calculate the ratio $F_2(\omega)/F_1(\omega)$ numerically. Discuss the detuning dependence of the different overtones relative to the fundamental, and draw a connection with the time–energy uncertainty principle.

d. Compare the accuracy of the short time formulas in parts a, b, and c with the exact harmonic oscillator results, based on Eq. 14.104 without any short time assumption. Show that they agree in the pre- and postresonant frequency regions. You will need to do a numerical calculation, using MATLAB, for example.

▶ Exercise 14.17

a. Show that in the case of multiple displaced modes one must replace σ in the above exercises by (Heller, 1982)

$$\sigma^2 = \sum_k \frac{V_k^2}{2\omega_{0k}}. \tag{14.110}$$

b. Next, show that in the short time limit the absorption spectrum is given by

$$\epsilon(\omega_{\mathrm{I}}) = \kappa \omega_{\mathrm{I}} \exp\left[-\frac{(\tilde{\omega}_{\mathrm{I}} - \Omega)^2}{2\sigma^2}\right], \tag{14.111}$$

where the usual absorption spectrum prefactors have been absorbed into the parameter κ.

c. Note that absolute resonance Raman intensities are very difficult to measure experimentally. The previous exercise showed how *relative* resonance Raman intensities can give *ratios* of the slope of the excited-state potential in different coordinates. Use the results of parts a and b of this exercise to describe how the absorption spectrum, Eq. 14.111 with Eq. 14.110, can be used to scale the slopes obtained from the preresonance Raman intensities to obtain an absolute calibration.

▶ **Exercise 14.18**   Expand the correlation function in Eq. 14.104 for small Δ. How does the ratio of two fundamentals in the Raman spectrum scale with Δ in this regime?

▶ **Exercise 14.19**   Now consider the case that $\phi_0(0)$ is a Gaussian centered on a parabolic barrier, so there is wavepacket spreading but no displacement of the center. Take the state ϕ_f again as the nth harmonic oscillator eigenstate.

a. Use the analytic solution for a Gaussian spreading on a parabolic barrier (see Exercise 3.19), together with the generating function for the Hermite polynomials, to calculate $C_{n0}(t)$. Show that $C_{n0}(t) = 0$ if n is odd. Compare your results with numerical calculations using MATLAB.

b. Expand the expression for $C_{n0}(t)$ in part a up to short times. Show that

$$|\alpha_{20}(\omega_{\mathrm{I}})|^2 = \frac{(\omega_0^2 - V_{xx})^2}{8\omega_0^2\sigma^4} F_1\left(\frac{\tilde{\omega}_{\mathrm{I}} - \Omega}{\sigma}\right). \tag{14.112}$$

c. From your numerical study of the ratio $F_2(\omega)/F_1(\omega)$ in Exercise 14.16, comment on the detuning dependence of the intensity of overtones arising from wavepacket spreading versus those arising from wavepacket displacement. Compare the analytical result in the pre- and postresonant frequency region with the numerical harmonic oscillator results from MATLAB without any short time assumption.

References

General

1. E. J.Heller, The semiclassical way to molecular spectroscopy, Acc. Chem. Res. 14, 368 (1981).
2. J. Franck, Elementary processes of photochemical reactions, Trans. Faraday Soc. 31, 536 (1925).
3. E. U. Condon, Nuclear motion associated with electron transitions in diatomic molecules, Phys. Rev. 32, 858 (1928).

Absorption

4. E. J. Heller, Quantum corrections to classical photodissociation models, J. Chem. Phys. 68, 2066 (1978).

5. R. T Pack, Simple theory of diffuse vibrational structure in continuous uv spectra of polyatomic molecules. I. Collinear photodissociation of symmetric triatomics, J. Chem. Phys. 65, 4765 (1976).

6. E. J. Heller, Photofragmentation of symmetric triatomic molecules: Time dependent picture, J. Chem. Phys. 68, 3891 (1978).

7. E. J. Heller and W. M. Gelbart, Normal mode spectra in pure local mode molecules, J. Chem. Phys. 73, 626 (1980).

8. E. J. Heller, E. B. Stechel and M. J. Davis, Molecular spectra, Fermi resonances, and classical motion, J. Chem. Phys. 73, 4720 (1980).

9. K. C. Kulander and E. J. Heller, Time dependent formulation of polyatomic photofragmentation: Application to H_3^+, J. Chem. Phys. 69, 2439 (1978).

Emission

10. L. Tutt, D. Tannor, E. J. Heller and J. I. Zink, The MIME effect: Absence of normal modes corresponding to vibronic spacings, Inorg. Chem. 21, 3858 (1982).

11. L. Tutt, D. Tannor, J. Schindler, E. J. Heller and J. I. Zink, Calculation of the missing mode effect frequencies from Raman intensities, J. Phys. Chem. 87, 3017 (1983).

Photoelectron Spectroscopy

12. A. J. Lorquet, J. C. Lorquet, J. Delwiche and M. J. Hubin-Franskin, Intramolecular dynamics by photoelectron spectrsocopy. I. Application to N_2^+, HBr^+, and HCN^+, J. Chem. Phys. 76, 4692 (1982).

Photodetachment/Transition State Spectroscopy

13. D. M. Neumark, Transition-state spectroscopy via negative ion photodetachment, Acc. Chem. Res. 26, 33 (1993).

14. R. B. Metz, S. E. Bradforth and D. M. Neumark, Transition state spectroscopy of bimolecular reactions using negative ion photodetachment, Adv. Chem. Phys. 81, 1 (1992).

15. D. E. Manolopoulos, K. Stark, H.-J. Werner, D. W. Arnold, S. E. Bradforth and D. M. Neumark, The transition state of the F + H_2 reaction, Science 262, 1852 (1993).

Resonance Raman Spectroscopy

16. H. A. Kramers and W. Heisenberg, Über die Streuung von Strahlen durch Atome, Z. Phys. 31, 681 (1925) (English translation, On the dispersion of radiation by atoms, in *Sources of Quantum Mechanics*, ed. B. L. van der Waerden (Dover, New York, 1968)).

17. P. A. M. Dirac, The quantum theory of dispersion, Proc. R. Soc. London 114, 710 (1927).

18. D. Lee and A. C. Albrecht, A unified view of Raman, resonance Raman, and fluorescence spectroscopy (and their analogues in two-photon absorption), in *Advances in Infrared and Raman Spectroscopy*, vol. 12, p. 179, ed. R. J. H. Clark and R. E. Hester (Wiley Heyden, London, 1985).

19. S.-Y. Lee and E. J. Heller, Time-dependent theory of Raman scattering, J. Chem. Phys. 71, 4777 (1979).

20. D. J. Tannor and E. J. Heller, Polyatomic Raman scattering for general harmonic potentials, J. Chem. Phys. 77, 202 (1982).

21. E. J. Heller, R. L. Sundberg and D. J. Tannor, Simple aspects of Raman scattering, J. Phys. Chem. 86, 1822 (1982).

22. D. E. Morris and W. H. Woodruff, Detailed aspects of Raman scattering: Overtone and combination intensities and prescriptions for determining excited-state structure, J. Phys. Chem. 89, 5795 (1985).

23. R. L. Sundberg and E. J. Heller, Preparation and dynamics of vibrational hot spots in polyatomics via Raman scattering, Chem. Phys. Lett. 93, 586 (1982).

24. S. O. Williams and D. G. Imre, Raman spectroscopy: Time dependent pictures, J. Phys. Chem. 92, 3363 (1988).

25. A. B. Myers and R. A. Mathies, Resonance Raman intensities: A probe of excited-state structure and dynamics, in *Biological Applications of Raman Spectroscopy*, ed. T. G. Spiro (Wiley-Interscience, New York, 1987), vol. 2, p. 1.

26. M. K. Lawless, S. D. Wickham and R. A. Mathies, Resonance Raman view of pericyclic photochemical ring-opening reactions: Beyond the Woodward-Hoffmann rules, Acc. Chem. Res. 28, 493 (1995).

27. D. Imre, J. Kinsey, R. Field and D. Katayama, Spectroscopic characterization of repulsive potential energy surfaces: Fluorescence spectrum of ozone, J. Phys. Chem. 86, 2564 (1982); D. Imre, J. L. Kinsey, A. Sinha and J. Krenos, Chemical dynamics studied by emission spectroscopy of dissociating molecules, J. Phys. Chem. 88, 3956 (1984).

28. D. J. Tannor, *Time Dependent Approach to Resonance Raman Scattering*, Ph.D. thesis, UCLA (1984).

29. D. J. Tannor, Anomalously polarized emission from molecules undergoing large amplitude motion, J. Phys. Chem. 92, 3341 (1988).

30. J. Z. Zhang, E. J. Heller, D. Huber and D. G. Imre, Spectroscopy and photodissociation dynamics of a 2-chromophore system: Raman scattering as a probe of non-adiabatic electronic coupling, J. Phys. Chem. 95, 6129 (1991).

Alkylbenzene Spectroscopy

31. J. B. Hopkins, D. E. Powers and R. E. Smalley, Vibrational relaxation in jet-cooled alkyl benzenes. I. Absorption spectra, J. Chem. Phys. 72, 5039 (1980); J. B. Hopkins, D. E. Powers, S. Mukamel and R. E. Smalley, Vibrational relaxation in jet-cooled alkyl benzenes. II. Fluorescence spectra, J. Chem. Phys. 72, 5049 (1980); J. B. Hopkins, D. E. Powers and R. E. Smalley, Vibrational relaxation in jet-cooled alkyl benzenes. III. Nanosecond time evolution, J. Chem. Phys. 73, 683 (1980).

32. D. J. Tannor, M. Blanco, and E. J. Heller, Simplification of the resonance fluorescence spectrum by detuning from absorption features, J. Phys. Chem. 88, 6240 (1984).

33. D. J. Tannor and E. J. Heller, Detuning from vibrational absorption features: A probe of intramolecular dynamical pathways, in *Time Resolved Vibrational Spectroscopy*, ed. G. H. Atkinson, (Gordon and Breach, 1987) p. 343.

Chapter 15

Strong Field Excitation

Until now we have dealt with the interaction of molecules only with weak fields, where perturbation theory is valid. In this chapter, we turn to the interaction with strong fields. The molecular dynamics becomes much more complicated in the presence of a strong field, and it is generally very difficult to have detailed intuition for what will happen. Nevertheless, for certain situations we are on solid ground. For structureless two-level systems in a single frequency field there is an (almost exact) analytic solution, known as the Rabi solution. The concept of Rabi cycling, and its dependence on intensity and detuning, provide the foundation for all other strong field behavior. Next we describe the Feynman–Vernon–Hellwarth (FVH) geometrical approach to the two-level system in the presence of a strong field. This approach gives intuition into a wide variety of processes involving two-level systems in the presence of fields with time-varying envelopes. We follow this with a discussion of dressed states, which are analyzed from several perspectives. In particular, the dressed states are eigenstates of the combined system + field Hamiltonian in an interaction picture. This viewpoint can be extended to include adiabatically changing frequencies and intensities, and the nonadiabatic corrections can be calculated. An important example is the STIRAP (Stimulated Raman Adiabatic Passage) method for robust population transfer in three-level systems, described in Section 15.4.2. Finally, we consider two variations of the Rabi and FVH solutions, extending the concepts from structureless two-level systems to coordinate-dependent potential energy surfaces.

15.1   Two-Level System

Consider a system with Hamiltonian H_0 having only two eigenstates, ψ_a and ψ_b, with energies $E_a = \hbar\omega_a$ and $E_b = \hbar\omega_b$. Define $\omega_0 = \omega_b - \omega_a$. The most general wavefunction for this system may be written as

$$\Psi(t) = a(t)e^{-i\omega_a t}\psi_a + b(t)e^{-i\omega_b t}\psi_b. \tag{15.1}$$

The coefficients $a(t)$ and $b(t)$ are subject to the constraint that $|a(t)|^2 + |b(t)|^2 = 1$. If this system is coupled via its transition dipole moment to a light field, with coupling represented as $V = -\mu_{ba}\varepsilon\cos(\omega t)$, then we may write the Time-Dependent Schrödinger Equation in matrix form as

$$i\hbar\frac{d}{dt}\begin{pmatrix} a(t)e^{-i\omega_a t} \\ b(t)e^{-i\omega_b t} \end{pmatrix} = \begin{pmatrix} E_a & -\mu_{ab}\varepsilon\cos(\omega t) \\ -\mu_{ba}\varepsilon\cos(\omega t) & E_b \end{pmatrix}\begin{pmatrix} a(t)e^{-i\omega_a t} \\ b(t)e^{-i\omega_b t} \end{pmatrix}. \tag{15.2}$$

This leads to a system of differential equations for the expansion coefficients:

$$\dot{a}(t) = \frac{i\mu\varepsilon}{2\hbar}(e^{+i(\omega-\omega_0)t} + e^{-i(\omega+\omega_0)t})b(t) \tag{15.3}$$

$$\dot{b}(t) = \frac{i\mu\varepsilon}{2\hbar}(e^{-i(\omega-\omega_0)t} + e^{+i(\omega+\omega_0)t})a(t), \tag{15.4}$$

where here and below we take $\mu_{ab} = \mu_{ba} = \mu$. Note that Eqs. 15.3–15.4 are equivalent to an interaction picture in which the free evolution of the two-level system has been removed (cf. Eq. 9.21). To continue we invoke the rotating wave approximation (RWA). Assume that ω is close to ω_0 and define a detuning, $\Delta = \omega - \omega_0$. If $|\Delta| \ll \omega_0$ then $\exp(-i(\omega - \omega_0)t)$ is slowly varying while $\exp(-i(\omega + \omega_0)t)$ is rapidly varying and therefore when integrated gives approximately zero. We therefore ignore this term. We have already come across the RWA in the weak field limit (cf. the discussion preceding Eq. 14.6). The equations of motion for the expansion coefficients then become

$$\dot{a}(t) = i\frac{\mu\varepsilon}{2\hbar}e^{+i\Delta t}b(t) \tag{15.5}$$

$$\dot{b}(t) = i\frac{\mu\varepsilon}{2\hbar}e^{-i\Delta t}a(t). \tag{15.6}$$

By taking another time derivative and substituting for $\dot{a}(t)$ we may solve these equations. The solution is

$$a(t) = -\left(\frac{\hbar}{\mu\varepsilon}\right)e^{\frac{i}{2}\Delta t}\left[(\Delta - \Omega)Ae^{\frac{i}{2}\Omega t} + (\Delta + \Omega)Be^{-\frac{i}{2}\Omega t}\right] \tag{15.7}$$

$$b(t) = e^{-\frac{i}{2}\Delta t}\left[Ae^{\frac{i}{2}\Omega t} + Be^{-\frac{i}{2}\Omega t}\right], \tag{15.8}$$

where the Rabi frequency, Ω, is defined as

$$\Omega = \sqrt{\Delta^2 + \left(\frac{\mu\varepsilon}{\hbar}\right)^2} = \sqrt{\Delta^2 + \bar{\Omega}^2}. \tag{15.9}$$

where we have defined $\bar{\Omega} = \mu\varepsilon/\hbar$. Notice that when $\Delta = 0$, $\Omega = \bar{\Omega}$.

If we choose as initial conditions $|a(0)|^2 = 1$ and $|b(0)|^2 = 0$ then $A = -B = \mu\varepsilon/(2\hbar\Omega)$ and the solution becomes

$$a(t) = e^{+\frac{i}{2}\Delta t}\left(\cos(\Omega t/2) - i\frac{\Delta}{\Omega}\sin(\Omega t/2)\right) \tag{15.10}$$

$$b(t) = e^{-\frac{i}{2}\Delta t}\left(\frac{\mu\varepsilon}{2\hbar\Omega}\right)(2i\sin(\Omega t/2)). \tag{15.11}$$

The populations as functions of time are then

$$|a(t)|^2 = \left(\frac{\Delta}{\Omega}\right)^2 + \left(\frac{\mu\varepsilon}{\hbar\Omega}\right)^2\cos^2(\Omega t/2) \tag{15.12}$$

$$|b(t)|^2 = \left(\frac{\mu\varepsilon}{\hbar\Omega}\right)^2\sin^2(\Omega t/2). \tag{15.13}$$

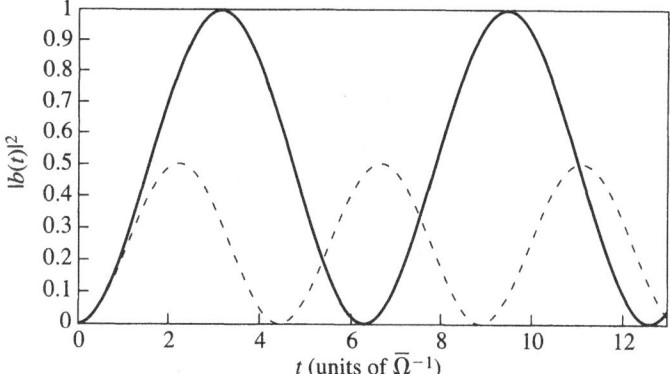

Figure 15.1 The population in the upper state as a function of time for resonant excitation (solid line) and for nonresonant excitation (dashed line).

The population in the upper state as a function of time is shown in Figure 15.1. There are several important things to note. At early times, resonant and nonresonant excitation produce the same population in the upper state because, for short times, the population in the upper state is independent of the detuning:

$$|b(t)|^2 = \left(\frac{\mu\varepsilon}{\hbar\Omega}\right)^2 \sin^2(\Omega t/2) \xrightarrow{t \text{ small}} \left(\frac{\mu\varepsilon}{2\hbar}\right)^2 t^2. \tag{15.14}$$

This result is consistent with the result of perturbation theory. One should also notice that resonant excitation completely cycles the population between the lower and upper state with a period of $2\pi/\bar{\Omega}$. Nonresonant excitation also cycles population between the states, with a period $2\pi/\Omega$, but never completely depopulates the lower state. Finally one should notice that since Ω for the nonresonant case is larger than $\bar{\Omega}$ (see Eq. 15.9), nonresonant excitation cycles population between the two states at a faster rate than resonant excitation.

By turning off the pulse $\varepsilon(t)$ at a specific final time one can control the nature of the final state produced. Consider specifically the resonance case, $\Delta = 0$, $\Omega = \bar{\Omega} = \frac{\mu\varepsilon}{\hbar}$. Equations 15.10–15.11 reduce to

$$a(t) = \cos\left(\frac{\bar{\Omega}t}{2}\right), \qquad b(t) = i \sin\left(\frac{\bar{\Omega}t}{2}\right). \tag{15.15}$$

For a pulse of duration $t = \pi/\bar{\Omega}$, at the final time all the amplitude is in the excited state; such a pulse is known as a π-pulse. For a pulse of duration $t = \pi/(2\bar{\Omega})$, at the final time the system is in a $50 - 50$ coherent superposition of the ground state and the excited states; such a pulse is known as a $\frac{\pi}{2}$-pulse.

▶ **Exercise 15.1**   Return to Eqs. 15.3–15.4. Assume that $\Delta = 0$ but that ε is time dependent, so that the equations take the form

$$\dot{a}(t) = i\frac{\bar{\Omega}(t)}{2}b(t), \qquad \dot{b}(t) = i\frac{\bar{\Omega}(t)}{2}a(t). \tag{15.16}$$

Show that the solutions of Eq. 15.16 are the following generalized form of Eq. 15.15:

$$a(t) = \cos\left(\int_0^t \frac{\bar{\Omega}(t')\, dt'}{2}\right), \qquad b(t) = i \sin\left(\int_0^t \frac{\bar{\Omega}(t')\, dt'}{2}\right). \tag{15.17}$$

Equation 15.17 indicates that for resonant excitation the population transfer depends only on the integral $\int_0^t \bar{\Omega}(t')\, dt'$ and not on the specific form of $\bar{\Omega}(t)$. This is known as the pulse area theorem.

15.2   The Feynman–Vernon–Hellwarth (FVH) Representation

A geometrical, and somewhat more general, approach to the study of two-level systems is provided by the Feynman–Vernon–Hellwarth picture. Recall that the Hamiltonian for a two-level system may be represented as a two-by-two matrix,

$$H = \begin{pmatrix} E_b & V_{ba} \\ V_{ab} & E_a \end{pmatrix}, \tag{15.18}$$

where we have now chosen a more general form for the system–field interaction:

$$V_{ba} = -\mu_{ba}\varepsilon(t). \tag{15.19}$$

The density matrix for the system may also be written as a two-by-two matrix:

$$\rho = \begin{pmatrix} bb^* & ba^* \\ ab^* & aa^* \end{pmatrix}, \tag{15.20}$$

where now a and b contain the bare system evolution phase factors. (Note that the ordering of the rows and columns in both the H and ρ matrices differs from that in the previous section; the present ordering leads to a more natural form of the geometrical representation below.) We can express the Hamiltonian and density matrices as linear combinations of the Pauli matrices, which form a complete basis for the space of two-by-two matrices. To simplify some of the results that follow, we define our basis as 1/2 times the usual Pauli matrices (Eq. 8.83). Thus,

$$\sigma_1 = \tfrac{1}{2}\begin{pmatrix} 0 & 1 \\ 1 & 0 \end{pmatrix}, \qquad \sigma_2 = \tfrac{1}{2}\begin{pmatrix} 0 & -i \\ i & 0 \end{pmatrix},$$

$$\sigma_3 = \tfrac{1}{2}\begin{pmatrix} 1 & 0 \\ 0 & -1 \end{pmatrix}, \qquad \sigma_4 = \tfrac{1}{2}\begin{pmatrix} 1 & 0 \\ 0 & 1 \end{pmatrix}. \tag{15.21}$$

We can therefore write

$$H = (V_{ab} + V_{ba})\sigma_1 + i(V_{ba} - V_{ab})\sigma_2 + (E_b - E_a)\sigma_3 + (E_b + E_a)\sigma_4, \tag{15.22}$$

$$\rho = (ab^* + a^*b)\sigma_1 + i(a^*b - ab^*)\sigma_2 + (bb^* - aa^*)\sigma_3 + (bb^* + aa^*)\sigma_4. \tag{15.23}$$

We now substitute Eqs. 15.22 and 15.23 into the quantum Liouville equation (Eq. 5.39):

$$i\hbar\dot{\rho} = [H, \rho]. \tag{15.24}$$

Note that the coefficient of σ_4 in the expansion of the density matrix is the total population and is therefore time independent. Furthermore, the σ matrices obey the commutation relations:

$$[\sigma_1, \sigma_2] = i\sigma_3 \qquad [\sigma_2, \sigma_3] = i\sigma_1 \qquad [\sigma_3, \sigma_1] = i\sigma_2 \qquad (15.25)$$

$$[\sigma_4, \sigma_i] = 0 \qquad \text{for } i = 1, 2, 3. \qquad (15.26)$$

The Hamiltonian and the density matrices can therefore be represented as expansions in terms of just σ_1, σ_2 and σ_3:

$$H = \overbrace{(V_{ab} + V_{ba})}^{\Omega_1} \sigma_1 + \overbrace{i(V_{ba} - V_{ab})}^{\Omega_2} \sigma_2 + \overbrace{(E_b - E_a)}^{\Omega_3} \sigma_3,$$

$$\rho = \underbrace{(ab^* + a^*b)}_{r_1} \sigma_1 + \underbrace{i(a^*b - ab^*)}_{r_2} \sigma_2 + \underbrace{(bb^* - aa^*)}_{r_3} \sigma_3.$$

The Liouville equation can then be written as

$$i\hbar(\dot{r}_1\sigma_1 + \dot{r}_2\sigma_2 + \dot{r}_3\sigma_3) = i(\Omega_2 r_3 - \Omega_3 r_2)\sigma_1 - i(\Omega_1 r_3 - \Omega_3 r_1)\sigma_2 + i(\Omega_1 r_2 - \Omega_2 r_1)\sigma_3.$$

$$(15.27)$$

If one now defines two three-dimensional vectors,

$$\boldsymbol{r} = (r_1, r_2, r_3), \qquad (15.28)$$

$$\boldsymbol{\Omega} = \frac{1}{\hbar}(\Omega_1, \Omega_2, \Omega_3), \qquad (15.29)$$

then one can rewrite the Liouville equation for the two-level system as

$$\frac{d}{dt}\boldsymbol{r} = \boldsymbol{\Omega} \times \boldsymbol{r}. \qquad (15.30)$$

Note that r_3 is the population difference between the upper and lower states. Having all the population in the lower state corresponds to $r_3 = -1$, while having a completely inverted population with no population in the lower state corresponds to $r_3 = +1$. In general, one can say that the z-component of $\boldsymbol{r}$ is then the population difference.

This representation is slightly inconvenient since Ω_1 and Ω_2 are explicitly time dependent. Specifically, if $\varepsilon(t) = \varepsilon \cos(\omega t)$, and making the rotating wave approximation as in the previous section, we find that $V_{ba} = -\mu_{ba}\frac{\varepsilon}{2}e^{-i\omega t}$, $V_{ab} = -\mu_{ab}\frac{\varepsilon}{2}e^{i\omega t}$, and hence

$$\boldsymbol{\Omega} = \left(-\frac{\mu\varepsilon}{\hbar}\cos(\omega t), -\frac{\mu\varepsilon}{\hbar}\sin(\omega t), \omega_0\right), \qquad (15.31)$$

where $\omega_0 = (E_b - E_a)/\hbar$. Here and below we assume that $\mu_{ab} = \mu_{ba} = \mu$. The time dependence of $\boldsymbol{\Omega}$ can be removed by transforming to a frame of reference rotating at the frequency of the light field. To see this, define a new population vector, $\boldsymbol{r}' = U^{-1}\boldsymbol{r}$, where

$$U^{-1} = \begin{pmatrix} \cos\omega t & \sin\omega t & 0 \\ -\sin\omega t & \cos\omega t & 0 \\ 0 & 0 & 1 \end{pmatrix}. \qquad (15.32)$$

The Liouville equation for the transformed vector, r', becomes

$$\frac{d}{dt}r' = \Omega' \times r',$$ (15.33)

where

$$\Omega' = (-\frac{\mu\varepsilon}{\hbar}, 0, \omega_0 - \omega)$$ (15.34)

is now time independent.

▶ Exercise 15.2   Derive Eq. 15.33.

Solution   First note that Eq. 15.30 can be rewritten

$$\frac{d}{dt}(Ur') = \Omega \times Ur'.$$ (15.35)

Expanding the derivative on the LHS, rearranging terms and left-multiplying by U^{-1} gives

$$\frac{d}{dt}r' = U^{-1}\Omega Ur' - U^{-1}\dot{U}r'.$$ (15.36)

Next note that Eq. 15.30 can be written in matrix form:

$$\begin{pmatrix} \dot{r}_1 \\ \dot{r}_2 \\ \dot{r}_3 \end{pmatrix} = \frac{1}{\hbar}\begin{pmatrix} 0 & -\Omega_3 & \Omega_2 \\ \Omega_3 & 0 & -\Omega_1 \\ -\Omega_2 & \Omega_1 & 0 \end{pmatrix}\begin{pmatrix} r_1 \\ r_2 \\ r_3 \end{pmatrix}.$$ (15.37)

Substituting the matrix form for Ω from Eq. 15.37 into Eq. 15.36 and using Eq. 15.32 for U gives

$$\begin{pmatrix} \dot{r}'_1 \\ \dot{r}'_2 \\ \dot{r}'_3 \end{pmatrix} = \begin{pmatrix} 0 & -(\omega_0 - \omega) & 0 \\ \omega_0 - \omega & 0 & \mu\varepsilon/\hbar \\ 0 & -\mu\varepsilon/\hbar & 0 \end{pmatrix}\begin{pmatrix} r'_1 \\ r'_2 \\ r'_3 \end{pmatrix}.$$ (15.38)

Expressing Eq. 15.38 in cross-product format yields Eq. 15.33 with Ω' given by Eq. 15.34.

The geometrical interpretation of Eq. 15.33 is that the state vector, r', precesses around the field vector, Ω', analogous to the precession of a classical gyromagnet in a magnetic field (Ω' plays the role of the magnetic field vector, and r' plays the role of the magnetic moment vector). This representation of the two-level system is called the Feynman–Vernon–Hellwarth, or FVH, representation; it gives a unified, pictorial view with which one can understand the effect of a wide variety of optical pulse effects in two-level systems. For example, the geometrical picture of Rabi cycling within the FVH picture is shown in Figure 15.2. Assuming that at $t = 0$ all the population is in the ground state, then the initial position of the r' vector is $(0, 0, -1)$, and so r' points along the negative z-axis. For a resonant field, $\omega_0 - \omega = 0$ and so the Ω' vector points along the x-axis. Equation 15.33 then says that the population vector simply precesses about the x-axis, at all times staying on the unit sphere. It then periodically points along the positive z-axis, which corresponds to having all the population in the upper state. If the field is nonresonant, then Ω' no longer points along the x-axis but along some other direction in the xz-plane. The population vector still precesses about the field vector, but now at some angle to the z-axis. Thus the projection onto

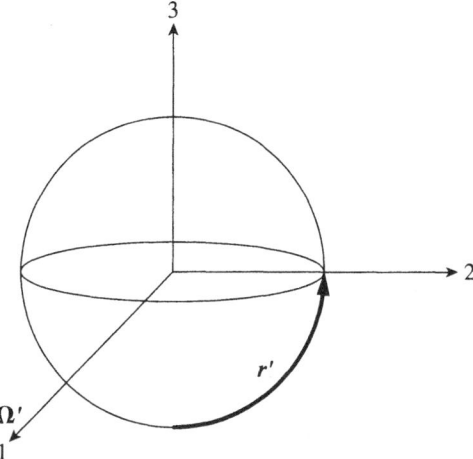

Figure 15.2 Feynman–Vernon–Hellwarth (FVH) diagram, exploiting the isomorphism between the two-level system and a state vector precessing on the unit sphere. The state vector, r', precesses around the field vector, Ω', according to the equation $\dot{r}' = \Omega' \times r'$. The z-component of the r' vector is the population difference between the two levels, while the x- and y-components refer to the polarization, namely, the real and imaginary part of the coherence between the amplitude in the two levels. In the frame of reference rotating at the carrier frequency, the z-component of the Ω' vector is the detuning of the field from resonance, while the x- and y-components indicate the field amplitude. In the rotating frame, the y-component of Ω' may be set equal to zero (since the overall phase of the field is irrelevant, assuming no coherence of the levels at $t = 0$), unless there is nonuniform change in phase in the field during the process.

the z-axis of r' never equals one and so there is never a complete population inversion. The FVH representation is the optical analog of a representation originally developed in nuclear magnetic resonance by Bloch; as such the state vector is sometimes called the "Bloch vector" and the unit sphere is sometimes called the "Bloch sphere."

The Feynman–Vernon–Hellwarth representation allows us to visualize the results of more complicated laser pulse sequences. A laser pulse that takes r' from $(0, 0, -1)$ to $(0, 0, 1)$ is called a π-pulse since the r' precesses π radians about the field vector. Similarly, a pulse that takes r' from $(0, 0, -1)$ to $(+1, 0, 0)$ is called a $\pi/2$-pulse. The state represented by the vector $(+1, 0, 0)$ is a coherent superposition of the upper and lower states of the system.

One interesting experiment is to apply a $\pi/2$-pulse followed by a $\pi/2$ phase shift of the field. This phase shift will bring Ω' parallel to r'. Since now $\Omega' \times r' = 0$, the population is fixed in time in a coherent superposition between the ground and excited states. This is called photon locking.

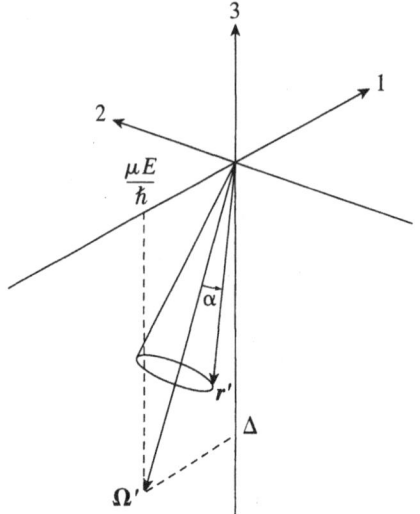

Figure 15.3 Feynman–Vernon–Hellwarth (FVH) diagram, showing the concept of adiabatic following. The Bloch vector, r', precesses in a narrow cone about the rotating frame torque vector, Ω'. As the detuning Δ changes from negative to positive, the field vector Ω' becomes inverted. If the change in Ω' is adiabatic the Bloch vector follows the field vector in this inversion process, corresponding to complete population transfer to the excited state.

A second interesting experiment is to begin with a pulse that is far below resonance and slowly and continuously sweep the frequency until the pulse is far above resonance. At $t = -\infty$ the field vector is pointing nearly along the $-z$-axis, and is therefore almost parallel to the state vector. As the field vector Ω' slowly moves from $z = -1$ to $z = +1$, the state vector r' adiabatically follows it, precessing about the instantaneous direction of the field vector. When, at $t \to +\infty$ the field vector is directed nearly along the $+z$-axis, the state vector is directed there as well, signifying complete population inversion. This scheme is known as "adiabatic following," and is represented geometrically in Figure 15.3. The remarkable feature of this method is its robustness—there is almost no sensitivity to field strength or the exact schedule of changing the frequency, provided the conditions for adiabaticity are met. For a quantitative treatment of adiabatic following, see Allen (1975), and references therein.

15.3   Dressed States

In this section we follow closely the treatment by Boyd (2003). We begin by returning to the Rabi solution of the two-level system and calculate the expectation value of the dipole moment for a system known to be in the ground state at $t = 0$. This quantity is given by

$$\langle \mu \rangle = \langle \Psi(t)|\mu|\Psi(t) \rangle, \tag{15.39}$$

with $\Psi(t)$ given by Eq. 15.1. We assume that $\langle \psi_a|\mu|\psi_a \rangle = \langle \psi_b|\mu|\psi_b \rangle = 0$, and $\mu_{ab} = \langle \psi_a|\mu|\psi_b \rangle = \langle \psi_b|\mu|\psi_a \rangle = \mu_{ba} = \mu$. Hence, the induced dipole moment is given by

$$\langle \mu \rangle = a^*(t)b(t)\mu e^{-i\omega_{ba}t} + \text{c.c.} \tag{15.40}$$

Using Eqs. 15.10–15.11 for $a(t)$ and $b(t)$ we have

$$\langle \mu \rangle = -\frac{1}{4\Omega^2}\frac{\mu\varepsilon}{\hbar}\left[2\Delta e^{-i\omega t} - (\Delta - \Omega)e^{-i(\omega+\Omega)t} - (\Delta + \Omega)e^{-i(\omega-\Omega)t}\right]\mu_{ab} + \text{c.c.}$$

$$\tag{15.41}$$

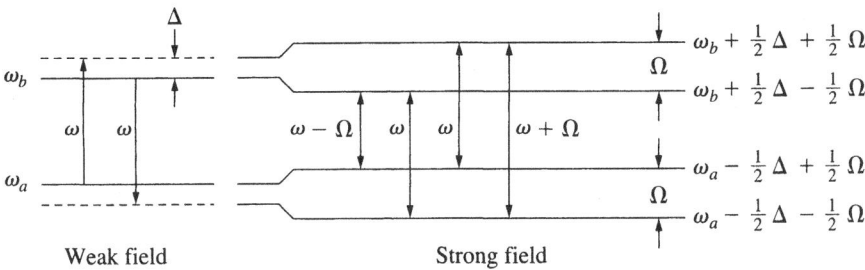

Figure 15.4 Frequency spectrum of the wavefunction given by Eq. 15.1, with $a(t)$ and $b(t)$ given by Eqs. 15.10 and 15.11, respectively. The figure assumes that the detuning $\Delta > 0$. Adapted from Boyd (2003).

This shows that the induced dipole oscillates not only at the driving frequency ω but also at the Rabi sideband frequencies $\omega + \Omega$ and $\omega - \Omega$. The presence of these frequencies is a result of the fact that all these frequencies appear in the time-dependent wavefunction, $\Psi(t)$, according to Eq. 15.1, where $a(t)$ contains frequencies $-\frac{1}{2}(\Delta \pm \Omega)$ and $b(t)$ contains frequencies $\frac{1}{2}(\Delta \pm \Omega)$. Figure 15.4 shows graphically the frequencies that are present in the wavefunction; the frequencies at which the dipole oscillates correspond to differences of the various frequency components of the wavefunction. In fact, in the presence of strong fields fluorescence is observed at all three of the transition frequencies. This is called the fluorescence triplet or Mollow triplet (Cohen-Tannoudji, 1992).

Another perspective on these combination frequencies in the induced dipole moment is that they are the difference frequencies between the "dressed states" of the system. The dressed states are stationary states of the Hamiltonian of the system + field. Since the dressed state is a superposition of the bare eigenstates ψ_a and ψ_b, the probability to be in level ψ_a or ψ_b is also constant in time. According to the general form of the solutions, Eqs. 15.7–15.8, the two dressed states correspond to two different pairs of integration parameters A and B that lead to time-independent probabilities for the occupancy of levels ψ_a and ψ_b. The first solution, which we designate as ψ_+, corresponds to the case in which

$$A = 0, \qquad B = 1 \qquad (\psi_+). \tag{15.42}$$

In this case,

$$a(t) = -\frac{\hbar}{\mu\varepsilon}(\Delta + \Omega)e^{\frac{i\Delta t}{2}}e^{-\frac{i\Omega t}{2}} \tag{15.43}$$

$$b(t) = e^{-\frac{i\Delta t}{2}}e^{-\frac{i\Omega t}{2}}. \tag{15.44}$$

Thus,

$$\Psi(t) = \psi_+(t) = a(t)e^{-i\omega_a t}\psi_a + b(t)e^{-i\omega_b t}\psi_b \tag{15.45}$$

$$= N_+\left[-\frac{\hbar}{\mu\varepsilon}(\Delta + \Omega)e^{\frac{i\Delta t}{2}}e^{-\frac{i\Omega t}{2}}e^{-i\omega_a t}\psi_a + e^{-\frac{i\Delta t}{2}}e^{-\frac{i\Omega t}{2}}e^{-i\omega_b t}\psi_b\right], \tag{15.46}$$

where N_+ is a normalization constant. The second solution, which we designate as ψ_-, corresponds to the case where

$$A = 1, \qquad B = 0 \qquad (\psi_-). \tag{15.47}$$

In this case,

$$a(t) = -\frac{\hbar}{\mu\varepsilon}(\Delta - \Omega)e^{\frac{i\Delta t}{2}}e^{\frac{i\Omega t}{2}} \tag{15.48}$$

$$b(t) = e^{-\frac{i\Delta t}{2}}e^{\frac{i\Omega t}{2}}. \tag{15.49}$$

Thus,

$$\Psi(t) = \psi_-(t) = a(t)e^{-i\omega_a t}\psi_a + b(t)e^{-i\omega_b t}\psi_b \tag{15.50}$$

$$= N_-\left[-\frac{\hbar}{\mu\varepsilon}(\Delta - \Omega)e^{\frac{i\Delta t}{2}}e^{\frac{i\Omega t}{2}}e^{-i\omega_a t}\psi_a + e^{-\frac{i\Delta t}{2}}e^{\frac{i\Omega t}{2}}e^{-i\omega_b t}\psi_b\right], \tag{15.51}$$

where N_- is a normalization constant. The normalization constants, $N_\pm$, are determined by the condition that

$$\langle\psi_\pm|\psi_\pm\rangle = 1 = N_\pm^2\left[\left(\frac{\hbar}{\mu\varepsilon}\right)^2(\Delta \pm \Omega)^2 + 1\right]. \tag{15.52}$$

Some rearrangement, using Eq. 15.9, gives

$$N_\pm = \left(\frac{\Omega \mp \Delta}{2\Omega}\right)^{1/2}, \tag{15.53}$$

and hence the explicit form for the normalized dressed states $\psi_\pm$ is

$$\psi_\pm = \mp\left[\frac{\Omega \pm \Delta}{2\Omega}\right]^{1/2}e^{\frac{i\Delta t}{2}}e^{-\frac{i\Omega t}{2}}e^{-i\omega_a t}\psi_a + \left[\frac{\Omega \mp \Delta}{2\Omega}\right]^{1/2}e^{\frac{-i\Delta t}{2}}e^{-\frac{i\Omega t}{2}}e^{-i\omega_b t}\psi_b. \tag{15.54}$$

▶ **Exercise 15.3**

a. Verify Eq. 15.53 for $N_\pm$.

b. Show that if the system is in a dressed state $\psi_\pm$, the probability for the system to be in level ψ_a is given by

$$|\langle\psi_a|\psi_\pm\rangle|^2 = \frac{\Omega \pm \Delta}{2\Omega}, \tag{15.55}$$

and the probability to be in level ψ_b is given by

$$|\langle\psi_b|\psi_\pm\rangle|^2 = \frac{\Omega \mp \Delta}{2\Omega}. \tag{15.56}$$

Note that these occupancies are constant in time, and hence the dressed states are stationary states of the coupled system + field Hamiltonian. Check that the occupancies sum to 1.

c. Show that the dressed states are orthogonal, that is,

$$\langle\psi_+|\psi_-\rangle = 0. \tag{15.57}$$

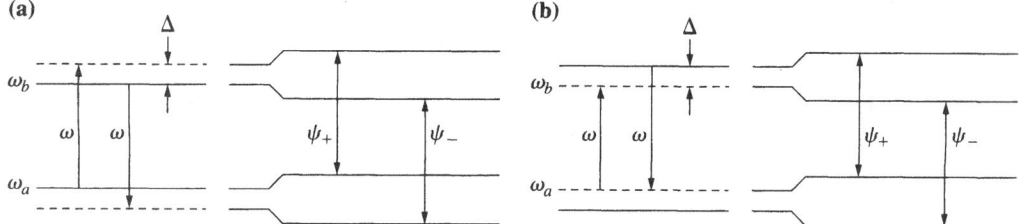

Figure 15.5 The dressed atomic states ψ_+ and ψ_- for (a) $\Delta > 0$, and (b) $\Delta < 0$. Adapted from Boyd (2003).

▶ Exercise 15.4

a. Show that the expectation value of the induced dipole for a system in a dressed state is given by

$$\langle \psi_\pm | \mu | \psi_\pm \rangle = \mp \frac{\mu \varepsilon}{2\hbar \Omega} \mu_{ba} e^{i\omega t} + \text{c.c.} \tag{15.58}$$

Note that the induced dipole moment in a dressed state oscillates only at the driving frequency.

b. Show that the transition dipole moment between two dressed states is given by

$$\langle \psi_\pm | \mu | \psi_\mp \rangle = \pm \mu_{ba} \left(\frac{\Omega \mp \Delta}{2\Omega} \right)^{1/2} e^{i\omega t} e^{\pm i\Omega t} \mp \mu_{ab} \left(\frac{\Omega \pm \Delta}{2\Omega} \right)^{1/2} e^{-i\omega t} e^{\pm i\Omega t}, \tag{15.59}$$

that is, the *transition* dipole moment between dressed states oscillates at the driving frequency but additionally has sidebands at $\omega \pm \Omega$.

c. Interpret the frequencies in part a and b in terms of the transition frequencies in the energy-level diagram in Figures 15.4 and 15.5. Note that although the dressed states correspond to initial conditions different from the Rabi solution, the same combination frequencies appear.

Consider the limiting form of the dressed states for the case of a weak applied field, that is, for $\frac{\mu \varepsilon}{\hbar} \ll |\Delta|$. In this limit, we can approximate the Rabi frequency Ω as

$$\Omega = |\Delta| \left(1 + \left(\frac{\mu \varepsilon}{\hbar \Delta} \right)^2 \right)^{1/2} \approx |\Delta| \left(1 + \frac{1}{2} \left(\frac{\mu \varepsilon}{\hbar \Delta} \right)^2 \right). \tag{15.60}$$

We treat the case of positive and negative Δ separately. For positive Δ we can approximate the dressed state wavefunctions of Eq. 15.54 as

$$\psi_+ = -\psi_a e^{-i\omega_a t} + \frac{\mu \varepsilon}{2\hbar \Delta} \psi_b e^{-i(\omega_b + \Delta)t} \tag{15.61}$$

$$\psi_- = \frac{\mu \varepsilon}{2\hbar \Delta} \psi_a e^{-i(\omega_a - \Delta)t} + \psi_b e^{-i\omega_b t}. \tag{15.62}$$

Note that in this limit ψ_+ has primarily the character of ψ_a while ψ_- is primarily ψ_b. The frequencies associated with the smaller contributions (ψ_b in ψ_+ and ψ_a in ψ_-) suggest the identification of these components with virtual levels induced by the transition (see Figure 15.5a). For the case of negative Δ we obtain

$$\psi_+ = -\frac{\mu\varepsilon}{2\hbar\Delta}\psi_a e^{-i(\omega_a-\Delta)t} - \psi_b e^{-i\omega_b t} \tag{15.63}$$

$$\psi_- = \psi_a e^{-i\omega_a t} - \frac{\mu\varepsilon}{2\hbar\Delta}\psi_b e^{-i(\omega_b+\Delta)t}. \tag{15.64}$$

Now ψ_+ is primarily ψ_b and ψ_- is primarily ψ_a. Again, the frequencies associated with the smaller contributions (now ψ_a in ψ_+ and ψ_b in ψ_-) suggest the identification of these components with virtual levels induced by the transition (see Figure 15.5b).

We end this section by discussing the connection between the treatment in this section and Floquet theory (Section 9.5). According to Floquet theory, the number of Floquet eigenvalues and eigenvectors is equal to the number of molecular states used in the basis. Since the treatment in this section deals with only two molecular states there can be only two Floquet states. In fact, the two Floquet states are just $\psi_+(t)$ and $\psi_-(t)$ (Eqs. 15.46 and 15.51, respectively). Indeed, ψ_+ is a superposition of the upper branches of the ground and excited doublet, while ψ_- is a superposition of the lower branches of the ground and excited doublet. To a certain extent, Figures 15.4–15.5 are misleading, giving the appearance that four dressed states emerged out of the two molecular basis states. Nevertheless, these pictures are very useful: they rationalize the experimental observation of the fluorescence (or Mollow) triplet of emission frequencies. In other words, although there are only two Floquet states, there are three transition frequencies, a fact that is corroborated by experiment.

▶ **Exercise 15.5** Rewrite the phases in the expressions for $\psi_+(t)$ and $\psi_-(t)$ (Eqs. 15.46 and 15.51) in terms of just ω and Ω. Show that, as required by Floquet theory, both $\psi_+(t)$ and $\psi_-(t)$ are products of a state vector that is periodic in time multiplied by an overall time-dependent phase factor. What are the Floquet eigenvalues? Compare your result with that obtained in Exercise 9.4. Where does the Floquet eigenvalue enter in that treatment?

15.4 Adiabatic Excitation with Strong Fields

In Section 15.2 we touched on the topic of adiabatically (slowly varying) Rabi frequencies. In Section 15.3 we introduced the dressed state picture, always assuming that the field carrier frequency and intensity were constant in time. In this section we combine the two ideas, generalizing the dressed state picture to allow for adiabatically changing frequencies and intensities. The dressed states emerge as the eigenvectors of the instantaneous Hamiltonian of the molecular system plus the field. In the first part of the section we specialize to the two-level system. For the case of adiabatic frequency changes we obtain a new perspective on the process of adiabatic passage, as motion on the adiabatically changing dressed state. In the second part we consider a three-level system, and focus just on adiabatic *intensity* changes. The dressed state picture suggests a remarkable, counterintuitive scheme known as Stimulated Raman Adiabatic Passage (STIRAP) for complete population transfer from level 1 to level 3 without populating level 2. Finally, we consider an approximation to the equations of motion for large detuning for two- and three-level systems. This approximation is often called "adiabatic elimination," although it should be distinguished clearly from the usual adiabatic approximation.

15.4.1   Adiabatically Changing Dressed States

Consider the Hamiltonian for a two-level system in the presence of a time-varying field. For generality, we will allow both the envelope and the carrier frequency of the field to vary in time. Thus we write

$$H(t) = \begin{pmatrix} E_1 & -\frac{\mu\varepsilon(t)}{2}e^{i\int_{-\infty}^{t}\omega(t')dt'} + \text{c.c.} \\ -\frac{\mu\varepsilon(t)}{2}e^{-i\int_{-\infty}^{t}\omega(t')dt'} + \text{c.c.} & E_2 \end{pmatrix}. \quad (15.65)$$

We now define an interaction picture where

$$U_0 = \begin{pmatrix} e^{-iE_1t/\hbar}e^{i\int_{-\infty}^{t}\Delta(t')dt'} & 0 \\ 0 & e^{-iE_2t/\hbar}e^{-i\int_{-\infty}^{t}\Delta(t')dt'} \end{pmatrix}, \qquad \Psi_I = U_0^{-1}\Psi, \quad (15.66)$$

where $\Delta(t) \equiv \omega(t) - \omega_0$. Then

$$H_I(t)$$

$$\equiv U_0^{-1}(t)H(t)U_0(t) - i\hbar U_0^{-1}(t)\dot{U}_0(t) \qquad (15.67)$$

$$= \begin{pmatrix} E_1 & -\frac{\mu\varepsilon(t)}{2} \\ -\frac{\mu\varepsilon(t)}{2} & E_2 \end{pmatrix} + \begin{pmatrix} -E_1 + \frac{\hbar\Delta(t)}{2} & 0 \\ 0 & -E_2 - \frac{\hbar\Delta(t)}{2} \end{pmatrix} = \frac{\hbar}{2}\begin{pmatrix} \Delta(t) & -\frac{\mu\varepsilon(t)}{\hbar} \\ -\frac{\mu\varepsilon(t)}{\hbar} & -\Delta(t) \end{pmatrix}.$$

$$(15.68)$$

Here and henceforth we neglect the counterrotating terms. (For the form of Eq. 15.67, see Exercise 9.3.) Physically speaking, the diagonal elements of $H_I(t)$ can be thought of in the following way: the lower level E_1 has been shifted up by the energy of the field, $\hbar\omega(t)$, and then the zero of energy has been redefined to be centered between E_2 and the shifted E_1. We may define a second transformation $U_1(t)$ that diagonalizes the matrix $H_I(t)$ at each instant of time:

$$U_1(t) = \begin{pmatrix} \cos(\theta(t)/2) & -\sin(\theta(t)/2) \\ \sin(\theta(t)/2) & \cos(\theta(t)/2) \end{pmatrix}, \quad (15.69)$$

where

$$\tan\theta(t) = \frac{\mu\varepsilon(t)}{\hbar\Delta(t)}. \quad (15.70)$$

The new Hamiltonian matrix, $H_I'(t)$, is given by

$$H_I'(t) = D(t) - i\hbar U_1^{-1}\dot{U}_1(t), \quad (15.71)$$

where

$$D(t) = U_1^{-1}(t)H_I(t)U_1(t) = \begin{pmatrix} E_-(t) & 0 \\ 0 & E_+(t) \end{pmatrix} \quad (15.72)$$

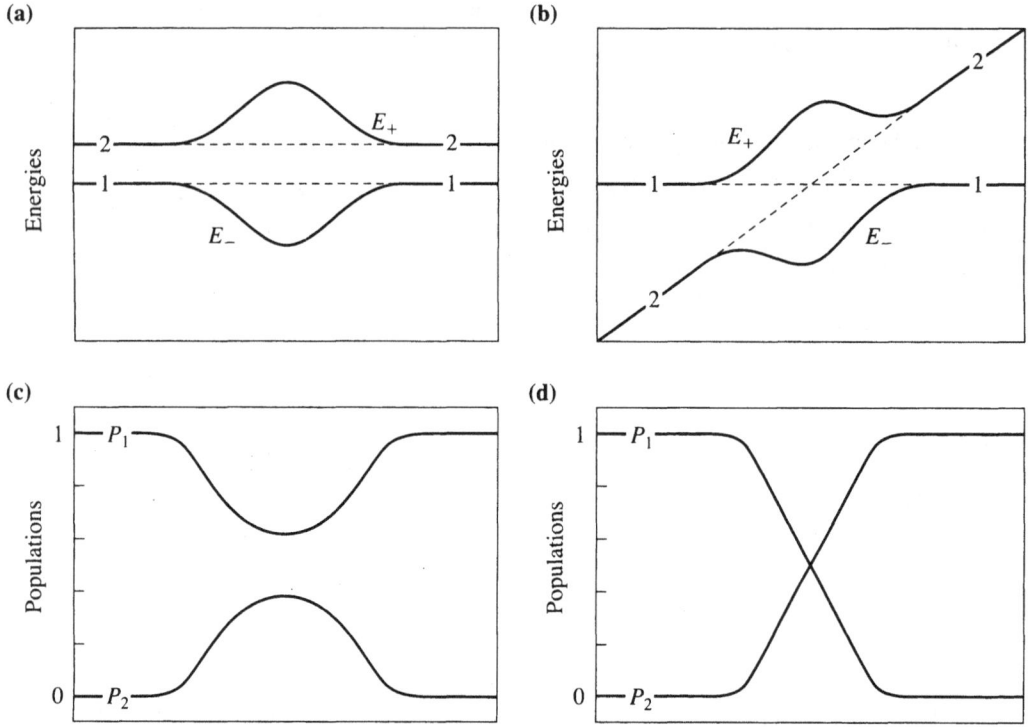

Figure 15.6 Time evolution of the adiabatic eigenvalues in a two-level system for (a) adiabatic intensity change with finite detuning, (b) adiabatic frequency sweep from below resonance to above resonance. The solid lines are the adiabatic energies while the dashed lines are the diabatic energies. (c) Population in the diabatic levels versus time for the adiabatic intensity change in (a). There is some exchange of population in the diabatic basis, but the exchange is reversed when the intensity is restored to its initial value. (d) Population in the diabatic levels versus time for the adiabatic frequency sweep in (b). There is complete transfer of population in the diabatic basis. Adapted from Vitanov (2001).

with the adiabatic eigenvalues, $E_\pm$, given by

$$E_\pm = \pm \frac{\hbar}{2}\sqrt{\Delta^2(t) + \left(\frac{\mu\varepsilon(t)}{\hbar}\right)^2} = \pm\frac{\hbar}{2}\Omega(t). \tag{15.73}$$

Figure 15.6 shows the adiabatic eigenvalues for two different cases: adiabatic intensity change (a) and adiabatic frequency sweep (b). We make three comments: 1) The adiabatic eigenvalues $E_\pm$ are the energies of the adiabatically changing field-dressed states, that is, the instantaneous eigenvalues of the molecular Hamiltonian + field in the interaction representation. 2) The adiabatic eigenvalues $E_\pm$ are essentially the Floquet eigenvalues (cf. Exercise 15.5), generalized to allow for adiabatic changes. 3) The difference between the eigenvalues $E_\pm$ and the diagonal elements of $H_I(t)$, $\pm\frac{\hbar}{2}\Delta(t)$, is known as the AC Stark shift. This represents the shift of the field-dressed eigenvalues of $H(t)$ due to their coupling. Note that the AC Stark shift depends on the field intensity.

▶ **Exercise 15.6**   Calculate the AC Stark shift in the limit that (a) $\Delta(t) \gg \mu\varepsilon(t)/\hbar$, and (b) $\mu\varepsilon(t)/\hbar \gg \Delta(t)$.

Solution

a.   For $\Delta(t) \gg \mu\varepsilon(t)/\hbar$ we have

$$E_{\pm}(t) = \pm\frac{\hbar}{2}\left(\Delta(t) + \frac{\left(\frac{\mu\varepsilon(t)}{\hbar}\right)^2}{2\Delta(t)} + \cdots\right) \tag{15.74}$$

and thus the AC Stark shift is

$$\Delta^{AC}(t) = \frac{(\mu\varepsilon(t))^2}{4\hbar\Delta(t)} \tag{15.75}$$

(cf. Eq. 15.60, taking into account Eq. 15.73). For $\mu\varepsilon(t)/\hbar \gg \Delta(t)$ we have

$$E_{\pm}(t) = \pm\frac{\mu\varepsilon(t)}{2} = \Delta^{AC}(t). \tag{15.76}$$

As an application of the adiabatic-dressed-states picture, consider an adiabatic frequency sweep from below resonance to above resonance (Figure 15.6b). As discussed at the end of Section 15.2, if the frequency sweep is adiabatic there will be complete population transfer from level 1 to level 2. There, the complete population transfer was visualized in terms in the tight precession of the Bloch vector around the field vector as the latter moved adiabatically from the south pole to the north pole of the Bloch sphere. Here, the same phenomenon is given a completely different representation, in terms of motion on a single adiabatic dressed state, the lower level in Figure 15.6b. At the beginning of the frequency sweep, $\Delta < 0$ and the lower adiabatic state corresponds to diabatic level 1, dressed by ω (LHS of Figure 15.6b). At the end of the frequency sweep, $\Delta > 0$ and now the upper adiabatic state corresponds to the diabatic level 1, dressed by ω (RHS of Figure 15.6b). In the middle of the frequency sweep, the adiabatic dressed states have an avoided crossing, corresponding to resonance between the two diabatic states. However, if the frequency sweep is slow enough the population will stay on the adiabatic state on which it started, leading to an overall process of complete population transfer from diabatic level 1 to diabatic level 2. The nonadiabatic corrections can be calculated with the aid of the Landau–Zener formula (Section 9.4.2) applied to the avoided crossing of the dressed-state levels, as shown in Exercise 15.7.

▶ **Exercise 15.7**   Find the nonadiabatic correction to complete adiabatic passage via a frequency sweep.

Solution   The TDSE in the interaction picture corresponding to Eq. 15.71 is

$$i\hbar\frac{d}{dt}\begin{pmatrix}\Psi_-\\\Psi_+\end{pmatrix} = \begin{pmatrix}E_- & 0\\0 & E_+\end{pmatrix}\begin{pmatrix}\Psi_-\\\Psi_+\end{pmatrix} - \frac{i\hbar\dot{\theta}}{2}\begin{pmatrix}0 & -1\\1 & 0\end{pmatrix}\begin{pmatrix}\Psi_-\\\Psi_+\end{pmatrix} \tag{15.77}$$

(cf. the treatment in Section 9.4.2). A calculation analogous to Eq. 9.113 gives

$$\frac{d}{dt}(\tan\theta) = \sec^2\theta\,\dot\theta = (1+\tan^2\theta)\dot\theta = \left(1+\frac{\bar\Omega^2}{\Delta^2}\right)\dot\theta = \frac{\dot{\bar\Omega}\Delta - \dot\Delta\bar\Omega}{\Delta^2}, \quad (15.78)$$

where the last two relations follow from Eq. 15.70 with $\bar\Omega \equiv \mu\varepsilon/\hbar$. Therefore,

$$\dot\theta = \frac{\dot{\bar\Omega}\Delta - \dot\Delta\bar\Omega}{\Omega^2}, \quad (15.79)$$

where for generality we have let both the intensity and the frequency change with time. The condition for adiabaticity is

$$|\dot\theta| = \left|\frac{\dot\Delta\bar\Omega - \dot{\bar\Omega}\Delta}{\Omega^2}\right| \ll \frac{E_+ - E_-}{\hbar} = \Omega(t) \quad (15.80)$$

or

$$\left|\dot\Delta\bar\Omega - \dot{\bar\Omega}\Delta\right| \ll \min_{\Delta(t)}\ \Omega^3 = \bar\Omega^3. \quad (15.81)$$

We now specialize to the case of an adiabatic frequency sweep. Neglecting the second term on the LHS of Eq. 15.81, the adiabaticity condition becomes

$$|\dot\Delta| \ll \bar\Omega^2. \quad (15.82)$$

The Landau–Zener formula (Eq. 9.119) estimates the total (time-integrated) probability for crossing between adiabatic states as

$$p = \exp\left[\frac{-2\pi V^2}{\hbar\,|\dot E|}\right] = \exp\left[\frac{-\pi\bar\Omega^2}{2\,|\dot\Delta|}\right], \quad (15.83)$$

where in the second equality we have identified $V = \hbar\bar\Omega/2$ and $|\dot E| = \hbar\,|\dot\Delta|$ (cf. Eqs. (9.102) and (15.2)). For slowly changing frequency, $\bar\Omega^2/|\dot\Delta| \gg 1$ and therefore $p \approx 0$, indicating that adiabatic passage is virtually complete.

15.4.2   Adiabatic Dressed States in a Three-Level System: Stimulated Raman Adiabatic Passage

In the previous section we discussed adiabatically changing dressed states in a two-level system. In this section we extend our treatment to the three-level system. We show how consideration of the adiabatically changing dressed states gives a simple recipe for transferring population completely from level 1 to level 3 without populating level 2. This method is known as Stimulated Raman Adiabatic Passage, or STIRAP.

Consider the three-level system shown in Figure 15.7. Such a system is called a Lambda (Λ) system because its structure resembles the form of the Greek letter Λ. The levels interact with a pair of pulses of frequency ω_p and ω_s, with slowly varying envelopes ε_p and ε_s, respectively. The pulse at frequency ω_p, called the "pump" pulse, is assumed to excite selectively the $|1\rangle \to |2\rangle$ transition, while the pulse at frequency ω_s, called the

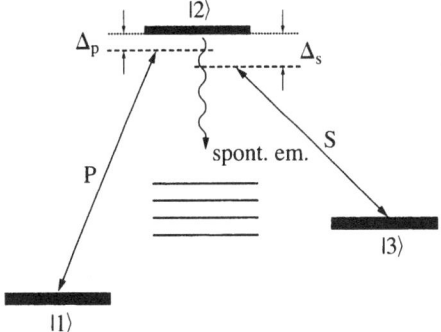

Figure 15.7 Three-level excitation scheme. The initially populated state $|1\rangle$ and the final state $|3\rangle$ are coupled by the pump laser, P, and the Stokes laser, S, via an intermediate state $|2\rangle$. The detuning of the pump and Stokes laser frequencies from the transition frequency to the intermediate state are Δ_p and Δ_s, respectively. From K. Bergmann et al. (1998).

"Stokes" pulse, is assumed to excite selectively the $|2\rangle \to |3\rangle$, transition, due to resonance considerations. Thus, the Hamiltonian is described by

$$H = \begin{pmatrix} E_1 & -\frac{\mu_{12}\varepsilon_p}{2}\left(e^{i\omega_p t} + \text{c.c.}\right) & 0 \\ -\frac{\mu_{21}\varepsilon_p}{2}\left(e^{-i\omega_p t} + \text{c.c.}\right) & E_2 & -\frac{\mu_{23}\varepsilon_s}{2}\left(e^{-i\omega_s t} + \text{c.c.}\right) \\ 0 & -\frac{\mu_{32}\varepsilon_s}{2}\left(e^{i\omega_s t} + \text{c.c.}\right) & E_3 \end{pmatrix}. \quad (15.84)$$

Note the sequential form of the coupling: level $|1\rangle$ is coupled to level $|2\rangle$ and $|2\rangle$ to $|3\rangle$ but $|1\rangle$ not directly to $|3\rangle$. The object is to find the envelopes ε_p and ε_s that transfer the population completely and robustly from level $|1\rangle$ where it starts at time 0, to level $|3\rangle$.

The analysis is facilitated by representing the evolution in an interaction picture (or equivalently, a rotating frame), using the free evolution operator,

$$U_0 = \begin{pmatrix} e^{-iE_2 t/\hbar}e^{i\omega_p t} & 0 & 0 \\ 0 & e^{-iE_2 t/\hbar} & 0 \\ 0 & 0 & e^{-iE_2 t/\hbar}e^{i\omega_s t} \end{pmatrix}. \quad (15.85)$$

Recalling the equation of motion in the interaction picture (Exercise 9.3),

$$i\hbar\dot{\psi}_I = U_0^{-1}HU_0\psi_I - i\hbar U_0^{-1}\dot{U}_0\psi_I = H_I\psi_I, \quad \text{with } \psi_I = U_0^{-1}\psi, \quad (15.86)$$

we find that

$$H_I = -\frac{\hbar}{2}\begin{pmatrix} -2\Delta_p & \Omega_p & 0 \\ \Omega_p & 0 & \Omega_s \\ 0 & \Omega_s & -2\Delta_s \end{pmatrix}. \quad (15.87)$$

We have defined $\Omega_p \equiv \mu_{21}\varepsilon_p/\hbar$, $\Omega_s \equiv \mu_{32}\varepsilon_s/\hbar$, $\Delta_p \equiv (E_1 + \omega_p - E_2)/\hbar$ and $\Delta_s = (E_3 + \omega_s - E_2)/\hbar$, and we have made the RWA.

We now specialize to the case of two-photon resonance, $\Delta_p = \Delta_s = \Delta$. For simplicity, we consider the case $\Delta = 0$ here and leave the case $\Delta \neq 0$ to Exercise 15.10. Setting $\Delta_p = \Delta_s = 0$, H_I takes the simple form

$$H_I = -\frac{\hbar}{2}\begin{pmatrix} 0 & \Omega_p & 0 \\ \Omega_p & 0 & \Omega_s \\ 0 & \Omega_s & 0 \end{pmatrix}. \quad (15.88)$$

This matrix can be diagonalized, yielding the eigenvalues

$$E^{0,\pm} \equiv \hbar\omega^{0,\pm} = 0, \pm\frac{\hbar}{2}\Omega, \qquad (15.89)$$

where we have defined (for the section on STIRAP only!)

$$\Omega \equiv (\Omega_p^2 + \Omega_s^2)^{1/2}. \qquad (15.90)$$

The corresponding eigenvectors are

$$|g_0\rangle = \left[\Omega_s|1\rangle - \Omega_p|3\rangle\right]\Omega^{-1} \qquad (15.91)$$

$$|g_\pm\rangle = \frac{1}{\sqrt{2}}\left[\left[\Omega_p|1\rangle + \Omega_s|3\rangle\right]\Omega^{-1} \mp |2\rangle\right]. \qquad (15.92)$$

Defining

$$\Theta = \arctan\left(\frac{\Omega_p}{\Omega_s}\right), \qquad (15.93)$$

the field-dressed eigenstates can be written as

$$|g_0\rangle = \cos\Theta|1\rangle - \sin\Theta|3\rangle \qquad (15.94)$$

$$|g_+\rangle = \sin\Theta|1\rangle - \frac{1}{\sqrt{2}}|2\rangle + \cos\Theta|3\rangle \qquad (15.95)$$

$$|g_-\rangle = \sin\Theta|1\rangle + \frac{1}{\sqrt{2}}|2\rangle + \cos\Theta|3\rangle. \qquad (15.96)$$

The eigenvectors $|g_0\rangle$, $|g_\pm\rangle$ are dressed states (cf. Section 15.3), that is, eigenfunctions of the instantaneous Hamiltonian in the presence of the fields. Note that $|g_0\rangle$ contains no contribution from level $|2\rangle$. This is the basis for the remarkable STIRAP (Stimulated Raman Adiabatic Passage) scheme, in which population is transferred from level $|1\rangle$ to level $|3\rangle$ with negligible intermediate population in level $|2\rangle$, as we now explain.

Assume that as $t \to -\infty$ all the population is in level $|1\rangle$. Inspection of Eq. 15.91 shows that this corresponds to all the population in the dressed state $|g_0\rangle$. Consider an adiabatic change in field intensities such that for $t \to -\infty$ $\Omega_s \gg \Omega_p$, while for $t \to \infty$ $\Omega_p \gg \Omega_s$ (see Figure 15.8a). Since the transformation is adiabatic, the population will remain in the dressed state $|g_0\rangle$ throughout the process, while changing its *character* from that of level $|1\rangle$ to that of level $|3\rangle$. Since the dressed state $|g_0\rangle$ contains no component of level $|2\rangle$, this adiabatic change leads to complete population transfer from level $|1\rangle$ to level $|3\rangle$, without populating level $|2\rangle$. This is the STIRAP scheme. Note the counterintuitive order of the pulses: Ω_s precedes Ω_p; naively, one would expect, given the sequential nature of the coupling $|1\rangle \to |2\rangle \to |3\rangle$, that Ω_p should precede Ω_s. Not only does the counterintuitive pulse sequence work, but this sequence is actually robust with respect to the exact shape of the envelope functions Ω_s and Ω_p, as long as three conditions are met: 1) the pump pulse and the Stokes pulse must have sufficient intensity to satisfy the adiabatic assumption (cf. Exercise 15.11); 2) the Stokes and pump pulses must have partial overlap for the scheme to work (cf. Exercise 15.8); 3) the relative phases of the Stokes and pump pulses must be kept extremely stable. In fact, Eq. 15.94 indicates that in the rotating frame, the phase of the

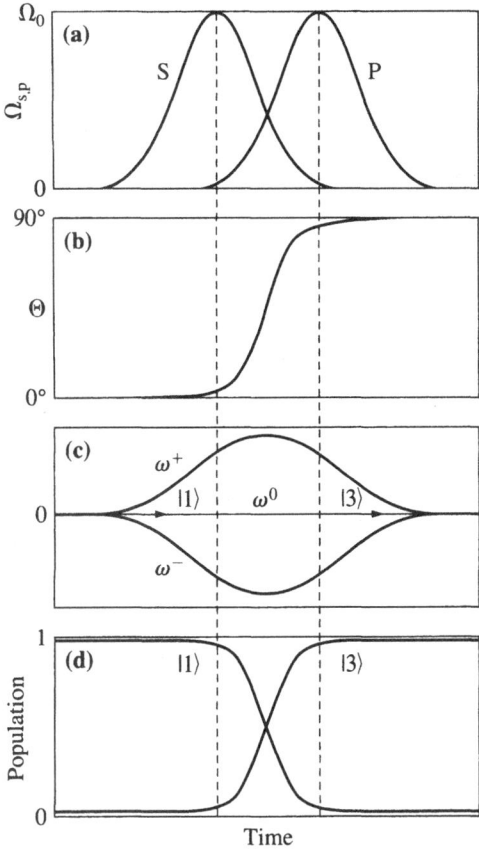

Figure 15.8 Time evolution of (a) the Rabi frequencies of the pump and Stokes laser; (b) the mixing angle Θ ($\tan(\Theta) \equiv \frac{\Omega_p(t)}{\Omega_s(t)}$); (c) the dressed-state eigenvalues ($\hbar\omega^{0,\pm} = E^{0,\pm}$); (d) the population of the initial level (starting at unity) and the final level (ending at unity). From K. Bergmann et al. (1998).

pump pulse must be precisely the opposite of the phase of the Stokes pulse. In a way that is not completely clear, this condition is achieved and maintained in the laboratory without any explicit control of the relative phase by the experimentalist.

▶ **Exercise 15.8**   Analyze the dressed states for the case where the Stokes and pump pulses do not overlap. Show why the STIRAP scheme breaks down in this case.

There is an interesting alternative perspective on the reason for the counterintuitive order of pulses in the STIRAP scheme (Malinovsky, 1997a). For simplicity, we specialize to $\Delta = 0$. From Eq. 15.87, the Schrödinger equation in the interaction picture takes the form

$$i\hbar \begin{pmatrix} \dot{a}_1 \\ \dot{a}_2 \\ \dot{a}_3 \end{pmatrix} = -\frac{\hbar}{2} \begin{pmatrix} 0 & \Omega_p & 0 \\ \Omega_p & 0 & \Omega_s \\ 0 & \Omega_s & 0 \end{pmatrix} \begin{pmatrix} a_1 \\ a_2 \\ a_3 \end{pmatrix}. \tag{15.97}$$

Note that a sufficient condition for the population in level $|2\rangle$ to remain constant is that the fields maintain the condition $\frac{d|a_2(t)|^2}{dt} = 0$. This condition can be reexpressed as

$$\frac{d|a_2(t)|^2}{dt} = 2\text{Re}\{a_2^*(t)\dot{a}_2(t)\} = \Omega_p(t)\text{Im}\{a_2^*(t)a_1(t)\} + \Omega_s(t)\text{Im}\{a_2^*(t)a_3(t)\} = 0. \tag{15.98}$$

Equation 15.98 is satisfied if

$$\Omega_p = -\Omega_0(t)\mathrm{Im}\{a_2^*(t)a_3(t)\} \tag{15.99}$$

$$\Omega_s = \Omega_0(t)\mathrm{Im}\{a_2^*(t)a_1(t)\}. \tag{15.100}$$

In physical terms, the time rate of change of $|a_2(t)|^2$ involves two contributions, that from level $|1\rangle$ and that from level $|3\rangle$. If these two contributions are equal and opposite, the population in level $|2\rangle$ will remain constant; in this case, level $|2\rangle$ corresponds to what is called a "dark" state, a state with no net absorption. From this perspective, STIRAP is an adiabatic passage through a dark state. At early times, when a_1 is large and a_3 is small, Ω_p must be small and Ω_s must be large in order for the contributions to cancel; at late times the situation is reversed: a_1 is small and a_3 is large, hence Ω_p must be large and Ω_s. This provides a second perspective on the counterintuitive ordering of the pulses in the STIRAP scheme.

15.4.3   Adiabatic Elimination: An Approximation for Large Detuning

In many cases a separation of time scales allows us to eliminate one or more of the equations of motion to obtain a simplification. The simplest example is a two-level system with off-resonant excitation:

$$i\hbar\frac{d}{dt}\begin{pmatrix} a \\ b \end{pmatrix} = -\frac{\hbar}{2}\begin{pmatrix} 0 & \bar{\Omega}e^{i\Delta t} \\ \bar{\Omega}e^{-i\Delta t} & 0 \end{pmatrix}\begin{pmatrix} a \\ b \end{pmatrix} \tag{15.101}$$

(cf. Eqs. 15.5–15.6). Integrating the equation for $\dot{b}$ and assuming that the envelope $\bar{\Omega}$ is slowly varying with respect to the oscillations $e^{-i\Delta t}$ so it can be taken out of the integral, we obtain

$$b \approx -\frac{\bar{\Omega}}{2\Delta}e^{-i\Delta t}a. \tag{15.102}$$

Substituting back into the equation for $\dot{a}$ we obtain

$$i\hbar\dot{a} = -\frac{\hbar}{2}\bar{\Omega}e^{i\Delta t}b \tag{15.103}$$

$$= \frac{\hbar\bar{\Omega}^2}{4\Delta}a. \tag{15.104}$$

Equation 15.104 provides an equation of motion for a from which b has been eliminated. Note that the effect of eliminating b is to provide an effective shift of $\frac{\hbar\bar{\Omega}^2}{4\Delta}$ in the energy of a, which is precisely the AC Stark shift described at Eq. 15.75.

Consider now a three-level system of the form shown in Figure 15.7. The equations of motion in the interaction picture with $U_0 = e^{-iH_0t/\hbar}$ are

$$i\hbar\frac{d}{dt}\begin{pmatrix} a_1 \\ a_2 \\ a_3 \end{pmatrix} = -\frac{\hbar}{2}\begin{pmatrix} 0 & \bar{\Omega}_pe^{i\Delta_pt} & 0 \\ \bar{\Omega}_pe^{-i\Delta_pt} & 0 & \bar{\Omega}_se^{-i\Delta_st} \\ 0 & \bar{\Omega}_se^{i\Delta_st} & 0 \end{pmatrix}\begin{pmatrix} a_1 \\ a_2 \\ a_3 \end{pmatrix}. \tag{15.105}$$

Integrating the equation of motion for $\dot{a}_2$ and assuming that $\bar{\Omega}_s$ and $\bar{\Omega}_p$ are slowly varying with respect to $e^{i\Delta_s t}$ and $e^{i\Delta_p t}$, respectively, so that they can be removed from the integral, we obtain

$$a_2 \approx \left[-\frac{\bar{\Omega}_p a_1}{2\Delta_p} e^{-i\Delta_p t} - \frac{\bar{\Omega}_s a_3 e^{-i\Delta_s t}}{2\Delta_s} \right]. \tag{15.106}$$

Substituting this expression for a_2 into Eq. 15.105 we obtain

$$i\hbar \frac{d}{dt} \begin{pmatrix} a_1 \\ a_3 \end{pmatrix} = \begin{pmatrix} \dfrac{\hbar\bar{\Omega}_p^2}{4\Delta_p} & \dfrac{\hbar\bar{\Omega}_p\bar{\Omega}_s}{4\Delta_s} e^{i\Delta_p t} e^{-i\Delta_s t} \\ \dfrac{\hbar\bar{\Omega}_p\bar{\Omega}_s}{4\Delta_s} e^{-i\Delta_p t} e^{i\Delta_s t} & \dfrac{\hbar\bar{\Omega}_s^2}{4\Delta_s} \end{pmatrix} \begin{pmatrix} a_1 \\ a_3 \end{pmatrix}, \tag{15.107}$$

indicating that the states (a_1, a_3) can be treated as an effective two-level system. The base energies of a_1 and a_3 are shifted by the AC Stark shifts $\frac{\hbar\bar{\Omega}_p^2}{4\Delta_p}$ and $\frac{\hbar\bar{\Omega}_s^2}{4\Delta_s}$ respectively.

15.5   Impulsive Excitation

When the two-level system in the Rabi solution is replaced by two displaced electronic states, with dependence on nuclear coordinates, there are a variety of new effects. One of these effects is the creation of a "hole" in the coordinate space wavefunction: there is a narrow region of coordinate over which the transition satisfies the resonance condition, and the portion of the wavefunction in this region will be preferentially promoted (although the ground state may be repopulated due to Rabi cycling, so that the instantaneous position of the hole depends on field strength, duration and frequency). The second effect has to do with momentum. If the excitation is short compared with the time scale for nuclear dynamics, the wavefunction at each different value of the coordinate may be considered to undergo independent Rabi cycling. A coordinate-dependent phase develops due to the different resonance offsets at different values of the coordinate. But a coordinate-dependent phase is momentum in quantum mechanics; thus the short pulse gives a momentum kick to the ground state wavepacket. The direction of the momentum kick again depends on field strength, duration and frequency.

In the general case, one requires the solution of the two-surface TDSE in the presence of the strong field. Specifically, if the pulse is comparable to or longer than the time scale for nuclear motion, the nuclear motion during the pulse mixes up the wavefunction at different values of the coordinate. However, if the excitation is impulsive (short compared with the time scale for nuclear motion), then the nuclear kinetic energy operator on both surfaces can be neglected during the pulse, each value of the nuclear coordinate is decoupled from the others, and the effect of the pulse can be obtained by considering the Rabi solution for each coordinate independently. Using Eqs. 15.10–15.11, we obtain the following equations for the wavepacket on the ground and excited electronic state surfaces, respectively:

$$\psi_a = a(t)e^{-i\omega_a t} = e^{i\Delta t/2}\left(\cos\left(\frac{\Omega t}{2}\right) - i\frac{\Delta}{\Omega}\sin\left(\frac{\Omega t}{2}\right) \right) e^{-i\omega_a t} \tag{15.108}$$

$$\psi_b = b(t)e^{-i\omega_b t} = e^{-i\Delta t/2}\left(\frac{\mu\varepsilon}{2\hbar\Omega} \right)\left(2i\sin\left(\frac{\Omega t}{2}\right) \right) e^{-i\omega_b t}. \tag{15.109}$$

Noting that $\omega_b + \Delta/2 = \omega_a + \omega - \Delta/2$ and making the replacements

$$\Delta = \omega - \omega_0 \quad \Longrightarrow \quad \Delta(R) = \omega - \frac{H_b - H_a}{\hbar} = \omega - \frac{V_b(R) - V_a(R)}{\hbar} \quad (15.110)$$

$$\Omega = \left(\Delta^2 + \left(\frac{\mu\varepsilon}{\hbar} \right)^2 \right)^{1/2} \quad \Longrightarrow \quad \Omega(R) = \left(\Delta^2(R) + \left(\frac{\mu\varepsilon}{\hbar} \right)^2 \right)^{1/2}, \quad (15.111)$$

we obtain

$$\psi_a(R, t) = e^{i\Delta(R)t/2} \left(\cos\left(\frac{\Omega(R)t}{2} \right) - i \frac{\Delta(R)}{\Omega(R)} \sin\left(\frac{\Omega(R)t}{2} \right) \right) e^{-i\omega_a t} \psi_a(R, 0) \quad (15.112)$$

$$\psi_b(R, t) = e^{i\Delta(R)t/2} \left(\frac{\mu\varepsilon}{2\hbar\Omega(R)} \right) \left(2i \sin\left(\frac{\Omega(R)t}{2} \right) \right) e^{-i\omega t} e^{-i\omega_a t} \psi_a(R, 0). \quad (15.113)$$

A somewhat more rigorous derivation of Eqs. 15.112–15.113 is given in Banin (1994); in comparing with their formulas note the difference in the sign of Δ and the factors of 2 in the definitions of Δ, ε and Ω.

The coordinate-dependent phase of the wavefunction is given by

$$\Phi_a(R) = -\omega_a t + \frac{\Delta(R)t}{2} + \arctan\left(-\frac{\Delta(R)}{\Omega(R)} \tan\left(\frac{\Omega(R)t}{2} \right) \right) \quad (15.114)$$

and

$$\Phi_b(R) = -(\omega_a + \omega)t + \frac{\Delta(R)t}{2} + \frac{\pi}{2}. \quad (15.115)$$

The momentum kick is given by

$$\delta P = \hbar \frac{d\Phi}{dR}. \quad (15.116)$$

Since in most photoexcitation processes $\frac{d\Delta}{dR} > 0$ the momentum kick on the excited state is positive. The ground state is more complicated. We consider three special cases.

1. At short times, the first nonzero term that enters into the phase is

$$\Phi_a = -\Delta(R) \left(\frac{\mu\varepsilon}{\hbar} \right)^2 \frac{t^3}{24}, \quad (15.117)$$

   which gives a negative momentum kick.

2. Far off resonance ($|\Delta| \gg \frac{\mu\varepsilon}{\hbar}$), there is rapid Rabi cycling. The average asymptotic momentum shift is

$$\delta P = \frac{(\mu\varepsilon)^2 t}{4\hbar\Delta^2} \frac{d\Delta}{dR}. \quad (15.118)$$

   Assuming $\frac{d\Delta}{dR} > 0$, this gives a positive momentum kick.

3. On resonance,

$$\delta P \approx \hbar \left. \frac{d\Phi_a}{dR} \right|_{R_h} = \frac{1}{2} \left(t - \frac{2\hbar}{\mu\varepsilon} \tan\left(\frac{\mu\varepsilon t}{2\hbar} \right) \right) \left. \frac{d\Delta}{dR} \right|_{R_h}, \quad (15.119)$$

   where R_h is the position of the hole. The term $\tan\left(\frac{\mu\varepsilon t}{2\hbar} \right)$ maximizes for $\frac{\mu\varepsilon t}{2\hbar}$ just below π, after which it changes sign. Assuming that $\frac{d\Delta}{dR} > 0$, this gives a negative momentum kick for pulses with $\frac{\mu \varepsilon t}{2\hbar}$ below π and a positive momentum kick for pulses with $\frac{\mu \varepsilon t}{2\hbar}$ above π. For further details the reader is referred to Smith (1994) and Banin (1994).

▶ Exercise 15.9   Derive Eqs. 15.117–15.119.

15.6   Optical Paralysis

The photon-locking mechanism, described in the earlier discussion of two-level systems, can be generalized to the case of two molecular electronic potential energy surfaces (Kosloff, 1992; Malinovsky, 1997b). The amplitude in each of the two levels becomes replaced by coherent wavepacket amplitudes on each of the two potential energy surfaces. The analog of the locking condition is that the total population in each of the two electronic levels is fixed; however, the population within each electronic manifold may be redistributed over any set of vibrational states. Thus, one can then excite or deexcite a vibrational wavepacket while keeping the electronic populations fixed. We term this "optical paralysis," or "excitation without demolition" since the electronic excitation that leads to sample degradation is turned off. To see how this works, we start from the matrix form of the Time-Dependent Schrödinger Equation for a two-surface system:

$$i\hbar \frac{\partial}{\partial t} \begin{pmatrix} \psi_a \\ \psi_b \end{pmatrix} = \begin{pmatrix} H_a & -\mu \varepsilon(t) \\ -\mu \varepsilon^*(t) & H_b \end{pmatrix} \begin{pmatrix} \psi_a \\ \psi_b \end{pmatrix}, \tag{15.120}$$

where H_a and H_b are the Hamiltonians for the lower and upper electronic potential energy surfaces, respectively. The rate of change of the ground state population is

$$\frac{d}{dt} \langle \psi_a | \psi_a \rangle = 2 \, \text{Re} \langle \psi_a | \dot{\psi}_a \rangle$$

$$= 2 \, \text{Re} \langle \psi_a | \left(-\frac{\mu \varepsilon(t)}{i\hbar} \right) | \psi_b \rangle$$

$$= -\frac{2}{\hbar} \text{Im} \left(\langle \psi_a | \mu | \psi_b \rangle \varepsilon(t) \right), \tag{15.121}$$

where we have used the Time-Dependent Schrödinger Equation to express $|\dot{\psi}_a\rangle$ in terms of $|\psi_b\rangle$. Notice that the structure of Eq. 15.121 is the same as that obtained in the two-level system. Specifically, compare the third component of Eq. 15.38:

$$\frac{d}{dt}(bb^* - aa^*) = -\frac{\mu \varepsilon}{\hbar}(i(a^*b - ab^*)) \implies \frac{d}{dt}(aa^*) = -\frac{2\mu \varepsilon}{\hbar} \text{Im}(a^*b). \tag{15.122}$$

Returning to Eq. 15.121, we now choose $\varepsilon(t) = \langle \psi_b | \mu | \psi_a \rangle$. Then

$$\frac{d}{dt} \langle \psi_a | \psi_a \rangle = -\frac{2}{\hbar} \text{Im} \left(\langle \psi_a | \mu | \psi_b \rangle \varepsilon(t) \right)$$

$$= -\frac{2}{\hbar} \text{Im} |\langle \psi_a | \mu | \psi_b \rangle|^2$$

$$= 0. \tag{15.123}$$

Thus the population in the ground electronic state and, by necessity, the population in the excited state, are fixed. More generally we can choose $\varepsilon(t) = C(t)\langle\psi_b|\mu|\psi_a\rangle$ for any real $C(t)$ and still maintain a fixed population difference. If we were to choose $\varepsilon(t) = iC(t)\langle\psi_b|\mu|\psi_a\rangle$ then we would cause population to be transferred to the excited electronic state, while choosing $\varepsilon(t) = -iC(t)\langle\psi_b|\mu|\psi_a\rangle$ would dump population from the excited state down to the ground state. So by controlling the phase of a laser, we can control the direction of population transfer between the two electronic states. This is a well-known property of two-level systems but only recently has it been appreciated in the context of potential energy surfaces.

Since the population is controlled by the phase of ε, we can try to use its amplitude, $C(t)$, to do something interesting. The change in the energy of the ground state under conditions of population locking is

$$\frac{d}{dt}E_a = \frac{d}{dt}\frac{\langle\psi_a|H_a|\psi_a\rangle}{\langle\psi_a|\psi_a\rangle}$$

$$= \frac{2}{\langle\psi_a|\psi_a\rangle}\,\mathrm{Re}\langle\psi_a|H_a|\dot{\psi}_a\rangle$$

$$= \frac{2}{\langle\psi_a|\psi_a\rangle}\,\mathrm{Re}\langle\psi_a|H_a\left(-\frac{\mu\varepsilon(t)}{i\hbar}\right)|\psi_b\rangle$$

$$= -\frac{2}{\hbar}\frac{1}{\langle\psi_a|\psi_a\rangle}\mathrm{Im}\{\langle\psi_a|H_a\mu|\psi_b\rangle C(t)\langle\psi_b|\mu|\psi_a\rangle\}. \tag{15.124}$$

We can therefore use the sign of $C(t)$ to control the sign of dE_a/dt and thus to "heat" or "cool" the ground state wavepacket. The magnitude of $C(t)$ controls the rate of "heating" (or "cooling"). In Figures 15.9–15.11 this heating strategy is applied to a harmonic model system ($m = \omega = 1$) to illustrate the basic soundness of the approach. Figure 15.9 shows

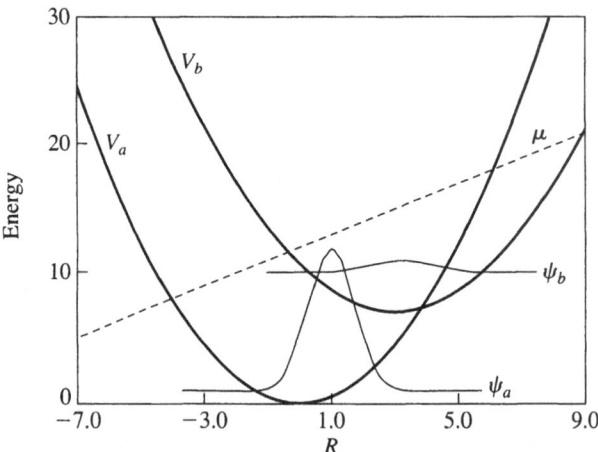

Figure 15.9 Potential energy curves and transition dipole moment of the model system for phase-locked vibrational heating. The ground and excited-state wavefunctions (absolute value) are shown after the first excitation pulse (not to scale). Adapted from Kosloff (1992).

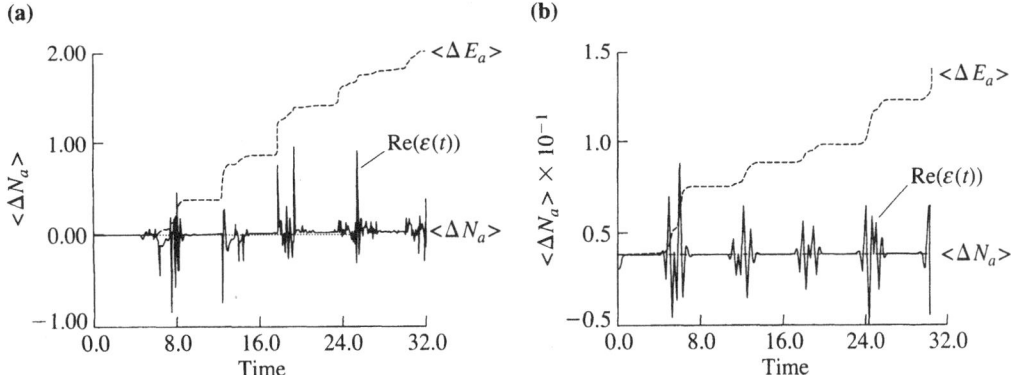

Figure 15.10 Change in population on the ground state $\langle \Delta N_a \rangle$ (dotted line), change in ground state energy $\langle \Delta E_a \rangle$ (dashed line) and the real part of the field ε as a function of time. (a) Strong field; (b) Weak field. The strong excitation pulse increases the ground state energy by a factor of 4 in approximately five vibrational periods. The units are $2\pi/\omega$. Adapted from Kosloff (1992).

the potential energy curves and transition dipole moment of the model system. Figure 15.10 shows, as a function of time, the real part of the field ε whose phase satisfies Eq. 15.123 and that makes the sign in Eq. 15.124 positive. Figure 15.10a shows a strong field, and Figure 15.10b shows a weak field. The plots show the change in population on the ground state $\langle \Delta N_a \rangle$ as a function of time, which is seen to be zero after an initial, weak seed pulse. Finally, the plots show the change in ground state energy $\langle \Delta E_a \rangle$ as a function of time, which is seen to increase monotonically in both cases, although about ten times faster for the strong field. An additional perspective on how the fields work is obtained by examining the power spectrum of the pulse of Figure 15.10a, shown in Figure 15.11. The field is seen to be highly structured, with a spacing of features approximately equal to the excited-state

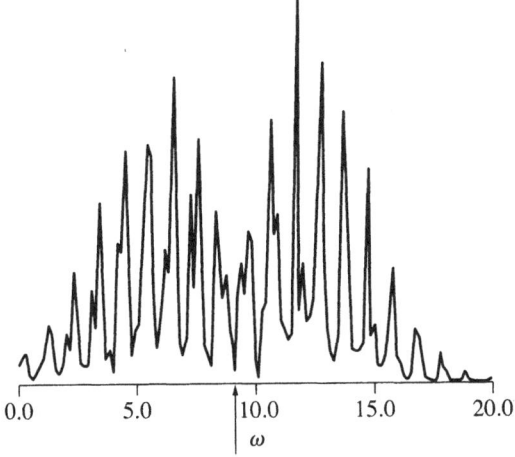

Figure 15.11 The power spectrum of the pulse of Figure 15.10a. The arrow indicates the frequency of the vertical transition. Adapted from Kosloff (1992).

vibrational frequency. The arrow indicates the frequency of the strongest vertical Franck–Condon transition: there is a conspicuous absence of intensity at this position, as well as at the position of the other vibrational resonances of the excited electronic state. This is entirely consistent with the zero population transfer condition: by exciting only at frequencies in between the excited-state vibrational resonances no population is actually transferred to the excited state. Yet, there is strong scattering, since the excitation is resonant, overall, with the excited electronic state. The heating from this perspective is a stimulated Raman process, in which virtual states are excited and emission is stimulated to higher and higher vibrational states of the ground electronic state, that is, Stokes transitions are systematically favored over anti-Stokes. If the phase relationship between the excitation frequencies were not controlled, the energy change in the ground electronic state would be expected to be random after a certain amount of time.

Problems

▶ **Exercise 15.10** In the text, we considered the form of the STIRAP eigenvalues and eigenvectors assuming that $\Delta_p = \Delta_s = 0$. In this exercise we return to the interaction picture Hamiltonian H_I in Eq. 15.87 and consider the case $\Delta_p = \Delta_s = \Delta \neq 0$.

a. Show that the eigenvalues of $H_I(t)$ are given by

$$E^{0,\pm} \equiv \hbar\omega^{0,\pm} = \hbar\Delta, \ \frac{\hbar\Delta}{2} \pm \frac{\hbar}{2}\sqrt{\Delta^2 + \Omega^2(t)}. \tag{15.125}$$

[Hint: the algebra simplifies if one adds the diagonal matrix $-\hbar\Delta\mathbf{1}$ to $H_I(t)$ and then adjusts the eigenvalues accordingly. Note that the addition of a diagonal matrix does not change the eigenvectors.]

b. Show that the eigenvectors are given by

$$|g_0\rangle = \left[\Omega_s|1\rangle - \Omega_p|3\rangle\right]\Omega^{-1} \tag{15.126}$$

$$|g_\pm\rangle = \left[\Omega_p|1\rangle + \Omega_s|3\rangle + 2\omega^{\mp}|2\rangle\right]N_{\mp}^{-1}, \tag{15.127}$$

where

$$N_{\pm} = \left((2\omega^{\pm})^2 + \Omega^2\right)^{1/2} \quad \text{and} \quad \Omega = \left(\Omega_p^2 + \Omega_s^2\right)^{1/2}, \tag{15.128}$$

and $\omega^{\pm}$ is defined by Eq. 15.125.

c. Define the angle

$$\Phi = \frac{1}{2}\arctan\left[\frac{\Omega}{-\Delta}\right]. \tag{15.129}$$

Show that the field-dressed eigenstates can be written as

$$|g_0\rangle = \cos\Theta|1\rangle - \sin\Theta|3\rangle \tag{15.130}$$

$$|g_+\rangle = \sin\Phi\sin\Theta|1\rangle - \cos\Phi|2\rangle + \sin\Phi\cos\Theta|3\rangle \tag{15.131}$$

$$|g_-\rangle = \cos\Phi\sin\Theta|1\rangle + \sin\Phi|2\rangle + \cos\Phi\cos\Theta|3\rangle. \qquad (15.132)$$

[Hint: use the formula $\tan 2A = 2\tan A/(1 - \tan^2 A)$.]

▶ **Exercise 15.11**

a. Use the form of the STIRAP eigenvalues and eigenvectors in Exercise 15.10 to define a transformation that diagonalizes the instantaneous, adiabatically changing STIRAP Hamiltonian. Show that this defines a new Hamiltonian

$$H_1' = D - i\hbar U_1^\dagger \dot{U}_1 \qquad (15.133)$$

$$= \begin{pmatrix} E^0(t) & 0 & 0 \\ 0 & E^+(t) & 0 \\ 0 & 0 & E^-(t) \end{pmatrix}$$

$$- i\hbar \left[\begin{pmatrix} 0 & \sin\Phi & \cos\Phi \\ -\sin\Phi & 0 & 0 \\ -\cos\Phi & 0 & 0 \end{pmatrix}\dot{\Theta} + \begin{pmatrix} 0 & 0 & 0 \\ 0 & 0 & -1 \\ 0 & 1 & 0 \end{pmatrix}\dot{\Phi} \right].$$

$$(15.134)$$

b. Show that Eq. 15.134 implies that, to satisfy the condition of adiabatic transfer of population, the Stokes and pump fields must be sufficiently strong so that the energy separations between the field-dressed states, $E^+ - E^0$ and $E^- - E_0$, are much larger than the rate of change of $|g_0\rangle$:

$$\left| \left\langle g_\pm \left| \frac{dg_0}{dt} \right\rangle \right| \ll \frac{|E^\pm - E^0|}{\hbar}, \qquad \left| \left\langle g_- \left| \frac{dg_+}{dt} \right\rangle \right| \ll \frac{|E^+ - E^-|}{\hbar}. \quad (15.135)$$

c. Show that Eq. 15.134 implies that the adiabaticity conditions

$$\left| \frac{d\Theta}{dt} \right| \ll \frac{|E^\pm - E^0|}{\hbar}, \qquad \left| \frac{d\Phi}{dt} \right| \ll \frac{|E^+ - E^-|}{\hbar}, \qquad (15.136)$$

must be satisfied for all t. ,

d. Show that Eq. 15.136 can be rewritten as

$$\left| \frac{\Omega_s(d\Omega_p/dt) - \Omega_p(d\Omega_s/dt)}{\Omega^2} \right| \ll \frac{|E^\pm - E^0|}{\hbar},$$

$$\left| \frac{\Delta(d\Omega/dt) - \Omega(d\Delta/dt)}{\Omega^2 + \Delta^2} \right| \ll \frac{|E^+ - E^-|}{\hbar}. \qquad (15.137)$$

References

Two-Level Systems, the Rabi Solution and the FVH Representation

1. I. I. Rabi, Space quantization in a gyrating magnetic field, Phys. Rev. 51, 652 (1937).
2. R. P. Feynman, F. L. Vernon, Jr. and R. W. Hellwarth, Geometrical representation of the Schrödinger equation for solving maser problems, J. Appl. Phys. 28, 49 (1957).
3. J. Steinfeld, *Molecules and Radiation* (MIT Press, Cambridge, MA 1986).

4. L. Allen and J. H. Eberly, *Optical Resonance in Two Level Atoms* (Wiley, New York: 1975).

5. M. Sargent III, M. O. Scully and W. E. Lamb, Jr., *Laser Physics* (Addison-Wesley, Reading, MA 1974).

6. M. D. Fayer, *Elements of Quantum Mechanics* (Oxford, New York, 2001).

Dressed States

7. R. W. Boyd, *Nonlinear Optics*, 2nd ed. (Academic Press, New York, 2003).

8. B. W. Shore, *The Theory of Coherent Atomic Excitation*, 2 vols. (Wiley, New York, 1990).

9. C. Cohen-Tannoudji, J. Dupont-Roc and G. Grynberg, *Atom-Photon Interactions: Basic Processes and Applications* (Wiley, New York, 1992).

10. P. R. Berman, Theory of fluorescence and probe absorption in the presence of a driving field using semiclassical dressed states, Phys. Rev. A 53, 2627 (1996).

Impulsive Excitation

11. U. Banin, A. Bartana, S. Ruhman and R. Kosloff, Impulsive excitation of coherent vibrational motion ground surface dynamics induced by intense short pulses, J. Chem. Phys. 101, 8461 (1994).

12. T. J. Smith, L. W. Ungar and J. A. Cina, Resonant short-pulse effects on nuclear motion in the electronic ground-state, *J. Lumin.* 58, 66 (1994).

Adiabatic Excitation

13. K. Bergmann, H. Theuer and B. W. Shore, Coherent population transfer among quantum states of atoms and molecules, Rev. Mod. Phys. 70, 1003 (1998).

14. V. Malinovsky and D. J. Tannor, Simple and robust extension of the stimulated Raman adiabatic passage technique to N-level systems, Phys. Rev. A 56, 4929 (1997a).

15. N. V. Vitanov, B. W. Shore and K. Bergmann, Adiabatic population transfer in multistate chains via dressed intermediate states, Eur. Phys. J. D 4, 15 (1998).

16. T. Nakajima, Population transfer in N-level systems by dressing fields, Phys. Rev. A 59, 559 (1999).

17. T. Halfmann, L. P. Yatsenko, M. Shapiro, B. W. Shore and K. Bergmann, Population trapping and laser induced continuum structures in He, Phys. Rev. A 58, R46 (1998).

18. M. N. Kobrak and S. A. Rice, Selective photochemistry via adiabatic passage: An extension of stimulated Raman adiabatic passage for degenerate final states, Phys. Rev. A 57, 2885 (1998).

19. S. Chelkowski and A. D. Bandrauk, Raman chirped adiabatic passage: A new method for selective excitation of high vibrational states, J. Raman Spect. 28, 459 (1997).

20. J. Oreg, F. T. Hioe and J. H. Eberly, Adiabatic following in multilevel systems, Phys. Rev. A 29, 690 (1984); J. Oreg, G. Hazak and J. H. Eberly, Multilevel inversion schemes in and beyond the adiabatic limit, Phys. Rev. A 32, 2776 (1985).

21. T. Nakajima, M. Elk, J. Zhang and P. Lambropoulos, Population transfer through the continuum, Phys. Rev. A 50, R913 (1994).

22. T. A. Laine and S. Stenholm, Adiabatic processes in three-level systems, Phys. Rev. A 53, 2501 (1996).

23. Y. Band and O. Magnes, Is adiabatic passage population transfer a solution to an optimal-control problem?, J. Chem. Phys. 101, 7528 (1994).

24. N. V. Vitanov, T. Halfmann, B. W. Shore and K. Bergmann, Laser-induced population transfer by adiabatic passage techniques, Annual Rev. Phys. Chem. 52, 763 (2001).

Optical Paralysis

25. R. Kosloff, A. D. Hammerich and D. Tannor, Excitation without demolition: Radiative excitation of ground-surface vibration by impulsive stimulated Raman scattering with damage control, Phys. Rev. Lett. 69, 2172 (1992).

26. A. Bartana, R. Kosloff and D. J. Tannor, Laser cooling of internal degrees of freedom. II., J. Chem. Phys. 106, 1435 (1997).

27. V. Malinovsky, C. Meier and D. J. Tannor, Optical paralysis in electronically congested systems: Application to large amplitude motion of ground state Na_2, Chem. Phys. 221, 67 (1997b).

Chapter 16

Design of Femtosecond Pulse Sequences to Control Chemical Reactions

Since the invention of the laser in 1960, the chemist's dream has been to take advantage of the unique properties of laser light—enormous intensities, precise frequency tunability and perhaps most importantly, phase coherence—to carry out photochemical reactions selectively and efficiently. Most industrial chemistry is performed using temperature and pressure control. Typically, a lot of energy is wasted and a lot of unwanted by-products are produced. Laser light is an ideal chemical reagent in that all its parameters can be completely controlled. It offers the promise of depositing energy in a molecule in a very nonstatistical fashion, and thereby performing chemistry cleanly and energetically efficiently.

In the period 1960–1985 many intuitive approaches to laser selective chemistry were tried. Most of these approaches focused on depositing energy in a sustained manner, using monochromatic radiation, into a particular state or mode of the molecule. For diatomic molecules this approach does not lead to fragmentation: a frequency that is resonant with vibrational transitions at the bottom of the well is off-resonant with vibrational transitions at higher energies, and it is impossible to reach the dissociation threshold, at least at moderate intensities. For polyatomic and larger molecules, the situation is reversed. There tends to be a very high density of vibrational/rotational states even at moderate energies, and there are always resonances. The problem is that many unwanted photoproducts are generally obtained and the distribution tends to be statistical and uncontrolled. The many photofragments that are produced can be traced to the nature of the molecular eigenstates at intermediate to high energies, where many degrees of freedom are simultaneously excited in a nonseparable way. This is often expressed by saying that in polyatomic molecules, intramolecular vibrational relaxation (IVR) is rapid compared with the time scale for breaking the desired chemical bond.

In the mid-1980s attention began to shift to more sophisticated ideas for laser control of chemical reactions. Both theory and experiment began to focus on the emerging capability of producing femtosecond laser pulses (10^{-15} s). What makes this time scale significant is that it is short compared with a vibrational period of motion (10^{-14} s–10^{-13} s). As a result, femtosecond pulses can prepare vibrational wavepackets that are localized to a portion of the vibrational orbit. Moreover, sequences of femtosecond pulses can "capture" the wavepacket in a precise phase of its vibrational motion and transfer it to a new electronic state; this can be carried out multiple times between multiple electronic states, with the time delays between the pulses controlling the properties of the wavepacket that is produced.

In addition to the delay between femtosecond pulses, in the mid-1980s both theory and experiment began to explore the consequences of *shaping* femtosecond pulses, in both amplitude and phase. Given the rate of experimental progress, by the mid-1980s it was already natural to ask: if any pulse shape or pulse sequence whatsoever could be crafted in the laboratory, what shape would be optimal for cleaving a particular bond in a particular molecule?

As we discuss below, the question of finding the *optimal* shape of laser pulses or pulse sequences to cleave a particular bond is naturally formulated as a problem in the calculus of variations. The calculus of variations is the mathematical apparatus for maximizing a *functional*: an objective function that depends on another function. This class of problems can be called "best shape" problems: find the best shape for a function, subject to constraints that will maximize or minimize a certain quantity. For example, the shape that encloses the maximum area for a given perimeter; the curve of minimum length that connects two points on a sphere, subject to the constraint that the curve lie on the sphere; the shape of a cable of fixed length and fixed endpoints that minimizes the potential energy; the trajectory of a fixed speed that goes from point a to point b in least time; the trajectory that goes from point a to point b in time t that minimizes the so-called action integral: all the above are searches for the best shape, and all are problems that are naturally formulated and solved using the calculus of variations. In our case, we are searching for the best shape of laser pulse (electric field amplitude vs. time) to maximize a certain chemical product yield. If we admit complex pulses this involves an optimization over the real and imaginary parts of the pulse shape. We may be interested in the optimal pulse subject to some constraints, for example, for a fixed total energy in the pulse.

There is another branch of mathematics, closely related to the calculus of variations, known as Optimal Control Theory (OCT). One may view Optimal Control Theory as the application of the calculus of variations to problems with differential equation constraints. Optimal Control Theory is used in chemical, electrical and aeronautical engineering, where the differential equation constraints may be chemical kinetic equations, electrical circuit equations, the Navier–Stokes equations for air flow, or Newton's equations. In our case, the differential equation constraint is the Time-Dependent Schrödinger Equation (TDSE) in the presence of the control, which is the electric field interacting with the dipole (permanent or transition dipole moment) of the molecule. From the point of view of control theory, this application presents many new features relative to conventional applications. Conceptually, the transfer of the mathematics of control theory to control the *microscopic* equations of motion is both a novel and potentially very important new direction.

This chapter begins by introducing a model triatomic ABC (masses H, H, D). This two-degrees-of-freedom system is the simplest paradigm for control of chemical products. The Tannor–Rice "pump–dump" scheme is reviewed briefly, to illustrate that some measure of control is indeed possible and to set the stage for the type of mechanisms that may emerge from the optimization procedure. This same example is then used to illustrate the formulation of the variational problem. The "seat of the pants" optimization procedure, developed by Kosloff et al., is then presented. Finally, the optimal control formalism is developed, leading to an expression for the optimal field that requires the solution to a set of coupled partial differential equations. The numerical solution of these coupled equations is then shown to involve many of the components from the "seat of the pants" procedure! The pulses that emerge from the optimization are analyzed using the Husimi transform technique, to learn about their mechanism. A general discussion of controllability theory is also provided.

The latter part of this chapter discusses additional topics relevant to quantum control. The first is an analytic formulation of the phase profile of a femtosecond pulse to obtain desired objectives in simple two- and three-level systems. Then, several examples of control of molecular dynamics via chirped pulses are presented. Finally, we give an introduction to learning algorithms, in which the iterative process of finding optimal pulse shapes can be conducted entirely in the laboratory through a feedback loop from the output back to the pulse shaper.

16.1   Intuitive Control Concepts

Consider the ground electronic state potential energy surface in Figure 16.1. This potential energy surface, corresponding to collinear ABC, has a region of stable ABC and two exit channels, one corresponding to A + BC. and one to AB + C. This system is the simplest paradigm for control of chemical product formation: a two-degrees-of-freedom system is the minimum that can display two distinct chemical products. The objective is, starting out in a well-defined initial state ($v = 0$ of the ABC molecule), to design an electric field as a function of time that twill steer the wavepacket out of channel 1, with no amplitude going out of channel 2, and vice versa.

We introduce a single excited electronic state surface at this point. The motivation is severalfold:

1. The interaction with the transition dipole moment $\mu(R)$ is generally much stronger than with the permanent dipole moment $d(R)$, since the constant part of $\mu(R)$ can promote transitions (Condon approximation), while only the coordinate dependence of $d(R)$ can promote pure vibrational transitions.

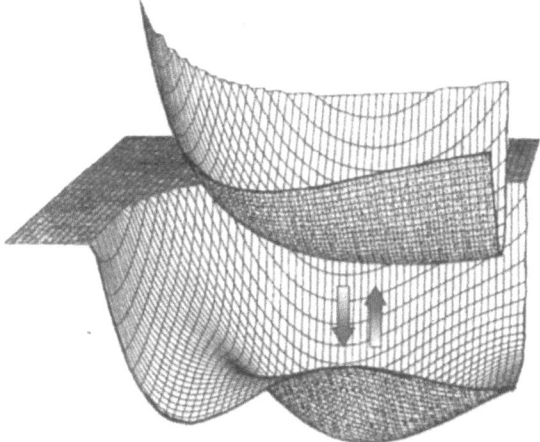

Figure 16.1 Stereoscopic view of the ground and excited-state potential energy surfaces for a model collinear ABC system with the masses of HHD. The ground state surface has a minimum, corresponding to the stable ABC molecule. This minimum is separated by saddle points from two distinct exit channels, one leading to AB + C the other to A + BC. The object is to use optical excitation and stimulated emission between the two surfaces to 'steer' the wavepacket selectively out of one of the exit channels.

2. The use of excited electronic states facilitates large changes in force on the molecule, effectively instantaneously, without necessarily using strong fields. In contrast, with a single surface one must make use of the forcing generated by the coordinate dependence of the dipole.

3. The technology for amplitude and phase control of optical pulses for many years was significantly ahead of the corresponding technology in the infrared.

The object now will be to steer the wavefunction out of a specific exit channel on the ground electronic state, using the excited electronic state as an intermediate.

Consider the following intuitive scheme, in which the timing between a pair of pulses is used to control the identity of products (Tannor, 1985, 1986). The scheme is based on the correspondence between the center of a wavepacket in time and that of a classical trajectory (Ehrenfest's theorem). The first pulse excites a vibrational wavepacket to the excited electronic state. A second pulse, incident after some time delay, stimulates emission of the vibrational wavepacket back to the ground electronic state. The time delay between the pulses controls the time that the vibrational wavepacket evolves on the excited-state potential energy surface. The Franck–Condon principle is used twice, once in the excitation stage and once in the stimulated emission stage. The role of the Franck–Condon principle in the excitation process is familiar by now: the vibrational wavepacket prepared by the first pulse on the excited electronic state is a replica of the initial vibrational state on the ground electronic state at $t = 0$. However, the role of the Franck–Condon principle in the stimulated emission process is new, and quite interesting: the vibrational wavepacket prepared by the second pulse on the *ground* electronic state has the same position and momentum as that of the *excited*-state wavepacket at the moment of the stimulated emission. The position and momentum of the wavepacket produced on the ground state give, in turn, some measure of control over fragmentation products on the ground electronic state. The classical version of this "pump–dump" scheme is illustrated in Figure 16.2, where a single classical trajectory

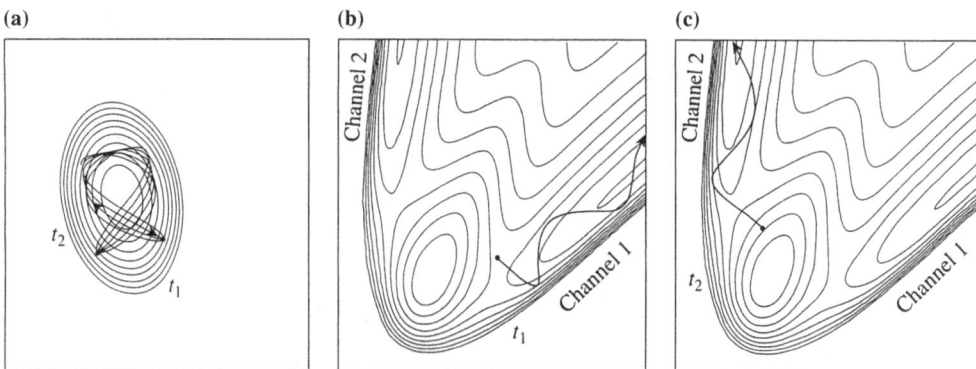

(a)   **(b)**   **(c)**

Figure 16.2 Equipotential contour plots of the ground and excited-state potential energy surfaces (here a harmonic excited state is used because that is the way the first calculations were done). (a) The classical trajectory that originates from rest on the ground state surface makes a vertical transition to the excited state, and subsequently undergoes Lissajous motion, which is shown superimposed. The arrows show the position and momentum of the trajectory at representative times t_1 and t_2. (b) For a vertical transition down at time t_1 (position and momentum conserved) the trajectory continues to evolve on the ground state surface and exits from channel 1 (masses H + HD). (c) If the transition down is at time t_2 the classical trajectory exits from channel 2 (masses HH + D). Adapted from Tannor (1986).

simulates the center of the quantum wavepacket. The trajectory originates at the ground state surface minimum (the equilibrium geometry). At $t = 0$ it is promoted to the excited-state potential surface (a two-dimensional harmonic oscillator in this model) where it originates at the Condon point—vertically above the ground state minimum. Since this position is displaced from equilibrium on the excited state, the trajectory begins to evolve, executing a two-dimensional Lissajous motion. After some time delay, the trajectory is brought down vertically to the ground state (keeping both the instantaneous position and momentum it had on the excited state) and allowed to continue to evolve on the ground state. Figure 16.2 shows that for one choice of time delay it will exit into channel 1; for a second choice of time delay it will exit into channel 2. Note how the position and momentum of the trajectory on the ground state, immediately after it comes down from the excited state, are consistent with the values it had when it left the excited state, and at the same time are ideally suited for exiting out of channel 1 (Figure 16.2b) or channel 2 (Figure 16.2c).

A full quantum mechanical calculation based on these classical ideas is shown in Figures 16.3–16.4. Parameters for the ground and excited-state potential energy surfaces are given by Tannor (1986). Note that the two exit channels differ by only an isotopic

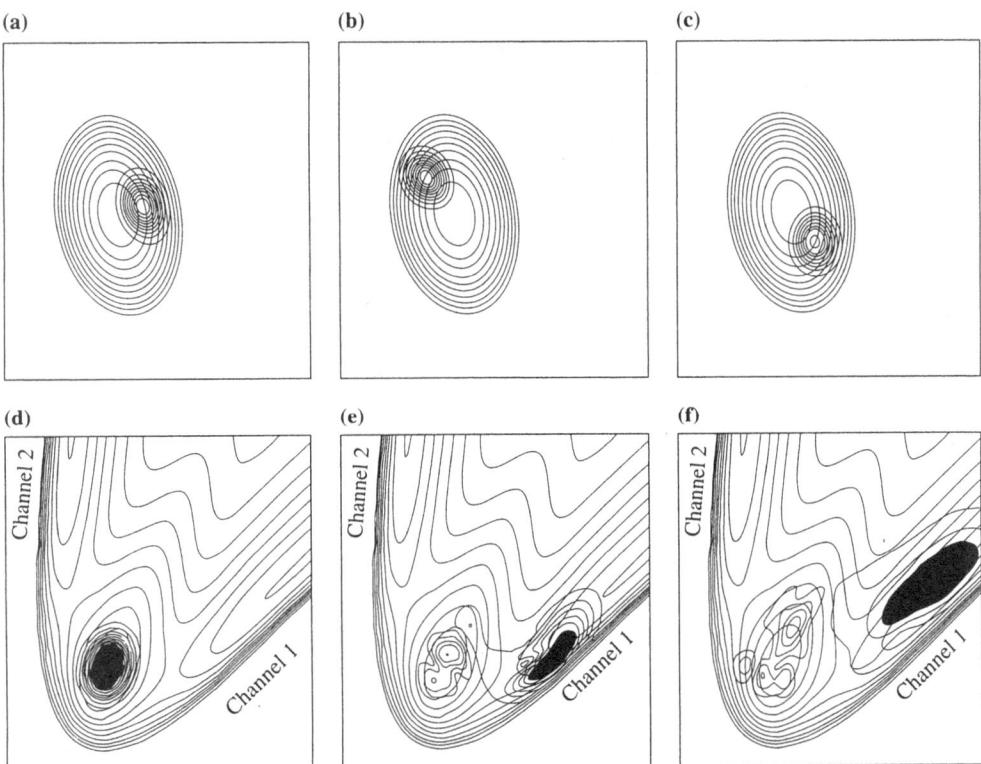

Figure 16.3 Top: Magnitude of the excited-state wavefunction for a pulse sequence of two Gaussians with time delay of 610 a.u. = 15 fs. (a) $t = 200$ a.u., (b) $t = 400$ a.u., (c) $t = 600$ a.u. Note the close correspondence with the results obtained for the classical trajectory (Figure 16.2a). Bottom: Magnitude of the ground state wavefunction for the same pulse sequence, at (d) $t = 0$, (e) $t = 800$ a.u., (f) $t = 1000$ a.u. Note the close correspondence with the classical trajectory of Figure 16.2b. Although some of the amplitude remains in the bound region, that which does exit does so exclusively from channel 1 (masses H + HD). Adapted from Tannor (1986).

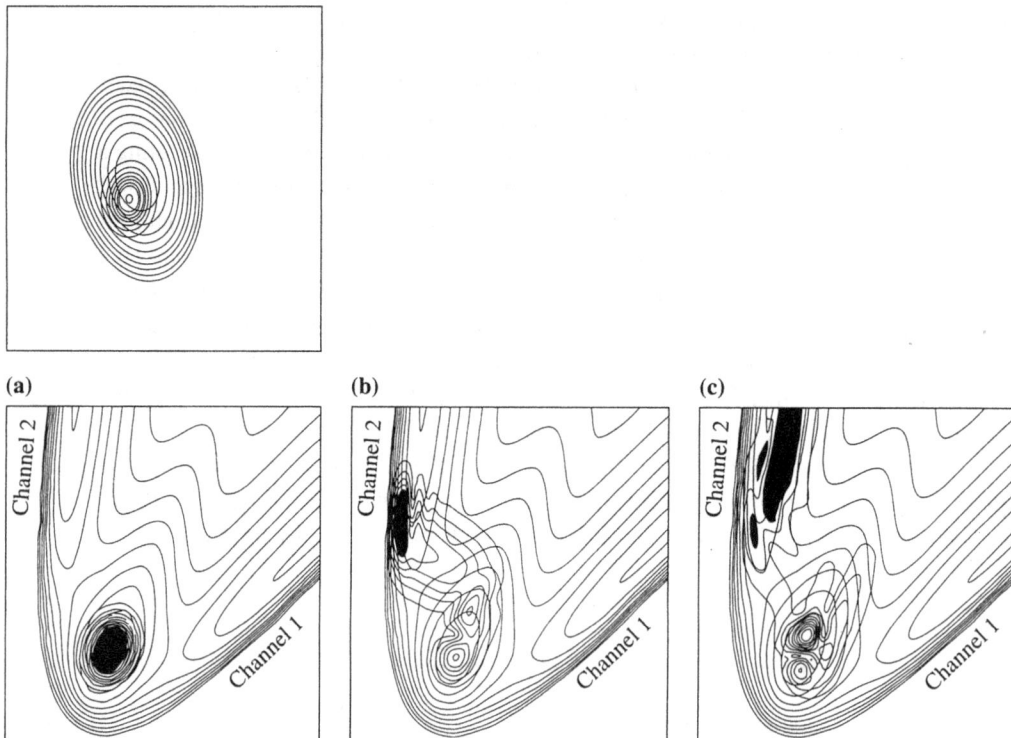

Figure 16.4 Magnitude of the ground and excited-state wavefunctions for a sequence of two Gaussian pulses with time delay of 810 a.u. Top: excited-state wavefunction at $t = 800$ a.u., before the second pulse. Bottom: Magnitude of the ground state wavefunction for the same pulse sequence, at (a) 0 a.u., (b) 1000 a.u., (c) 1200 a.u. Although not all the amplitude exits, the amplitude that does exit does so exclusively from channel 2 (masses HH + D). Note the close correspondence with the classical trajectory of Figure 16.2c. Adapted from Tannor (1986).

substitution (channel 1 corresponds to masses H + HD while channel 2 corresponds to masses HH + D. The dynamics of the two-electronic-state model is solved, starting in the lowest vibrational eigenstate of the ground electronic state, in the presence of a pair of femtosecond pulses that couple the states. Because the pulses are taken to be much shorter than a vibrational period, the effect of the pulses is to prepare a wavepacket on the excited/ground state that is almost an exact replica of the instantaneous wavefunction on the other surface. Thus, the first pulse prepares an initial wavepacket that is almost a perfect Gaussian, and that begins to evolve on the excited-state surface. The second pulse transfers the instantaneous wavepacket at the arrival time of the pulse back to the ground state, where it continues to evolve on the ground state surface, with initial conditions determined by its position and momentum at the time of arrival from the excited state. For one choice of time delay the exit out of channel 1 is almost completely selective (Figure 16.3), while for a second choice of time delay the exit out of channel 2 is almost completely selective (Figure 16.4). Note the close correspondence with the classical model: the wavepacket on the excited state is executing a Lissajous motion almost identical with that of the classical trajectory (the wavepacket is a nearly Gaussian wavepacket on a two-dimensional harmonic oscillator). On the ground state, the wavepacket becomes spatially extended but its exit channel, as well as the partitioning of energy into translation and vibration (i.e., parallel and perpendicular to the exit direction), is seen to be in close agreement with the corresponding classical trajectory.

Before turning to pulse shaping and optimization, we point out an interesting alternative perspective on pump–dump control. A central tenet of Feynman's path integral approach to quantum mechanics is to think of quantum interference as arising from multiple dynamical paths that lead from the same initial to the same final state. The simplest example of this interference involves an initial state, two intermediate states and a single final state, although if the objective is to control some branching ratio at the final energy then at least two final states are necessary. By controlling the phase with which each of the two intermediate states contributes to the final state, one may control constructive versus destructive interference in the final states. This is the basis of the Brumer–Shapiro approach to coherent control (Brumer, 1992, 1995). It is interesting to note that pump–dump control can be viewed entirely from this perspective. Now, however, instead of two intermediate states there are many, corresponding to the vibrational levels of the excited electronic state. The control of the phase that determines how each of these intermediate levels contributes to the final state is achieved via the time delay between the excitation and the stimulated emission pulse. This "interfering pathways" interpretation of pump–dump control is shown in Figure 16.5.

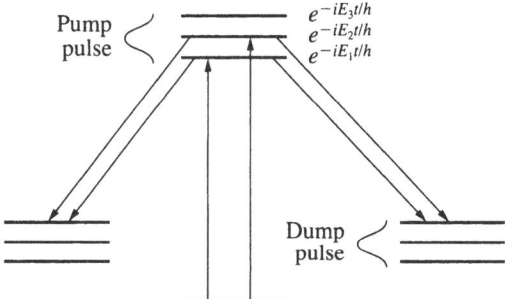

Figure 16.5 Multiple pathway interference interpretation of pump–dump control. Since each of the pair of pulses contains many frequency components, there are an infinite number of combination frequencies that lead to the same final energy state, which generally interfere. The time delay between the pump and dump pulses controls the relative phase among these pathways, and hence determines whether the interference is constructive or destructive. The frequency-domain interpretation highlights two important features of coherent control. First, if final products are to be controlled there must be degeneracy in the dissociative continuum. Second, a single interaction with the light, no matter how it is shaped, cannot produce control of final products; at least two interactions with the field are needed to produce interfering pathways.

16.2   Variational Formulation of Control of Product Formation

The pump–dump scheme is significant for three reasons: it shows that control is possible; it gives a starting point for the design of optimal pulse shapes; and it gives a framework for interpreting the action of two-pulse and more complicated pulse sequences. Nevertheless, the approach is limited: in general with the best choice of time delay and central frequency of the pulses one may achieve only partial selectivity. In addition, this scheme does not exploit the optical phase of the light. Intuition breaks down for more complicated processes and classical pictures cannot adequately describe the role of the phase of the light and the wavefunction. The next step is therefore to address the question, How is it possible to take advantage of the many additional available parameters (pulse shaping, multiple pulse sequences and so on)—in general an $\varepsilon(t)$ with arbitrary complexity—to maximize and perhaps obtain perfect selectivity?

Posing the problem mathematically, one seeks to maximize (Tannor, 1985, 1988)

$$J \equiv \lim_{T \to \infty} \langle \psi(T) | P_\alpha | \psi(T) \rangle, \tag{16.1}$$

where P_α is a projection operator for chemical channel α (here, α takes on two values, referring to arrangement channels A + BC and AB + C; in general, in a triatomic molecule ABC, α takes on three values, 1, 2, 3, referring to arrangement channels A + BC, AB + C, and AC + B). For photodissociation problems, the time T is understood to be longer than the duration of the pulse sequence, $\varepsilon(t)$; the yield, J, is defined as $T \to \infty$, that is, after the wavepacket amplitude has time to reach its asymptotic arrangement. For other types of problems, the time T can be some finite time at which the target state is desired. The key observation is that the quantity J is a *functional* of $\varepsilon(t)$: J is a function of a function, because $\psi(T)$ depends on the whole history of $\varepsilon(t)$. To make this dependence on $\varepsilon(T)$ explicit we may write

$$J[\varepsilon(t)] \equiv \lim_{T \to \infty} \langle \psi([\varepsilon(t)], T) | P_\alpha | \psi([\varepsilon(t)], T) \rangle, \tag{16.2}$$

where square brackets are used to indicate functional dependence. The problem of maximizing a function of a function has a rich history in mathematical physics, and falls into the class of problems belonging to the calculus of variations.

In the perturbative regime it is necessary to introduce a constraint on the norm of the field; otherwise, the larger the field the larger the yield. This constraint takes the form

$$E \equiv \int_0^T dt |\varepsilon(t)|^2, \tag{16.3}$$

where E is a constant. This leads to a maximization problem for the modified objective functional (Tannor, 1985, 1988),

$$\bar{J} \equiv \lim_{T \to \infty} \langle \psi(T) | P_\alpha | \psi(T) \rangle - \lambda \int_0^T dt |\varepsilon(t)|^2, \tag{16.4}$$

where λ is a Lagrange multiplier. Alternatively, it is possible to let the norm of the field vary, but put a penalty on the size of the norm. In this case, one treats λ as fixed: the higher the value of λ the higher the penalty. In the presence of strong fields it is not clear that such a constraint or penalty is necessary, although it does serve to keep the field intensity within physically reasonable limits. However, if this constraint is abandoned the resulting optimal control problem is "singular," and somewhat more sophisticated numerical methods are required.

16.2.1   Perturbative Formulation: Calculus of Variations

Within first-order perturbation theory we have

$$\psi(T) = \frac{1}{i\hbar} \int_0^T e^{-iH_b(T-t)/\hbar}\{-\mu\varepsilon(t)\}e^{-iH_a t/\hbar}\psi_i dt. \tag{16.5}$$

where ψ_i is the initial vibrational state. As a preliminary example, we will choose the objective to be a specially shaped wavepacket, $\tilde{\phi}$, on the excited electronic state (subscript b) at the finite target time T, and therefore the projection operator P is given by

$$P = |\tilde{\phi}_b\rangle\langle\tilde{\phi}_b|. \tag{16.6}$$

Substituting Eqs. 16.5 and 16.6 into Eq. 16.4, we obtain

$$\bar{J} = \left| \frac{1}{i\hbar} \int_0^T \langle\tilde{\phi}|e^{-iH_b(T-t)/\hbar}\{-\mu\varepsilon(t)\}e^{-iH_a t/\hbar}|\psi_i\rangle dt \right|^2 - \lambda \int_0^T |\varepsilon(t)|^2 dt. \tag{16.7}$$

Assuming that ψ_i is an eigenstate of H_a with eigenvalue E_i, we can replace the factor $e^{-iH_a t/\hbar}$ by $e^{-iE_i t/\hbar}$, which cancels in the absolute value; we will henceforth drop this factor. Defining

$$C(t) = \langle\tilde{\phi}_b|e^{-iH_b(T-t)/\hbar}|\phi_i\rangle, \tag{16.8}$$

where $\phi_i = \mu\psi_i$, we have

$$\bar{J} = \frac{1}{\hbar^2} \int_0^T \int_0^T C^*(t)\varepsilon^*(t)C(t')\varepsilon(t')\,dt\,dt' - \lambda \int_0^T |\varepsilon(t)|^2 dt. \tag{16.9}$$

The condition for an extremum (maximum, minimum or generalized saddle point) of the functional is that

$$\frac{\delta\bar{J}}{\delta\tilde{\varepsilon}} = 0, \tag{16.10}$$

where $\tilde{\varepsilon}$ is the optimal field. This equation requires some explanation. $\bar{J}$ is a real quantity, while ε is complex. If we take $\mathrm{Re}\varepsilon$ and $\mathrm{Im}\varepsilon$ as independent variables, the condition $\frac{\delta\bar{J}}{\delta\tilde{\varepsilon}} = 0$ is interpreted as

$$\frac{\delta\bar{J}}{\delta\mathrm{Re}\tilde{\varepsilon}} = 0 \quad \text{and} \quad \frac{\delta\bar{J}}{\delta\mathrm{Im}\tilde{\varepsilon}} = 0. \tag{16.11}$$

Alternatively, ε and ε^* can be viewed as the independent variables, and thus the two conditions for an extremum are

$$\frac{\delta\bar{J}}{\delta\tilde{\varepsilon}} = 0 \quad \text{and} \quad \frac{\delta\bar{J}}{\delta\tilde{\varepsilon}^*} = 0. \tag{16.12}$$

Often, as in Eq. 16.9, $\bar{J}$ will have a symmetry with respect to ε and ε^*; then the two conditions in Eq. 16.12 tend to be redundant and one condition suffices. For our present purposes it is slightly more convenient to take the derivative with respect to ε^*, since from the structure of Eq. 16.3 that gives directly a relation for ε. Therefore,

$$\delta \bar{J} = \frac{1}{\hbar^2} \int_0^T \int_0^T C(t') \varepsilon(t') C^*(t) \, dt' \delta \varepsilon^*(t) \, dt - \lambda \int_0^T \varepsilon(t) \delta \varepsilon^*(t) \, dt = 0. \quad (16.13)$$

Since the RHS must be zero for all $\delta \varepsilon^*(t)$, we have

$$\frac{\delta \bar{J}}{\delta \varepsilon^*} = \frac{1}{\hbar^2} \int_0^T C(t') \varepsilon(t') C^*(t) \, dt' - \lambda \varepsilon(t) = 0. \quad (16.14)$$

The solution of this equation is easily checked to be (Tannor, 1985; Cao, 1997)

$$\varepsilon(t) = \alpha C^*(t) = \alpha \langle \phi_i | e^{i H_b (T-t)/\hbar} | \tilde{\phi} \rangle = \alpha \langle \phi_i | \tilde{\phi}_b(t) \rangle, \quad (16.15)$$

where α is an arbitrary (complex but time-independent) multiplicative factor and

$$\lambda = \frac{1}{\hbar^2} \int_0^T |C(t)|^2 dt = \frac{1}{\hbar^2 |\alpha|^2} \int_0^T |\varepsilon(t)|^2 dt. \quad (16.16)$$

Equation 16.15 has the appealing physical interpretation that the optimal field is determined by propagating the target state $\tilde{\phi}_b$ backward in time under H_b from T to t and overlapping it with the initial state; the amplitude and phase of the overlap at time t determine the amplitude and phase of the field at time t.

Before leaving the subject of optimization of first-order processes, we present a simple, alternative approach (Uberna, 1998). The target state on the excited electronic state is expressed as a superposition of excited-state vibrational eigenstates:

$$\tilde{\phi}_b(T) = \sum_n a_n e^{-i\phi_n(T)} \psi_n. \quad (16.17)$$

If we separate out the preparation stage from the subsequent dynamics we obtain

$$\tilde{\phi}_b(T) = \sum_n a_n e^{-i E_n T/\hbar} e^{-i\phi_n} \psi_n, \quad (16.18)$$

where $a_n e^{-i\phi_n}$ is determined by the preparation process. In particular,

$$a_n e^{-i\phi_n} = \tilde{\varepsilon}(\omega_n - \omega_i) \langle \psi_n | \mu | \psi_i \rangle, \quad (16.19)$$

where $\tilde{\varepsilon}(\omega_n - \omega_i)$ is the Fourier component of the field at the transition frequency, $\omega_n - \omega_i$, and $\langle \psi_n | \mu | \psi_i \rangle$ is the Franck–Condon factor. Rearranging Eq. 16.19 we find

$$\tilde{\varepsilon}(\omega_n - \omega_i) = \frac{a_n e^{-i\phi_n}}{\langle \psi_n | \mu | \psi_i \rangle}, \quad (16.20)$$

which gives a prescription for the amplitude and phases of the optimal field at the transition frequencies $\omega_n - \omega_i$.

The first-order treatment given here provides a nice introduction to the variational method. However, in some sense it does not achieve true coherent control. For example, if the excited electronic state is dissociative with an intrinsic branching ratio of different photochemical products, first-order excitation with any shaped pulse simply produces the intrinsic branching ratio at each energy E, multiplied by the power spectrum of the excitation at that energy. This is not to say that first-order excitation cannot produce interestingly shaped wavepackets; but it cannot control quantities like photochemical yields that involve asymptotic time properties. However, within second-order perturbation theory, control of asymptotic time properties is indeed possible, as we have already seen in the pump–dump scheme. The second-order variational treatment has been worked out by several groups. It is more involved than the first-order treatment and we will omit a discussion of it here, going directly to the complete strong field formulation.

16.2.2   Optimal Control Theory

In the previous section we described how the calculus of variations can be used, in the weak field regime, to find optimal pulse shapes and pulse sequences that in principle can be of arbitrary complexity. If the field is not weak, one cannot write an equation like Eq. 16.5, in which the field appears in a simple way in the expression for the wavefunction; the wavefunction at time T will in general have a complicated nonlinear dependence on the entire history of the field at previous times. The weak field limitation can be overcome by using Optimal Control Theory (OCT), where the complicated dependence of $\psi(T)$ on $\varepsilon(t)$ is effectively removed by *deconstraining* the problem using a Lagrange multiplier.

As an illustration of the OCT formalism, we apply it to the two-electronic-state problem of control of reaction products described in Section 16.1. In this example, the Time-Dependent Schrödinger Equation written as a 2×2 matrix in a Born–Oppenheimer basis set, Eq. 16.22, is introduced into the objective functional with a Lagrange multiplier, sometimes called the "dual" or "adjoint" wavefunction, $\chi(x, t)$. The modified objective functional may now be written as (Peirce, 1988; Kosloff, 1989)

$$\bar{J} \equiv \lim_{T \to \infty} \langle \psi(T)|P_\alpha|\psi(T)\rangle + 2\text{Re}\left\{\int_0^T dt \langle \chi(t)| - \frac{\partial}{\partial t} + \frac{H}{i\hbar}|\psi(t)\rangle\right\} - \lambda \int_0^T dt |\varepsilon(t)|^2,$$

(16.21)

where

$$H \equiv \begin{pmatrix} H_a & -\mu\varepsilon^*(t) \\ -\mu\varepsilon(t) & H_b \end{pmatrix}$$

(16.22)

and

$$\psi \equiv \begin{pmatrix} \psi_a \\ \psi_b \end{pmatrix} \qquad \chi \equiv \begin{pmatrix} \chi_a \\ \chi_b \end{pmatrix}.$$

(16.23)

It is clear that as long as ψ satisfies the Time-Dependent Schrödinger Equation the new term in $\bar{J}$ will vanish for any $\chi(x, t)$. The function of the new term is to make the variations of $\bar{J}$ with respect to ε and with respect to ψ independent, to first order in $\delta\varepsilon$, that is, to "deconstrain" ψ and ε. It turns out that because ε and ε^* enter in different ways in Eq. 16.22 we will need to take the derivative with respect to these two variables independently.

Integrating Eq. 16.21 by parts we obtain

$$\bar{J} \equiv \lim_{T \to \infty} \langle \psi(T)|P_\alpha|\psi(T)\rangle - 2\text{Re}\langle \chi|\psi\rangle|_0^T$$

$$+ 2\text{Re}\int_0^T dt \left\{\langle \chi(t)|\frac{H}{i\hbar}|\psi(t)\rangle + \langle \dot{\chi}|\psi\rangle\right\} - \lambda \int_0^T dt |\varepsilon(t)|^2. \quad (16.24)$$

Because the functional can now be varied without an auxiliary constraint equation, the variations of $\psi(t)$, $\psi(T)$, $\varepsilon(t)$ and $\varepsilon^*(t)$ may now be taken independently. The conditions for a stationary point are

$$\frac{\delta \bar{J}}{\delta \psi(t)} = 0 \qquad \frac{\delta \bar{J}}{\delta \psi(T)} = 0 \qquad \frac{\delta \bar{J}}{\delta \varepsilon} = 0 \qquad \frac{\delta \bar{J}}{\delta \varepsilon^*} = 0. \tag{16.25}$$

Thus,

$$
\begin{aligned}
\delta \bar{J} = \lim_{T \to \infty} & \; 2\mathrm{Re}\langle \psi(T)|P_\alpha|\delta\psi(T)\rangle - 2\mathrm{Re}\langle \chi(T)|\delta\psi(T)\rangle \\
& + 2\mathrm{Re} \int_0^T dt \left\{ \left\langle \frac{H}{-i\hbar}\chi(t) \,\Big|\, \delta\psi(t) \right\rangle + \langle \dot{\chi}(t)|\delta\psi(t)\rangle \right\} \\
& + 2\mathrm{Re} \int_0^T dt \left\{ \left\langle \chi(t) \,\Big|\, \frac{1}{i\hbar}\frac{\delta H}{\delta \varepsilon^*} \,\Big|\, \psi(t) \right\rangle \delta\varepsilon^* + \left\langle \chi(t) \,\Big|\, \frac{1}{i\hbar}\frac{\delta H}{\delta \varepsilon} \,\Big|\, \psi(t) \right\rangle \delta\varepsilon \right\} \\
& - \lambda \int_0^T dt\, \varepsilon(t)\delta\varepsilon^*(t) - \lambda \int_0^T dt\, \varepsilon^*(t)\delta\varepsilon(t), \tag{16.26}
\end{aligned}
$$

where

$$\frac{\delta H}{\delta \varepsilon^*} = \begin{pmatrix} 0 & -\mu \\ 0 & 0 \end{pmatrix}, \qquad \frac{\delta H}{\delta \varepsilon} = \begin{pmatrix} 0 & 0 \\ -\mu & 0 \end{pmatrix}. \tag{16.27}$$

In order to satisfy the requirement that $\frac{\delta \bar{J}}{\delta \psi(t)} = 0$ it is convenient to choose χ to satisfy the equation

$$i\hbar \frac{\partial \chi}{\partial t} = H\chi, \tag{16.28}$$

that is, to have χ satisfy the Time-Dependent Schrödinger Equation. Similarly, in order to satisfy the requirement that $\frac{\delta \bar{J}}{\delta \psi(T)} = 0$ it is convenient to choose the boundary condition of χ at the *final* time, T, to be

$$\chi(x, T) = P_\alpha \psi(x, T); \tag{16.29}$$

in other words, at the final time T, χ is equal to the projection operator operating on the Schrödinger wavefunction at the final time. These conditions "conspire," so that a change in ε, which would ordinarily change $\bar{J}$ through the dependence of $\psi(T)$ on ε, does not do so to first order in the field.

For a physically meaningful solution it is *required* that

$$i\hbar \frac{\partial \psi}{\partial t} = H\psi \tag{16.30}$$

$$\psi(x, 0) = \psi_0(x). \tag{16.31}$$

Some workers prefer to derive this latter equation from the condition that $\frac{\delta \bar{J}}{\delta \chi} = 0$. The physical interpretation is that if one removes the constraint that ψ must satisfy the Time-Dependent Schrödinger Equation, one can attain only a maximum value of J higher than or equal to the physical maximum. If one then proceeds to minimize this "fictitious" $\bar{J}$ with respect to χ one can hope to recover the global maximum of J corresponding to the physical problem, but only if Eqs. 16.30–16.31 are satisfied; otherwise the condition $\frac{\delta \bar{J}}{\delta \chi} = 0$

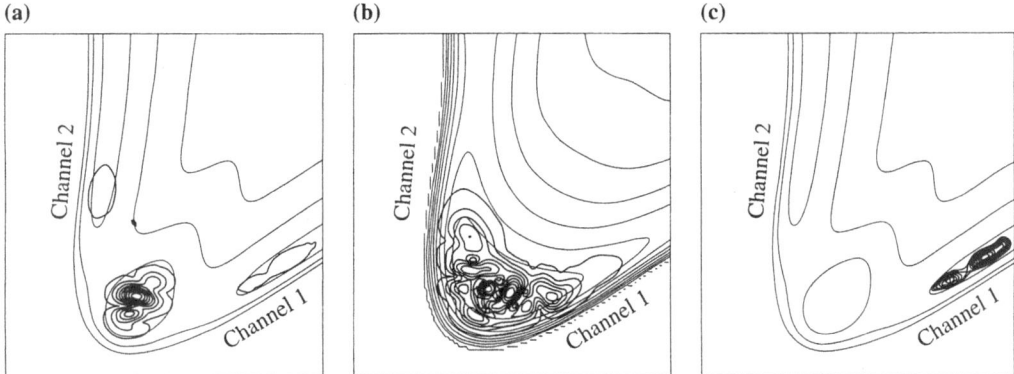

Figure 16.6 Pictorial explanation of an iterative method for optimizing the amount of wavepacket amplitude that exits out the desired chemical channel. (a) The ground state wavefunction $\psi_a(T)$ at the final time T. Note that there is amplitude in *both* exit channels, which is the generic result obtained with an initial guess at the optimal pulse. (b) The excited-state wavefunction at the final time, $\psi_b(T)$. (c) The effect of applying a projection operator $P_\alpha = P_1$ to the ground state wavefunction at the final time, which sets to zero all amplitude that is not in channel 1. This new wavefunction is called χ and the figure therefore illustrates $\chi(T)$. Adapted from Kosloff (1989).

provides a least upper bound to the physical J. This forms the basis for a powerful, but unconventional, approach to functional maximization.

We now turn to the equation for the optimal field. Taking the complex conjugate of the term in Eq. 16.26 that is proportional to $\delta\varepsilon$ we can express the integrand entirely in terms of $\delta\varepsilon^*$. The optimal $\varepsilon(t)$ must satisfy the condition that $\frac{\delta \bar{J}}{\delta\varepsilon^*(t)} = 0$, which leads to the equation

$$\varepsilon(t) = \frac{i}{\hbar\lambda}[\langle\chi_a(t)|\mu|\psi_b(t)\rangle - \langle\psi_a(t)|\mu|\chi_b(t)\rangle], \tag{16.32}$$

where we have used Eq. 16.27. Equation 16.32 is our central equation for the optimal field.

Equations 16.28–16.32 form the basis for a double-ended boundary value problem. ψ is known at $t = 0$; taking a guess for $\varepsilon(t)$ one can propagate ψ forward in time to obtain $\psi(t)$. At time T the projection operator P_α may be applied to obtain $\chi(T)$, which may be propagated backward in time to obtain $\chi(t)$. Note, however, that the above description is not self-consistent: the guess of $\varepsilon(t)$ used to propagate $\psi(t)$ forward in time and to propagate $\chi(t)$ backward in time is not, in general, equal to the value of $\varepsilon(t)$ given by Eq. 16.32. Thus, in general one has to solve these equations iteratively. This iterative scheme is described pictorially in Figures 16.6–16.7.

Equation 16.29 for χ has the following physical interpretation: at the final time T, after the pulse at the current iteration has finished, all amplitude that is not in the desired chemical channel is discarded (compare Figure 16.6a with Figure 16.6c). The remaining amplitude is $\chi(x, T)$; it is propagated back to $t = 0$ (Eq. 16.28), essentially to find out what were the components of the field that gave rise only to this portion of amplitude. According to Eq. 16.32, the first term in the new field is given by the instantaneous dipole moment corresponding to the *filtered* wavefunction on the ground state, χ_a, and the *unfiltered* wavefunction on the excited state, ψ_b, multiplied by i. This is the correct phase to make $\psi(T)$ at the next iteration as close as possible to $\chi(T)$ at the previous iteration, in other

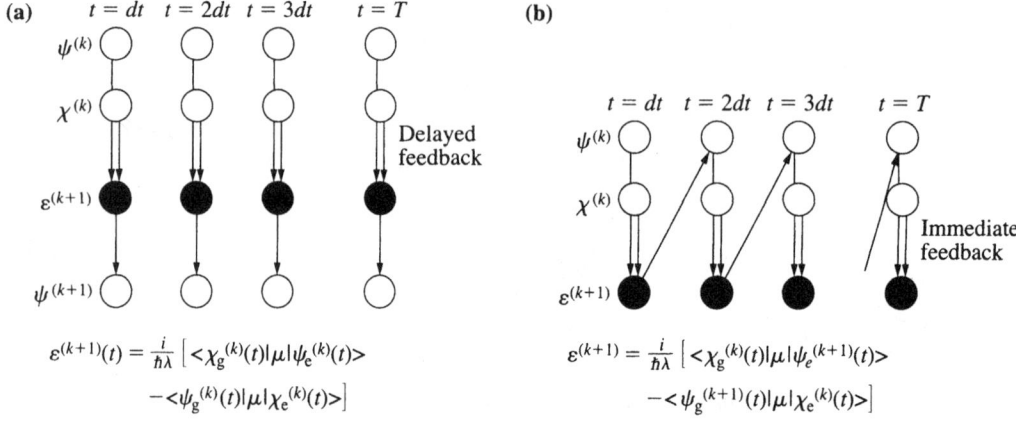

Figure 16.7 In the optimization procedure, $\chi(T)$ is propagated backward to $t = 0$; χ and ψ are then propagated forward in parallel. (a) In the gradient method both χ and ψ are propagated under the old field; the new field $\varepsilon^{(k+1)}(t)$ is calculated from the overlap of $\psi^{(k)}(t)$ and $\chi^{(k)}(t)$. (b) In the Krotov method (Krotov, 1984) χ is propagated under the old field and ψ is propagated under the new field $\varepsilon^{(k+1)}(t)$, which is calculated "on the fly" from the overlap between $\psi^{(k+1)}(t)$ and $\chi^{(k)}(t)$. In both methods, at the final time T the projection operator is applied to the new $\psi(T)$ to construct a new $\chi(T)$. These steps are repeated until convergence is obtained. For more details see Tannor (1992). Adapted from Somlói (1993).

words, to increase the fraction of amplitude in the desired exit channel. The second term in Eq. 16.32 corresponds to higher-order processes, as seen in Exercise 16.1.

▶ **Exercise 16.1**   Show the relationship between Eq. 16.32 and the first-order perturbation theory result, Eq. 16.15.

Solution   If the objective is a target state on the excited-state potential, as in the first-order treatment, then $\chi_a(T)$ vanishes by definition. To lowest order, $\chi_a(t)$ is therefore zero at all times, and the first term vanishes; similarly, to lowest order, $\psi_a(t)$ in the second term is replaced by ψ_i. Therefore, $\varepsilon(t)$ is proportional to $\langle \phi_i | \tilde{\phi}_b(t) \rangle$, where $\phi_i = \mu \psi_i$, in agreement with Eq. 16.15.

16.2.3   Hamiltonian Formulation of the Optimal Control Equations

There is an alternative formulation of the set of Eqs. 16.28–16.31 that bears the same relationship to the original formulation as Hamiltonian mechanics does to Lagrangian mechanics (Bryson, 1975). A control Hamiltonian, $\mathbf{H}$, is constructed and the optimal pulse is that for which the Hamiltonian is an extremal, so that $\frac{\partial \mathbf{H}}{\partial \varepsilon} = 0$. Algorithmically, the search for the optimal pulse can then be performed as a maximization of $\mathbf{H}$. One should not confuse the optimization Hamiltonian $\mathbf{H}$ with the physical Hamiltonian H.

The optimization Hamiltonian is defined as $\mathbf{H} \equiv 2\mathrm{Re}\langle \chi | H/(i\hbar) | \psi \rangle - \lambda |\varepsilon|^2$. This is the Legendre transform of the original objective, neglecting the boundary term $\langle \psi(T) | P | \psi(T) \rangle$. Hamilton's equations take the form

$$\dot{\chi} = -\frac{\partial \mathbf{H}}{\partial \psi} = \frac{H}{i\hbar}\chi \tag{16.33}$$

$$\dot{\psi} = \frac{\partial \mathbf{H}}{\partial \chi} = \frac{H}{i\hbar}\psi \tag{16.34}$$

in agreement with Eqs. 16.28 and 16.30 above. In taking the derivatives with respect to ψ and χ in Eqs. 16.33–16.34, we are treating ψ and χ at each value of x as independent variables. Analogous equations exist with $\psi \to \psi^*$ and $\chi \to \chi^*$, but they are redundant. The condition for the optimal field is

$$\tilde{\varepsilon} = \arg\max_{\varepsilon} \mathbf{H}(t, \varepsilon, \psi(\varepsilon)) = \frac{i}{\lambda}\left[\left\langle \chi_a \left|\frac{\mu}{\hbar}\right| \psi_b \right\rangle - \left\langle \psi_a \left|\frac{\mu}{\hbar}\right| \chi_b \right\rangle\right] \tag{16.35}$$

in agreement with Eq. 16.32.

It is straightforward to show that $\frac{d\mathbf{H}}{dt} = 0$ for $\varepsilon = \tilde{\varepsilon}(t)$, that is, the control Hamiltonian is time independent on the optimal trajectory if $\mathbf{H}$ does not depend *explicitly* on time (i.e., other than through its dependence on ψ, χ and ε). This relationship can be used as a diagnostic of how close a particular control is to one that satisfies the necessary conditions for optimality. Note the analogy to the relationship in classical mechanics, that if $\frac{\partial H}{\partial t} = 0$ then $\frac{dH}{dt} = 0$ (i.e., energy is conserved) on a trajectory for which the integral of the Lagrangian is an extremum (i.e., any physical trajectory).

▶ **Exercise 16.2**   Show that $\frac{d\mathbf{H}}{dt} = 0$ along the optimal trajectory.

Solution

$$\frac{d\mathbf{H}}{dt} = \frac{\partial \mathbf{H}}{\partial t} + \frac{\partial \mathbf{H}}{\partial \psi}\dot{\psi} + \frac{\partial \mathbf{H}}{\partial \varepsilon^*}\dot{\varepsilon}^* + \frac{\partial \mathbf{H}}{\partial \chi}\dot{\chi} \tag{16.36}$$

$$= 2\mathrm{Re}\left\{\left\langle \chi \left|\frac{H}{i\hbar}\right| \dot{\psi} \right\rangle + \left\langle \dot{\chi} \left|\frac{H}{i\hbar}\right| \psi \right\rangle\right\} + \frac{\partial \mathbf{H}}{\partial \varepsilon^*}\dot{\varepsilon}^* \tag{16.37}$$

$$= 2\mathrm{Re}\left\{\left\langle \frac{H}{-i\hbar}\chi + \dot{\chi} \left|\frac{H}{i\hbar}\right| \psi \right\rangle\right\} + \frac{\partial \mathbf{H}}{\partial \varepsilon^*}\dot{\varepsilon}^* \tag{16.38}$$

$$= \frac{\partial \mathbf{H}}{\partial \varepsilon^*}\dot{\varepsilon}^*. \tag{16.39}$$

Since the last equation is zero along the optimal trajectory we have $\frac{d\mathbf{H}}{dt} = 0$.

16.2.4   Controllability Theory

It is natural to ask, Are there general criteria to decide if, given initial state ψ_i at time 0, it is possible to reach *any* final state ψ_f at time T? The question of controllability has a rich history. The main results have antecedents in the theory of differential equations associated with the names Frobenius and Chow, but were translated to the context of control theory by Brockett (1972, 1973a) and Jurdjevic and Sussman (1972). The theory of controllability as formulated in the general context of dynamical systems may be applied to the control of finite-dimensional quantum systems without the need for any significant modification (Butkovskiy, 1990; Ramakrishna, 1995), although the infinite-dimensional case is still incompletely understood (Huang, 1983). Some striking new ideas exploiting a geometric analysis of controllability have recently been introduced in the context of N-spin systems (Khaneja, 2001a, 2001b), with relevance to optimizing gates in quantum computing.

Linear Control

Consider first a linear control system

$$\dot{x}(t) = Ax(t) + Bu(t), \tag{16.40}$$

where u is the control (M components) and x is the state (N components). (This may be augmented by the equation $y(t) = Cx(t)$, where instead of x, y is the output (p components).) Equation 16.40 is said to be controllable if for every x_0 and x_1 and every $T > 0$ there exists a $u(t)$ such that if $x(0) = x_0$ then $x(T) = x_1$. A necessary and sufficient condition for controllability is that (see, for example, Brockett, 1970)

$$\text{Rank}(B, AB, \ldots, A^{N-1}B) = N, \tag{16.41}$$

where "," indicates a column partition and Rank indicates the number of linearly independent columns.

The proof of Eq. 16.41 goes as follows. The formal solution of Eq. 16.40 is

$$x(t) = \int_{t_0}^{t} e^{A(t-t')} Bu(t') \, dt' + e^{A(t-t_0)} x(t_0). \tag{16.42}$$

For simplicity, we will take $x(t_0) = 0$. From the Cayley–Hamilton theorem (see Exercises 8.44, 8.77), $e^{A(t-t')}$ can be written as

$$e^{A(t-t')} = \sum_{i=0}^{N-1} \alpha_i A^i (t - t')^i, \tag{16.43}$$

where the $\{\alpha_i\}$ are some scalar coefficients. Hence

$$x(t) = \int_{t_0}^{t} \sum_{i=0}^{N-1} \alpha_i A^i (t - t')^i Bu(t') \, dt'. \tag{16.44}$$

The vectors of the form $A^i Bu$, $i = 0, \ldots N - 1$, correspond to directions in the Hilbert space. Dividing the time interval $[t_0, t]$ into N subdivisions and viewing the value of $u(t')$ at these intermediate times as N independent controls, in principle one can create any state spanned by the set $\{A^i B\}$. Therefore, if $\text{Rank}(B, AB, \ldots, A^{N-1}B) = N$, then the reachable space is the full Hilbert space, and the N independent controls $u(t')$ allow any state in this space to be accessed. The usual claim in linear controllability theory is even stronger, that all states in this span can be created in an arbitrarily short time. This formal result can be understood by noting that in principle, no matter how short the time interval $t - t_0$ is, it can be subdivided into N subintervals giving N independent controls. However, this result has to be treated with caution since going to extremely short times requires extremely strong controls, for which the linear equations of motion are almost certainly no longer physically valid.

▶ **Exercise 16.3**   The derivation of the rank condition in Eq. 16.41 assumes that x, A and B are real. Generalize the derivation of Eq. 16.41 to the case that x, A and B are complex.

Solution   If x, A and B are complex we may separate the real and imaginary parts as follows:

$$(x_r \overset{\cdot}{+} i x_i) = (A_r + i A_i)(x_r + i x_i) + (B_r + i B_i)(u_r + i u_i), \tag{16.45}$$

which can be rearranged to give

$$\begin{pmatrix} \dot{x}_r \\ \dot{x}_i \end{pmatrix} = \begin{pmatrix} A_r & -A_i \\ A_i & A_r \end{pmatrix} \begin{pmatrix} x_r \\ x_i \end{pmatrix} + \begin{pmatrix} B_r & -B_i \\ B_i & B_r \end{pmatrix} \begin{pmatrix} u_r \\ u_i \end{pmatrix} \tag{16.46}$$

$$\equiv \dot{\tilde{x}} = \tilde{A}\tilde{x} + \tilde{B}\tilde{u}, \tag{16.47}$$

where $\tilde{x}$ is a vector of length $2N$, $\tilde{u}$ is a vector of length $2M$, $\tilde{A}$ is a matrix of dimension $2N \times 2N$ and $\tilde{B}$ is a matrix of dimension $2N \times 2M$. The rank condition can be applied to Eq. 16.47 with the result that for complete controllability we need

$$\text{Rank}(\tilde{B}, \tilde{A}\tilde{B}, \cdots, \tilde{A}^{N-1}\tilde{B}) = 2N. \tag{16.48}$$

Note that Eq. 16.40 is not homogeneous in x. Therefore, it is not directly applicable to the full Schrödinger equation, which is homogeneous in ψ. However, Eq. 16.40 can be applied to the TDSE within the context of time-dependent perturbation theory, where one obtains a linear, inhomogeneous equation for $\psi^{(1)}$. For instance, in the case of two Born–Oppenheimer potentials we had

$$\psi_b^{(1)}(t) = \frac{1}{i\hbar} \int_{t_0}^{t} e^{-iH_b(t-t')/\hbar} \{-\mu\varepsilon(t')\} e^{-iE_i(t'-t_0)/\hbar} \psi_i(t_0) dt' \tag{16.49}$$

(cf. Eq. 13.10). Consider representing the excited-state Hilbert space in a discrete basis of dimension N. The rank condition tells us that if

$$\text{Rank}(\tilde{\psi}_i, \tilde{H}_b\tilde{\psi}_i, \ldots, \tilde{H}_b^{N-1}\tilde{\psi}_i) = 2N \tag{16.50}$$

then the system is completely controllable. This implies that any final state $\psi_b^{(1)}(t)$ can be created from ψ_i in an arbitrarily short time, which contradicts physical intuition in that it may take a finite amount of time for evolution under H_b to reach the desired final state. The formal resolution to this paradox is that even if the dynamics under H_b is very short, it can produce N basis functions that are marginally linearly independent. Formally, these N basis functions, combined with N independent values of $u(t')$ in the time interval $[t_0, t]$, is enough to create any desired state in the N-dimensional Hilbert space. However, in practice, if the propagation time under H_b is very short, the basis will be so close to linearly dependent that expanding a distant state will be essentially impossible. Moreover, as discussed above, the control fields become larger and larger as the time allowed to reach the target is taken to zero, and the linear equations that arise from perturbation theory are no longer physically appropriate. Thus, the statement that if a linear system is completely controllable the target state can be reached arbitrarily quickly must be taken with caution.

Bilinear Control

Now consider control equations of the form

$$\dot{x}(t) = \left(A + \sum_{i=1}^{M} u_i(t) B_i \right) x(t). \tag{16.51}$$

This family of control problems is bilinear (because the product of u and x enters in the equation of motion). In quantum mechanics, the role of A is played by the operator $-iH_0/\hbar$ and B by $-iH_1/\hbar$, and hence we will focus on the case that A and B are skew-Hermitian matrices ($A = iH$, where H is Hermitian), which ensures that the norm of $x(t)$ is independent of time. This is a subtle, but significant, difference between $x(t)$ in Eq. 16.40 and in Eq. 16.51. The requirement of norm preservation in Eq. 16.51 implies that $x(t)$ does not belong to a linear vector space (since generally $\alpha x_1 + \beta x_2$ will not be of unit norm and hence is not an admissible state); as a result, for the bilinear equation, Eq. 16.51, there is no direct analog of the rank condition for linear systems. The next best thing is to consider the controllability of the *evolution operator*, rather than the state. As we have seen in the context of the TDSE, the evolution operator (here a unitary $N \times N$ matrix) obeys the same equation of motion as the state vector

$$\dot{X}(t) = \left(A + \sum_{i=1}^{M} u_i(t) B_i \right) X(t), \qquad X(0) = \mathbf{1}. \tag{16.52}$$

If the evolution operator is completely controllable (within the space of unitary matrices of dimension $N \times N$) then it can take any initial state into any final state, and it follows that there is complete controllability of the state space. However, controllability of the evolution operator is a much stricter requirement than complete state-to-state control because it requires simultaneous control of all N^2 parameters of the evolution operator, whereas to move any state to any other state requires controlling only one row or one column of the $N \times N$ evolution operator at a time (N parameters). Thus, controllability of X is sufficient but not necessary for controllability of x. Since X is just the propagator, the reachable set of evolution operators $X(T)$ determines the reachable set of $x(T)$ via matrix-vector multiplication.

The analysis of the controllability of X involves three major threads of reasoning:

1. As we have seen in other contexts, every evolution operator can be written in the form $X(t) = e^{Ct}$, where C is a skew-Hermitian matrix. An arbitrary $N \times N$ skew-Hermitian matrix has N^2 independent parameters (N real diagonal elements and $N(N-1)/2$ complex off-diagonal elements), and therefore any $N \times N$ unitary matrix X can be represented as

$$X = e^{\sum_{i=1}^{N^2} \alpha_i C_i}, \tag{16.53}$$

where the C_i are linearly independent skew-Hermitian matrices. Thus the problem of being able to synthesize any unitary operator reduces to the problem of being able to generate and control independently the coefficients of the N^2 *generators* $\{C_i\}$ that sit in the exponent.

2. Consider the case that there are only two generators, B_1 and B_2, and that they can be controlled independently. By choosing appropriate piecewise continuous

controls, one can concatenate the evolution operators associated with the individual generators to obtain the following composite evolution operator:

$$e^{B_1 t} e^{B_2 t} e^{-B_1 t} e^{-B_2 t} = 1 + [B_1, B_2] t^2 + O(t^3) = e^{[B_1, B_2] t^2} + O(t^3). \quad (16.54)$$

The physical significance of Eq. 16.54 is that by concatenating the evolution operators associated with B_1 and B_2 in the proper order one can generate a new generator $[B_1, B_2]$. This new generator may be given an abstract geometric interpretation as a new "direction" in the linear vector space of the generators. By extension, all iterated commutators of B_1 and B_2 lead to new generators and hence new allowed directions.

▶ **Exercise 16.4**   Verify Eq. 16.54 by expanding both sides of the equation in a Taylor series and comparing terms of the same order.

---

3. If B_1 and B_2 can be independently controlled, and $\{C_i\}$, $i = 1, \ldots, i_{max}$, belong to the algebra generated by B_1 and B_2 (that is, C_i is a linear combination of the iterated commutators of B_1 and B_2), then any unitary operator that can be written in the form

$$X = e^{\sum_{i=1}^{i_{max}} \alpha_i C_i} \quad (16.55)$$

can be synthesized by the proper choice of controls of B_1 and B_2 (Brockett, 1972). To a certain extent, this result is a close cousin of the Wei–Norman factorization, but applied in the reverse direction: rather than trying to express the exponential of a sum as a product of exponentials, as in Wei–Norman, here we are claiming that the product of exponentials can be represented as the exponential of a sum. Comparing Eq. 16.55 with Eq. 16.53 we see that if $i_{max} = N^2$, B_1 and B_2 can be used to generate any desired $N \times N$ unitary matrix. More generally, given a set of $\{B_i\}$, $i = 1, \ldots, M$, then the set $\{C_i\}$ in Eq. 16.55 consists of all the iterated commutators of the $\{B_i\}$, and i_{max} is the dimension of this iterated commutator algebra. The one case that requires some special care is the generator associated with the free evolution, A (Jurdjevic, 1972). Since evolution under A is only forward and not backward, we cannot apply Eq. 16.54 to argue that the direction generated by $[A, B_i]$ can be generated. However, for a wide class of systems the motion under A is quasiperiodic. In this case, simply by waiting long enough one can effectively have the direction $(-A)$ participate in the algebra. For these systems (technically called compact Lie groups) the algebra is determined by considering the iterated commutators of A along with $\{B_i\}$. For compact Lie groups (which is the case for unitary evolution of N-level quantum systems), if the combined algebra generated by $\{A, B_1, \ldots\}$ is of dimension N^2 then X is completely controllable. There is another important difference between the generator A and the $\{B_i\}$. As opposed to the $\{B_i\}$, the strength of A is not under our control. As a result, there are restrictions on the time required to reach a matrix X (or a state x) that relies on evolution under A.

In summary, for unitary evolution of N-level quantum systems, if the combined algebra generated by $\{A, B_1, \ldots\}$ is of dimension N^2 then X is completely controllable. An explicit algorithm for determining whether the algebra generated by $\{A, B_1, \cdots\}$ is of dimension N^2

is given in Schirmer (2001). A MATLAB code implementing this algorithm can be found at the web site for this book.

▶ Exercise 16.5

    a. Show that the criterion for nonlinear controllability reduces to that of linear controllability in the appropriate limit.

    b. Show why allowing for nonlinear control will always give a value of the linear rank condition equal to or greater than that predicted by the linear test.

16.3 Applications of the Variational Formulation

We now turn to several applications of the variational formulation to control of fragmentation in molecular systems. The systems to be described are simplified theoretical models, but the qualitative physics is expected to be very similar in typical molecular systems.

16.3.1 Model ABC System

The first example of the variational formulation is the model ABC system used earlier to illustrate the pump–dump control. The objective, J, is given by Eq. 16.1, with P_α a projection operator for chemical arrangement channel A + BC or AB + C (right- or left-hand exit channels in Figure 16.2b–c). A penalty, λ, is imposed on the integrated field intensity, and hence the modified objective functional $\bar{J}$ is given exactly by Eq. 16.21.

The pulses presented below will be analyzed using a Husimi transform in a time-frequency representation. The Husimi transform is proportional to the absolute value squared of

$$\tilde{Q}(\omega_0, t_0) = \int_{-\infty}^{\infty} \left(e^{-\alpha(t-t_0)^2 + i\omega_0(t-t_0)} \right)^* \varepsilon(t) \, dt. \tag{16.56}$$

The quantity $\tilde{Q}(\omega_0, t_0)$ is the overlap of the pulse with a family of Gaussians whose centers vary continuously in both time and frequency. It transforms a time (or frequency) distribution into a two-dimensional time–frequency distribution, allowing one to visualize both the time delay and the frequency offset of the pump and dump pulses simultaneously. It is very similar to a Wigner transform, in that it transforms a one-dimensional function into a two-dimensional function of the original variable and its conjugate. However, there are two reasons to prefer the Husimi transform over the Wigner transform to analyze shaped pulses and pulse sequences. First, the Husimi distribution by its very definition decomposes the pulse into a sum of Gaussians, which is conducive to analyzing the mechanism of the pulse, as well as representing the pulse in terms of a small number of subpulses. Second, for the simplest case of two Gaussian pulses with a time delay, the Husimi distribution produces simply two Gaussians in time–frequency space, while the Wigner distribution gives oscillatory amplitude between the two pulses. The former is a simpler and arguably a more physically meaningful representation of the pulse, facilitating the unravelling of the mechanism by which the pulse produces its control. There are times, however, when the

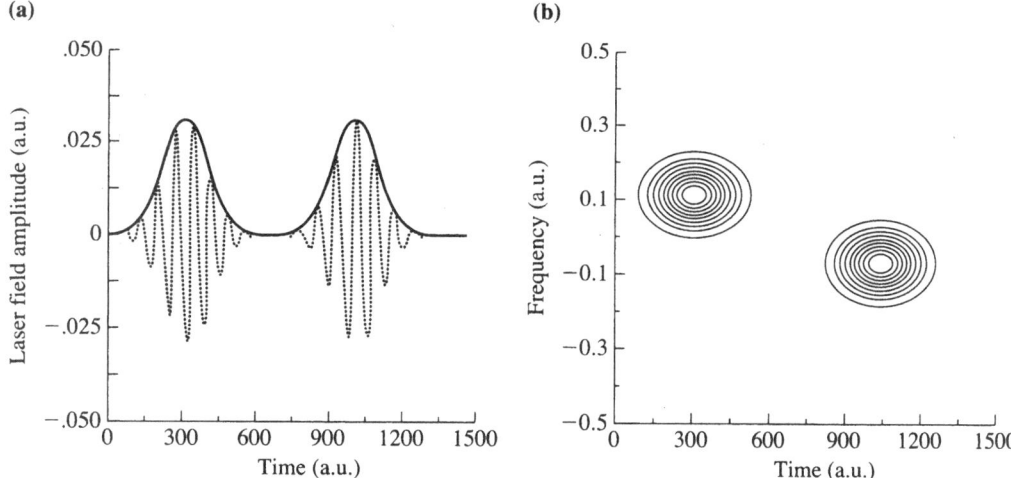

Figure 16.8 (a) Two–pulse sequence used in the pump–dump scheme. (b) The Husimi time–frequency distribution corresponding to the two pulse sequence in (a), constructed by taking the overlap of the pulse sequence with a two-parameter family of Gaussians, characterized by different centers in time and carrier frequency, and plotting the overlap as a function of these two parameters. Note that the Husimi distribution allows one to visualize both the time delay and frequency offset of the pump and dump pulses simultaneously. Adapted from Kosloff (1989).

Wigner representation is preferred, the oscillations in the Wigner distribution serving as a diagnostic of phase coherence between subpulses.

Figure 16.8a shows the initial guess for the pulse sequence (two Gaussians), and Figure 16.8b shows its Husimi transform. Figure 16.9a shows the pulse obtained after about 50 iterations, designed to yield high probability in arrangement channel A + BC, and Figure 16.9b shows its Husimi transform. Figure 16.9c shows the norm on the ground and excited-state surfaces as a function of time, under the influence of the pulse in Figure 16.9a. Figure 16.9d shows the wavepacket on the ground state potential surface after 1500 a.u. (37.5 fs), under the influence of the pulse in Figure 16.9a. Note the high degree of selectivity for exit into the A + BC channel. Figure 16.10a shows the pulse designed to yield high probability in the AB + C channel, and Figure 16.10b shows its Husimi transform. Figure 16.10c shows the norm on the two surfaces as a function of time, under the influence of the pulse in Figure 16.10a. Figure 16.10d shows the wavepacket on the ground state potential surface after 1500 a.u. (37.5 fs), under the influence of the pulse in Figure 16.10a.

It is clear that these pulses are remarkably effective in directing wavepacket amplitude out one channel or the other. Equally intriguing is the superficial similarity between the two pulses, and between both pulses and the initial guess. The pulses in Figures 16.9a and 16.10a occupy identical windows in time, in contrast with the simple pump–dump control scheme of Tannor and Rice, which makes use of different windows in time to achieve different products. The Husimi plots show that the two pulses are even similar in their joint time–frequency distributions. There are significant differences, however. The pulse in Figure 16.9a is still primarily a two-pulse, pump–dump sequence. The pulse in Figure 16.10a, however, has substantial amplitude between the pump and dump, and is much closer to being continuous excitation/de-excitation. This is confirmed by inspection of the norm in Figures 16.9b and 16.10b. The norm in Figure 16.9b changes once from the

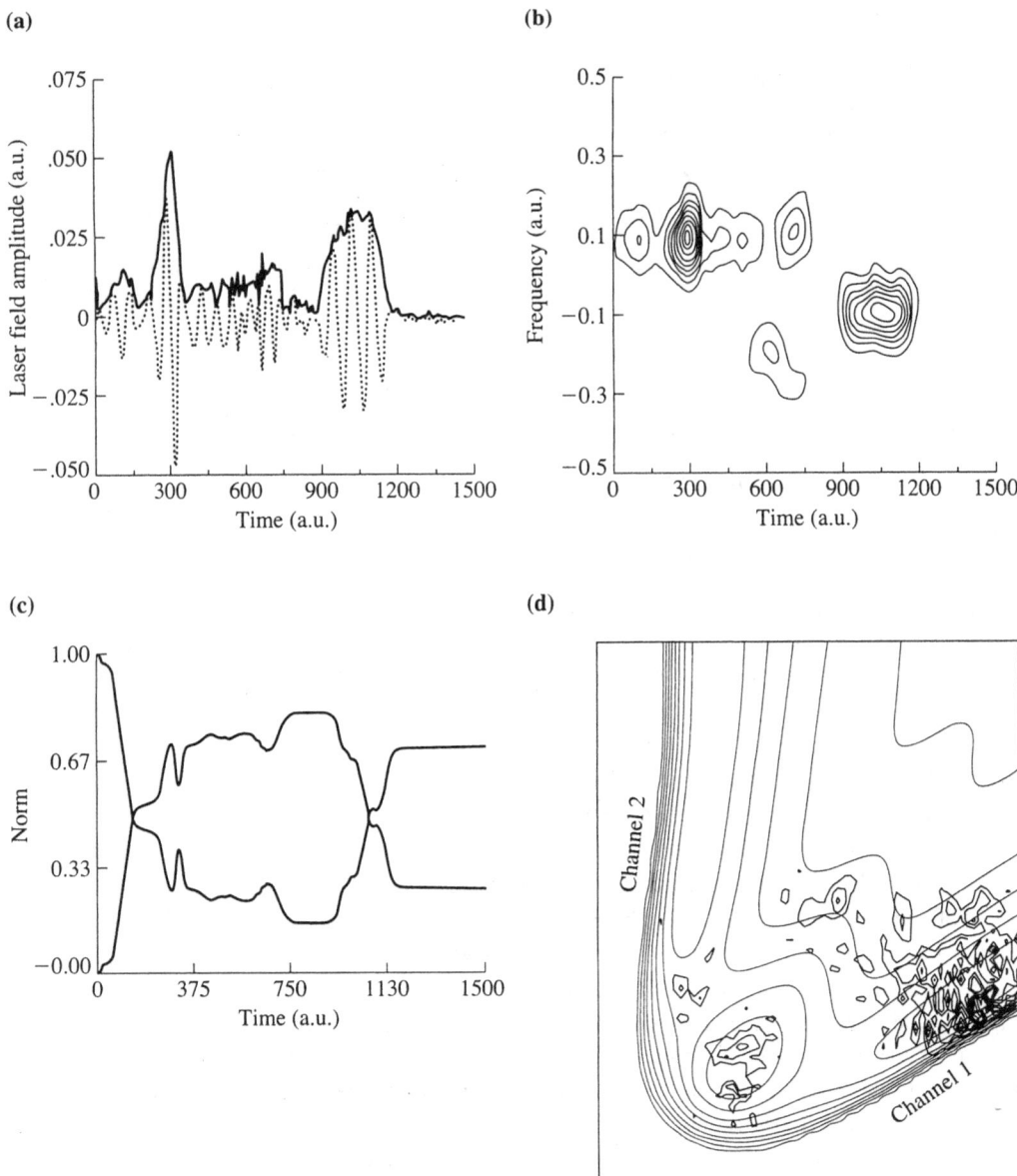

Figure 16.9 (a) The pulse sequence resulting from the optimization procedure described above, with the objective of directing amplitude out channel 1. (b) The Husimi distribution corresponding to the pulse sequence in (a). Note that the qualitative feature of a pump and a dump pulse survives. (c) Norm of the ground and excited electronic state populations versus time for evolution under the pulse sequence shown in (a). (d) Absolute value of the ground state wavefunction at 1500 a.u. (37.5 fs), propagated under the pulse sequence shown in (a), superimposed on equipotential contours of the ground state surface. Adapted from Tannor (1991).

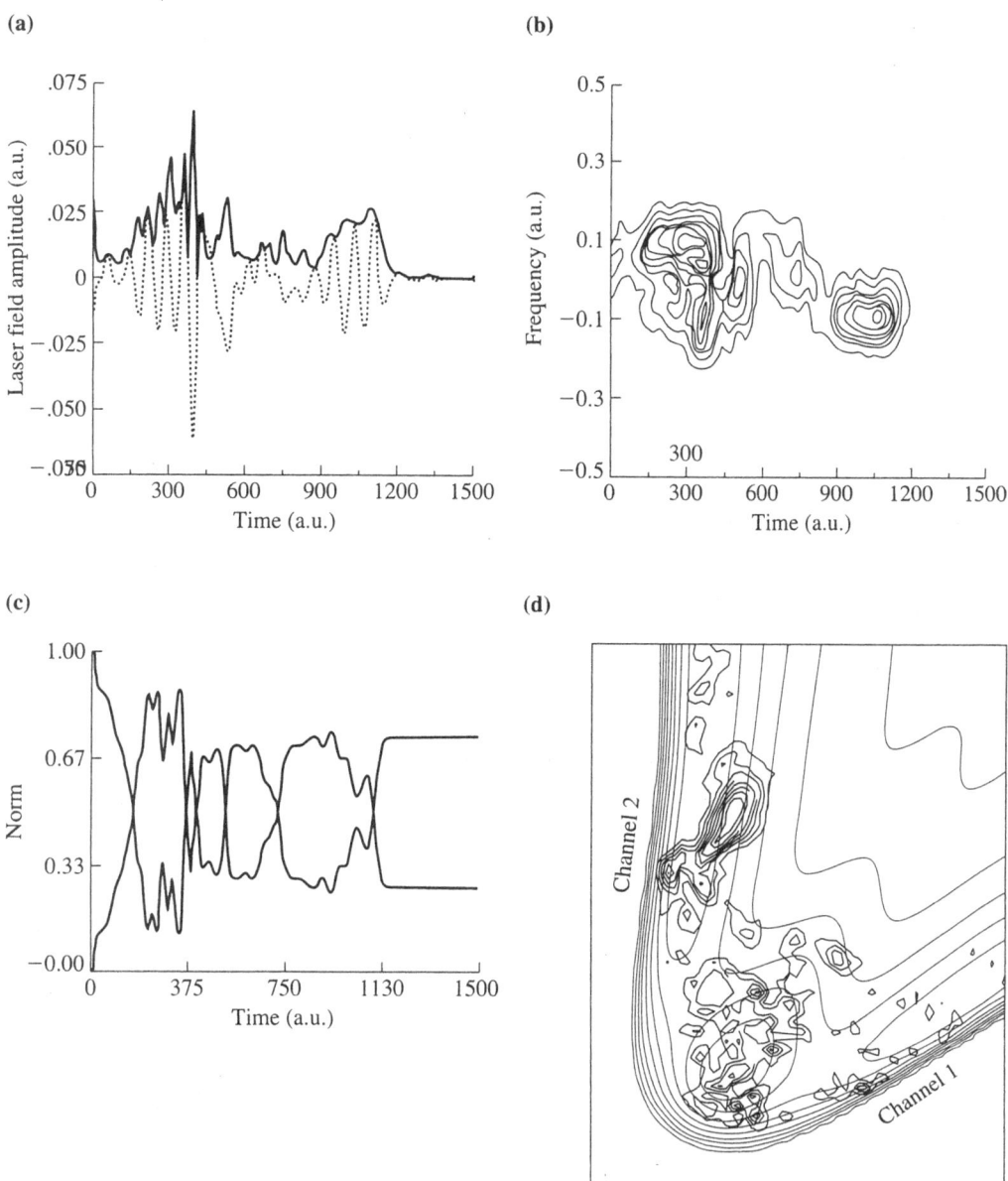

(a)

(b)

(c)

(d)

Figure 16.10 (a) The pulse sequence resulting from the same iterative optimization procedure used in Figure 16.9, but with the objective of directing amplitude out channel 2 (i.e., the projection operator for channel 2, rather than channel 1 is used). (b) The Husimi distribution corresponding to the pulse sequence in (a). Note that some of the character of pump and dump pulses is still present, but there is a significant amount of pulse amplitude at times in between. This pulse sequence, viewed in its entirety, is actually closer to a periodic frequency sweep. (c) Norm of the ground and excited electronic state populations versus time for evolution under the pulse sequence shown in (a). (d) Absolute value of the ground state wavefunction at 1500 a.u. (37.5 fs), propagated under the pulse sequence shown in (a), superimposed on equipotential contours of the ground state surface. Adapted from Tannor (1991).

pump pulse and once again from the dump pulse. The norm in Figure 16.10b, corresponding to the pulse in Figure 16.10a, undergoes three full cycles. It should be noted that there may be significant phase differences between the two pulses in Figures 16.9a and 16.10a that are not evident in the Husimi plots, the latter showing only the absolute value of the distribution.

The pulses in Figures 16.9a and 16.10a provide examples of how simple mechanisms may underlie the pulses calculated using the TDSE/OCT machinery. The pulse sequence in Figure 16.9a is qualitatively a simple pump–dump form: the time delay between the two pulses is such that the wavepacket on the excited-state surface has the correct position and momentum to exit out of channel 1 on the ground state surface. The pulse sequence in Figure 16.10a reflects a different mechanism, closer to a cyclically repeating pump and dump. The cyclical pump–dump pulse sequence provides a natural mechanism for preparing highly excited symmetric stretch vibration, since this is the coordinate with greatest difference in forces on the two potential surfaces and thus this motion is amplified on each cycle. This can be seen in pictures of the wavepacket motion (not shown). If amplified sufficiently this motion will lead to selective three-body fragmentation (A + B + C). In the present study, however, with the objective of A + BC fragmentation, the symmetric stretch motion is amplified only in the first stage of the pulse sequence. In the second stage of the pulse sequence the energy in the symmetric stretch is channeled into the AB + C channel.

16.3.2 A System with Realistic Parameters: I_2

The I_2 system, which has been the subject of a substantial amount of femtosecond work, is studied theoretically as a prospect for experimental control. The I_2 system lacks the complexity of a polyatomic system, with the prospect for selective bond breaking. Nevertheless, its optical properties make it well suited for study with available femtosecond technology, and the vibrational time scale is sufficiently long for currently achievable pulse sequences to affect the dynamics on the time scale of a fraction of a period. Furthermore, because its spectroscopy and potential energy curves are well known, it is an ideal system for initial comparisons between theory and experiment.

Several theoretical studies have been undertaken to explore the use of femtosecond pulse sequences to achieve dissociation of I_2. Dissociation on the B state is taken as the objective because of its experimental accessibility, although dissociation on any other electronic state (including the X state) could have been chosen as well. The scientific objective of the study is to find one or several *interesting* pulse sequences (i.e., those whose mechanisms for achieving the objective are neither overly trivial nor so complex as to defy a simple interpretation). If the pulse sequences are sufficiently interesting it is anticipated that they would stimulate experiments, despite the absence of issues of chemical branching in this simple system.

In work by Somlói et al. (1993), an unconstrained optimization of the electric field was performed, designed to achieve maximal photodissociation yield on the B state. This was to give the field the maximum freedom, so that unexpected mechanisms might be discovered. Analysis of the optimal pulse sequence indicated that it was exploiting a quasi-CW mechanism: the frequency of the excitation is resonant with the B state continuum, and produces efficient dissociation, albeit in an uninteresting way.

To find a different mechanism from the quasi continuous-wave (CW) excitation, the electric field was restricted to the frequency interval from 0 to 19752 cm^{-1}. The optimal

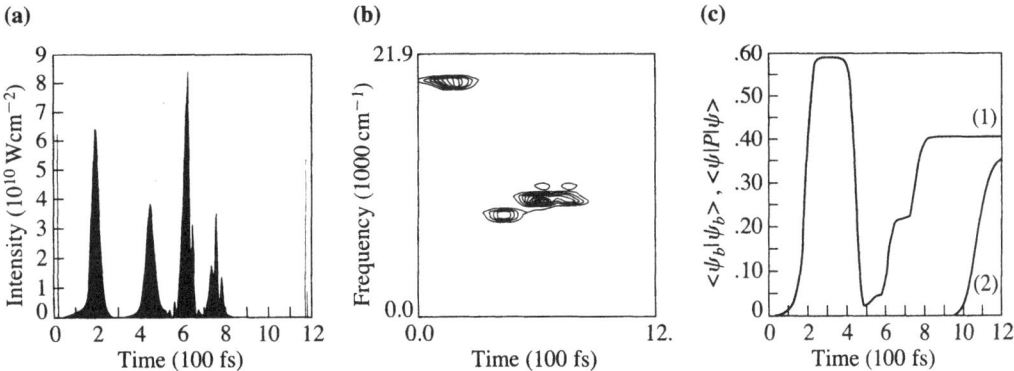

Figure 16.11 (a) Optimal pulse for dissociation on the B electronic state of I_2, where frequencies corresponding to direct excitation from the ground vibrational state into the continuum have been filtered out. (b) Husimi distribution for the pulse sequence in (a). (c) B state population (1) and probability for exit on the B state (2) as a function of time for the pulse sequence in (a). Adapted from Somlói (1993).

pulse sequence under these conditions takes the form of a "pump-dump-pump" sequence (Figure 16.11). (Although at first glance the pulse sequence looks like a four-pulse sequence, the third and fourth pulses are both performing part of the same function. This point will be expanded below.) This pulse sequence is a natural extension of the "pump–dump" scheme (Tannor, 1985, 1986). The first "photon" excites the molecule, which then evolves on the excited electronic state potential surface until the second "photon" dumps it by stimulated emission. Since the ground state potential energy surface is steeper than the excited state, the kinetic energy gained by the wavefunction on the ground state potential energy surface is enough to generate dissociation on the excited-state potential energy surface after the molecule interacts with the third "photon."

As mentioned above, although the pulse sequence consists of four pulses, the third and fourth pulses are performing the same function, namely, the final pump. This statement is supported by the Husimi distribution, Figure 16.11b, which indicates that the third and fourth pulses are at the same frequency. Moreover, the plot of the excited-state population versus time, Figure 16.11c, shows clearly that both the third and fourth pulses are serving to pump amplitude to the excited state. Extensive tests removing the third and fourth pulses separately indicate that their effect is simply additive: they are not using interference to increase the excited-state population. With the help of the Husimi transforms of the optimal electric field a sequence of four Gaussian-shape pulses was constructed as a possible candidate for experimental application. This approximating field gave 11% for dissociation probability, and the wavefunction is well localized at the final time T. With higher values of the amplitudes the dissociation probability can be increased further, since no final optimization was done on the Gaussians with respect to their intensity. Both with respect to intensity and pulse duration these Gaussian pulses should be producible in the laboratory with current technology.

The remaining question is, How good is the physical model for I_2? A similar model was able to reproduce the qualitative features of the experimental wavepacket interferometry signal shown in Fig. 13.4 (see Chapter 13, ref. 6). That fact is encouraging, and suggests that the qualitative effects described here should persist in the real system. Nevertheless, a more sophisticated model—including a more accurate potential, initially hot vibrational states of the ground electronic state and inclusion of the rotational degree of freedom (in particular, averaging over orientations)—is required to give completely accurate experimental predictions.

16.4   Multiple Pathway Interference

As mentioned in Section 16.1, Feynman's path integral approach to quantum mechanics leads naturally to thinking of quantum interference as arising from multiple dynamical paths that lead from the same initial to the same final state. The simplest example of this interference involves an initial state, two intermediate states, and a single final state, although if the objective is to control some branching ratio at the final energy then at least two final states are necessary. By controlling the phase with which each of the two intermediate states contributes to the final state, one may control constructive versus destructive interference in the final states. This is the basis of the Brumer–Shapiro approach to coherent control (Brumer, 1992, 1995).

   With advances in the technology of amplitude and phase shaping of femtosecond pulses, one can have a virtual infinity of interfering pathways, all ending up at the same final state. Figure 16.12 shows a schematic energy-level diagram of nonresonant two-photon processes. A cesium atom is irradiated using a single broadband laser pulse whose carrier frequency is one-half the $6S_{1/2} \to 8S_{1/2}$ transition frequency. Note that there is a continuum of pairs of frequencies under the pulse envelope such that $\omega_1 + \omega_2 = \omega_0$, the two-photon resonance frequency. The existence of multiple paths to the same final state that can all be excited by the same pulse envelope gives rise to the possibility of quantum interference and a rich assortment of coherent control strategies. In the most general case, the amplitude and phase of each frequency contained under the pulse envelope can be independently controlled. Then the amplitude for the transition from the initial to the final state via two-photon resonance is determined by taking into account all pairwise frequencies that satisfy the two-photon resonance condition, and adding (integrating) them with the appropriate amplitude and phase. Figure 16.13 shows the experimental arrangement for controlling the amplitude and phase of each frequency component under the pulse envelope independently.

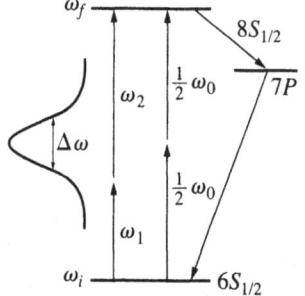

Figure 16.12 Schematic energy-level diagram of the $6S_{1/2}$-$8S_{1/2}$ two-photon absorption in Cs. The power spectrum of the pulse is shown at the left. Note that there is a continuum of pairs of frequencies under the pulse envelope such that $\omega_1 + \omega_2 = \omega_0$, the two photon resonance frequency. The existence of multiple paths to the same final state gives rise to quantum interference and the possibility of coherent control.

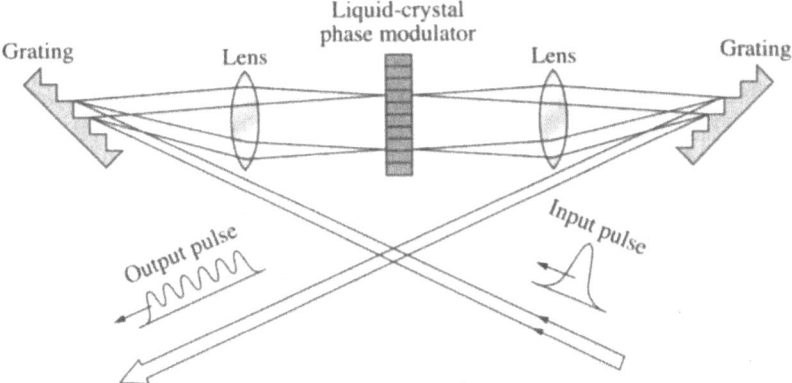

Figure 16.13 Experimental arrangement for phase and amplitude shaping of pulses.
There are a symmetrically arranged pair of diffraction gratings and achromatic lenses.
In the center is a programmable one-dimensional liquid-crystal spatial light modulator
(SLM array) composed of 128 discrete computer-controlled elements. In the absence
of the SLM array, the output pulse would be identical in amplitude and phase to the
input pulse; the SLM array allows the independent control of the phase and amplitude
of 128 different component frequencies of the pulse. Adapted from Weiner (1990).

The notion of interference of all the two-photon pathways can be put on a quantitative
basis using second-order time-dependent perturbation theory. The projection of the second-
order amplitude on the final state ψ_f can be written as

$$\langle \psi_f | \psi^{(2)}(t) \rangle = -\frac{1}{4\hbar^2} \sum_n \mu_{fn}\mu_{ni} \int_{-\infty}^{t} \int_{-\infty}^{t_1} \varepsilon(t_1)\varepsilon(t_2)e^{i\omega_{fn}t_1}e^{i\omega_{ni}t_2}\, dt_2\, dt_1, \quad (16.57)$$

where $\omega_{ni} = \frac{E_n - E_i}{\hbar}$, and so on, and the summation is performed over all possible intermediate states n of the unperturbed atom. Equation 16.57 can be derived from Eq. 14.47 by
assuming a coordinate-independent transition dipole moment in the latter equation and inserting two complete sets $\{\psi_n\}$ of eigenstates of H_b. Formally, the sum over n in Eq. 16.57
includes all excited electronic states. If we assume that all the intermediate states are far from
resonance, their contribution will add coherently only for a very short duration. Defining
this effective time duration as $\bar{\omega}^{-1}$, we may take the transition amplitude as being constant
during this interval. Hence (see Meshulach (1999)),

$$\sum_n \mu_{fn}\mu_{ni}e^{iE_n(t_2-t_1)/\hbar} \approx \begin{cases} \langle \psi_f | \mu^2 | \psi_i \rangle, & |t_1 - t_2| \leq \bar{\omega}^{-1} \\ 0, & |t_1 - t_2| \geq \bar{\omega}^{-1} \end{cases}. \quad (16.58)$$

Assuming that $\bar{\omega}^{-1} \ll (\Delta\omega)^{-1}$, where $\Delta\omega$ is the spectral bandwidth, we obtain

$$P_{i \to f} = \lim_{t \to \infty} \langle \psi^{(2)}(t) | \psi^{(2)}(t) \rangle \quad (16.59)$$

$$= \frac{1}{16\hbar^4} \left| \frac{\langle \psi_f | \mu^2 | \psi_i \rangle}{\bar{\omega}} \right|^2 \left| \int_{-\infty}^{\infty} \varepsilon^2(t)e^{i\omega_0 t}\, dt \right|^2, \quad (16.60)$$

where $\omega_0 = (E_f - E_i)/\hbar$. Denoting the last factor in Eq. 16.60 as S_2 and rewriting this factor in the frequency domain we obtain (Meshulach, 1999)

$$S_2 = \left| \int_{-\infty}^{\infty} \varepsilon^2(t) e^{i\omega_0 t} dt \right|^2 \tag{16.61}$$

$$= \left| \int_{-\infty}^{\infty} \tilde{\varepsilon}(\omega) \tilde{\varepsilon}(\omega_0 - \omega) \, d\omega \right|^2 \tag{16.62}$$

$$= \left| \int_{-\infty}^{\infty} \tilde{\varepsilon}(\omega_0/2 + \Omega) \tilde{\varepsilon}(\omega_0/2 - \Omega) \, d\Omega \right|^2 \tag{16.63}$$

$$= \left| \int_{-\infty}^{\infty} A(\omega_0/2 + \Omega) A(\omega_0/2 - \Omega) e^{i[\Phi(\omega_0/2+\Omega)+\Phi(\omega_0/2-\Omega)]} \, d\Omega \right|^2, \tag{16.64}$$

where $\tilde{\varepsilon}(\omega) = A(\omega) e^{i\Phi(\omega)}$ is the Fourier transform of $\varepsilon(t)$, and $A(\omega)$ and $\Phi(\omega)$ are the spectral amplitude and the spectral phase, respectively. Equation 16.62 follows from the Fourier convolution theorem, and Eq. 16.63 follows from the change of variables $\Omega = \omega - \omega_0/2$. Equation 16.64 indicates that two-photon transitions occur for all pairs of photons with frequencies ω_1 and ω_2, with $\omega_1 + \omega_2 = \omega_0$, and ω_1 and ω_2 lying within the spectrum of the exciting pulse. The two-photon transition probability can therefore be controlled by tailoring the spectral phase within a single pulse. The change of variables in Eq. 16.63 is not necessary, but is convenient for thinking about the pairs of photons as being symmetrically arranged around $\omega_0/2$.

One can easily show that S_2, and hence the two-photon transition probability, is maximized, for a given energy and power spectrum $A^2(\omega)$, by a transform-limited pulse, that is, $\Phi(\Omega) = 0$. This pulse has the minimum time duration. However, consider a pulse with the same energy and power spectrum, that is, with fixed $A(\omega)$ but having any *antisymmetric* spectral phase distribution around the two-photon transition frequency $\omega_0/2$, that is, $\Phi(\omega_0/2 + \Omega) = -\Phi(\omega_0/2 - \Omega)$. Figure 16.14a shows $\tilde{\varepsilon}$ for a transform-limited pulse and for a pulse with an antisymmetric phase profile. The absolute values of both are the same (shown as the envelope) and hence the two pulses have the same power spectrum. For a pulse with an antisymmetric phase profile, the phase terms in Eq. 16.64 cancel each other pairwise, so that S_2 and therefore the two-photon transition probability is identical with that of the transform-limited pulse. Figure 16.14d shows the power spectrum of $\varepsilon^2(t)$ for both pulses: although the two power spectra differ at most frequencies, at $\omega = 0$, which corresponds here to the two-level transition frequency ω_0, the two-photon power spectra are identical. Figure 16.14b–c shows $\varepsilon(t)$ and $\varepsilon^2(t)$, respectively, for the two pulses. $\varepsilon(t)$ and $\varepsilon^2(t)$ for the transform-limited pulse have narrow peaks at the origin. $\varepsilon(t)$ and $\varepsilon^2(t)$ for the pulse with the antisymmetric phase profile consist of a sequence of pulses, and correspondingly the peak power is lower; this could have significant advantages for efficient excitation of a sample while minimizing damage due to multiphoton ionization or dissociation.

While antisymmetric phase profiles do not affect the two-photon transition, other profiles can change the intensity of the two-photon transition. In fact, some pulses can completely turn off the two-photon transition; such pulses may be called "dark pulses." One may view these dark pulses, according to Eq. 16.60, as having a vanishing power spectrum

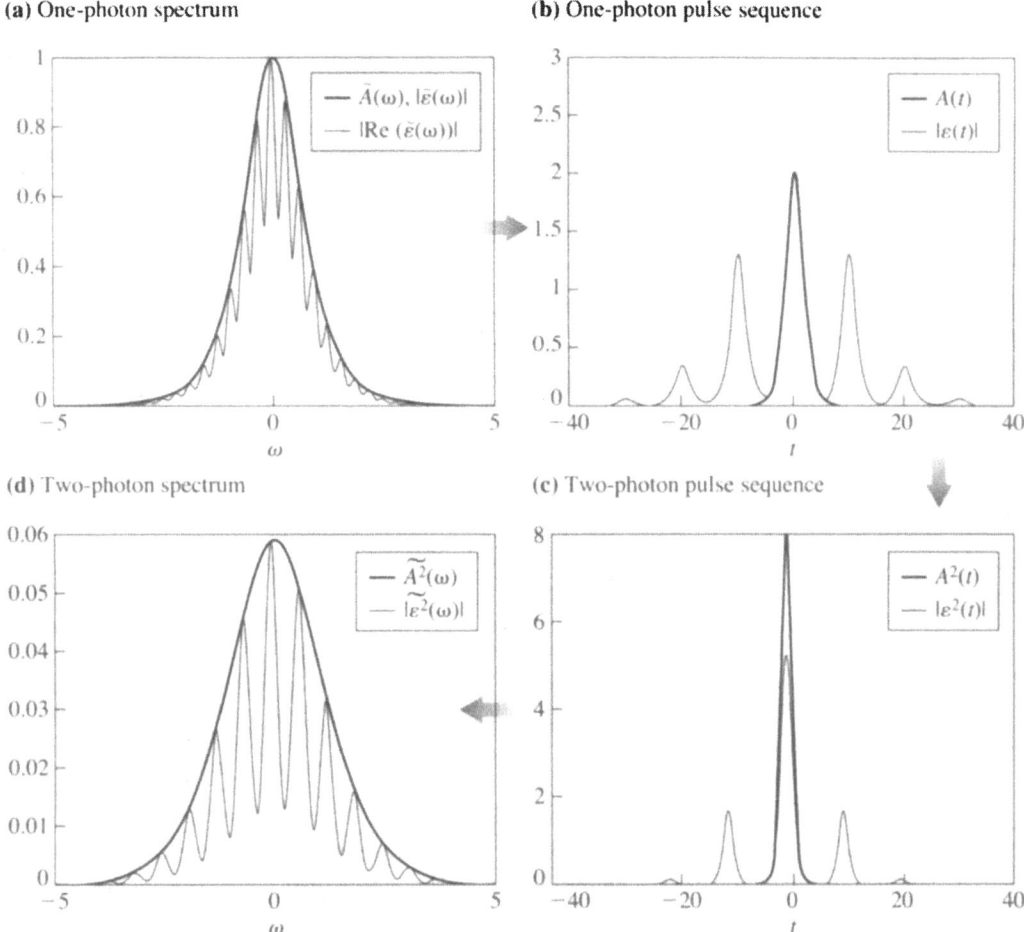

(a) One-photon spectrum

(b) One-photon pulse sequence

(d) Two-photon spectrum

(c) Two-photon pulse sequence

Figure 16.14 (a) One-photon spectrum for a transform-limited pulse ($\tilde{\varepsilon}(\omega) = \tilde{A}(\omega)$) and for a phase-modulated pulse ($\tilde{\varepsilon}(\omega) = A(\omega)\exp[i\alpha\sin(\beta\omega)]$ for $\alpha = 1$, $\beta = 10$). The envelope shows $|\tilde{\varepsilon}(\omega)|$, which is the same for both pulses. (b) $\varepsilon(t)$ for the two pulses in (a). The transform-limited pulse is a single peak centered at $t = 0$, while the phase modulated pulse in frequency leads to a pulse sequence in time. (c) $|\varepsilon^2(t)|$ for the two pulses in (a). (d) The two-photon power spectrum, $|\tilde{\varepsilon^2}(\omega)|$, for the two pulses in (a). The solid curve is for the transform-limited pulse while the dashed curve is for the phase-modulated pulse. Note that there are points at which the two power spectra are equal, in particular at $\omega = 0$, which corresponds here to the two-photon transition frequency, as well as at all $\omega = n2\pi/\beta$ (see text).

of $\varepsilon^2(t)$ at $\omega = \omega_0$. Consider a pulse with a single, discontinuous phase step of π at some value of Ω. The phase function may be written

$$\Phi = \frac{\pi}{2}s(\Omega - \delta), \tag{16.65}$$

where $s(\omega) = \pm 1$ according to the sign of ω. For $\delta = 0$ we obtain an antisymmetric phase profile around $\Omega = 0$ of the type discussed above. The two-photon absorption as a function of δ is shown in Figure 16.15 (upper panel). When $\delta \gg \Delta\omega$, the phase profile is effectively flat over the power spectrum of the pulse, and the two-photon transition is the same as

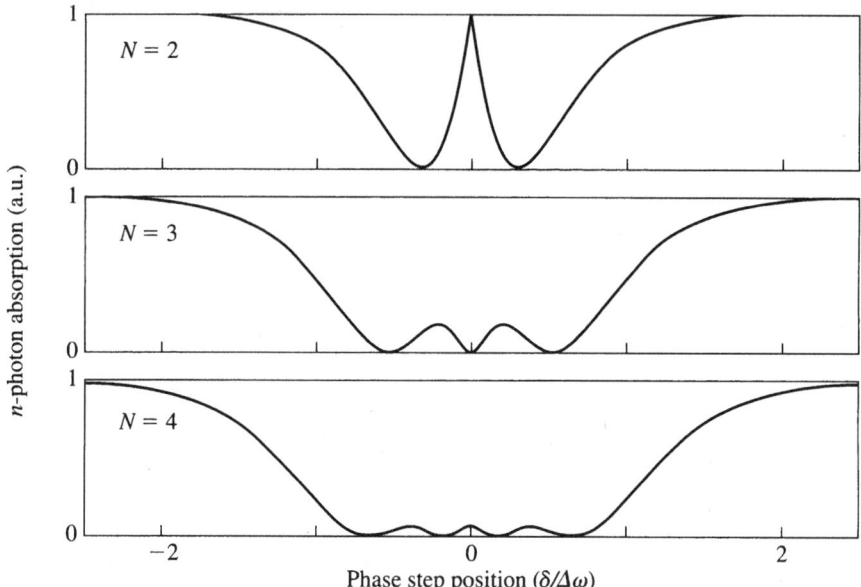

Figure 16.15 Coherent quantum control of N-photon transition in a two-level system for $N = 2, 3, 4$. The transition probability for excitation by a pulse with a π spectral step is shown as a function of the normalized step position $\delta/\Delta\omega$. The transition probabilities are normalized to the probability of excitation by a transform-limited pulse. Adapted from Meshulach (1999).

for a transform-limited pulse. For $\delta = 0$, the spectrum has an antisymmetric phase profile and again the two-photon transition is the same as for the transform-limited pulse, by the arguments of the previous paragraph. However, for intermediate values of δ the two-photon absorption can vanish. These phase profiles correspond to dark pulses.

Equation 16.61 can be rewritten as

$$S_2 = \left| \int_{-\infty}^{\infty} \varepsilon^2(t) e^{i\omega_0 t} \, dt \right|^2$$

$$= \left| \int_{-\infty}^{\infty} \tilde{\varepsilon}(\omega) \tilde{\varepsilon}(\omega_0 - \omega) \, d\omega \right|^2$$

$$= \left| \int_{-\infty}^{\infty} \tilde{\varepsilon}(\omega) \tilde{\varepsilon}^*(\omega - \omega_0) \, d\omega \right|^2 \tag{16.66}$$

$$= \left| \int_{-\infty}^{\infty} A(\omega) A^*(\omega - \omega_0) e^{i[\Phi(\omega) - \Phi(\omega - \omega_0)]} \, d\omega \right|^2, \tag{16.67}$$

where we have used the fact that since $\varepsilon(t)$ is real $\tilde{\varepsilon}(\omega) = \tilde{\varepsilon}^*(-\omega)$. Equation 16.67 is convenient for analyzing Raman excitations: this equation has the interpretation that the Raman transition is excited by all pairs of frequencies that are separated by the Raman frequency. It is interesting to consider what happens when the phase function $\Phi(\omega)$ in Eq. 16.67 is periodic,

$$\Phi(\omega) = \alpha \cos(\beta\omega + \phi). \tag{16.68}$$

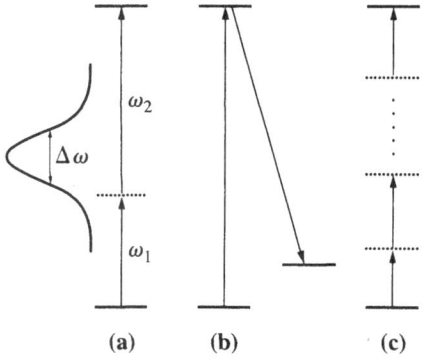

Figure 16.16 Schematic energy-level diagrams of nonresonant multiphoton processes. (a) Two-photon absorption, with a schematic representation of the pulse spectrum. (b) The analogous energy-level diagram for the stimulated Raman process. (c) The analogous diagram for multiphoton absorption. Adapted from Meshulach (1999).

We have already noted that S_2 takes on its maximum value for a transform-limited pulse, $\Phi = 0$. However, for any value of β satisfying $\beta\omega_0 = n2\pi$, where n is an integer, S_2 will be equal to its transform-limited value. This can be observed in Figure 16.14d. Such periodic phase structures will split an initial transform-limited pulse into a pulse sequence with repetition rate $\beta = 2\pi/\omega_0$. A pulse sequence of this type reduces the peak power of the pulse and narrows down the response of the Raman excitation, to cancel all but a desired transition.

We consider now the extension of the discussion above to N-photon processes (see Figure 16.16). The generalization of Eq. 16.60 to N-photon processes is obtained by considering the power spectrum of $\varepsilon^N(t)$ at $\omega = \omega_0$ (Meshulach, 1999):

$$P_{i\to f}^N \propto \left| \int_{-\infty}^{\infty} \varepsilon^N(t)e^{i\omega_0 t} dt \right|^2 . \tag{16.69}$$

The second and third panels in Figure 16.15 show the N-photon absorption as a function of δ for $N = 3$ and $N = 4$. Note the following differences from the two-photon case: a) the width of the region of reduced N-photon absorption increases with N; b) the absorption at $\delta = 0$ is much reduced from that of the transform-limited pulse and even vanishes for N odd; c) the number of values of δ for which the pulse is perfectly dark is precisely equal to N.

16.5   Chirped Pulse Excitation

One of the simplest forms of pulse shaping is chirping, in which the central frequency of the pulse changes with time. Consider the simplest case of linear chirping:

$$\varepsilon(t) = \varepsilon_0 \exp\left[-\frac{(t-t_0)^2}{2\tau^2} - i\omega_0(t-t_0) - ic\frac{(t-t_0)^2}{2} \right] \tag{16.70}$$

$$= \varepsilon_0 \exp\left[-\frac{(t-t_0)^2}{2\tau^2} - i(t-t_0)\left(\omega_0 + c\frac{t-t_0}{2}\right) \right], \tag{16.71}$$

where τ is the pulse duration and $c/2$ is the linear chirp rate in time. Since $\varepsilon(t)$ is a generalized Gaussian it can be Fourier transformed analytically to yield

$$\tilde{\varepsilon}(\omega) = \int_{-\infty}^{\infty} \varepsilon(t) e^{i\omega t} \, dt \tag{16.72}$$

$$= \varepsilon_0 \left(\frac{2\pi}{\gamma^2} \right)^{1/2} e^{-(\omega-\omega_0)^2/2\gamma^2} \tag{16.73}$$

$$= \varepsilon_0 \left(\frac{2\pi\tau^2}{1 + ic\tau^2} \right)^{1/2} e^{-(\omega-\omega_0)^2/2\Gamma^2 + ic'(\omega-\omega_0)^2/2}, \tag{16.74}$$

where we have defined $\gamma^2 = 1/\tau^2 + ic$, $\Gamma^2 = 1/\tau^2 + c^2\tau^2$ and $c' = c\tau^2/\Gamma^2$. Γ is the frequency bandwidth of the pulse and $c'/2$ is the frequency chirp. Note the following symmetric relations between bandwidth and chirp in the time and frequency domains:

$$c'\Gamma^2 = c\tau^2 \tag{16.75}$$

$$\Gamma^2 = 1/\tau^2 + c^2\tau^2 \iff \tau^2 = 1/\Gamma^2 + c'^2\Gamma^2. \tag{16.76}$$

▶ **Exercise 16.6**   Verify the RHS of Eq. 16.76.

Solution   Solving Eq. 16.75 for c and substituting into the LHS of Eq. 16.76 we obtain

$$\Gamma^2 = (1 + c'^2\Gamma^4)/\tau^2 \tag{16.77}$$

and therefore

$$\tau^2 = 1/\Gamma^2 + c'^2\Gamma^2. \tag{16.78}$$

One distinguishes an upwardly chirped pulse, in which the central frequency increases with time, from a downwardly chirped pulse, in which the central frequency decreases with time. Figure 16.17a and c shows an upwardly and a downwardly chirped pulse, respectively. The chirp is conveniently visualized in a Husimi time–frequency representation. In the Husimi representation, an upward chirp corresponds to a positive slope of the major axis of the Husimi distribution, and a downward chirp corresponds to a negative slope of the major axis.

▶ **Exercise 16.7**

    a. Show that the Wigner transform of Eq. 16.71 for $\varepsilon(t)$ to the time–frequency domain is

$$F(t, \omega) = \varepsilon_0^2 \exp\left[-\frac{(t - t_0)^2}{\tau^2} - \tau^2[\omega - (\omega_0 + c(t - t_0))]^2 \right]. \tag{16.79}$$

    b. Show that this agrees with the form of the distributions in Figure 16.17.

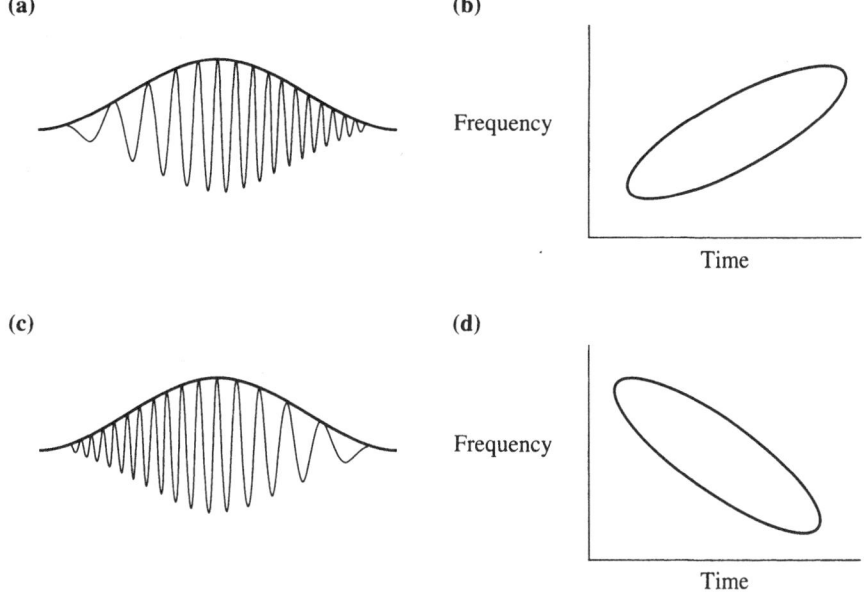

(a)

(b)

Frequency

Time

(c)

(d)

Frequency

Time

Figure 16.17 (a) Amplitude versus time for an upwardly chirped pulse. (b) Time–frequency Husimi distribution for the pulse in (a). (c) Amplitude versus time for a downwardly chirped pulse. (d) Time–frequency Husimi distribution for the pulse in (c).

Excitation of a molecule with a chirped pulse can lead to many different, even opposite physical effects, depending on the nature of the molecular Hamiltonian, the nature of the field–matter coupling (e.g., permanent vs. transition dipole moment) and the strength of the field. In this section we will touch on only two of the effects, both occuring in optical transitions coupled by the transition dipole moment interaction.

The first effect is wavepacket *focusing* using chirped excitation (Kohler, 1995; Cao, 1997). This effect can be described within first-order perturbation theory, treating the chirped field and the anharmonic wavepacket evolution on the excited electronic state within a single coherent framework. Following Wilson and coworkers, we distinguish the case where the wavepacket focusing occurs *before* the classical turning point is reached (a "cannonball") and the case where the focusing occurs *after* the turning the point is reached and the wavepacket is already on its way back to the Franck–Condon region (a "reflectron").

Consider the objective of producing a reflectron. The simplest way of understanding why a chirped pulse will produce the desired focusing is to consider the first-order analysis in Eqs. 16.17–16.19:

$$\tilde{\phi}(T) = \sum_n a_n e^{-iE_n T/\hbar} e^{-i\phi_n} \psi_n, \qquad (16.80)$$

with

$$a_n e^{-i\phi_n} = \tilde{\varepsilon}(\omega_n - \omega_i)\langle \psi_n|\mu|\psi_i\rangle. \qquad (16.81)$$

For a Morse oscillator, the spacing between vibrational levels, $E_{n+1} - E_n$, gets smaller with higher energy, and therefore the higher energy components effectively have a longer

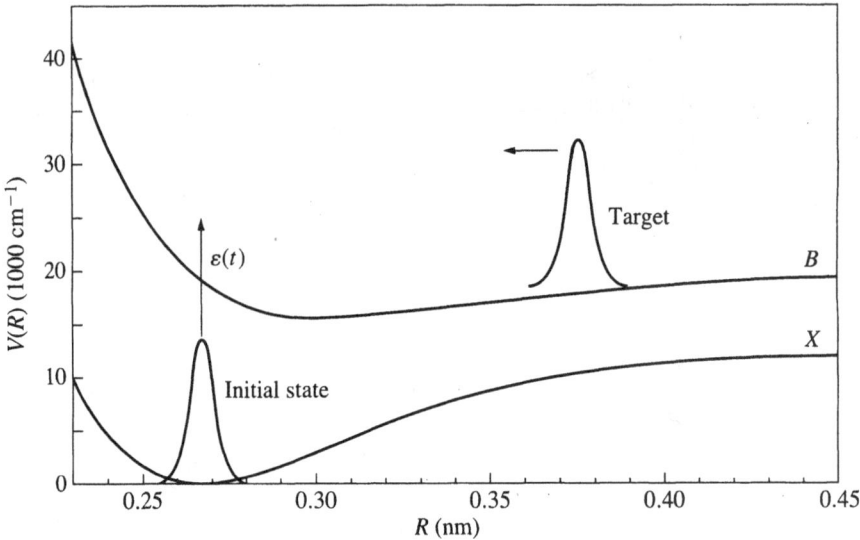

Figure 16.18 The initial and target states for chirped control of wavepacket evolution. The initial state is the ground vibrational state of the ground electronic state, while the target state is a "reflectron": a maximally narrow wavepacket with negative momentum, having reflected off the outer turning point of the excited-state potential. Adapted from Kohler (1995).

period than the lower energy components. To get the entire wavepacket to come back to the Franck–Condon region at the same time one should excite the slower, higher energy components before the faster, lower energy components. This suggests a downward chirp, which indeed is successful in full simulations as well as in experiments in producing a reflectron. The shortcoming of this argument is that the spacing between the energy levels of the Morse oscillator is relevant for predicting the dynamics on the time scale of a period or so; this qualitative argument is therefore of limited use in predicting the chirp that will produce a cannonball. Equation 16.106 (Exercise 16.13) gives a more quantitative treatment, and shows that a positive chirp will maximize the overlap with the cannonball state, and a negative chirp will maximize the overlap with the reflectron state.

Figure 16.18 shows the initial and target states for chirped control of wavepacket evolution. The initial state is the ground vibrational state of the ground electronic state, while the target state is a maximally narrow wavepacket with negative momentum, having reflected off the outer turning point of the excited-state potential, a "reflectron." Figure 16.19 shows the Wigner time–frequency distribution of the pulse that creates the reflectron in Figure 16.18. The negative chirp appears as the downward rotation of the axes of the time–frequency distribution. The negative chirp is in agreement with the prediction of Eq. 16.106.

The second effect of chirp that we will describe is its influence on wavepacket population transfer between electronic states. Ruhman and Kosloff (Ruhman, 1990) developed a simple physical picture, showing how a *downwardly* chirped pulse could efficiently lead to intrapulse impulsive stimulated Raman scattering. This effect is shown pictorially in Figure 16.20: as the wavepacket moves downhill on the excited-state potential surface, the energy gap between the ground and excited potentials decreases. If the frequency of the pulse decreases with time in such a way as to match the potential energy gap at the instantaneous center of the wavepacket, the later portions of the excitation pulse will be

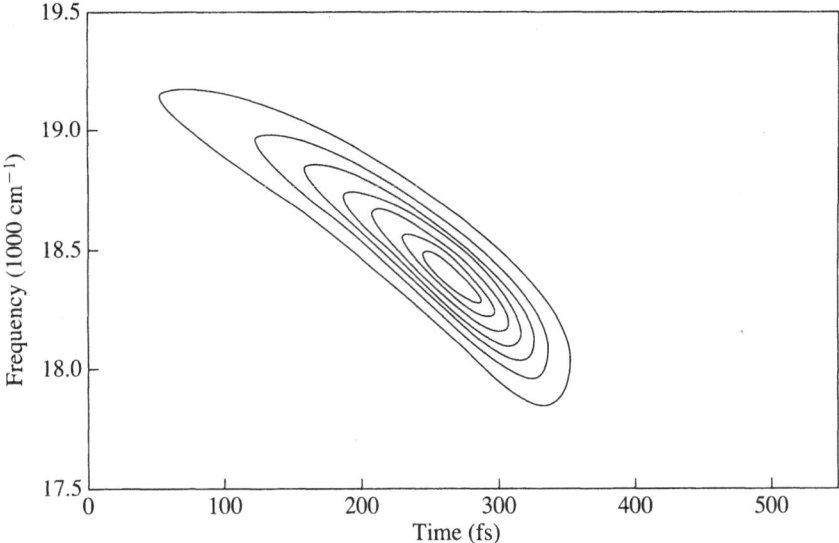

Figure 16.19 Wigner time–frequency distribution of the pulse that creates the reflectron in Figure 16.18. The negative chirp shows up as the downward rotation of the axes of the time–frequency distribution. Adapted from Kohler (1995).

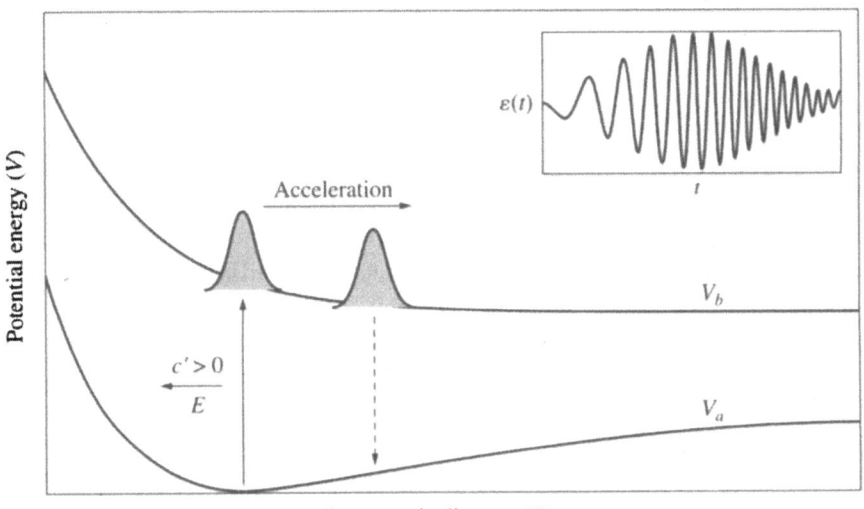

Figure 16.20 Schematic diagram of intrapulse pump–dump mechanism. Such a mechanism is favored by a *downward* chirp, so that the instantaneous frequency is matched to the instantaneous energy separation of the two potential energy curves at the center of the wavepacket. The inset shows a pulse with an *upward* chirp; such a pulse is as *inefficient* as possible in dumping amplitude back to the ground state, and hence is optimal for achieving complete population inversion to the excited electronic state. This pulse is called a "molecular π pulse." Adapted from Cao (1998).

efficient at stimulating amplitude back to the ground electronic state. A few years later, the intrapulse pump–dump effect was inverted by Wilson and coworkers (Cao, 2000). Those workers reasoned that if a downwardly chirped pulse is efficient at transferring amplitude back to the ground electronic state, an *upwardly* chirped pulse should turn off the stimulated Raman scattering, leading to efficient population transfer to the excited electronic state. Indeed, they showed first theoretically and then experimentally that they could achieve nearly 100% population transfer to the excited electronic state using an upwardly chirped pulse (Figure 16.20). Numerical calculations (Cao, 2000) show that an intense, positively chirped pulse is robust in inverting the full thermal distribution of rotational–vibrational states over a wide range of chirp rates, pulse intensities, temperatures and even electronic dephasing times.

The effect of chirp on population transfer is a strong field effect, although it is already discernable at second-order perturbation theory, as illustrated in Exercise 16.14. In that exercise it is shown that negative chirp maximizes the intrapulse impulsive Raman scattering back to the ground state potential. This suggests that positive chirp will minimize impulsive Raman scattering back to the ground state potential, consistent with the numerical calculations showing that positive chirp *maximizes* population transfer to the excited electronic state surface.

16.6 Learning Algorithms

The use of optimal control theory to design optimal laser pulse shapes suffers from a number of problems when the computed pulse shapes are tested in the lab. First, for all but the smallest molecules, we do not know the Hamiltonian (i.e., the multidimensional potential energy surfaces) accurately enough to predict the pulses that will work for the real molecule. Second, the laser pulses in the lab inevitably have a certain amount of noise, as well as uncertainty in amplitude and phase. This error in the pulse can lead to failure of the control mechanism. Finally, there are effects that are generally neglected in model calculations, such as additional potential energy surfaces, spin–orbit coupling, propagation effects in thick samples, rotational levels, finite temperature and so forth that can also lead to the failure of the calculated optimal pulse to perform as desired when applied to the real system.

All these problems can be circumvented using a variant strategy for iteratively finding optimal pulses that leaves theory out of the loop completely. These strategies are sometimes called "feedback" or "adaptive" control. In a seminal paper, Judson and Rabitz suggested letting the molecule solve its own Schrödinger equation, which it can do much faster and much more accurately than any computer, automatically including all potential energy surfaces, laser uncertainties, pulse propagation effects and so on (Judson, 1992). The idea is to set up an input–output feedback loop such that a given input produces a certain output, which is fed back into a computer that evaluates the merit of the input. Based on the merit, the input is then modified in the next iteration. The simplest scheme to imagine is a Metropolis Monte Carlo scheme. An initially random set of amplitudes and phases is chosen, and a merit is assigned to the output. Then a random change, in either amplitude or phase, is made; if the change increases the merit, the change is kept. If the change decreases the merit, the change is kept with a probability $e^{-\beta E}/Q$, where E measures the merit, β is a parameter that determines the probability of merit-decreasing transitions and Q is a normalization factor.

Note that in most engineering applications of control theory the time scale for the process is usually seconds or longer; as a result, the measurement-feedback loop can be incorporated into the process while it is going on. In contrast, in quantum control the typical time scales of the processes of interest, $10^{-15} - 10^{-12}$ s, are much shorter than the time scale under which the control can be updated (10^{-3} s). This implies that the measurement-feedback loop is of no value for the original system (i.e., molecule) on which the measurement was performed; instead, the feedback is introduced in the input control for an identical but distinct system (i.e., molecule). This type of strategy is sometimes called "adaptive control," to distinguish it from bona fide feedback control, which feeds back on the original system in real time.

A more sophisticated type of adaptive control is based on analogies with Darwinian evolution. Specifically, one starts with a random assortment of initial pulse sequences. The merit that is assigned to the output of a particular pulse sequence is considered its "fitness." Each segment of each pulse sequence is considered to be a "gene." The strategy is to let the pulse sequences reproduce, via crossover of "genetic material" as well as via random mutations, creating a new "generation" of pulse sequences. At each generation, only those pulse sequences with the highest values of the fitness criterion are kept. By this combination of crossover of genetic material, mutation and survival of the fittest, the biological process of evolution is simulated. After many generations, the new pulse sequences are expected to have a higher fitness than the original ensemble. A schematic diagram of experimental adaptive feedback is shown in Figure 16.21.

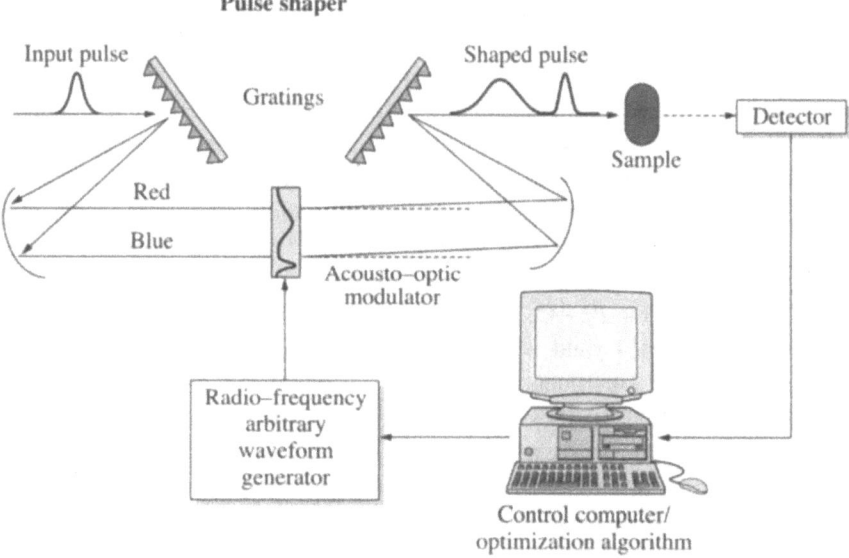

Figure 16.21 Schematic diagram showing closed-loop feedback control. An acousto–optic modulator (in other schemes, a liquid-crystal mask) imprints a phase and amplitude profile on the pulse. The pulse impinges on a sample, and some experimental observable that reflects the merit of the pulse is measured. The merit of one pulse (or many pulses) is used by the computer to determine a new pulse (or family of pulses) for the next iteration. The term "adaptive" control is to be preferred over the term feedback control, since the time scale to apply the feedback is longer than the time scale for an individual experiment; the new pulse calculated by the computer is applied to a new molecule. Adapted from Assion (1998).

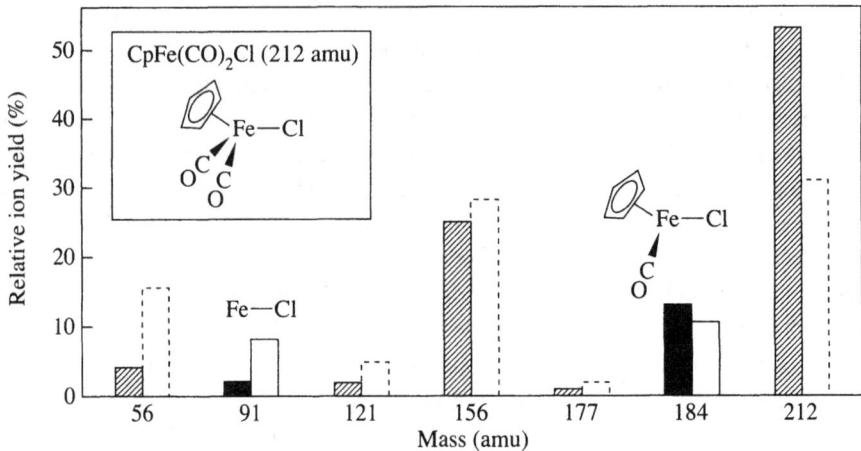

Figure 16.22 Mass spectrum of CpFe(CO)$_2$Cl. The peak at 91 amu represents an FeCl fragment, while that at 184 amu represents CpFeCOCl. The white boxes represent the fragment yields when a pulse optimized for FeCl production is applied, while the black boxes and cross-hatched boxes represent the fragment yields when a pulse optimized for CpFeCOCl production is applied. The contrast ratio that can be achieved is about a factor of 5. Adapted from Assion (1998).

Adaptive feedback has led to extremely impressive examples of experimental quantum control in the last few years, including control of fluorescence, laser pulse duration, high harmonic generation, chemical branching ratios and energy transfer in a light-harvesting complex. We show here two examples of control of chemical branching ratios. The first is the landmark set of experiments by Gerber et al. showing control over the fragmentation of CpFe(CO)$_2$Cl to CpFeCOCl versus FeCl (Assion, 1998). Figure 16.22 shows the mass spectrum of fragments. The peak at 91 amu represents an FeCl fragment, while that at 184 amu represents CpFeCOCl. The contrast ratio that can be achieved is about a factor of 5. Note that the Fe—Cl bond is the weakest in the molecule, and hence the enhancement of the FeCl yield is particularly interesting. The second example is the beautiful set of recent experiments by Levis et al. on control of the fragmentation of acetophenone (C$_6$H$_5$COCH$_3$) (Levis, 2001). Figure 16.23a shows the mass spectrum of C$_6$H$_5$COCH$_3$. Figure 16.23b shows the relative ion yield for C$_6$H$_5$CO$^+$, C$_6$H$_5^+$ and the C$_6$H$_5$CO$^+$/C$_6$H$_5^+$ ratio as a function of generation when maximization of this ratio is the objective in the closed-loop experiment. Figure 16.23c is the same as Figure 16.23b except that the ratio C$_6$H$_5^+$/C$_6$H$_5$CO$^+$ is to be maximized. Figure 16.23d shows the average signal for toluene (C$_6$H$_5$CH$_3^+$) as a function of generation when maximization of this ion signal is the objective. The corresponding chemical branching processes are shown on the right of Figure 16.23. Note that the production of toluene requires the extrusion of the CO moiety and the joining of C$_6$H$_5$ with the CH$_3$ group; this requires the breaking of two chemical bonds and the formation of a new one, and is thus a highly counterintutitive photochemical product. Moreover, toluene is completely absent from the electron impact ionization of C$_6$H$_5$COCH$_3$, indicating that toluene production is really active control of chemical reactions in the fullest sense of the word.

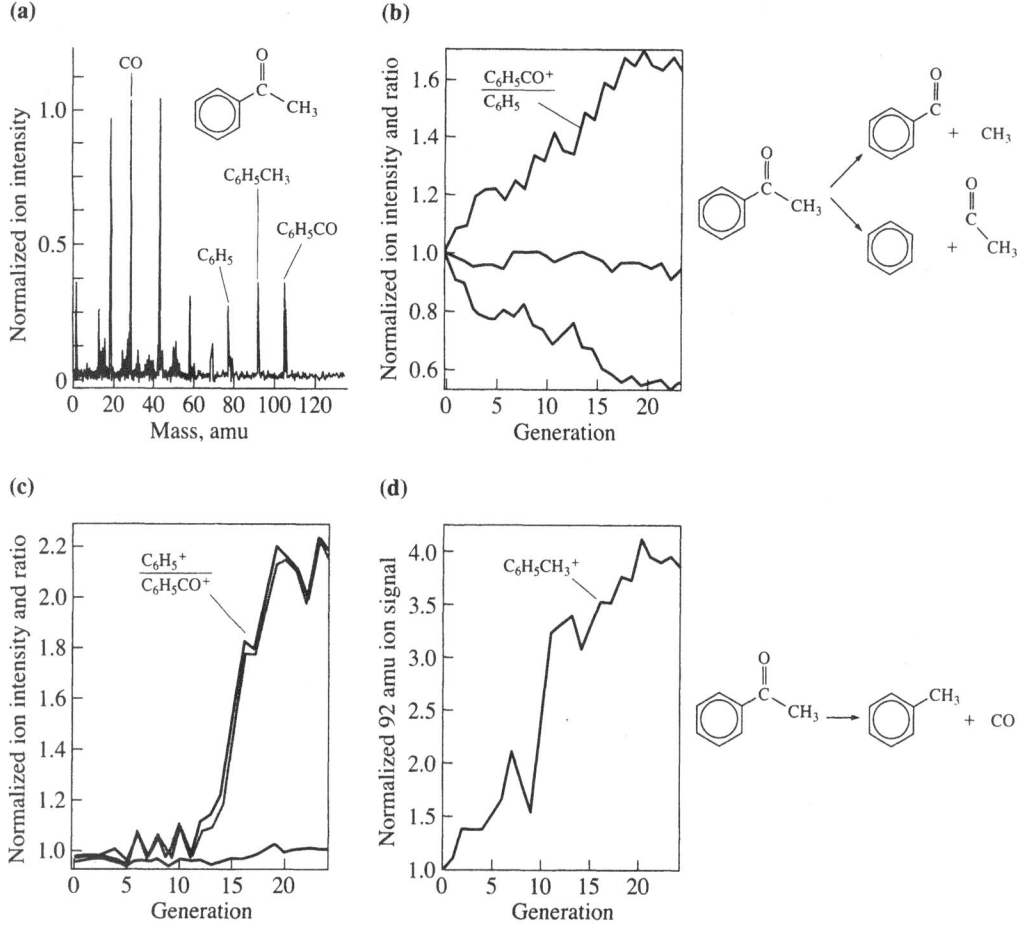

Figure 16.23 Left: (a) Mass spectrum of acetophenone ($C_6H_5COCH_3$). (b) The relative ion yield for $C_6H_5CO^+$, $C_6H_5^+$ and the $C_6H_5CO^+/C_6H_5^+$ ratio as a function of generation when maximization of this ratio is the merit in the closed-loop experiment. (c) The relative ion yield for $C_6H_5CO^+$, $C_6H_5^+$ and the $C_6H_5^+/C_6H_5CO^+$ ratio as a function of generation when maximization of this ratio is the merit. (d) The average signal for $C_6H_5CH_3^+$ as a function of generation when maximization of this ion signal is the merit. Right: The chemical branching processes corresponding to the figures on the left. Note that the formation of $C_6H_5CH_3^+$ requires the extrusion of the CO moiety, which requires the breaking of two chemical bonds and the formation of a new one. Adapted from Levis (2001).

Problems

▶ **Exercise 16.8** Consider the objective of a maximum total excited-state population at time T:

$$J = \langle \psi^{(1)}(T) | \psi^{(1)}(T) \rangle. \tag{16.82}$$

a. Show that J can be written as

$$J = \frac{1}{\hbar^2} \int_0^T \int_0^T \langle \phi_i | \varepsilon^*(t) e^{-i\tilde{H}_b(t-t'')/\hbar} \varepsilon(t'') | \phi_i \rangle dt\, dt'', \tag{16.83}$$

where $\tilde{H}_b = H_b - E_i$ with H_b and E_i defined as in Section 16.2.1.

   b. Define the modified objective as in Eq. 16.4 with J given by Eq. 16.83. Show that the optimal field is given by

$$\tilde{\varepsilon}(t) = \frac{1}{\lambda} \int_0^T M(t, t') \tilde{\varepsilon}(t') \, dt', \tag{16.84}$$

where

$$M(t, t') = \frac{1}{\hbar^2} \langle \phi_i | e^{-i\bar{H}(t-t')/\hbar} | \phi_i \rangle. \tag{16.85}$$

   c. Equation 16.84 is an integral equation for $\varepsilon(t)$. Show that by discretizing time this equation can be written as

$$\lambda \varepsilon(t_k) = \sum_{j=j_{min}}^{j_{max}} M(t_k, t_j) \varepsilon(t_j). \tag{16.86}$$

Describe how this equation could be solved numerically, and which value of λ is optimal (Yan, 1993).

▶ **Exercise 16.9**

   a. Show that if the $\{B_i\}$ have zero trace, the algebra generated by them consists only of matrices with zero trace.

   b. Show that if A is included in the algebra, $\text{Tr}(A) \neq 0$, then the only traceless matrix in the extended algebra is A.

   c. Show that the dimension of the algebra of all traceless $n \times n$ matrices is $n^2 - 1$.

   d. Finally, show that with the set of all traceless $n \times n$ matrices one can generate any $n \times n$ unitary operator within an overall phase.

▶ **Exercise 16.10**

   a. (3-level system) Consider the TDSE

$$i\hbar \frac{da}{dt} = (H_0 + H_1)a \tag{16.87}$$

with

$$H_0 = \begin{pmatrix} E_1 & 0 & 0 \\ 0 & E_2 & 0 \\ 0 & 0 & E_3 \end{pmatrix} \qquad H_1 = \begin{pmatrix} 0 & \alpha & \beta \\ \alpha & 0 & 0 \\ \beta & 0 & 0 \end{pmatrix}. \tag{16.88}$$

Show that this system of equations is controllable.

   b. Consider the system in **a.** except with

$$H_0 = \begin{pmatrix} E_1 & 0 & 0 \\ 0 & E_2 & 0 \\ 0 & 0 & E_2 \end{pmatrix}. \tag{16.89}$$

Show that this system is not controllable.

c. Consider the system in **a.** except with

$$H_1 = \begin{pmatrix} 0 & \alpha & \alpha \\ \alpha & 0 & 0 \\ \alpha & 0 & 0 \end{pmatrix}. \tag{16.90}$$

Is it controllable?

▶ Exercise 16.11

a. (4-level system) Consider the TDSE

$$i\hbar \frac{da}{dt} = (H_0 + H_1)a \tag{16.91}$$

with

$$H_0 = \begin{pmatrix} E_1 & 0 & 0 & 0 \\ 0 & E_2 & 0 & 0 \\ 0 & 0 & E_3 & 0 \\ 0 & 0 & 0 & E_4 \end{pmatrix} \qquad H_1 = \begin{pmatrix} 0 & \alpha & \beta & 0 \\ \alpha & 0 & 0 & \gamma \\ \beta & 0 & 0 & \delta \\ 0 & \gamma & \delta & 0 \end{pmatrix}. \tag{16.92}$$

Show that this system of equations is controllable.

b. Consider the system in **a.** except with

$$H_0 = \begin{pmatrix} E_1 & 0 & 0 & 0 \\ 0 & E_2 & 0 & 0 \\ 0 & 0 & E_2 & 0 \\ 0 & 0 & 0 & E_4 \end{pmatrix}. \tag{16.93}$$

Is it controllable?

c. Consider the system in **b.** with each of the following choices for H_1:

$$H_1 = \begin{pmatrix} 0 & \alpha & \beta & 0 \\ \alpha & 0 & 0 & \alpha \\ \beta & 0 & 0 & \beta \\ 0 & \alpha & \beta & 0 \end{pmatrix}; \qquad H_1 = \begin{pmatrix} 0 & \alpha & \beta & 0 \\ \alpha & 0 & 0 & \beta \\ \beta & 0 & 0 & \alpha \\ 0 & \beta & \alpha & 0 \end{pmatrix};$$

$$H_1 = \begin{pmatrix} 0 & \alpha & \beta & 0 \\ \alpha & 0 & 0 & \beta \\ \beta & 0 & 0 & -\alpha \\ 0 & \beta & -\alpha & 0 \end{pmatrix}. \tag{16.94}$$

Which of these systems is controllable?

▶ Exercise 16.12   In Section 16.4, we implicitly considered the two-photon absorption assuming there is no intermediate resonance. If there is an intermediate resonance (see Figure 16.24a), its contribution to the two-photon absorption is

$$a^r \approx -\frac{1}{i\hbar^2}\mu_{fn}\mu_{ni}\left[i\pi\varepsilon(\omega_{ni})\varepsilon(\omega_{fi}-\omega_{ni}) + \mathcal{P}\int_{-\infty}^{\infty}\frac{\varepsilon(\omega)\varepsilon(\omega_{fi}-\omega)}{\omega_{ni}-\omega}d\omega\right] \tag{16.95}$$

where $\mathcal{P}$ is the Cauchy principal value (cf. Eq. A.22) and ω_{ni}, $\omega_{fi} - \omega_{ni} = \omega_{fn}$ are the resonance frequencies. (Do not confuse $i = \sqrt{-1}$ with the label i, which appears only as a subscript.) The second term is known as the "dispersion term," and is of the form of

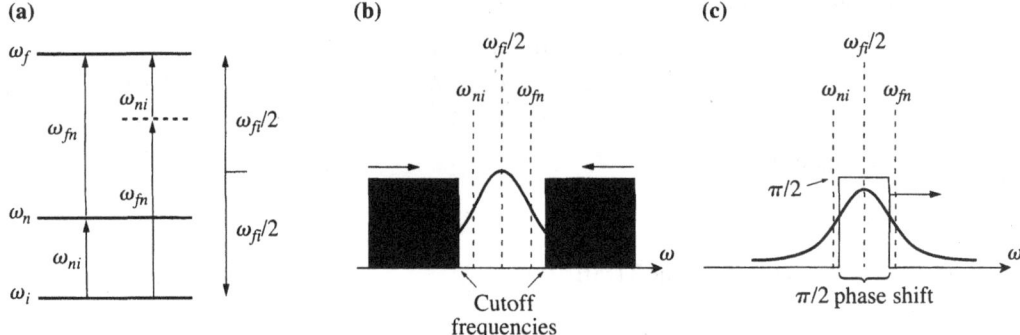

Figure 16.24 (a) Energy-level diagram for the initial state, i, the intermediate state, n, and the final state, f. (b) An adjustable slit used to block spectral bands of the exciting pulse symmetrically around $\omega_{fi}/2$. (c) The window for the optimal $\pi/2$ phase mask, its leading and trailing edges located at the resonance frequencies ω_{ni} and ω_{fn}, respectively. The phase mask ensures that two-photon transitions whose intermediate energies are above the energy of the resonant intermediate state contribute with the opposite sign from two-photon transitions whose intermediate energies are below the intermediate resonance. Adapted from Dudovich et al. (2001).

the KHD resonance denominator of Eq. 14.3. Assume that the pulse bandwidth, $\Delta\omega$, is large compared with the frequency separation of the resonance from the two-photon carrier frequency, that is, $\Delta\omega \gg |\omega_{fi}/2 - \omega_{ni}|$. Note the change in sign of the denominator of the dispersion term as ω passes from below to above ω_{ni}.

   a. Show that the contribution of the dispersion term to the two-photon absorption can be increased by blocking the red detuned ($\omega < \omega_{ni}$) and blue detuned ($\omega > \omega_{fn}$) frequency components of the pulse. We have assumed for definiteness that $\omega_{ni} < \omega_{fn/2}$; see Figure 16.24b.

   b. Show that the contribution of the dispersion term to the two-photon absorption can be further increased if, instead of blocking the red and blue detuned frequencies, the phase of these two sets of detuned frequencies is given a pairwise π-phase shift relative to the frequencies in between. Show that this may be accomplished by imposing a phase shift of $\pi/2$ over the spectral region $\omega_{ni} < \omega < \omega_{fn}$; see Figure 16.24c (Dudovich, 2001).

▶ **Exercise 16.13**   Show that for an initial wavepacket on the inner turning point of a Morse potential, the optimal field to create a cannonball is upwardly chirped while the optimal field to create a reflectron is downwardly chirped. Use the first-order perturbation theory result, Eq. 16.15, that the optimal field for a specified target is given by

$$\varepsilon(t) = \langle \phi_f | e^{-iH_b\tau/\hbar} | \phi_i \rangle^* \tag{16.96}$$

where ϕ_i is the initial state, ϕ_f is the target state, t_f is the final time, and $\tau = t_f - t$. Assume that $\phi_i(x)$ and $\phi_f(x)$ are Gaussian of the form

$$\phi_i(x) = N e^{-\alpha_i(x-x_i)^2 + ip_i(x-x_i)/\hbar} \tag{16.97}$$

$$\phi_f(x) = N e^{-\alpha_f(x-x_f)^2 + ip_f(x-x_f)/\hbar}. \tag{16.98}$$

The cannonball and reflectron are special cases of $\phi_f(x)$ with large α_f and $p_f > 0$ or $p_f < 0$, respectively, where $p_f = p_i(\tau_0)$, $x_f = x_i(\tau_0)$. Assume that the classical trajectory associated with the center of ϕ_i gets to the center of ϕ_f in time t.

a. Substitute Eqs. 16.97–16.98 into Eq. 16.96 and use the semiclassical expression, Eq. 10.107, to obtain

$$\varepsilon(t) = N^*(\tau) e^{-iS(\tau)/\hbar} e^{-\frac{1}{4}[\Delta p(\tau)] \cdot \Pi^* \cdot [\Delta p(\tau)]}, \tag{16.99}$$

where

$$\Delta p(\tau) = [p_f(\tau) - p_f, \, p_i(\tau) - p_i], \tag{16.100}$$

$p_f(\tau_0) = p_f$, $p_i(\tau_0) = p_i$ and Π is given by Eq. 10.108.

b. Expand $S(\tau)$, $p_f(\tau)$ and $p_i(\tau)$ around $\tau = \tau_0$,

$$S(\tau) = S(\tau_0) + \frac{\partial S}{\partial \tau_0}(\tau - \tau_0) + \frac{1}{2}\frac{\partial^2 S}{\partial \tau_0^2}(\tau - \tau_0)^2 + \cdots \tag{16.101}$$

$$p_f(\tau) - p_f = \frac{\partial p_f}{\partial \tau_0}(\tau - \tau_0) \qquad p_i(\tau) - p_i = \frac{\partial p_i}{\partial \tau_0}(\tau - \tau_0), \tag{16.102}$$

and substitute these expansions into Eq. 16.99. Match terms with Eq. 16.71 to show that the temporal center is

$$t_0 = t_f - \tau_0, \tag{16.103}$$

the carrier frequency is

$$\hbar\omega_0 = -\partial S(\tau)/\partial\tau_0 = E, \tag{16.104}$$

the pulse duration is

$$\frac{1}{\tau^2} = [\partial p(\tau_0)] \cdot \mathrm{Re}\Pi \cdot [\partial p(\tau_0)] \tag{16.105}$$

and the linear chirp rate is

$$c = \frac{\partial^2 S}{\partial \tau_0^2} - \frac{1}{2}[\partial p(\tau_0)] \cdot \mathrm{Im}\Pi \cdot [\partial p(\tau_0)]. \tag{16.106}$$

We have used the shorthand notation $[\partial p(\tau_0)]$ for the two-component vector $[\partial p_f/\partial \tau_0, \partial p_i/\partial \tau_0]$.

c. If we neglect the second term in Eq. 16.106 we find that $c = \partial^2 S/\partial \tau_0^2 = -\partial E/\partial \tau_0$. For a Morse oscillator, as the energy increases the potential becomes harder near the inner turning point and softer near the outer turning point (cf. the excited-state potential in Figure 16.18). For target time $\tau < t_{\text{outer}}$, where t_{outer} is the time for the wavepacket to reach the outer turning point, $\partial\tau/\partial E$ is dominated by the inner turning point, and hence $\partial\tau/\partial E < 0$. For target time $\tau > t_{\text{outer}}$, $\partial\tau/\partial E$ is dominated by the outer turning point (recall that overall $\partial T/\partial E > 0$ for a Morse oscillator, where T is the classical period at energy E), and hence $\partial \tau / \partial E > 0$. Use the above considerations to show that Eq. 16.106 predicts that a positive chirp will maximize the overlap with the cannonball state, and a negative chirp will maximize the overlap with the reflectron state (Cao, 1997).

▶ **Exercise 16.14** Consider the second-order expression for the ground state wavefunction

$$\psi^{(2)}(T) = \left(\frac{1}{i\hbar}\right)^2 \int_{-\infty}^{T} dt_2 \int_{-\infty}^{t_2} dt_1 \hat{G}(t_2, t_1) \varepsilon^*(t_2) \varepsilon(t_1) \psi_0, \tag{16.107}$$

where

$$\hat{G}(t_2, t_1) = e^{-i H_a (T - t_2)/\hbar} e^{-i H_b (t_2 - t_1)/\hbar} e^{-i H_a t_1/\hbar} \tag{16.108}$$

and we assume that the transition dipole moment μ is a constant.

a. Show that if the kinetic energy is neglected, the second-order expression for the population on the ground electronic state can be written as

$$N_0^{(2)} = \left| \frac{\mu^2}{\hbar^2} \int_{-\infty}^{\infty} dt_2 \int_{-\infty}^{t_2} dt_1 G(t_2, t_1) \varepsilon(t_2) \varepsilon(t_1) \right|^2, \tag{16.109}$$

where

$$G(t_2, t_1) = e^{-i\omega_1(x_0)t_2 + i\omega_1(x_0)t_1}, \tag{16.110}$$

$$\hbar\omega_1(x) = V_b(x) - V_a(x), \tag{16.111}$$

and x_0 is the center of ψ_0. We have assumed that T is longer than the duration of $\varepsilon(t)$.

b. Consider now the effect of the kinetic energy. Show that to first order in the kinetic energy

$$\hat{G}(t_2, t_1) \approx e^{i\omega_2(\hat{x})t_2} e^{-i\eta(x)\hat{p}/\hbar} e^{i\omega_1(x)t_1}, \tag{16.112}$$

where $s = t_2 - t_1 \geq 0$ and $\eta(x) = f_b(x)s^2/2m$, which is the displacement of the wavepacket on the excited state due to the force $f_b(x) = -V_b'(x)$.

c. Show that Eq. 16.112 leads to an expression for the second-order population on the ground electronic state of the form of Eq. 16.109, with

$$G(t_2, t_1) = e^{-i\omega_1(x_1)t_2 + i\omega_1(x_0)t_1}, \tag{16.113}$$

where $x_1 = x_0 + \eta(x_0)$ is the center of the displaced wavepacket on the excited-state surface.

d. Expand $\omega_1(x_1)$ to linear order in x:

$$\omega_1(x_1) \approx \omega_1(x_0) - \alpha(x_0)s^2, \tag{16.114}$$

where $\alpha(x_0) = \omega_1' V_1'/2m = (V_1')^2/2m\hbar \geq 0$, to obtain

$$G(t_2, t_1) \approx e^{-i\omega_1(x_0)t_2 + i\alpha s^2 t_2 + i\omega_1(x_0)t_1}. \tag{16.115}$$

e. Substitute Eq. 16.115 into Eq. 16.109 and take $\varepsilon(t)$ to be a chirped pulse of the form in Eq. 16.71. Show that

$$N_0^{(2)} = P^2 |C(\Delta)|^2, \tag{16.116}$$

where

$$C(\Delta) = \int_0^\infty ds \, \exp\left[-\frac{s^2}{4\tau^2} - \frac{\tau^2 s^2}{4}(c + \alpha s)^2 - is\Delta + i\frac{\alpha}{2}s^3 \right], \tag{16.117}$$

$P = \int_{-\infty}^\infty \left| \frac{\mu\varepsilon(t)}{\hbar} \right|^2 dt = \pi^{1/2} \left(\frac{\mu\varepsilon_0}{\hbar} \right)^2 \tau$ and $\Delta = \omega_1(x_0) - \omega_0$ is the detuning.

f. Maximize $N_0^{(2)}$ by minimizing the square term in the exponent of $C(\Delta)$ with respect to c (i.e., $(c + \alpha s)^2 \approx 0$). Since the average value of s is on the order of the pulse duration (i.e., $s \approx \tau$) show that an estimate for the optimal linear chirp is $c \approx -\alpha\tau$ (Cao, 1998).

References

General

1. S. A. Rice and M. Zhao, *Optical Control of Molecular Dynamics* (Wiley, New York, 2000).

2. D. J. Tannor, Design of femtosecond optical pulse sequences to control photochemical products, in *Molecules in Laser Fields*, ed. A. Bandrauk (Dekker, New York, 1994).

3. Controlling chemical reactions with laser light, Science 255, 1643 (1992).

4. S. A. Rice, New ideas for guiding the evolution of a quantum system, Science 258, 412 (1992).

5. W. S. Warren, H. Rabitz and M.Dahleh, Science 259, 1581 (1993).

6. P. Brumer and M. Shapiro, Laser control of molecular processes, Annu. Rev. Phys. Chem. 43, 257 (1992).

7. P. Brumer and M. Shapiro, Laser control of molecular motions, Scientific American 272, 34 (March, 1995).

8. B. Kohler, J. L. Krause, F. Raksi, K. R. Wilson, V. V. Yakovlev, R. M. Whitnell and J. J. Yang, Controlling the future of matter, Acc. Chem. Res. 28, 133 (1995).

9. R. J. Gordon and S. A. Rice, Active control of the dynamics of atoms and molecules, Annu. Rev. Phys. Chem. (1997).

10. T. Brixner and G. Gerber, Quantum control of gas-phase and liquid-phase femtochemistry, Chem. Phys. Phys. Chem. 4, 418 (2003).

11. M. Shapiro and P. Brumer, *Principles of the Quantum Control of Molecular Processes* (Wiley-Interscience, Hoboken, NJ, 2003).

General Theory of Optimal Control

12. A. E. Bryson, Jr. and Y.-C. Ho, *Applied Optimal Control* (Hemisphere, New York, 1975).

13. V. F. Krotov, *Global Methods in Optimal Control Theory* (Dekker, New York, 1996).

14. D. E. Kirk, *Optimal Control Theory: An Introduction* (Prentice Hall, Englewood Cliffs, NJ, 1970).

15. D. P. Bertsekas, *Dynamic Programming and Optimal Control* (Athena Scientific, Belmont, MA, 1987).

Controllability

16. R. W. Brockett, *Finite Dimensional Linear Systems* (Wiley, New York, 1970).

17. R. W. Brockett, System theory on group manifolds and coset spaces, SIAM J. Control 10, 265 (1972).

18. R. W. Brockett, Lie theory and control systems defined on spheres, SIAM J. Appl. Math. 25, 213 (1973a).

19. R. Brockett, Beijing lectures on nonlinear control systems, AMS/IP Studies in Advanced Mathematics, v. 17 (2000).

20. V. Jurdjevic and H. J. Sussman, Control systems on Lie groups, J. Diff. Eqns. 12, 313 (1972).

21. G. M. Huang, T. J. Tarn and J. W. Clark, On the controllability of quantum-mechanical systems, J. Math. Phys. 24, 2608 (1983).

22. V. Ramakrishna, M. V. Salapaka, M. Dahleh, H. Rabitz and A. Peirce, Controllability of molecular systems, Phys. Rev. A 51, 960 (1995).

23. S. G. Schirmer, H. Fu and A. I. Solomon, Complete controllability of quantum systems, Phys. Rev. A 63, 063410 (2001).

24. A. G. Butkovskiy and Yu. I. Samoilenko, *Control of Quantum-Mechanical Processes and Systems* (Kluwer, 1990).

25. S. P. Shah, D. J. Tannor and S. A. Rice, Controllability of population transfer to degenerate states: Analytical and numerical results for a four-level system, Phys. Rev. A 66, 033405 (2002).

26. G. Turinici and H. Rabitz, Quantum wavefunction controllability, Chem. Phys. 267, 1 (2001).

27. F. Albertini and D. D'Alessandro, Notions of controllability for bilinear multilevel quantum systems, IEEE Transactions on Automatic Control, 48, 1399 (2003).

28. C. Rangan and A. M. Bloch, Control of finite-dimensional quantum systems: Application to a spin-1/2 particle coupled with a finite quantum harmonic oscillator, J. Math. Phys. 46, 032106 (2005).

29. N. Khaneja, R. Brockett and S. J. Glaser, Time optimal control in spin systems, Phys. Rev. A 63, 032308 (2001a).

30. N. Khaneja and S. J. Glaser, Cartan decomposition of $SU(2^n)$ and control of spin systems, Chem. Phys. 267, 11 (2001b).

31. N. Khaneja, T. Reiss, B. Luy and S. J. Glaser, Optimal control of spin dynamics in the presence of relaxation, J. Mag. Res. 162, 311 (2003).

Pump–Dump or Pump–Pump Control

32. D. J. Tannor and S. A. Rice, Control of selectivity of chemical reaction via control of wavepacket evolution, J. Chem. Phys. 83, 5013 (1985).

33. D. J. Tannor, R. Kosloff and S. A. Rice, Coherent pulse sequence induced control of selectivity of reactions: Exact quantum mechanical calculations, J. Chem. Phys. 85, 5805 (1986).

34. D. J. Tannor and S. A. Rice, Coherent pulse sequence control of product formation in chemical reactions, Adv. Chem. Phys. 70, 441 (1988).

35. Y. Yan, J. Che and J. L. Krause, Optimal pump-dump control, Chem. Phys. 217, 297 (1997).

36. R. Mitrić, M. Hartmann, J. Pittner and V. Bonačić-Koutecký, New strategy for optimal control of femtosecond pump-dump processes, J. Phys. Chem. A 106, 10477 (2002).

37. T. Baumert, J. Helbing and G. Gerber, Coherent control with femtosecond laser pulses, Adv. Chem. Phys. 101, 47 (1997).

38. E. D. Potter, J. L. Herek, S. Pedersen, Q. Liu and A. H. Zewail, Femtosecond laser control of a chemical reaction, Nature 355, 66 (1992).

39. G. K. Paramonov and P. Saalfrank, A new pump & dump strategy to control chemical reactivity at surfaces: Application to photoisomerization of adsorbates, Chem. Phys. Lett. 301, 509 (1999).

Optimal Control of Molecular Dynamics

40. D. J. Tannor and S. A. Rice, Control of selectivity of chemical reaction via control of wavepacket evolution, J. Chem. Phys. 83, 5013 (1985).

41. D. J. Tannor and S. A. Rice, Coherent pulse sequence control of product formation in chemical reactions, Adv. Chem. Phys. 70, 441 (1988).

42. A. P. Peirce, M. A. Dahleh and H. Rabitz, Optimal control of quantum mechanical systems: Existence, numerical approximations, and applications, Phys. Rev. A37, 4950 (1988).

43. R. Kosloff, S. A. Rice, P. Gaspard, S. Tersigni and D. J. Tannor, Wavepacket dancing: Achieving chemical selectivity by shaping light pulses, Chem. Phys. 139, 201 (1989).

44. D. J.Tannor and Y. Jin, Design of femtosecond pulse sequences to control photochemical products, in *Mode Selective Chemistry*, eds. B. Pullman, J. Jortner, and R. D. Levine (Kluwer, Dordrecht, 1991).

45. V. F. Krotov and I. N. Fel'dman, An iterative method for solving optimal control problems, Engrg. Cybernetics 21, 123 (1984).

46. W. Jakubetz and J. Manz, Theory of optimal laser pulses for selective transitions between molecular eigenstates, Chem. Phys. Lett. 165, 100 (1990).

47. S. Shi and H. Rabitz, Optimal control of bond selectivity in unimolecular reactions, Comp. Phys. Comm. 63, 71 (1991).

48. Y. J. Yan, R. E. Gillilan, R. M. Whitnell, K. R. Wilson and S. Mukamel, Optimal control of molecular dynamics: Liouville space theory, J. Chem. Phys. 97, 2320 (1993).

49. D. J. Tannor, V. A. Kazakov, and V. Orlov, Control of photochemical branching: Novel procedures for finding optimal pulses and global upper bounds, in *Time-Dependent Quantum Molecular Dynamics*, eds. J. Broeckhove and L. Lathouwers, NATO ASI Series B: Physics vol. 299 (Plenum, New York, 1992).

50. J. Somlói, V. A. Kazakov and D. J. Tannor, Controlled dissociation of I_2 via optical transitions between the X and B electronic states, Chem. Phys. 172, 85 (1993).

51. A. Bartana, R. Kosloff and D. J. Tannor, Laser cooling of internal degrees of freedom. II., J. Chem. Phys. 106, 1435 (1997).

52. Y. Ohtsuki, W. S. Zhu and H. Rabitz, Monotonically convergent algorithm for quantum optimal control with dissipation, J. Chem. Phys. 110, 9825 (1999).

53. L. Liu and J. T. Muckerman, Strong-field optical control of vibrational dynamics: Vibrational Stark effect in planar acetylene, J. Chem. Phys. 110, 2446 (1999).

Femtosecond Pulse Shaping—Technology and Applications

54. A. M. Weiner and J. P. Heritage, Picosecond and femtosecond Fourier pulse shape synthesis, Revue de Physique Appliquée 22, 1619 (1987).

55. A. M. Weiner, D. E. Leaird, J. S. Patel and J. R. Wullert, Programmable femtosecond pulse shaping by use of a multielement liquid-crystal phase modulator, Opt. Lett. 15, 326 (1990).

56. C. W. Hillegas, J. X. Tull, D. Goswami, D. Strickland and W. S. Warren, Femtosecond laser pulse shaping by use of microsecond radio-frequency pulses, Opt. Lett. 19, 737 (1994).

57. M. M. Wefers and K. A. Nelson, Generation of high-fidelity programmable ultrafast optical waveforms, Opt. Lett. 20, 1047 (1995).

58. A. M. Weiner, D. E. Leaird, G. P. Wiederrecht and K. A. Nelson, Femtosecond multiple-pulse impulsive stimulated Raman scattering spectroscopy, J. Opt. Soc. Am. B 8, 1264 (1991).

59. T. C. Weinacht, J. Ahn and P. H. Bucksbaum, Controlling the shape of a quantum wavefunction, Nature 397, 233 (1999).

Husimi Distribution

60. K. Husimi, Some formal properties of the density matrix, Proc. Phys. Math. Soc. Jpn. 22, 264 (1940).

61. Y. Kano, A new phase-space distribution function in the statistical theory of the electromagnetic field, J. Math. Phys. 6, 1913 (1965).

62. W. Schleich, *Quantum Optics in Phase Space* (Wiley-VCH, New York, 2001).

Coherent Control via Interfering Pathways: Frequency Domain

63. P. Brumer and M. Shapiro, Laser control of molecular processes, Annu. Rev. Phys. Chem. 43, 257 (1992).

64. L. Zhu, V. Kleiman, X. Li, S.-P. Lu, K. Trentelman and R. J. Gordon, Coherent laser control of the product distribution obtained in the photoexcitation of HI, Science 270, 77 (1995).

65. F. Wang, C. Chen and D. S. Elliott, Product state control through interfering excitation routes, Phys. Rev. Lett. 77, 2416 (1996).

66. E. Charron, A. Giusti-Suzor and F. H. Mies, Coherent control of isotope separation— Separation in HD^+ photodissociation by strong fields, Phys. Rev. Lett. 71, 692 (1993).

67. B. Sheehy, B. Walker and L. F. DiMauro, Phase-control in the 2-color photodissociation of HD^+, Phys. Rev. Lett. 74, 463 (1996).

68. G. Kurizki, M. Shapiro and P. Brumer, Phase-coherent control of photocurrent directionality in semiconductors, Phys. Rev. B 39, 3435 (1989).

69. E. Dupont, P. B. Corkum, H. C. Liu, M. Buchanan and Z. R. Wasilewski, Phase-controlled currents in semiconductors, Phys. Rev. Lett. 74, 3596 (1995).

70. T. Seideman, The role of a molecular phase in two-pathway excitation schemes, J. Chem. Phys. 108, 1915 (1998).

Coherent Control via Interfering Pathways: Time Domain

71. R. Uberna, M. Khalil, R. M. Williams, J. M. Papanikolas and S. R. Leone, Phase and amplitude control in the formation and detection of rotational wavepackets in the $E^1\sigma_g^+$ state of Li_2, J. Chem. Phys. 108, 9259 (1998).

72. D. Meshulach and Y. Silberberg, Coherent quantum control of two-photon transitions by a femtosecond laser pulse, Nature 396, 239 (1998).

73. P. H. Bucksbaum, Photonics: An atomic dimmer switch, Nature 396, 217 (1998).

74. D. Meshulach and Y. Silberberg, Coherent quantum control of multiphoton transitions by shaped ultrashort optical pulses, Phys. Rev. A 60, 1287 (1999).

75. N. Dudovich, B. Dayan, S. M. Gallagher Faeder and Y. Silberberg, Transform limited pulses are not optimal for resonant multiphoton transitions, Phys. Rev. Lett. 86, 47 (2001).

Chirping

76. B. Kohler, J. L. Krause, F. Raksi, K. R. Wilson, V. V. Yakovlev, R. M. Whitnell and Y. Yan, Controlling the future of matter, Acct. Chem. Res. 28, 133 (1995).

77. J. Cao and K. R. Wilson, A simple physical picture for quantum control of wavepacket localization, J. Chem. Phys. 107, 1441 (1997).

78. M. Sterling, R. Zadoyan and V. A. Apkarian, Interrogation and control of condensed phase chemical dynamics with linearly chirped pulses: I_2 in solid Kr, J. Chem. Phys. 104, 6497 (1996).

79. J. Cao and K. R. Wilson, Detecting wavepacket motion in pump-probe experiments: Theoretical analysis, J. Chem. Phys. 106, 5062 (1997).

80. S. Chelkowski and A. D. Bandrauk, Raman chirped adiabatic passage: A new method for selective excitation of high vibrational states, J. Raman Spect. 28, 459 (1997).

81. D. M. Villeneuve, S. A. Aseyev, P. Dietrich, M. Spanner, M. Y. Ivanov and P. B. Corkum, Forced molecular rotation in an optical centrifuge, Phys. Rev. Lett. 85, 542 (2000).

82. G. K. Paramonov, Coherent control of linear and nonlinear mulitphoton excitation of molecular vibrations, Chem. Phys. 177, 169 (1993); J. Manz and G. K. Paramonov, Laser control scheme for state-selective ultrafast vibrational-excitation of the HOD molecule, J. Phys. Chem. 97, 12625 (1993).

83. S. Ruhman and R. Kosloff, Application of chirped ultrashort pulses for generating large-amplitude ground-state vibrational coherence—A computer-simulation, J. Opt. Soc. Am. B 7, 1748 (1990).

84. G. Cerullo, C. J. Bardeen, Q. Wang and C. V. Shank, High-power femtosecond chirped pulse excitation of molecules in solution, Chem. Phys. Lett. 262, 362 (1996).

85. J. Cao, J. Che and K. R. Wilson, Intrapulse dynamical effects in multiphoton processes: Theoretical analysis, J. Phys. Chem. A 102, 4284 (1998).

86. J. Cao, C. Bardeen and K. R. Wilson, Molecular π pulses: Population inversion with positively chirped short pulses, J. Chem. Phys. 113, 1898 (2000).

Learning Algorithms

87. D. E. Goldberg, *Genetic Algorithms in Search, Optimization, and Machine Learning* (Addison-Wesley, Reading, MA, 1989).

88. L. Davis, *Handbook of Genetic Algorithms* (van Nostrand Reinhold, New York, 1991).

89. R. S. Judson and H. Rabitz, Teaching lasers to control molecules, Phys. Rev. Lett 68, 1500 (1992).

90. C. J. Bardeen, V. V. Yakovlev, K. R. Wilson, S. D. Carpenter, P. M. Weber and W. S. Warren, Feedback quantum control of molecular electronic population transfer, Chem. Phys. Lett 280, 151 (1997).

91. A. Assion, T. Baumert, M. Bergt, T. Brixner, B. Kiefer, V. Seyfried, M. Strehle and G. Gerber, Control of chemical reactions by feedback-optimized phase-shaped femtosecond laser pulses, Science 282, 919 (1998).

92. T. C. Weinacht, J. L. White and P. H. Bucksbaum, Toward strong field mode-selective chemistry, J. Phys. Chem. A 103, 10166 (1999).

93. R. J. Levis, G. M. Menkir and H. Rabitz, Selective bond dissociation and rearrangement with optimally tailored, strong-field laser pulses, Science 292, 709 (2001).

94. S. Vajda, A. Bartelt, E. C. Kaposta, T. Leisner, C. Lupulescu, S. Minemoto, P. Rosendo-Francisco and L. Wöste, Feedback optimization of shaped femtosecond laser pulses for controlling the wavepacket dynamics and reactivity of mixed alkaline clusters, Chem. Phys. 267, 231 (2001).

95. R. Bartels, S. Backus, E. Zeek, L. Misoguti, G. Vdovin, I. P. Christov, M. M. Murnane and H. C. Kapteyn, Shaped-pulse optimization of coherent emission of high-harmonic soft X-rays, Nature 406, 164 (2000).

96. J. L. Herek, W. Wohlleben, R. J. Cogdell, D. Zeidler and M. Motzkus, Quantum control of energy flow in light harvesting, Nature 417, 533 (2002).

97. M. Q. Phan and H. Rabitz, Learning control of quantum-mechanical systems by laboratory identification of effective input-output maps, Chem. Phys. 217, 389 (1997); M. Q. Phan and H. Rabitz, A self-guided algorithm for learning control of quantum-mechanical systems, J. Chem. Phys. 110, 34 (1999).

98. J. Botina and H. Rabitz, Reduced control dynamics for complex quantum systems, Phys. Rev. A 55, 1634 (1997); J. Botina and H. Rabitz, Learning control algorithm for nonlinear maps, Phys. Rev. E 55, 5338 (1997).

99. G. T. Töth, A. Lörincz and H. Rabitz, The effect of control field and measurement imprecision on laboratory feedback control of quantum systems, J. Chem. Phys. 101, 3715 (1994).

100. T. Hornung, M. Motzkus and R. de Vivie-Riedle, Adapting optimal control theory and using learning loops to provide experimentally feasible shaping mask patterns, J. Chem. Phys. 115, 3105 (2001).

101. J. V. Tietz and S.-I. Chu, Energy deposition in SO_2 via intense infrared laser multiphoton excitation, Chem. Phys. Lett. 101, 446 (1983).

102. J. Chang and R. E. Wyatt, Preselecting paths for multiphoton dynamics using artificial intelligence, J. Chem. Phys. 85, 1826 (1986).

Chapter 17

Wavepacket Approach to Photodissociation

17.1 Introduction

This chapter deals with the wavepacket formulation of photodissociation. Photodissociation is a rich area of chemistry, of practical importance to organic chemists on one hand, and amenable to detailed experimental study by physical chemists on the other. The central theme of this chapter is the time-dependent formulation of both the partial and the total photodissociation cross sections in terms of the dynamics of a wavepacket initiated in the Franck–Condon region of the excited state potential.

Despite our ultimate interest in the time-dependent formulation of photodissociation, we introduce the time-*independent* formulation first. This is for two reasons. The first is to provide exposure to another, more conventional perspective on photodissociation. But more importantly, it is to lay the groundwork for the derivation of the time-dependent expressions. In particular, in our treatment the time-independent scattering eigenstates are formulated as the Fourier transform of wavepackets with appropriately chosen boundary conditions. These boundary conditions are then expressed naturally using the Møller operators, which are closely related to the interaction picture. The Møller operators then facilitate the derivation of the time-dependent formulation of the partial photodissociation cross section.

This chapter on photodissociation, and the next chapter on reactive scattering, are inextricably intertwined. Photodissociation is often called a "half-collision" process, since the process begins, as it were, from the middle—from the parent complex—and proceeds to products, whereas in reactive scattering the process proceeds from reactants, to complex, to products. It is therefore natural that we begin with photodissociation before proceeding to the more complicated full-collision process. Indeed, virtually all the concepts and mathematical methods developed in this chapter—asymptotic Hamiltonians, scattering eigenstates, and Møller operators—will be used again in the next chapter, but *doubled*—once for reactants and once for products.

This chapter and the next chapter are to some extent the climax of the book, both physically and mathematically. Three of the cornerstones of the earlier chapters—the interaction picture (Chapter 9), the spectral method for constructing eigenstates (Chapter 6), and the Fourier transform of time-correlation functions (Chapters 6 and 14)—will all come together in these chapters.

17.2   The Eigenstates of an Asymptotic Hamiltonian

In any scattering event it is assumed that at large enough separation of the fragments (either reactants or products) the interaction between the fragments vanishes. We therefore write, in general,

$$H = H_0 + V, \tag{17.1}$$

where H is the full Hamiltonian, H_0 is the asymptotic Hamiltonian and V is the difference between the two—the interaction potential between the fragments, which is zero at large distances but nonzero at finite distances.

The asymptotic Hamiltonian, H_0, is by definition separable in the translational (R) and the fragment internal degrees of freedom (r):

$$H_0 = T + h, \tag{17.2}$$

where h is the fragment Hamiltonian and T is the kinetic energy operator for relative translation of the fragments. For definiteness we choose

$$T(R) = -\frac{\hbar^2}{2m}\frac{\partial^2}{\partial R^2}, \tag{17.3}$$

where m is the reduced mass associated with the relative translation of fragments. Since H_0 is separable, the asymptotic eigenstates, satisfying the equation

$$H_0\psi_E = E\psi_E, \tag{17.4}$$

can be chosen as a product of an eigenstate of the internal Hamiltonian and a translational eigenstate,

$$\psi_E = \psi_{n,\pm k}(E) = \chi_n\sqrt{\frac{m}{\hbar k}}\,\psi_{\pm k}, \tag{17.5}$$

where the χ_n are orthonormal eigenstates of the internal Hamiltonian, h,

$$h\chi_n(r) = E_n\chi_n(r) \tag{17.6}$$

$$\langle\chi_{n'}|\chi_n\rangle = \delta_{n'n}, \tag{17.7}$$

and the $\psi_{\pm k}$ are the momentum-normalized eigenstates of $T(R)$,

$$T(R)\psi_{\pm k}(R) = \frac{\hbar^2 k^2}{2m}\psi_{\pm k}(R) \tag{17.8}$$

$$\psi_{\pm k}(R) = \frac{e^{\pm ikR}}{\sqrt{2\pi}} \tag{17.9}$$

$$\langle\psi_{\pm k'}|\psi_{\pm k}\rangle = \delta_{\pm\pm}\delta(k - k'), \tag{17.10}$$

where $\delta_{\pm\pm} = 1$ if the sign preceding k and k' is the same and is zero otherwise. The total energy is a sum of the relative translational energy and internal energy of the fragments:

$$E = E_n + \frac{\hbar^2 k^2}{2m}. \tag{17.11}$$

The asymptotic TISE, Eq. 17.4, can now be rewritten as

$$H_0\psi_{n,\pm k}(E) = E\psi_{n,\pm k}(E), \tag{17.12}$$

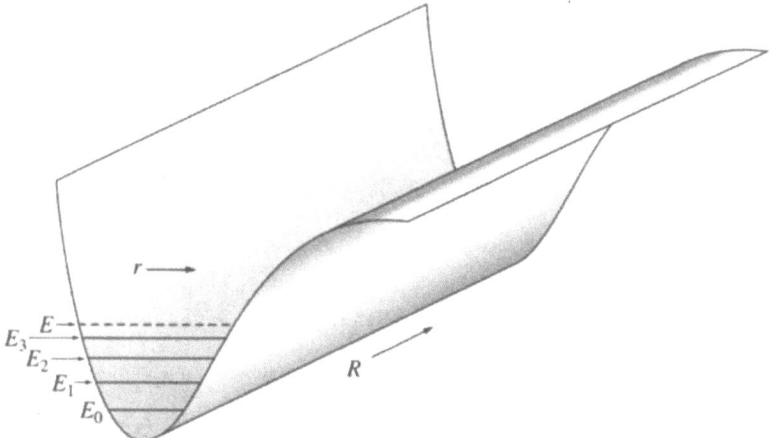

Figure 17.1 Schematic picture of the potential energy function for the separable Hamiltonian, H_0, corresponding to the asymptotic region of the potential. The potential is a sum of an internal fragment potential in r and free particle motion in the interfragment translational coordinate, R.

with $\psi_{n,\pm k}(E)$ given by Eq. 17.5 and E given by Eq. 17.11. The factor $\sqrt{\frac{m}{\hbar k}} = \frac{1}{\sqrt{v}}$ was introduced into Eq. 17.5 to make the full eigenstate, $\psi_{n,\pm k}(E)$, energy normalized:

$$\langle \psi_{n',\pm k'}(E')|\psi_{n,\pm k}(E)\rangle = \delta_{n'n}\delta_{\pm\pm}\delta(E - E'). \tag{17.13}$$

Note that the degeneracy of the states $\{\psi_{n,\pm k}(E)\}$ is equal to $2N$, where N is the number of internal states (i.e., eigenstates of h) accessible at energy E, and the factor of 2 comes from the two possible signs of k. The number N is referred to as the number of "open channels." For example, at the energy E shown in Figure 17.1 there are four accessible values of E_n. For each of these values of E_n, the deficit $E - E_n$ can be made up in translational energy. Associated with this translational energy there are two possible signs of the momentum k. Because of this degeneracy, any linear combination of these $2N$ eigenstates is also an eigenstate of H_0, but the states with well-defined labels $(n, \pm k)$ provide a convenient basis for understanding any more general wavefunction, as well as for counting the degeneracy.

17.3   The Eigenstates of a Scattering Hamiltonian

17.3.1   Correlation with Separable States

Now consider the eigenstates of the full Hamiltonian, H, the so-called scattering eigenstates. In the asymptotic region of the potential these eigenstates must be identical to the eigenstates of H_0, since $H = H_0$ in that region and therefore, locally, they must satisfy the same TISE. Although the scattering eigenstates can become very complicated in the interaction region, they can be labelled by the separable state $\psi_{n,k}(E)$ to which they correlate in the asymptotic region. For example, the scattering eigenstate labelled $\psi_n^+(E)$ has the same spatial dependence in the asymptotic region as $\psi_{n,k}(E)$. Similarly, the scattering eigenstate labelled $\psi_n^-(E)$ has the same spatial dependence in the asymptotic region as $\psi_{n,-k}(E)$ (see Figure 17.2). To avoid unnecessary subscripts, we leave off the subscript $\pm k$ from the

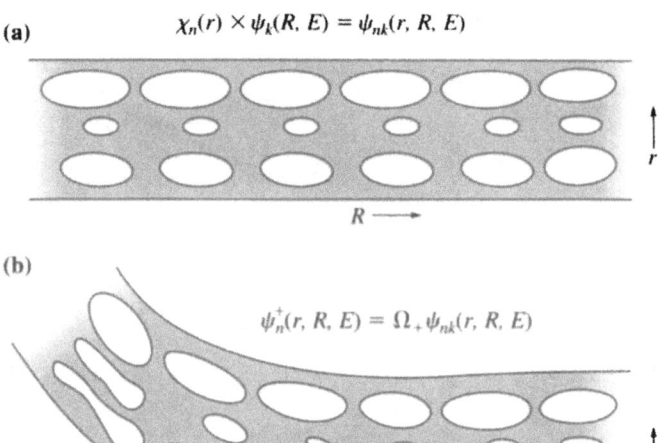

Figure 17.2 (a) Separable wavefunction $\psi_{n,k}(E) = \chi_n(r) \sqrt{\frac{m}{\hbar k}} \psi_k(R)$, eigenstate of H_0. (b) Scattering eigenstate $\psi_n^+(E)$, which correlates with the incoming separable state $\psi_{n,k}(E)$ in the asymptotic region. The outgoing portion of the wavefunction would be quite complicated and is not shown.

scattering eigenstates; the magnitude of k is determined by conservation of energy from Eq. 17.11 and the sign is given by the superscript $\pm$.

The scattering eigenstates solve the TISE

$$H\psi_n^{\pm}(E) = E\psi_n^{\pm}(E), \tag{17.14}$$

with the normalization conditions

$$\langle \psi_{n'}^+(E')|\psi_n^+(E)\rangle = \langle \psi_{n'}^-(E')|\psi_n^-(E)\rangle = \delta_{n'n}\delta(E - E'). \tag{17.15}$$

Note that the factor $\delta_{\pm\pm}$ in Eq. 17.13 is absent here. The overlap of an incoming and an outgoing scattering eigenstate is in general nonzero, and in fact such overlaps will be the main topic of Chapter 18.

It is worth adding a few more comments to sharpen our description of the scattering states. For simplicity, we will assume that we are dealing with only a single arrangement channel, and thus all amplitude ultimately exits through the same region where it came in.

1. Consider the scattering state $\psi_n^+(E)$ in Figure 17.2. In saying that $\psi_n^+(E)$ has the same asymptotic spatial dependence as $\psi_{n,k}(E)$ we are comparing just the positive k part. In addition to the separable, positive k part in this region, there is in general a complicated, nonseparable superposition of negative k states. This outgoing portion of the scattering state corresponds to reflection of the wavefunction from the interaction region. Similarly, the scattering state $\psi_n^-(E)$ has the asymptotic spatial dependence of $\psi_{n,-k}(E)$ in its negative k component, but in addition has a

complicated, nonseparable superposition of positive k states that can be considered as arising from reflection of the separable outgoing state backward in time.

2. Recall that the degeneracy of the eigenstates of H_0 was $2N$, N being the number of internal states available at energy E ("open channels") and the factor of 2 coming from the two possible signs of the momentum. For the scattering states, since the incoming and outgoing states are not independent, this degeneracy is reduced to N. One way of understanding this reduction in degeneracy is to note that each of the sets of states $\{\psi_n^+(E)\}$ and $\{\psi_n^-(E)\}$ forms a complete basis, the former decomposing an arbitrary wavefunction in terms of its incoming components and the latter decomposing an arbitrary wavefunction in terms of its outgoing components. In the presence of multiple arrangement channels, the degeneracy is the total number of open channels over all arrangements. This corresponds to the observation that the union of just incoming states from *all* arrangement channels constitutes a complete basis, and similarly the union of just outgoing states from all arrangement channels constitutes a complete basis. Thus, an arbitrary scattering wavefunction may be analyzed completely in terms of its past or in terms of its future, as discussed for one-dimensional systems in Section 7.2.

3. For photodissociation we will need only the $\psi_n^-(E)$ states. What is measured in photodissociation experiments is the "branching ratio" of population into each of the accessible internal states of products. Since products correspond to the future of the molecule, the appropriate theoretical description of these experiments is to decompose the initial parent wavefunction in terms of a basis of scattering states with well-defined internal states in the future; as discussed above, these are the $\psi_n^-(E)$ states. Although we need only the $\psi_n^-(E)$ states for this chapter, it is nevertheless convenient to introduce the incoming and outgoing scattering eigenstates together, since the formulas are nearly identical and both will be needed in the next chapter, on reactive scattering.

17.3.2 Spectral Method for Constructing Scattering Eigenstates

Now that we understand something of the nature and the degeneracy of the eigenstates in the continuum, we turn to the time-domain represention of these eigenstates. We have discussed earlier that it is possible to view eigenstates as a superposition of wavepackets, via the relationship

$$\psi_E \propto \int_{-\infty}^{\infty} e^{-iHt/\hbar}\phi e^{iEt/\hbar}\, dt, \tag{17.16}$$

and that one may in fact use this relationship to select specific eigenstates from the same wavepacket ϕ. In Chapter 6, the discussion was in the context of bound systems, and assumed no degeneracy; in that case, the frequency (i.e., the energy) of the Fourier transform of the wavepacket uniquely determines the eigenstate that is constructed. The question arises: in the continuum, where in more than one dimension there is virtually always a massive degeneracy, how does one go about using the spectral method to select a particular eigenstate from within this degenerate manifold? The answer is that the particular degenerate eigenstate that is produced is controlled by the choice of the initial wavepacket used to produce it. To construct the eigenstate $\psi_n^+(E)$, for example, one needs to choose the wavepacket

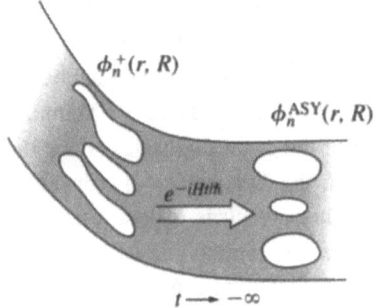

Figure 17.3 The figure shows two wavefunctions. On the left is the asymptotic separable state $\phi_n^{ASY}(R, r) = \chi_n(r)g^{ASY}(R)$, used to prepare the eigenstate $\psi_n^{\pm}(E)$. On the right is the nonseparable state $\phi_n^{+}(R, r)$ defined by Eq. 17.20. Because this state evolves backward in time into $\chi_n(r)g^{ASY}(R)$, and because Eq. 17.16 involves the integral over $(-\infty, \infty)$, the nonseparable wavefunction on the right will generate the same scattering eigenstate as the separable eigenstate on the left.

$\phi = \phi_n^{+}$; the exact properties of ϕ_n^{+} will be discussed below but the operational definition is any choice of ϕ that produces $\psi_n^{+}(E)$. The simplest choice of ϕ_n^{+} is

$$\phi_n^{+}(R, r) = \phi_n^{ASY}(R, r) = \chi_n(r)g^{ASY}(R), \tag{17.17}$$

which is a product of an *eigenstate* $\chi_n(r)$ of the internal Hamiltonian h in the internal degrees of freedom and a *wavepacket*, $g^{ASY}(R)$, located in the asymptotic region of the potential with incoming momentum in the translational coordinate (see Figure 17.3). However, more generally, since the expression in Eq. 17.16 includes propagation backward in time to $t \rightarrow -\infty$, one needs to choose $\phi = \phi_n^{+}$ only to *correlate* (i.e., evolve) into a separable state of the type described above as $t \rightarrow -\infty$ (again, see Figure 17.3). This can be expressed quantitatively by the condition

$$\lim_{t \rightarrow -\infty} e^{iH_0t/\hbar}e^{-iHt/\hbar}\phi_n^{+}(R, r) = \chi_n(r)g^{+}(R). \tag{17.18}$$

Note that the factor $e^{-iHt/\hbar}$ for negative times corresponds to propagation backward in time. Why don't we simply write $\lim_{t \rightarrow -\infty} e^{-iHt/\hbar}\phi_n^{+}(R, r)$? The extra factor of $e^{iH_0t/\hbar}$ is introduced to make the limit exist. For suitably chosen $\phi_n^{+}(R, r)$, as $t \rightarrow -\infty$ the wavepacket reaches the asymptotic region of reactants and becomes separable in R and r; but there is continued free propagation backward in time, and hence there is no well-defined limit to the form of the wavepacket. The factor $e^{iH_0t/\hbar}$ cancels this free propagation in the asymptotic region, and ensures the existence of a limiting form of the wavepacket. The sequence of these propagations is shown in Figure 17.4a. Note that we have written $g^{+}(R)$ in Eq. 17.18 instead of $g^{ASY}(R)$, since the translational wavepacket $g^{+}(R)$ is generally not located in the asymptotic region after the action of $e^{-iH_0t/\hbar}$.

The use of $e^{iH_0t/\hbar}$ to obtain a limit was discussed previously in Section 9.1 on the interaction picture (see Eq. 9.13 and Figure 9.1). In fact we recognize that

$$e^{iH_0t/\hbar}e^{-iHt/\hbar} = U_1(t), \tag{17.19}$$

the propagator in the interaction picture.

We now rearrange Eq. 17.18 as follows:

$$\phi_n^{+}(R, r) = \lim_{t \rightarrow -\infty} e^{iHt/\hbar}e^{-iH_0t/\hbar}\left\{\chi_n(r)g^{+}(R)\right\}. \tag{17.20}$$

The action of H_0 on the separable wavefunction is particularly simple: $\chi_n(R)$ is invariant while $g^{+}(R)$ evolves into $g^{ASY}(R)$. This asymptotic wavepacket is then propagated into

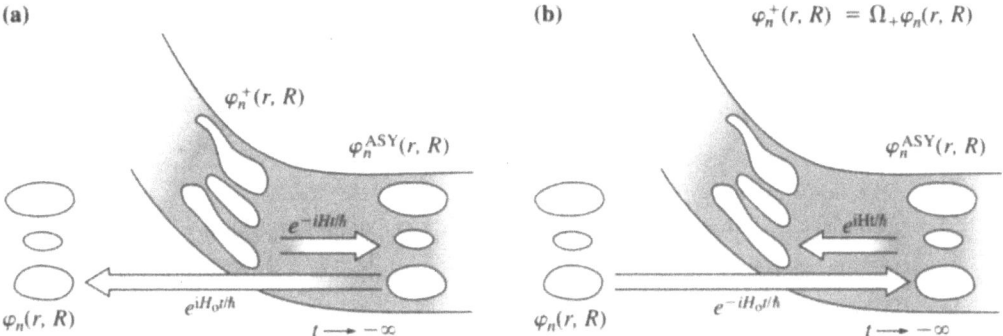

Figure 17.4 (a) This figure should be compared with Figure 17.3. Backward evolution in time of $\phi_n^+(r, R)$ under H gives $\phi_n^{ASY}(r, R) = \chi_n(R)g^{ASY}(R)$ located in the asymptotic region. Subsequent forward evolution under H_0 gives the separable state $\phi_n(r, R) = \chi_n(R)g^+(R)$ located in the interaction region. (b) This is the time reversal of (a). Backward evolution of $\phi_n(r, R)$ under H_0 gives the state $\phi_n^{ASY}(r, R)$; subsequent forward evolution under H gives the state $\phi_n^+(r, R)$. The Møller operator Ω_+ is the sequential operation of backward propagation under H_0 and forward propagation under H in the limit $t \to \infty$, and thus $\phi_n^+(r, R) = \Omega_+\phi_n(r, R)$. Alternatively, the Møller operator can be viewed as *mapping* $\phi_n(r, R)$ to $\phi_n^+(r, R)$.

the interaction region under the full H for the same amount of time. This action distorts the wavepacket; it does not leave it separable. The sequence of these propagations is shown in Figure 17.4b. Mathematically, the moving of the limit in Eq. 17.18 from one side of the equation to the other requires care. On physical grounds the limit exists, because additional time will simply increase propagation into the reactant channel, which is precisely cancelled in the forward propagation under the full H.

The previous paragraph provides a prescription for obtaining a wavepacket ϕ_n^+ that will generate the eigenstate $\psi_n^+(E)$. However, there is still an undetermined multiplicative factor, where the factor depends on the component of $\psi_n^+(E)$ contained in the wavepacket ϕ_n^+. To construct energy-normalized eigenfunctions we will therefore need at hand the coefficient of $\psi_n^+(E)$ contained in ϕ_n^+. The wavepacket ϕ_n^+ may be expanded in terms of the energy-normalized eigenstates of the full Hamiltonian:

$$\phi_n^+ = \int_0^\infty \eta(E)\psi_n^+(E)\, dE \tag{17.21}$$

with coefficients $\eta(E)$ given by

$$\eta(E) = \langle \psi_n^+(E)|\phi_n^+\rangle \tag{17.22}$$

$$= \langle \psi_{n,k}(E)|\phi_n\rangle \tag{17.23}$$

$$= \left\langle \chi_n\sqrt{\frac{m}{\hbar k}}\,\psi_k \,\middle|\, \chi_n g^+ \right\rangle \tag{17.24}$$

$$= \sqrt{\frac{m}{2\pi\hbar k}} \int e^{-ikR}g^+(R)\, dR. \tag{17.25}$$

The transition from Eq. 17.22 to Eq. 17.23, which shows that the overlap of the + states is identical to that of the corresponding states without the +, is a remarkable result, which will be proved in the next section. The final result, Eq. 17.25, indicates that $\eta(E)$ is simply related to the k component of $g^+(R)$ at the value $k = \frac{\sqrt{2m(E-E_n)}}{\hbar}$, and hence is computed trivially from a k Fourier transform of the initial wavepacket. Substituting Eq. 17.21 into Eq. 17.16 we obtain an expression for the energy-normalized eigenstate $\psi_n^+(E)$:

$$\psi_n^+(E) = \frac{1}{2\pi\hbar\eta(E)} \int_{-\infty}^{\infty} e^{-iHt/\hbar} \phi_n^+ e^{iEt/\hbar}\, dt. \tag{17.26}$$

Similarly, to obtain the eigenstate $\psi_n^-(E)$, one should choose $\phi = \phi_n^-$ to be in the asymptotic region of the potential with quantum numbers n and momentum such that the wavepacket is outgoing (this will actually take the packet into the interaction region when time is reversed). Analogous to Eq. 17.20, we have

$$\lim_{t\to+\infty} e^{iHt/\hbar} e^{-iH_0 t/\hbar} \left\{ \chi_n(r)g^-(R) \right\} = \phi^-(R, r), \tag{17.27}$$

where $g^-(R)$ is a wavepacket located in the asymptotic region of the potential with outgoing momentum in the translational coordinate. The energy-normalized eigenstate is

$$\psi_n^-(E) = \frac{1}{2\pi\hbar\mu(E)} \int_{-\infty}^{\infty} e^{-iHt/\hbar} \phi_n^- e^{iEt/\hbar} dt, \tag{17.28}$$

where $\mu(E)$ is the coefficient of the energy-normalized eigenstate $\psi_n^-(E)$ contained in ϕ_n^-:

$$\phi_n^- = \int_0^{\infty} \mu(E)\psi_n^-(E)\, dE. \tag{17.29}$$

The quantity $\mu(E)$ is given by

$$\mu(E) = \langle \psi_n^-(E)|\phi_n^-\rangle \tag{17.30}$$

$$= \langle \psi_n(E)|\phi_n\rangle \tag{17.31}$$

$$= \left\langle \chi_n\sqrt{\frac{m}{\hbar k}}\psi_{-k} \,\middle|\, \chi_n g^- \right\rangle \tag{17.32}$$

$$= \sqrt{\frac{m}{2\pi\hbar k}} \int e^{ikR} g^-(R) dR, \tag{17.33}$$

where Eq. 17.31 will be proved with the aid of the Møller operators in the next section.

Equations 17.26 and 17.28 are the central results of this section. They give a precise relation for the energy-normalized scattering eigenstates in terms of the Fourier transform of a wavepacket with suitable boundary conditions.

17.4   Møller Operators

In this section we will define the Møller operators, Ω_+ and Ω_-, and describe some of their most important properties. The Møller operators are central objects in the formal theory of scattering, and their use enormously simplifies certain derivations, such as Eqs. 17.23 and 17.31.

In Section 17.3.2 (Eqs. 17.20 and 17.27) we had the formulas

$$\lim_{t\to\mp\infty} e^{iHt/\hbar} e^{-iH_0 t/\hbar} \left\{ \chi_n(r)g^\pm(R) \right\} = \phi^\pm(R, r). \tag{17.34}$$

Defining the Møller operators as

$$\Omega_{\pm} = \lim_{t \to \mp\infty} e^{iHt/\hbar} e^{-iH_0 t/\hbar}, \tag{17.35}$$

we may rewrite these equations as

$$\Omega_{\pm} \left\{ \chi_n(r) g^{\pm}(R) \right\} = \phi_n^{\pm}(R, r). \tag{17.36}$$

The effect of the Møller operators on an initial state ϕ can be viewed in two different ways. The first perspective is sequential: the Møller operators propagate the initial state backward (forward) to time $t = -\infty$ ($t = +\infty$) under H_0 and then forward (backward) to time $t = 0$ using the full H (see Figure 17.4b). In practice, some large but finite value of τ is chosen for the backward propagation. As the propagation time τ is increased the states $\phi^{\pm}$ reach an asymptotic limit. Note that the backward propagation under H_0 can often be done analytically. The second perspective on the Møller operators is nonsequential: the combined action of the two parts of the Møller operator provides a *mapping* of the set of separable states $\{\phi_n = \chi_n g^{\pm}\}$ to the set of "Møller states," $\{\phi_n^{\pm}\}$ (cf. Eq. 17.36). Because the evolution under H and under H_0 largely cancels out, the mapping, although it distorts the shape of the wavefunction, often produces minimal displacement or spreading (compare ϕ_n and ϕ_n^+ in Figure 17.4). Many important properties of the Møller operators are statements about the properties of this map. There is a numerical counterpart to this nonsequential perspective on the Møller operators. The Møller operator is equivalent to the infinite time limit of the adjoint of the evolution operator in the interaction picture:

$$\Omega_{\pm} = \lim_{t \to \pm\infty} U_I^{\dagger}(t; 0) = \lim_{t \to \pm\infty} U_I(0; t). \tag{17.37}$$

The numerical counterpart of the nonsequential perspective is to propagate ϕ_n in the interaction picture. As the propagation time t increases, the interaction picture wavepacket evolves from ϕ_n to $\phi_n^{\pm}$ monotonically; the change at each time step becomes smaller and smaller as the interaction wavepacket approaches its asymptotic limit. Each step in the interaction picture propagation can be viewed as a mapping, producing only modest changes in the interaction wavepacket, with no need to consider the intermediate Schrödinger wavefunction. The concatenation of these mappings produces the Møller operator as $t \to \infty$. The time τ required for the sequence of interaction picture mappings to converge to the Møller operator is just the same as the time τ required for backward propagation to reach the asymptotic region in the sequential approach.

Because Møller operators are a concatenation of evolution operators, it is natural that they preserve the norm of the wavefunction on which they operate. Operators having this property are called "isometric," and satisfy the property

$$\Omega_{\pm}^{\dagger} \Omega_{\pm} = \mathbf{1}. \tag{17.38}$$

However, in general the Møller operators are not unitary:

$$\Omega_{\pm}^{\dagger} \neq \Omega_{\pm}^{-1}. \tag{17.39}$$

Unitarity is satisfied only if there are no bound states supported by the Hamiltonian. For an operator to be unitary it must be isometric, but it must also satisfy the property that the *range* of functions produced by the operator spans the same Hilbert space as the *domain* of allowed input functions. The domain of the Møller operators is the full Hilbert space—any initial state is an acceptable function on which the Møller operator can act. However, the range is the space of scattering functions—that is, the output state must lie in the space orthogonal to that of the bound states. For an extensive discussion, see Taylor (1972).

▶ Exercise 17.1

  a. Sketch the action of (i) $\Omega_\pm$, (ii) $\Omega_\pm^\dagger$ and (iii) $\Omega_\pm^{-1}$ on a bound state by considering the sequential action of propagation under H_0 and H, or H and H_0.

  b. Show that $\Omega_\pm^\dagger \neq \Omega_\pm^{-1}$ when both operators are applied to a bound state.

  c. Try to find an initial state such that the output under the action of Ω_+ (or Ω_-) is a bound state.

The eigenstates of the full H satisfy the equation

$$\Omega_\pm \psi_{n,\pm k}(E) = \psi_n^\pm(E), \tag{17.40}$$

a relationship analogous to Eq. 17.36 derived above for the wavepacket. This is a remarkable result that allows one to manipulate the nonsquare normalizable scattering eigenstates, $\psi_n^\pm(E)$, in much the same way that one manipulates the square integrable wavefunctions, $\phi_n^\pm$. For example, in Section 17.3.2 we wrote without proof that $\langle \psi_n^\pm(E)|\phi_n^\pm \rangle = \langle \psi_{n,\pm k}(E)|\phi_n \rangle$ (Eqs. 17.23, 17.31). Substituting Eqs. 17.36 and 17.40 into Eq. 17.22, and using Eq. 17.38, we immediately obtain Eq. 17.23. Similarly, substituting into Eq. 17.30 we obtain Eq. 17.31.

▶ Exercise 17.2   Derive Eq. 17.40.

Solution   Consider the expansion of the translational wavepacket in terms of the asymptotic translational eigenstates:

$$g^\pm = \int_0^\infty \eta'(E)\sqrt{\frac{m}{\hbar k}}\,\psi_{\pm k}(E)\,dE. \tag{17.41}$$

Multiplying both sides by $\chi_n(r)$, applying $\Omega_\pm$ and using Eq. 17.36 we obtain

$$\phi^\pm = \int_0^\infty \eta'(E)\Omega_\pm \psi_{n,\pm k}(E)\,dE, \tag{17.42}$$

where we have used Eq. 17.5. Compare Eq. 17.42 with Eqs. 17.21 and 17.29. Since the two integrals on the RHS hold for arbitrary $\eta(E)$ and $\eta'(E)$, setting $\eta'(E) = \eta(E)$ we must have $\psi_n^\pm(E) = \Omega_\pm \psi_n(E)$ for all E.

In the next exercise we establish certain further properties of the Møller operators.

▶ Exercise 17.3

  a. Show that

$$e^{iH\tau/\hbar}\Omega_\pm = \Omega_\pm e^{iH_0\tau/\hbar}. \tag{17.43}$$

b. Take the derivative with respect to τ of both sides of Eq. 17.43 and set $\tau = 0$ to show that

$$H\Omega_\pm = \Omega_\pm H_0. \tag{17.44}$$

The two forms of Eq. 17.44, with the $+$ and $-$ sign, respectively, are called the intertwining relations.

c. Use the result of part b to show that

$$H_0 = \Omega_\pm^\dagger H \Omega_\pm. \tag{17.45}$$

17.5   Wavepacket Formulation of Photodissociation

17.5.1   Photodissociation Partial Cross Sections

The photodissociation partial cross section, $\sigma_n(E)$, is given by the overlap of the Franck–Condon wavefunction, ϕ_0, with the energy-normalized outgoing scattering state $\psi_n^-(E)$ (see, for example, Kulander (1978), Shapiro (2003)):

$$\sigma_n(E) = \frac{4\pi^2\omega}{3\hbar c}|\langle\psi_n^-(E)|\phi_0\rangle|^2 \equiv \frac{4\pi^2\omega}{3\hbar c}\bar\sigma_n(E), \tag{17.46}$$

where $\phi_0 = \mu\psi_0$, and μ is the transition dipole moment connecting the ground and dissociative electronic states. The expression for $\bar\sigma$ can be rewritten in the following way (Kulander, 1978):

$$\bar\sigma_n(E) = |\langle\psi_n^-(E)|\phi_0\rangle|^2 \tag{17.47}$$

$$= \left|\int_{-\infty}^{\infty}\left(\lim_{t\to\infty} e^{iHt/\hbar}e^{-iH_0t/\hbar}\chi_n(r)\sqrt{\frac{m}{2\pi\hbar k}}e^{ikR}\right)^* \phi_0(R,r)\,dR\,dr\right|^2 \tag{17.48}$$

$$= \lim_{t\to\infty}\left|\int_{-\infty}^{\infty}(e^{-iH_0t/\hbar}\chi_n(r)e^{ikR})^*e^{-iHt/\hbar}\phi_0(R,r)\,dR\,dr\right|^2\frac{m}{2\pi\hbar k} \tag{17.49}$$

$$= \lim_{t\to\infty}\left|\int_{-\infty}^{\infty}\chi_n(r)e^{-ikR}\phi_0(R,r,t)\,dR\,dr\right|^2\frac{m}{2\pi\hbar k}, \tag{17.50}$$

where $\phi_0(R,r,t) = e^{-iHt/\hbar}\phi_0(R,r)$. In Eq. 17.50 we have used the fact that $e^{-iH_0t/\hbar}$ operating on an eigenstate of H_0 gives just a simple phase factor, which cancels out when the absolute value of the integral is taken. The mathematical manipulations in Eqs. 17.48–17.50 correspond to a profound shift in perspective on the photodissociation process and how to calculate it. In the initial expression, Eq. 17.46, the initial state ϕ_0 is stationary; all the "dynamics" is in the functional form of the scattering eigenstates, $\psi_n^-(E)$. These scattering eigenstates have asymptotic boundary conditions, meaning they must be propagated *inward* from infinity to the Franck–Condon region (corresponding to backward in time), where they are overlapped with the initial state. In contrast, in the final formula, Eq. 17.50, all the dynamics is in the *initial* state, corresponding to the intuitive idea that in photodissociation there is a wavepacket born in the Franck–Condon region of an unstable potential that proceeds to evolve *outward* to the asymptotic region on that unstable potential. Once in the asymptotic region, the overlap is calculated between the nonseparable asymptotic wavepacket and the separable, asymptotic energy eigenstates.

Equation 17.50 is convenient for numerical application. One needs only know the *separable* asymptotic states, not the scattering states, to evaluate the partial cross section. By moving the evolution operator from the bra to the ket, only a single wavepacket—the one uniquely defined by the vertical transition from the initial state—needs to be propagated to obtain all the partial cross sections. Heller has compared the economy of using Eq. 17.50 versus Eq. 17.46 to that of calculating a cannonball trajectory from the cannon to its destination versus calculating cannonball trajectories that start at all destinations and finding those that enter the cannon!

17.5.2   Total Photodissociation Cross Sections

The total photodissociation cross section may be expressed as the sum of all the partials:

$$\sigma(E) = \sum_n \sigma_n(E) = \frac{4\pi^2\omega}{3\hbar c} \sum_n \bar{\sigma}_n(E) \tag{17.51}$$

$$= \frac{4\pi^2\omega}{3\hbar c} \sum_n \int_0^\infty \langle\phi_0|\psi_n^-(E')\rangle\langle\psi_n^-(E')|\phi_0\rangle\delta(E - E')\,dE' \tag{17.52}$$

$$= \frac{2\pi\omega}{3\hbar c} \sum_n \int_0^\infty \int_{-\infty}^\infty \langle\phi_0|\psi_n^-(E')\rangle\langle\psi_n^-(E')|e^{-iHt/\hbar}|\phi_0\rangle e^{iEt/\hbar}\,dt\,dE', \tag{17.53}$$

where Eq. 17.52 follows from Eq. 17.46 together with the definition of the δ function. Because the scattering states, $\{\psi_n^-(E')\}$, form a complete set of outgoing states, they represent a complete basis for the outgoing wavepacket. We can therefore remove the projection onto these states and write

$$\sigma(E) = \frac{2\pi\omega}{3\hbar c} \int_{-\infty}^\infty \langle\phi_0|e^{-iHt/\hbar}|\phi_0\rangle e^{iEt/\hbar}\,dt. \tag{17.54}$$

Notice that the final expression, Eq. 17.54, expresses the total photodissociation cross section as the Fourier transform of the autocorrelation of the initial wavepacket, ϕ_0, propagated on the excited-state potential energy surface. We have seen this form before; in fact Eq. 17.54 is virtually identical to the expression for the electronic absorption/emission spectrum discussed in great detail in Chapter 14, Eq. 14.21. All the qualitative pictures of excited-state wavepacket motion discussed in Chapter 14, and the relationship between features in the time and frequency domain (see also Chapter 6) hold equally well for photodissociation as for absorption and emission. Indeed, several of the examples in Chapter 14 involved dissociative processes.

The equality of the photoabsorption spectrum and the photodissociation spectrum is true under two conditions:

1. That there are no radiative or radiationless processes that lead to channels other than the dissociative one;

2. That the scattering states must form a complete set of outgoing states. The latter is true only if the excited electronic state has no true bound states; otherwise, the absorption spectrum will include a bound state component in addition to the dissociative component, the latter being equal to the sum of the partials.

17.6   Applications

We now discuss several representative examples, which were historically among the first applications of the wavepacket correlation function formulas given above. The first is $CH_3I \rightarrow CH_3 + I$, which has been a benchmark for quantum mechanical calculations of photodissociation. In the parent molecule, before dissociation, the carbon atom is at the center of what is essentially a tetrahedral geometry; the C atom is not coplanar with the three H atoms. However, after dissociation, the free CH_3 radical is planar (see Figure 17.5 (left)). As a result, the CH_3 radical is expected to be vibrationally excited in the out-of-plane, or "umbrella," bending mode. Shapiro and Bersohn (Shapiro, 1980) developed a reduced two-dimensional model of the dissociation in which only the CH_3—I stretch (the dissociation coordinate, Q_3) and the CH_3 umbrella bending mode (Q_2) are included. The potential energy surface of Shapiro and Bersohn is shown in Figure 17.5 (right). A schematic trajectory is shown superimposed, exiting along the dissociation coordinate but then sensing the change in geometry in the umbrella vibration and beginning to move in the perpendicular coordinate. Figure 17.6a shows the total absorption spectrum calculated using Eq. 17.54, and the sum of the partial cross sections to individual vibrational states of the CH_3 radical (in principle the two should be identical). The individual partial cross sections, calculated using Eq. 17.50, are shown in Figure 17.6b.

Our next example involves chemical branching, in which there is more than one possible set of fragments in dissociation. The example we choose is $HOD \rightarrow H + OD$, $D + OH$. From the point of view of wavepacket dynamics we can make an interesting and general observation about systems that exhibit chemical branching. Each different set of chemical products corresponds to a distinct and orthogonal region of configuration space. These orthogonal regions are called arrangement channels; we have already met them in Chapter 16 on control and they will be discussed more formally in the next chapter on chemical reactions. Since we know that wavepacket amplitude ends up in more than one of these disjoint regions, there must be a wavepacket bifurcation process along the way.

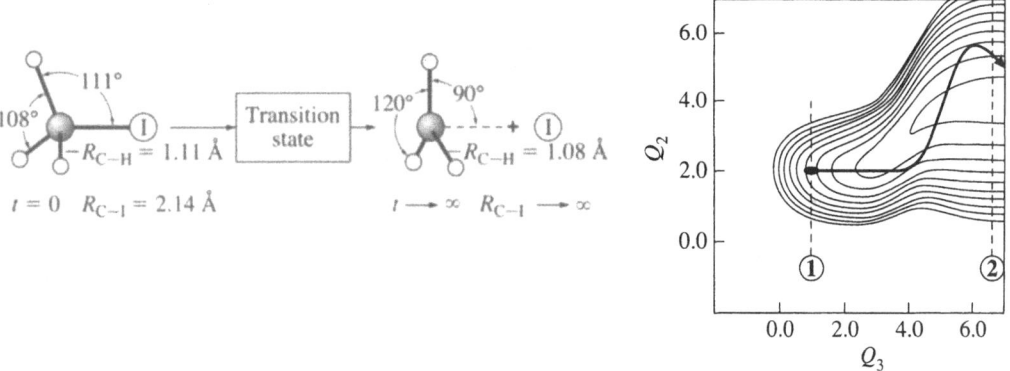

Figure 17.5 Left: Structural transformation in CH_3I photodissociation. Right: Schematic illustration of the CH_3I excited electronic state potential energy surface in the dissociation coordinate (Q_3) and the umbrella bending mode (Q_2). Adapted from D. Imre et al. (1984).

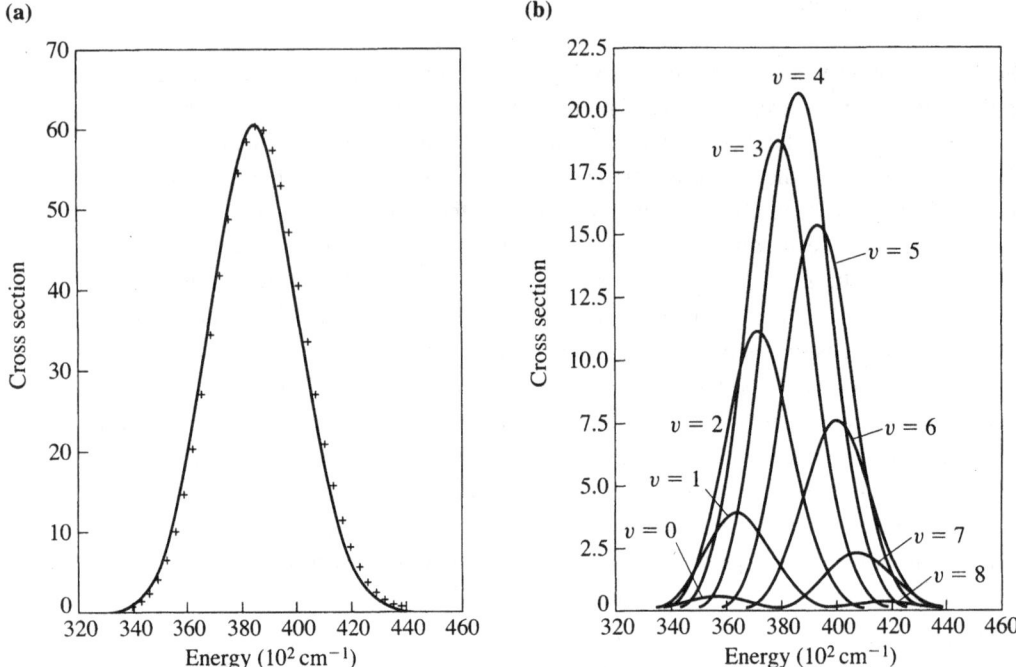

Figure 17.6 (a) Total photodissociation cross section of CH_3I as a function of frequency. The solid line is the sum over partial cross sections and the +'s are the Fourier transform of the wavepacket correlation function. (b) Partial photodissociation cross sections for CH_3I as a function of frequency, calculated by exact wavepacket propagation, for $v = 0, 1, \ldots, 8$. Adapted from Lee (1982).

Figure 17.7 shows a two-dimensional potential energy surface for the HOD electronically excited state, in the two bond-stretching coordinates. Superimposed is the ground vibrational state of the ground electronic state ($v'' = 0$). The total photodissociation cross section, shown in Figure 17.7b, was calculated according to Eq. 17.54. We may define a chemical partial cross section for each *arrangement channel* (H + OD vs. D + OH), in which we sum over all vibrational states consistent with a particular chemical identity of fragments. The calculation of these chemical partial cross sections can be performed without calculating the individual vibrational state partial cross sections in the following way. One propagates the initial wavepacket until it has completed bifurcating into components with distinct chemical outcomes; the Fourier transform of the autocorrelation of these individual wavepacket components gives the chemical partial cross sections of interest. Figure 17.7 also shows the chemical partial cross sections for H + OD and D + OH, which are seen to follow the total absorption spectrum fairly well. The branching ratio of H + OD/D + OH as a function of frequency is also shown. Note that this branching ratio goes to infinity at the low-frequency end of the absorption spectrum. This is because of a zero-point energy effect: the zero-point energy of OH is higher than OD, and thus there is a small range of frequencies that are above the threshold for OD production but below the threshold for OH production.

It is tantalizing to inquire if one can control the branching ratio of the fragmentation pathways by varying the initial vibrational state in the ground electronic state. The

(a) **(b)**

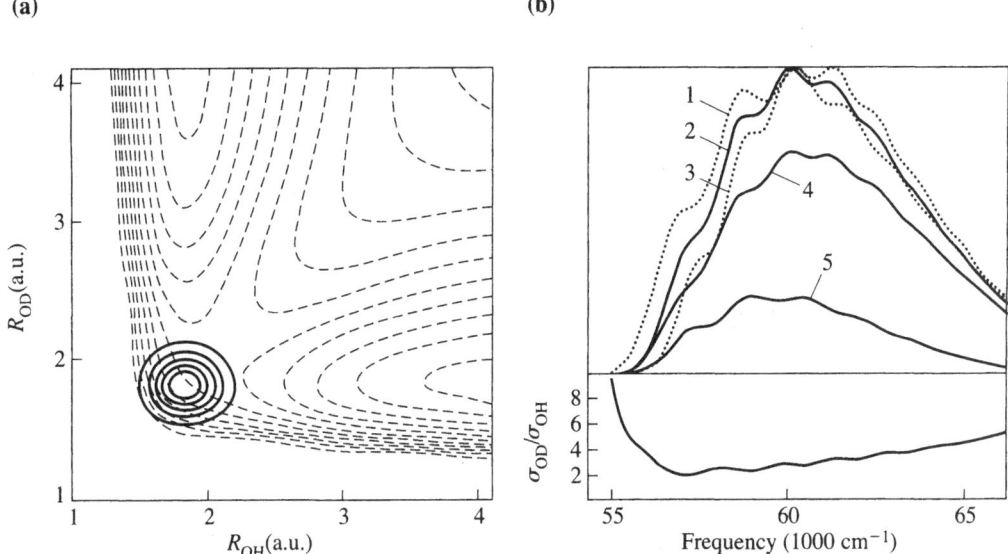

Figure 17.7 (a) Contour plot of the first excited PES (dashed lines) of HOD for H—O—D bond angle 104.52°. The solid contours represent the initial wavepacket prepared on this surface starting from $v'' = 0$. (b) Absorption spectrum (dotted lines) for (1) H_2O and (3) D_2O starting from $v'' = 0$, and partial absorption spectra (solid lines) for (4) H + OD and (5) D + OH for transition starting from $v'' = 0$ in HOD. The solid line (2) is the absorption spectrum of HOD, which is the sum of the two partial spectra 4 and 5. The bottom part of this figure shows the branching ratio (H + OD)/(D + OH) as a function of frequency. Reprinted from Zhang et al. (1989).

O—H and O—D vibrations in HOD are known to be good "local modes"; vibrational energy deposited in one of these modes essentially never gets to the other mode. This suggests that exciting one or several quanta of OH (OD) vibration before the electronic excitation occurs will prepare an initial wavepacket with energy already in one of the modes and will therefore predispose the system to fragment into D + OH (H + OD). This is a form of active control of chemical reactions, although it is somewhat less sophisticated than the optimal control framework discussed in Chapter 16, and is generally called "vibrationally mediated photodissociation."

Figure 17.8 shows contour plots of some of the vibrationally excited eigenstates of the ground electronic state, superimposed on the excited electronic state surface. The local mode character of the ground state is apparent in the orientation of the vibrational wavefunctions along either the OH or OD coordinates. The shaded regions of the wavefunctions correspond to the low-energy excitation; it is clear that these portions of the wavefunction are predisposed to exit selectively into one particular arrangement channel.

Figure 17.9 shows that indeed, by a combination of initial state preparation and electronic excitation at the low-frequency end of the absorption spectrum, almost complete control over the branching ratio is possible.

Our final example is the photodissociation of ICN → I(I*) + CN(j). This example introduces two new features. The first is the electronic branching ratio between I and I*, which is associated with an electronically nonadiabatic interaction during the dissociation.

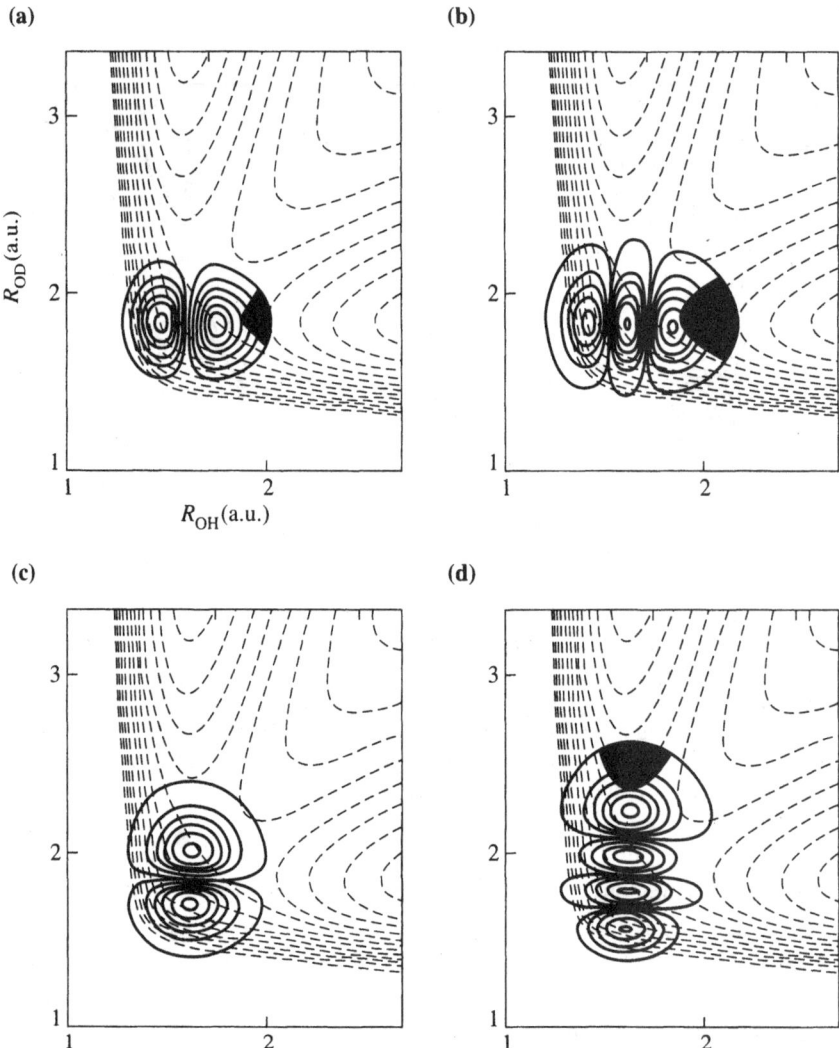

Figure 17.8 Initial wavepackets (solid contours) prepared on the excited-state PES (dashed lines) of HOD for transitions starting from (a) $v''_{OH} = 1$, (b) $v''_{OH} = 2$, (c) $v''_{OD} = 1$, (d) $v''_{OD} = 3$. The shaded regions are used to indicate the low-energy portions of each wavepacket. Reprinted from Zhang et al. (1989).

The second is the calculation of the partial cross section to the different *rotational* states of the CN fragment. The model consists of two potential energy surfaces, one correlating with I + CN, the other with I* + CN; while both are dissociative the former is linear (i.e., stable in the bending coordinate) and the latter is bent (i.e., unstable in the bending coordinate). Experimentally, no excitation of the CN vibration is observed, and this coordinate is neglected in the model.

The initial wavepacket is prepared entirely on the linear state. Since the ground state is linear, the amplitude that exits on this surface is expected to remain close to linear

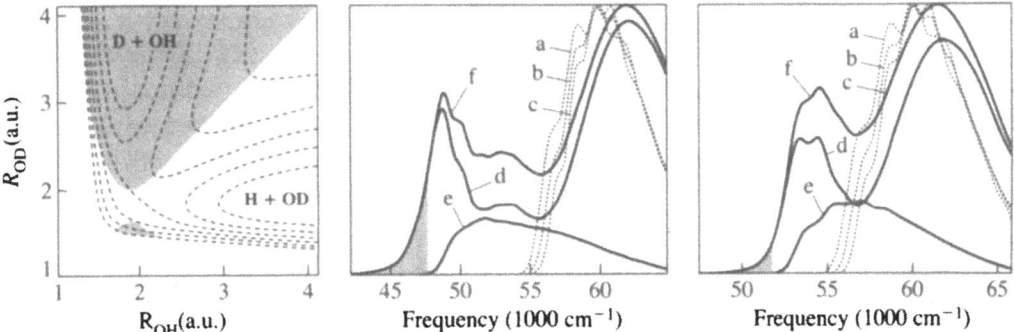

Figure 17.9 Left: Diagram showing the regions of the excited-state potential surface of HOD that lead to D + OH production (shaded) and H + OD production (white). Middle: Absorption spectrum for (a) H_2O, (b) HOD and (c) D_2O, starting from $v'' = 0$. Chemical partial absorption spectra (solid lines) for (d) H + OD and (e) D + OH for transition starting from $v''_{OH} = 2$ in HOD. The solid line f is the sum of the two partial spectra d and e. The shaded area marks the bond-selective region of frequency. Right: Absorption spectrum for (a) H_2O, (b) HOD and (c) D_2O, starting from $v'' = 0$. Chemical partial absorption spectra (solid lines) for (d) H + OD and (e) D + OH for transition starting from $v''_{OD} = 3$ in HOD. The solid line f is the sum of the two partial spectra d and e. The shaded area marks the bond-selective region of frequency. Reprinted from Zhang et al. (1989).

geometry at all times and therefore to be rotationally cold. However, amplitude that crosses nonadiabatically to the bent surface is expected to exit rotationally hot. Thus, the I fragment is expected to correlate with a cold rotational distribution, I* with hot. Figure 17.10a, b shows the amplitude on both the linear and bent surfaces, after the wavepacket on both surfaces has reached the asymptotic region. The wavepacket amplitude is shown not in the coordinate variables (R, θ), but in the conjugate variable (k, j). As in the usual case with representation in k-space, the wavefunction in these variables has the same information content as in (R, θ); however, there are two advantages to this representation. The first is that the amplitude in these coordinates (more precisely, its absolute value) becomes time independent when the wavepacket reaches the asymptotic region, since there is no longer any acceleration in translation or torque in the bend (mathematically, both $\hat{p}$ and $\hat{j}$ commute with the asymptotic Hamiltonian). The second advantage is that the partial cross section, $\sigma_j(E)$, can be directly related to the amplitude in this representation: since $E = \hbar^2 k^2/2\mu + \hbar^2 j(j + 1)/2\mu r^2$, the choice of a particular j determines k (the equation is approximately an ellipse). Thus, taking the intersection of different ellipsoidal curves with the wavefunctions in Figure 17.10a, b and squaring the amplitude gives the spectrum $\sigma_j(E)$, after taking into account energy normalization.

Note the low rotational excitation of the wavepacket on the linear surface (Figure 17.10a) and the high rotational excitation on the bent surface (Figure 17.10b). The partial cross sections are shown in Figure 17.10c, d, e for three different excitation energies, corresponding to different values of E. Note that although there are some differences in the spectra at different energies, the basic pattern of a rotationally cold distribution in the linear state and a rotationally hot distribution in the bent state persists.

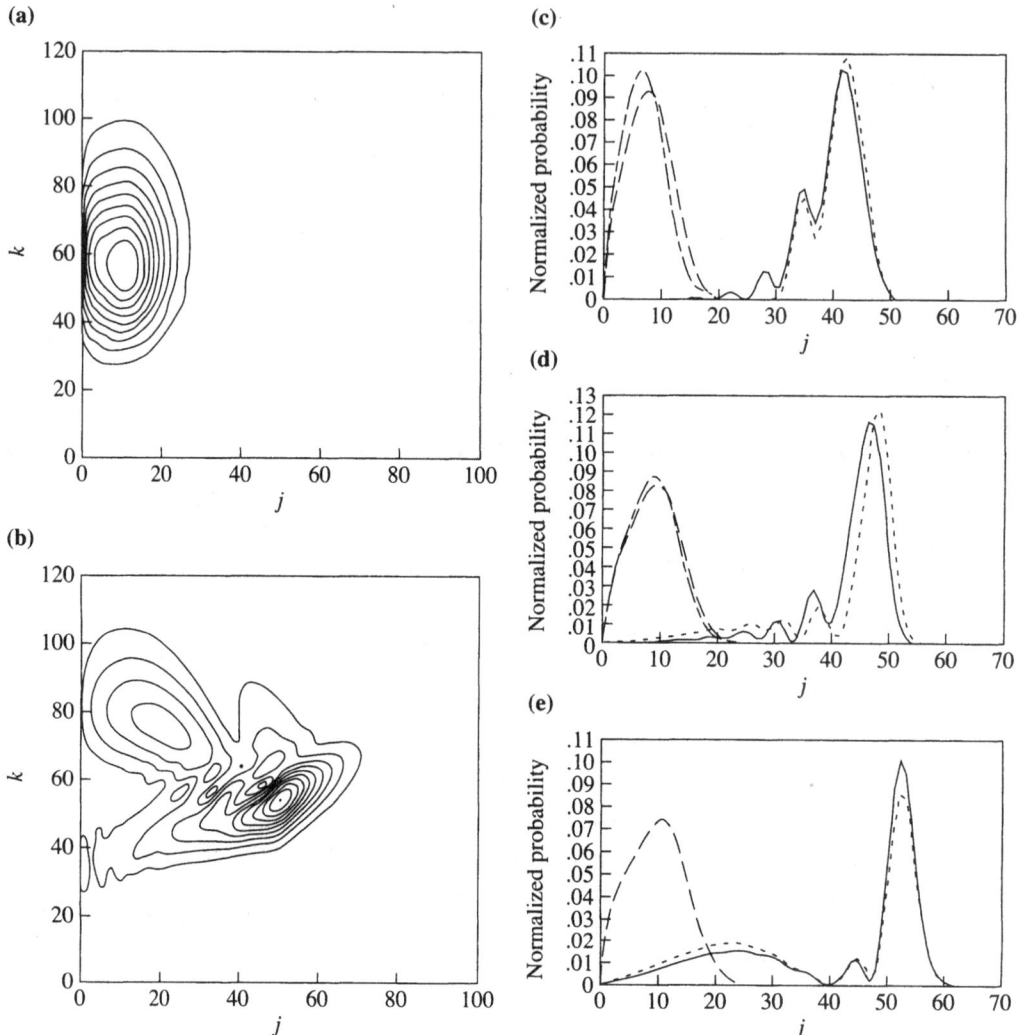

Figure 17.10 Amplitude of the wavepacket for ICN photodissociation at $t = 50.8$ fs in the conjugate momentum variables $k = k_R$ and j, on the (a) linear and (b) bent surfaces, respectively. (c, d, e) Rotational distribution calculated from the wavepackets in (a) and (b) when it reaches its asymptotic form (≈ 100 fs). The rotational distributions are shown at (c) 280 nm, (d) 266 nm, and (e) 248 nm. The long dashed and solid lines are the results of the time-dependent calculation on the linear and bent surfaces, respectively, while the dot-dashed and short dashed lines are the rotational distributions for the linear and bent surfaces from a time-independent calculation. Reprinted from Qian (1992).

References

General

1. R. Schinke, *Photodissociation Dynamics*, (Cambridge University Press, Cambridge, 1993).
2. L. Butler, Chemical reaction dynamics beyond the Born–Oppenheimer approximation, Ann. Rev. Phys. Chem. 49, 125 (1998).
3. J. P. Simons, Stereochemistry and control in molecular reaction dynamics, Faraday Disc. 113, 1 (1999).
4. F. F. Crim, Vibrationally mediated photodissociated: Exploring excited-state surfaces and controlling decomposition pathways, Ann. Rev. Phys. Chem. 44, 397 (1993).

Dissociation to One Arrangement Channel

5. K. C. Kulander and E. J. Heller, Time dependent formulation of polyatomic photofragmentation: Application to H_3^+, J. Chem. Phys. 69, 2439 (1978).
6. R. T Pack, Simple theory of diffuse vibrational structure in continuous UV spectra of polyatomic molecules. I. Collinear photodissociation of symmetric triatomics, J. Chem. Phys. 65, 4765 (1976).
7. E. J. Heller, Photofragmentation of symmetric triatomic molecules: Time dependent picture, J. Chem. Phys. 68, 3891 (1978).
8. S.-Y. Lee and E. J. Heller, Exact time-dependent wavepacket propagation: Application to the photodissociation of methyl iodide, J. Chem. Phys. 76, 3035 (1982).
9. N. E. Henriksen and E. J. Heller, Quantum dynamics for vibrational and rotational degrees of freedom using Gaussian wavepackets: Application to the three-dimensional photodissociation dynamics of ICN, J. Chem. Phys. 91, 4700 (1989).
10. D. Chasman, D. J. Tannor and D. G. Imre, Photoabsorption and photoemission of ozone in the Hartley band, J. Chem. Phys. 89, 6667 (1988).
11. B. R. Johnson and J. L. Kinsey, Phys. Rev. Lett. 62, 1607 (1989); J. Chem. Phys. 91, 7638 (1989).
12. F. LeQuere and C. Leforestier, Hyperspherical formulation of the photodissociation of ozone, J. Chem. Phys. 94, 1118 (1991).
13. G. G. Balint-Kurti, R. N. Dixon, C. C. Marston and A. J. Mulholland, The calculation of product quantum state distributions and partial cross-sections in time-dependent molecular collision and photodissociation theory, Comp. Phys. Comm. 63, 126 (1991).
14. S. K. Gray and C. E. Wozny, Fragmentation mechanisms from three-dimensional wavepacket studies: Vibrational predissociation of $NeCl_2$, $HeCl_2$, NeICl, and HeICl, J. Chem. Phys. 94, 2817 (1991).
15. D. Imre, J. Kinsey, A. Sinha and J. Krenos, Chemical dynamics studied by emission spectroscopy of dissociating molecules, J. Phys. Chem. 88, 3956 (1984).

Chemical Branching Ratios in Photodissociation

16. D. Krajnovich, L. Butler and Y. T. Lee, UV photodissociation of C_2F_5Br, C_2F_5I, and 1,2-C_2F_4BrI, J. Chem. Phys. 81, 3031 (1984).
17. S. Das and D. J. Tannor, Chemical branching in two-chromophore systems: Application to the photodissociation of C_2F_4IBr, J. Chem. Phys. 91, 2324 (1989).
18. K. Weide, S. Hennig and R. Schinke, Photodissociation of vibrationally excited water in the first absorption band, J. Chem. Phys. 91, 7630 (1989); R. L. Vander Wal, J. L. Scott, F. F. Crim, K. Weide and R. Schinke, J. Chem. Phys. 94, 3548 (1991).

19. J. Zhang, D. G. Imre and J. H. Frederick, HOD spectroscopy and photodissociation dynamics, J. Phys. Chem. **93**, 1840 (1989).

20. I. Bar, Y. Cohen, T. Arusi-Parpar, S. Rosenwaks and J. J. Valentini, J. Chem. Phys. **95**, 3341 (1991).

Electronic Branching Ratios in Photodissociation

21. L. Butler, Chemical reaction dynamics beyond the Born-Oppenheimer approximation, Ann. Rev. Phys. Chem. **49**, 125 (1998).

22. N. Delaney, J. Faeder, P. E. Maslen and R. Parson, Photodissociation, recombination, and electron transfer in cluster ions: A nonadiabatic molecular dynamics study of $I_2^-(CO_2)_n$, J. Phys. Chem. A **101**, 8147 (1997).

23. J. Zhang, E. J. Heller, D. Huber, D. G. Imre and D. Tannor, CH_2I_2 photodissociation: Dynamical modeling, J. Chem. Phys. **89**, 3602 (1988).

24. X.-P. Jiang, R. Heather and H. Metiu, Time dependent calculations of the absorption spectrum of a photodissociating system with two interacting excited electronic states, J. Chem. Phys. **90**, 2555 (1989).

25. J. Qian, C. J. Williams and D. J. Tannor, Understanding the origin of rotational distributions in triatomic photodissociation: A k-j wavepacket study of ICN, J. Chem. Phys. **97**, 6300 (1992).

26. J. Qian, D. J. Tannor, Y. Amatatsu and K. Morokuma, Ab initio structure and wavepacket dynamics of ICN photodissociation, J. Chem. Phys. **101**, 9597 (1994).

27. J. M. Bowman, R. C. Mayrhofer and Y. Amatatsu, Coupled-channel scattering calculations of ICN $\tilde{A} \leftarrow \tilde{X}$ photodissociation using ab initio potentials, J. Chem. Phys. **101**, 8564 (1994).

28. H. Wei and T. Carrington, A time-dependent calculation of the alignment and orientation of the CN fragment of the photodissociation of ICN, J. Chem. Phys. **105**, 141 (1996).

29. J. F. Black, J. R. Waldeck and R. N. Zare, Evidence for three interacting potential-energy surfaces in the photodissociation of ICN at 249 nm, J. Chem. Phys. **92**, 3519 (1990).

30. Z. H. Kim, A. J. Alexander, S. A. Kandel, T. P. Rakitzis and R. N. Zare, Orientation as a probe of photodissociation dynamics, Faraday Disc. **113**, 27 (1999).

Time-Independent Methods in Photodissociation

31. M. Shapiro and R. Bersohn, Vibrational energy distribution of the CH_3 radical photodissociated from CH_3I, J. Chem. Phys. **73**, 3810 (1980).

32. G. G. Balint-Kurti and M. Shapiro, Photofragmentation of triatomic molecules: Theory of angular and state distribution of product fragments, Chem. Phys. **61**, 137 (1981).

33. M. Shapiro and R. Bersohn, Theories of the dynamics of photodissociation, Annu. Rev. Phys. Chem. **33**, 409 (1982).

34. M. Shapiro and P. Brumer, *Principles of the Quantum Control of Molecular Processes* (Wiley-Interscience, Hoboken, NJ, 2003).

35. D. E. Manolopoulos and M. H. Alexander, Quantum flux redistribution during molecular photodissociation, J. Chem. Phys. **97**, 2527 (1992).

36. H. M. Lambert, P. J. Dagdigian and M. H. Alexander, Spin-orbit branching in the photofragmentation of HCl at long wavelength, J. Chem. Phys. **108**, 4460 (1998).

Chapter 18

Wavepacket Correlation Function Formulation
of Reactive Scattering

In this chapter we discuss the quantum theory of chemical reactions, the theory of which is based on a formalism called quantum reactive scattering. Scattering theory is traditionally considered one of the most difficult topics in quantum mechanics, and within scattering theory the theory of reactive scattering is considered especially difficult because of the complications of dealing with the rearrangement of particles. Yet, it is precisely this rearrangement that is at the heart of chemical reactions—chemical reactions involve an approach of reactants, a rearrangement of the atoms, and a separation of the products—and thus reactive scattering is too important to neglect in a basic chemical physics training.

Fortunately, scattering theory in general, and reactive scattering in particular, are very easy to grasp intuitively using a time-dependent approach. A wavepacket corresponding to a well-defined internal state of reactants enters the interaction region and bifurcates: part returns to reactants and part proceeds to (one or more) different product arrangements. The projection of the product wavepacket onto the asymptotic eigenstates of products gives the various transition amplitudes from the initial state to all final states—a column of the scattering or S-matrix. Despite the great intuitive appeal of this description, this is actually not the approach we adopt in this chapter. Instead, we adopt an approach that treats reactants and products symmetrically: a wavepacket corresponding to a well-defined internal state of reactants is propagated forward in time, while a wavepacket corresponding to a well-defined internal state of products is propagated backward in time, and the cross-correlation function between these wavepackets is calculated (see Figure 18.3 below). The Fourier transform of the cross-correlation between these wavepackets, properly normalized, gives the scattering amplitude, or S-matrix element, from the selected initial state to the selected final state. Although one normally thinks of a chemical reaction as proceeding forward in time—from reactants, to complex, to products—on further reflection there must be a symmetry between reactants and products: there can be no absolute definition of reactants and products since, depending on the experiment, today's reactants can be tomorrow's products. Besides highlighting this formal symmetry, known as time-reversal symmetry, the symmetric treatment of reactants and products we adopt here avoids to a great extent the numerical difficulties posed by the existence of multiple arrangement channels and multiple coordinate systems.

As in the previous chapter, before deriving the time-dependent formulation we first present the time-independent treatment. The time-independent expression for the S-matrix is given by the overlap between degenerate scattering eigenstates—one corresponding to a well-defined internal state of reactants and one corresponding to a well-defined internal state of products. In the previous chapter we learned how to express the scattering eigenstates in terms of the Fourier transform of "Møller wavepackets" (Eqs. 17.26 and 17.28). In this chapter we will use this relation twice: once for reactants and once for products. Expressing both reactant and product eigenstates as the Fourier transform of wavepackets, we derive our central formula for the individual S-matrix elements as the Fourier transform of the cross-correlation between a reactant and product Møller wavepacket.

As mentioned earlier, this chapter is in some sense the climax of the book. Three major ideas from the earlier chapters come together in our central formula for the S-matrix elements—the interaction representation (Chapter 9), the spectral method for constructing eigenstates (Chapter 6), and the Fourier transform of time-correlation functions (Chapters 6 and 14).

18.1   The Concept of an Arrangement Channel and the Problem of Coordinate Systems

In Chapter 16, we considered a model two-degree-of-freedom system, corresponding to collinear ABC, which could fragment in two ways: ABC → A + BC, or ABC → AB + C. The different sets of fragmentation products, A + BC and AB + C, are termed "arrangement channels." More generally, a triatomic ABC can fragment into A + BC, AB + C, AC + B or A + B + C. A chemical reaction corresponds to a transition from an initial state that is in one arrangement channel to a final state that is in another arrangement channel, such as A + BC → ABC → AB + C.

Each arrangement channel, γ, corresponds to a different partitioning of the full set of atoms in the system into fragments. As in the single arrangement case, at large enough separation of the fragments the interaction between fragments vanishes and one can define an asymptotic Hamiltonian, which is the sum of an internal Hamiltonian of fragments h^γ and the relative kinetic energy between fragments T^γ:

$$H_0^\gamma = T^\gamma + h^\gamma. \tag{18.1}$$

Note that in the multiarrangement case both the asymptotic Hamiltonian and its partitioning into different degrees of freedom depend on the arrangement channel, γ.

For a triatomic molecule, it is convenient to express H_0 in Jacobi coordinates (see Section 12.5.2). In Jacobi coordinates, the three internal coordinates are r, R and θ, where r is the separation between atoms 2 and 3, R is the separation of atom 1 from the 2–3 center of mass, and θ is the angle between the r and R vectors. Clearly, the Jacobi coordinates corresponding to A + BC are different from that corresponding to AB + C or AC + B. This is the famous coordinate problem in reactive scattering.

Henceforth, we will specialize to collinear triatomics. In Jacobi coordinates the Hamiltonian H_0^γ for a collinear triatomic takes the form

$$H_0^\gamma(R_\gamma, r_\gamma) = T^\gamma(R_\gamma) + h^\gamma(r_\gamma) = -\frac{\hbar^2}{2m_\gamma}\frac{\partial^2}{\partial R_\gamma^2} + h^\gamma(r_\gamma), \tag{18.2}$$

where m_γ is the reduced mass associated with the relative translation of fragments in arrangement channel γ.

In general, we may express the total Hamiltonian as

$$H = H_0^\gamma(R_\gamma, r_\gamma) + V^\gamma(R_\gamma, r_\gamma). \tag{18.3}$$

Although H does not depend on γ, V^γ does. At large enough separation of the fragments, the interaction between the fragments vanishes, that is, $V^\gamma(R, r) \to 0$ as $R_\gamma \to \infty$. This vanishing of V^γ manifests itself in the form of the potential energy surface, the equipotential contours becoming parallel at large R^γ. This applies to large separation in all arrangement channels, γ, and hence the potential energy contour maps for reactive scattering typically have multiple regions of parallel equipotentials, corresponding to the different values of γ.

18.2   The Eigenstates of a Scattering Hamiltonian with Multiple Arrangement Channels

In the previous chapter we introduced the incoming and outgoing scattering eigenstates. However, the discussion in the previous chapter was restricted to a single arrangement channel. In reactive scattering, since the final chemical arrangement of the atoms in the products is different from that in the reactants, we will need to consider scattering eigenstates corresponding to multiple chemical arrangements. We will be interested in incoming eigenstates with well-defined boundary conditions in the reactant arrangement channel, and outgoing eigenstates with well-defined boundary conditions in the product arrangement channel. The S-matrix will then be defined in terms of the overlap of degenerate scattering eigenstates with these different boundary conditions. This is the time-*independent* formulation of reactive scattering.

18.2.1   The Eigenstates of the Asymptotic Hamiltonians

Consider the reaction $A + BC \to AB + C$. For any particular choice of γ ($A + BC$ and $AB + C$ in Figure 18.1), the properties of the eigenstates of the asymptotic Hamiltonian are essentially identical to those of the eigenstates of the asymptotic Hamiltonian for the single arrangement channel case discussed in the previous chapter, Eqs. 17.4–17.13, with the addition of a label for the arrangement channel. For example, the asymptotic eigenstates of arrangement channel γ satisfy the equation

$$H_0^\gamma \psi_{\gamma E} = E \psi_{\gamma E}, \tag{18.4}$$

where

$$\psi_{\gamma E} = \chi_{\gamma n}(r_\gamma) \sqrt{\frac{m_\gamma}{\hbar k_\gamma}} \psi_{\pm k_\gamma}(R_\gamma) = \psi_{\gamma n, \pm k}(E) \tag{18.5}$$

and

$$E = E_{\gamma n} + \frac{\hbar^2 k_\gamma^2}{2m_\gamma}. \tag{18.6}$$

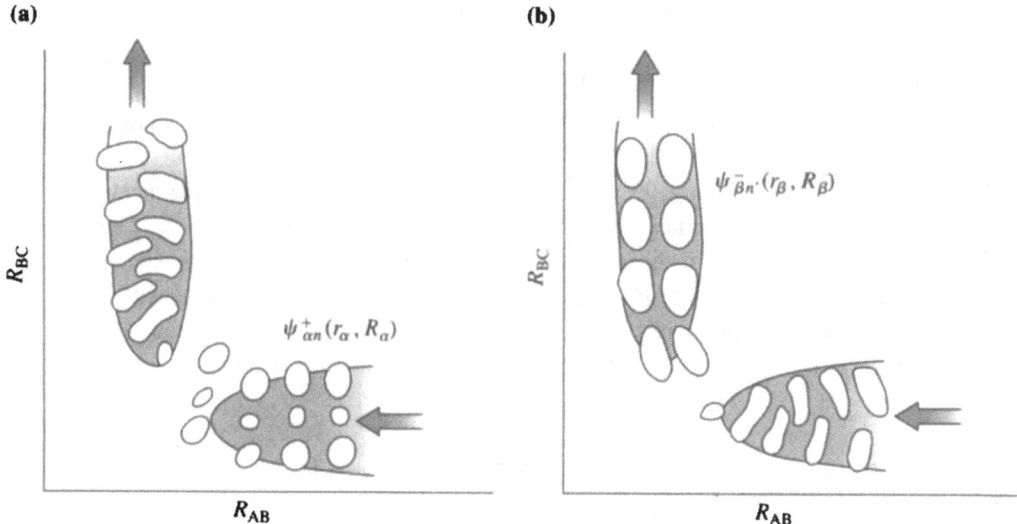

(a)

R_{BC}

$\psi^+_{\alpha n}(r_\alpha, R_\alpha)$

R_{AB}

(b)

R_{BC}

$\psi^-_{\beta n'}(r_\beta, R_\beta)$

R_{AB}

Figure 18.1 Equipotential energy contours for collinear ABC as a function of R_{AB} and R_{BC}. (a) A scattering eigenstate that correlates with a precise internal state of reactants, $\psi^+_{\alpha n}(r_\alpha, R_\alpha)(E)$. (b) A scattering eigenstate that correlates with a specific internal state of products, $\psi^-_{\beta n'}(r_\beta, R_\beta)(E')$. (These pictures are meant to be schematic. For clarity of presentation, we have not portrayed the reflected portions of both eigenstates.) The overlap between the scattering eigenstates in (a) and (b) gives the S-matrix element $S_{\beta n', \alpha n}\delta(E, E')$. The δ function comes from the fact that the scattering eigenstates extend to infinity, hence if their overlap exists it is infinite.

The eigenstates of the asymptotic Hamiltonian corresponding to a particular arrangement channel γ satisfy the orthogonality relation

$$\langle\psi_{\gamma n', \pm k'}(E')|\psi_{\gamma n, \pm k}(E)\rangle = \delta_{n'n}\delta_{\pm\pm}\delta(E - E'), \tag{18.7}$$

but there is no requirement of orthogonality of the asymptotic eigenstates in different arrangements.

18.2.2 The Eigenstates of the Scattering Hamiltonian

Now consider the eigenstates of the full Hamiltonian, H. In the asymptotic regions of the potential these eigenstates satisfy the same TISE as the eigenstates of H_0^γ. Recall that for a single arrangement channel the scattering eigenstates can correlate with a separable incoming or outgoing state but not both; so in the multiple arrangement channel case the scattering eigenstates can correlate with a separable incoming or outgoing state in only one arrangment channel. For example, the scattering eigenstate labelled $\psi^+_{\alpha n}(E)$ has the same spatial dependence in the asymptotic region α as $\psi_{\alpha n k}(E)$, but no simple dependence in asymptotic region β. Similarly, the scattering eigenstate labelled $\psi^-_{\beta n}(E)$ has the same spatial dependence in the asymptotic region as $\psi_{\beta n, -k}(E)$, but no simple dependence in asymptotic region α. This idea is shown schematically in Figure 18.1. (a) shows a scattering eigenstate with incoming boundary conditions from arrangement channel α having two vibrational quanta (as reflected in the two nodes perpendicular to the reaction coordinate in the asymptotic reactant region); (b) shows a scattering eigenstate with outgoing boundary conditions to arrangement channel β having one vibrational quantum (as reflected in the one node perpendicular to the reaction coordinate in the product asymptotic region). The simple nodal structure of the former distorts as it enters the interaction region; likewise, the nodal structure of the latter distorts as it enters the interaction region backward in time.

The orthogonality conditions on the scattering eigenstates for multiple arrangements are an extension of those for the single arrangement case. Recall that for a single arrangement, incoming scattering eigenstates corresponding to different initial internal states are orthogonal; similarly, outgoing scattering eigenstates corresponding to different final internal states are orthogonal. In the multiarrangement case, incoming scattering states corresponding to different initial internal states in the same arrangement are still orthogonal; but in addition, incoming scattering states originating from different arrangement channels are always orthogonal, regardless of their internal state. This condition is expressed as follows:

$$\langle \psi^+_{\gamma'n'}(E')|\psi^+_{\gamma n}(E)\rangle = \delta_{\gamma'\gamma}\delta_{n'n}\delta(E - E'). \tag{18.8}$$

Similarly, outgoing scattering states corresponding to different final internal states in the same arrangement channel are orthogonal; and in addition, outgoing scattering states correlating to different arrangement channels are orthogonal for any internal state:

$$\langle \psi^-_{\gamma'n'}(E')|\psi^-_{\gamma n}(E)\rangle = \delta_{\gamma'\gamma}\delta_{n'n}\delta(E - E'). \tag{18.9}$$

What is the degeneracy of scattering eigenstates at energy E? One might think that the total degeneracy is the sum over the degeneracy of each of the asymptotic Hamiltonians. Since the degeneracy of H_0^γ is $2N_\gamma$, this would predict a degeneracy of $2\sum_\gamma N_\gamma$. However, as in the single arrangement case, the incoming and outgoing eigenstates of the full Hamiltonian are not independent: the incoming scattering states from all arrangements are enough to span the space (dimension $N_\alpha + N_\beta$ for two arrangement channels), as are the outgoing scattering states to all arrangments (also of dimension $N_\alpha + N_\beta$ for two arrangement channels). The number of independent eigenstates at energy E is therefore actually smaller by a factor of 2: the degeneracy is $N_\alpha + N_\beta$ for two arrangement channels, or $\sum_\gamma N_\gamma$ for more arrangement channels. This number, $\sum_\gamma N_\gamma$, is called the number of "open channels" at energy E. The concept of two different complete sets of basis functions, one based on the *history* of the states and one based on the *destiny* of the states, appeared first in Section 7.2 and was seen again in Chapter 17.

18.2.3   The S-Matrix: Time-Independent Viewpoint

In the previous section, we saw that scattering eigenstates corresponding to incoming boundary conditions from different arrangement channels are orthogonal, and similarly, scattering eigenstates corresponding to outgoing boundary conditions to different final arrangement channels are orthogonal. However, degenerate eigenstates—one associated with incoming boundary conditions and one associated with outgoing boundary conditions—are in general not orthogonal. In fact, it is the overlap between them that defines the transition amplitude from an incoming state with one set of internal quantum numbers to an outgoing state with a new set of internal quantum numbers:

$$S_{\beta n',\alpha n}(E)\delta(E - E') = \langle \psi^-_{\beta n'}(E')|\psi^+_{\alpha n}(E)\rangle. \tag{18.10}$$

Here and below we use α to represent the initial arrangement channel and β to represent the final arrangement channel. One can visualize the quantity $S_{\beta n',\alpha n}(E)$ as being formed from the overlap of degenerate scattering eigenstates at energy E of the form shown in

$$
\begin{array}{c}
\overbrace{\begin{array}{cc} N_r & \quad N_p \end{array}} \\[6pt]
\begin{array}{c} N_r \\[18pt] N_p \end{array}
\left(
\begin{array}{c|c}
\mathbf{S}_{rr} & \mathbf{S}_{rp} \\
\hline
\mathbf{S}_{pr} & \mathbf{S}_{pp}
\end{array}
\right)
\end{array}
$$

Figure 18.2 Block structure of the scattering correlation matrix.

Figure 18.1. Note that the δ function comes from the fact that the scattering eigenstates extend to infinity; hence if their overlap exists it is infinite.

Clearly, one can construct a matrix of such overlaps of scattering eigenstates, for all initial levels αn going to all final levels $\beta n'$. This matrix is called the scattering or S-matrix, and is the central object in scattering theory. For reactive collisions, $\alpha \neq \beta$; however, it is convenient to define the S-matrix to include nonreactive collisions as well, $\alpha = \beta$. When we want to emphasize that we are dealing with reactive collisions we will use r for the initial arrangement channel and p for the final arrangement channel.

Let us focus first on the reactive block of the S-matrix. If we choose some specific energy E, the dimension of the reactive matrix will be $N_p \times N_r$, where N_p is the number of internal levels of product and N_r is the number of internal levels of reactant available at energy E (product and reactant "open channels," respectively). Similarly, the nonreactive block has dimension $N_r \times N_r$. The nonreactive transitions can be further broken down into elastic scattering, if $n' = n$, and inelastic scattering, if $n' \neq n$. Combined with the reactive S-matrix we obtain a matrix of dimension $(N_r + N_p) \times N_r$, corresponding to N_r reactant states (columns) going to $N_r + N_p$ final states (rows). It turns out that it is useful to extend the S-matrix to include transitions from level n' of *products* and going to either level n of reactants (dimension $N_r \times N_p$) or to level n'' of products (dimension $N_p \times N_p$). The combination of these two latter matrices gives a matrix of dimension $(N_r + N_p) \times N_p$, corresponding to N_p product states (columns) going to $N_r + N_p$ final states (rows). Putting all four of these matrices together, the overall S-matrix is square, with dimension $(N_r + N_p) \times (N_r + N_p)$ (Figure 18.2). Thus, the dimension of the S-matrix is equal to the number of open channels at energy E, reflecting the fact that the S-matrix is a measure of overlap between two complete sets of bases for the degenerate manifold at energy E, the incoming and outgoing states, both bases being of dimension $N_r + N_p$. We now revert to the notation α and β for the initial and final arrangement channels, regardless of whether the process is reactive or nonreactive. To minimize the number of subscripts we follow conventional usage and will use α and β to label all the internal state variables as well as the arrangement channel.

One of the important properties of the S-matrix is that it is symmetric:

$$S_{\beta\alpha}(E) = S_{\alpha\beta}(E). \tag{18.11}$$

This is a consequence of time-reversal symmetry. This symmetry holds for both the non-reactive blocks ($\beta = \alpha$) as well as the reactive blocks ($\alpha \neq \beta$). We will derive this relationship below when we develop the time-dependent formulation of the S-matrix.

There is an interesting relationship between the S-matrix and the Møller operators. In Chapter 17 we showed that the scattering eigenstate $\psi_n^{\pm}$ could be expressed as the

Møller operator $\Omega_\pm$ operating on the asymptotic eigenstate $\psi_{nk}(E)$ (Eqs. 17.35 and 17.40). For reactive scattering we must generalize this idea to multiple arrangement channels, the key difference now being that H_0 in the definition of the Møller operator must carry an arrangement channel label, γ, where $\gamma = \alpha, \beta$. Thus,

$$\psi_\alpha^+(E) = \Omega_+^\alpha \psi_{\alpha k}(E) \qquad \Omega_+^\alpha = \lim_{t \to -\infty} e^{iHt/\hbar} e^{-iH_0^\alpha t/\hbar} \tag{18.12}$$

and

$$\psi_\beta^-(E') = \Omega_-^\beta \psi_{\beta k}(E') \qquad \Omega_-^\beta = \lim_{t \to \infty} e^{iHt/\hbar} e^{-iH_0^\beta t/\hbar}. \tag{18.13}$$

Substituting Eq. 18.12 into Eq. 18.10 we obtain

$$S_{\beta\alpha}(E)\delta(E - E') = \langle \psi_\beta(E')|\Omega_-^{\beta\dagger}\Omega_+^\alpha|\psi_\alpha(E)\rangle = \langle \psi_{\beta k'}(E')|S|\psi_{\alpha k}(E)\rangle, \tag{18.14}$$

where we have defined the scattering operator S, which is related to the Møller operators by the equation (Taylor, 1987)

$$S = \Omega_-^{\beta\dagger}\Omega_+^\alpha. \tag{18.15}$$

Note that we use boldface notation for the scattering *matrix* and plain italic notation for the scattering *operator*. Using the explicit form for the Møller operators, the scattering operator can be written as

$$S = \lim_{t \to \infty} e^{iH_0^\beta t/\hbar} e^{-iHt/\hbar} e^{iH_0^\alpha t/\hbar}. \tag{18.16}$$

One of the key properties of the S-operator is that it is unitary. Equation 18.16 is enough to show that the S-operator is norm preserving or "isometric," but to show that it is unitary we need the additional property that it describes a one-to-one mapping. Consider first the case of single arrangement scattering. The Møller operators, $\Omega_\pm$, map the full set of states in the Hilbert space $\mathcal{H}$ into a subspace of the full Hilbert space ($\mathcal{R}_+$ for Ω_+ and $\mathcal{R}_-$ for Ω_-). Because the Møller operators do not provide a one-to-one mapping they are isometric and not unitary. The proof that the S-operator is unitary is based on the observation that both $\mathcal{R}_+$ and $\mathcal{R}_-$ are the orthogonal complement of the subspace $\mathcal{B}$ of all bound states:

$$\mathcal{R}_+ + \mathcal{B} = \mathcal{R}_- + \mathcal{B} = \mathbf{1}, \tag{18.17}$$

and hence,

$$\mathcal{R}_+ = \mathcal{R}_- = \mathcal{R}. \tag{18.18}$$

It follows that the composite mapping $\Omega_\pm^\dagger\Omega_\pm$ is from $\mathcal{H}$ to $\mathcal{R}$ and then from $\mathcal{R}$ back to $\mathcal{H}$. Since the overall domain and range are equal to $\mathcal{H}$, the mapping is one-to-one.

For the multiarrangement case the proof is more subtle, and we only sketch the argument here. The Møller operator Ω_+^α maps the space $\mathcal{P}_\alpha$ into the space $\mathcal{R}_\alpha^+$, while the Møller operator $\Omega_-^{\beta\dagger}$ maps the space $\mathcal{R}_\beta^-$ into the space $\mathcal{P}_\beta$; neither the final space nor the intermediate space is identical to the initial space. To show unitarity, one must generalize the S-operator by writing

$$S = \lim_{t \to \infty} (P_\alpha + P_\beta + \cdots)e^{iH_0t/\hbar}e^{-iHt/\hbar}e^{iH_0t/\hbar}(P_\alpha + P_\beta + \cdots) \tag{18.19}$$

$$= \lim_{t \to \infty} (P_\alpha e^{iH_0^\alpha t/\hbar} + P_\beta e^{iH_0^\beta t/\hbar} + \cdots)e^{-iHt/\hbar}(e^{iH_0^\alpha t/\hbar}P_\alpha + e^{iH_0^\beta t/\hbar}P_\beta + \cdots), \tag{18.20}$$

where P_α is a projection operator for the space $\mathcal{P}_\alpha$, and so forth. Now the mapping is from $\sum_\gamma \mathcal{P}_\gamma$ to $\sum_\gamma \mathcal{R}_\gamma^+$ and from $\sum_\gamma \mathcal{R}_\gamma^-$ back to $\sum_\gamma \mathcal{P}_\gamma$. Because

$$\sum_\gamma \mathcal{R}_\gamma^+ + \mathcal{B} = \sum_\gamma \mathcal{R}_\gamma^- + \mathcal{B} = 1$$

(since both sums span the space orthogonal to the bound states), it follows that the overall mapping is one-to-one. Combined with the property that the multiarrangement S-operator is linear and norm preserving, this is sufficient to guarantee unitarity. For an extensive discussion of the unitarity of the S-matrix the reader is referred to Taylor (1987).

The next exercise deals with some additional properties of the S-operator.

▶ Exercise 18.1

    a. Show that

$$e^{iH_0^\beta \tau} S = S e^{iH_0^\alpha \tau}. \tag{18.21}$$

    b. Use the result of part a to show that

$$H_0^\beta S = S H_0^\alpha. \tag{18.22}$$

    c. Derive the result of part b using the intertwining relation for the Møller operators (Eq. 17.44) twice.

18.3   Wavepacket Cross-Correlation Function Formulation of $S_{\beta\alpha}(E)$

We opened this chapter with a heuristic description of reactive scattering transition amplitudes in terms of the Fourier transform of the cross-correlation between a reactant wavepacket going forward in time and a product wavepacket going backward in time. We will now show that an expression of this type can be developed that is rigorously equivalent to the conventional definition of the S-matrix elements, Eq. 18.10. The key to the derivation is the spectral formula expressing eigenstates as a superposition of wavepackets. The strategy of the derivation is to use Eqs. 17.26 and 17.28 to construct $\psi_\alpha^+(E)$ and $\psi_\beta^-(E')$, and then simply construct $S_{\beta\alpha}$ from the overlap, using Eq. 18.10. Since the bra and the ket have both been expressed in terms of wavepacket motion, we expect the resulting expression for $S_{\beta\alpha}(E)$, although it is energy resolved, to be an expression whose fundamental constituents are time-dependent wavepackets.

18.3.1   Derivation of the Cross-Correlation Function Expression for $S_{\beta\alpha}(E)$

The scattering eigenstates in the case of multiple arrangement channels can be generated using the spectral method, entailing only a slight generalization of the treatment in Chapter 17, Section 17.3.1. To obtain the incoming eigenstate $\psi_\alpha^+(E)$, one must choose $\phi = \phi_\alpha^+$ to correlate with a wavepacket in the asymptotic region of reactant channel α with internal quantum numbers n and momentum such that the wavepacket is incoming. The energy-normalized incoming scattering eigenstate then takes the form

$$\psi_\alpha^+(E) = \frac{1}{2\pi\hbar\eta_\alpha(E)} \int_{-\infty}^{\infty} e^{-iHt/\hbar}\phi_\alpha^+ e^{iEt/\hbar}\, dt, \tag{18.23}$$

where $\eta_\alpha(E)$ is the coefficient of the energy-normalized eigenstate $\psi_\alpha^+(E)$ contained in ϕ_α^+. Similarly, to obtain the outgoing eigenstate $\psi_\beta^-(E')$, one must choose $\phi = \phi_\beta^-$ to correlate with a wavepacket in the asymptotic region of product channel β with internal quantum numbers n' and momentum such that the wavepacket is outgoing. The energy-normalized outgoing eigenstate is then given by

$$\psi_\beta^-(E') = \frac{1}{2\pi\hbar\mu_\beta(E')} \int_{-\infty}^{\infty} e^{-iHt/\hbar}\phi_\beta^- e^{iE't/\hbar}\, dt, \tag{18.24}$$

where $\mu_\beta(E')$ is the coefficient of the energy-normalized eigenstate $\psi_\beta^-(E')$ contained in ϕ_β^-.

From Eq. 18.10, the S-matrix is then simply related to the overlap between these two energy-normalized eigenfunctions:

$$S_{\beta\alpha}(E)\delta(E - E') = \langle\psi_\beta(E')|S|\psi_\alpha(E)\rangle = \langle\psi_\beta^-(E')|\psi_\alpha^+(E)\rangle. \tag{18.25}$$

Substituting Eqs. 18.23 and 18.24 into Eq. 18.25, we obtain

$$S_{\beta\alpha}(E)\delta(E - E')$$

$$= \frac{(2\pi\hbar)^{-2}}{\mu_\beta^*(E')\eta_\alpha(E)} \int_{-\infty}^{\infty}\int_{-\infty}^{\infty} \langle\phi_\beta^-|e^{-iH(t'-t'')/\hbar}|\phi_\alpha^+\rangle e^{iE_+(t'-t'')/\hbar} e^{iE_-(t'+t'')/\hbar}\, dt'\, dt'', \tag{18.26}$$

where $E_+ \equiv \frac{E+E'}{2}$ and $E_- \equiv \frac{E-E'}{2}$. With the definitions $t = t' - t''$, $s = t' + t''$, the above equation becomes

$$S_{\beta\alpha}(E)\delta(E - E') = \frac{(2\pi\hbar)^{-2}}{2\mu_\beta^*(E')\eta_\alpha(E)} \int_{-\infty}^{\infty}\int_{-\infty}^{\infty} \langle\phi_\beta^-|e^{-iHt/\hbar}|\phi_\alpha^+\rangle e^{iE_+t/\hbar}\, dt\, e^{iE_- s/\hbar}\, ds \tag{18.27}$$

$$= \frac{(2\pi\hbar)^{-2}}{\mu_\beta^*(E')\eta_\alpha(E)} \int_{-\infty}^{\infty} \langle\phi_\beta^-|e^{-iHt/\hbar}|\phi_\alpha^+\rangle e^{iE_+t/\hbar}\, dt\, 2\pi\hbar\delta(E - E'), \tag{18.28}$$

where the factor $\frac{1}{2}$ in Eq. 18.27 comes from the Jacobian associated with the change of variables, and is cancelled by a factor of 2 coming from $\delta(\frac{E-E'}{2})$.

Comparing both sides of Eq. 18.28 we find (Tannor, 1993)

$$S_{\beta\alpha}(E) = \frac{(2\pi\hbar)^{-1}}{\mu_\beta^*(E)\eta_\alpha(E)} \int_{-\infty}^{\infty} \langle\phi_\beta^-|e^{-iHt/\hbar}|\phi_\alpha^+\rangle e^{iEt/\hbar}\, dt, \tag{18.29}$$

the desired expression.

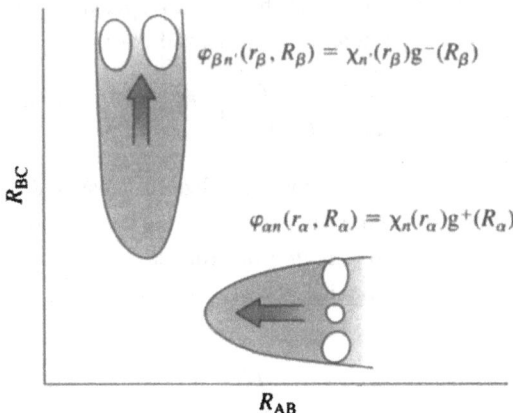

$$\varphi_{\beta n'}(r_\beta, R_\beta) = \chi_{n'}(r_\beta)g^-(R_\beta)$$

$$\varphi_{\alpha n}(r_\alpha, R_\alpha) = \chi_n(r_\alpha)g^+(R_\alpha)$$

Figure 18.3 Equipotential energy contours for collinear ABC as a function of R_{AB} and R_{BC}. Superimposed are a separable wavepacket in the reactant asymptotic region, $\phi_{\alpha n}^+(r_\alpha, R_\alpha) = \chi_{\alpha n}(r_\alpha)g^+(R_\alpha)$, and a separable product wavepacket in the product asymptotic region, $\phi_{\beta n'}^-(r_\beta, R_\beta) = \chi_{\beta n'}(r_\beta)g^-(R_\beta)$. (Strictly speaking, the asymptotic wavepackets are separable in Jacobi, not bond coordinates.)

Figure 18.3 shows a separable wavepacket in the reactant asymptotic region, $\phi_{\alpha n}^+(r_\alpha, R_\alpha) = \chi_{\alpha n}(r_\alpha)g^+(R_\alpha)$, and a separable product wavepacket in the product asymptotic region, $\phi_\beta^-(r_\beta, R_\beta) = \chi_{\beta n'}(r_\beta)g^-(R_\beta)$. For clarity, we now specify the internal (vibrational) quantum number explicitly. If the reactant wavepacket is propagated forward in time and the product wavepacket is propagated backward in time to the central interaction region, these packets will in general have some nonzero overlap. (Note that the product wavepacket going backward in time still has its momentum pointing toward the product region.) It is intuitively reasonable that this overlap is a measure of the probability amplitude for making the transition from vibrational state n of reactants to vibrational state n' of products. However, the reactant and product wavepackets contain many energy components, and hence this overlap will contain the amplitude for the $n \to n'$ transition only in an energy-averaged sense. Nevertheless, the *time-correlation function* between the reactant and product wavepackets contains all the information on the $n \to n'$ transition amplitude: through the Fourier transform of the time-correlation function, all the energy-resolved information on the $n \to n'$ transition can be extracted. As we have seen in several other places, the Fourier transform essentially acts as a filter, selecting just a single energy component out of the time-correlation function that matches the energy of the Fourier transform.

18.3.2 Time-Reversal Symmetry of the S-Matrix

A well-known property of the S-matrix is its symmetry under time reversal (Taylor, 1987):

$$S_{\beta'\alpha'} = S_{\alpha\beta}, \tag{18.30}$$

where α and β label the arrangement channel and internal quantum numbers of the initial and final states, respectively, and the primes indicate the time-reversed value of the quantum numbers. Time reversal changes the sign of all the momenta and the spin in the system, but leaves other quantum numbers, for example, vibrations, unchanged. The physical meaning of this symmetry, which is sometimes called microscopic reversibility or detailed balance, is that the probability amplitude for scattering $\phi_{\beta'} \leftarrow \phi_{\alpha'}$ (for the process in which initial and final states are time reversed) is the same as the probability of scattering $\phi_\alpha \leftarrow \phi_\beta$ (for the process in which the roles of initial and final states are exchanged). An appealing feature of the correlation function expression for $S_{\beta\alpha}$, Eq. 18.29, is that this symmetry is immediately apparent. Moreover, the correlation function has the same symmetry under time reversal as the S-matrix. We now proceed to demonstrate these statements.

The time-reversal operator, T, operating on the state ϕ_γ gives the complex conjugate:

$$T\phi_\gamma \equiv \phi_{\gamma'} = \phi_\gamma^*. \tag{18.31}$$

The time-reversal operator is therefore *antilinear*:

$$Tc\phi_\gamma = c^* T\phi_\gamma. \tag{18.32}$$

The adjoint of an antilinear operator is defined according to the following equation:

$$\langle \phi_\beta | T\phi_\alpha \rangle = \langle T^\dagger \phi_\beta | \phi_\alpha \rangle^*. \tag{18.33}$$

Since T is norm preserving and antilinear it is an *antiunitary* operator, and satisfies the equation

$$T^\dagger T = T T^\dagger = 1. \tag{18.34}$$

The effect of time reversal on the Møller operators is to interchange Ω_+ and Ω_- (Taylor, 1987):

$$T^\dagger \Omega_\pm^\gamma T = \Omega_\mp^\gamma. \tag{18.35}$$

We now have all the relations necessary to examine the behavior of the correlation function under time reversal of the incoming and outgoing states:

$$C_{\beta'\alpha'} \equiv \langle \phi_{\beta'} | \Omega_-^{\beta\dagger} e^{-iHt/\hbar} \Omega_+^\alpha | \phi_{\alpha'} \rangle \tag{18.36}$$

$$= \langle \phi_\beta | T^\dagger \Omega_-^{\beta\dagger} T T^\dagger e^{-iHt/\hbar} T T^\dagger \Omega_+^\alpha T | \phi_\alpha \rangle^* \tag{18.37}$$

$$= \langle \phi_\beta | \Omega_+^{\beta\dagger} e^{iHt/\hbar} \Omega_-^\alpha | \phi_\alpha \rangle^* \tag{18.38}$$

$$= \langle \phi_\alpha | \Omega_-^{\alpha\dagger} e^{-iHt/\hbar} \Omega_+^\beta | \phi_\beta \rangle \tag{18.39}$$

$$= C_{\alpha\beta}. \tag{18.40}$$

Using Eq. 18.40 we find that the S-matrix element between the time-reversed states is given by

$$S_{\beta'\alpha'} = \frac{(2\pi\hbar)^{-1}}{\mu_{\beta'}^* \eta_{\alpha'}} \int_{-\infty}^\infty \langle \phi_{\beta'} | \Omega_-^\dagger e^{-iHt/\hbar} \Omega_+ | \phi_{\alpha'} \rangle e^{iEt/\hbar} \, dt \tag{18.41}$$

$$= \frac{(2\pi\hbar)^{-1}}{\mu_\beta \eta_\alpha^*} \int_{-\infty}^\infty \langle \phi_\alpha | \Omega_-^\dagger e^{-iHt/\hbar} \Omega_+ | \phi_\beta \rangle e^{iEt/\hbar} \, dt \tag{18.42}$$

$$= S_{\alpha\beta}. \tag{18.43}$$

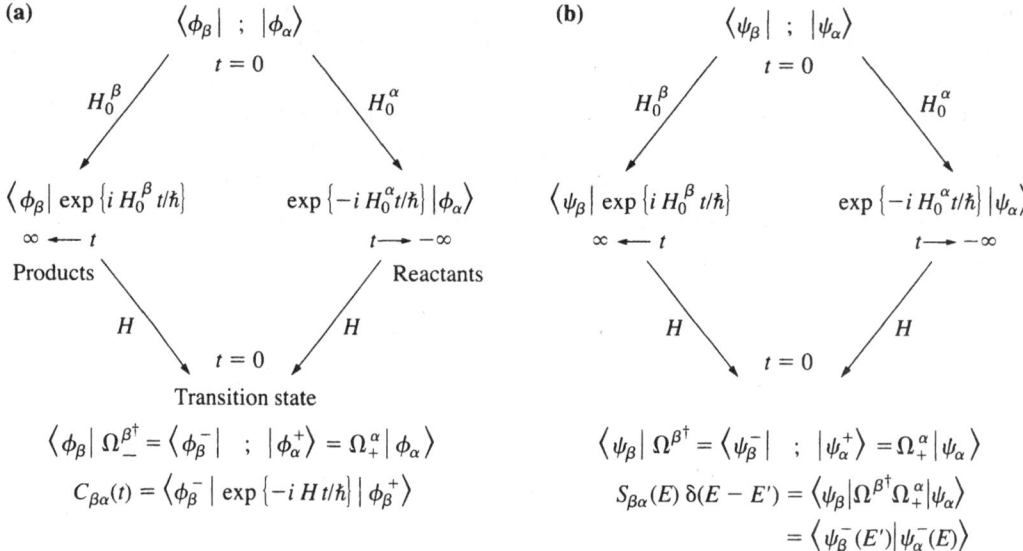

Figure 18.4 (a) Schematic diagram of the computational scheme for constructing S-matrix elements described in the text. Note the symmetry with respect to reactants and products, and the overlapping of states in the transition state region. Reprinted from D. J. Tannor and D. E. Weeks (1993). (b) The corresponding diagram for the *time-independent* formulation of the S-matrix. The symmetry with respect to reactants and products has always been a cornerstone of the time-independent formulation and provided motivation for developing the time-dependent formulation represented by (a).

This completes the proof of the two assertions. The key to the derivation is that time reversal in quantum mechanics is equivalent to complex conjugation. Therefore, if we do both operations (Eq. 18.38) the new result must equal the original quantity.

For elastic scattering $\beta = \alpha$; in this case it is possible to *choose* the wavepackets in Eq. 18.29 in such a way that $\phi_\beta = \phi_\alpha^*$. The correlation function now has an additional symmetry: it is not quite an autocorrelation function, because the bra is the complex conjugate of the ket; we shall refer to it as an "anti-autocorrelation function," because it is related to an autocorrelation function through the operation of the antiunitary time-reversal operator on the bra. The simple relationship of bra and ket implies that in practice the operation of only one of the Møller operators needs to be calculated.

The dynamical components of the expression for the S-matrix are summarized in Figure 18.4. The symmetry of the figure highlights the time-reversal symmetry of both the correlation function and the S-matrix. Thus, the time-dependent formulation of reactive scattering involves three dynamical parts: the application of the reactant Møller operator on a separable reactant wavepacket, the application of the product Møller operator on a separable product wavepacket, and the calculation of the time-correlation function between these two Møller wavepackets. In contrast, photodissociation requires the calculation of the product Møller wavepacket. Thus, in the time-dependent formulation, photodissociation corresponds to just *one-third* of the full-collision process!

18.4   Application to Collinear $H + H_2 \rightarrow H_2 + H$

In Section 7.2 we presented a simple application of the cross-correlation function formulation of the S-matrix to one-dimensional barrier reflection and transmission coefficients (Figure 7.6). Equation 7.45, which was used there, is the one-dimensional version of Eq. 18.29. In this section, we illustrate the ideas using a two-dimensional system, collinear $H + H_2$ reactive scattering. The two-dimensional case introduces the new aspect of an internal degree of freedom of reactants and products that couples to the reaction coordinate. In the general N-dimensional case, there are more internal degrees of freedom, but the basic elements of interplay of internal and translational degrees of freedom are already contained in the two-dimensional case.

The reaction $H + H_2 \rightarrow H_2 + H$ in collinear geometry is the simplest model of a chemical reaction, and has been well studied by theoreticians since the early 1970s. This system has served as a traditional benchmark for testing new methods for calculating reactive scattering probabilities. Here we show the calculation of state-to-state reaction probabilities for this reaction using the wavepacket correlation function formalism presented above, using the so-called LSTH potential surface (Weeks, 1994).

Figure 18.5a shows the initial state of the reactants at $t = 0$. It is a product of the $v = 0$ eigenfunction of the asymptotic reactant vibrational Hamiltonian with a Gaussian in

(a)

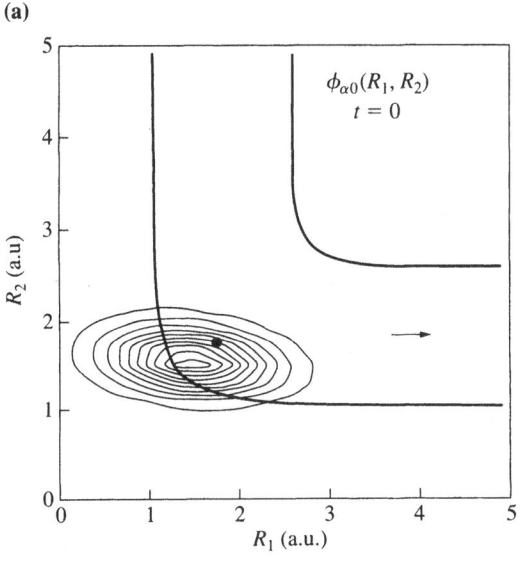

(b)

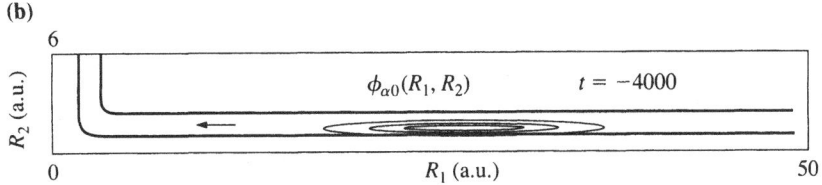

Figure 18.5 Calculating S-matrix elements for the collinear $H + H_2 \rightarrow H_2 + H$ reaction using the LSTH potential surface. (a) Initial state of reactants. (b) The reactant channel packet at $t = -4000$ a.u. The arrow indicates the direction of the following stage of the propagation. Reprinted from D. E. Weeks and D. J. Tannor (1994).

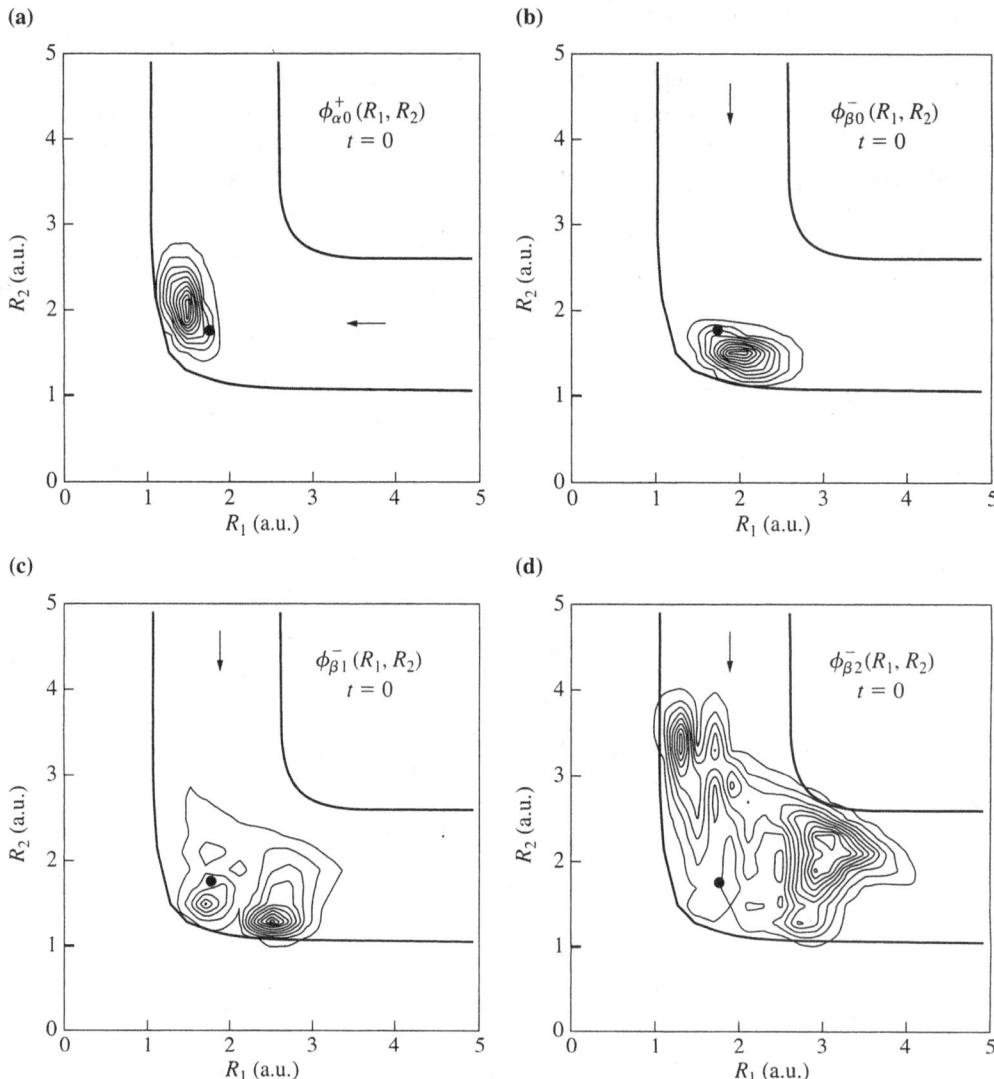

Figure 18.6 Møller wavepackets for collinear $H + H_2$. (a) $v = 0$ reactants; (b) $v = 0$ products; (c) $v = 1$ products; (d) $v = 2$ products. Reprinted from D. E. Weeks and D. J. Tannor (1994).

the translational coordinate, but located in the interaction region of the potential. Figure 18.5b shows the reactant channel packet at $t = -4000$ a.u. It is the state in Figure 18.5a, propagated backward in time under the Hamiltonian H_0^α (the asymptotic Hamiltonian corresponding to reactants). Since this Hamiltonian is separable in vibration and translation it does not introduce any correlation in these degrees of freedom into the wavepacket. Figure 18.6a shows the Møller wavepacket corresponding to $v = 0$ reactants: the state in Figure 18.5b propagated forward in time back to $t = 0$, but under the influence of the full H. Note that the net effect of the backward and forward propagations is to distort the initial wavepacket (i.e., to introduce correlations between vibration and translation) but to leave it localized in approximately the same region as the initial state.

Figure 18.6b–d shows the Møller wavepacket corresponding to $v = 0$, 1 and 2 of products, respectively. These states are the result of the same type of backward/forward propagation as in Figure 18.6a, but with initial states that are a product of $v = 0, 1, 2$ eigen-functions of the asymptotic *product* vibrational Hamiltonian with a Gaussian in the *product* translational coordinate of products, and the backward propagation performed under H_0^β, which is the asymptotic Hamiltonian corresponding to products. The cross-correlation functions necessary to compute the S-matrix elements $0 \to 0$, $0 \to 1$ and $0 \to 2$ are obtained by propagating the incoming Møller wavepacket, Figure 18.6a, and computing the time-dependent overlap with each of the three outgoing Møller wavepackets, Figure 18.6b–d. These cross-correlation functions are shown in Figure 18.7a–c. The reaction probabilities, $|S_{\beta\alpha}(E)|^2$, for the transitions $0 \to 0$, $0 \to 1$ and $0 \to 2$, computed using the cross-correlation functions in Figure 18.7a–c, are shown in Figure 18.7d–f. These reaction probabilities are in excellent agreement with calculations performed using a wide variety of other methods (see, for example, Bondi, 1982).

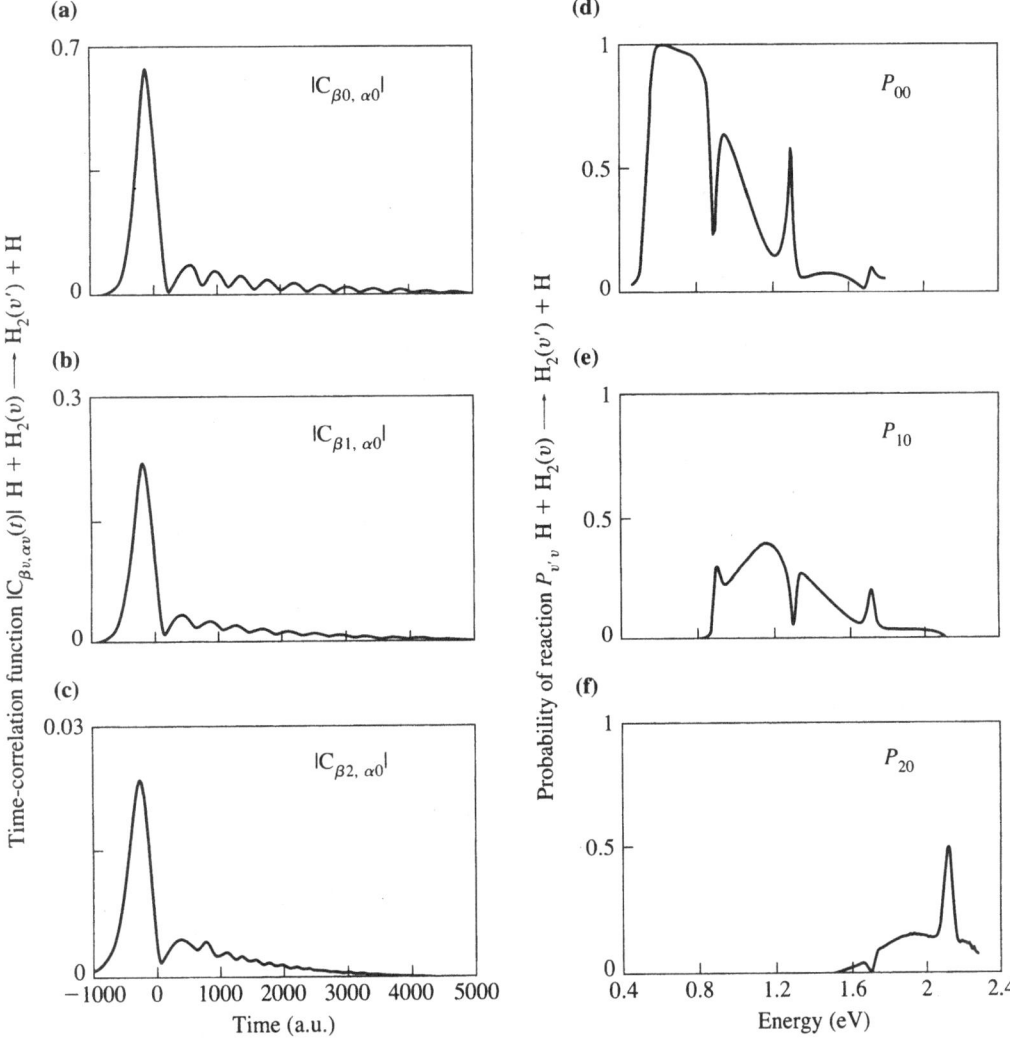

Figure 18.7 Cross-correlation functions for collinear H + H₂. (a) $0 \to 0$; (b) $0 \to 1$; (c) $0 \to 2$. State-to-state scattering probability for collinear H + H₂, corresponding to the collinear functions in (a)–(c). (d) $0 \to 0$; (e) $0 \to 1$; (f) $0 \to 2$. Reprinted from D. E. Weeks and D. J. Tannor (1994).

18.5   Cumulative Reaction Probability

The reaction probability may be defined at different levels of detail. Complete information about the scattering process requires knowledge of all the individual S-matrix elements, $\{S_{\beta\alpha}(E)\}$. Often, however, only some state-averaged quantities are of relevance, such as the initial (final) state-selected *total* reaction probability

$$N_{\alpha(\beta)}(E) = \sum_{\beta(\alpha)} |S_{\beta\alpha}(E)|^2, \tag{18.44}$$

or the *cumulative* reaction probability

$$N(E) = \sum_{\alpha\beta} |S_{\beta\alpha}(E)|^2. \tag{18.45}$$

It is the cumulative reaction probability, averaged over the energy distribution, that gives the thermal reaction rate constant

$$k(T) = \frac{1}{2\pi\hbar Q(T)} \int_0^\infty N(E) e^{-\frac{E}{k_B T}} dE, \tag{18.46}$$

where $Q(T)$ is the reactant partition function. From the practical point of view it is advantageous to find the total or cumulative reaction probabilities directly without explicit reference to the state-to-state probabilities. An example of such a formulation is the expression of Miller, Schwartz and Tromp (Miller, 1983):

$$N(E) = \frac{(2\pi\hbar)^2}{2} \mathrm{Tr}\left(\bar{F}\delta(E-H)\bar{F}\delta(E-H)\right), \tag{18.47}$$

where H is the Hamiltonian of the system and $\bar{F}$ is the symmetrized flux operator. Formally, the trace form of $N(E)$ can be evaluated in any complete basis. Since Eq. 18.47 does not refer to the asymptotic states explicitly, $N(E)$ can be calculated using only dynamics in the transition state region (i.e., interaction region) (see Figure 18.8).

A variety of improvements to Eq. 18.47 have been proposed and tested by Miller's group (Seideman, 1992; Manthe, 1993). These improvements exploit absorbing boundary conditions to minimize the space of the calculation, and the recognition that only a small set of eigenfunctions of the argument of the trace in Eq. 18.47 have any amplitude. Zhang and Light have developed what they call a wavepacket transition state theory for calculating $N(E)$, in which a small number of wavepackets are propagated at the transition state to construct scattering eigenfunctions, from which the flux operators in Eq. 18.47 can be evaluated efficiently (Zhang, 1996).

Given the averaging that occurs in $N(E)$, this seems like a particularly natural quantity to calculate semiclassically. Yet, it was found that semiclassical calculations of $N(E)$ can be expensive even in one dimension (Spath, 1996), and difficult or impossible to converge at all for larger systems. Given the conceptual and numerical advantages of the wavepacket correlation function formulation of the S-matrix, the question naturally arises whether it is possible to use ideas in the same spirit to calculate state-selected total reaction probabilities, $N_\alpha(E)$, and cumulative total reaction probabilities, $N(E)$. A brute force way to calculate $N(E)$ is to calculate the S-matrix elements for every state-to-state transition and then sum the squares of the absolute values. However, as discussed above, the individual S-matrix

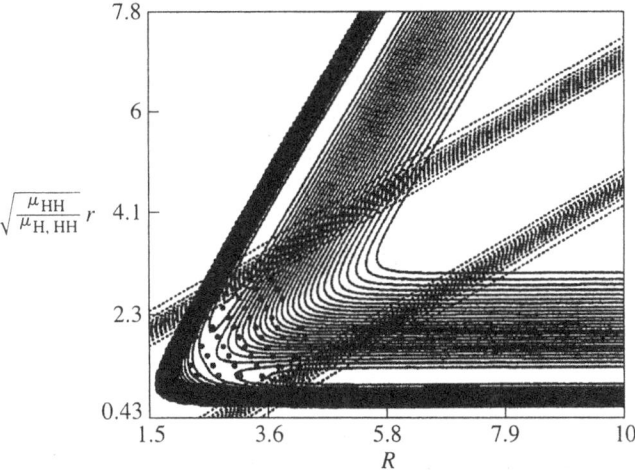

Figure 18.8 Contour diagram of the potential energy surface for the collinear $H + H_2 \rightarrow H_2 + H$ reaction. The dashed lines are contours of an absorbing potential that does not affect the cumulative reaction probability. Notice that the absorbing potential extends into the interaction region, an indication that the cumulative reaction probability is completely determined in the region around the transition state and does not require propagation to the asymptotic region. Adapted from Miller (1998).

elements are defined in terms of specific internal states in both the reactant and product arrangement channels, which in turn implies that the calculation requires precise asymptotic boundary conditions for both reactant and product wavepackets. Yet, as we have seen, the cumulative reaction probability is in principle independent of the asymptotic basis and can be calculated in a relatively small region around the transition state without knowledge of the asymptotic boundary conditions. We describe below two key formulas that maintain the spirit of the correlation function approach but remove, wholly or partially, the need for asymptotic boundary conditions. The first, for the state-selected total reaction probability, $N_\alpha(E)$, is (Garashchuk, 1998)

$$N_\alpha(E) = \frac{\int_{-\infty}^{\infty} dt \langle \hat{P}^+ \Phi_\alpha^+ | e^{-iHt/\hbar} | \hat{P}^+ \Phi_\alpha^+ \rangle e^{iEt/\hbar}}{\int_{-\infty}^{\infty} dt \langle \Phi_\alpha^+ | e^{-iHt/\hbar} | \Phi_\alpha^+ \rangle e^{iEt/\hbar}}. \tag{18.48}$$

This expression has the appealing property that both numerator and denominator are Fourier transforms of wavepacket *auto*correlation functions. $\hat{P}^+$ is a projection operator that formally removes the nonreactive portions from the initial wavepacket Φ_α^+. The physical interpretation of the expression is that the reaction probability from state α to all states β is the ratio of the spectrum of the reactive part of the wavepacket normalized by the total spectrum of the wavepacket.

The second key formula is for the cumulative reaction probability, $N(E)$ (Garashchuk, 1999a, 1999b):

$$N(E) = \mathrm{Tr}\left(\mathbf{A}_{\mathrm{pr}}(E) \mathbf{A}_{\mathrm{r}}^{-1}(E) \mathbf{A}_{\mathrm{pr}}^{\dagger}(E) \mathbf{A}_{\mathrm{p}}^{-1}(E) \right). \tag{18.49}$$

Here, the $\mathbf{A}(E)$ are matrices that are constructed via the Fourier transform of matrices of the form $\mathbf{C}(t)$ element by element. The latter is an array of reactant–product wavepacket cross-correlation functions in the case of $\mathbf{A}_{pr}(\mathbf{C}_{pr})$, and reactant–reactant or product–product cross-correlation functions in the case of $\mathbf{A}_r(\mathbf{C}_r)$ and $\mathbf{A}_p(\mathbf{C}_p)$, respectively. The elements in these matrices connect all initial and all final states, in an arbitrary (even nonorthogonal) internal basis. Equations 18.48 and 18.49 are derived in Sections 18.5.1 and 18.5.2, respectively.

As we show below, Eq. 18.49 combines beautifully with the HHKK propagator (Section 10.5.3), giving a semiclassical $N(E)$ that is in semiquantitative agreement with the converged quantum result. The semiclassical calculation of $N(E)$ is in fact more efficient and robust than the calculation of the individual state-to-state probabilities. In contrast, Eq. 18.48 for $N_\alpha(E)$ surprisingly does not combine well with semiclassical implementation.

18.5.1 Wavepacket Correlation Expression for the Initial State-Selected Reaction Probability

The goal of this section is to find reaction probabilities averaged over the final (or initial) distribution of internal states,

$$N_{\alpha(\beta)}(E) = \sum_{\beta(\alpha)} |S_{\beta\alpha}(E)|^2.$$

This quantity does not necessarily require the knowledge of the internal eigenstates of the product (or reactant) asymptotic Hamiltonian, nor even of the dynamics in the exit (or entrance) region of the potential, where the distribution between the internal eigenstates of fragments is still changing but the total reaction probability is not. In this section we take advantage of this idea to develop a variety of simple expressions for $N_{\alpha(\beta)}(E)$ that are in the spirit of the wavepacket correlation function approach of the previous section but do not require information on the asymptotic product (or reactant) asymptotic internal states, or the exit (entrance) channel dynamics.

The starting point of the derivations is the definition of the S-matrix:

$$S_{\beta\alpha}(E)\delta(E - E') = \langle \psi_{\beta,E'}^- | \psi_{\alpha,E}^+ \rangle. \tag{18.50}$$

The reaction probability from the initial state with internal quantum numbers α to a specific final state of the reactants with the internal quantum numbers β is proportional to

$$S_{\beta\alpha}^*(E)\delta(E - E')S_{\beta\alpha}(E'')\delta(E' - E'') = \langle \psi_{\alpha,E}^+ | \psi_{\beta,E'}^- \rangle \langle \psi_{\beta,E'}^- | \psi_{\alpha,E''}^+ \rangle. \tag{18.51}$$

Integration over E' and summation over all β of the product channel gives

$$\sum_{\beta} S_{\beta\alpha}^*(E)S_{\beta\alpha}(E'')\delta(E - E'') = \langle \psi_{\alpha,E}^+ | \hat{P}^+ | \psi_{\alpha,E''}^+ \rangle, \tag{18.52}$$

where $\hat{P}^+$ denotes the operator projecting onto products

$$\hat{P}^+ \equiv \sum_{\beta} \int_0^\infty |\psi_{\beta,E'}^-\rangle \langle \psi_{\beta,E'}^-| dE'. \tag{18.53}$$

Using the relationship

$$\psi_{\alpha,E}^{+} = \frac{1}{\eta_{\alpha}(E)} \delta(E - H)\Phi_{\alpha}^{+}, \tag{18.54}$$

which is an expression of the spectral representation of the scattering eigenstate, we have

$$\sum_{\beta} S_{\beta\alpha}^{*}(E) S_{\beta\alpha}(E'') \delta(E - E'') = \frac{1}{\eta_{\alpha}^{*}(E)\eta_{\alpha}(E'')} \langle \Phi_{\alpha}^{+} | \delta(E - H)\hat{P}^{+}\delta(E'' - H) | \Phi_{\alpha}^{+} \rangle$$

$$= \frac{1}{\eta_{\alpha}^{*}(E)\eta_{\alpha}(E'')} \langle \Phi_{\alpha}^{+} | \delta(E - H)\hat{P}^{+} | \Phi_{\alpha}^{+} \rangle \delta(E'' - E). \tag{18.55}$$

Integrating Eq. 18.55 over E'' and replacing $\delta(E - H)$ by its integral expression we obtain

$$N_{\alpha}(E) = \frac{(2\pi\hbar)^{-1}}{|\eta_{\alpha}(E)|^{2}} \int_{-\infty}^{\infty} dt \langle \Phi_{\alpha}^{+} | \hat{P}^{+} e^{-iHt/\hbar} | \Phi_{\alpha}^{+} \rangle e^{iEt/\hbar}. \tag{18.56}$$

Equation 18.56 expresses $N_{\alpha}(E)$ as the cross-correlation function of the incoming wavepacket $|\Phi_{\alpha}\rangle$ with its reactive part $\hat{P}^{+}|\Phi_{\alpha}^{+}\rangle$.

Using the properties of the projection operator $(\hat{P}^{+})^{2} = \hat{P}^{+}$ and $(\hat{P}^{+})^{\dagger} = \hat{P}^{+}$ and the spectral representation of the normalization coefficient η_{α} we can rewrite Eq. 18.56 in a symmetrized way:

$$N_{\alpha}(E) = \frac{\int_{-\infty}^{\infty} dt \langle \hat{P}^{+}\Phi_{\alpha}^{+} | e^{-iHt/\hbar} | \hat{P}^{+}\Phi_{\alpha}^{+} \rangle e^{iEt/\hbar}}{\int_{-\infty}^{\infty} dt \langle \Phi_{\alpha}^{+} | e^{-iHt/\hbar} | \Phi_{\alpha}^{+} \rangle e^{iEt/\hbar}}. \tag{18.57}$$

This expression is manifestly real since it involves the Fourier transform only of autocorrelation functions. It expresses the state-selected total reaction probability as a ratio of two spectra, that of the reactive wavepacket divided by that of the initial wavepacket, which has both reactive and nonreactive components. The latter spectrum normalizes the former, ensuring that the total reaction probability is between 0 to 1 and that the result at each energy is independent of the choice of energy distribution in the initial translational wavepacket.

All expressions for the state-selected total reaction probability can be reversed to treat the case of a single well-defined internal state of products, summed over all internal states of reactants. To do so, the labels α and β must be interchanged, and the projection operator must be redefined to project onto reactants.

18.5.2 $N(E)$ Based on Reactant–Product Cross-Correlation Functions

The derivations of Sections 18.3.1 and 18.5.1 for state-to-state and initial-state-selected total reaction probabilities, respectively, involve wavepackets with well-defined internal quantum numbers α for reactants and/or β for products. In this section we show how the previous expressions for $|S_{\beta\alpha}(E)|^{2}$ and $N_{\alpha}(E)$ in terms of wavepacket correlation functions can be generalized to calculate the *cumulative* reaction probability with an arbitrary complete set of internal functions. Note that in order to obtain the cumulative reaction probability by brute force, as a sum of the state-to-state probabilities, one has to define and propagate as many wavepackets as there are internal states. Thus, for a system with N_{r} internal eigenstates in the asymptotic channel of reactants and N_{p} internal eigenstates in the asymptotic channel of products, one needs a set $\{\Phi_r\}$ of N_r wavepackets on the reactant side that correlate with the internal eigenstates of reactants, and a set of N_p wavepackets $\{\Phi_p\}$ on the product side correlating with internal product eigenstates. Now, let us consider wavepackets that are arbitrary in the internal degrees of freedom, instead of being eigenstates. These wavepackets are some unknown linear combinations of the energy eigenstates with different internal quantum numbers, with incoming boundary conditions for reactant wavepackets and with outgoing boundary conditions for product wavepackets. The vector of reactant and product wavepackets can be written as

$$\left(|\Phi_r^1\rangle \ldots |\Phi_r^{N_r}\rangle \right) = \int_0^\infty dE \left(|\psi_{1,E}^+\rangle \ldots |\psi_{N_r,E}^+\rangle \right) \mathbf{M}_r(E) \tag{18.58}$$

for reactants and

$$\left(|\Phi_p^1\rangle \ldots |\Phi_p^{N_p}\rangle \right) = \int_0^\infty dE \left(|\psi_{1,E}^-\rangle \ldots |\psi_{N_p,E}^-\rangle \right) \mathbf{M}_p(E) \tag{18.59}$$

for products. We emphasize that the matrices $\mathbf{M}_{r/p}$ are unknown and will not appear in our final working expression. Here and below we represent the wavepackets as vectors explicitly to avoid additional indices required for tensor notation.

Now we propagate the reactant wavepackets and calculate all the reactant–reactant cross-correlation functions; similarly we calculate all the product–product cross-correlation functions:

$$C_r^{ij}(t) = \langle \Phi_r^i | \exp(-iHt/\hbar) | \Phi_r^j \rangle \quad \text{and} \quad C_p^{ij}(t) = \langle \Phi_p^i | \exp(-iHt/\hbar) | \Phi_p^j \rangle$$

for $\{i, j\} = 1 \ldots N_r$ for reactants and $\{i, j\} = 1 \ldots N_p$ for products. This defines an $N_r \times N_r$ matrix for reactants and an $N_p \times N_p$ matrix for products. Calculating the Fourier transform of each element,

$$A_r^{ij}(E) = \int_{-\infty}^\infty dt \, C_r^{ij}(t) e^{iEt/\hbar} \quad \text{and} \quad A_p^{ij}(E) = \int_{-\infty}^\infty dt \, C_p^{ij}(t) e^{iEt/\hbar},$$

we form new matrices, $\mathbf{A}_r(E)$ and $\mathbf{A}_p(E)$. It is readily verified that $\mathbf{A}_r(E)$ and $\mathbf{A}_p(E)$ may be written as

$$\mathbf{A}_r(E) = 2\pi \mathbf{M}_r^\dagger(E)\mathbf{M}_r(E), \qquad \mathbf{A}_p(E) = 2\pi \mathbf{M}_p^\dagger(E)\mathbf{M}_p(E). \tag{18.60}$$

Equations 18.60 provide a strategy for eliminating the unknown coefficients $\mathbf{M}_r(E)$ and $\mathbf{M}_p(E)$ in terms of the known matrices $\mathbf{A}_r(E)$ and $\mathbf{A}_p(E)$, provided that the matrix $\mathbf{M}_{r/p}$ and its adjoint always come together; below we will show that this is indeed the case.

Finally, we define the reactant–product cross-correlation functions

$$C_{pr}^{ij}(t) = \langle \Phi_p^i | \exp(-iHt/\hbar) | \Phi_r^j \rangle$$

and their Fourier transforms $A_{pr}^{ij}(E)$, $i = 1 \ldots N_p$, $j = 1 \ldots N_r$. From Eqs. 18.58 and 18.59 the matrix $\mathbf{A}_{pr}(E) = \{A_{pr}^{ij}(E)\}$ can be written as

$$\mathbf{A}_{pr}(E) = 2\pi \mathbf{M}_p^\dagger(E)\mathbf{S}_{pr}(E)\mathbf{M}_r(E), \tag{18.61}$$

where $\mathbf{S}_{\mathrm{pr}}(E)$ is a block of the scattering matrix that connects the chosen reactants and products. Note that the product of the matrices in Eq. 18.61 produces the matrix $\mathbf{S}_{\mathrm{pr}}(E)$ of dimension $N_{\mathrm{p}} \times N_{\mathrm{r}}$.

Now, realizing that the cumulative reaction probability can be expressed as a *trace*

$$N(E) = \mathrm{Tr}\left(\mathbf{S}_{\mathrm{pr}}(E)\mathbf{S}_{\mathrm{pr}}^{\dagger}(E)\right) \tag{18.62}$$

and using Eqs. 18.61 and 18.60 we obtain

$$N(E) = \mathrm{Tr}\left(\mathbf{A}_{\mathrm{pr}}(E)\mathbf{A}_{\mathrm{r}}^{-1}(E)\mathbf{A}_{\mathrm{pr}}^{\dagger}(E)\mathbf{A}_{\mathrm{p}}^{-1}(E)\right). \tag{18.63}$$

This is our final expression for $N(E)$ in terms of the dynamics of incoming and outgoing wavepackets in an *arbitrary basis of internal functions*. Note that all dependence on the matrices $\mathbf{M}_{\mathrm{r}}$ and $\mathbf{M}_{\mathrm{p}}$, which would require knowledge of the asymptotic states, has disappeared.

If the wavepackets are initially located in the asymptotic region of the potential and they each correspond to a single internal eigenstate α and β, then Eq. 18.63 reduces to $N(E) = \sum_{\alpha\beta} |S_{\beta\alpha}(E)|^2$. Each S-matrix element in this sum takes the form of Eq. 18.29.

Equation 18.63 was implemented for the collinear $H + H_2 \rightarrow H_2 + H$ reaction on the Wall–Porter potential surface, using both a fully quantum mechanical calculation and using the semiclassical propagator of Heller–Herman–Kay–Kluk, described in detail in Section 10.5.3. As shown in Figure 18.9, the cumulative reaction probability for the HHKK and quantum mechanical propagation are in semiquantitative agreement.

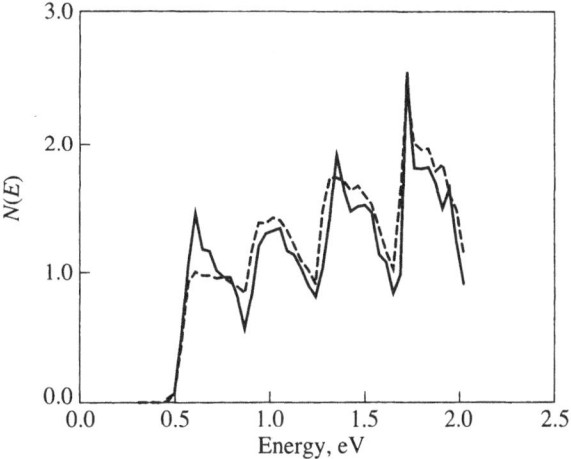

Figure 18.9 The cumulative reaction probability, obtained from the semiclassical propagation of wavepackets (solid line) and the quantum mechanical result (dashed line). The wavepackets were initially located at $R_0 = 4.7$ a.u. in the translational coordinate and were constructed as the harmonic oscillator eigenstates in the vibrational coordinate. Adapted from Garashchuk (1999a).

We have reached the end of our journey. The calculation of cumulative reaction probabilities has brought together almost all the major themes of this book: wavepacket correlation functions, chemical reaction dynamics and semiclassical approximation methods. It is my fervent hope that the reader will have obtained some appreciation of the fundamental unity and simplicity of time-dependent quantum mechanics, and will be inspired to use the concepts and methods presented here in original and imaginative new contexts.

References

General

1. J. Taylor, *Scattering Theory: The Quantum Theory of Nonrelativistic Collisions* (Krieger, Malabar, FL, 1987).
2. R. G. Newton, *Scattering Theory of Waves and Particles* (Springer-Verlag, New York, 1982).
3. R. D. Levine, *Quantum Mechanics of Molecular Rate Processes* (Clarendon Press, Oxford, 1969).
4. M. Baer, ed., *Theory of Chemical Reaction Dynamics*, 4 vols. (CRC Press, Boca Raton, FL, 1985).
5. J. Z. H. Zhang, *Theory and Application of Quantum Molecular Dynamics* (World Scientific, Singapore, 1999).
6. D. Manolopoulos, State-to-state reactive scattering, in *Encyclopedia of Computational Chemistry*, ed. P. Von R. Schleyer (Wiley, Chichester, 1999).
7. D. J. Tannor and S. Garashchuk, Semiclassical calculation of chemical reaction dynamics via wavepacket correlation functions, Annu. Rev. Phys. Chem. 51, 553 (2000).
8. G. C. Schatz and M. A. Ratner, *Quantum Mechanics in Chemistry* (Prentice Hall, Englewood Cliffs, NJ, 1993).

Spectral Method

9. M. J. Davis and E. J. Heller, Multidimensional wave functions from classical trajectories, J. Chem. Phys. 75, 3916 (1981).
10. M. D. Feit and J. A. Fleck, Jr., Solution of the Schrödinger equation by a spectral method II: Vibrational energy levels of triatomic molecules, J. Chem. Phys. 78, 301 (1983).
11. R. Viswanathan, S. Shi, E. Villalonga and H. Rabitz, Calculation of scattering wave functions by a numerical procedure based on the Møller wave operator, J. Chem. Phys. 91, 2333 (1989).

Reactive Scattering Calculations Using Wavepackets

12. E. A. McCullough, Jr. and R. E. Wyatt, Quantum dynamics of the collinear (H, H_2) reaction, J. Chem. Phys. 51, 1253 (1969); E. A. McCullough Jr. and R. E. Wyatt, Dynamics of the collinear $H+H_2$ reaction. I. Probability density and flux, J. Chem. Phys. 54, 3578 (1971).
13. E. J. Heller, Time-dependent approach to semiclassical dynamics, J. Chem. Phys. 62, 1544 (1975); E. J. Heller, Classical S-matrix limit of wavepacket dynamics, J. Chem. Phys. 65, 4979 (1976).
14. R. C. Mowrey and D. J. Kouri, On a hybrid close-coupling-wavepacket approach to molecular scattering, Chem. Phys. Lett. 119, 285 (1985); J. Z. H. Zhang and D. J. Kouri, Phys. Rev. A 34, 2687 (1986).

15. D. Neuhauser, M. Baer, R. S. Judson and D. J. Kouri, The application of time-dependent wavepacket methods to reactive scattering, Comp. Phys. Comm. 63, 460 (1991).

16. G. G. Balint-Kurti, R. N. Dixon, C. C. Marston and A. J. Mulholland, The calculation of product quantum state distributions and partial cross-sections in time-dependent molecular collision and photodissociation theory, Comp. Phys. Comm. 63, 126 (1991).

17. D. H. Zhang, M. A. Collins and S.-Y. Lee, First principles theory for the $H+H_2O$, D_2 reactions, Science 290, 961 (2000).

18. F. Huarte-Larranaga and U. Manthe, Full dimensional quantum calculations of the $CH_4+H \rightarrow CH_3+H_2$ reaction rate, J. Phys. Chem. A 105, 2522 (2001).

19. S. C. Althorpe, Quantum wavepacket method for state-to-state reactive cross-sections, J. Chem. Phys. 114, 1601 (2001).

20. S. C. Althorpe, F. Fernandez-Alonso, B. D. Bean, J. D. Ayers, A. E. Pomerantz, R. N. Zare and E. Wrede, Observation and interpretation of a time-delayed mechanism in the hydrogen exchange reaction, Nature 416, 67 (2002).

Wavepacket Correlation Function Formulation of the S-matrix

21. D. J. Tannor and D. E. Weeks, Wavepacket correlation function formulation of scattering theory: The quantum analog of classical S-matrix theory, J. Chem. Phys. 98, 3884 (1993); D. E. Weeks and D. J. Tannor, A time-dependent formulation of the scattering matrix using Møller operators, Chem. Phys. Lett. 207, 301 (1993).

22. D. E. Weeks and D. J. Tannor, A time-dependent formulation of the scattering matrix for the collinear reaction $H+H_2(v) \rightarrow H_2(v')+H$, Chem. Phys. Lett. 224, 451 (1994).

23. D. K. Bondi and J. N. L. Conner, A new numerical method for collinear quantum reactive scattering using Delves' coordinates: Application to the $H + H_2$ $(n \leq 7) \rightarrow H_2$ $(m \leq 7) + H$ chemical reaction, Chem. Phys. Lett. 92, 570 (1982).

24. A. Jäckle and H.-D. Meyer, Reactive scattering using the multiconfiguration time-dependent Hartree approximation: General aspects and application to the collinear $H + H_2 \rightarrow H_2+H$ reaction, J. Chem. Phys. 102, 5605 (1995).

25. J. Dai and J. Z. H. Zhang, Time-dependent wavepacket approach to state-to-state reactive scattering and application to $H+O_2$ reaction, J. Phys. Chem. 100, 6898 (1996).

26. D. J. Kouri, Y. Huang, W. Zhu and D. K. Hoffman, Variational principles for the time-independent wavepacket-Schrödinger and wavepacket-Lippmann-Schwinger equations, J. Chem. Phys. 100, 3662 (1994).

27. N. Rom, J. W. Pang and D. Neuhauser, Scattering matrix elements by a time independent wavepacket complex scaling formalism, J. Chem. Phys. 105, 10436 (1996).

28. F. Grossmann and E. J. Heller, A semiclassical correlation function approach to barrier tunneling, Chem. Phys. Lett. 241, 45 (1995).

29. S. Garashchuk and D. J. Tannor, Wavepacket correlation function approach to $H_2(v)+H \rightarrow H+H_2(v')$: Semiclassical implementation, Chem. Phys. Lett. 262, 477 (1996).

30. S. Garashchuk, F. Grossmann and D. J. Tannor, Semiclassical approach to the hydrogen exchange reaction: Reactive and transition state dynamics, J. Chem. Soc., Faraday Trans. 93, 781 (1997).

Cumulative Reaction Probability

31. W. H. Miller, S. D. Schwartz and J. W. Tromp, Quantum mechanical rate constants for bimolecular reactions, J. Chem. Phys. 79, 4889 (1983).

32. T. Seideman and W. H. Miller, Calculation of the cumulative reaction probability via a discrete variable representation with absorbing boundary conditions, J. Chem. Phys. 96, 4412 (1992).

33. U. Manthe and W. H. Miller, The cumulative reaction probability as eigenvalue problem, J. Chem. Phys. 99, 3411 (1993).

34. D. H. Zhang and J. C. Light, Cumulative reaction probability via transition state wavepackets, J. Chem. Phys. 104, 6184 (1996).

35. B. Spath and W. H. Miller, Semiclassical calculation of cumulative reaction probabilities, J. Chem. Phys. 104, 95 (1996).

36. W. H. Miller, "Direct" and "correct" calculation of canonical and microcanonical rate constants for chemical reactions, J. Phys. Chem. A 102, 793 (1998).

37. S. Garashchuk and D. J. Tannor, Cumulative reaction probability in terms of reactant-product wavepacket correlation functions, J. Chem. Phys. 110, 2761 (1999a).

38. S. Garashchuk and D. J. Tannor, Semiclassical calculation of cumulative reaction probabilities, Phys. Chem. Chem. Phys. 1, 1081 (1999b).

Appendix A

The Dirac Delta Function and the Cauchy Principal Value

In this Appendix we discuss the definition and properties of the Dirac delta function and the Cauchy principal value.

A.1   The Dirac Delta Function

Consider the function $R_a(x)$:

$$R_a(x) = \begin{cases} \frac{1}{2a} & -a < x < a \\ 0 & x < -a, \quad x > a. \end{cases} \tag{A.1}$$

$R_a(x)$ is a rectangular function with area $\int_{-\infty}^{\infty} R_a(x)\, dx = 1$. We define the Dirac delta function as the limit of $R_a(x)$ as $a \to 0$:

$$\delta(x) = \lim_{a \to 0} R_a(x) \tag{A.2}$$

Pictorially, the Dirac delta function is an infinitely high and infinitely narrow rectangle, defined such that the integrated area is equal to 1:

$$\int_{-\infty}^{\infty} \delta(x)\, dx = \int_{-\infty}^{\infty} \lim_{a \to 0} R_a(x)\, dx = \lim_{a \to 0} \int_{-\infty}^{\infty} R_a(x)\, dx = \lim_{a \to 0} 1 = 1. \tag{A.3}$$

A basic property of the delta function is that when it multiplies any smooth function $f(x)$, $f(x)$ can be replaced with its value at the delta function:

$$f(x)\delta(x - x_0)\, dx = f(x_0)\delta(x - x_0). \tag{A.4}$$

This result can be rationalized as follows. Since the delta function vanishes everywhere except at $x = x_0$, it kills $f(x)$ everywhere except in the infinitesimal neighborhood of $x = x_0$. In the infinitesimal neighborhood of $x = x_0$, $f(x) \approx f(x_0)$. An immediate consequence of Eq. A.4 is that when the delta function appears inside an integral along with a smooth function, the value of the integral is simply the value of the *integrand* at the position of the delta function. That is,

$$\int_{-\infty}^{\infty} f(x)\delta(x - x_0)dx = f(x_0). \tag{A.5}$$

▶ **Exercise A.1**   Show that the integral of the product of two delta functions is

$$\int_{-\infty}^{\infty} \delta(x - x_0)\delta(x - x_1)dx = \delta(x_0 - x_1). \tag{A.6}$$

▶ **Exercise A.2**   Show that

$$\int_{x_1}^{x_2} \delta(x - x_0)dx = \begin{cases} 1 & \text{if } x_1 \leq x_0 \leq x_2 \\ 0 & x_0 < x_1 \text{ or } x_0 > x_2 \end{cases} ; \tag{A.7}$$

that is, the integral of a delta function over a finite interval gives 1 if the delta function lies within that interval and gives zero otherwise. [Hint: consider the integral $\int_{-\infty}^{\infty} f(x)\delta(x - x_0)\, dx$, where $f(x) = 1$ for $x_1 \leq x \leq x_2$ and $f(x) = 0$ outside the interval $[x_1, x_2]$.]

An important property of the Dirac delta function is

$$\delta(ax) = \frac{1}{|a|}\delta(x). \tag{A.8}$$

More generally,

$$\delta(f(x)) = \sum_i \frac{1}{|f'(x_i)|}\delta(x - x_i), \tag{A.9}$$

where $f'(x_i)$ is the derivative of $f(x)$ with respect to x evaluated at the positions x_i where $f(x_i)$ vanishes, provided that $f'(x_i) \neq 0$. The summation is over all such positions x_i.

▶ **Exercise A.3**

   a. Derive Eq. A.8 by considering the expression A.1 and changing variables.
   b. Derive Eq. A.9 under the assumption that $f(x) \approx f'(x_i)(x - x_i)$ for $x \approx x_i$, and using the result of (a).

▶ **Exercise A.4**   Show that if $a \neq b$,

$$\delta((x - a)(x - b)) = \frac{1}{|a - b|} \left[\delta(x - a) + \delta(x - b) \right]. \tag{A.10}$$

We defined the Dirac delta function in terms of the limiting case of a rectangular area that gets taller and narrower with its area preserved at 1. However, the specific choice of a rectangular shape is arbitrary, and the Dirac delta function can be defined as the limit of a wide variety of other functions. For example,

$$\lim_{\alpha \to \infty} \left(\frac{\alpha}{\pi}\right)^{1/4} e^{-\alpha x^2} \tag{A.11}$$

also becomes infinitely narrow and tall while keeping the area constant and equal to 1.

One of the most interesting and important representations of the Dirac delta function is

$$\delta(x) = \frac{1}{2\pi} \int_{-\infty}^{\infty} e^{ikx}\, dk. \tag{A.12}$$

What is striking about this representation is that it is not immediately obvious that it describes a function that is infinitely narrow and infinitely high. However, this representation can be rationalized as follows. First, note that if $x = 0$ then the integrand is equal to 1 and hence the RHS is infinite, consistent with the rectangular definition of the delta function. Second, note that for any $x \neq 0$ the integrand oscillates periodically and hence in integrating these oscillations from $(-\infty, \infty)$ the positive and negative contributions cancel and one obtains zero. More generally,

$$\delta(x - x_0) = \frac{1}{2\pi} \int_{-\infty}^{\infty} e^{-ikx_0} e^{ikx} \, dk. \tag{A.13}$$

The significance of Eq. A.13 is that both e^{-ikx_0} and e^{ikx} are periodically oscillating functions. If x and x_0 are different, even ever so slightly, then over the infinite range of integration the functions e^{ikx_0} and e^{ikx} will continue to get in phase and out of phase, leading to an overall cancellation of position and negative contributions in the product. Only if x and x_0 are exactly equal will the two functions continue to oscillate in phase for the entire range of integration, in this case returning an infinite value for the integral. Equation A.13 can be viewed as the fundamental equation of Fourier analysis, and indeed we use it throughout the book to calculate the inverse Fourier transform.

▶ **Exercise A.5** Derive each of the following representations of the Dirac delta function by viewing Eq. A.12 as the limit of another expression and performing the integral analytically. (The other expression may involve rewriting the limits of integration or inserting a factor in the integrand that goes to 1 in the limit.)

a. $\delta(x) = \frac{1}{\sqrt{\pi}} \lim_{\epsilon \to 0} \frac{1}{\sqrt{\epsilon}} e^{-x^2/\epsilon}$

b. $\delta(x) = \frac{1}{\pi} \lim_{\epsilon \to 0} \frac{\epsilon}{x^2 + \epsilon^2}$

c. $\delta(x) = \frac{1}{\pi} \lim_{N \to \infty} \frac{\sin Nx}{x}$

d. Show that the representation $\delta(x) = \frac{1}{2} \frac{d^2}{dx^2} |x|$ satisfies the conditions of Eq. A.7.

A.2   The Cauchy Principal Value

In this section we study the quantity $\int_0^{\infty} e^{ikx} \, dk$. We shall see that under appropriate conditions we can write

$$\int_0^{\infty} e^{ikx} \, dk = i\mathcal{P}\left(\frac{1}{x}\right) + \pi\delta(x), \tag{A.14}$$

where $\mathcal{P}$ will be defined immediately below. As this point we simply draw attention to the fact that, as opposed to $\int_{-\infty}^{\infty} e^{ikx} \, dk$, $\int_0^{\infty}$ has a finite imaginary part in addition to the delta function real part.

We begin our analysis by reconsidering the integral $\int_{-\infty}^{\infty} e^{ikx} \, dk$. We do not a priori know how to evaluate the integrand of this integral as $k \to \infty$ or $k \to -\infty$. We adopt the convention that the integrand is understood to have a small but nonvanishing factor $e^{\pm \epsilon k}$ that *regularizes* the integral so that the value of the integrand tapers off to 0 at each of the two infinite limits. In Exercise A.5 we added regularizing factors in order to derive alternative expressions for the delta function; here we are adding these same factors to legitimize the mathematics. Thus,

$$\int_{-\infty}^{\infty} e^{ikx}\,dk = \lim_{\epsilon \to 0} \int_{-\infty}^{\infty} e^{ikx} e^{-\epsilon|k|}\,dk \tag{A.15}$$

$$= \lim_{\epsilon \to 0} \int_{-\infty}^{0} e^{ikx} e^{+\epsilon k}\,dk + \lim_{\epsilon \to 0} \int_{0}^{\infty} e^{ikx} e^{-\epsilon k}\,dk \tag{A.16}$$

$$= \lim_{\epsilon \to 0} \left\{ \frac{1}{ix + \epsilon} - \frac{1}{ix - \epsilon} \right\} \tag{A.17}$$

$$= \lim_{\epsilon \to 0} \frac{2\epsilon}{x^2 + \epsilon^2} = 2\pi\delta(x). \tag{A.18}$$

In the last step we are simply using one of the representations of the delta function that we have already seen (Exercise A.5b).

Let us now consider the half integral $\int_0^\infty e^{ikx}dk$:

$$\int_{0}^{\infty} e^{ikx}\,dk = \lim_{\epsilon \to 0} \int_{0}^{\infty} e^{ikx - \epsilon k}\,dk \tag{A.19}$$

$$= \lim_{\epsilon \to 0} \frac{1}{ix - \epsilon} \tag{A.20}$$

$$= \lim_{\epsilon \to 0} \frac{ix}{x^2 + \epsilon^2} + \lim_{\epsilon \to 0} \frac{\epsilon}{x^2 + \epsilon^2}. \tag{A.21}$$

Comparing with Eq. A.18 we see that the second term in Eq. A.21 is equal to $\pi\delta(x)$. The first term goes to $\frac{i}{x}$ as $\epsilon \to 0$. The one point that requires special attention is $x = 0$. It turns out that in spite of the divergence of the function $\frac{i}{x}$ at this point, if $\frac{i}{x}$ appears inside an integral multiplied by a smooth function $f(x)$ there is no contribution from the neighborhood of the point $x = 0$. This is because the function $\frac{1}{x}$ changes sign infinitely rapidly in this neighborhood and hence the positive and negative contributions to the integral cancel:

$$\int_{-\infty}^{\infty} f(x)\frac{i}{x}\,dx = \int_{-\infty}^{-\epsilon} f(x)\frac{i}{x}\,dx + \int_{\epsilon}^{\infty} f(x)\frac{i}{x}\,dx \equiv i\mathcal{P}\int_{-\infty}^{\infty} f(x)\frac{1}{x}\,dx. \tag{A.22}$$

$\mathcal{P}$ is called the Cauchy principal value of the function $\frac{1}{x}$. $\mathcal{P}\left(\frac{1}{x}\right)$ is defined to be identical with $\frac{1}{x}$ except at the singular point $x = 0$, which is excised from the domain. The ability to replace $\frac{1}{x}$ with $\mathcal{P}\left(\frac{1}{x}\right)$ is a formal way of saying that there is no contribution to the integral from the neighborhood of $x = 0$.

Thus, provided that the expression is in an integrand together with a well-behaved function of $f(x)$ we have the *effective* equality

$$\int_{0}^{\infty} e^{ikx}\,dk = \lim_{\epsilon \to 0} \frac{1}{ix - \epsilon} = i\mathcal{P}\left(\frac{1}{x}\right) + \pi\delta(x). \tag{A.23}$$

We make several comments about this final result:

1. The real part of this integral is 1/2 times the real part of $\int_{-\infty}^{\infty} e^{ikx}\, dx$. This makes sense since the range of integration is only $[0, \infty)$ instead of $(-\infty, \infty)$.

2. The imaginary part of the integral comes from the truncation of the lower limit of the integral at 0. This can be seen by treating $\int_0^{\infty} e^{ikx}\, dk$ as an ordinary exponential integral and considering the contribution from the integration limit 0.

3. The function $\mathcal{P}\left(\frac{1}{x}\right)$ excises the singular point $x = 0$ from the domain of the function $\frac{1}{x}$. Note that this is the one point for which the delta function is nonvanishing. In a way that can be made rigorous using the tools of complex analysis, the delta function in fact fills in for the missing point $x = 0$, with the phase difference of $\pi/2$ a consequence of deforming the integration along the real axis such that the contour circumvents the singular point $x = 0$.

The next two Exercises establish some useful properties of the integral of $e^{\pm ikx}$ when only one of the limits goes to $\pm\infty$. These relations are used in Section 7.4.1 to combine various terms to obtain a δ function, hence we will refer to these relations as "rules for combining partial δ functions."

▶ **Exercise A.6**   Show that $\int_{-\infty}^{0} e^{-ikx}\, dk = \int_0^{\infty} e^{ikx}\, dk$.

Solution   As above, we adopt the convention that each integral is understood to have a regularizing factor such that the integrand vanishes as $t \to -\infty$ or $t \to \infty$. Thus

$$\int_{-\infty}^{0} e^{-ikx}\, dk = \lim_{\epsilon \to 0} \int_{-\infty}^{0} e^{-ikx+\epsilon k}\, dk = \lim_{\epsilon \to 0} \frac{1}{-ix+\epsilon}$$

$$= \lim_{\epsilon \to 0} -\frac{1}{ix-\epsilon} = \lim_{\epsilon \to 0} \int_0^{\infty} e^{ikx-\epsilon k}\, dk = \int_0^{\infty} e^{ikx}\, dk. \quad \text{(A.24)}$$

▶ **Exercise A.7**

a. Consider the four expressions

$$(i)\ \int_{-\infty}^{0} e^{ikx}dk, \qquad (ii)\ \int_{-\infty}^{a} e^{ikx}dk, \qquad (iii)\ \int_0^{\infty} e^{ikx}dk,$$

$$(iv)\ \int_a^{\infty} e^{ikx}dk. \qquad\qquad\qquad\qquad\qquad\qquad\quad \text{(A.25)}$$

Rewrite each of these expressions in the form $A + B\delta(x)$, and show that B is identical in all four cases. [Hint: you will need Eq. A.4.]

b. Generalize your results from **a.** from e^{ikx} to $e^{ikf(x)}$. For each of the four cases in **a**, write the expression in the form $A + B\delta(x)$, and show that B is identical in all four cases.

References

1. P. A. M. Dirac, *The Principles of Quantum Mechanics* (Oxford University Press, Oxford, 1958).

2. C. Cohen-Tannoudji, B. Diu and F. Laloë, *Quantum Mechanics* vol II (Wiley, New York, 1977), Appendix II.

3. F. Reif, *Fundamentals of Statistical and Thermal Physics* (McGraw-Hill, New York, 1965), Appendix 7.

4. E. Merzbacher, *Quantum Mechanics* (Wiley, New York, 1970), 2nd ed.

5. M. D. Fayer, *Elements of Quantum Mechanics* (Oxford University Press, New York, 2001).

6. M. J. Lighthill, *Fourier Analysis and Generalized Functions* (Cambridge University Press, Cambridge, 1958).

Appendix B

Composite Systems

In this Appendix we consider composite systems, that is, systems consisting of several degrees of freedom. We introduce many of the most important concepts such as entanglement and separability by focusing on the simplest example, of a composite system consisting of two two-level systems. We also use this system to introduce the concept of the *reduced* density matrix, that is, the density matrix of a subsystem, obtained by tracing out the remaining degree(s) of freedom. As we shall see, even if the composite system corresponds to a pure state, the reduced density matrix is generally a probabilistic mixture of pure states of the remaining Hilbert space. In fact, the concept of the density operator is sometimes motivated in this way: that the probabilistic mixture of pure states associated with the density operator reflects the loss of information associated with the tracing out of part of a larger system. In this sense, this Appendix complements and adds to the contents of Chapter 5.

B.1 Wavefunction of a Composite System: Separability and Entanglement

Consider a Hilbert space corresponding to two two-level systems. We can define a basis for system 1 as $|1:+\rangle$, $|1:-\rangle$ and a basis for system 2 as $|2:+\rangle$, $|2:-\rangle$. A general state of system 1 is

$$|\phi(1)\rangle = \alpha_1|1:+\rangle + \beta_1|1:-\rangle \tag{B.1}$$

and similarly a general state of system 2 is

$$|\chi(2)\rangle = \alpha_2|1:+\rangle + \beta_2|2:-\rangle. \tag{B.2}$$

A basis for the composite system may be defined as

$$|++\rangle = |1:+\rangle|2:+\rangle \tag{B.3}$$
$$|+-\rangle = |1:+\rangle|2:-\rangle \tag{B.4}$$
$$|-+\rangle = |1:-\rangle|2:+\rangle \tag{B.5}$$
$$|--\rangle = |1:-\rangle|2:-\rangle. \tag{B.6}$$

A product state of the composite system takes the form

$$|\phi(1)\rangle|\chi(2)\rangle = \alpha_1\alpha_2|++\rangle + \alpha_1\beta_2|+-\rangle + \alpha_2\beta_1|-+\rangle + \beta_1\beta_2|--\rangle. \tag{B.7}$$

But note that this is not the most general state of the composite system: the most general state has the form

$$|\psi\rangle = \alpha|++\rangle + \beta|+-\rangle + \gamma|-+\rangle + \delta|--\rangle, \tag{B.8}$$

with the only constraint being

$$|\alpha|^2 + |\beta|^2 + |\gamma|^2 + |\delta|^2 = 1. \tag{B.9}$$

For Eq. B.8 to be of the form of Eq. B.7 we must have

$$\frac{\alpha}{\beta} = \frac{\gamma}{\delta}. \tag{B.10}$$

Composite states that are of the form Eq. B.7 are called "separable." Composite states that are of the form of Eq. B.8 but not of the form of Eq. B.7 are called "entangled."

There is an important result that holds for wavefunctions of bipartite systems known as "Schmidt decomposition," which states that an arbitrary wavefunction for a bipartite system can be written in the form

$$|\psi^{AB}\rangle = \sum_i \lambda_i |\tilde{\psi}_i^A\rangle |\tilde{\psi}_i^B\rangle \tag{B.11}$$

with $\lambda_i > 0$ and $|\tilde{\psi}_i^A\rangle$, $|\tilde{\psi}_i^B\rangle$ orthonormal vectors in the Hilbert space of A and B, respectively. Equation B.11 holds true even if the dimension of the Hilbert space of A and B is different. The number of nonzero λ_i is called the "Schmidt number." If the Schmidt number is greater than one, the system is entangled. The largest possible value of the Schmidt number is the dimension of the smaller of the two Hilbert spaces A and B, $\min(\dim(A), \dim(B))$.

▶ **Exercise B.1**   Find the Schmidt decomposition and Schmidt number for each of the four Bell states,

$$(i) \quad \frac{1}{\sqrt{2}}(|++\rangle + |--\rangle), \qquad (ii) \quad \frac{1}{\sqrt{2}}(|++\rangle - |--\rangle)$$

$$(iii) \quad \frac{1}{\sqrt{2}}(|+-\rangle + |-+\rangle), \qquad (iv) \quad \frac{1}{\sqrt{2}}(|+-\rangle - |-+\rangle). \tag{B.12}$$

B.2   Density Matrix of a Composite System

We now consider the density matrix of composite systems. We need to define the Kronecker product of two arrays. Following standard notation, the Kronecker product of an m by n matrix A with a p by q matrix B is an mp by nq matrix C:

$$C = A \otimes B \equiv \begin{bmatrix} A_{11}B & A_{12}B & \dots & A_{1n}B \\ A_{21}B & A_{22}B & \dots & A_{2n}B \\ \vdots & \vdots & \vdots & \vdots \\ A_{m1}B & A_{m2}B & \dots & A_{mn}B \end{bmatrix}, \tag{B.13}$$

or in component notation,

$$C_{(i-1)p+\alpha,(j-1)q+\beta} = A_{ij}B_{\alpha\beta} \equiv c_{i\alpha, j\beta}, \tag{B.14}$$
$$i = 1, \dots, m; \quad j = 1, \dots, n; \quad \alpha = 1, \dots, p; \quad \beta = 1, \dots, q,$$

where we use Roman letters to label A and Greek letters to label B. On the RHS we have introduced the symbol lowercase c, defined by Eq. B.14, which provides a convenient and intuitive labelling scheme for the matrix elements of the matrix C. Note that since the density matrix is square we will be interested below in the case $mp = nq$ and particularly $m = n$ and $p = q$.

One way in which composite density matrices arise is from the tensor product of density matrices that are defined on different spaces. For example, if

$$\rho^A = \frac{1}{2}\begin{pmatrix} 1 & 1 \\ 1 & 1 \end{pmatrix} \text{ and } \rho^B = \frac{1}{2}\begin{pmatrix} 1 & -1 \\ -1 & 1 \end{pmatrix} \tag{B.15}$$

one may represent the composite density matrix ρ^{AB} using the Kronecker product as

$$\rho^{AB} = \rho^A \otimes \rho^B = \frac{1}{4}\left(\begin{array}{cc|cc} 1 & -1 & 1 & -1 \\ -1 & 1 & -1 & 1 \\ \hline 1 & -1 & 1 & -1 \\ -1 & 1 & -1 & 1 \end{array}\right), \tag{B.16}$$

where the horizontal and vertical lines in the ρ^{AB} matrix are just for pedagogical purposes, to highlight the way in which the ρ^B block structure enters into ρ^{AB} in this example. A second way in which composite density matrices arise is from pure states of composite systems. If the latter are entangled then ρ^{AB} cannot be reduced to the form $\rho^A \otimes \rho^B$. For example, the composite density matrix associated with the first Bell state is

$$\rho^{AB} = \frac{1}{\sqrt{2}}\begin{pmatrix} 1 \\ 0 \\ 0 \\ 1 \end{pmatrix} \cdot \frac{1}{\sqrt{2}}(1\,0\,0\,1) = \frac{1}{2}\begin{pmatrix} 1 & 0 & 0 & 1 \\ 0 & 0 & 0 & 0 \\ 0 & 0 & 0 & 0 \\ 1 & 0 & 0 & 1 \end{pmatrix}, \tag{B.17}$$

which cannot be factored into the form $\rho^A \otimes \rho^B$.

From the bookkeeping perspective, notice that in the first example the composite density matrix was formed from the *tensor* product of two 2×2 matrices while in the second example the composite density matrix was formed from the *outer* product of a 4×1 and a 1×4 vector, but either way the result is a 4×4 density matrix.

B.3   Reduced Density Matrix

We now consider the reverse process, of passing from a density matrix of the composite system to a reduced density matrix of one of the subsystems. This is accomplished by the partial trace,

$$\rho^A = \text{Tr}_B(\rho^{AB}) \qquad \rho^B = \text{Tr}_A(\rho^{AB}). \tag{B.18}$$

The partial trace is defined in two stages. First, we consider the partial trace of a single element of the composite density matrix

$$\text{Tr}_B(|a_i\rangle\langle a_j| \otimes |b_\alpha\rangle\langle b_\beta|) \equiv |a_i\rangle\langle a_j|\text{Tr}(|b_\alpha\rangle\langle b_\beta|), \tag{B.19}$$

where $|a_i\rangle$ and $|a_j\rangle$ are any two vectors in the state space of A, and $|b_\alpha\rangle$ and $|b_\beta\rangle$ are any two vectors in the state space of B. The trace operation on the RHS is the usual trace for system B, so that $\text{Tr}(|b_\alpha\rangle\langle b_\beta|) = \langle b_\beta|b_\alpha\rangle$. In the second stage, the definition of the partial trace is extended by linearity to the full composite density matrix.

▶ **Exercise B.2**   Extend the definition of the partial trace to the full composite density matrix

$$\rho^{AB} = \sum_{ij\alpha\beta} c_{i\alpha,j\beta}|a_i\rangle\langle a_j| \otimes |b_\alpha\rangle\langle b_\beta|. \tag{B.20}$$

Derive general expressions for ρ^A and for ρ^B in terms of $c_{i\alpha,j\beta}$.

Solution

$$\rho^A = \text{Tr}_B(\rho^{AB}) = \sum_{i,j=1}^{m,n} |a_i\rangle\langle a_j| \sum_{\alpha,\beta=1}^{p,q} c_{i\alpha,j\beta}\langle b_\beta|b_\alpha\rangle \equiv \sum_{i,j=1}^{m,n} c'_{ij}|a_i\rangle\langle a_j|. \tag{B.21}$$

$$\rho^B = \text{Tr}_A(\rho^{AB}) = \sum_{\alpha,\beta=1}^{p,q} |b_\alpha\rangle\langle b_\beta| \sum_{i,j=1}^{m,n} c_{i\alpha,j\beta}\langle a_j|a_i\rangle \equiv \sum_{\alpha,\beta=1}^{p,q} c''_{\alpha,\beta}|b_\alpha\rangle\langle b_\beta|. \tag{B.22}$$

For $\langle b_\beta|b_\alpha\rangle = \delta_{\alpha\beta}$ we have

$$c'_{ij} = \sum_{\alpha=1}^{p} c_{i\alpha,j\alpha} \tag{B.23}$$

while for $\langle a_j|a_i\rangle = \delta_{ij}$ we have

$$c''_{\alpha\beta} = \sum_{i=1}^{m} c_{i\alpha,i\beta}. \tag{B.24}$$

The partial trace has a graphical interpretation, which depends on whether the trace is over B or A. Consider the partial trace over B of ρ^{AB} in Eq. B.16:

$$\rho^A = \text{Tr}_B \frac{1}{4}\begin{pmatrix} 1 & -1 & 1 & -1 \\ -1 & 1 & -1 & 1 \\ 1 & -1 & 1 & -1 \\ -1 & 1 & -1 & 1 \end{pmatrix} = \frac{1}{4}\left(\begin{array}{cc|cc} 1 & & 1 & \\ & + & & + \\ & 1 & & 1 \\ \hline 1 & & 1 & \\ & + & & + \\ & 1 & & 1 \end{array}\right) = \frac{1}{2}\begin{pmatrix} 1 & 1 \\ 1 & 1 \end{pmatrix}. \tag{B.25}$$

The partial trace over B has the interpretation of adding the diagonal elements in each block to form the "block traces," and then combining the block traces into a new matrix of dimension $\dim(A)$. Now consider the partial trace over A of ρ^{AB}:

$$\rho^B = \text{Tr}_A \frac{1}{4}\begin{pmatrix} 1 & -1 & 1 & -1 \\ -1 & 1 & -1 & 1 \\ 1 & -1 & 1 & -1 \\ -1 & 1 & -1 & 1 \end{pmatrix} = \frac{1}{4}\left(\begin{bmatrix} 1 & -1 \\ -1 & 1 \end{bmatrix} + \begin{bmatrix} 1 & -1 \\ -1 & 1 \end{bmatrix}\right) = \frac{1}{2}\begin{pmatrix} 1 & -1 \\ -1 & 1 \end{pmatrix}. \tag{B.26}$$

The partial trace over A has the interpretation of discarding the off-diagonal blocks and adding the diagonal blocks element by element to form a new matrix of dimension dim(B).

Consider now the partial trace of ρ^{AB} given by Eq. B.17, corresponding to the first Bell state (cf. Exercise B.1). The partial trace over B gives

$$\mathrm{Tr}_B \frac{1}{2} \begin{pmatrix} 1 & 0 & 0 & 1 \\ 0 & 0 & 0 & 0 \\ 0 & 0 & 0 & 0 \\ 1 & 0 & 0 & 1 \end{pmatrix} = \frac{1}{2} \left(\begin{array}{cc|cc} 1 & & 0 & \\ & + 0 & & + 0 \\ \hline 0 & & 0 & \\ & + 0 & & + 1 \end{array} \right) = \frac{1}{2} \begin{pmatrix} 1 & 0 \\ 0 & 1 \end{pmatrix}$$

(B.27)

while the partial trace over A gives

$$\mathrm{Tr}_A \frac{1}{2} \begin{pmatrix} 1 & 0 & 0 & 1 \\ 0 & 0 & 0 & 0 \\ 0 & 0 & 0 & 0 \\ 1 & 0 & 0 & 1 \end{pmatrix} = \frac{1}{2} \left(\begin{bmatrix} 1 & 0 \\ 0 & 0 \end{bmatrix} + \begin{bmatrix} 0 & 0 \\ 0 & 1 \end{bmatrix} \right) = \frac{1}{2} \begin{pmatrix} 1 & 0 \\ 0 & 1 \end{pmatrix}. \quad \text{(B.28)}$$

Similarly, the other three Bell states give reduced density matrices of the form

$$\rho^{A,B} = \frac{1}{2} \begin{pmatrix} 1 & 0 \\ 0 & 1 \end{pmatrix}$$

(B.29)

whether the partial trace is performed over B or over A.

▶ **Exercise B.3**   Verify that this is the form of the reduced density matrix for the other three Bell states.

B.4   Purity of the Reduced Density Matrix

The example of the Bell states shows a remarkable property: that even if the wavefunction of the composite system is pure, the density matrix of the subsystem after the partial trace operation may be mixed. Some reflection reveals that if the wavefunction of the composite system is pure and *separable* the reduced density matrix must correspond to a pure state. Thus, entanglement is necessary in order to obtain a mixed reduced density matrix starting from a pure composite system.

As we shall now prove, entanglement is also *sufficient* for the reduced density matrix to be mixed. Since the Schmidt number is an indicator of entanglement in a composite wavefunction, one might guess that the Schmidt coefficients $\{\lambda_i\}$ of the composite wavefunction are related to the purity of the reduced density matrices that derive from the composite density matrix. To establish this connection we first note that there is a simple relationship between the Schmidt number of the composite wavefunction and the *eigenvalues* of the reduced density matrices. Using Eq. B.11, we have

$$\rho^{AB} = \sum_{i,j=1}^{n} \lambda_i \lambda_j |\tilde{\psi}_i^A\rangle\langle\tilde{\psi}_j^A| \otimes |\tilde{\psi}_i^B\rangle\langle\tilde{\psi}_j^B| \tag{B.30}$$

so that

$$\rho^A = \text{Tr}_B(\rho^{AB}) = \sum_{i=1}^{n} \lambda_i^2 |\tilde{\psi}_i^A\rangle\langle\tilde{\psi}_i^A| \tag{B.31}$$

$$\rho^B = \text{Tr}_A(\rho^{AB}) = \sum_{i=1}^{n} \lambda_i^2 |\tilde{\psi}_i^B\rangle\langle\tilde{\psi}_i^B|. \tag{B.32}$$

Equations B.31–B.32 can be viewed as diagonal representations of ρ^A and ρ^B, respectively. Viewed this way, it is apparent that the eigenvalues of ρ^A and ρ^B are both given by

$$p_i = \lambda_i^2. \tag{B.33}$$

Equation B.33 expresses the remarkable property that given a pure, bipartite wavefunction the eigenvalues of both reduced density matrices are equal. This holds even if the size of the Hilbert space of the two subsystems is different; in that case, the additional eigenvalues of the larger system must be zero. Since the eigenvalues of both reduced density matrices, ρ^A and ρ^B, are identical so must be their purities:

$$\text{Tr}((\rho^A)^2) = \text{Tr}((\rho^B)^2) = \sum_{i=1}^{n} p_i^2 = \sum_{i=1}^{n} \lambda_i^4. \tag{B.34}$$

Only if the composite wavefunction is separable is there a single nonzero value of λ; in this case and only in this case we have $\text{Tr}((\rho^A)^2) = \text{Tr}((\rho^B)^2) = 1$, corresponding to a pure state of the subsystems. If the composite system is not separable we necessarily have $\text{Tr}((\rho^A)^2) = \text{Tr}((\rho^B)^2) < 1$, expressing the fully general result that entanglement of a composite system implies a lack of purity of the reduced density matrices of the subsystems.

We have seen that entangled pure states of composite systems give reduced density matrices that are mixed. The converse is perhaps even more remarkable: given *any* density matrix ρ one can always find a pure state of a composite system that gives ρ as the reduced density matrix after performing a partial trace. In fact, there is a school of thought in which the probabilistic mixture of the density operator should always be viewed as arising from the tracing out of part of a larger system.

▶ **Exercise B.4**   Given a density matrix $\rho^A = \sum_i p_i |\psi_i^A\rangle\langle\psi_i^A|$ show that ρ^A can be "purified" by introducing a system B that has the same state space as system A with orthogonal basis states $|\psi_i^B\rangle$ and defining a pure state for the composite system

$$|AB\rangle = \sum_i \sqrt{p_i} |\psi_i^A\rangle |\psi_i^B\rangle. \tag{B.35}$$

Solution   The reduced density operator for system A corresponding to the state $|AB\rangle$ is

$$\text{Tr}_B(|AB\rangle\langle AB|) = \sum_{ij} \sqrt{p_i p_j} |\psi_i^A\rangle \langle \psi_j^A| \text{Tr}(|\psi_i^B\rangle\langle \psi_j^B|) \tag{B.36}$$

$$= \sum_{ij} \sqrt{p_i p_j} |\psi_i^A\rangle \langle \psi_j^A| \delta_{ij} = \sum_i p_i |\psi_i^A\rangle \langle \psi_i^A| = \rho^A. \tag{B.37}$$

In the general case, a bipartite density matrix is called separable if it can be written in the form (Werner, 1989)

$$\rho^{AB} = \sum_{i=1}^N p_i \rho_i^A \otimes \rho_i^B. \tag{B.38}$$

Entangled pure states (states of the form of Eq. B.11 with more than one nonzero λ) generate a composite density matrix ρ^{AB} that cannot be written in the form of Eq. B.38, and hence ρ^{AB} is not separable. However, in the general case no simple sufficiency condition is known to determine if a bipartite density matrix is separable, except in special cases (low-dimensional Hilbert spaces). This is an active area of research.

References

General

1. M. A. Nielsen and I. L. Chuang, *Quantum Computation and Quantum Information*, (Cambridge University Press, Cambridge, 2000).

2. A. Peres, *Quantum Theory: Concepts and Methods*, (Kluwer Academic, Dordrecht, 1993).

3. J. Preskill, *Physics 229: Advanced Mathematical Methods of Physics: Quantum Computation and Information*, Caltech, 1998. URL: http://www.theory.caltech.edu/people/preskill/ph229/.

4. K. Kraus, *States, Effects and Operations: Fundamental Notions of Quantum Theory*, Lecture Notes in Physics, Vol. 190, (Springer-Verlag, Berlin, 1983).

5. J. von Neumann, *Mathematical Foundations of Quantum Mechanics* (Princeton University Press, Princeton, 1955).

Entanglement and Measurement

6. C. Cohen-Tannoudji, B. Diu and F. Laloë, *Quantum Mechanics* vol. I (Wiley, New York, 1977), Complement DIV.

7. A. Einstein, B. Podolsky and N. Rosen, Can quantum-mechanical description of physical reality be considered complete?, Phys. Rev. 47, 777 (1935).

8. J. S. Bell, *Speakable and Unspeakable in Quantum Mechanics* (Cambridge University Press, Cambridge, 1987).

9. D. M. Greenberger, M. A. Horne, A. Shimony and A. Zeilinger, Bell's theorem without inequalities, Am. J. Phys. 58, 1131 (1990).

10. N. D. Mermin, Quantum mysteries revisited, Am. J. Phys. 58, 731 (1990).

Density Matrices of Composite Systems

11. R. F. Werner, Quantum states with Einstein-Podolsky-Rosen correlations admitting a hidden-variable model, Phys. Rev. A 40, 4277 (1989).

12. A. Peres, Separability criterion for density matrices, Phys. Rev. Lett. 77, 1413 (1996).

13. M. Horodecki, P. Horodecki and R. Horodecki, Separability of mixed states: Necessary and sufficient conditions, Phys. Lett. A 223, 1 (1996).

Appendix C

Normalization and Orthogonality of Scattering Eigenstates

The primary purpose of this Appendix is to derive the normalization and orthogonality relations among the scattering eigenstates that are introduced in Section 7.4. However, in the course of the derivation we obtain a general equation relating the asymptotic coefficients of any two scattering eigenstates (Eq. C.6). This equation includes as special cases the four symmetry relations among r, t, r' and t' that are usually derived from the method of conservation of flux, as well as the four S-matrix elements for one-dimensional barrier scattering.

Our approach is to first derive the normalization and orthogonality for a potential with a single step. We then show that the results apply equally well to an arbitrary number of steps. Since an arbitrary one-dimensional scattering potential can be viewed as the limit of concatenating a finite number of potential steps, the results are fully general.

In Section 7.4 we give a simplified version of this Appendix. There, the normalization and orthogonality relations are not derived, but assumed. It is shown there that the four symmetry relations follow just from these two assumptions. Moreover, the four S-matrix elements follow immediately by examining the nonorthogonal overlaps.

C.1 Single Step

We begin by reconsidering the problem of scattering from a single step treated in Section 7.4, but now we take a general form for the incoming and outgoing waves. Consider a general wavefunction of the form appearing in Figure C.1a. Let $x = a$ be the boundary between region I and region III (region II is infinitely narrow in this example). Continuity of the wavefunction and its first derivative at $x = a$ gives two equations. These equations may be arranged in the form

$$\mathbf{m}(k) \begin{pmatrix} A \\ B \end{pmatrix} = \mathbf{m}(\kappa) \begin{pmatrix} C \\ D \end{pmatrix}. \tag{C.1}$$

We define a matrix $\mathbf{M}(kK) \equiv \mathbf{m}^{-1}(k)\mathbf{m}(\kappa)$. It is easy to check (Exercise C.2) that

$$\begin{pmatrix} A \\ B \end{pmatrix} = \begin{pmatrix} M_{11} & M_{12} \\ M_{21} & M_{22} \end{pmatrix} \begin{pmatrix} C \\ D \end{pmatrix} = \frac{1}{2k} \begin{pmatrix} (k+\kappa)e^{-i(k-\kappa)a} & (k-\kappa)e^{-i(k+\kappa)a} \\ (k-\kappa)e^{i(k+\kappa)a} & (k+\kappa)e^{i(k-\kappa)a} \end{pmatrix} \begin{pmatrix} C \\ D \end{pmatrix}. \tag{C.2}$$

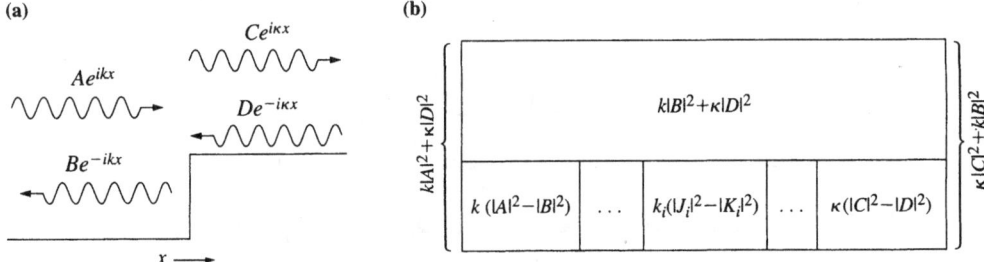

Figure C.1 (a) General form of a scattering eigenstate for a potential with a single step, with incoming and outgoing waves in each direction. (b) Schematic representation of the amplitudes in each zone of a general scattering potential, showing the equality $k(|A|^2 - |B|^2 = k_i(|J_i|^2 - |K_i|^2) = \kappa(|C|^2 - |D|^2)$. Adding $k|B|^2 + \kappa|D|^2$ to all regions gives $k|A|^2 + \kappa|D|^2 = \cdots = \kappa|C|^2 + k|B|^2$ (see text).

Consider now the overlap of two independent scattering eigenstates, $\tilde{\psi}$ and $\tilde{\psi}'$, neither of which is necessarily incoming or outgoing:

	Region I	Region III
$\tilde{\psi}$	$A(k)e^{ikx} + B(k)e^{-ikx}$	$C(\kappa)e^{i\kappa x} + D(\kappa)e^{-i\kappa x}$
$\tilde{\psi}'$	$A'(k')e^{ik'x} + B'(k')e^{-ik'x}$	$C'(\kappa')e^{i\kappa'x} + D'(\kappa')e^{-i\kappa'x}$ (C.3)

The fact that $\tilde{\psi}$ and $\tilde{\psi}'$ are independent implies that A', B', C', D' are in general different from A, B, C, D. For example, for the overlap of $\psi_\alpha^+(k)$ with $\psi_\beta^+(k)$ we have ($A = 1$, $B = r$, $C = t$, $D = 0$) and ($A' = 0$, $B' = t'$, $C' = r'$, $D' = 1$). The general form for the overlap is

$$\langle \tilde{\psi}'(k')|\tilde{\psi}(k)\rangle = \frac{1}{2\pi} \left\{ \int_{-\infty}^{0} (A'^*(k')e^{-ik'x} + B'^*(k')e^{ik'x})(A(k)e^{ikx} + B(k)e^{-ikx})\, dx \right.$$

$$\left. + \int_{0}^{\infty} (C'^*(\kappa')e^{-i\kappa'x} + D'^*(\kappa')e^{i\kappa'x})(C(\kappa)e^{i\kappa x} + D(\kappa)e^{-i\kappa x})\, dx \right\}. \quad (C.4)$$

For simplicity we have taken the boundary to be at $x = 0$ although as discussed following Eq. C.8 the final result is independent of the position of the boundary. To evaluate this overlap we proceed as follows:

1. We first express $A(k)$, $B(k)$ in terms of the matrix $\mathbf{M}(k\kappa)$ times $C(\kappa)$, $D(\kappa)$, and $A'^*(k')$, $B'^*(k')$ in terms of $\mathbf{M}(k'\kappa')$ times $C'^*(\kappa')$, $D'^*(\kappa')$.

2. We substitute the expressions for $A(k)$, $B(k)$, $A'^*(k')$ and $B'^*(k')$ in item 1 into Eq. C.4 and multiply out the terms in parentheses. We then group the terms into four sets, according to whether they contain the factor $C'^*(\kappa')C(\kappa)$, $C'^*(\kappa')D(\kappa)$, $D'^*(\kappa')C(\kappa)$ or $D'^*(\kappa')D(\kappa)$.

3. Each of the factors, for example, $C'^*(\kappa')C(\kappa)$, has associated with it five terms. The x dependence of each of these terms appears only in the complex exponential, which takes either the rotating form, $e^{\pm i(k-k')x}$ or $e^{\pm i(\kappa-\kappa')x}$, or the counterrotating form, $e^{\pm i(k+k')x}$ or $e^{\pm i(\kappa+\kappa')x}$. Performing the integral over x, the rotating terms lead to $\pm i(k - k')$ in the denominator while the counterrotating terms lead to $\pm i(k + k')$ in the denominator (the contribution to the integral coming from the limits $-\infty$ and

∞ is taken to be 0, consistent with the discussion of the regularizing function in Appendix A).

4. Substituting the explicit expressions for the elements of **M** from Eq. C.2, we collect the five terms associated with each factor into a single term. It turns out that the numerator of this term *always* vanishes. This cancellation of terms requires the inclusion of both the rotating and the counterrotating terms.

5. Since the numerator vanishes quite generally, the coefficient of each factor as determined by (numerator)/(denominator) will vanish unless the denominator vanishes. Taking into account that k, $k' > 0$, the denominators of the counterrotating terms can never vanish; however, the denominators of the rotating terms do vanish for $k = k'$, giving 0/0. We set $k' = k$ in the arguments of A'^*, B'^*, C'^*, D'^* but not in the complex exponentials, and calculate the value of the integral using relations for "partial δ functions" derived in Appendix A.

We find that

$$\langle \tilde{\psi}'(k')|\tilde{\psi}(k)\rangle = \frac{1}{2\pi}\left\{ \int_{-\infty}^{0} \left\{ A'^*A + \frac{\kappa}{k}D'^*D \right\} e^{i(k-k')x}dx \right.$$

$$\left. + \int_{0}^{\infty} \left\{ \frac{\kappa}{k}C'^*C + B'^*B \right\} e^{i(k-k')x}dx \right\}. \qquad \text{(C.5)}$$

Expressing A, B, A'^*, B'^* in terms of C, D, C'^*, D'^* multiplied by elements of the matrix $\mathbf{M}(k\kappa)$ and using the symmetry properties of $\mathbf{M}(k\kappa)$ (Exercise C.3), we find that the coefficients of each factor, for example, C'^*C or C'^*D, are equal in region I and region III. This implies that

$$A'^*(k)A(k) + \frac{\kappa}{k}D'^*(k)D(k) = \frac{\kappa}{k}C'^*(\kappa)C(\kappa) + B'^*(k)B(k), \qquad \text{(C.6)}$$

where $\kappa = \kappa(k)$.

▶ **Exercise C.1**   Identify (A, B, C, D) and (A', B', C', D') for each of the four overlaps in Table 7.3. Show that the four relations among r, t, r' and t' in Table 7.3 are all special cases of Eq. C.6.

The special case $A' = A$ in Eq. C.6 gives the relation

$$k|A|^2 + \kappa|D|^2 = \kappa|C|^2 + k|B|^2. \qquad \text{(C.7)}$$

Equation C.7 can be interpreted in terms of the equality of the norm of the total incoming probability and the total outgoing probability. This equation can be rearranged into the form

$$k(|A|^2 - |B|^2) = \kappa(|C|^2 - |D|^2). \qquad \text{(C.8)}$$

Equation C.8 (and Eq. C.6 as well, for that matter) can also be obtained by applying the method of conservation of flux (Section 7.3.3) to the wavefunctions in Eq. C.3.

In the discussion above, we chose the position of the potential step to be at 0. However, the discussion leading to Eqs. C.5 and C.6 is valid for any position a of the potential step. Specifically, for the case $k \neq k'$ the exponential phase containing the position of the step,

a, drops out. For the case $k = k'$, the coefficients of the δ functions are independent of a (Exercise A.7). In fact, the $k = k'$ term is unaffected if the limits of the first integral are $(-\infty, a_L]$ and the limits of the second integral are $[a_R, \infty)$, that is, even if the boundaries of the integrals do not join continuously. This will be important for the concatenation of multiple steps, discussed below.

C.2   Concatenation of Steps

It is now relatively straightforward to extend the procedure above to an arbitrary number of steps. Consider first the case $k \neq k'$. Notice that the contributions to the integrals in Eq. C.4 came only from the boundary between regions, in other words, the contribution from the upper limit of the left integral cancelled the contribution from the lower limit of the right integral. The same will hold true no matter how many steps there are: the contribution from the upper limit of one region will always cancel with the contribution from the lower limit of the next integral. Therefore, the net contribution from all integrals vanishes unless $k = k'(\kappa')$. Consider now the case $k = k'$, and suppose that there are many intermediate regions with wavefunctions of the form $J_i e^{ik_i x} + K_i e^{-ik_i x}$. Extending the treatment of the single step potential, we have

$$k(A'^*A - B'^*B) = \ldots = k_i(J_i'^*J_i - K_i'^*K_i) = \ldots = \kappa(C'^*C - D'^*D), \quad \text{(C.9)}$$

and therefore

$$kA'^*A + \kappa D'^*D = k_i(J_i'^*J_i - K_i'^*K_i) + kB'^*B + \kappa D'^*D = \ldots = \kappa C'^*C + kB'^*B. \quad \text{(C.10)}$$

A pictorial representation of this rearrangement of terms is shown in Figure C.1b. Equation C.10 indicates that the coefficient of $e^{i(k-k')x}$ in the left asymptotic region is equal to the coefficient of $e^{i(k-k')x}$ in the right asymptotic region. The discussion following Eq. C.8 shows that combining the two portions of the integral of $e^{i(k-k')x}$ gives

$$\langle \tilde{\psi}' | \tilde{\psi} \rangle = (kA'^*A + \kappa D'^*D)\delta(k - k') = (\kappa C'^*C + kB'^*B)\delta(k - k') \quad \text{(C.11)}$$

despite the finite intervening barrier region. Equating the coefficients of the δ functions in Eq. C.11, we conclude that Eq. C.6 is valid for arbitrary piecewise-continuous potentials.

Problems

▶ **Exercise C.2**   Verify Eq. C.2 for the M-matrix.

▶ **Exercise C.3**

a.   Verify the following symmetries of $M(k\kappa)$:

$$\text{(i) } M_{11}^*M_{11} - M_{21}^*M_{21} = \frac{\kappa}{k}$$

$$\text{(ii) } M_{22}^*M_{22} - M_{12}^*M_{12} = \frac{\kappa}{kt}$$

(iii) $M_{12}^* M_{11} - M_{22}^* M_{21} = 0$

(iv) $M_{11}^* M_{12} - M_{21}^* M_{22} = 0$ $\qquad$ (C.12)

Show that if an arbitrary number of M matrices are multiplied together, the resulting matrix has the same symmetries as the individual M matrices.

b. Calculate the determinant of M. Find a simple expression for the determinant of a product of an arbitrary number of M matrices.

c. We may define the S-matrix in terms of the following equation:

$$\begin{pmatrix} B/\sqrt{k} \\ C/\sqrt{\kappa} \end{pmatrix} = \begin{pmatrix} S_{11} & S_{12} \\ S_{21} & S_{22} \end{pmatrix} \begin{pmatrix} A/\sqrt{k} \\ D/\sqrt{\kappa} \end{pmatrix}. \qquad (C.13)$$

Whereas the M-matrix is a relationship between amplitude on the left and amplitude on the right, the S-matrix is a relationship between incoming and outgoing amplitude. Use Eqs. C.2 and C.13 to find an expression for each of the elements of M in terms of the elements of S and vice versa.

d. Show that each of the symmetries of the M-matrix in part a corresponds to a symmetry of the S-matrix.

References

1. L. D. Landau and E. M. Lifshitz, *Quantum Mechanics: Non-relativistic Theory* (Pergamon, New York, 1977), p. 63.
2. L. D. Fadeev and O. A. Yakubovsky, *Lectures on Quantum Mechanics for Students–Mathematicians* (Leningrad University Publishing House, Leningrad, 1980) (in Russian).
3. E. Merzbacher, *Quantum Mechanics* (Wiley, New York, 1970), 2nd ed., Chapter 6.

Appendix D

Units and Conversions

D.1 Energy Conversions and Time Scales

Table D.1 Conversion between different energy units

	a.u.	eV	cm^{-1}	kcal/mol	Hz
a.u.	1	27.2107	2.19474×10^5	6.27503×10^2	6.57966×10^{15}
eV	3.67502×10^{-2}	1	8.06573×10^3	23.0609	2.41804×10^{14}
cm^{-1}	4.55633×10^{-6}	1.23981×10^{-4}	1	2.85911×10^{-3}	2.99793×10^{10}
kcal/mol	1.59362×10^{-3}	4.33634×10^{-2}	3.49757×10^2	1	1.04854×10^{13}
Hz	1.51983×10^{-16}	4.13558×10^{-15}	3.33565×10^{-11}	9.53702×10^{-14}	1

Table D.2 Atomic units

Quantity	Unit	Physical significance	Value in cgs units
Charge	e	Electron charge	4.80298×10^{-10} esu
Mass	m or m_e	Electron mass	9.1091×10^{-28} gm
Length	$a_0 = \hbar^2/me^2$	Bohr radius	0.529167×10^{-8} cm
Velocity	$v_0 = e^2/\hbar$	Electron velocity in first Bohr orbit	2.18765×10^8 cm sec^{-1}
Time	a_0/v_0	Time required for electron in first Bohr orbit to travel one Bohr radius	2.41888×10^{-17} sec
Energy	e^2/a_0	Twice the IP for hydrogen (infinite nuclear mass)	4.35942×10^{-11} erg

Table D.3 Conversion between time, frequency and energy for 1 fs

Time	Frequency $= \nu = 1/T$	Energy $= h\nu$
1 fs $= 10^{-15}$ s	10^{15} Hz	10^{15} s$^{-1} \times \frac{s}{2.99792 \times 10^{10} cm} = 3.33564 \times 10^4$ cm^{-1}
41.3414 a.u.	$1/41.3414$ a.u. $= .0241888$ a.u.	$2\pi/41.3414$ a.u. $= .151983$ a.u. *

* Note the factor of 2π in the conversion from frequency to energy in a.u., since $\hbar = 1 = \frac{h}{2\pi}$

Table D.4 Time and frequency scales for some representative processes

	Time (in periods)	Frequency (1/period)
Electron in the first Bohr orbit	1.52×10^{-16} s	$\left(\frac{1}{1.52\times10^{-16}\text{s}}\right) \times \left(\frac{\text{s}}{3\times10^{10}\text{cm}}\right) = 2.19 \times 10^5 \text{ cm}^{-1}$
Vibration of H_2	7.58×10^{-15} s	$\left(\frac{1}{7.58\times10^{-15}\text{s}}\right) \times \left(\frac{\text{s}}{3\times10^{10}\text{ cm}}\right) = 4.40 \times 10^3 \text{ cm}^{-1}$
Vibration of I_2	1.67×10^{-13} s	$\left(\frac{1}{1.67\times10^{-13}\text{s}}\right) \times \left(\frac{\text{s}}{3\times10^{10}\text{cm}}\right) = 2.14 \times 10^2 \text{ cm}^{-1}$
Rotation of H_2	2.74×10^{-13} s	$\left(\frac{1}{2.74\times10^{-13}\text{s}}\right) \times \left(\frac{\text{s}}{3\times10^{10}\text{cm}}\right) = 2 \times 60.8 \text{ cm}^{-1}$
Rotation of I_2	4.50×10^{-10} s	$\left(\frac{1}{4.50\times10^{-10}\text{s}}\right) \times \left(\frac{\text{s}}{3\times10^{10}\text{cm}}\right) = 2 \times .037 \text{ cm}^{-1}$

D.2   Units of Electric Field Amplitude and Intensity

In Chapter 13 we used the relation (Eq. 13.51)

$$E = \int_V \frac{\mathbf{E}^2(t) + \mathbf{B}^2(t)}{8\pi} dV. \tag{D.1}$$

Clearly, the electric field has units of (energy/volume)$^{1/2}$. To convert from the electric field in atomic units to cgs units (the unit of the electric field in cgs units is called the esu = statvolt/cm = statC/cm^2) we note that

$$\left(\frac{\text{energy}}{\text{volume}}\right)^{1/2} (\text{a.u.}) = \left(\frac{4.35942 \times 10^{-11}\text{erg}}{(0.529167 \times 10^{-8}\text{cm})^3}\right)^{1/2} = 1.7152 \times 10^7 \text{esu.} \tag{D.2}$$

We also used the expression (Eq. 13.53)

$$E = \int \frac{|\varepsilon(t)^2|}{2\pi} Ac \, dt. \tag{D.3}$$

According to this expression, the energy of the electric field per unit area per unit time, called the "irradiance" or "power density," is

$$\frac{1}{A}\frac{dE}{dt} = \frac{|\varepsilon(t)|^2 c}{2\pi}. \tag{D.4}$$

The irradiance is often specified in W/cm^2. To convert from atomic units to W/cm^2 we note that

$$\frac{\text{energy}}{(\text{area} \times \text{time})} (\text{a.u.}) = \frac{4.35942 \times 10^{-11} \text{ erg}}{(0.529167 \times 10^{-8} \text{ cm})^2 \times (2.41888 \times 10^{-17} \text{ sec})} \times \frac{10^{-7} \text{ J}}{\text{erg}}$$

$$= 6.436 \times 10^{15} \frac{\text{W}}{\text{cm}^2}. \tag{D.5}$$

Index

Key to index with Examples: Throughout this index, the following abbreviations have been used: A = Appendix A; 8 = Chapter 8; 2.1.3 = Section 2.1.3; ex. 15.10 = Exercise 15.10; 12H = history section.